W9-BWO-464

Annual Review of Ecology and Systematics

ANNUAL REVIEW OF ECOLOGY AND SYSTEMATICS

VOLUME 31, 2000

DAPHNE GAIL FAUTIN, *Editor*
University of Kansas

DOUGLAS J. FUTUYMA, *Associate Editor*
State University of New York at Stony Brook

FRANCES C. JAMES, *Associate Editor*
Florida State University

www.AnnualReviews.org science@AnnualReviews.org 650-493-4400

ANNUAL REVIEWS
4139 El Camino Way • P.O. BOX 10139 • Palo Alto, California 94303-0139

ANNUAL REVIEWS
Palo Alto, California, USA

International Standard Serial Number: 0066-4162
International Standard Book Number: 0-8243-1431-X
Library of Congress Catalog Card Number: 71-135616

⊗ The paper used in this publication meets the minimum requirements of American National Standards for Information Sciences—Permanence of Paper for Printed Library Materials. ANSI Z39.48-1992.

TYPESET BY TECHBOOKS, FAIRFAX, VA
PRINTED AND BOUND IN THE UNITED STATES OF AMERICA

Annual Review of Ecology and Systematics
Volume 31, 2000

CONTENTS

Related Articles

From the ***Annual Review of Earth and Planetary Sciences***, Volume 28 (2000):

Palynology after Y2K–Understanding the Source Area of Pollen in Sediments, Margaret B. Davis

Dinosaur Reproduction and Parenting, John R. Horner

Chemical Weathering, Atmospheric CO_2, and Climate, Lee Kump, Susan L. Brantley, and Michael A. Arthur

From the ***Annual Review of Energy and Environment***, Volume 24 (1999)

On the Road to Global Ecology, H. A. Mooney

Harmful Algal Blooms: An Emerging Public Health Problem with Possible Links to Human Stress on the Environment, J. Glenn Morris, Jr.

Non-CO_2 Greenhouse Gases in the Atmosphere, M. A. K. Khail

From the ***Annual Review of Entomology***, Volume 45 (2000)

The Current State of Insect Molecular Systematics: A Thriving Tower of Babel, Michael S. Caterino, Soowon Cho, and Felix A. H. Sperling

Control of Insect Pests with Entomopathogenic Nematodes: The Impact of Molecular Biology and Phylogenetic Reconstruction, Jie Liu, G. O. Poinar, Jr., and R. E. Berry

Evolutionary Ecology of Progeny Size in Arthropods, Charles W. Fox and Mary Ellen Czesak

Phylogenetic System and Zoogeography of the Plecoptera, Peter Zwick

From the ***Annual Review of Microbiology***, Volume 54 (2000)

Oral Microbial Communities: Biofilms, Interactions, and Genetic Systems, Paul E. Kolenbrander

Bacterial Virulence Gene Regulation: An Evolutionary Perspective, Peggy A. Cotter and Victor J. DiRita

Pathogenicity Islands and the Evolution of Microbes, Jörg Hacker and James B. Kaper

From the ***Annual Review of Phytopathology***, Volume 38 (2000)

The Ecology and Biogeography of Microorganisms on Plant Surfaces, John H. Andrews and Robin F. Harris

Genetic Diversity and Evolution of Closteroviruses, Alexander V. Karasev

Role of Horizontal Gene Transfer in the Evolution of Fungi, U. Liane Rosewich and H. Corby Kistler

Phellinus weirii and Other Native Root Pathogens as Determinants of Forest Structure and Process in Western North America, E. M. Hansen and Ellen Michaels Goheen

ANNUAL REVIEWS is a nonprofit scientific publisher established to promote the advancement of the sciences. Beginning in 1932 with the *Annual Review of Biochemistry*, the Company has pursued as its principal function the publication of high-quality, reasonably priced *Annual Review* volumes. The volumes are organized by Editors and Editorial Committees who invite qualified authors to contribute critical articles reviewing significant developments within each major discipline. The Editor-in-Chief invites those interested in serving as future Editorial Committee members to communicate directly with him. Annual Reviews is administered by a Board of Directors, whose members serve without compensation.

Annual Reviews of

Anthropology
Astronomy and Astrophysics
Biochemistry
Biomedical Engineering
Biophysics and Biomolecular Structure
Cell and Developmental Biology
Earth and Planetary Sciences
Ecology and Systematics
Energy and the Environment
Entomology
Fluid Mechanics
Genetics
Genomics and Human Genetics
Immunology
Materials Science
Medicine
Microbiology
Neuroscience
Nuclear and Particle Science
Nutrition
Pharmacology and Toxicology
Physical Chemistry
Physiology
Phytopathology
Plant Physiology and Plant Molecular Biology
Political Science
Psychology
Public Health
Sociology

SPECIAL PUBLICATIONS
Excitement and Fascination of Science, Vols. 1, 2, 3, and 4

Annu. Rev. Ecol. Syst. 2000. 31:1–7

PREFACE: A Millennial View of Ecology and Systematics, and *ARES* at Age 30

Richard F. Johnston
Natural History Museum, 602 Dyche Hall, University of Kansas, Lawrence, Kansas 66045-2454; e-mail: rfj@falcon.cc.ukans.edu

INTRODUCTION

The convention of a temporal millennium reflects, among other matters, the historical decision of the West to use a decimal arithmetic. Arbitrary and artificial as observance of a millennial ending may be, it nevertheless provides an occasion for celebrations, as well as invitations for examining the past. I am pleased to respond to an invitation to discuss some of our better efforts in past time, although social and political incompetence in the second Christian millennium also invites comment. Selected fragments of the history of biology, especially of ecology and systematics, are my focus; ultimately I aim at a paternal assessment of this *Annual Review*.

SYSTEMATICS AND ECOLOGY

Achievements in systematics and ecology occurred late in the second Christian millennium. Earlier developments were indifferent, which is true for much early science. For most of the past thousand years in the West (even currently in some intellectual deserts) the supernatural, magic, and superstition were more important than science (6, 17). Significantly, theology was long considered to be a fundamental part of science, with all other sciences generally subordinate to it (17). Nevertheless, theologians as well as scientists had difficulties as a result of, or were punished for, their work in the millennium, because novel points of view or new knowledge could contradict received wisdom and easy answers or would threaten vested interests. That modern science actually developed in the last millennium is one of the triumphs of human understanding (14). Other branches of science had some early successes, but biology was slow to develop, perhaps because it gradually included study of humans. For whatever reasons, biology was "a catastrophe" (20) and poorly pursued through most of those thousand years. We really have had only 200 years of scientific biology, or, for that matter, science in general (18). In medicine, an applied field in which the need for information, as distinguished from opinion, was of immediate practical importance, some practitioners were

0066-4162/00/1120-0001$14.00

still bleeding people for a variety of ailments as late as the nineteenth century. Perversely, some patients actually survived and, the placebo effect being unrecognized, supported the status quo.

The development of systematics required accumulation of a great deal of descriptive information about organisms of the world. Although that remains critical, the discipline could barely define itself, much less flourish, before an evolutionary biology existed. Earlier efforts at systematics were on record since at least the time of Aristotle (1), one of whose contributions was to show how difficult such study was going to be for most of the succeeding two millennia. Later, at the end of the eighteenth century, an evolutionary view of life gradually became possible through the work of Lamarck (e.g. 13). Then, after another 50 years, systematics was fundamentally reorganized by the Darwin-Wallace hypothesis of evolution by natural selection (2). The hypothesis, the Big Bang of theoretical biology, reorganized the science and made an informed systematics a necessity (e.g. 12). Such reorganization also required a change in social and political thinking, a change not yet complete at this writing (e.g. 4).

With a workable theory of evolution, the paleontologic record became understandable, and systematics gradually was transformed into a discipline with promise of becoming a science. Its continued development, however, depended on the incorporation of genetics, which required another 50 to 60 years (3, 14). In the interim, systematists split into a number of irreconcilable cliques seemingly dedicated to producing irreproducible phylogenies, programs that persisted into the twentieth century. Later, toward mid-century, Hennig (7) and Sokal & Sneath (21) showed that reproducible phylogenies could be made by using either of two, quite different, assumptions. Systematics, no longer an idiosyncratic art, was largely restructured (12), and its current condition, augmented by molecular biology (8) and electronic data processing, is more robust than at any previous time. It is orders of magnitude more complex and demanding a field of study than anyone could have imagined at the opening of the past century.

Ecology, only the faint outlines of which are detectable in ancient time (22), has no 2000-year, or even millennial, history to live down. Ecology is more complicated, or has to consider more variables, than other sciences. At least for this reason hard-core ecology developed notably late, and we were into the twentieth century before ecologists could identify their concern as scientific natural history. British ecologists formed a scientific society in 1912 and Americans in 1915. But it took another five years before the Americans announced, "...the future is in our own hands" (16). The subsequent proliferation of the formal literature of ecology has been astounding, as your library stacks attest.

Evolutionary biology was naturally introduced into ecology through the subfield of population biology, and today the whole of ecology presents an evolutionary face. Thus, the science has been good. The social ramifications of ecologic thought nevertheless (or perhaps therefore) have generated persistent political resistance. This almost wholly reflects social and political irresponsibility; the general public has been slow to comprehend the significance of conclusions framed within probability theory, conclusions that reflect the inordinately complex and multivariate nature of ecological processes (and are responsible for the canard that ecologists cannot communicate). It is nevertheless clear that ecology, to its everlasting credit, will almost certainly continue to show that easy answers, and those based upon greed, are not answers at all.

THE REVIEW

Significance of the Series

The *Annual Review of Ecology and Systematics* (*ARES*) has been with us for the last third of the last century, precisely the period in which the most important developments in ecology and systematics have occurred. Ecology and systematics are today at levels of sophistication and substantive accomplishment that were not anticipated when *ARES* was conceived in 1968. It therefore seems fair to ask whether or to what extent *ARES* has participated in this development. I must note that, as the original Editor of *ARES*, I have a conflict of interest in making this assessment.

The role assumed by *ARES* from the beginning has been largely reactive. A review, by definition, reacts, or is responsible, to the knowledge within the boundaries of its concern, tracking developments but not causing them. Perceptive tracking alone could have kept *ARES* acceptably current. But, in addition, *ARES* has always included, directly or implicitly, critical commentary resulting from input of advisory committees by what they recommend, of editors by what they include, and, ultimately, of authors by what they write. We also hoped to direct interest to those fields of study in which potential for advance seemed likely. And, because the advancement of science is largely the purview of working scientists who publish in research journals, members of that population were solicited to do the reviews.

Responsive tracking of the field and knitting up threads of independent lines of research by competent members of the guild are ample evidence of good intentions. But what we really wish to know is whether such intentions have been realized—how well has the review done its job? Aside from Volume 1, critical reviewers have not said a lot about *ARES*, and mostly they have noted that it covers an enormous field of study, worrying that specialists—a grasslands botanist, an arachnologist, a zooplankton ecologist—might find nothing of specialty interest in a given volume of *ARES*. The concern is real enough, and there is not much of an explanatory response to it. I note, however, that this problem accompanies the territory: even specialty journals regularly cover phenomena or taxa peripheral to a reader's interest or competence.

Fortunately, and independent of opinion, some quantitative measures of the value of at least part of what has appeared in *ARES* actually exist, because the Institute for Scientific Information (ISI) produces the *Science Citation Index* (SCI; e.g. 19), an annual catalog over a great range of sciences that notes which articles are cited, who cited them, and where and when they were cited. Most published papers are never cited, so one that is can be taken to be a paper of above-average quality, and its long-term citation record probably is an index to the level of that quality. These assumptions are supported by a recent analysis (5), which shows, among other matters, that authors tend to cite papers that enhance their own, thus usually selecting papers of high quality. Beyond this, ISI provides some descriptive statistics: review articles tend to be cited more frequently than articles reporting new research; the average cited item is cited about two times per year, and the average cited author about eight times. Papers three to six years old are cited most frequently, and in most years about half the citations are seven years old or less. After ten years an article has a low probability of being cited—papers 25 years old receive three quarters of one per cent of all citations, and those 30 years old one half of one per cent.

Under such a strict regime of selection, the citation record of a paper is clearly a meaningful measure of its quality and utility. I examined the SCI for 1998 (19) to see if any of the 15 reviews in Volume 1 of *ARES* were cited after 28 years and found that nearly half of them had at least one citation. Since we know that the probability of finding an unspecified 30-year-old article out of a set of 200 citations chosen at random is 0.005, the probability of finding citations to seven 28-year-old articles from a specific volume of one of the hundreds of journals examined by ISI is certainly represented by a smaller number.

Articles from other volumes of *ARES* also show substantial lifespans, and for most volumes of *ARES*, the average half-lives of cited articles exceed ten years. Long half-lives are characteristic of review articles; research reports have lesser citation lifespans. As an example, authors of articles appearing in *ARES* for 1998 deal mostly with the research literature, and they cite papers with half-lives of 7.1 years.

ARES can be accessed on the World Wide Web, so it has a record of its utility there (9). JSTOR is a nonprofit organization offering electronic access to a wide

TABLE 1 Most frequently accessed articles from *ARES* (JSTOR ranking by total viewings plus total printings as of June 28, 1999)

1. Human Population Growth and Global Land-Use/Cover Change, WL, Meyer, BL Turner II; vol. 23, 1992.
2. Global Change and Coral Reef Ecosystems, SV Smith, RW Buddemeier; vol. 23, 1992.
3. Landscape Ecology: The Effect of Pattern on Process, MG Turner; vol. 20, 1989.
4. Global Environmental Change: An Introduction, PM Vitousek; vol. 23, 1992.
5. Front Matter and Preface, the Editors and the Editorial Committee; vol. 1, 1970.
6. The Role of Disturbance in Natural Communities, WP Sousa; vol. 15, 1984.
7. Global Change and Freshwater Ecosystems, SR Carpenter, SG Fisher, NB Grimm, JF Kitchell; vol. 23, 1992.
8. Biological Invasions by Exotic Grasses, the Grass/Fire Cycle, and Global Change, CM D'Antonio, PM Vitousek; vol. 23, 1992.
9. Multivariate Analysis in Ecology and Systematics: Panacea or Pandora's Box?, FC James, CE McCulloch; vol. 21, 1990.
10. Population Viability Analysis, MS Boyce; vol. 23, 1992.

variety of scholarly journals; it has included *ARES* since January 1998. From Volumes 1 through 24, 527 articles from *ARES* have been stored and are available for viewing and/or printing. Of these, 98.1% had been viewed, and 85.2% printed, as of early summer 1999. At that time, JSTOR actually offered a total of 21,294 articles from *ARES* and four additional journals in ecology; of these, 65.8% had been viewed and 41.2% had been printed. The differences in usage of articles in *ARES* and the larger sample are probably owing to the review function of *ARES*.

The top ten articles in *ARES* (see Table 1), based on viewed plus printed scores, include six from 1992, and one each from 1970, 1984, 1989, and 1990. The 25 articles most frequently requested for viewing or printing refer to 11 different volumes.

Additional information on the point of concern can be found in the *Journal Citation Report* (*JCR;* 11), also published annually by ISI. The *JCR* computes an "impact factor" for journals (as well as the "publication half-life" noted earlier). The impact factor is a number indicating a ratio of the number of citations in the scientific literature of articles from a journal divided by the number of citable items published in that journal, over a two-year period. The impact factor is generally taken to represent the significance or clout of a publication. ISI notes that the impact factor tends to discount the advantages of a large journal over smaller ones, of frequently published journals over those of lesser frequency, and of older versus newer journals.

In ISI's category "ecology," *ARES* began recording high impact factor scores in the mid-1970s and was ranked first for a long time. Scores have remained high, and *ARES* has regularly been among the top four journals in the world in ISI's ecological impact factor. The other journals of the top four in recent years are *Trends in Ecology and Evolution* (*TREE*), *Advances in Ecological Research*, *Evolution, Wildlife Monographs*, and *Ecological Monographs*. The last three are research journals; the others are reviews but differ from *ARES* by covering only the field of ecology. *Advances* is a review annual, featuring about four long reviews per volume. *TREE* is a monthly review. The schedule permits *TREE* to secure articles of some immediacy on hot research areas and to be more proactive than other journals. One possible result of this is that *TREE* currently ranks first in impact factor in ISI's ecology category. Another result can be seen in the citation-life of articles, the average for which in *TREE* runs from three to four years; those in *Advances*, *Monographs*, and *ARES* are from eight to more than ten years. The journals are topically distinct, their preferred numbers and lengths of articles differ, and their temporal appearance varies; they provide broadly intersecting but generally different services to the scientific community.

If we examine journals that have scored in the top ten in *JCR* impact factor in ISI's ecology category in recent years, some 15 publications are included. The journals show reasonably broad niche diversification but are fairly readily identifiable as ecological: *TREE*, *Advances in Ecological Research*, *Ecology*, *Ecological Monographs*, *Journal of Animal Ecology*, *Evolutionary Ecology*, *Journal of Ecology*, *Molecular Ecology*, *Advances in Microbial Ecology*, *Ecological Applications*, and *Microbial Ecology*. The remaining four are broader in topical content: *Evolution*, *Wildlife Monographs*, *The American Naturalist*, and *ARES*. None other than

ARES regularly addresses systematics, although *The American Naturalist* and *Evolution* occasionally do so. To the point of this assessment, judging by citation-life of articles and impact factor scores, *ARES* stands in good company and clearly has played a significant role in the recent development of ecology. The assessment of course cannot be applied to systematics.

Current Developments

The short review emphasized by *TREE* has recently become important in *Science*, a weekly journal that now includes commentaries on research papers, each being a mini-review of a part of the scientific endeavor. Other journals also include reviews as regular items—*The Auk*, a specialized research journal in ornithology, already features at least one review in each quarterly number. And the current reorganization of publications of the Ecological Society of America actually includes plans for an American *TREE*. The expansion of knowledge in all biological specialties generates need for more reviews on ever more tightly focused information subsets. Users are now required to spend ever more time on journals in their specialties. That requirement must dilute readership for journals of heterogeneous scope.

Because impact factor scores have achieved widespread attention and significance, the way in which the scores are computed is obviously also of significance. Currently, the computation of factor scores for "ecological" journals is inadequate, because for *ARES*, *Evolution*, and *The American Naturalist*, to consider three publications in the ecological top ten, undefined fields (systematics, behavior, genetics, non-ecologic theory) are included in ISI's statistical manipulations for the defined field (ecology). What this can mean to impact factor scores can be judged by some history of *ARES*. In the beginning, pages for systematics in *ARES* ran at less than 30% of the total. At that time, annual impact factor scores for *ARES* in ISI's ecology were the highest. More recently, systematics has secured about half the pages in *ARES*, reflecting the increase of topical range in, as well as the remarkable quality of, recent developments in systematics. Relative to the numbers in ecology, there are fewer systematists, fewer journals devoted to systematics, and fewer citations of articles in systematics; as a consequence, *ARES* now ranks third in annual "ecological" impact factor score. Were ISI to remove non-ecological articles in computing ecological impact factor scores, a more realistic ranking of journals publishing ecological studies would result. If such a move were made, working out impact factor scores in systematics would then become a responsible endeavor for ISI.

The Future

ISI's impact factor scores lead to annual bragging rights, and publications ranking high now include their ranks in advertising circulars. This recognizes the meeting of the academy with the marketplace, which is as it should be—it surely cannot be avoided. This is one reason I hope ISI can provide some intelligence in the allocation of articles to its category of ecology. Additionally, bragging rights aside, users or potential users deserve to be informed, and current practice by ISI in treating all articles in *ARES* as ecological is misleading.

Some readers may remember that Charles Michener and I went to some lengths to ensure that *ARES* would concern both ecology and systematics. This was not because the two disciplines are interchangeable but because they are mutually supportive—the systematic (evolutionary) play occurs in the ecologic theater (10). I think the relationship is worth emphasizing even now, and it is good to see that the disciplines (which ought to be separated by ISI) are still steadily united by *ARES*.

ACKNOWLEDGMENTS

Thanks are owing to Ike Burke, NancyLee Donham, and Daphne G Fautin for help in preparing this preface.

Visit the Annual Reviews home page at www.AnnualReviews.org

LITERATURE CITED

1. Aristotle. [1984.] *The Complete Works of Aristotle: the Revised Oxford Translation.* Princeton, NJ: Princeton Univ. Press
2. Darwin C, Wallace AR. 1858. [Evolution by natural selection.] *Linnaean Soc.*, London
3. Dobzhansky T. 1937. *Genetics and the Origin of Species.* New York: Columbia Univ. Press
4. Dobzhansky T. 1962. *Mankind Evolving.* New Haven/London: Yale Univ. Press
5. Franck G. 1999. Scientific communication–a vanity fair? *Science* 286:53–55
6. Hanson RB, Bloom FE. 1999. Fending off furtive strategists. *Science* 278:1847
7. Hennig W. 1966. *Phylogenetic Systematics.* Urbana: Univ. Ill. Press
8. Hillis D, Moritz C. 1990. *Molecular Systematics.* Sunderland, MA: Sinauer
9. http://ecolsys.AnnualReviews.org/ Also: http://www.jstor.org/
10. Hutchinson GE. 1965. *The Ecological Theater and the Evolutionary Play.* New Haven, CT: Yale Univ. Press
11. Institute For Scientific Information, Inc. 1997. *Journal Citation Reports.* Philadelphia, PA: ISI Press
12. Kauffman SA. 1993. *The Origins of Order.* New York/Oxford: Oxford Univ. Press
13. Lamarck JB. 1809. *Zoological Philosophy.* ed. H Elliott, 1963. New York: Hafner
14. Mayr E. 1991. *One Long Argument.* Cambridge, MA: Harvard Univ. Press
15. Miller D. 1999. Being an absolute skeptic. *Science* 284:1625–26
16. Moore B. 1920. The scope of ecology. *Ecology* 1:3–5
17. Sarton G. 1927. *Introduction to the History of Science.* Vol. 1. Baltimore: Williams & Wilkins
18. Sarton G. 1937. *Introduction to the History of Science.* Vol. 3, Pt. 1. Baltimore: Williams & Wilkins
19. Institute for Scientific Information. 1998. *Science Citation Index.* Philadelphia, PA: ISI
20. Silver BL. 1998. *The Ascent of Science.* New York: Oxford Univ. Press
21. Sokal R, Sneath PHA. 1973. *Numerical Taxonomy.* San Francisco: Freeman
22. Theophrastus of Eresus. [1916.] *Enquiry into Plants.* Vol. 1. London: Heinemann/ New York: G. P. Putnam's Sons

Annu. Rev. Ecol. Syst. 2000. 31:9–32

The Kinship Theory of Genomic Imprinting

David Haig
Department of Organismic and Evolutionary Biology, Harvard University, 26 Oxford Street, Cambridge, Massachusetts 02138; e-mail: dhaig@oeb.harvard.edu

Key Words genetic conflict, inclusive fitness, parent-offspring conflict, kin selection, sex chromosomes

■ **Abstract** The inclusive fitness effect attributable to an allele can be divided into an effect on matrilineal kin when the allele is maternally derived and an effect on patrilineal kin when paternally derived. However, the allele is not subject to selection on its effects on patrilineal kin when maternally derived nor on its effects on matrilineal kin when paternally derived. As a result, natural selection may favor alleles with effects that differ, depending on the allele's parental origin. At autosomal loci, this process is predicted to lead to the silencing of alleles when inherited from one or the other parent. At X-linked loci subject to random X inactivation, the process is predicted to lead to quantitative differences of expression between maternal and paternal alleles but not to complete silencing of one allele. The implications of this theory and some challenges to the theory are reviewed.

INTRODUCTION

Mendel observed that the progenies of reciprocal crosses appeared identical whether a dominant character was transmitted by the seed or pollen parent (56). Although reciprocal crosses do not always yield similar progenies, most exceptions can be explained by subsidiary hypotheses (sex-linkage, cytoplasmic inheritance, apomixis, maternal effects) that do not challenge the basic hypothesis that the phenotypic expression of a gene is unchanged by the sex of the transmitting parent. The general validity of this hypothesis has been overwhelmingly supported by classical and molecular genetics. Nevertheless, the phenomenon of genomic imprinting has shown that the hypothesis is not universally true. The *GNAS1* locus on human chromosome 20, for example, shows a complex pattern of expression in which some transcripts are expressed from both copies of the locus, some transcripts are expressed only from the paternally derived allele and other transcripts only from the maternally derived allele (34). Thus, a past environment—whether a gene was present in a male or female germ line in the previous generation—can affect how the gene is expressed in the current generation. Because an allele that is maternally derived in one generation may be paternally derived in the next, the two alleles at the locus must be distinguished by some difference (an imprint) that is perpetuated through multiple cell divisions but that can be erased and reset.

The current paper addresses the question of why subtle differences between alleles passing through male and female germ lines have been elaborated by natural selection to become a mechanism of transcriptional control for some genes but not others, in some organisms but not others. The paper focuses on the hypothesis that such parent-of-origin effects are the outcome of conflicting selective forces acting on maternally derived and paternally derived alleles at loci that influence interactions among kin. This hypothesis has been known by various names (including the conflict hypothesis and tug-of-war hypothesis), of which I use the kinship theory of imprinting (85) because of the unique role the hypothesis ascribes to interactions among kin. The large literature on the nature of the marks that record parental origin and the mechanisms by which these differences affect transcription is reviewed in References 1, 5, 47, and 63 and provides examples of the expanding list of imprinted genes and parent-of-origin effects.

THE KINSHIP THEORY OF IMPRINTING

Symmetric and Asymmetric Kin

Descendants of the two alleles at a locus are potential competitors for future domination of the gene pool. Despite this divergence in long-term interests, the alleles have a common short-term interest in increasing the number of successful gametes produced by their shared individual. How then can maternally derived and paternally derived alleles be selected to express conflicting interests? An answer to this conundrum is provided by the observation that an individual may be more closely related to other individuals via his father than via his mother, or the reverse. If the individual's actions have fitness consequences for such relatives, the symmetry between the short-term interests of maternally derived and paternally derived alleles is broken.

Two different factors of one half have entered into traditional calculations of relatedness (23, 24, 85). The first arises when calculating forward from parent to offspring and reflects the random nature of meiotic segregation. The second arises when calculating backward from offspring to parent and reflects uncertainty about whether a randomly chosen allele entered a zygote in an egg or sperm. If this information were provided, the probability of one half that an allele entered the offspring via an egg would decompose into probabilities of one for maternally derived alleles and zero for paternally derived alleles (and the reverse for an allele that entered via a sperm). The forward calculation is unaffected by whether parental origin is specified, but the backward calculation is not. When one individual is related to another via a backward step, imprinting may change the coefficients of relatedness that are required in applications of Hamilton's Rule.

Consider the expression of an allele that causes a benefit (B) to the individual in whom it is expressed at a cost (C) to the individual's mother (where costs and benefits are measured as differences relative to some alternative allele). An allele with these effects would be expressed when favored by natural selection if $B - rC > 0$, where r is a measure of how costs to mothers are weighted relative to benefits to offspring. If the implicit comparison were between two alleles, both of whose effects were independent of parental origin, 50% of the time the allele would be expressed when maternally derived (in which case costs to mothers should be given equal weight to effects on self), and 50% of the time the allele would be expressed when paternally derived (in which case costs to mothers should be given zero weight). The appropriate value of r would be the average of these weights ($r = 1/2$). By contrast, if the implicit comparison were between two alleles that were expressed only when maternally derived, an allele would be without cost or benefit when paternally derived and the appropriate weight would be one, whereas the weight would be zero if the comparison were between two alleles that were expressed only when paternally derived.

Suppose instead that the cost were experienced by a maternal half-sister of the offspring's mother. The traditional value of r for such a relative is one eighth, calculated as the product of two backward steps (from offspring to mother to grandmother) and one forward step (from grandmother to aunt). However, if alleles at the locus were expressed only when maternally derived, the appropriate value of r would be one quarter, whereas if alleles were expressed only when inherited from a maternal grandmother, the appropriate value would be one half. There is no evidence as yet that grandparental origin influences gene expression, and subsequent discussion assumes that all backward steps in calculations of relatedness contribute a factor of one half, except for the initial step from offspring to parent. If so, the traditional coefficient of relatedness (r) can be viewed as an average of distinct coefficients of matrilineal (m) and patrilineal (p) relatedness: $r = (m + p)/2$.

Kin can be classified as either symmetric ($m = p$) or asymmetric ($m \neq p$). An individual's symmetric kin include herself, her direct descendants, and her full sibs. Most other relatives are asymmetric kin, including her parents, grandparents, aunts, uncles, and cousins. The kinship theory proposes that genomic imprinting has evolved as a mechanism of transcriptional control at loci whose expression has fitness consequences for asymmetric kin. The theory was first developed in the context of postzygotic maternal care (mothers are asymmetric kin of their offspring) but was subsequently generalized to all kinds of asymmetric kin. For clarity, the general theory is presented first, with discussion of the special case of parent-offspring relations postponed to a subsequent section.

Matrilineal and Patrilineal Inclusive Fitness

Haig (21) proposed a partition of an allele's inclusive fitness effect (δW) into an effect on matrilineal kin (δW_m) and an effect on patrilineal kin (δW_p):

$$\delta W = \frac{1}{2}(\delta W_m + \delta W_p) = \frac{1}{2}\left(\sum_{i=0} m_i \delta a_i + \sum_{j=0} p_j \delta b_j\right).$$

Here, δa_i is the effect on individual i (when the allele is maternally derived); δb_j is the effect on individual j (when the allele is paternally derived); and m_i and p_j are the corresponding coefficients of matrilineal and patrilineal relatedness of these individuals for individual 0 (self). In this partition, kin are divided into a matriline (individuals with $m > 0$) and a patriline (individuals with $p > 0$). Some classes of relatives may belong to both matriline and patriline (e.g., self and self's symmetric kin). Effects on mothers are given equal weight to effects on self (both $m = 1$) in calculations of δW_m. Effects on fathers are given equal weight to effects on self (both $p = 1$) in calculations of δW_p.

An allele's effects on matrilineal kin when paternally derived (δV_p) and its effects on patrilineal kin when maternally derived (δV_m) are not included in δW, but they can be used to define what one might call the allele's excluded fitness effect:

$$\delta V = \frac{1}{2}(\delta V_m + \delta V_p) = \frac{1}{2}\left(\sum_{i=0} p_i \delta a_i + \sum_{j=0} m_j \delta b_j\right).$$

On average, an allele at an unimprinted locus has the same effects on patrilineal kin when it is maternally derived as when it is paternally derived. In other words, the allele's excluded fitness effect when paternally derived (δV_p) has the same expectation as its inclusive fitness effect when maternally derived (δW_m). Substituting δV_m for δW_m yields

$$\delta W = \frac{1}{2}(\delta V_p + \delta W_p) = \sum_{j=0} \left(\frac{m_j + p_j}{2}\right) \delta b_j.$$

Therefore, natural selection at unimprinted loci acts to increase average inclusive fitness (an equivalent result can be obtained by substituting δV_m for δW_p). This provides a justification for the standard practice of using coefficients of average relatedness rather than coefficients of parent-specific relatedness, but only at unimprinted loci. An allele at an unimprinted locus will not increase in frequency if its benefit to matrilines is outweighed by its cost to patrilines, or vice versa.

Monoallelic expression uncouples inclusive and excluded fitness effects. At maternally silent loci, all δa_i are zero ($\delta V_m = \delta W_m = 0$). Therefore, natural selection acts to increase patrilineal inclusive fitness (δW_p) without regard for effects on matrilines (δV_p). An allele at a maternally silent locus can increase in frequency even if its cost to matrilines greatly exceeds its benefit to patrilines. At paternally silent loci, all δb_j are zero ($\delta V_p = \delta W_p = 0$), and natural selection acts to increase matrilineal inclusive fitness (δW_m) without regard for effects on patrilines (δV_m).

Quantitative Expression

Simple models suggest that imprinting of autosomal loci will usually be an all-or-none phenomenon. Suppose that an allele's strategy can be represented by a vector $\{x, y\}$, where x is the allele's level of expression when maternally derived and y its level of expression when paternally derived. Further suppose that each

δa_i and δb_j can be represented by a differentiable function of the total level of gene expression X. Then the kinship theory predicts that the evolutionarily stable strategy (ESS) at the locus will be either 'symmetric' or 'asymmetric' (21).

A symmetric ESS occurs when maternally derived and paternally derived alleles favor the same total level of gene expression. At such an ESS, matrilineal and patrilineal inclusive fitness would be decreased by mutant alleles that cause either small increases or decreases of expression. The ESS is described as symmetric because perturbations have the same effect on matrilineal and patrilineal inclusive fitness, not because the levels of expression of maternally derived and paternally derived alleles are necessarily equal. If both alleles favor X^*, any strategy $\{x^*, y^*\}$ for which $x^* + y^* = X^*$ is an ESS, including the unimprinted strategy $\{X^*/2, X^*/2\}$ and the imprinted strategies $\{X^*, 0\}$ and $\{0, X^*\}$. Of these, the unimprinted strategy appears the most likely to be observed in nature because it minimizes costs associated with deleterious mutations (58). However, imprinted strategies cannot be formally excluded, especially if maternally derived and paternally derived alleles have been previously subject to selection for different levels of total expression (61).

An asymmetric (or parentally antagonistic) ESS occurs when maternally derived and paternally derived alleles favor different total levels of gene expression. In the absence of imprinting, this conflict is resolved by a compromise, with the ESS level of production intermediate between the two parental optima. Small perturbations of expression in the neighborhood of the ESS would cause an increase in patrilineal inclusive fitness and a decrease in matrilineal inclusive fitness, or the reverse. In the presence of imprinting, the conflict is resolved by a fait accompli (14); the allele that favors the higher amount produces this amount and the other allele is silent (29). The strategy is stable because the silent allele cannot reduce its own production below zero. This form of conflict resolution has been called the loudest-voice-prevails principle (20). If maternally derived alleles favor a higher level of gene expression than paternally derived alleles, the paternal allele is silent. At such an ESS, small increases in gene expression would result in decreases of patrilineal and matrilineal inclusive fitness, whereas small decreases in gene expression would result in increases of patrilineal inclusive fitness but decreases of matrilineal inclusive fitness. If paternally derived alleles favor a higher level of gene expression than maternally derived alleles, the maternal allele is silent and the previous conditions are reversed (21).

Qualitative Effects

The models discussed in the previous section considered only quantitative mutations that change an allele's level of expression. However, the loudest-voice-prevails principle has important consequences for the kinds of qualitative mutations that can succeed at an imprinted locus. Suppose that it is paternally derived alleles that favor the higher level of gene product; then maternally derived alleles are predicted to be silent at the ESS. Once the established allele at a locus is silent when maternally derived, any mutation that does not reactivate maternal expression—for example, a mutation that changes the coding sequence or causes the allele to be expressed in a new tissue—is subject to selection solely on its effects on patrilineal inclusive fitness. As a consequence, qualitative mutations with strongly deleterious effects for matrilines can become fixed at a maternally silent locus. Alleles at imprinted loci will therefore tend to accumulate parentally antagonistic effects. This long-term evolutionary process will reinforce imprinted expression because it results in the pleiotropic association of traits that enhance matrilineal interests at paternally silent loci and of traits that enhance patrilineal interests at maternally silent loci.

Why So Few Imprinted Genes?

A symmetric, unimprinted ESS appears balanced on a knife-edge. If maternally derived or paternally derived alleles favor different amounts of gene product—no matter how small the difference—simple models predict an asymmetric ESS at which one allele is silent. Despite this prediction, the vast majority of genes have biallelic expression. A number of suggestions have been made as to why this should be the case.

The principal effects of most genes may be to increase or decrease the fitness of the individual in which the gene is expressed, with minimal consequences for asymmetric kin. Even if a gene has effects on asymmetric kin, these effects must be dosage-sensitive for natural selection to favor changes in expression levels. At loci where loss-of-function mutations are recessive, inactivation of one allele has little discernible effect on phenotype, and selection in favor of imprinted alleles will be weak. Therefore, few genes may have the kind of dosage-sensitive effects on asymmetric kin that would favor the evolution of imprinting. Moreover, imprinting cannot evolve if there is no variation on which to select. The paucity of imprinted genes could partially be explained if mutant alleles with parent-specific expression are rare (21).

Imprinted expression of a locus would not be expected if the selective forces favoring monoallelic expression were outweighed by countervailing costs. The most obvious cost is increased exposure to the effects of deleterious recessives when one allele is silent (69), but there may be others. Mochizuki and coworkers showed that the cost of deleterious recessives could favor biallelic expression of a fetal growth enhancer despite multiple paternity of a female's offspring (58). This is likely to be an important consideration only at loci where parentally antagonistic effects are weak because the costs of deleterious mutations at an imprinted locus are small. At equilibrium there is only one selective death for each new deleterious mutation (25).

Imprinting may be rare because of conflicts between "imprinter" genes expressed in parents and imprinted genes expressed in offspring (3). For example, genes expressed in fathers will favor lower demands on mothers than will paternally expressed genes in offspring (see below). Therefore, genes expressed in the paternal germ line might be selected to erase any gametic marks responsible for imprinted expression of paternally derived alleles in offspring.

PARENT-OFFSPRING RELATIONS

Genes Expressed in Offspring

The matrilineal and patrilineal inclusive fitness effects of an allele (expressed in offspring) that modulates offspring demands on mothers are $\delta W_m = \delta a_o + \delta a_m$ and $\delta W_p = \delta b_o + \delta b_f$, respectively. Here, δa_o is the allele's effect on offspring when maternally derived; δb_o its effect on offspring when paternally derived; δa_m its effect on the residual reproductive value (RRV) of mothers when maternally derived; and δb_f its effect on the RRV of fathers when paternally derived. All four effects are associated with parent-specific relatednesses of one. An allele's effect on mothers when paternally derived (δb_m) and its effect on fathers when maternally derived (δa_f) are associated with zero relatedness and do not appear in δW_m and δW_p. Paternally derived alleles in offspring, however, need not always be selected to maximize benefits to offspring without regard to costs to mothers because costs to mothers may be associated with correlated costs to fathers. For example, a cost to a mother's RRV will be associated with an equal cost to her partner's RRV (and vice versa) if females and males have all of their offspring with a single partner.

Most previous formulations of the kinship theory (15, 27, 29, 60) have side-stepped the complication that costs to mothers may be correlated with costs to fathers by defining the cost to a mother's RRV as a cost to the mother's other offspring. The rate of multiple paternity appeared in these formulations as a discounting factor in the patrilineal relatedness of these other offspring. In the current formulation, the rate of multiple paternity appears as a discounting factor in δb_f. The new method of accounting is more easily extended to conflicts between maternally derived and paternally derived alleles that would arise if females are monandrous but interfere with their partners' ability to sire offspring with other females (3, 49). This method would extend as well to the absence of conflict that would occur, despite frequent partner change and half-sib families, if each and every cost of parental care were shared equally by an offspring's parents (49, 67).

When maternal care imposes greater costs on the RRVs of mothers than of fathers, the kinship theory predicts that alleles at paternally expressed loci of offspring will have been selected to make greater demands on mothers than will alleles at unimprinted loci, which will have been selected to make greater demands on mothers than will alleles at maternally expressed loci of offspring (3, 15). Therefore, if females have offspring by more than one male, fetal growth enhancers are predicted to be paternally expressed and maternally silent at evolutionary equilibrium whereas fetal growth inhibitors are predicted to be maternally expressed and paternally silent (58).

Haig (20) modeled the quantitative expression of a placental hormone (secreted into the maternal bloodstream) that increased nutrient supplies for all members of the current litter at the expense of members of future litters. In this model, multiple paternity within litters and changes of paternity between litters had opposite effects on the level of hormone production. Multiple paternity within litters reduced the expression of paternally derived alleles because benefits were then shared with a larger proportion of potential freeloaders of zero patrilineal relatedness. Paternity change between litters reduced the patrilineal relatedness of future litters and thereby favored increased hormone production by members of the current litter.

Genes Expressed in Parents

Although parents are asymmetric kin of offspring, offspring are symmetric kin of parents. If the fitness effects of the previous section were caused by alleles expressed in mothers rather than offspring, $\delta W_m = \delta a_o/2 + \delta a_m$ and $\delta W_p = \delta b_o/2 + \delta b_m$ (for alleles expressed in fathers, $\delta W_m = \delta a_o/2 + \delta a_f$ and $\delta W_p = \delta b_o/2 + \delta b_f$). At unimprinted loci expressed in parents, $\delta W_m = \delta W_p$ because $\delta a_o = \delta b_o$, $\delta a_m = \delta b_m$, $\delta a_f = \delta b_f$. Therefore, loci responsible for parental care are not predicted to be imprinted (with the proviso that δW_m and δW_p may differ if they contain additional nonzero terms for asymmetric kin of the parent (25).

When genes that modulate offspring demands are expressed in parents rather than offspring, benefits to offspring are discounted by a relatedness of one half. That is, genes expressed in mothers are selected to favor a lower level of maternal investment than are maternally expressed genes in offspring (15). Similarly, genes expressed in fathers are selected to favor a lower level of maternal investment in offspring than are paternally derived genes expressed in offspring, if the father has some chance of having other offspring by the same mother. Genes of parents express different interests from genes of offspring because of asymmetric information. Once a gene finds itself in offspring the gene "knows" the outcome of one toss of the meiotic coin, but the outcome remains "unknown" for genes in the parent. Burt & Trivers suggested (3) that this difference in information can result in conflicts between imprinter genes of parents and imprinted genes of offspring.

OTHER ASYMMETRIC RELATIONS

The logic of the kinship theory applies to all interactions with asymmetric kin, not just with parents and half-sibs (23, 85). Nevertheless, the selective forces favoring imprinting are likely to be weaker when an allele's expression affects other kinds of asymmetric kin because asymmetries of relatedness are maximal for relations with a parent ($m = 1$ versus $p = 0$ for a mother, the reverse for a father) but become progressively weaker for more distant kin. Moreover, there is no selection for imprinting if an allele's effects on asymmetric kin are unbiased with respect to matrilines and patrilines. Biased effects require either direct recognition of matrilineal and patrilineal kin or an asymmetry in social relations that ensures individuals interact preferentially with one side of the family.

The different parental roles of mothers and fathers provide a reason why an allele will often have disproportionate effects on mothers compared to fathers (and on maternal half-sibs compared to paternal half-sibs) but a question remains, how could an allele discriminate in its effects between maternal and paternal first half-cousins? Two social asymmetries have been discussed in this context (23, 85). First, if the variance of reproductive success is greater for males than for females, a population will contain more paternal half-sibs and their descendants than maternal half-sibs and their descendants. Second, if there is preferential dispersal of one sex, an individual may interact predominantly with kin of the non-dispersing parent. These factors can interact to produce complex asymmetries of relatedness. For example, if male offspring disperse, female offspring remain in their natal group, and paternity within the group is dominated by a single male immigrant until he is supplanted by a new unrelated male, then an individual will often have higher patrilineal than matrilineal relatedness to members of her own age class and their offspring, but higher matrilineal than patrilineal relatedness to members of older age classes and their offspring (23).

CHALLENGES TO THE THEORY

One of the most effective ways to clarify the predictions of a theory is to show how it would explain what appears, at first sight, to be contradictory evidence. In this section, I outline some challenges to the kinship theory and how these challenges can be rebutted. Presenting a case for the defense seems preferable to maintaining a pretense of impartiality in a debate in which I have been an active participant. Although there is no reason why a single hypothesis should explain all examples of imprinting, it is desirable to minimize superfluous hypotheses and expand the explanatory domain of an already successful theory.

Diallelic Models

The models discussed above find an ESS $\{x^*, y^*\}$ from among an infinite set of alleles $\{x, y\}$ in which maternal expression x and paternal expression y are allowed to take any non-negative value (26). A different approach has been taken by Spencer and coworkers (76, 77), who presented a series of models in which there are two alleles: an unimprinted allele $\{z, z\}$ and an imprinted allele $\{0, z\}$ or $\{z, 0\}$. In their models, the level of expression z is implicitly a constant that does not evolve. Contrary to predictions of ESS models, these authors found that imprinted alleles can invade in the absence of multiple paternity, that multiple paternity has no effect on the dynamics of models with maternal-silencing, and that stable polymorphisms of imprinted and unimprinted alleles are possible.

When models have such different structures, it is hardly surprising that they make different predictions. A choice between the models' predictions therefore devolves upon which set of simplifying assumptions are deemed more relevant to the question of interest. Haig (26) argued that the diallelic models of Spencer and colleagues ignore the effects of ongoing mutation and therefore describe a process of short-term rather than long-term evolution (see 10, 30 for discussion of this distinction). From this perspective, the demonstration that an imprinted allele can displace an unimprinted allele in the absence of multiple paternity is merely a demonstration that a total level of expression z is sometimes superior to $2z$ when no constraints are placed on the evolutionary plausibility of z. Consistent with this interpretation, a dominant genetic modifier that alters expression from $\{z, z\}$ to $\{z/2, z/2\}$ can invade a population fixed for $\{z, z\}$ under the same conditions as can imprinted alleles $\{z, 0\}$ or $\{0, z\}$ if multiple paternity is absent, whereas if multiple paternity is present, the imprinted alleles can invade under a subset of conditions for which the modifier cannot invade, but the modifier can never invade under conditions for which neither imprinted allele can invade (39).

Reverse Imprinting

Some loss-of-function mutations of imprinted genes and some uniparental disomies—i.e., an individual with both copies of a chromosome derived from one parent—have phenotypes that have been interpreted as contradicting the kinship theory (40, 41). *Mash2*, for example, is a paternally silent locus that is strongly expressed in early mouse trophoblast. Mutational inactivation of the maternally derived allele results in embryonic death with major placental defects, specifically, absence of spongiotrophoblast and poor development of labyrinthine trophoblast, but excess development of trophoblast giant cells (13, 83). *Mash2* has thus been interpreted as a paternally silent enhancer of placental growth, whereas the kinship theory predicts that placental growth enhancers will be maternally silent. Similarly, paternal disomies in mice of proximal chromosome 7 and distal chromosome 17 are associated with deleterious postnatal effects that have been interpreted as contradicting the kinship theory (4).

For the most part, these criticisms appear to result from a simple misunderstanding of the nature of an asymmetric ESS. If maternally derived alleles favor a level of gene product X_m and paternally derived alleles favor a level of gene product X_p, where $X_m < X_p$, then the theory predicts that the paternal allele will produce X_p and the maternal allele will be silent at the ESS. Between X_m and X_p, changes in the level of gene product are predicted to have opposite effects on patrilineal and matrilineal inclusive fitness, but increases above X_p and decreases below X_m will be detrimental to both (21). A maternal disomy at this locus, or knockouts of the paternally derived allele, would result in zero gene product, whereas a paternal disomy would result in $2X_p$. Therefore, both kinds of perturbation would result in levels of gene expression that lie outside the zone of conflict and would be associated with phenotypes that are detrimental to both patrilineal and matrilineal inclusive fitness. Nevertheless, uniparental disomies and loss-of-function mutations can provide evidence for testing the kinship hypothesis in the clues they provide about the phenotypic effect of changes in gene expression within the zone of conflict.

Iwasa and coworkers (42, 44) have developed models to explain how the anomalous cases of *Mash2* and of paternal disomies with retarded embryonic growth can be made compatible with the kinship theory. In the case of paternal disomies, they considered expression of a locus that increased an offspring's relative allocation of resources to placental, rather than embryonic, growth. Paternally derived alleles were shown to favor greater proportional allocation to placental growth because this increased the total uptake of maternal resources. The ESS at such a locus would therefore have the form $\{0, y^*\}$, with paternal disomies producing $2y^*$, a level of production that could result in excessive allocation to the placenta at the expense of the embryo's own growth ("overshoot"). Thus, their model is a special case of the general principle that disomies will result in phenotypes that are detrimental to both matrilineal and patrilineal inclusive fitness. In the case of *Mash2*, these authors argued that preferential maternal expression of an embryonic growth enhancer would be predicted if high growth rates were associated with an increased risk of early abortion (44). Although the model is ingenious, it seems simpler to explain the phenotype of *Mash2* knockouts as an example of "overshoot" in the allocation of various cell types during placental development (21).

Imprinting Where Not Predicted

Imprinting in Oviparous Vertebrates The kinship theory posits that the principal selective force favoring the major effects of genomic imprinting on mammalian development has been conflict between the maternally and paternally derived genomes of offspring over the level of maternal investment. As a corollary, genomic imprinting is not expected to have major developmental effects in oviparous taxa because an offspring's paternally derived genome can do nothing to influence the level of maternal investment (with the caveat that post-hatching interactions ammong asymmetric kin could favor imprinting of genes affecting social behaviors). Therefore, the observation of methylation differences between maternally and paternally transmitted transgenes in zebrafish (54)—a species without postzygotic parental care—has been interpreted as inconsistent with the theory (55).

The supposed inconsistency with the kinship theory appears to be a misinterpretation of the theory's domain of explanation. In fact, the theory presupposes the existence of differences between maternal and paternal alleles; otherwise there would be nothing to select upon. What the theory does claim is that given a mechanism that causes alleles at some loci to have different levels of maternal and paternal expression—even if these differences are initially small—the cumulative processes of natural selection and new mutation will result in qualitative differences in gene expression resulting in major phenotypic effects when there is a conflict of interests between maternal and paternal genomes, but not when such conflicts are absent. Thus, the existence of differential methylation in zebrafish adds weight to the kinship theory because it provides evidence of a pre-existing mechanism for generating parent-specific differences, but only so long as differential methylation does not have major phenotypic effects. The kinship theory therefore receives support from the observation that androgenetic and gynogenetic zebrafish are phenotypically normal (8, 78). The adaptive function of methylation and why it should differ between male and female germlines are important questions upon which the kinship theory is silent.

Imprinting in Monogamous Species In the context of parent-offspring relations, the kinship theory predicts that the "interests" of maternally derived and paternally derived alleles are identical when all costs to a mother's RRV are associated with an equal cost to the father's RRV, and vice versa. This would be the case, for example, if individuals of both sexes were constrained to have all of their offspring with a single partner. Therefore, there would be no selective force favoring the origin of imprinted expression in a species with strict lifetime monogamy. For this reason, Hurst (38, 41) argued that the existence of imprinting in "monogamous" *Peromyscus polionotus* (88) and in predominantly self-fertilizing *Arabidopsis thaliana* (70) adds to accumulating evidence against the kinship theory.

The kinship theory can parry this thrust in two ways. The first is to question whether *P. polionotus* and *A. thaliana* are truly monogamous (25). The rate of partner change between successive litters of *P. polionotus* is substantial (20% in one study; 11) and the rate of outcrossing in *A. thaliana* probably exceeds the per locus mutation rate (74). Thus, serial monogamy in *P. polionotus* is consistent with continuing selection for imprinting (25), and the same may be true of the mating system of *A. thaliana*, although this case is less strong. The second is to note that if maternal and paternal alleles agree on the same level of combined gene expression X^*, the unimprinted strategy $\{X^*/2, X^*/2\}$ is the midpoint on a continuum of possible symmetric ESSs from $\{0, X^*\}$ to $\{X^*, 0\}$. If a locus evolved imprinted expression because maternal and paternal alleles previously favored different levels of expression—but the mating system changed so that maternal and paternal alleles favor the same level—there are many ways to adjust gene expression to achieve the new consensus that do not involve the loss of imprinting (61).

For the above reasons, the kinship theory does not predict a rapid loss of imprinting when a species' mating system shifts toward greater monogamy. However, the theory does predict that the shift will result in reduced expression of paternally expressed genes and reduced conflict costs. Thus, the kinship theory is supported by the observation that in both *P. polionotus and A. thaliana* the growth-promoting effects of paternal alleles appear attenuated relative to related taxa with a higher incidence of partner change (70, 88).

Imprinting of Genes Affecting Maternal Behavior Natural selection for imprinted expression is not expected at loci whose effects are limited to symmetric kin. Offspring are symmetric kin of their mothers (i.e., a mother's maternally derived and paternally derived alleles are equally likely to be transmitted to each of her offspring). Therefore, the kinship theory does not predict imprinting of loci affecting maternal behavior unless maternal care has fitness consequences for other (asymmetric) kin of the mother (such as the female's own mother or her matrilineal half-sisters). However, null mutations of two paternally expressed loci in mice result in impaired maternal care (48, 50). The implications for the kinship theory are ambiguous because both null mutations also cause prenatal growth retardation. Imprinting of these loci could therefore be explained by their effects on growth, without requiring a separate explanation for the parent-of-origin effects on maternal behavior. The more interesting possibility is that the promotion of prenatal growth and postnatal maternal behavior are pleiotropically associated because both serve patrilineal interests. If so, this would imply that increased care for a female mouse's own offspring has occurred at the expense of investment in other matrilineal kin (25).

Other Hypotheses

Alternative explanations for the evolution of imprinting have continued to proliferate since the reviews of Haig & Trivers (28) and Hurst (37), but I do not attempt a new review here. Rather, I limit myself to some brief comments on the ovarian time-bomb hypothesis (87) because this is perhaps the most commonly cited alternative to the kinship theory, and the minimization of variance hypothesis (37) because this has been claimed to explain many of the same phenomena as the kinship theory.

Varmuza & Mann proposed (87) that genomic imprinting is an adaptation to protect eutherian females from the development of invasive trophoblast in ovarian germ cell tumors (87; see 19, 59, 75 for critiques). My previous argument that this hypothesis could not explain the existence of paternally silent growth inhibitors (19, 28) has been shown to be fallacious (44). Nevertheless, I believe that the kinship theory—with its emphasis on the conflicting interests of maternal and paternal genomes—provides a more satisfying general explanation for the evolution of imprinting because it can explain many more phenomena, including the imprinting of genes affecting seed development where there is no risk of germ cell tumors (29, 30). Moreover, theories based on genetic conflicts can explain why trophoblast is often invasive (16), whereas this is merely accepted without explanation in Varmuza & Mann's hypothesis.

Hurst & McVean (41, p. 702) claimed that most of the features of imprinting that are explained by the kinship theory are also explained by a model in which "imprinting is an adaptation to control growth rates in embryos in which the uptake of resources is continuous and flexible over time." This appears to be Hurst's (40) proposal that imprinting is a means whereby cooperative offspring minimize the variance in resource extraction from their mother. The central assumption of this hypothesis is that a lower variance in the rate of transcription can be achieved with monoallelic expression than with biallelic expression. This assumption is questionable. One could argue instead that coefficients of variation will be lower in a biallelic system because stochastic processes occur independently at the two loci (6).

There is an important sense in which neither of these hypotheses—nor any other hypothesis that attempts to explain the origin of imprinting—is an alternative to the kinship theory. An allele's effects when paternally derived are subject to selection solely on their consequences for patrilines, whereas an allele's effects when maternally derived are subject to selection solely on their consequences for matrilines, whether or not its locus is imprinted, and whether or not other factors have played a role in the origin of imprinting. Therefore, if imprinted expression evolves—for whatever reason—the logic of the kinship theory will apply if an allele's expression has consequences for asymmetric kin. Trophoblast growth in Varmuza &

Mann's hypothesis and resource extraction in Hurst's hypothesis clearly have consequences for at least one asymmetric relative (an offspring's mother). Therefore, formal models of these hypotheses will need to take account of the different selective forces acting on alleles of maternal and paternal origin.

SEX CHROMOSOMES

X-Linked Relatedness

Mammalian fathers transmit an X chromosome to their daughters but a Y chromosome to their sons. Thus, an offspring's sex reveals the outcome of meiosis for sex-linked loci of males: forward steps from fathers to daughters (and backward steps from sons to mothers) contribute a factor of one to coefficients of X-linked relatedness, whereas forward steps from fathers to sons (and backward steps from sons to fathers) contribute a factor of zero. As a consequence, coefficients of relatedness will differ for autosomal and X-linked loci if individuals are related via a father-to-offspring link in which the offspring's sex is specified, and any individual that is related to another via a chain that contains a father-to-son link is a nonrelative from the perspective of X-linked loci. As a corollary, X-chromosomal matrilines and patrilines contain fewer individuals than autosomal matrilines and patrilines, but some of the included individuals have higher X-linked than autosomal relatedness. For example, at an autosomal locus, patrilineal relatedness is one half for a paternal half-sib of either sex, whereas at an X-linked locus of a female, patrilineal relatedness is one for a paternal half-sister (all of a father's daughters receive an identical X chromosome) but zero for a paternal half-brother.

Imprinting at X-linked loci will be restricted to females because males lack paternally derived alleles (the distinction between symmetric and asymmetric kin is meaningless for X-linked loci of males). The kinship theory predicts that natural selection favors imprinting when expression of alleles at a locus has fitness consequences for kin with different degrees of matrilineal and patrilineal relatedness at that locus. Symmetric kin at autosomal loci may be asymmetric kin at X-linked loci, or the reverse (the latter requires some degree of inbreeding). For example, a female's full-sibs and their descendants are symmetric autosomal kin, but they are asymmetric X-chromosomal kin whenever sibs are distinguished by sex. That is, a female's full-sisters have higher patrilineal than matrilineal relatedness at X-linked loci ($m = 1/2, p = 1$), whereas her full-brothers are matrilineal kin but patrilineal non-kin ($m = 1/2; p = 0$). Maternally silent X-linked alleles are therefore predicted to favor full-sisters without regard for costs to brothers (3, 23), whereas paternally silent X-linked alleles are predicted to oppose these effects.

X-Linked Inclusive Fitness

Over the course of several generations, an average X-linked allele spends one third of its time as a maternally derived allele in males, one third as a maternally

TABLE 1 Invasion criteria for new alleles at an X-linked locus

Type of X-linked locus	Expression[a]	Mutant alleles favored if
Biallelic expression, not sex-limited	$\{x, y, z\}$	$\delta W_x + \delta W_y + \delta W_z > 0$
Biallelic expression, female-limited	$\{0, y, z\}$	$\delta W_x + \delta W_z > 0$
Male-limited	$\{x, 0, 0\}$	$\delta W_y > 0$
Paternally silent	$\{x, y, 0\}$	$\delta W_x + \delta W_y > 0$
Maternally silent	$\{0, 0, z\}$	$\delta W_z > 0$
Paternally silent and female-limited	$\{0, y, 0\}$	$\delta W_x > 0$

[a]An allele's strategy is represented by the triplet $\{x, y, z\}$, where x is the allele's level of expression when maternally derived in males; y, its expression when maternally derived in females; and z, its expression when paternally derived in females.

derived allele in females, and one third as a paternally derived allele in females. Therefore, the allele's inclusive fitness effect, δW, will be an equally weighted sum of its effects in each of these circumstances (δW_x, δW_y, δW_z respectively);

$$\delta W = \frac{1}{3}(\delta W_x + \delta W_y + \delta W_z) = \frac{1}{3}\left(\sum_{i=0} m_i \delta a_i + \sum_{j=0} m_j \delta b_j + \sum_{k=0} p_k \delta c_k\right)$$

where m_i, m_j, p_k are the appropriate coefficients of X linked relatedness.

Conditions favorable to the invasion of a new allele are summarized in Table 1 for different patterns of expression at an X-linked locus. Monoallelic and/or sex-limited expression reduce one or more of δW_x, δW_y, δW_z to zero. At paternally silent loci, δW_z is zero. Therefore, natural selection favors alleles that increase matrilineal inclusive fitness ($\delta W_x + \delta W_y > 0$) without regard for effects on patrilines. As a corollary, experimental or mutational reactivation of paternally silent loci is predicted to be particularly detrimental to patrilines. At maternally silent loci, δW_x and δW_y are zero. Natural selection favors alleles that increase patrilineal inclusive fitness of females ($\delta W_z > 0$) without regard for effects on matrilines.

Haplodiploidy and Other X-Linked Genomes

The entire genome of haplodiploid taxa is effectively X-linked: haploid males develop from unfertilized eggs and lack paternally derived alleles; diploid females develop from fertilized eggs and receive alleles from both parents. These genetic systems thus provide a "natural experiment" in which the selective forces acting on X chromosomes are not masked by selection acting on autosomes.

Hamilton proposed that the repeated evolution of nonreproductive helpers in the haplodiploid Hymenoptera was a consequence of increased relatedness among sisters: daughters of a singly mated female share three quarters of their genes by descent, rather than half, because all of their haploid father's sperm carry an identical genome (31). The boost to relatedness among sisters caused by haplodiploidy (and X-linkage) results solely from increased patrilineal relatedness; $r = 3/4$ is an average of $m = 1/2$ and $p = 1$ (14). From this perspective, the daughters of a singly mated female are a matrilineal sibship but a patrilineal clone. Thus, the kinship theory predicts that paternally derived alleles will be under stronger selection than maternally derived alleles to promote behaviors that benefit sisters if single-mating is common. If a female mates with many males, most of her daughters will be maternal half-sibs, and the relatedness asymmetries are reversed; it is maternally derived alleles that are more strongly predisposed to favor sisters. Recent evidence for the existence of imprinting in a parasitoid wasp (9) strengthens the possibility that imprinted genes will also be found to play an important role in the control of hymenopteran social behaviors (14, 22).

Many of the kinship theory's predictions about X-linked loci and haplodiploidy also apply to parahaplodiploid systems of paternal genome loss that occur in coccoid scale insects and sciarid flies, among other taxa (35). In these groups, males develop from fertilized eggs, but only a male's maternally derived alleles are transmitted to his offspring. The similarity to haplodiploidy is particularly close for scale insects because the paternal genome of males is inactivated or eliminated during early development (64). Therefore, paternally derived alleles of males are expected to have minimal phenotypic effects. In sciarid flies, on the other hand, paternally derived alleles of males are expressed (57) and are therefore subject to selection on their effects on patrilineal kin. Paternally derived alleles of males might therefore be selected to promote reproduction by sisters in their own sibship at the expense of the males' own reproduction. This possibility does not arise in most sciarids because females produce offspring of a single sex only, either all sons or all daughters (12).

The elimination of the paternal genome in scale insects and sciarid flies provides a dramatic demonstration that maternally derived and paternally derived genomes are not equivalent in these taxa, but this form of imprinting does not directly compare with the locus-specific imprinting of mammals in which alleles of both parental origins are transmitted to offspring. I have argued that these genetic systems have evolved as the outcome of a system of meiotic drive in which parental origin marks one set of chromosomes for elimination (17, 18).

X Inactivation

Two patterns of X inactivation occur in female mammals. Paternal X inactivation is observed in the somatic cells of female marsupials (7) and in trophoblast and yolk sac of mice (66, 82). Random X inactivation occurs in the somatic cells of female eutherians and (probably) in human trophoblast (51); only a single X is active in any given cell, but the paternal and maternal X are active in different cells of the same female (52). This section discusses some of the implications of X inactivation for the kinship theory and briefly comments on possible implications of the theory for understanding the evolution of paternal X inactivation.

Paternal X Inactivation Natural selection at paternally silent loci favors alleles that increase matrilineal inclusive fitness without regard for effects on patrilines (Table 1). Therefore, X-linked genes expressed in female marsupials and in trophoblast and yolk sac of female mice are predicted to evolve strong biases in favor of matrilineal interests. In the specific context of maternal-embryo relations in mice, reactivation of the inactive paternal X chromosome is predicted to retard embryonic growth.

Evidence that murine X chromosomes harbor inhibitors of placental growth comes from a recent knockout of *Xist* (53). When a disabled copy of *Xist* was inherited maternally, female offspring were viable because the paternal X chromosome (with an intact copy of *Xist*) was inactivated in extraembryonic membranes. By contrast, there was a profound failure of placental development, associated with early death, when an embryo's paternal copy of *Xist* was disabled, presumably because both copies of the X chromosome remained active in placental tissues. The lethal effect could be ascribed to the increased number of active X chromosomes (rather than to the expression of imprinted genes on the paternal X) because an XO mouse that inherited a disabled paternal copy of *Xist* was viable.

Moore and colleagues (60, 62) have suggested that paternal X inactivation may be the outcome of an evolutionary conflict between maternal and paternal interests. In this view, inactivation of the paternal X was initiated by maternally derived genes because the paternal X of an ancestral mammal carried imprinted growth enhancers that benefited patrilines at the expense of matrilines. An alternative hypothesis would view the evolution of paternal X inactivation as a response to a matrilineal bias in the effects of X-linked genes. In this scenario, paternally derived alleles would have gained an advantage from shutting down their own chromosome because ancestral X chromosomes carried unimprinted growth inhibitors that benefited the matriline at the expense of the patriline.

Why might unimprinted X-linked loci have possessed a matrilineal bias in their effects? At an autosomal locus with biallelic expression in both sexes, a new allele cannot increase in frequency if its costs when present in females outweigh its benefits when present in males (or vice versa), nor can it increase in frequency if its costs to matrilineal kin when maternally derived exceed its benefits to patrilineal kin when paternally derived (or vice versa). Therefore, evolution at autosomal loci is not expected to systematically favor females over males nor matrilines over patrilines. By contrast, at an X-linked locus with biallelic expression in females and hemizygous expression in males, a new allele's effects when present in females are given twice the weight of its effects when present in males, and its effects on matrilines when maternally derived are given twice the weight of its effects on patrilines when paternally derived. Therefore, allelic substitutions at such loci will tend to favor females at the expense of males and matrilines at the expense of patrilines.

Random X Inactivation At an autosomal locus, maternally derived and paternally derived alleles are expressed in the same cells and contribute their gene products to a single pool. Whichever allele favors the higher level of gene product is predicted to produce its favored amount with the other allele silent. By contrast, at a locus subject to random X inactivation, maternally derived and paternally derived alleles contribute their gene products to different pools: Each allele can seemingly produce its favored amount in different cells. Thus, the logic of the loudest-voice-prevails principle does not apply at imprinted loci subject to random X inactivation (particularly at loci with cell-autonomous effects). Therefore, if both alleles are expressed, but at different levels, imprinting may be more difficult to detect at X-linked loci than at autosomal loci. Moreover, if both alleles are active, "qualitative" mutations at imprinted X-linked loci will be subject to natural selection on their effects for both matrilines and patrilines.

Sex-Specific Expression Hypothesis

Paternally derived alleles at X-linked loci are restricted to females. Therefore, it has been proposed that there will be selection for imprints on the paternal X which are specifically favorable to females (45); that genomic imprinting may function as a mechanism of haplodiploid sex determination (68); and that imprinting of X-linked loci may be a hormone-independent mechanism of achieving sexual dimorphism during mammalian development (43, 72).

Could imprinting be favored at X-linked loci independently of effects on asymmetric kin? There is no reason in principle why not. Indeed, the presence of a paternally derived genome appears to determine female development in *Nasonia vitripennis* (9). Once imprinting has evolved at an X-linked locus, for whatever reason, the selective forces acting on an allele's effects when paternally derived will be female-specific and specific to patrilineal kin; both sets of forces must be considered in evolutionary models, and their relative importance becomes an empirical question.

Iwasa & Pomiankowski (43) have argued that the pattern of X-linked imprinting in humans and mice contradicts predictions of the kinship theory. They proposed instead that imprinting has evolved to control sex-specific gene expression in early embryos before gonadal sex determination. Two lines of evidence are claimed to be inconsistent with the kinship theory. First, experimental data from mice show that X^p inhibits embryonic growth relative to X^m (superscripts refer to the maternal or paternal origin of the X chromosome). Second, X^p0 humans have greater social skills than X^m0 humans (73).

The claim that X^p inhibits embryonic growth relative to X^m is based on the observation that X^mY and X^m0 embryos are larger at 10.5 days post coitum than X^mX^p embryos, which, in turn, are larger than X^p0 embryos (84). In comparisons of X^p0 with X^m0 embryos and X^mX^p with X^m0 embryos, X^p is indeed associated with poorer growth. However, in comparisons of $X^mX^mX^p$ with X^mX^p embryos (71) and X^mX^mY embryos with X^mX^pY embryos (79), X^m is associated with poorer growth. A plausible interpretation of these data is that trophoblast develops very poorly if its only X chromosome is a normally inactive X^p (46), but that the presence of two active X^m chromosomes also inhibits development of trophoblast (80). The slow growth of X^mX^p relative to X^m0 embryos appears to be an effect of having two (rather than one) active X chromosomes during early development (2, 81). These interpretations are consistent with predictions of the kinship theory that the effects of unimprinted loci on the X chromosome will show a matrilineal bias and that X-linked loci expressed in tissues with paternal X inactivation will have been selected to favor matrilineal interests without regard for costs to patrilines.

For Iwasa & Pomiankowski (43), the observed differences between X^p0 and X^m0 humans suggested sexual dimorphism in the adaptive value of social skills (43). These authors did not discuss effects on kin, but if their hypothesis is to be made maximally distinct from the kinship theory, such sex differences would be reflected solely in the individual fitness component of inclusive fitness. From the perspective of the kinship theory, the observed differences would be interpreted as evidence that the expression of social skills has had fitness implications for asymmetric kin of females. I suspect that not enough is currently known about the context of human social evolution, nor about the properties of imprinted loci on the X chromosome, to make predictions that discriminate between the hypotheses.

On a final note, marsupials show substantial sexual differentiation before gonadal sex determination (65, 86), yet inactivation of the paternal X has the effect that the single active X of females is maternally derived (as is the single X of males)—not what one would expect if the principal function of X-linked imprinting were to achieve sex-specific expression. A similar argument applies to paternal X inactivation in mouse trophoblast.

Y Chromosomes

X and Y chromosomes segregate at male meiosis I and have opposite patterns of inheritance with respect to the sex of a male's offspring. Genes on the Y chromosome are transmitted to all of his sons but none of his daughters, and from these sons to their sons, and so on, in an unbroken chain of male-to-male transmission. Because Y-linked genes are restricted to males and are always paternally derived, they are expected to favor the growth of their own embryo at the expense of the mother (36) and to favor the growth of full-brothers at the expense of full-sisters and maternal half-siblings of either sex (85).

PROSPECTS

During the life of an individual organism, gene expression responds adaptively to information from the organism's internal and external environment. These responses are not limited to effects of the contemporaneous environment but include effects of past environments. If such evolved responses are possible within a generation, there seems no reason in principle why they could not also occur between generations. Until recently, however, it was generally believed that each individual starts life with a blank slate; historical (epigenetic) information was not transferred between generations. Somatic cells, it was believed, might retain information acquired during the current generation, but either germ cells were quarantined from such effects or all information was wiped clean during gametogenesis.

Genomic imprinting conclusively demonstrates that at least one bit of information (maternal versus paternal origin) can be transmitted epigenetically from one generation to the next. If so, could not other useful information be similarly transmitted? For example, the optimal allocation of resources in good times between fat stores and growth might differ for individuals in environments with, on average, a famine every two generations versus a famine every ten generations. If past famines could leave an epigenetic trace in the germ line, these modifications could adaptively modulate gene expression in the current generation. Many similar examples can be envisioned.

The initial selective advantage favoring the evolution of contingent responses to information from past environments need have nothing to do with effects on asymmetric kin, but there are at least two reasons why such information would not be symmetrically transmitted by both sexes. First, males and females may have different information. The nondispersing sex, for example, would have better information about local conditions. Second, male and female germ lines are so different that it seems unlikely that identical DNA modifications would occur in both. Because of these profound differences between the biochemical environments of male and female germ lines—one actively dividing in adult life, the other arrested at mid-meiosis since early development—parental origin is perhaps the simplest information that could be transmitted to the next generation. Future work will show whether it is an exception to a general rule that information is not transmitted or is just one piece of information among many.

ACKNOWLEDGMENTS

Austin Burt, Glenn Herrick, Rolf Ohlsson, Jon Seger, and Robert Trivers have all made helpful comments on the manuscript.

LITERATURE CITED

1. Bartolomei MS, Tilghman SM. 1997. Genomic imprinting in mammals. *Annu. Rev. Genet.* 31:493–525
2. Burgoyne PS, Thornhill AR, Boudrean SK, Darling SM, Bishop CE, Evans EP. 1995. The genetic basis of XX-XY differences present before gonadal sex differentiation in the mouse. *Philos. Trans. R. Soc. London B* 350:253–61
3. Burt A, Trivers R. 1998. Genetic conflicts in genomic imprinting. *Proc. R. Soc. London B* 265:2393–97
4. Cattanach BM, Barr JA, Evans EP, Burtenshaw M, Beechey CV, et al. 1992. A candidate mouse model for Prader-Willi syndrome which shows an absence of Snrpn expression. *Nat. Genet.* 2:270–74
5. Constância M, Pickard B, Kelsey G, Reik

W. 1998. Imprinting mechanisms. *Genome Res.* 8:881–900
6. Cook DL, Gerber AN, Tapscott SJ. 1998. Modeling stochastic gene expression: implications for haploinsufficiency. *Proc. Natl. Acad. Sci. USA* 95:15641–46
7. Cooper DW, Johnston PG, Watson JM, Graves JAM. 1993. X-inactivation in marsupials and monotremes. *Semin. Devel. Biol.* 4:117–28
8. Corley-Smith GE, Lim CJ, Brandhorst BP. 1996. Production of androgenetic zebrafish (*Danio rerio*). *Genetics* 142:1265–76
9. Dobson S, Tanouye M. 1996. The paternal sex ratio chromosome induces chromosome loss independently of *Wolbachia* in the wasp *Nasonia vitripennis*. *Dev. Genes Evol.* 206:207–17
10. Eshel I. 1996. On the changing concept of evolutionary population stability as a reflection of a changing point of view in the quantitative theory of evolution. *J. Math. Biol.* 34:485–510
11. Foltz DW. 1981. Genetic evidence for long-term monogamy in a small rodent, *Peromyscus polionotus*. *Am. Nat.* 117:665–75
12. Gerbi SA. 1986. Unusual chromosome movements in sciarid flies. In *Germ Line – Soma Differentiation, Results and Problems in Cell Differentiation*, ed. W Hennig, 13:71–104. Berlin: Springer Verlag
13. Guillemot F, Caspary T, Tilghman SM, Copeland NG, Gilbert DJ, et al. 1995. Genomic imprinting of *Mash2* required for trophoblast development. *Nat. Genet.* 9:235–41
14. Haig D. 1992. Intragenomic conflict and the evolution of eusociality. *J. Theor. Biol.* 156:401–3
15. Haig D. 1992. Genomic imprinting and the theory of parent-offspring conflict. *Semin. Devel. Biol.* 3:153–60
16. Haig D. 1993. Genetic conflicts in human pregnancy. *Q. Rev. Biol.* 68:495–532
17. Haig D. 1993. The evolution of unusual chromosomal systems in sciarid flies: intragenomic conflict and the sex ratio. *J. Evol. Biol.* 6:249–61
18. Haig D. 1993. The evolution of unusual chromosomal systems in coccoids: extraordinary sex ratios revisited. *J. Evol. Biol.* 6:69–77
19. Haig D. 1994. Refusing the ovarian time bomb. *Trends Genet.* 10:346–47
20. Haig D. 1996. Placental hormones, genomic imprinting, and maternal-fetal communication. *J. Evol. Biol.* 9:357–80
21. Haig D. 1997. Parental antagonism, relatedness asymmetries, and genomic imprinting. *Proc. R. Soc. London B* 264:1657–62
22. Haig D. 1998. Mother's boy or daddy's girl? Sex determination in Hymenoptera. *Trends Ecol. Evol.* 13:380–81
23. Haig D. 1999. Genomic imprinting, sex-biased dispersal, and social behavior. *Ann. NY Acad. Sci.* 907:149–63
24. Haig D. 1999. Asymmetric relations: internal conflicts and the horror of incest. *Evol. Hum. Behav.* 20:83–98
25. Haig D. 1999. Genomic imprinting and the private life of *Peromyscus polionotus*. *Nat. Genet.* 22:131
26. Haig D. 1999. Multiple paternity and genomic imprinting. *Genetics* 151:1229–31
27. Haig D, Graham C. 1991. Genomic imprinting and the strange case of the insulin-like growth factor-II receptor. *Cell* 64:1045–46
28. Haig D, Trivers R. 1995. The evolution of parental imprinting: a review of hypotheses. See Ref. 65a, pp. 17–28
29. Haig D, Westoby M. 1989. Parent-specific gene expression and the triploid endosperm. *Am. Nat.* 134:147–55
30. Haig D, Westoby M. 1991. Genomic imprinting in endosperm: its effects on seed development in crosses between species and between different ploidies of the same species, and its implications for the evolution of apomixis. *Philos. Trans. R. Soc. London B* 333:1–13
31. Hamilton WD. 1964. The genetical

evolution of social behaviour. II. *J. Theor. Biol.* 7:17–52
32. Hamilton WD. 1972. Altruism and related phenomena, mainly in social insects. *Annu. Rev. Ecol. Syst.* 3:193–232
33. Hammerstein P. 1996. Darwinian adaptation, population genetics and the streetcar theory of evolution. *J. Math. Biol.* 34:511–32
34. Hayward BE, Moran V, Strain L, Bonthron DT. 1998. Bidirectional imprinting of a single gene: *GNAS1* encodes maternally, paternally, and biallelically derived proteins. *Proc. Natl. Acad. Sci. USA* 95:15475–80
35. Herrick G, Seger J. 1999. Imprinting and paternal genome elimination in insects. See Ref. 65a, pp. 41–71
36. Hurst LD. 1994. Embryonic growth and the evolution of the mammalian Y chromosome. I. The Y as an attractor for selfish growth factors. *Heredity* 73:223–32
37. Hurst LD. 1997. Evolutionary theories of genomic imprinting. In *Genomic Imprinting*, ed. W Reik, A Surani, pp. 211–237. Oxford, UK: Oxford Univ. Press
38. Hurst LD. 1998. *Peromysci*, promiscuity and imprinting. *Nat. Genet.* 20:315–16
39. Hurst LD. 1999. Is multiple paternity necessary for the evolution of genomic imprinting? *Genetics* 153:509–12
40. Hurst LD, McVean GT. 1997. Growth effects of uniparental disomies and the conflict theory of genomic imprinting. *Trends Genet.* 13:436–43
41. Hurst LD, McVean GT. 1998. Do we understand the evolution of genomic imprinting? *Curr. Opin. Genet. Dev.* 8:701–8
42. Iwasa Y. 1998. The conflict theory of genomic imprinting: How much can be explained? *Curr. Top. Devel. Biol.* 40:255–93
43. Iwasa Y, Pomiankowski A. 1999. Sex specific X chromosome expression caused by genomic imprinting. *J. Theor. Biol.* 197:487–95
44. Iwasa Y, Mochizuki A, Takeda Y. 1999. The evolution of genomic imprinting: abortion and overshoot explain aberrations. *Evol. Ecol. Res.* 1:129–50
45. Jablonka E, Lamb MJ. 1990. The evolution of heteromorphic sex chromosomes. *Biol. Rev.* 65:249–76
46. Jamieson RV, Tan SS, Tam PPL. 1998. Retarded postimplantation development of X0 mouse embryos: impact of the parental origin of the monosomic X chromosome. *Dev. Biol.* 201:13–25
47. Latham KE. 1999. Epigenetic modification and imprinting of the mammalian genome during development. *Curr. Top. Dev. Biol.* 43:1–49
48. Lefebvre L, Viville S, Barton SC, Ishino F, Keverne EB, Surani MA. 1998. Abnormal maternal behaviour and growth retardation associated with loss of the imprinted gene *Mest. Nat. Genet.* 20:163–69
49. Lessels CM, Parker GA. 1999. Parent-offspring conflict: the full-sib–half-sib fallacy. *Proc. R. Soc. London B* 266:1637–43
50. Li LL, Keverne EB, Aparicio SA, Ishino F, Barton SC, Surani MA. 1999. Regulation of maternal behavior and offspring growth by paternally expressed *Peg3*. *Science* 284:330–33
51. Looijenga LHJ, Gillis AJM, Verkerk AJMH, van Putten WLJ, Oosterhuis JW. 1999. Heterogenous X inactivation in trophoblastic cells of human full-term female placentas. *Am. J. Hum. Genet.* 64:1445–52
52. Lyon MF. 1999. Imprinting and X-chromosome inactivation. See Ref. 65a, pp. 73–90
53. Marahrens Y, Panning B, Dausman J, Strauss W, Jaenisch R. 1997. *Xist*-deficient mice are defective in dosage compensation but not spermatogenesis. *Genes Dev.* 11:156–166
54. Martin CC, McGowan R. 1995. Parent-of-origin specific effects on the methylation of a transgene in the zebrafish, *Danio rerio*. *Dev. Genet.* 17:233–39
55. McGowan R, Martin CC. 1997. DNA methylation and genome imprinting in the

zebrafish, *Danio rerio*: some evolutionary ramifications. *Biochem. Cell Biol.* 75:499–506
56. Mendel G. 1909 [1865]. Experiments in plant hybridisation. In *Mendel's Principles of Heredity*, transl. W Bateson. Cambridge, UK: Cambridge Univ. Press. (From German)
57. Metz CW. 1938. Chromosome behavior, inheritance and sex determination in *Sciara*. *Am. Nat.* 72:485–520
58. Mochizuki A, Takeda Y, Iwasa Y. 1996. The evolution of genomic imprinting. *Genetics* 144:1283–95
59. Moore T. 1994. Refusing the ovarian time bomb. *Trends Genet.* 10:347–48
60. Moore T, Haig D. 1991. Genomic imprinting in mammalian development: a parental tug-of-war. *Trends Genet.* 7:45–49
61. Moore T, Mills W. 1999. Imprinting and monogamy. *Nat. Genet.* 22:130–31
62. Moore T, Hurst LD, Reik W. 1995. Genetic conflict and evolution of mammalian X-chromosome inactivation. *Dev. Genet.* 17:206–11
63. Morison IM, Reeve AE. 1998. A catalogue of imprinted genes and parent-of-origin effects in humans and animals. *Hum. Mol. Genet.* 7:1599–1609
64. Nur U. 1989. Reproductive biology and genetics. In *Armoured Scale Insects, Their Biology, Natural Enemies and Control*, ed. D Rosen, A:A179–90. Amsterdam: Elsevier
65. O WS, Short RV, Renfree MB, Shaw G. 1988. Primary genetic control of somatic sexual differentiation in a mammal. *Nature* 331:716–17
65a. Ohlsson R, Hall K, Ritzen M. 1995. *Genomic Imprinting: Causes and Consequences*. Cambridge, UK: Cambridge Univ. Press
66. Papaioannou VE, West JD. 1981. Relationship between the parental origin of the X chromosomes, embryonic cell lineage and X chromosome expression in mice. *Genet. Res.* 37:183–97
67. Parker GA. 1985. Models of parent-offspring conflict. V. Effects of the behaviour of the two parents. *Animal Behav.* 33:519–33
68. Poirié M, Périquet G, Beukeboom L. 1992. The hymenopteran way of determining sex. *Semin. Dev. Biol.* 3:357–61
69. Sapienza C. 1989. Genome imprinting and dominance modification. *Ann. NY Acad. Sci.* 564:24–38
70. Scott RJ, Spielman M, Bailey J, Dickinson HG. 1998. Parent-of-origin effects on seed development in *Arabidopsis thaliana*. *Development* 125:3329–41
71. Shao C, Takagi N. 1990. An extra maternally derived X chromosome is deleterious to early mouse development. *Development* 110:969–75
72. Skuse DH. 1999. Genomic imprinting of the X chromosome: a novel mechanism for the evolution of sexual dimorphism. *J. Lab. Clin. Med.* 133:23–32
73. Skuse DH, James RS, Bishop DVM, Coppin B, Dalton P, et al. 1997. Evidence from Turner's syndrome of an imprinted X-linked locus affecting cognitive function. *Nature* 387:705–8
74. Snape JW, Lawrence MJ. 1971. The breeding system of *Arabidopsis thaliana*. *Heredity* 27:299–302
75. Solter D. 1994. Refusing the ovarian time bomb. *Trends Genet.* 10:346
76. Spencer HG, Clark AG, Feldman MW. 1999. Genetic conflicts and the evolutionary origin of genomic imprinting. *Trends Ecol. Evol.* 14:197–201
77. Spencer HG, Feldman MW, Clark AG. 1998. Genetic conflicts, multiple paternity and the evolution of genomic imprinting. *Genetics* 148:893–904
78. Streisinger G, Walker C, Dower N, Knauber D, Singer F. 1981. Production of clones of homozygous diploid zebra fish (*Brachydanio rerio*). *Nature* 291:293–96
79. Tada T, Takagi N, Adler ID. 1993. Parental imprinting on the mouse X chromosome: effects on the early development of

XO, XXY and XXX embryos. *Genet. Res.* 62:139–48
80. Takagi N. 1991. Abnormal X-chromosome dosage compensation as a possible cause of early developmental failure in mice. *Dev. Growth Different.* 33:429–35
81. Takagi N, Abe K. 1990. Detrimental effects of two active X chromosomes on early mouse development. *Development* 109:189–201
82. Takagi N, Sasaki M. 1975. Preferential inactivation of the paternally derived X chromosome in the extraembryonic membranes of the mouse. *Nature* 256:640–42
83. Tanaka M, Gertsenstein M, Rossant J, Nagy A. 1997. *Mash2* acts cell autonomously in mouse spongiotrophoblast development. *Dev. Biol.* 190:55–65
84. Thornhill AR, Burgoyne PS. 1993. A paternally imprinted X chromosome retards the development of the early mouse embryo. *Development* 118:171–74
85. Trivers R, Burt A. 1999. Kinship and genomic imprinting. See Ref. 65a, pp. 1–21
86. van der Schoot P, Payne AP, Kersten W. 1999. Sex difference in target seeking behavior of developing cremaster muscles and the resulting first visible sign of somatic sexual differentiation in marsupial mammals. *Anat. Rec.* 255:130–41
87. Varmuza S, Mann M. 1994. Genomic imprinting—defusing the ovarian time bomb. *Trends Genet.* 10:118–23
88. Vrana PB, Guan XJ, Ingram RS, Tilghman SM. 1998. Genomic imprinting is disrupted in interspecific *Peromyscus* hybrids. *Nature Genet.* 20:362–65
89. Whitney G. 1976. Genetic substrates for the initial evolution of human sociality. I. Sex chromosome mechanisms. *Am. Nat.* 110:867–75

Annu. Rev. Ecol. Syst. 2000. 31:33–59

Cenozoic Mammalian Herbivores from the Americas: Reconstructing Ancient Diets and Terrestrial Communities

Bruce J. MacFadden
Florida Museum of Natural History, University of Florida, Gainesville, Florida 32611; e-mail: bmacfadd@flmnh.ufl.edu

Key Words browsers, grazers, isotopes, teeth, morphology, paleoclimate

■ **Abstract** Herbivory first evolved in terrestrial mammals during the late Cretaceous, ~100 million years ago (Mya). Of the ~35 ordinal-level clades of extinct or extant eutherian mammals from the New World, ~24 have been adapted to herbivory in one form or another. Dental adaptations for specialized terrestrial browsing are first recognized during the early Cenozoic (Paleocene-Eocene). Mammalian herbivores adapted for grazing did not become widespread in the New World until the middle Cenozoic; it seems that this adaptation and the spread of grasslands occurred during the late Oligocene (30 Mya) in South America ~10 million years earlier than in North America (20 Mya). Carbon isotopic evidence from fossil herbivore teeth indicates that C3 plants predominated until the late Miocene (~8 Mya). Thereafter, C3 and C4 terrestrial communities diversified. Late Pleistocene extinctions ~10,000 years ago decimated the diversity of mammalian herbivores, particularly those of larger body size.

INTRODUCTION

As primary consumers of plant biomass, herbivores represent the majority of diversity in ancient mammalian radiations. The fossil record of mammalian herbivores in North and South America is relatively well represented over the past 65 million years (My). During this time there have been considerable changes in climate and plant diversity that affected the structure and distribution of mammalian herbivore communities. In the past several decades, some important factors have influenced our understanding of, and our ability to reconstruct, ancient mammalian herbivore communities. Paleontological discoveries continuously improve our knowledge of the fossil record and oftentimes fill in critical gaps. New techniques, such as analyses of stable isotopes and enamel microwear, have advanced our ability to make paleodietary interpretations. Continuous refinements in dating techniques allow a better understanding of the time sequence of mammalian herbivore community evolution.

0066-4162/00/1120-0033$14.00

The fossil record reveals a time dimension not available to modern ecologists. Paleontologists can track discrete communities through millions of years during which basic community structure is preserved but new taxa originate and fill ecological niches vacated by taxa that became extinct. This is called a chronofauna, a term originally proposed by Olson (64, 65).

This paper reviews the fossil record of terrestrial mammalian herbivore communities over the past 65 my, the Cenozoic Era in North and South America. Whenever possible, emphasis is placed on recent discoveries and new techniques that enhance understanding of this subject. The Cenozoic was a time of great global change, and it is somewhat artificial to devote this review to the New World, when parallel faunal changes occurred in the Old World. However, this focus is determined by the available space; the interested reader can also consult previous reviews on this general subject (31, 36, 94, 95).

BACKGROUND AND METHODS

In 1873, the Russian paleontologist Kowalevsky published a classic paper (38) describing fossil horses from Europe. He asserted that the evolution of horses with short-crowned teeth to those with high-crowned teeth during the Miocene indicated a corresponding change in diet from browsing to grazing. Since that time, the tooth crown height of fossil herbivores has been used to interpret the diets of ancient mammals (Figure 1). Within the past few decades, other techniques have added independent evidence to an understanding of ancient herbivore diets. Complementing gross tooth morphology, studies of skull morphology, and enamel microwear, stable carbon isotopes now can be used to reconstruct Cenozoic herbivore communities.

Morphology

Modern herbivorous mammals with short-crowned (brachyodont) teeth, e.g., the tapir (*Tapirus*) or deer (*Odocoileus*), are generally adapted to feeding on soft, leafy vegetation and hence are browsers. In contrast, modern herbivorous mammals with high-crowned teeth, e.g., the zebra (*Equus*) and bison (*Bison*), are generally grazers. High-crowned is defined as unworn premolar or molar teeth in which the height exceeds the occlusal length of the tooth (Figure 2*A*). High-crowned teeth are either of finite growth, like those of horses, or ever-growing (hypselodont) during the animal's lifetime, like those of some grazing rodents. The adaptive significance of high-crowned teeth is, in most cases, related to the abundance of phytoliths in grasses. Phytoliths are microscopic bodies of silica (SiO_2, the same compound as glass) within grasses that tend to wear teeth down and are an adaptation of the plant against herbivory (59). There are, however, some exceptions to this general short-crowned browser/high-crowned grazer pattern. For example, short-crowned llamas (*Lama*) are principally grazers (29, 52), whereas some extinct high-crowned

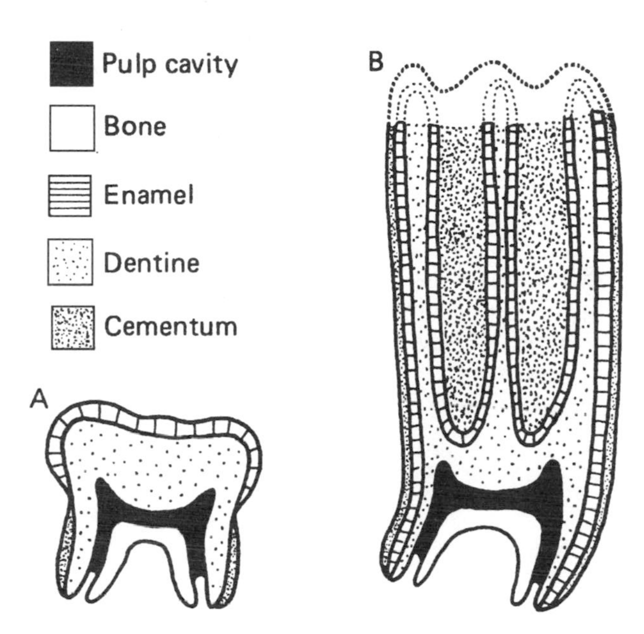

Figure 1 Comparison of short-crowned (brachyodont) human tooth (*a*) versus high-crowned horse tooth (*b*) showing expansion of the crown relative to the root area. From Ref. 33 and reproduced with permission of Cambridge University Press.

horses were principally browsers (53). Nevertheless, tooth morphology serves as a general model to interpret extinct herbivore diets.

There also is a correlation between high-crowned teeth and open-country habitats. Thus, some high-crowned herbivores may incorporate considerable amounts of grit into their diets from dust on the plant foods that they eat close to the land surface (29). Terrestrial grazers are generally considered to include animals that feed predominantly (>90%) on grasses (29), although a species is also a grazer if it crops plants other than grasses (e.g., forbs) that form the low ground cover in some biomes.

Certain cranial characters are also highly correlated to diet in extant herbivorous mammals. For example, the muzzle shape and incisor width of modern ungulates distinguish browsers from grazers (Figure 2). Grazers tend to have more transversely straight muzzles and incisors of generally similar size, which together form a functional cropping mechanism to procure grass and other plants near the ground surface. In contrast, browsers have more rounded muzzles and differentiated incisors, which together form a cropping mechanism for selective feeding from trees and shrubs (32, 84). These differences are also apparent in fossil herbivores, e.g., Oligocene notoungulates in South America (78) and Miocene sympatric horse species in North America (47). Other characters, like the depth of the jaw, development of the bony masseter prominence on the cheek, and position of the orbit, also indicate browsing versus grazing diets (85).

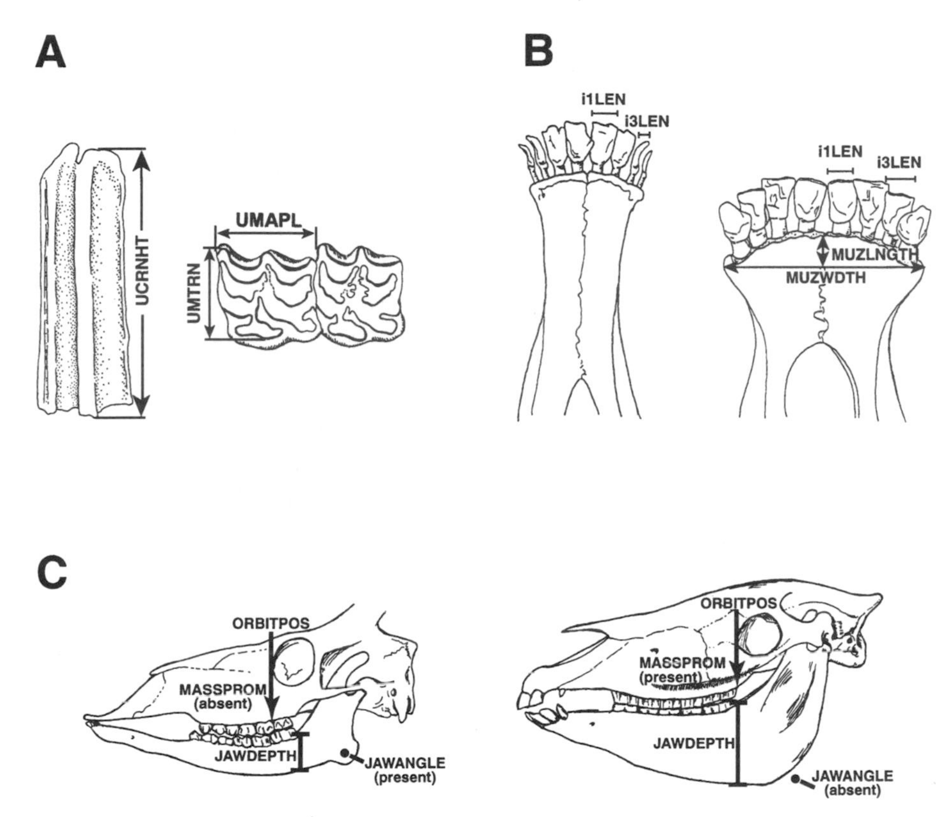

Figure 2 Dental and skull characters used to assess browsing versus grazing. (A) crown height, or hypsodonty index (HI) is the ratio of unworn crown height (UCRNHT) to molar anteroposterior length (UMAPL); in this case the HI is ~2.6, indicating a grazer. (B) Relative incisor lengths (i1LEN and i3LEN) and muzzle width (MUZWDTH) indicate a browser (narrow, left) or a grazer (broad, right). (C) The position of the orbit (ORBITPOS), presence or absence of the masseter prominence (MASSPROM), and depth and posterior angle of the jaw (JAWDEPTH, JAWANGLE) indicate a browser (left) or a grazer (right). Modified from Ref. 52 and reproduced with permission of the Paleontological Society.

The mastication of various foodstuffs imparts distinctive microscopic wear patterns to the tooth enamel of modern herbivorous mammals with known diets (83, 86). Using modern analogs, extinct herbivore diets can be interpreted using dental microwear. These enamel-microwear patterns can be quantified by determining the proportion of scratches versus pits on the enamel occlusal surface. Thus the enamel of a browser like the black-fronted duiker (*Cephalophus niger*) will have a large number of pits, which are caused by percussing leaves, in contrast to a grazer like the bison (*Bison bison*), which has many scratches caused by chewing abrasive grasses (Figure 3).

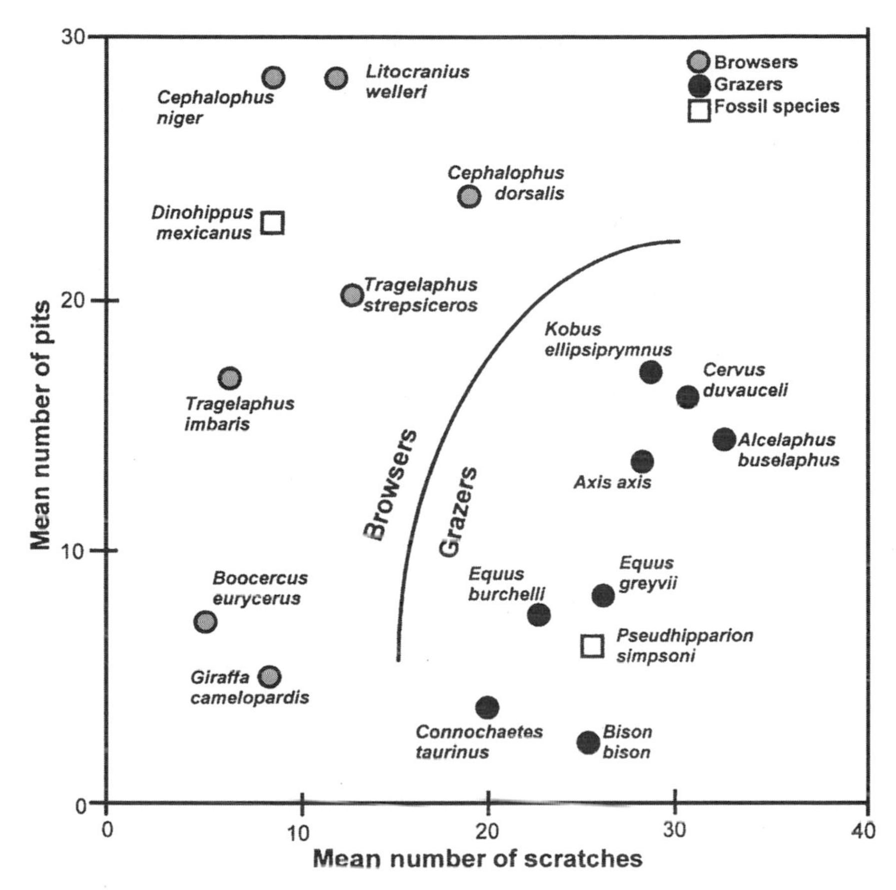

Figure 3 Enamel microwear studies the proportion of pits versus scratches per unit area (in this case 0.5 mm^2) for modern mammalian herbivores with known diets. Browsers have a high proportion of pits, whereas grazers have a high proportion of scratches. These modern analogs are used as a model to interpret the diets of fossil mammalian herbivores. Of the late Miocene fossil horses depicted here, *Dinohippus mexicanus* is a browser and *Pseudhipparion simpsoni* is a grazer. Modified from Ref. 53 and reproduced with permission of the AAAS.

Carbon Isotopes

Stable carbon isotopes have recently been used to interpret the diets and ecology of fossil mammals. For extinct herbivores, analysis of tooth enamel carbonate indicates whether the animal was feeding on C3 plants, C4 grasses, or a mixture of these foodstuffs (e.g., 37, 41, 52, 74). This technique can be used to discriminate diets because plants photosynthesize carbon by two principal pathways. The dominant photosynthetic pathway for terrestrial plants is the Calvin cycle, in which carbon is first incorporated into 3-carbon compounds, hence the term C3. About 85% of terrestrial plant biomass, including trees, most shrubs, forbs, and high-latitude or high-elevation grasses, use the Calvin cycle. In contrast, about 10% of terrestrial plant biomass photosynthesizes carbon using the Hatch-Slack cycle, in which carbon is first incorporated into 4-carbon compounds, hence C4. (The third photosynthetic pathway, CAM, or Crassulacean acid metabolism, is used by a small fraction of terrestrial plant biomass such as succulents. It is of minor relevance in the current study.) Present-day tropical and temperate grasses are predominantly C4, and they are adapted to more highly seasonal and arid climates than are C3 plants (17, 18). Not only do C3 and C4 photosynthetic pathways incorporate carbon into different compounds, they also fractionate the stable isotopes of carbon (^{12}C and ^{13}C) in different proportions. The ratio of $^{13}C/^{12}C$ is conventionally expressed as

$$\delta^{13}C(\text{in parts per mil, ‰}) = (R_{sample}/(R_{standard} - 1) \times 1000,$$

where $R = {}^{13}C/^{12}C$. All measurements of an unknown (such as fossil-tooth enamel) are compared to the standard PDB (a Cretaceous marine mollusk PeeDee belemnite, which has a $\delta^{13}C$ of 0 ‰). C3 plants characteristically have $\delta^{13}C$ values ranging from −34 ‰ to −23 ‰, with a mean of −27 ‰; C4 plants characteristically have $\delta^{13}C$ values ranging from −17 ‰ to −9 ‰, with a mean of −13 ‰ (3, 16, 19). When mammals eat plants, the $\delta^{13}C$ is enriched in their skeletal tissues, so the $\delta^{13}C$ of tooth enamel is ~12–14 ‰ more positive than the corresponding plant foods (8, 37; Figure 4).

The use of carbon isotopes for dietary and community reconstructions of mammalian herbivores is best suited to the late Cenozoic, after the diversification of C3 and C4 plant communities ~8 Mya. Prior to this time, most of the dietary reconstructions presented below rely principally on morphological interpretations.

TERRESTRIAL HERBIVORE PHYLOGENY, ORIGINS, AND DISTRIBUTION

Herbivory is a widespread adaptation in the history of mammals. Of the ~35 ordinal-level clades of extinct and extant mammals from the New World, two thirds contain herbivorous species and half are exclusively herbivores (33; Figure 5). These clades include such diverse specializations as folivores, frugivores, granivores, succulent feeders, mixed-plant feeders, and grazers.

During the late Triassic through early Cretaceous (225 to 120 Mya), all mammals were small-bodied (<5 kg) and lacked the distinctive dental adaptations that indicate plant-eating. Although some of these primitive mammals probably were omnivores, with some percentage of plants in their diets, it is difficult to make meaningful dietary interpretations of most Mesozoic mammals (12).

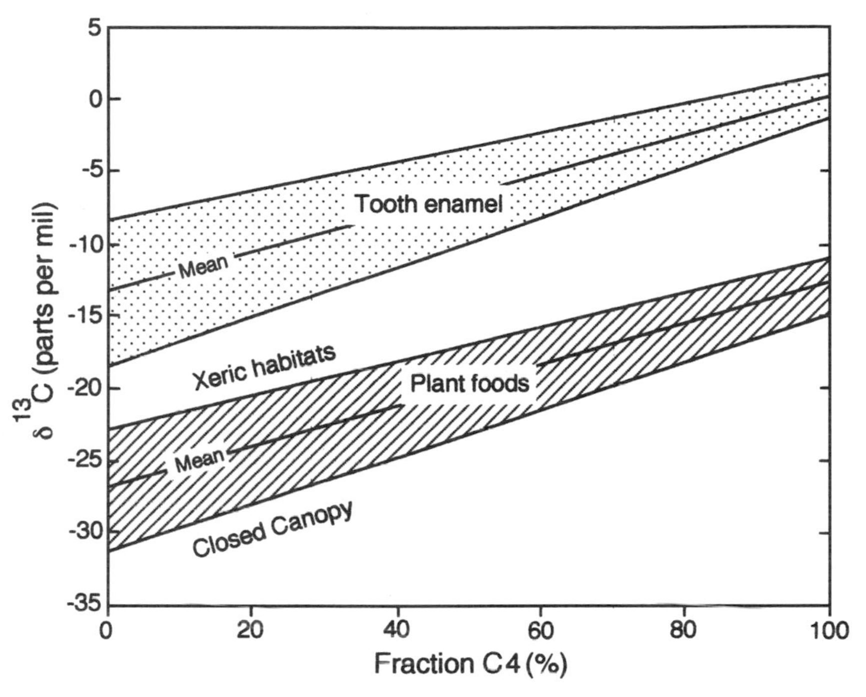

Figure 4 Carbon isotopic values, expressed in δ^{13} C_{PDB} of end-member C3 plants (left side, bottom = 0% C4) and C4 grasses (right side, bottom = 100% C4) as compared to δ^{13}C values of tooth enamel of C3 feeders (left side, top) versus C4 grazers (right side, top). Relative to plant foods, tooth enamel carbonate is enriched by about 12–14 ‰ (8, 37).

Eutherian dental evolution is marked by a key innovation during the early Cretaceous, ~100 Mya, the tribosphenic tooth (4, 69). Prior to this time, mammalian teeth were relatively simple and probably were principally sectorial (adapted for cutting), with a lesser crushing or grinding function. The tribosphenic tooth is characterized by an internal principal cusp, the protocone, in the uppers, which occludes with a talonid basin in the lowers. The evolution of the protocone and talonid basin, the function of which has been likened to that of a mortar and pestle, resulted in a morphologically complex tooth with increased surface area for mastication.

Some of the eutherian mammal clades that later developed highly specialized adaptations for herbivory originated during the late Cretaceous, 100 to 65 Mya. In North America during the late Cretaceous, some eutherian "ungulate" and condylarth clades had dentitions that were apparently adapted for masticating fibrous plant foods of low nutritive value and/or seeds and nuts of higher energy (1). The fossil record of Mesozoic mammals in South America is relatively fragmentary (11, 13), although some notable recent discoveries have been made (68). Given the overall paucity of well-preserved material, little can be said about the dental

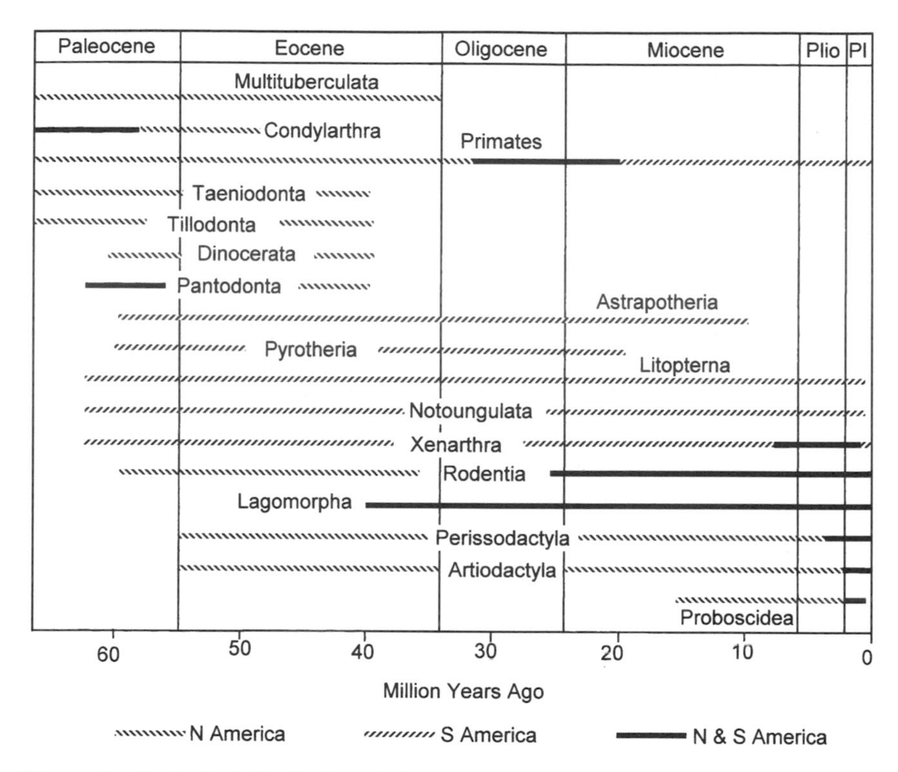

Figure 5 Geological distribution of the major ordinal-level groups of terrestrial mammalian herbivores from the Cenozoic of the Americas. Other groups with a low percentage of herbivores (Marsupialia, Chiroptera) or of low diversity that are not discussed in text (Xenungulata) are not shown here. Compiled from numerous sources, e.g. Refs. 11, 34, 67, 68.

function and diet of South American Mesozoic mammals. Nevertheless, in the New World (i.e., mostly North America), we have evidence of the primitive mammal clades that later developed specialized herbivorous adaptations.

EARLY TERTIARY MAMMALIAN HERBIVORES AND CLIMATE

The first half of the Cenozoic, represented by the Paleocene and Eocene epochs (65 to 34 Mya), represents a time of major global change. Stable isotopic evidence from both marine and terrestrial sequences at the Paleocene/Eocene boundary (~55 Mya), indicates there was a relatively abrupt global warming event resulting in much warmer "hothouse" conditions during the Eocene (37, 71). The mean annual temperature in low and middle latitudes during the early Eocene is estimated to have been between 25°C and 30°C (Figure 6), ~15°C to 20°C warmer

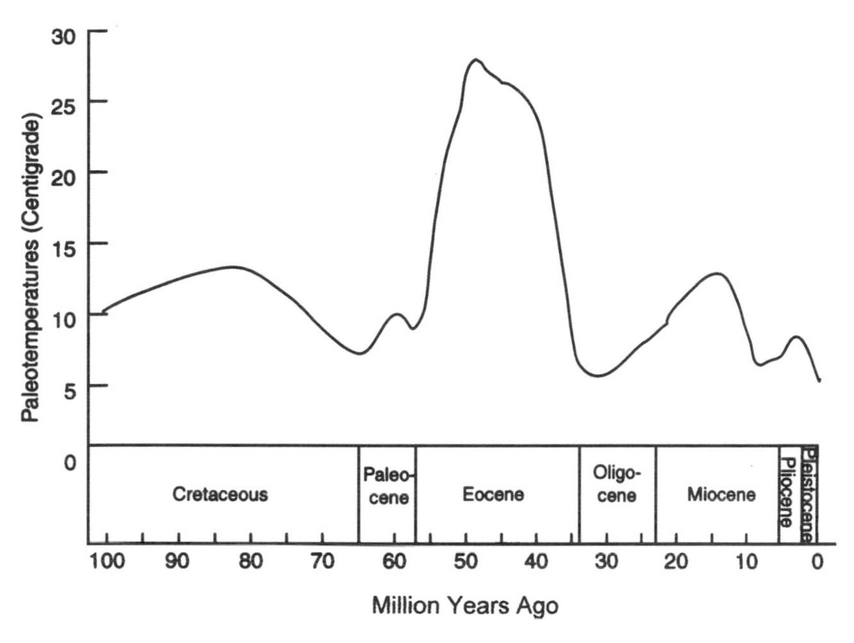

Figure 6 Change in mean annual temperature over the past 100 My as reconstructed from deep-sea oxygen isotope and paleobotanical data. From Ref. 63 and reproduced with permission of the Missouri Botanical Garden.

than today. This was the most significant global warming event during the Cenozoic, and it had a profound effect on the terrestrial biota. For example, in the well-sampled Paleocene/Eocene sequence in northern Wyoming, plant macrofossil diversity doubled (25 to 50 recognizable species) between 55 and 53 Mya (100; Figure 7). Thus, the fossil record indicates a dramatic shift from Paleocene mesic woodlands to Eocene subtropical vegetation throughout much of North America. The presence of cold-intolerant turtles, tortoises, crocodylians, and primates throughout North America, including at extreme northern latitudes (e.g., in the Canadian Arctic at 80° N; 24), indicates global expansion of subtropical and some temperate belts during this time (71). Although grasses originated during the early Cenozoic (the oldest reported fossil grass is from the Paleoocene/Eocene of Tennessee, 14), there were no extensive grasslands then as there are today.

By the Paleocene, faunal interchange between the Americas, as represented by the pan-American groups Condylarthra and Pantodonta, had ended. Herbivore diversification thereafter until the late Cenozoic was represented by separate adaptive radiations of endemic groups in South and North America. The isolation of South America during most of the Cenozoic gives us a unique opportunity to compare radiations of mammalian herbivores that originated from different clades and to observe parallel evolution of adaptations. Similarly, the reconnection of the

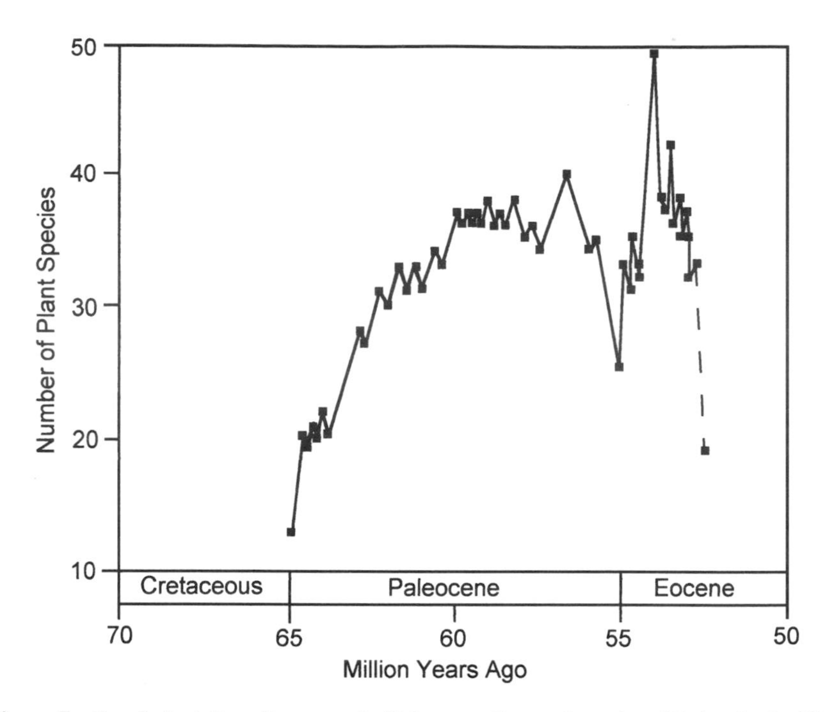

Figure 7 Fossil plant diversity across the Paleocene-Eocene boundary. Notice the doubling in the early Paleocene and subsequent drop from 40 to 25 species within about 2 My (from 57 to 55 Mya). From Ref. 100 and reproduced with permission of Elsevier.

Americas at the end of the Cenozoic had a profound impact on the native biotas. These "natural evolutionary experiments" are discussed below.

Morphological evidence from both North and South American mammalian faunas during the Paleocene and Eocene reveals considerable body size diversification (particularly increase; 11) and dental specializations. Dental morphology of numerous clades indicates a complex tooth apparently specialized for herbivory, including increased tooth area for mastication of plant foodstuffs. In the smaller range of body sizes (<5 kg), the abundance of forested biomes during this time yielded a correspondingly diverse array of arboreally adapted herbivores, including folivores, frugivores, and granivores (71).

In the New World, many of the modern groups of terrestrial herbivores originated and/or radiated during this time. The orders include Xenarthra (edentates), Lagomorph, Rodentia, Perissodactyla, and Artiodactyla (Figure 5). Now-extinct orders of New World terrestrial herbivorous mammals alive at that time included the Tillodontia, Taeniodontia, Condylarthra, Dinocerata, Astrapotheria, Pyrotheria, Pantodonta, and non-therian Multituberculata. With a few exceptions, all of these Paleocene and Eocene herbivores had relatively brachyodont dentitions, indicating a principally browsing diet. They exploited the niches created in open, nonarboreal, habitats. Taeniodonts and tillodonts from North America were high-crowned exceptions. Their diets have been reconstructed as having contained a considerable proportion of underground roots and tubers. They probably contended with grit on their foods, which explains the hypsodont teeth (42, 43).

The earliest horse *Hyracotherium* ("eohippus"; family Equidae) is known from the early Eocene of Holarctica. Because of its widespread abundance, and its importance near the base of both the equid and perissodactyl diversifications, *Hyracotherium* is an instructive example of an early Cenozoic mammalian herbivore. Traditionally, because of its short-crowned dentition, *Hyracotherium* was considered a forest-dwelling browser (e.g., 62, 80, 81). However, an extraordinary population of 24 individuals from a single quarry in Colorado is interpreted to represent an open-country form, which fed in early savanna-like, woodland mosaics (22). But instead of grass, this horse may have fed on soft ground vegetation, including herbaceous dicots and dry-adapted ferns, as well as some low shrubs (e.g. hackberry—*Celtis*). Fossilized brain endocasts of *Hyracotherium* indicate a relatively advanced neocortex, suggesting increased tactile sensitivity of the lips for selective feeding (75). Dental microwear studies of *Hyracotherium* show mostly pits, indicating a mixed-browsing diet, the animals having fed primarily on fruits, bushes, seeds, ferns, and other leaves (5). The mean $\delta^{13}C$ of *Hyracotherium* teeth is -12.5 ‰, indicating a diet of C3 plants (91).

The early diversification of the orders Perissodactyla and Artiodactyla occurred during the Eocene. By the Oligocene, ~34 to 30 Mya, more advanced artiodactyls diversified, particularly within the ruminants. It has been suggested that ruminant digestion provided a competitive advantage relative to hind-gut perissodactyl fermentation. Perissodactyls remained of modest diversity while the artiodactyls increased dramatically in diversity later during the Cenozoic, and some workers (81) have suggested a causal interdependence between these two orders. It seems, however, that this is an oversimplification and actually climatic and vegetative changes during this time are also causative factors in the relative diversities of the Perissodactyla and Artiodactyla (10, 30).

In South America, the fossil record of herbivorous mammals is predominantly represented by brachyodont forms, indicating principally browsing (70). An exception, the Archaeohyracidae, a family of small-bodied notoungulates, had ever-growing teeth. This may have been an early (Eocene) shift to grazing (82), an adaptation that greatly expanded during the later Cenozoic. Another group, the polydolopid marsupials, was rodent-like, an adaptation also seen in the multituberculates in North America during the Paleocene and Eocene (31, 67, 68).

EOCENE GREENHOUSE TO OLIGOCENE ICEHOUSE

For almost a century paleontologists have recognized a major change in Eocene to Oligocene mammalian faunas from Europe, termed the Grand Coupure ("big cut") (71, 73). A series of climatic events (including the Grand Coupure) occurred during a period of ~7 My from ~39 to 32 Mya (71, 73). Climatic reconstructions indicate significant global change from the peak Eocene "greenhouse" conditions to early Oligocene colder "icehouse" conditions. Temperatures are estimated to have dropped ~20°C from 50 to 35 Mya (Figure 6). Profound faunal and floral changes, including extinctions, originations, and geographic range shifts, during the late Eocene and into the early Oligocene occurred in such diverse groups as plant macrofossils, marine microfossils, marine invertebrates, amphibians, reptiles, and land mammals. Recent analysis of the ecology across the Eocene/Oligocene boundary indicates a change from principally warm, humid forest types during the Eocene to more zonal climates during the Oligocene that included arid, colder, and more open savanna-like habitats (71, 73).

The shift from greenhouse to icehouse conditions coincided with sweeping changes in New World herbivore communities. Archaic groups such as the condylarths, primitive primates, tillodonts, taeniodonts, and uintatheres (Dinocerata) became extinct by the end of the Eocene. As global climate shifted toward more glacial conditions, sea levels dropped, providing opportunities for intercontinental dispersal routes, such as the Bering land bridge connecting Holarctica. Immigrant herbivores from the Old World that first appeared during the Eocene in North America include lagomorphs, rhinoceroses, and several clades of artiodactyls (e.g., camels). With the possible exception of the lagomorphs, the other groups of mammalian herbivores had short-crowned teeth and were presumably browsers (31).

Some of the best early Tertiary mammal faunas of South American endemic groups are known from classic early and middle Eocene localities in Argentina, ~53 to 45 Mya (21). These faunas include xenarthrans, notoungulates, and litopterns, which together represent the majority of the herbivore diversity on that continent at that time (several other ordinal-level groups that also existed in South America during this time are either poorly represented by fossils or were of low diversity and enigmatic phylogenetic relationship). Before the late Eocene, 80% of the herbivorous mammals were short-crowned and thus are interpreted to have been browsers, and 10% were either partially or fully high-crowned and thus are interpreted either to have been early grazers or to have fed on gritty foods (67, 70). Fossil phytoliths first become common in the late Eocene of South America, suggesting the spread of early grassland biomes (27, 87). As we will see below, the adaptive radiation of high-crowned grazers and presumed spread of grasslands occurred at different times in North and South America.

PRECOCIOUS HYPSODONTY AND GRAZERS IN SOUTH AMERICA

Patterson & Pascual (70, also 67) noticed a fundamental change in dental morphology and presumed diets of middle Cenozoic mammalian herbivores in South America. The early and middle Eocene (45 Mya) herbivores are principally short-crowned, presumed to be browsers. In younger faunas of late Eocene/early Oligocene "Tinguirican" (~35 Mya; 101) and late Oligocene/early Miocene

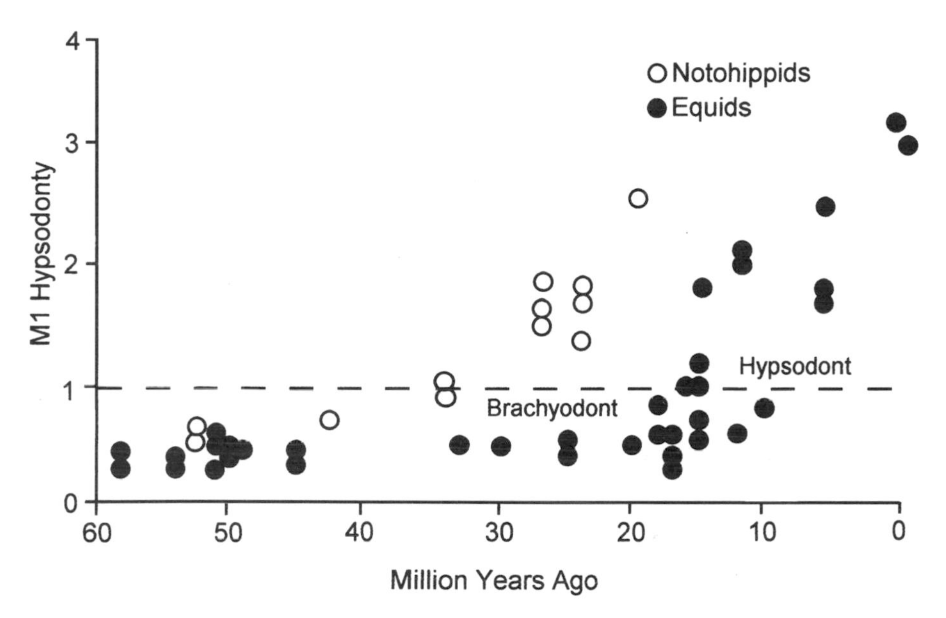

Figure 8 Advent of "precociously hypsodont" South American notohippid notoungulates at ~25 to 30 Mya compared to hypsodonty in North American horses at ~15 to 20 Mya. Modified from Ref. 78 and reproduced with permission of the Society of Vertebrate Paleontology.

Deseadan (~30 to 20 Mya) ages, the dominant small- to medium-sized herbivores are represented by the Notoungulata, which include a striking diversity of high-crowned forms—with hypsodonty indices between 1.5 and 2 (Figure 8). The teeth are characteristically evergrowing (hypselodont).

The abundance of South American mammals with high-crowned teeth (70) in the middle Tertiary has been suggested to indicate the advent of grassland communities, but this advanced morphology occurs as much as 15 My earlier in South America than in North America. This asynchrony has led to the concept of "precocious hypsodonty" (70) for South American notoungulates relative to, for example, horses (family Equidae) in North America (Figure 8). Carbon isotopic evidence from both short- and high-crowned herbivores from Deseadan localities in Argentina and Bolivia uniformly yields δ^{13}C values more negative than ~−10‰ (50, 54), indicating that these presumed grazers were feeding on C3 grasses. This indicates a major difference from modern terrestrial grasslands, which are predominantly C4. There are both climatic and physiological explanations for this difference, as is discussed below.

The deseadan Salla Beds of Bolivia, which contain precociously hypsodont herbivores, have been precisely dated between 28 and 25 Mya (35). The Salla fauna includes a diversity of short-crowned herbivores (*Pyrotherium*, astrapotheres, litopterns), small- to medium-sized high-crowned notoungulates, early caviomorph rodents, and the primate *Branisella*. Modern primates occur in rain forests, so by analogy it might be suggested that Salla was tropical. However, given the

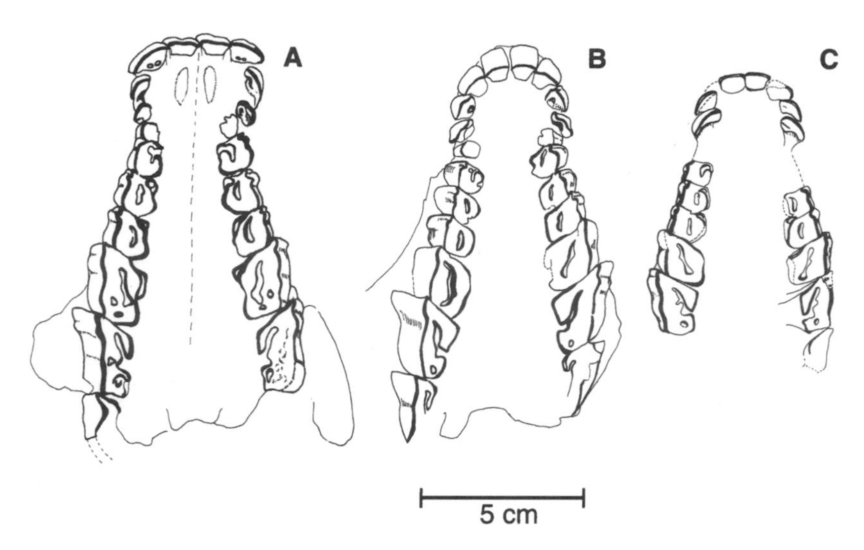

Figure 9 Ventral views of upper dentitions of notoungulates from the late Oligocene of Salla, Bolivia, showing differences in incisor series. In *Pascualihippus* (*a*), the incisor series is transversely linear, indicating a grazer. In *Eurygenium* (*b*) and *Rhynchippus* (*c*), the incisor series is rounded, indicating a browser, or more selective feeder. From Ref. 78 and reproduced with permission of the Society of Vertebrate Paleontology.

preponderance of high-crowned herbivores, and based on associated evidence from the sediments, Salla was reconstructed as a relatively arid grassland habitat (46). The Salla fauna contains three notoungulates, *Pascualihippus*, *Eurygenium*, and *Rhynchippus*, all of which have high-crowned teeth (HIs of 1.5, 1.8, and 1.4, respectively; also see Figure 2*a*), but they have different incisor morphologies. *Pascualihippus* has a broad, transversely linear incisor series, indicating a grazer (Figure 9*A*). In contrast, *Eurygenium* and *Rhynchippus* have rounded muzzles (Figure 9*B*,*C*), which otherwise might suggest a browser (84). The incisor morphology of *Eurygenium* and *Rhynchippus* suggest browsing, whereas their high-crowned teeth suggest grazing. How then does high-crowned dental morphology fit with the rounded incisor series? *Eurygenium* and *Rhynchippus* are both interpreted as selective feeders, and thus with niches more specialized than that of *Pascualihippus* (78). *Eurygenium* and *Rhynchippus*, although of similar incisor and cheek tooth morphologies, differed in size, which apparently facilitated their niche separations within the same community, perhaps corresponding to "Hutchinsonian ratios" (26, 77).

In roughly equivalent 30 Mya Oligocene sediments from North America, as represented by the classic Badlands of South Dakota and Nebraska, the terrestrial mammalian herbivore community was relatively primitive in its morphological and presumed feeding adaptations. Although there was a significant diversification of the principal ungulate groups Artiodactyla and Perissodactyla in the North American Oligocene, they had uniformly short-crowned teeth with rounded incisors, indicating a browsing diet. Enamel-microwear studies of the common Oligocene horse *Mesohippus* indicate that it was a browser and fed on a varied diet of fruits, shoots, and leaves (5). Thus, in contrast to their North America counterparts, the South American mammalian herbivores were indeed precociously hypsodont.

MIOCENE SAVANNAS OF NORTH AMERICA

The extensive North American fossil record reveals major floral and faunal changes during the late Oligocene and early Miocene, ~20 to 25 Mya. Although fossil grasses occur in earlier Cenozoic sediments in North America (14), the first evidence of abundant fossil grasslands is from the early Miocene (55, 87). The adaptive radiation of hypsodont herbivorous mammals occurred during the middle Miocene, ~15 to 20 Mya. Members of the Rodentia, Proboscidea, Perissodactyla, and Artiodactyla all developed hypsodont clades. This resulted in a striking array of herbivores, including browsers, mixed feeders, and the grazing guild (48), with a diversity and community structure similar to that of the modern African savanna (94, 96). Simpson (80) called this the Great Transformation, a time when mammals invaded a new adaptive zone (81) that allowed them to exploit grassland resources. The evolutionary "cost" of becoming grazers was rapidly accelerated tooth wear caused by feeding on abrasive grasses containing phytoliths. The response in grazers was the evolution of hypsodonty. This time of fundamental morphological change (76), as groups of herbivorous mammals in North America evolved from primarily browsers to primarily grazers, resulted in the terrestrial grazing guild.

Horses (family Equidae) diversified rapidly during the Great Transformation with pronounced morphological reorganization of the skull and dentition, resulting in a total of 13 genera in late Miocene faunas (Figure 10). Evidence from

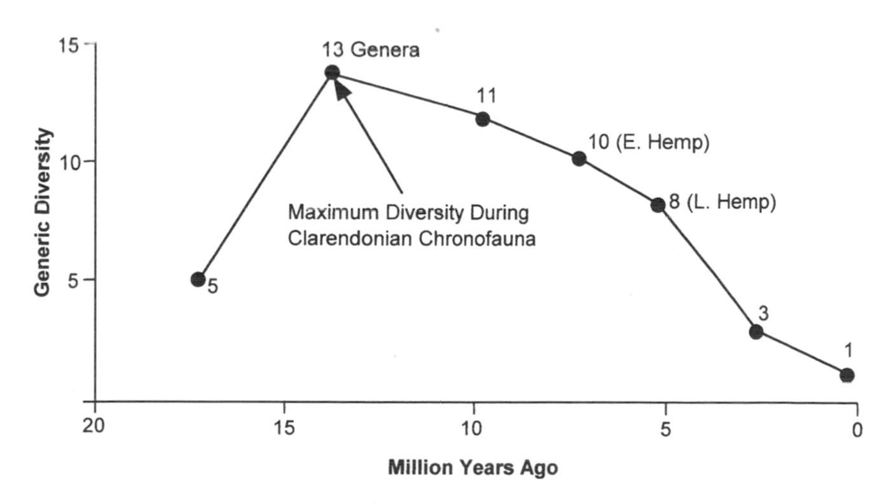

Figure 10 Generic diversity of Miocene and Pliocene horses (Equidae) from North America showing decline after about 15 Mya. Hemp = Hemphillian land-mammal age. From Ref. 47 and reproduced with permission of Cambridge University Press.

Figure 11 SEM showing enamel microwear in modern and fossil horses. Top: *Equus burchelli*, the modern Burchell's zebra, showing a predominance and density of elongated scratches characteristic of a grazer. Bottom: Late Miocene horse *Cormohipparion* from North America showing a lower density of scratches and the presence of pits; this pattern is characteristic of a mixed feeder. From Ref. 23 and reproduced with permission of the Finnish Zoological Publishing Board.

muzzle morphology and hypsodonty indicates that these coexisting horses spanned a broad range of diets (47). Enamel microwear of North American *Cormohipparion* (23; Figure 11) and Old World *Hipparion* (5) indicates that these horses fed on, respectively, a mixed diet and principally grass. (23). Evidence of grazing also comes from the transversely linear incisor morphology shape of the late Miocene equid *Calippus* (25). During the height of the Clarendonian chronofauna ~10 Mya, horses reached their peak diversity of 13 genera, declining in diversity thereafter. The Clarendonian chronofauna is a North American mammalian assemblage from ~12 to 7 Mya (92, 94, 96). The first horses to become extinct during this time were short-crowned, browsing anchitheres, presumably in response to the decline of forested communities. Thereafter, during the late Miocene, diversity dropped to eight genera by the late Hemphillian, 5 Mya (Figure 8).

Evidence of diverse herbivorous adaptations in the Clarendonian chronofauna comes from a variety of Miocene mammals. At the height of the Clarendonian chronofauna in Nebraska, mammalian herbivore taxa were represented by 87% grazers, 10% mixed feeders, and 3% browsers (96). Aepycameline camels evolved stilt-legs, elongated cervical vertebrae, and body heights approaching 6 m, thereby occupying a niche similar to modern-day giraffes (96). Gomphothere proboscideans, with elongated, flattened, and spatulate lower incisor tusks, were previously thought to have scooped up aquatic vegetation, but more recently have been reinterpreted as mixed feeders, with their incisors performing a variety of food-procuring roles (39). Definitive paleontological evidence of grazing is represented by the presence of fossil grass in the dental cavities of the late Miocene rhinoceros *Teleoceras* from the Ashfall Fossil Beds in Nebraska (90).

Is there evidence of carbon isotopic discrimination like that seen during the transition from browsers with short-crowned teeth to grazers with high-crowned teeth during the middle Miocene? Although horses became hypsodont between 15 and 20 Mya, there is no corresponding shift of $\delta^{13}C$ in the Equidae (9, 91). Mammalian herbivores had tooth enamel carbonate $\delta^{13}C$ values characteristically less than -10% (7, 49, 50, 54, 91). Thus, mammal herbivores in the New World prior to the late Miocene lived in C3 grass biomes that were widespread throughout temperate and tropical climes, a situation very different from modern-day ecosystems.

LATE MIOCENE GLOBAL CARBON SHIFT

Prior to the late Miocene, carbon isotopic evidence from North and South America indicate that mammalian herbivores existed in a C3 world, with tooth enamel carbonate $\delta^{13}C$ values characteristically <10 ‰ (7, 49, 50, 54, 91). The dominance of C3 plants in terrestrial ecosystems throughout the early and middle Cenozoic relates to atmospheric CO_2 concentrations. C3 plants are favored in regimes of elevated levels of atmospheric CO_2, as has been modeled for most of the Cenozoic (2). During the late Miocene, however, major global change reduced atmospheric CO_2 to below a critical threshold and increased aridity and seasonality, all of which favor C4 photosynthesis (17, 18). Carbon isotopic values from fossil mammals, ancient soils, and deep-sea cores indicate a major global change ~7–8 Mya (6, 7, 40, 74). $\delta^{13}C$ values of post–late-Miocene fossil grazing mammals and grassland paleosol carbonates have $\delta^{13}C$ values around 0 ‰ indicating C4 terrestrial communities. (Figure 12).

This late Miocene global carbon shift had a profound effect on the history of Cenozoic ecosystems: It resulted in a latitudinal gradient of C3/C4 grasses, with C3 grasses predominating in colder, more polar regions, and C4 grasses predominating in temperate and tropical regions (51, 89). C4 grasses are generally found in ecosystems of lower productivity, which in turn support a lower overall biomass and diversity.

After 7 Mya, horses continued to decline in diversity. It is plausible that horses were part of the overall drop in diversity supported by lower productivity steppe-like C4 grasslands. Which equid taxa survived this extinction? Isotopic and

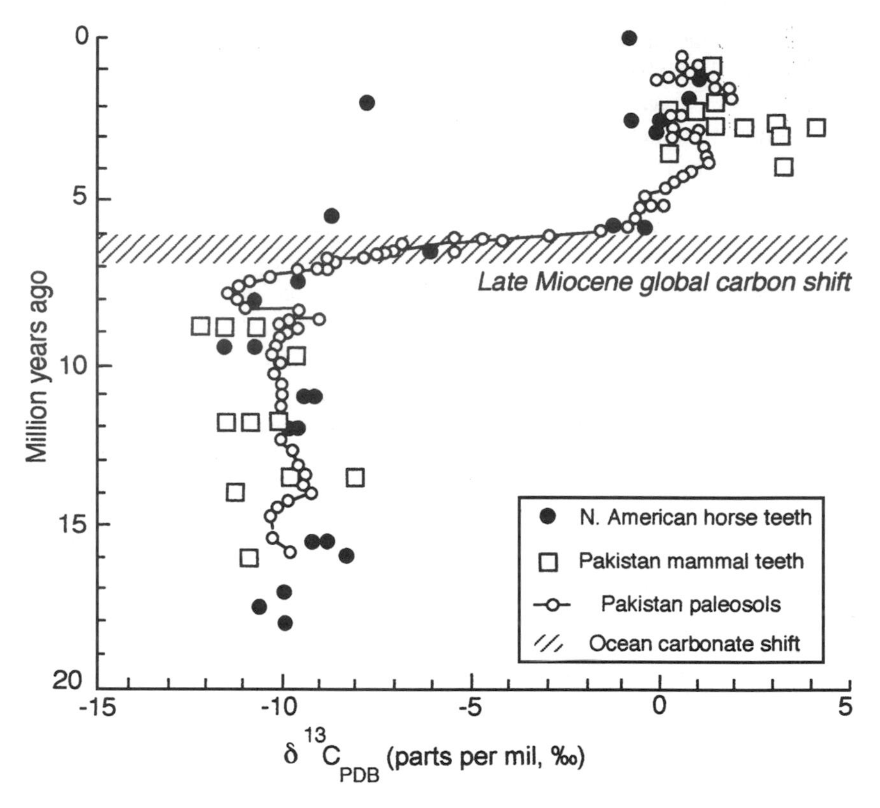

Figure 12 Isotopic evidence for the late Miocene global carbon shift between ~7 and 8 Mya. Modified from Ref. 6 and reproduced with permission of Macmillan.

microwear studies of the six coexisting high-crowned equid genera from 5-My-old sediments in Florida indicate that these horses included grazers, mixed feeders, and C3 browsers (53). Two grazing and one browsing species became extinct during the latest Miocene at 5 Mya. Other taxa of North American mammalian herbivores also experienced extinctions at 5 Mya, including the family Rhinocerotidae, and genera and species of Rodentia, Artiodactyla, and Proboscidea (96). The corresponding South American record at 5 Mya is not well known. Although C4 grasses are known to have been part of South American grazing diets (50, 54), it is difficult to determine if extinctions were above background levels at this time.

GREAT AMERICAN INTERCHANGE AND PLEISTOCENE MEGAHERBIVORES

For most of the Cenozoic, North and South America were geographically isolated from one another, and unique mammalian herbivore communities evolved on each of these continents. During the Pliocene (5 to 2 Mya), however, this faunal isolation changed with the formation of a dry-land connection by the closing of the Isthmus of Panama ~3 Mya. This closure resulted in a land-bridge dispersal corridor for the various immigrants. Paleontologists have studied this "Great American Interchange" (GAI; 88), the height of which occurred during the middle Pleistocene, ~1 Mya. The community structure and faunal equilibrium were disrupted by the addition of immigrant mammals into the resident biotas. Immigrant herbivores from North American that dispersed into South America included phyllotine rodents, mastodons, tapirs, horses, peccaries, llamas, and deer. Herbivores from South America that immigrated into North America included caviomorph rodents and the edentates, including armadillos, glyptodonts, and giant ground sloths.

It has been suggested that during times of faunal interchange, the immigrants, or "invaders," generally have a competitive advantage over the native, or resident biota, as Simpson (79, 82) asserted was the case during the GAI. Although the concept of competition is difficult to quantify and test in the fossil record, nevertheless, some general patterns of the GAI are instructive here. First, in South America (Figure 13), the native fauna of ungulate mammals consisted of 12–13 genera before the formation of the land bridge, and decreased to five and then three genera during the GAI. Correspondingly, the North American invaders

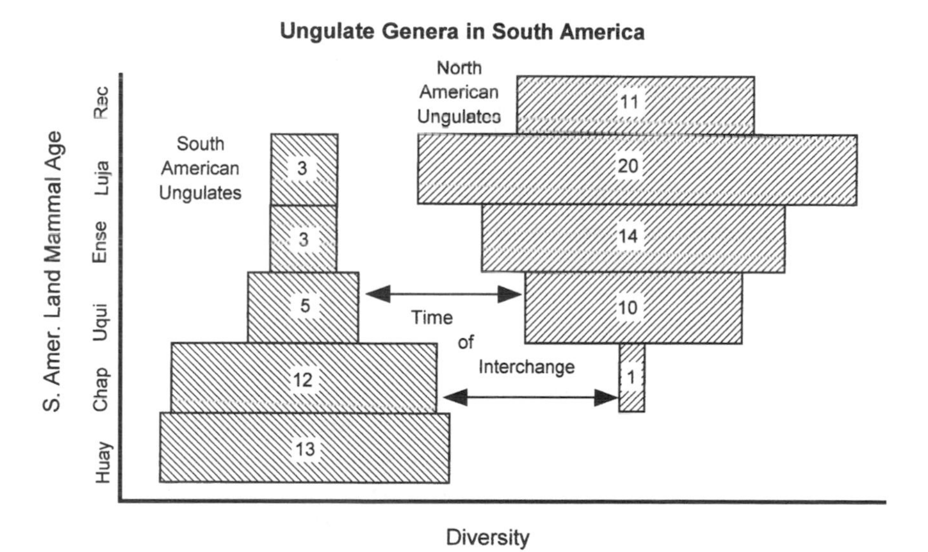

Figure 13 Turnover of ungulate genera in South America during the Plio-Pleistocene Great American Interchange. Modified from Ref. 92 and reproduced with permission of the Paleontological Society. Land-mammal abbreviations and boundaries are as follows (Ref. 21): Huay, Huayquerian, Miocene, 7 to 9 Mya; (hiatus); Chap, Chapadmalalan, Pliocene, 3.4 to 4 Mya; Uquian, Pliocene–early Pleistocene, 1.5 to 3 Mya; Ense, Ensenadan, middle Pleistocene, 0.5 to 1.5 Mya; Luja, Lujanian, late Pleistocene, 10,000 years ago to 0.5 Mya.

went from one to 10 to 14 to 20 genera before the late Pleistocene extinctions (93, 95). Second, the pattern of extinctions during the GAI is asymmetric; that is, the immigrants apparently affected the native herbivores in South America, but the reverse was not true, i.e. in North America the immigrant taxa from South America did not seem to have a competitive advantage over the native herbivores (82, 98, 99). It is tempting to suggest simple explanations (e.g. competition) for complex biological phenomena such as the GAI. Other factors, however, including climate change and loss of specialized habitats, almost certainly were related to the demise of the native herbivore faunas throughout the Americas during the Pleistocene.

The Tarija fauna of southern Bolivia is a classic middle Pleistocene GAI locality. It serves as a good example from which mammalian herbivore community structure can be examined. The Tarija medium- and large-bodied herbivores comprise two dozen species within the endemic orders Edentata, Notoungulata, Litopterna, and caviomorph Rodentia and to these were added the North American immigrant orders Proboscidea, Perissodactyla, and Artiodactyla. A multivariate study of the carbon isotopes, cranial morphology, and hypsodonty of these Tarija herbivores (except for the edentates, which lack enamel for carbon isotopic analysis) revealed some interesting patterns within the herbivore community (52). Of the 13 herbivore species analyzed, three were browsers, five were mixed feeders, and five were grazers. The grazers span the largest range of body size, from *Vicugna* to the mastodon *Cuvieronius*. The larger body sizes (to the right on the first principal component axis, Figure 14) supports the idea that grazers need longer retention times for processing low-nutritive-value forage in their gut (58, 66). Within families consisting of more than one herbivore species at Tarija, there is evidence of niche differentiation. For example, of the three sympatric llamas, two are grazers (*Lama, Vicugna*) and one is a mixed feeder (*Palaeolama*), and of the horses, two species are mixed feeders (*Hippidion, Onohippidium*) and one is a pure C4 grazer (*Equus*; Figure 14).

Large-bodied herbivores (megaherbivores) have a significant impact on terrestrial communities (66). In North America, the ecological impact of megaherbivore immigrations was profound during the GAI. Sloths were consummate browsers, as has been documented by the presence of low shrub twigs and roots in the dung of the late Pleistocene sloth *Nothriotheriops* from Arizona (44). Glyptodonts have heavily infolded, high-crowned teeth, and were probably mixed feeders with a significant amount of grass in their diets. An added component of biogeographic complexity resulted from immigrations across Holarctica during the Pleistocene. Two Old World immigrant megaherbivores, in particular, played a role in the hypergrazer niche, which during the Pliocene was occupied by *Equus* (49). The mammoth, *Mammuthus*, and bison, *Bison*, dispersed into North America at respectively 1.5 and 0.5 Mya (45). Studies of the carbon isotopic values of the teeth of mammoth and bison indicate that they were principally grazers, and *Equus* was also a grazer at this time, although the horse had more of a grazing/mixed feeding component to its diet during this time (15, 20, 36).

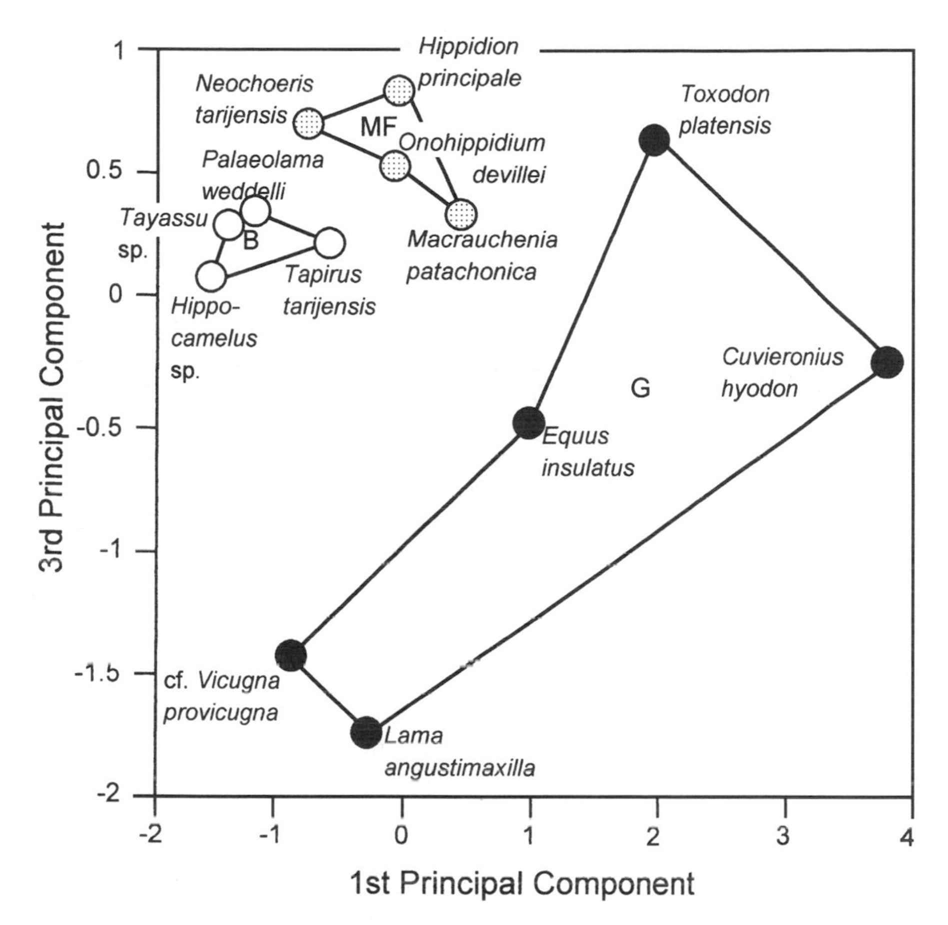

Figure 14 Principal components analysis of 13 herbivore taxa from the middle Pleistocene of Tarija, Bolivia, using carbon isotopes, molar length, muzzle width, and hypsodonty index. The grazers (G, black circles), mixed feeders (MF, shaded circles), and browsers (B, white circles) fall into discrete groups in multivariate space. From Ref. 52 and reproduced with permission of the Paleontological Society.

Only a fraction of the vast herbivore diversity that existed throughout the Americas during the Pleistocene persists today. In a very short period of time, from ~15,000 to 10,000 years ago, megaherbivores such as mammoths, mastodons, horses, sloths, glyptodonts, and giant armadillos became extinct throughout the Americas, and other taxa including llamas, tapirs, and bison have greatly reduced ranges today relative to those before 15,000 years ago (97, 99). Two principal hypotheses have explained these late Pleistocene megaherbivore extinctions. The first is climate and habitat change; the other is that humans rapidly populated the Americas, resulting in "Pleistocene overkill" of large prey species (60, 61). As with all complex natural systems, these two explanations undoubtedly both played roles as causal factors of late Pleistocene extinctions.

CONCLUDING COMMENTS

The analysis of Cenozoic herbivore communities in the Americas illustrates numerous evolutionary and ecological principles and patterns. We see the coevolution of grass and grazers, the evolution of similar adaptations (e.g., hypsodonty) in different groups on different continents, the biotic reaction to major climate perturbations, the change in diversity of browsing and grazing herbivore groups with corresponding changes in vegetation, and the extinctions resulting from climate and/or biotic factors. These examples are more clearly interpreted using a combination of morphological and isotopic techniques now available to paleontologists.

The fossil record of Cenozoic mammalian herbivores reveals both change and stability of overall community structure through time, the idea inherent in the concept of the chronofauna. The fundamental question about long-term changes in community evolution is whether biotic equilibrium and change are affected principally by external, climatic factors, or by internal factors such as species interactions and competition (28, 31, 72). Natural systems are rarely simple and, in reality, long-term changes are almost certainly structured by both climatic and biotic factors. The fossil record provides definitive and unique evidence from which we can understand the history of evolving communities of Cenozoic herbivores of the Americas.

ACKNOWLEDGMENTS

I thank RM Hunt and SD Webb for making helpful comments on this manuscript. This paper was written during support from NSF grants EAR-9506550, IBN-9528020, and EAR-9909186. This is University of Florida Contribution to Paleobiology no. 515.

LITERATURE CITED

1. Archibald JD. 1998. Archaic ungulates ("Condylarthra"). In *Evolution of Tertiary Mammals of North America*. Vol. 1: *Terrestrial Carnivores, Ungulates, and Ungulate-like Mammals*, ed. CM Janis, KM Scott, LL Jacobs, pp. 292–331. New York: Cambridge Univ. Press
2. Berner R. 1994. GEOCARB II: a revised model of atmospheric CO_2 over Phanerozoic time. *Am. J. Sci.* 294:56–94
3. Boutton TW. 1991. Stable carbon isotope ratios of natural minerals: II. Atmospheric, terrestrial, marine, and freshwater environments. In *Carbon Isotope Techniques*, ed. DC Coleman, B Fry, pp. 173–95. San Diego, CA: Academic Press
4. Bown TM, Kraus MJ. 1979. Origin of the tribosphenic molar and metatherian

and eutherian dental formulae. In *Mesozoic Mammals: The First Two-Thirds of Mammalian History*, ed. JA Lillegraven, Z Kielan-Jaworoska, WA Clemens, pp. 172–81. Berkeley: Univ. Calif. Press

5. Caprini A. 1998. The food habits of some Eocene to present-day Equidae deduced from observation of the teeth under the S.E.M. *Palaeontolograph. Italica* 85:153–76
6. Cerling TE, Wang Y, Quade J. 1993. Global ecological change in the late Miocene: expansion of C4 ecosystems. *Nature* 361:344–45
7. Cerling TE, et al. 1997. Global vegetation through the Miocene/Pliocene boundary. *Nature* 389:153–58
8. Cerling TE, Harris JM. 1999. Carbon isotope fractionation between diet and bioapatite in ungulate mammals and implications for ecological and paleoecological studies. *Oecologia* 120:347–63
9. Cerling TE, Harris JM, MacFadden BJ. 1998. Carbon isotopes, diets of North American equids, and the evolution of C4 grasslands. In *Stable Isotopes: Integration of Biological, Ecological, and Geochemical Processes*, ed. H Griffiths, pp. 363–79. Oxford, UK: Bios Sci. Pub.
10. Cifelli RL. 1982. Patterns of evolution among Artiodactyla and Perissodactyla. *Evolution* 35:433–40
11. Cifelli RL. 1985. South American ungulate evolution and extinction. In *The Great American Biotic Interchange*, ed. FG Stehli, SD Webb, pp. 249–66. New York: Plenum
12. Clemens WA, Keilan-Jaworoska Z. 1979. Multituberculata. In *Mesozoic Mammals: The First Two-Thirds of Mammalian History*, ed. JA Lillegraven, Z Kielan-Jaworoska, WA Clemens, pp. 99–149. Berkeley: Univ. Calif. Press
13. Clemens WA, Lillegraven JA, Lindsay EH, Simpson GG. 1979. Where, when, and what—a survey of known Mesozoic mammal distribution. In *Mesozoic Mammals: The First Two-Thirds of Mammalian History*, ed. JA Lillegraven, Z Kielan-Jaworoska, WA Clemens, pp. 7–58. Berkeley: Univ. Calif. Press
14. Crepet WL, Feldman GD. 1991. The earliest remains of grasses in the fossil record. *Am. J. Bot.* 78:1010–14
15. Cronin SL, Betancourt J, Quade J. 1998. Late Pleistocene C4 plant dominance and summer rainfall in the southwestern United States from isotopic study of herbivore teeth. *Q. Res.* 50:179–93
16. Deines P. 1980. The isotopic composition of reduced organic carbon. In *Handbook of Environmental Isotope Chemistry*. Vol. 1: *The Terrestrial Environment*, ed. P Fritz, JC Fontes, pp. 329–406. Amsterdam: Elsevier
17. Ehleringer JR, Sage RF, Flanagan LB, Pearcy RW. 1991. Climate change and the evolution of C_4 photosynthesis. *Trends Ecol. Evol.* 6:95–99
18. Ehleringer JR, Cerling TE, Helliker BR. 1997. C_4 photosynthesis, atmospheric CO_2, and climate. *Oecologia* 112:285–99
19. Farquahar GD, Ehleringer JR, Hubick KT. 1989. Carbon isotopic discrimination and photosynthesis. *Annu. Rev. Plant Physiol. Plant Mol. Biol.* 40:503–37
20. Feranec RS, MacFadden BJ. 2000. Evolution of the grazing niche in Pleistocene mammals: evidence from stable isotopes. *Palaeogeogr. Palaeoclimatol. Palaeoecol.* In press
21. Flynn JJ, Swisher CC. 1995. Cenozoic South American land mammal ages: correlation to global chronologies. In *Geochronology, Time Scales and Global Stratigraphic Correlation*, ed. WA Berggren, DV Kent, MP Aubry, J Hardenbol, pp. 317–33. Tulsa, OK: Soc. Sed. Geol. Publ. 54.
22. Gingerich PD. 1981. Variation, sexual dimorphism, and social structure in the early Eocene horse *Hyracotherium* (Mammalia, Perissodactlya). *Paleobiology* 7:443–55
23. Hayek LA, Bernor RL, Solounias N,

Steigerwald P. 1991. Preliminary studies of hipparionine horse diet as measured by tooth microwear. *Ann. Zool. Fennici* 28:187–200.
24. Hickey LJ, West RM, Dawson MR, Choi DK. 1983. Arctic terrestrial biota: paleomagnetic evidence of age disparity with mid-northern latitudes during the late Cretaceous and early Tertiary. *Science* 221:1153–56
25. Hulbert RC. 1988. *Calippus* and *Protohippus* (Mammalia, Perissodactyla, Equidae) from the Miocene (Barstovian–early Hemphillian) of the Gulf Coastal Plain. *Bull. Florida Mus. Nat. Hist., Biol. Sci.* 33:229–338
26. Hutchinson GE 1959. Homage to Santa Rosalia, or why are there so many kinds of animals? *Am. Nat.* 93:145–59
27. Jacobs BF, Kingston JD, Jacobs LL. 1999. The origin of grass-dominated ecosystems. *Ann. Missouri Bot. Gard.* 86:590–643
28. Janis CM. 1984. The use of fossil ungulate communities as indicators of climate and environment. In *Fossils and Climate*, ed. P Brenchley pp. 85–104. New York: Wiley
29. Janis CM. 1988. An estimation of tooth volume and hypsodonty indices in ungulate mammals, and the correlation of these factors with dietary preference. In *Teeth Revisited: Proc. VIIth Int. Symp. Dental Morphology*, ed. DE Russell, JP Santoro, D Sigogneau-Russell, pp. 367–87. *Mém. Mus. Nat. Hist. Nat., Paris, Sér.* C, 53
30. Janis CM. 1989. A climatic explanation for patterns of evolutionary diversity in ungulate mammals. *Palaeontology* 32:463–81
31. Janis CM. 1993. Tertiary mammal evolution in the context of changing climates, vegetation, and tectonic events. *Annu. Rev. Ecol. Syst.* 24:467–500
32. Janis CM, Ehrhardt D. 1988. Correlation of relative muzzle width and relatve incisor width with dietary preference in ungulates. *Zool. J. Linnean Soc.* 92:267–84
33. Janis CM, Fortelius M. 1988. On the means whereby mammals achieve functional durability of their dentitions, with special reference to limiting factors. *Bio. Rev.* 63:197–230
34. Janis CM, Scott KM, Jacobs LL, eds. 1998. *Evolution of Tertiary Mammals of North America.* Vol. 1: *Terrestrial Carnivores, Ungulates, and Ungulatelike Mammals*. New York: Cambridge Univ. Press. 691 pp.
35. Kay RF, MacFadden BJ, Madden RH, Sandeman H, Anaya F. 1997. Revised age of the Salla Beds, Bolivia, and its bearing on the age of the Deseadan South American land mammal "age." *J. Vert. Paleontol.* 18:189–99
36. Koch PL. 1998. Isotopic reconstruction of past continental environments. *Annu. Rev. Earth Planet. Sci.* 26:573–613
37. Koch PL, Zachos JC, Gingerich PD. 1992. Correlation between isotope records in marine and continental carbon reservoirs near the Palaeocene/Eocene boundary. *Nature* 358:319–22
38. Kowalevsky V. 1873. Sur *L'Anchitherium aurelianense* Cuv. et sur l'histoire paléntologique des chevaux. *Mém. L'Acad. Impériale Sci. St. Pétersbourg, 7th* Sér. 20(5):1–73, 3 plates
39. Lambert WD. 1992. The feeding habits of the shovel-tusked gomphotheres: evidence from tusk wear patterns. *Paleobiology* 18:132–47
40. Latorre C, Quade J, McIntosh WC. 1997. The expansion of C4 grasses and global change in the late Miocene: evidence from the Americas. *Earth Planet. Sci. Lett.* 146:83–96
41. Lee-Thorp JA, Van der Merwe NJ. 1987. Carbon isotope analysis of fossil bone apatite. *S. Afr. J. Sci.* 83:712–75
42. Lucas SG, Schoch RG, Williamson TE. 1998. Taeniodonta. In *Evolution of Tertiary Mammals of North America*, ed. CM Janis, KM Scott, LL Jacobs, pp. 260–67. New York: Cambridge Univ. Press
43. Lucas SG, Schoch RG. 1998. Tillodonta. In *Evolution of Tertiary Mammals of North*

America, ed. CM Janis, KM Scott, LL Jacobs, pp. 268–73. New York: Cambridge Univ. Press

44. Lull RS. 1929. A remarkable ground sloth. *Mem. Peabody Mus. Nat. Hist.* 3:1–39
45. Lundelius EL, et al. 1987. The North American Quaternary sequence. In *Cenozoic Mammals of North America: Geochronology and Biostratigraphy*, ed. MO Woodburne, pp. 211–35. Berkeley: Univ. Calif. Press
46. MacFadden BJ. 1990. Chronology of Cenozoic primate localities in South America. *J. Hum. Evol.* 19:7–21
47. MacFadden BJ. 1992. *Fossil Horses: Systematics, Paleobiology, and Evolution of the Family Equidae*. New York: Cambridge Univ. Press 369 pp.
48. MacFadden BJ 1997. Origin and evolution of the grazing guild in New World terrestrial mammals. *Trends Ecol. Evol.* 12:182–87
49. MacFadden BJ, Cerling TE. 1996. Mammalian herbivore communities, ancient feeding ecology, and carbon isotopes: a 10-million-year sequence from the Neogene of Florida. *J. Vert. Paleontol.* 16:103–15
50. MacFadden BJ, Cerling TE, Prado J. 1996. Cenozoic terrestrial ecosystem evolution in Argentina: evidence from stable isotopes of carbon isotopes of fossil mammal teeth. *Palaios* 319–327
51. MacFadden BJ, Cerling TE, Harris JM, Prado J. 1999. Ancient latitudinal gradients of C_3/C_4 grasses interpreted from stable isotopes of New World Pleistocene horse (*Equus*) teeth. *Global Ecol. Biogeogr.* 8:137–49
52. MacFadden BJ, Shockey BJ. 1997. Ancient feeding ecology and niche differentiation of Pleistocene mammalian herbivores from Tarija, Bolivia: morphological and isotopic evidence. *Paleobiology* 23:77–100
53. MacFadden BJ, Solounias N, Cerling TE. 1999. Ancient diets, ecology, and extinction of 5-million-year-old horses from Florida. *Science* 283:824–27
54. MacFadden BJ, Wang Y, Cerling TE, Anaya F. 1994. South American fossil mammals and carbon isotopes: a 25-million-year sequence from the Bolivian Andes. *Palaeogeogr. Palaeoclimatol. Palaeoecol.* 107:257–68
55. MacGinitie HD. 1962. The Kilgore flora: a late Miocene flora from northern Nebraska. *Univ. Calif. Pub. Geol. Sci.* 35:67–158
56. McKenna MC. 1972. Was Europe connected directly to North America prior to the middle Eocene? In *Evolutionary Biology*, ed. T Dobzhansky, MK Hecht, WC Steere, pp. 179–88. New York: Appleton-Century-Crofts
57. McKenna MC. 1975. Fossil mammals and the early Eocene land continuity. *Ann. Missouri. Bot. Gard.* 62:335–53
58. McNaughton SJ. 1991. Evolutionary ecology of large tropical herbivores. *In Plant-Animal Interactions: Evolutionary Ecology in Tropical and Temperate Regions*, ed. PW Price et al, pp. 509–22. New York: Wiley
59. McNaughton SJ, Tarrants JL, McNaughton MM, Davies RH. 1985. Silica as a defense against herbivory and as a growth promoter in African grasses. *Ecology* 66:528–35
60. Martin PS. 1973. The discovery of America. *Science* 179:969–74
61. Martin PS. 1984. Prehistoric overkill: the global model. In *Quaternary Extinctions: A Prehistoric Revolution*, ed. PS Martin, RG Klein, pp. 354–403. Tucson: Univ. Ariz. Press
62. Matthew WD. 1926. The evolution of the horse: a record and its interpretation. *Q. Rev. Biol.* 1:139–85
63. Novacek MJ. 1999. 100 million years of land vertebrate evolution: the Cretaceous–early Tertiary transition. *Ann. Missouri. Bot. Gard.* 86:230–58
64. Olson EC. 1952. The evolution of a Permian vertebrate chronofauna. *Evolution* 6:181–96
65. Olson EC. 1966. Community evolution and

the origin of mammals. *Ecology* 47:291–302

66. Owen-Smith N. 1988. *Megaherbivores: The Influence of Very Large Body Size on Ecology*. Cambridge, UK: Cambridge Univ. Press. 369 pp.
67. Pascual R, Ortiz Jaureguizar EO. 1990. Evolving climates and mammal faunas in Cenozoic South America. *J. Hum. Evol.* 19:23–60
68. Pascual R, Ortiz Jaureguizar EO. 1993. Evolutionary pattern of land mammal faunas during the late Cretaceous and Paleocene in South America: a comparison with the North American pattern. *Ann. Zool. Fennici* 28:245–52
69. Patterson B. 1956. Early Cretaceous mammals and the evolution of mammalian molar teeth. *Fieldiana Geol.* 13:1–105
70. Patterson B, Pascual, R. 1972. The fossil mammal faunas of South America. In *Evolution, Mammals, and Southern Continents*, ed. A Keast, FC Erk, B Glass, pp. 247–309. Albany: State Univ. NY Press
71. Prothero DR. 1994. *The Eocene-Oligocene Transition: Paradise Lost*. New York: Columbia Univ. Press. 291 pp.
72. Prothero DR. 1999. Does climatic change drive mammalian evolution? *GSA Today* 9:1–7
73. Prothero DR, Berggren WA, eds. 1992. *Eocene-Oligocene Climate and Biotic Evolution*. Princeton, NJ: Princeton Univ. Press. 568 pp.
74. Quade J, et al. 1992. A 16 million year record of paleodiet from Pakistan using carbon isotopes in fossil teeth. *Chem. Geol. (Isotope Geosci. Sec.)* 94:183–92
75. Radinsky LR. 1976. Oldest horse brains: more advanced than previously realized. *Science* 194:626–27
76. Radinsky LR. 1983. Allometry and reorganization in horse skull proportions. *Science* 221:1189–91
77. Roth VL. 1981. Constancy in the size rations of sympatric species. *Am. Nat.* 118:394–404
78. Shockey BJ. 1997. Two new notoungulates (family Notohippidae) from the Salla Beds of Bolivia (Deseadan: late Oligocene): systematics and functional morphology. *J. Vert. Paleontol.* 17:584–99
79. Simpson GG. 1950. The Nhistory of the fauna of South America. *Am. Sci.* 38:261–89
80. Simpson GG. 1951. *Horses*. Oxford, UK: Oxford Univ. Press. 247 pp.
81. Simpson GG 1953. *The Major Features of Evolution*. New York: Columbia Univ. Press. 434 pp.
82. Simpson GG. 1980. *Splendid Isolation: The Curious History of South American Mammals*. New Haven: Yale Univ. Press. 266 pp.
83. Solounias N, Hayek LA. 1993. New methods of tooth microwear analysis and application to dietary determination of two extinct antelopes. *J. Zool.* 229:142–55
84. Solounias N, Moelleken SMC. 1993. Dietary adaptation of some ruminants determined by premaxillary shape. *J. Mamm.* 74:1059–71
85. Solounias N, Moelleken SMC, Plavcan JM. 1995. Predicting the diet of extinct bovids using masseteric morphology. *J. Vert. Paleontol.* 15:795–805
86. Solounias N, Teaford M, Walker A. 1988. Interpreting the diet of extinct ruminants: The case of a non-browsing giraffid. *Paleobiology* 14:287–300
87. Stebbins GL. 1981. Coevolution of grasses and herbivores. *Ann. Missouri Bot. Gard.* 68:75–86
88. Stehli FG, Webb SD, eds. 1985. *The Great American Biotic Interchange*. New York: Plenum. 532 pp.
89. Terri JA, Stowe LG. 1976. Climatic patterns and distribution of C_4 grasses in North America. *Oecologia* 23:1–12
90. Voorhies MR, Thomasson JR. 1979. Fossil grass antothecia within Miocene rhinoceros skeletons: diet of an extinct species. *Science* 206:331–33
91. Wang Y, Cerling TE, MacFadden BJ. 1994.

Fossil horses and carbon isotopes: new evidence for Cenozoic dietary, habitat, and ecosystem changes in North America. *Palaeogeogr. Palaeoclimatol. Palaeoecol.* 107:269–79
92. Webb SD. 1969. The Burge and Minnechaduza Clarendonian mammalian faunas of north-central Nebraska. *Univ. Calif. Pub. Geol. Sci.* 78:1–191
93. Webb SD. 1976. Mammalian faunal dynamics of the great American interchange. *Paleobiology* 2:220–34
94. Webb SD. 1977. A history of savanna vertebrates in the New World. Part I. North America. *Annu. Rev. Ecol. Syst.* 8:355–80
95. Webb SD. 1978. A history of savanna vertebrates in the New World. Part II. South America and the Great Interchange. *Annu. Rev. Ecol. Syst.* 9:393–426
96. Webb SD. 1983. The rise and fall of the late Miocene ungulate fauna in North America. In *Coevolution*, ed. MH Nitecki, pp. 267–306. Chicago: Univ. Chicago Press
97. Webb SD. 1984. Ten million years of mammal extinctions in North America. In *Quaternary Extinctions: A Prehistoric Revolution*, ed. PS Martin, HE Wright, pp. 189–210. Tucson: Univ. Ariz. Press
98. Webb SD. 1991. Ecogeography and the Great American Interchange. *Paleobiology* 17:266–80
99. Webb SD, Barnosky A. 1989. Faunal dynamics of Pleistocene mammals. *Annu. Rev. Earth Planet. Sci.* 17:413–38
100. Wing SL, Alroy J, Hickey LJ. 1995. Plant and mammal diversity in the Paleocene to early Eocene of the Bighorn Basin. *Palaeogeogr. Palaeoclimatol. Palaeoecol.* 115:117–55
101. Wyss AR, et al. 1994. Paleogene mammals from the Andes of central Chile: a preliminary taxonomic, biostratigraphic, and geochronological assessment. *Am. Mus. Novitates* 3098:1–31

Annu. Rev. Ecol. Syst. 2000. 31:61–78

CONSERVATION ISSUES IN NEW ZEALAND

John Craig, Sandra Anderson, Mick Clout, Bob Creese, Neil Mitchell, John Ogden, Mere Roberts, and Graham Ussher
School of Environmental and Marine Sciences, University of Auckland, Auckland, New Zealand; e-mail: j.craig@auckland.ac.nz

Key Words sustainable management, representativeness, invasive species, marine, Maori

■ **Abstract** Conservation in New Zealand is failing to halt an ongoing decline in biodiversity. Classical problems of ecosystem loss and fragmentation have largely been countered in some regions by reservation of 30% of total land area. Unsustainable harvesting of native biodiversity has stopped; indeed harvesting of terrestrial species is rare. In contrast, marine reserves do not cover even 1% of the managed area, and harvest of native species, some of it unsustainable, are a major industry. Introduced pests, especially mammals, are the overwhelming conservation problem. Legislation, management, and considerable public opinion is based on preservationist ideals that demand the sanctity of native land biodiversity. Considerable success in threatened species management, island eradications, and mainland control of pests is increasing opportunities for restoration. New legislation is increasingly built on concepts of sustainability and offers the opportunity for integrating conservation, use, and development. Realization of these opportunities requires greater understanding of the relative merits of preservation versus sustainability, the dynamics and costs of pest control, the need for ecosystem processes in addition to individual species, and the involvement of people, especially the rights of indigenous Maori. Understanding marine environments and linking attitudes to land and sea is also a challenge.

INTRODUCTION

World conservation is at a crossroads of decision making (37, 53, 77), and New Zealand provides an extreme example of the issues and possible solutions (4). On the positive side, greater than 30% of New Zealand's total land area has been reserved (4, 79), there is a single government agency responsible for most conservation activities (45), single-species recovery programs are largely successful (26, 78, 79), and there is widespread public support for biodiversity conservation (4, 36, 79). This, however, contrasts sharply with ongoing biodiversity decline (4): For example, 50% of bird species are threatened, introduced pests are degrading all protected areas, the legal framework is based on preservationist ideals (16),

0066-4162/00/1120-0061$14.00

and government funding allows less than 5% of the protected lands to be managed sustainably (45). In addition Maori and local communities have little influence.

New Zealand is heavily dependent on natural renewable resources for much of its income (79), yet the consideration of sustainability is slow and hampered by a management dichotomy based on extremes of market forces versus preservation. Seventy percent of the land and most of the sea is managed unsustainably, while the remaining 30% of the land is locked up in reserves (81). Major gains in conservation have been made since the 1970s, when the rampant mining of natural capital (79) incited calls for protection and preservation (86). Moving New Zealand to sustainable management of all land and sea is the current challenge (4, 35, 47, 79, 81, 87, 95). Without it, New Zealand's biodiversity on the main islands will resemble that of other devastated islands such as Guam (48).

History and Status of New Zealand Biodiversity

The evolutionary history of New Zealand diverged markedly from the rest of the world's about 65–80 million years ago (31, 79), when it separated from the southern continent of Gondwanaland and became a separate island archipelago. While land area has varied with changes in sea level, New Zealand totals 26 million ha in three main islands plus another 700 islands greater than 5 ha. These stretch from the subtropics to the subantarctic (29°S to 52°S). Spanning two tectonic plates results in a diverse landscape dominated by mountains (including volcanoes), rolling hills, and river flats. The marine environment under New Zealand's jurisdiction is an order of magnitude greater than the land mass. Marine environments include long and highly indented coastlines, deep ocean plains and trenches, and underwater seamounts.

Whereas the rest of the world's biota was influenced by the evolution of mammals, in New Zealand birds were the largest animals in all terrestrial ecosystems (8), and there are high levels of endemism in most taxa (14). Ratites were common, and large size and flightlessness or much reduced flight ability were frequent (8). Sea birds comprised 30% of the avifauna. The dominance of birds in turn influenced the form and distribution of some trees and shrubs (7, 94). The reptiles included the tuatara (*Sphenodon* spp.), as well as geckos and skinks but not snakes and crocodiles. Added to these Gondwana relics are a range of invertebrate, reptile, bird, and plant species that have arrived by the westerly wind drift and putative island hopping from the north (55, 64, 108).

Throughout the Holocene (last 10,000 years) most of New Zealand's land surface was evergreen rain forest of differing structure and composition (128): Beech (*Nothofagus*) dominate montane and southern forests, whereas lowlands have conifer/broadleaf forests. South Island mountain beech (*N. solandri*) forests are the least diverse forests. In contrast, kauri forests of northern New Zealand have the highest biodiversity—averaging 18 tree species per hectare including many endemics (74, 98). Many canopy trees are light-demanding: Large-scale regeneration occurs only after widespread disturbance (99, 101). Disturbances include volcanism and earthquakes, and resultant fires, land-slips, and floods, as well as occasional tropical cyclones in the north and exceptional snowfalls or avalanches in the south. Massive exogenous disturbances with long return times have shaped the forests. Faunal communities of these forests included many large flightless birds (8), most of which are now extinct. Lizard and invertebrate fauna are poorly known with many species undescribed (65, 121).

Effects of People

New Zealand was the last major land mass to be colonized by people (2, 79, 99, 109). Polynesian predecessors of the modern Maori arrived some 700–1000 years ago; Europeans some 200 years ago. Together, people and their companion animals have greatly modified the landscape through ecosystem and species loss. Birds, especially land birds, are the worst affected (75, 76, 79).

Causal agents have been human hunting, ecosystem loss and fragmentation, and predation and competition from introduced pests (4, 45, 79). Rain forests have been reduced from an original 78% of land area to approximately 23% (79). Wetlands have been reduced by over 90% of their pre-European area of 700,000–1,000,000 ha (79, 91). Native grasslands initially increased from 1.5 million ha to some 8 million ha as a result of Maori fires, but degradation by burning, oversowing with European pasture grasses, and poor land management rapidly allowed invasion by introduced rabbits and weeds, making these lands ecological and agricultural deserts (132). Although large tracts of connected forest can still be found on the west coast of the South Island, much of the remainder is fragmented. Introduced pests are universally distributed through most areas. In addition, New Zealand has more introduced vascular plants (2400 species) than there are native species (2300 species, 86% endemic) (79, 102). Arrivals continue naturally, supplemented by the enormous influx brought, intentionally and unintentionally, by people. Six food plants were introduced by Maori, whereas the rest, introduced by Europeans, dominate the production landscape. They continue to enter the country, mostly as horticultural plants, at the rate of ca. 11 species per year (6).

Maori hunting eliminated 26 species (30%) of endemic land birds and 4 (18%) of the endemic sea birds, while ecosystem loss and companion animals eliminated a further 8 land birds. Many other species were reduced to localized populations. Tuatara, some lizards, and some invertebrates were also eliminated from the main islands. Seals and sea lions disappeared from northern areas.

European colonization started with the further reduction of seals and sea lions even in the subantarctic islands. More rapid ecosystem destruction to provide timber and allow pastoral agriculture reduced forest cover and saw the extinction of a further 16 land birds as well as a bat, a fish, and a number of invertebrates and plants (79). An overwhelming majority of the remaining fauna have declined, and many are continuing to do so. Nationally, birds, bats, lizards, frogs, and invertebrates are characterized by low population density or local extinction, range contraction, and severe population fragmentation (122). Offshore islands provide sole refuge for both species of tuatara and 37% of lizard species (121) and many birds.

Reductions have been greatest in the north, where most people live. For example, historic northern North Island reptile fauna comprising tuatara, 6 species of gecko, and 11 skinks has been reduced by over 50% to 4 species of gecko and 4 or 5 skinks (121).

Introduced Pests

Introduced weeds clearly pose a threat to indigenous communities, especially in small lowland remnants with relatively large edge:core ratios. Among the 230 species of environmental weeds (131), five types can be emphasized: (*a*) climbers (e.g. *Clematis vitalba*) which smother natives; (*b*) shade-tolerant herbs (e.g. ginger *Hedychium* spp.) form extensive clones, suppressing the natural regeneration of the understory species; (*c*) species with long-lived seed banks (e.g. *Lycesteria formosa*), which can colonize natural gaps and subvert the normal gap-regeneration processes; (*d*) woody seral species (e.g. *Hakea* and *Acacia* spp.); and (*e*) bird-dispersed trees (e.g. *Acmena smithii*). Most weeds appear unable to invade intact forest, but others such as *Acmena* do.

Weeds are not the only alien biota affecting landscapes and are rarely the chief concern of conservation managers (25). Polynesian colonists introduced the Pacific rat (*Rattus exulans*) and the dog (*Canis familiaris*). European explorers and colonists introduced more species. Since 1769 over 80 species have become established, including 34 mammals (63). Predatory mammals, including rats, mustelids, and cats, have eliminated or continue to reduce many animal species including those responsible for pollination and seed dispersal. Herbivorous mammals such as brushtail possums (*Trichosurus vulpecula*), deer (*Cervus* spp.), and goats (*Capra hircus*) are nearly ubiquitous, markedly altering the structure and composition of native plant communities (23, 32, 63). Brushtail possums are also significant nest predators of threatened birds (18). In addition to alien plants and mammals, many species of alien insects, birds, and fish have also become established (6, 73). Social wasps (*Vespula vulgaris*) are especially problematic in beech (*Nothofagus*) forests (13, 83, 119).

A government analysis (84) lists 403 New Zealand taxa as threatened, including 159 plants, 98 invertebrates, and 146 vertebrates. Recent reappraisal lists 511 plants (22% of endemic flora) as threatened (43). New Zealand birds include a particularly high proportion of threatened species (25). Forty-one of the 45 threatened birds (10) are endemic, and many now occur only on mammal-free islands (26). Despite greater diversity than vascular plants and terrestrial vertebrates, invertebrate taxa comprise less than 0.1% of all threatened taxonomic groups, a disparity more related to knowledge than reality (65). Large (>10 mm body length), flightless, nocturnal, litter-dwelling invertebrates have been particularly affected.

CONSERVATION ISSUES

Preservation or Sustainable Management

The debate between preservation and sustainable management as alternatives currently polarizes conservation. The Conservation Act (1987) has the overarching goal of "the preservation and protection of natural and historic resources for the purpose of maintaining their intrinsic values, providing for their appreciation and recreational enjoyment by the public, and safeguarding the options of future generations." Largely seen and interpreted as championing the preservation philosophy, this Act sets conservation apart from other human activities, rather than viewing it as an integral part of sustainable management and of managing people as part of functioning ecosystems.

The Resource Management Act (1991), despite its name, is a sustainability law aimed at advancing ecological, social, and economic goals jointly through use, development, and conservation. The requirement to "sustain the life supporting capacity of air, water, soil and ecosystems" sets a clear ecological goal that embodies effective conservation ideals. This requirement allows more effective interpretation of conservation into economic and social understanding than the oxymoron of "maintaining intrinsic values" (34, 38).

Most New Zealanders (85%) live in towns and cities (79) removed from day-to-day contact with functioning native ecosystems. For many, the presence of massive old trees, possibly pre-dating Maori arrival, suggests that areas of "primeval" forest remain. The ancient geological lineage of some genera reinforces this concept of a primeval unchanging forest, deserving of preservation, and as a model for restoration. However, a history of continued vegetation change plus the loss of many animals suggests this is not a long-term reality.

By definition, under preservation ideals, native biodiversity cannot be used and hence has no recognized economic value. Indeed Parliament has ruled that access to public lands is free, effectively making use and abuse also free (34). Such rulings preclude ecotourism (28), as the costs of use or ecosystem maintenance such as pest control cannot be internalized. Government is the sole funder (35, 56), thus ensuring that conservation is influenced more by politics than sustainability. Despite strong public interest (4, 36) and calls by government for greater public participation (4, 45, 79), government publications record ongoing biodiversity decline (4) with this approach. Indeed it is a tribute to individuals that declines have been reversed in some areas and for some species.

Moreover, since it cannot be used, native biodiversity on private land (70%) receives no economic recognition for the ecosystem services that it provides to society. In the absence of realizable value, land owners carry all the costs of pest control, and it is economically rational to replace natives with exotic species that have economic value. Not surprisingly, introduced grasses, sheep, cattle, and pine trees dominate most landscapes. Indeed 50% of New Zealand has been converted to pasture compared with a world average of only 25% (79). Exotic forest monocultures relatively low in native biodiversity (100) cover 6% of the land and are increasing, whereas less than 3% of the land is in indigenous forestry and none is being planted (79).

Monetary valuation of native biodiversity is abhorrent to most conservationists, yet without realistic valuation, native biodiversity will not be considered in most decision making (30). This will also entrench conservation as the exception across production landscapes, further reinforcing the dichotomy between preservation and sustainable management (81, 93). This issue was recently highlighted in plans for sustainable management of beech forests, where political expediency prevailed. The belief among the largely urban conservation movement is that managers of the productive landscape—foresters, fishers, and farmers—should become more conservation orientated, rather than that any of the conservation estate should be physically used. The abundant possibilities for integrating production and conservation are hindered when these are seen as mutually exclusive land uses (81).

Representativeness

Although 30% of the land area is in reserves, there are ongoing calls for more. Most conservation land is either in the super-humid regions or the uplands and montane areas that are not useful for production (75, 103, 123). In contrast protected areas in the fertile lowlands tend to be small, fragmented, isolated, extensively modified (88%), and generally poorly managed. Yet some 50% of New Zealand is below 300 m elevation. This is clearly illustrated in Northland (80, 82), where ecosystem loss is especially concentrated in the lowlands (Table 1, page 72) and reserves are more common on cooler, less productive slopes (Table 2, page 72).

This pattern conflicts with the requirement of the Reserves Act (1977) and the Conservation Act (1987) that reserves should represent the "original" character of the country. Such views deny the dynamic nature of ecosystems as confirmed from historic ecology (59, 74). Few areas of New Zealand have had a stable forest composition for more than a few tree generations (59). Change, rather than stability, in composition appears to be the rule, at all scales.

This is not to say that there are no relict areas of formerly more widespread vegetation types, nor that preservation of forests that have remained relatively unchanged since before European colonization is not a laudable goal in some areas. However, it is not possible to maintain representative areas unchanged in the long term, nor is it possible to define a primeval restoration goal, except in a very general sense. Moreover, the dynamics of historical ecosystem change also weakens the dichotomy between what is regarded as original (and therefore desirable) versus modified. It provides the basis for a new paradigm in conservation that would accept greater merging of the indigenous and exotic biota and provide a class of conservation land in which management goals would recognize the need to integrate the protected and productive components of the landscape (81). Such a change would also be more accepting of people as part of the landscape and would more closely align with Maori values.

The main response of the Department of Conservation to the ongoing declines in native ecosystems has been to concentrate management in areas where pests can be most easily eliminated or controlled. Thus, as with the production–protection dichotomy, limited resources and a protectionist philosophy may be leading to a division of the public conservation estate into priority areas of intensive management, where the aim is to restore and maintain a largely pre-European biota, and the remainder. In the latter—by far the greatest area—the interaction between the indigenous community and introduced pests is tackled with varying degrees of enthusiasm and effectiveness by inadequately funded conservancies and regional councils.

Intensively managed reserves are necessary to sustain some of the endangered fauna and some forest areas where compositional gradients representative of formerly extensive landscapes can be seen. However, it is clear that ecosystems covering most of the landscape have changed, are changing, and will continue to do so. How these areas are managed requires reasoned debate. They could provide for multiple uses, as envisaged in the Resource Management Act. For example, selective timber yield and recreational hunting in such areas could remove the pressure for these activities in fully protected reserves. They could also be moved to community or even privatized management. Recent reactions to the proposal for sustainable beech forest management that included minimal logging suggest that New Zealand society is not yet prepared to accept these options.

Management of Alien Species

In New Zealand, as elsewhere in the world, the response to invasive species that threaten native biodiversity has been minimal action. New legislation, however, requires action. For most mainland sites, control is the only option, but eradication is possible on islands. Eradication of introduced mammals from islands has been a major advance in New Zealand conservation in recent years (125, 126). Ungulates have been eradicated from islands up to 12,000 ha, cats from areas up to 3000 ha, and brushtail possums and rats from areas up to 2000 ha. These and other successes have resulted from the availability of single-dose anticoagulant poisons, such as brodifacoum, and the development of bait stations and aerial application (115, 126). Rodents have now been eradicated from 60 islands (126; CR Veitch, personal communication). Islands cleared of introduced mammals are now routinely used as important conservation sites for threatened species management (26, 124). Eradication of alien plants from New Zealand islands has been attempted less often, but intensive campaigns are under way, including an attempt to eradicate exotic estuarine grasses (*Spartina* spp.) (92, 102).

Biological control has been tried for a small number of plant and animal species, whereas poisoning is common. The most extensive chemical control of an invasive species in New Zealand is the ongoing campaign against the brushtail possum to protect native ecosystems and to prevent the transmission of bovine tuberculosis to livestock. This typically involves the aerial distribution of baits containing 1080 poison (sodium monofluoroacetate) over large areas of native forest or the use of anticoagulant toxin in bait stations. Such control also reduces populations of other invasive mammals such as rats, pigs, and deer. Secondary poisoning of feral cats and mustelids feeding on poisoned rodents can also occur (1, 50, 89), although, less beneficial, diet switching to birds by surviving stoats can occur after rats decline (88).

Temporary control of possums and other mammals to enhance the breeding success of native birds is now routine (27, 61, 96). The concept of "mainland islands" has resulted. These are defined areas of natural habitat on the New Zealand mainland, selected for permanent, intensive pest control and ecological restoration (113). Within these areas, invasive species (especially mammals) are reduced to minimum densities to permit recovery of threatened native species and ecosystem processes. Early results are promising (25, 28, 60, 113) although high costs, accumulation of toxins, nontarget game mammal poisoning (46), short-term declines in rare species (107, 112), and public wariness of poisons are concerns.

Reintroductions

New Zealand has a long history of conducting species reintroductions on land as a method of ecological restoration, especially of islands following pest eradications, but comparable experiments in the sea are only just being considered. Since 1960, ncarly 260 species transfers involving at least 66 animal species have been documented (51, 124). Numbers of transfers and the diversity of taxa have steadily increased since 1980. Birds dominated initially; thereafter transfers have included frogs, invertebrates, and especially reptiles. More effective predator control has increased the number of sites available for ecosystem restoration. As a result a number of species previously known only as single small populations now have multiple populations and some have a less threatened status (70). Numbers of founders used are typically lower than those used elsewhere (54), but success rates match or exceed overseas levels (5), possibly because New Zealand wildlife is less susceptible to inbreeding depression (33), and managers tend to select destinations in which species are known to do well (5).

Most threatened species in New Zealand are inaccessible to the general public. Current "mainland islands" are distant from major population centers, and most species translocations take place to remote mainland sites or to islands where people are denied access. The exceptions are Tiritiri Matangi Island (36) near the largest city (Auckland), to which eight threatened or rare birds have been translocated, and Matiu/Somes Island, in Wellington, which recently received the highly endangered Brother's Island tuatara (*Sphenodon guntheri*). New initiatives such as the urban wildlife refuge surrounded by a mammal-proof fence, in Wellington, (71) and Wenderholm Regional Park, near Auckland, partly redress the current imbalance between actual and desired access by the public to native wildlife, especially threatened species (36), but much more could be done.

A limit to suitable restoration sites is an emerging problem for species translocations. Predator-free islands are limited, as is knowledge. Reintroductions have traditionally been conducted as one-off, nonreplicated events (trials) to locations that mirror source habitats. Reintroductions designed as well-planned experimental comparisons will more rapidly advance knowledge (5, 124).

Conservation of Functional Diversity

Although most biodiversity conservation in New Zealand has focused on individual species, there is an increasing awareness of the importance of functional diversity and not just species richness in maintaining the integrity of ecosystems (38, 123). Ensuring pollination and seed dispersal (19) highlights this issue. Some protected forests, especially those in northern regions, lack many of their original pollinators and seed dispersers; hence their future is unknown. Poor understanding hinders management.

Early studies based on floral characteristics (syndromes) suggested that, with few exceptions, the largely generalized native flora was pollinated by a range of unspecialized insects (58, 117). A later review conceded bird and bat visitation but argued this was probably incidental and most likely to result in self-pollination (52). This view remains influential (24). More recent work on island refuges, where predator eradication and species translocations have restored some trophic links, has provided new information (3, 22). Honeyeaters are observed regularly and persistently visiting a variety of apparently "entomophilous" flowers in the cooler months when insect activity is reduced, making these birds the likely pollinators (3, 22, 66). Pollinator limitation attributed to a decline in honeyeaters has been documented for mainland populations of some species (e.g. 44). Native lizards and the endangered short-tailed bat (*Mystacina tuberculata*) are also potential pollinators. Existence of multiple floral visitors belies the fact that only one of these may be an effective pollinator (129). Depletion of the array of endemic pollinators on mainland New Zealand is now cause for concern. Paralleling this loss is the introduction of floral visitors, either intentionally to pollinate the flowers of imported fruit and crop plants (e.g. bees *Apis mellifera* and *Bombus* spp.), or as self-introductions (e.g. silvereye *Zosterops lateralis*). The status of these recent arrivals as pollinators of native flora is uncertain.

New Zealand forests have a high percentage (70%) of woody plants with fleshy fruits suited to vertebrate dispersal (21). While lizards (130), native bats (69), seabirds, and seals (3, 94) disperse some seeds, most fruits are probably dispersed by forest birds (28). Likely impacts of their reduction in variety and abundance in the recent past is recognized (8, 21, 24, 67). Several large-fruited native plants now depend almost entirely on the native pigeon or kereru (*Hemiphaga novaeseelandiae*) for their dispersal, a precarious situation given the current declines of kereru populations (24, 68).

Introduced seed dispersers may compensate to some extent (3, 21), but the extreme fragmentation of lowland forests makes it difficult for widespread seed dispersal to occur. As a result, remnant forests are showing a changing and reducing species diversity (133).

Maori and Conservation

An international consensus is developing, among those concerned with the long-term future of protected areas, that the long-term goal of sustainability is more likely where local communities are involved. Biodiversity benefits, as well the social, cultural, and economic well-being of local people (49). The Convention on Biological Diversity ratified by the New Zealand government affirms such rights for indigenous and local peoples (106). Recent legislation including the Conservation Act, the Resource Management Act, and the Fisheries Act recognize the rights of Maori as recorded in the 1840 Treaty of Waitangi. Article Two guarantees Maori rights over their natural resources including fishing, forests, traditional foods, etc. Yet practical recognition of traditional ecological knowledge (15), environmental responsibilities (111), and resource rights is minimal.

Some agreement has been reached for joint control over the harvesting of traditional resources such as whale bone, pounamu or jade, freshwater fish, and titi or mutton birds (72, 90, 110). Joint management of traditional fishing grounds (taiapure) has also been provided for in legislation (116). But in virtually all cases, ultimate control and decision making still reside with government. True partnerships, involving the application of traditional knowledge in the sustainable management of the resources and biodiversity within protected areas, has been successfully implemented in Australia and Canada. These demonstrate a win-win outcome for both partners and for the environment. Despite these examples, and our own Treaty obligations, co-management of the public conservation estate is still unknown in Aotearoa/New Zealand.

Recognition of intellectual property rights, and the equity issues that arise from them, similarly requires government attention. This has resulted in the 1993 Mataatua declaration and a claim to the Waitangi Tribunal (Wai 262) seeking recognition, restoration, and protection of Maori cultural and intellectual property rights over flora and fauna (127).

Despite these initiatives and some major legislative advances both internationally and nationally, Maori still remain on the periphery of efforts aimed at the conservation and management of biodiversity. Greater effort is needed to identify and remove the individual and institutional barriers that prevent empowerment of Maori and other community groups in achieving sustainable biodiversity outcomes in Aotearoa/New Zealand.

Marine Conservation

Sustainable management and conservation in the sea lags that on land (120). Moreover, the difficulty of marine research (11, 40, 104) means that scientific knowledge of the sea is relatively poor. Marine habitats are poorly described, their extent uncertain, and their resilience to anthropogenic disturbance unknown. Unlike in land management, reserves are insignificant, and harvest of native species is the basis of a major export industry. In line with concepts of maximum harvest rather than sustainable management, stocks of most commercial species have been "mined" to between 25% and 70% of original levels (79).

Fishing impacts both the species being targeted and the surrounding marine environment. Traditional fisheries management is concerned solely with controlling the catch of economically valuable species. New Zealand in 1986 was the first country to implement an extensive system of individual transferable quotas (ITQs) as its primary fisheries management tool. This system requires the government to annually set a total allowable catch (TAC), which includes catch for commercial, recreational, and traditional (Maori) fishers as well as illegal take (poaching). The commercial component of the TAC (the TACC) is then allocated proportionally to each quota holder as a fully transferable property right.

The New Zealand ITQ system is viewed as an economic success because it optimizes fishing effort, despite substantial costs (17). In conservation terms, success is harder to judge. Catch has been stabilized for many harvested species, but poor knowledge of the ecology of most species involved severely limits the ability to estimate sustainable TACCs. In addition, 1999 amendments to the Fisheries Act (1996), driven by the political ideology of deregulation, allow the New Zealand fishing industry a much greater role in the actual stock assessments. Although this internalizes costs, without independent audit public confidence in sustainability is low. Further difficulties arise for many inshore fisheries where the recreational component is relatively large and control difficult. Claims of public rights for everyone to harvest have resulted in classic examples of "the tragedy of the commons," especially with intertidal shellfish. Estimates of recreational take are poor (62) and recovery after depletion slow (85). More worrying is the effect of a species-based management system on the overall structure of marine communities (9, 104), including habitat degradation (118) and seabird and mammal bycatch (114).

New Zealand's 35 cetaceans and 6 pinnipeds are protected within the 200-mile EEZ by the New Zealand Marine Mammal Protection Act (1978). The Department of Conservation administers two marine mammal sanctuaries to protect a population of the endangered Hector's dolphin and the remaining population of Hooker's sea lions (114). Entanglement in fishing nets is a major problem for marine mammals that are the objects of conservation, and the southern trawl fishery for squid has been closed early on several occasions in recent years when the total allowable catch of endemic sea lions was exceeded.

Of 159 foreign marine species known to have entered New Zealand, 130 subsequently became established (39) although their effects are little known (41, 57). Of most recent concern is the brown seaweed (*Undaria pinnatifida*), which has become established in southern ports. The risks of importation of additional marine invaders on the hulls of boats or in ballast water are now controlled by the Ministry of Fisheries under the Biosecurity Act (1993), but effective control of existing marine exotics is nearly impossible.

The Marine Reserves Act (1971), introduced to provide protected areas for scientific study (42), has been used to create 16 reserves ranging in size from 748 km^2 to less than 1 km^2 (40). Under this legislation, all marine life is protected from all forms of disturbance. Past action has been largely ad hoc, and there are calls for a broader network (11, 12) to increase representation. As on land, evaluation of the

TABLE 1 The conversation status of forests at different altitudes in the Northland region, New Zealand, based on a survey of 1500 sites >5 ha

Altitude	Area (ha)	% of land area	% loss of forests	Mean forest area (ha)	% area in protection
<100 m	324,330	59.0%	79.3%	35	15.6%
100 m–200 m	140,751	25.6%	66.7%	68	17.7%
200 m–300 m	54,193	9.9%	66.0%	258	17.6%
>300 m[a]	30,538	5.5%	17.0%	910	53.0%

[a]The highest point in the region is 774 m, but for this study the data above 300 m has been aggregated.

TABLE 2 The distribution by aspect of protected sites in a survey of 900 sites in Rodney District (Auckland/Northland)

	North	North-east	East	South-east	South	South-west	West	North-west	Flat
Proportion of slopes	20	13	6	16	7	10	8	12	8
Proportions of protected sites	19	4	26	11	26	9	1	4	0

Note: At these latitudes sourtherly aspects are cool and easterly most exposed to severe storms.

effectiveness of these reserves in enhancing fish stocks and restoring ecosystem functioning awaits planned experimental comparisons.

CONCLUSIONS

New Zealand's biodiversity has been severely affected by the actions of people. These changes are relatively recent and ongoing. Increased understanding and action plus legislative changes have all assisted, but recognition that the declines are ongoing requires more radical and urgent action. More open debate of the ecological, social, and economic realities of New Zealand's environmental management is required to sustain both biodiversity and the economy in the long term.

LITERATURE CITED

1. Alterio N, Brown K, Moller H. 1997. Secondary poisoning of mustelids in a New Zealand *Nothofagus* forest. *J. Zool.* (*Lon.*) 243:863–69
2. Anderson A. 1991. The chronology of colonisation of New Zealand. *Antiquity* 65: 767–95
3. Anderson SH. 1997. *Changes in ecosystem*

processes: the dynamics of pollination and seed dispersal in New Zealand forests. MSc thesis. Univ. Auck., NZ. 171 pp.

4. Anon. 1998. *New Zealand's biodiversity strategy: our chance to turn the tide*. Draft. Wellington, N.Z.: Dep. Conserv. Ministry for the Environ. 142 pp.
5. Armstrong DP, McLean IG. 1995. New Zealand translocations: theory and practice. *Pac. Cons. Biol.* 2:39–54
6. Atkinson IAE, Cameron EK. 1993. Human influence on the terrestrial biota and biotic communities of New Zealand. *TREE* 8:447–51
7. Atkinson IAE, Greenwood RM. 1989. Relationship between moas and plants. *N.Z. J. Ecol.* 12:S67–S95
8. Atkinson IAE, Millener PR. 1991. An ornithological glimpse into New Zealand's pre-human past. *Acta XX Congr. Int. Ornithol.* 1:129–94
9. Babcock RC, Kelly S, Shears NT, Walker JW, Willis TJ. 1999. Changes in community structure in temperate marine reserves. *Mar. Ecol. Prog. Ser.* 189:125–34
10. Baillie J, Groombridge B. 1996. *1996 IUCN Red List of Threatened Animals*. Gland/Washington, DC: IUCN & Conserv. Int.
11. Ballantine WJ. 1991. Marine reserves for New Zealand. *Leigh Lab. Bull. No. 25*. 195 pp.
12. Ballantine WJ. 1999. *Marine reserves in New Zealand: the development of the concepts and the principles*. Oral presentation to workshop on Marine Protected Areas, Kordi, Korea, Nov. 1999
13. Beggs JR, Wilson PR. 1991. The kaka *Nestor meridionalis*, a New Zealand parrot endangered by introduced wasps and mammals. *Biol. Conserv.* 56:23–38
14. Bell. 1991. Recent avifaunal changes and the history of ornithology in New Zealand. *Acta XX Ornithol. Int.* 1:195–230
15. Berkes F. 1999. *Sacred Ecology: Traditional Ecological Knowledge and Resource Management*. Philadelphia: Taylor & Francis. 209 pp.
16. Bosselmann K, Taylor P. 1995. The New Zealand law and conservation. *Pac. Conserv. Biol.* 2:113–20
17. Boyd RO, Dewees CM. 1992. Putting theory into practice: individual transferable quotas in New Zealand fisheries. *Soc. Nat. Resourc.* 5:179–98
18. Brown K, Innes J, Shorten R. 1993. Evidence that possums prey on and scavenge birds' eggs, birds and mammals. *Notornis* 40:1–9
19. Buchmann SL, Nabhan GP. 1996. *The Forgotten Pollinators*. Washington: Island Press. 292 pp.
20. Burrows CJ. 1994. The seeds always know best. *N.Z. J. Bot.* 32:349–63
21. Burrows CJ. 1994. Fruit types and seed dispersal modes of woody plants in Ahuriri Summit Bush, Port Hills, Western Banks Peninsula, Canterbury, New Zealand. *N.Z. J. Bot.* 32:169–81
22. Castro I. 1997. Honeyeaters and the New Zealand forest flora: the utilisation and profitability of small flowers. *N.Z. J. Ecol.* 21:169–79
23. Challies C. 1990. Red deer. See Ref 63, pp. 436–48
24. Clout MN. 1989. The importance of birds as browsers, pollinators and seed dispersers in New Zealand forests. *N.Z. J. Ecol.* 12: 27–33
25. Clout MN. 1999. Biodiversity conservation and the management of invasive animals in New Zealand. In *Invasive Species and Biodiversity Management*, ed. OT Sandlund, PJ Schei, A Viken, pp. 349–59. London: Kluwer
26. Clout MN, Craig JL. 1995. The conservation of critically endangered flightless birds in New Zealand. *Ibis* 137:S181–90
27. Clout MN, Denyer K, James RE, McFadden IG. 1995. Breeding success of New Zealand pigeons (*Hemiphaga novaeseelandiae*) in relation to control of introduced mammals. *N.Z. J. Ecol.* 19:209–12

28. Clout MN, Saunders AJ. 1995. Conservation and ecological restoration in New Zealand. *Pac. Conserv. Biol.* 2:91–98
29. Commonwealth of Australia 1994. *National Ecotourism Strategy.* Canberra: Aust. Gov. 68 pp.
30. Constanza R, d'Arge R, de Groot R, Farber S, Grasso M, et al. 1997. The value of the world's ecosystem services and natural capital. *Nature* 387:253–60
31. Cooper RA, Milliner PR. 1993. The New Zealand biota: historical background and new research. *TREE* 8:429–33
32. Cowan PE. 1990. Brushtail possum. See Ref. 62, pp. 68–98
33. Craig JL. 1991. Are small populations viable? *Acta XX Congr. Int. Ornithol.* pp. 2546–52
34. Craig JL. 1997. Managing bird populations: for whom and at what cost? *Pac. Conserv. Biol.* 3:172–82
35. Craig JL. 1998. An economic analysis of New Zealand's conservation strategies. *Agenda* 5:311–22
36. Craig JL, Craig CJ, Murphy BD, Murphy AJ. 1995. Community involvement for effective conservation: What does the community want? In *Nature Conservation 4: The Role of Networks*, ed. DA Saunders, JL Craig, EM Mattiske, pp. 187–94. Chipping Norton, Austral.: Surrey Beatty & Sons
37. Craig JL, Mitchell NM, Saunders DA. 2000. *Nature Conservation 5: Conservation in Production Environments: Managing the Matrix.* Chipping Norton, Austr.: Surrey Beatty & Sons
38. Craig JL, Stewart AM. 1994. Conservation: a starfish without a central disk? *Pac. Conserv. Biol.* 1:163–68
39. Cranfield HJ, Gordon DP, Willam RC, Marshall BA, Battershill CN, et al. 1998. Adventive marine species in New Zealand. *NIWA Tech. Rep. No. 34.* 48 pp.
40. Creese RG, Cole RG. 1995. Marine conservation in New Zealand. *Pac. Conserv. Biol.* 2:55–63
41. Creese RG, Hooker SH, De Luca S, Wharton Y. 1997. The ecology and environmental impact of the introduced Asian date mussel, *Musculista senhousia. N.Z. J. Mar. Freshwater Res.* 31:225–36
42. Creese RG, Jeffs A. 1993. Biological research in New Zealand Marine Reserves. In *Proc. Int. Temperate Reef Symp., 2nd*, ed. C Battershill, D Schiel, G Jones, R Creese, A MacDiarmid, pp. 15–22. Wellington, N.Z.: NIWA Marine
43. de Lange PJ, Heenan PB, Given DR, Norton DA, Ogle CC, et al. 1999. Threatened and uncommon plants of New Zealand. *N.Z. J. Bot.* 37:603–28
44. de Lange PJ, Norton DA, Molloy BPJ. 1996. *Ecology and Conservation of New Zealand's* Loranthaceous *Mistletoes.* Wellington, N.Z.: Dep. Conserv.
45. Department of Conservation. 1998. *Restoring the Dawn Chorus: the Department of Conservation Strategic Plan 1998–2002.* Wellington, N.Z.: Dep. Conserv. 64 pp.
46. Eason CT, Milne L, Potts M, Morriss G, Wright GRG, et al. 1999. Secondary and tertiary risks associated with brodifacoum. *N.Z. J. Ecol.* 23:219–24
47. Fisher DE. 1991. *The Resource Management Legislation of 1991: A Judical Analysis of its Objectives. Resource Management.* Wellington, N.Z.: Brooker & Friend
48. Fritts TH, Rodda GH. 1998. The role of introduced species in the degradation of island ecosystems. *Annu. Rev. Ecol. Syst.* 29:113–40
49. Furze B, De Lacy T, Birckhead J. 1996. *Culture, Conservation and Biodiversity.* Chichester, UK: John Wiley & Sons. 269 pp.
50. Gillies CA, Pierce RJ. 1999. Secondary poisoning of mammalian predators during possum and rodent control operations at Trounson Kauri Park, Northland, New Zealand. *N.Z. J. Ecol.* 23:183–92
51. Girardet S. 2000. *Tools for saving endangered species: eradication, translocation, triangulation.* PhD thesis. Auckland Univ., Auckland, N.Z. 252 pp.

52. Godley EJ. 1979. Flower biology in New Zealand. *N.Z. J. Bot.* 17:441–66
53. Goodland R. 1995. The concept of environmental sustainability. *Annu. Rev. Ecol. Syst.* 26:1–24
54. Griffith B, Scott JM, Carpenter JW, Reed C. 1989. Translocation as a species conservation tool: status and strategy. *Science* 245:477–80
55. Hardy GS. 1977. The New Zealand Scincidae (Reptilia: Lacertilia): a taxonomic and zoogeographic study. *N.Z. J. Zool.* 4:221–325
56. Hartley P. 1997. *Conservation Strategies for New Zealand.* Wellington, N.Z.: Business Roundtable
57. Hayward BW. 1997. Introduced marine organisms in New Zealand and their impact in the Waitemata Harbour, Auckland. *Tane* 36:197–223
58. Heine E. 1938. Observations on the pollination of New Zealand flowering plants. *Trans. R. Soc. N.Z.* 67:133–48
59. Horrocks M, Ogden J. 2000. Evidence of late glacial and Holocene tree-line fluctuations from pollen diagrams from the sub-alpine zone on Mt. Hauhungatahi, Tongoriro National Park, New Zealand. *Holocene* 10:61–73
60. Innes J, Barker G. 1999. Ecological consequences of toxin use for mammalian pest control in New Zealand—an overview. *N.Z. J. Ecol.* 23:111–27
61. Innes J, Hay JR, Flux I, Bradfield P, Speed H, et al. 1999. Successful recovery of North Island kokako (*Callaeas cinerea wilsoni*) populations, by adaptive management. *Biol. Cons.* 87:210–14
62. Kearney MB. 1999. *Ecology and Management of* Austrovenus stutchburyi *in the Whangateau Harbour.* MSc thesis. Univ. Auck., Auckland, N.Z. 132 pp.
63. King CM, ed. 1990. *The Handbook of New Zealand Mammals.* Oxford, UK: Oxford Univ. Press
64. Klimaszewski J, Watt JC. 1997. *Coleoptera: Family-Group Review and Keys to Identification. Fauna of New Zealand No. 37.* Canterbury, N.Z.: Manaaki Whenua Press. 199 pp.
65. Kuschel G. 1990. *Beetles in a Suburban Environment: A New Zealand Case Study. DSIR Plant Protection Report No. 3.* DSIR. 119 pp.
66. Ladley JJ, Kelly D, Robertson AW. 1997. Explosive flowering, nectar production, breeding systems and pollinators of New Zealand mistletoes (Loranthaceae). *N.Z. J. Bot.* 35:345–60
67. Lee W. 1988. Fruit colour in relation to the ecology and habit of the *Coprosma* species in New Zealand. *Oikos* 53:325–31
68. Lee WG, Clout MC, Robertson HA, Wilson JB. 1991. Avian dispersers and fleshy fruits in New Zealand. *Acta XX Congr. Int. Ornithol.* 3:1617–23
69. Lord JM. 1991. Pollination and seed dispersal in *Freycinetia baueriana*, a dioecious liane that has lost its bat pollinator. *N.Z. J. Bot.* 29:83–86
70. Lovegrove T. 1996. Island releases of saddlebacks *Philesturnus carunculatus* in New Zealand. *Biol. Conserv.* 77:151–57
71. Lynch J. 1995. Back to the future: Karori—form reservoir to wildlife sanctuary. *F & B* 275:12–19
72. Lyver P, Moller H. 1999. *Titi harvest by Rakiura Maori: a case study of the use of Maori traditional environmental knowledge for sustainable resource management.* Presented at Manaaki Whenua conf., Te Papa, Wellington, April 21–23, 1999
73. McDowall RM. 1990. *New Zealand Freshwater Fishes: A Natural History and Guide.* Auckland, N.Z.: Heinemann Reed. 553 pp.
74. McGlone MS. 1985. Plant biogeography and the late Cenozoic history of New Zealand. *N.Z. J. Bot.* 23:723–49
75. McGlone MS. 1989. The Polynesian settlement of New Zealand in relation to environmental and biotic changes. *N.Z. J. Ecol.* 12:S115–29
76. McGlone MS, Anderson AJ, Holdaway

RN. 1994. An ecological approach to the Polynesian settlement of New Zealand. In *The Origins of the First New Zealanders*, ed. DG Sutton, pp. 136–63. Auckland, N.Z.: Auckland Univ. Press
77. Meffe GK, Carroll CR. 1994. *Principles of Conservation Biology*. Sunderland, MA: Sinauer Assoc.
78. Merton D. 1992. The legacy of old blue. *N.Z. J. Zool.* 16:65–68
79. Ministry for the Environment. 1997. *The State of New Zealand's Environment*. Wellington: Min. Environ. 802 pp.
80. Mitchell ND, Campbell GH, Cutting M, Ayres B. 1992. *The Protected Natural Areas Programme—Rodney Ecological District*. Auckland, N.Z.: Dep. Conserv. 193 pp.
81. Mitchell ND, Craig JL. 2000. Managing the matrix: realigning paradigms toward sustainability. See Ref. 37, pp. 26–34
82. Mitchell ND, Park GN. 1983. Indigenous forest map of Whangaroa-Kaikohe. NZMS 290 sheet P 04/05. Wellington, N.Z.; Gov. of N.Z.
83. Moller H, Plunkett GM, Tilley JAV, Toft RJ, Beggs, JR. 1990. Establishment of the wasp parasitoid, *Sphecophaga vesparum* (Hymneoptera: Ichneumonidae), in New Zealand. *N.Z. J. Zool.* 18:199–208
84. Molloy J, Davis A. 1994. *Setting Priorities for the Conservation of New Zealand's Threatened Plants and Animals*. Wellington, N.Z.: Dep. Conserv. 2nd ed.
85. Morrison MA, Browne GN. 1999. Intertidal shellfish population surveys in the Auckland region, 1998–1999, and associated yield estimates. *N.Z. Fisheries Assessment Res. Doc. 99/43*. 21 pp.
86. Morton J, Ogden J, Hughes T. 1984. *To Save a Forest: Whirinaki*. Auckland, N.Z.: Bateman. 111 pp.
87. Morton J. 1995. The future of New Zealand conservation: ethics and politics. *Pac. Conserv. Biol.* 2:2–6
88. Murphy EC, Bradfield P. 1992. Change in diet of stoats following poisoning of rats in a New Zealand forest. *N.Z. J. Ecol.* 16:137–40
89. Murphy EC, Robbins L, Young JB, Dowding JE. 1999. Secondary poisoning of stoats after an aerial 1080 poison operation in Pureora Forest, New Zealand. *N.Z. J. Ecol.* 23:175–82
90. New Zealand Government 1997. *Crown Settlement Offer: Consultation Document from the Ngai Tahu Negotiating Group*. Christchurch, N.Z.: Ngai Tahu Publ. Ltd. 68 pp.
91. Newsome PFJ. 1987. *The Vegetation Cover of New Zealand*. Wellington, N.Z.: Natl. Water & Soil Conserv. Authority, 153 pp.
92. Nichols P. 1999. *Further investigations into the environmental impacts of Spartina grass and its control*. Unpublished MSc thesis, Univ. Auckland, Auckland, N.Z. 125 pp.
93. Norton DA. 1999. Sand plain forest fragmentation and residential development, Invercargill City, New Zealand. See Ref. 37, pp. 157–65
94. Norton DA, de Lange PJ, Garnock-Jones PJ, Given DR. 1997. The role of seabirds and seals in the survival of coastal plants: lessons from New Zealand *Lepidium* (Brassicaceae). *Biol. Conserv.* 6:765–85
95. Norton DA, Miller CJ. 2000. Some issues and options for the conservation of native biodiversity in rural New Zealand. *Ecol. Manage. Rest.* 1:In press
96. O'Donnell CFJ, Dilks PJ, Elliott GP. 1992. Control of a stoat population irruption to enhance yellowhead breeding success. *Sci. Res. Internal Rep. No. 124*, Dep. Conserv., Wellington, N.Z.
97. Ogden J. 1988. Forest dynamics and stand-level dieback in New Zealand's *Nothofagus* forests. *Geojournal* 17:225–30
98. Ogden J. 1995. The long-term conservation of forest diversity in New Zealand. *Pac. Conserv. Biol.* 2:77–90
99. Ogden J, Basher L, McGlone M. 1998. Fire, forest regeneration and links with

early human habitation: evidence from New Zealand. *Ann. Bot.* 81:687–96

100. Ogden J, Braggins J, Stretton K, Anderson S. 1997. Plant species richness under *Pinus radiata* stands on the central North Island volcanic plateau, New Zealand. *N.Z. J. Ecol.* 21:17–29
101. Ogden J, Stewart GH, Allen RB. 1996. Ecology of New Zealand *Nothofagus* forest. In *The Ecology and Biogeography of* Nothofagus *Forests*, ed. TT Veblen, RS Hill, J Read, pp. 25–82. New Haven, CT: Yale Univ. Press
102. Owen SJ. 1998. *Department of Conservation Strategic Plan for Managing Invasive Weeds*. Wellington, N.Z.: Dep. Conserv.
103. Park GN. 1995. *Nga Uruora*. Wellington, N.Z.: Victoria Univ. Press. 376 pp.
104. Pauly D, Christensen V, Dalsgaard J, Froese R, Torres F. 1998. Fishing down marine food webs. *Science* 279:860–63
105. Pauly D, Christensen V, Froese R, Palomares ML. 2000. Fishing down aquatic food webs. *Am. Sci.* 88:46–51
106. Posey DA. 1996. *Traditional Resource Rights*. Switzerland: IUCN. 221 pp.
107. Powlesland RG, Knegtmans JW, Marshall ISG. 1999. Costs and benefits of aerial 1080 possum control operations using carrot baits to North Island robins (*Petroica australis longipes*), Pureora Forest Park. *N.Z. J. Ecol.* 23:149–60
108. Raven PH. 1973. Evolution of subalpine and alpine plant groups in New Zealand. *N.Z. J. Bot.* 11:177–200
109. Roberts M. 1991. Origin, dispersal routes and geographic distribution of *Rattus exulans* with special reference to New Zealand. *Pac. Sci.* 45:123–30
110. Roberts M, ed. 1998. *Protocol for the management of whale strandings in Ngatiwai rohe*. Auckland, N.Z.: Dep. Conserv.
111. Roberts M, Haami B. 1999. Science and other knowledge systems: coming of age in the new millennium. *Pac. World* 54:16–22
112. Robertson HA, Colbourne RM, Graham PJ, Miller PJ, Pierce, RJ. 1999. Survival of brown kiwi (*Apteryx mantelli*) exposed to brodifacoum poison in Northland, New Zealand. *N.Z. J. Ecol.* 23:225–32
113. Saunders AJ. 1998. Vertebrate pest control at mainland islands. *Proc. 11th Aust. Vertebrate Pest Conf., 11th*, pp. 61–68
114. Slooten E, Dawson S. 1995. Conservation of marine mammals in New Zealand. *Pac. Conserv. Biol.* 2:64–76
115. Taylor RH, Thomas BW. 1989. Eradication of Norway rats (*Rattus norvegicus*) from Hawea Island, Fiordland, using brodifacoum. *N.Z. J. Ecol.* 12:23–32
116. Te Puni Kokiri. 1993. *Mataatua Declaration on Cultural and Intellectual Property Rights of Indigenous Peoples Recommendation 2:10*. Wellington, N.Z.: Te Puni Kokiri
117. Thomson G. 1927. The pollination of New Zealand flowers by birds and insects. *Proc. N.Z. Inst.* 57:106–25
118. Thrush SF, Hewitt JE, Cummings VJ, Dayton PK, Cryer M, et al. 1998. Disturbance of the marine benthic habitat by commercial fishing: impacts at the scale of the fishery. *Ecol. Appl.* 8:866–79
119. Toft RJ, Beggs JR. 1995. Seasonality of crane flies (Diptera: Tipulidae) in South Island beech forest in relation to the abundance of *Vespula* wasps (Hymneoptera: Vespidae). *N.Z. Entomol.* 18:37–43
120. Towns DR, Ballantine WJ. 1993. Conservation and restoration of New Zealand island ecosystems. *TREE* 8:452–57
121. Towns DR, Daugherty CH, Cree A. 1999. Raising the prospects for a forgotten fauna: a review of ten years of conservation effort for New Zealand reptiles. *Biol. Conserv.* In press
122. Towns DR, Daugherty CH. 1994. Patterns of range contractions and extinctions in the New Zealand herpetofauna following human colonisation. *N.Z. J. Zool.* 21:325–39

123. Towns DR, Williams M. 1993. Conservation in New Zealand: towards a redefined approach. *J. R. Soc. N.Z.* 23:61–78
124. Ussher GT. 1999. *Restoration of threatened species populations: tuatara rehabilitations and reintroductions*. PhD thesis. Auckland Univ., Auckland, N.Z. 214 pp.
125. Veitch CR, Bell BD. 1990. Eradication of introduced animals from the islands of New Zealand. In *Ecological Restoration of New Zealand Islands*, ed. DR Towns, CH Daugherty, IAE Atkinson, pp. 137–46, *Conserv. Sci. Publ.* 2. Wellington: Dep. Conserv.
126. Veitch CR. 1994. Habitat repair: a necessary prerequisite to translocation of threatened birds. In *Reintroduction Biology of Australian and New Zealand Fauna*, ed. M Serena, pp. 97–104. Chipping Norton, Aust.: Surrey Beatty & Sons
127. Waitangi Tribunal. 1991. Claim 262—relating to the protection, control, conservation, management, treatment, propagation, sale, dispersal, utilisation and restriction on the use of and transmission or knowledge of New Zealand indigenous flora and fauna and the genetic resources contained therein.
128. Wardle P. 1991. *Vegetation of New Zealand*. Cambridge, UK: Cambridge Univ. Press. 672 pp.
129. Webb CJ. 1994. Pollination, self-incompatibility and fruit production in *Corokia cotoneaster* (Escalloniaceae). *N.Z. J. Bot.* 32:385–92
130. Whitaker AH. 1987. The roles of lizards in New Zealand plant reproductive strategies. *N.Z. J. Bot.* 25:315–27
131. Williams PA. 1997. Ecology and management of invasive weeds. *Conserv. Sci. Publ. No. 7*. Wellington, N.Z.: Dep. Conserv.
132. Working Party on Sustainable Land Management. 1994. *South Island High Country Review*. Wellington, N.Z.: Rep. Ministers of Conserv., Agric. Environ. 184 pp.
133. Young A, Mitchell NM. 1994. Microclimate and vegetational edge effects in a fragmented podocarp-broadleaf forest in New Zealand. *Biol. Conserv.* 67:63–72

Annu. Rev. Ecol. Syst. 2000. 31:79–105

THE EVOLUTION OF PREDATOR-PREY INTERACTIONS: Theory and Evidence

Peter A. Abrams
Department of Zoology, University of Toronto, 25 Harbord Street, Toronto, Ontario M5S 3G5 Canada; e-mail: abrams@zoo.utoronto.ca

Key Words coevolution, predation, stability

■ **Abstract** Recent theories regarding the evolution of predator-prey interactions is reviewed. This includes theory about the dynamics and stability of both populations and traits, as well as theory predicting how predatory and anti-predator traits should respond to environmental changes. Evolution can stabilize or destabilize interactions; stability is most likely when only the predator evolves, or when traits in one or both species are under strong stabilizing selection. Stability seems least likely when there is coevolution and a bi-directional axis of prey vulnerability. When population cycles exist, adaptation may either increase or decrease the amplitude of those cycles. An increase in the defensive ability of prey is less likely to produce evolutionary countermeasures in its partner than is a comparable increase in attack ability of the predator. Increased productivity may increase or decrease offensive and defensive adaptations. The apparent predominance of evolutionary responses of prey to predators over those of predators to prey is in general accord with equilibrium theory, but theory on stability may be difficult to confirm or refute. Recent work on geographically structured populations promises to advance our understanding of the evolution of predator-prey interactions.

INTRODUCTION

In this article, the term "predation" is used to describe an interaction in which individuals of one species kill and are capable of consuming a significant fraction of the biomass of individuals of another species. This definition includes finches that consume seeds, and the interaction between insect parasitoids and their hosts. However, it does not include most disease organisms and also does not include many herbivores. This definition was chosen because the interactions that fall under the definition can be modeled using a common mathematical framework. Most parasite-host relationships (which fall under some definitions of predation) require models where the number of infected hosts is the index of the abundance of the natural enemy. It is also necessary to take the longevity of infected hosts into account to determine parasite birth (transmission) rates. These properties lead to

0066-4162/00/1120-0079$14.00

quite different mathematical representations than those used for predators under the current definition. Even within this somewhat narrow definition of predation, however, there are very few species that are not engaged in some form of predator-prey interaction.

Given their major effects on fitness, traits that affect the ability to accomplish or avoid predation should therefore be under strong selection. Many current-day evolutionary biologists believe that predation has played a major role in determining patterns in the history of life on this planet, such as the increase in maximum complexity of organisms (32, 49, 66, 98). At the same time, the evolution of traits related to predation in both predator and prey has proven to be difficult to understand in theory, and difficult to study in the field. Many textbooks fail even to mention the evolution of traits related to predation, or at most, they devote a couple of pages to the subject (e.g. 45). The 33rd symposium of the British Ecological Society treated the subject of "Genes in Ecology" with only a passing reference in one chapter to the subject of predator-prey coevolution (22). Interpretations of temporal changes in predation-related characteristics of species have often been controversial (e.g. 61 vs. 76).

These difficulties have not prevented the growth of theory predicting the potential evolutionary trajectories of predator-prey interactions. Darwin (31) proposed that selection for catching various types of prey could lead to diversification of geographical races of predators. However, he did not discuss the simultaneous evolution of both species. Cott (30) may have been the first person (in 1940) to present the currently popular view of predator-prey coevolution as an arms race. In the following decades, most biologists who worked on the problem reverted to thinking about one-way evolutionary interactions between predator and prey. In the late 1960s and early 1970s, many biologists were concerned with explaining why predators did not evolve such a high efficiency that they drove their prey extinct (see Slobodkin's discussion of "prudent predation" in 90). Coevolution began to be mentioned more frequently beginning in the late 1960s. Some key works were Pimentel's verbal descriptions and laboratory experiments suggesting the possibility of a "genetic feedback mechanism" (74, 75) and a set of theoretical models of coevolution (including 65, 82, 83, 89). Dawkins & Krebs's (33) description of predator-prey coevolution as an arms race also helped to revive interest in the field.

Nevertheless, the empirical problems alluded to above have not yielded significantly to any advances in concepts or technology, and most of what we know comes from theoretical studies. This article reviews the questions asked, and both the assumptions and predictions made by previous theoretical treatments of the evolution of predator-prey interactions. It then reviews empirical evidence relating to both the assumptions and the predictions of the theory. In previous reviews of the evolution of predator-prey systems, there has been an excessive amount of interest in whether changes in the interaction of predator and prey represent coevolution in the narrow sense of Janzen's definition (60). Here evolutionary change in either species must evoke an evolutionary change in the other, which then changes the original trait value of the first species. This complicated scenario can best be understood by building upon the simpler cases in which only one of the two interacting species undergoes significant evolutionary change. Most interactions in nature are asymmetrical, and there is some evidence that predator-prey interactions are often characterized by greater responses of prey to predators than vice versa. Vermeij (97, 98), for example, has argued forcefully that predators affect the evolution of their prey, but prey do not significantly affect the evolution of predators. Nevertheless, it is both unlikely and difficult to prove that two-way effects are nonexistent. Therefore, the question 'Is it coevolution?' is not considered in any detail here. This article considers the evolution of one party in the predatory interaction, as well as coevolution of both species.

A REVIEW OF PREDATOR-PREY MODELS

The evolution of traits related to predation cannot be understood without some understanding of the dynamics of interacting populations that do not exhibit any significant evolutionary change. The two standard models for predator-prey and parasitoid-host interactions are the Lotka-Volterra and Nicholson-Bailey models. In their simplest forms both models lack any density dependence in the growth of the victim population, which leads to nonpersistent dynamics (neutrally stable cycles in the case of Lotka-Volterra, and diverging cycles ending in extinction for the Nicholson-Bailey system). The density-dependent versions of these two models are thus the minimal representations of a predator-prey system. The Lotka-Volterra model for a prey population of size N and a predator population of size P may be written,

$$\frac{dN}{dt} = rN\left(1 - \frac{N}{K}\right) - CNP$$

$$\frac{dP}{dt} = P(BCN - D) \qquad \text{1a,b.}$$

where r and K are the maximum per capita growth rate and carrying capacity of the prey; C is the capture rate per unit time per unit prey density by an average predator; B is the conversion efficiency of ingested prey into new predators; and D may be interpreted either as a per capita death rate or a per capita food requirement for maintenance and replacement of predators. The product, CN, represents the predator's functional response, i.e. the relationship between prey density, N, and the amount ingested by an average predator. The most common variant of equations (1a,b) is to replace this linear functional response with a type-2 (58) response. The latter is often described by the disk equation, $CN/(1 + ChN)$, where h is the amount of time required to capture a prey individual, during which further search for, or capture of prey is impossible. Equations (1a,b) always have a locally stable two-species equilibrium, provided the prey's carrying capacity is large enough

($K > D/(BC)$). If the type-2 response is substituted for the linear response in Equations (1a,b), then predator-prey cycles are possible when the predator is efficient (i.e. when the equilibrium prey density is low relative to its carrying capacity). Three other common modifications of these equations are: (*a*) a nonlinear relationship between amount eaten and per capita predator reproductive rate (i.e. a nonlinear numerical response); (*b*) a negative effect of predator density on the numerical or functional responses; and (*c*) nonlinear density-dependence of prey population growth.

The Nicholson-Bailey model and variants of it are difference equations that characterize population densities only at discrete intervals. Use of such models is most appropriate when reproduction occurs seasonally. The simplest case assumes that prey are eaten by predators continuously throughout the interval (i.e. season), but that predator numbers do not change within the interval. Then prey reproduction is based on the number of surviving prey multiplied by a per-individual birth rate that is reduced appropriately according to the density of prey (either at the beginning of the interval or averaged over the interval). The reproductive output of the predators is directly proportional to the number of prey eaten in the simplest case. A simple model for prey and predator numbers at time $t + 1$, given N_t prey and P_t predators at time t, is:

$$N_{t+1} = N_t Exp\left(r\left(1 - \frac{N_t}{K}\right) - CP_t\right)$$

$$P_{t+1} = BN_t(1 - Exp(-CP_t)) \qquad \text{2a,b.}$$

Again, it is possible to modify these equations to incorporate other forms of density dependent prey growth, nonlinear predator functional or numerical responses, and direct effects of predator density on the predator's per capita growth rate. The dynamics of equations (2) are still not fully understood, although conditions for the existence of a locally stable equilibrium were published over 20 years ago (20). Relatively high values of r and C can produce cycles or chaotic dynamics (12, 62).

The time course of evolutionary change in any trait that affects parameters in the above equations (or analogous parameters in more detailed equations) will usually depend on the population densities and patterns of change in population densities of one or both species. Thus, there is no way to study the evolution of such traits independently of the population dynamics that the ecological interaction implies.

TRAITS AND EVOLUTIONARY MODELS OF THEIR DYNAMICS

Many traits affect prey mortality and predator population growth rates in a predator-prey interaction. These may be classified according to which parameters of a population-dynamical model they influence. A few basic parameters appear in almost all such models. For the predator, the universally present parameters potentially affected by one or more traits are: (*a*) the individual's maximum capture rate of prey (C in Equations 1 and 2), and (*b*) the individual's per capita intake rate of prey required for zero population growth of the predator (B/D in Equation 1). In most cases, the relationship between prey abundance and the predator's intake rate (its functional response) is nonlinear, and predator traits may affect the shape of this relationship, for example, by changing handling time.

The shape of the relationship between intake rate and per capita growth rate (the predator's numerical response) may also be affected by the predator's traits. This relationship is assumed to be linear in both Equations 1 and 2, but it need not be. Any characteristics that increase the efficiency of conversion of food into new predators will affect the initial slope of the numerical response and may affect other aspects of its shape. A particular trait will often affect more than one parameter of the predator's population growth function. For example, larger jaw muscles in a predator that captures prey using its mouth may have the following effects: (*a*) greater maximum capture rate, because fewer prey escape following initial contact; (*b*) shorter handling time, because prey can be subdued or ingested more rapidly; (*c*) greater intake required for zero population growth, because larger muscles are energetically expensive, or offspring size or gestation period must increase. When predator abilities change with age, traits that affect life history parameters of necessity also affect predation-related parameters.

The traits of prey species may also be classified by the parameter(s) of a population dynamics model that are affected by those traits. The single parameter that is always present in predator-prey models is the predator's maximum per capita capture rate of prey (C in Equations 1 and 2). This is, of course, a function of prey traits as well as those of the predator. A lower maximum capture rate may be caused by traits that reduce the prey's chance of encountering a predator, of being detected if encounter occurs, or of escape following detection. Escape can often be brought about if the prey can discourage the predator from attacking by appearing dangerous, non-nutritious or unpalatable, or so proficient at escape that the attack would be futile. Life-history adaptations of prey to deal with high risk of predation may include faster progress through vulnerable stages or more general adaptations to high mortality (e.g. earlier reproduction and greater reproductive investment).

Most evolutionary models assume that there are pleiotropic effects of any trait that has an effect on capture rates. For the predator, traits that increase capture rates reduce some other component of fitness; for the prey, traits that decrease predation rates have costs. This assumption of tradeoffs or costs has empirical support from several detailed studies (for example, 25, 63). It is also a logical necessity in many (but not all) models, because without it traits would evolve to infinite values. In fact, some models (80, 89, and some models in 81) have solutions in which traits continue to increase indefinitely, although the system persists because adaptations in one species offset those in the other. Such models are not realistic (as Rosenzweig et al point out in 81). The exception to this argument for the necessity of costs is when an intermediate value of the trait produces the greatest value of a predation-rate-enhancing trait in the predator, or of an anti-predator trait in the prey.

Understanding the evolution of a predator-prey interaction requires a description of the potential dynamics of one or more traits in one or both species through time. Because the genetics of such traits are generally largely or completely unknown, most recent models have adopted approaches that are largely phenotypically based. Here, no restrictions are placed on the values of the predation-related traits, and the strength of selection determines the rate of change of average trait values. Three approaches fall under this general description. The first are models in which the traits change according to quantitative genetic recursion equations (64; applied in 84) or approximation to these equations (15, 91; applied in 13, 14). The second approach assumes that populations consist of asexual clones, and mutations with small effects produce new clones having slightly different phenotypes. If the new clone has a higher fitness than the resident, it will invade and either replace it or coexist with it (73). This approach is used by Dieckmann & Law (36), Dieckmann et al (37), and Marrow et al (70). The third approach is one that can be applied only when the equilibrium is stable; it simply looks for strategies that cannot be invaded by other strategies (27). All of these approaches are often well approximated by the same model in which the mean values of continuous traits change at rates proportional to the derivative of individual fitness with respect to the individual's trait value (15, 36, 99). Predictions are usually (but not always) very similar to those of multilocus models in which many loci have similar, small, additive effects on a character (38). The preceding discussion and the rest of this chapter ignores the role of evolutionary forces other than selection (and to a limited extent, mutation). This is more a reflection of the lack of knowledge about the roles of gene flow and drift than evidence that they are unimportant. Although most of the models have been based on extremely simple population models like Equations 1 and 2, there have now been enough studies of different models that some of the general features of these simplified scenarios are becoming clear.

This review tries to synthesize the findings of these previous models and focuses on two types of questions. The first is how evolution or coevolution affects the stability of predator-prey systems. The second question asks what is the response of predation-related traits in one species of the predator-prey pair to an environmental change or a change in the characteristics of its partner in the interaction? These are the questions that have most commonly been addressed in studies of predator-prey systems.

Population cycles are a potential consequence of predator-prey interactions without any evolution, and both predatory and anti-predator traits have been shown to affect the stability of the ecological interaction (7, 9, 13, 14, 79, 88). If the evolutionary dynamics of the traits that determine predation rates are unstable, this will drive population cycles. The previous discussions of the ability of evolutionary change in predator-prey systems to drive cycles have ranged from the conclusion that cycles occur very seldom (88) to the conclusion that cycles are almost inevitable (69). Neither of these extreme positions applies to the range of biologically plausible models that have been analyzed to date.

The second question concerned the response of traits to environmental change. The topics that have attracted the most attention to date are the responses of traits in one species to a change in the other species (2, 4, 5, 33, 34), and the response of traits to factors external to the predator and prey, such as enrichment of the prey's food source or changes in the predator's mortality rate (4, 56, 80). Here again, a variety of results have been obtained.

EVOLUTION AND STABILITY

Perhaps the most interesting questions about the effects of evolution of predation-related traits on stability are: Can evolution produce cycles in otherwise stable systems? (If so, under what conditions?); Can evolution dampen or eliminate cycles that occur in the absence of evolutionary change? (Again, under what conditions?); Do population cycles significantly change the mean values of adaptively evolving traits?

To answer the first question, we must start with a stable version of a system like Equations 1 or 2, add dynamics of one or more of the parameters in that population dynamical model, and determine whether the resulting expanded system is still stable. This has been done for the case of predator evolution in (7), for prey evolution in (71) and (14), and for coevolution of both species in (13, 36, 37, 46, 56, 69–71, 85, 89, 94, 96). The theoretical results obtained thus far suggest that, although evolution in one species can cause cycles, the most likely source of evolutionary instability is the interaction between evolutionary variables in both species. However, some rather special conditions must be met for this to occur.

Predator Evolution

Predator evolution alone is able to drive cycles in otherwise stable systems (7), but it appears to be relatively unlikely to do so. Cycles require that higher prey densities select for lower predator consumption rates and that the response to selection be very rapid. The optimal capture-rate parameter (C in Equations (1) or (2)) should often decrease in response to increased prey abundance. This is favored by a predator numerical response (per capita growth rate) that increases at a decreasing rate as prey densities become high. Under these conditions, it is advantageous to reduce costly traits that increase capture when prey are abundant or easily caught because increased food intake increases population growth by only a small amount. This decrease in predation when prey are abundant has a destabilizing influence on the system because it allows the prey population to continue growing when it is abundant. To produce population cycles, however, the evolutionary decrease in capture rates must be quite rapid, and no empirical examples give evidence of this mechanism.

Evolutionary stabilization of predator-prey systems with unstable population dynamics is also possible. When constraints on the predator's capture rates or reproductive rates are minimal, then increases in prey abundance should select for investment in costly capture-related traits; when prey are rare, the costs are likely to exceed the potential benefits of such traits. The result will be greater values of capture-related traits when prey are common and lower values of these traits when prey are rare. Both of these outcomes favor increased stability, because both push prey densities toward an equilibrium with intermediate densities.

Prey Evolution

The evolution of prey defensive traits is more likely to be a source of instability than is evolution of capture-related traits in the predator. Instability can take the form of population and trait cycles (as in 14) or of evolution that leads to extinction of the prey species (72). Each of these outcomes depends on the presence of a saturating functional response on the part of the predator (absent from many previous models of evolution in predator prey systems). The cyclic outcome occurs via the following mechanism. As predators reduce the number of prey, they become less satiated, raising the risk of capture for prey individuals. This increases the selection for resistance on the part of prey, which leads to a reduced predator population and selection to reduce costly defensive traits. Given the appropriate time scale of responses, the time lag between changes in predator population and prey vulnerability will result in cycles. This requires relatively slow predator population dynamics and relatively rapid evolutionary change on the part of the prey, in addition to the nonlinear predator functional response. If this process occurs, then there is a positive feedback between cycle amplitude and relaxed selection on the prey's defensive trait. Greater vulnerability results in great amplitude population cycles, which imply that the predator is usually either rare or satiated. This further relaxes selection for defense.

If the process of positive feedback described in the preceding paragraph is not halted, population cycles continue to grow in amplitude, ending in extinction of one or both species (14). This runaway process requires that the prey be able to gain a growth rate advantage by decreasing defense, even when defense is already low. In the examples considered by Matsuda & Abrams (72), the predator population was assumed to be constant as the result of alternative resources other than the focal prey. A constant predator population prevents evolutionary or population cycles. In this case, prey evolution could cause a steady decrease in the prey population without any cycles. This again depends on a type-2 functional response and occurs because predators become less satiated as prey become more difficult to capture. This increases the risk, selecting for even greater defense. If the greater defense requires a reduced capture rate of the prey's own food (e.g. if vulnerability is based on time spent foraging), then the outcome can be evolution to vanishingly small population sizes of the prey species. In a stochastic system, extinction occurs rapidly under this scenario.

Prey evolution is not always destabilizing. As Abrams & Matsuda (14) noted, a linear predator functional response ensures that evolution of costly defensive traits in the prey will always promote stability. Stabilization is also possible with nonlinear functional responses (14, 59), provided that the nonlinearity is not too pronounced. Stabilization of an otherwise cyclic predator-prey system via evolution of the prey is also possible (14). The most important condition for this outcome is that the relationship between vulnerability and prey growth rate be negatively accelerating (i.e. have a negative second derivative). This is a very reasonable assumption for many prey traits. Predators have maximum capture rates, which can prevent capture rates from increasing further once prey are very easy to capture. Furthermore, prey are likely to have maximum growth rates that cannot be exceeded by becoming very vulnerable to predators. Given this sort of vulnerability-growth rate relationship, greater predator density in the course of a population cycle is likely to select for greater defense, which will dampen or eliminate the cycles.

Many possible scenarios for evolutionary change in predator or prey have yet to be examined in any detail. Different scenarios in the simple sorts of models discussed above can be produced by assuming that traits affect different pairs (or trios, etc.) of population dynamics parameters. For example, evolution in a predator with a type-2 functional response may produce both greater values of the per-individual capture rate C and higher values of the handling time, h. Alternatively, a tradeoff relationship could exist between handling time and per capita death rate or between handling time and efficiency of conversion of prey into predators. Abrams (2, 4) presented a fairly extensive list of possible relationships but did not explore these in the context of unstable population dynamics.

Coevolution

The vast majority of the studies touching on stability have assumed that both predator and prey are capable of significant evolutionary change. Most of these have found that predator-prey coevolution can lead to cycles in both traits and population densities. There have been enough studies with different assumptions that it is now possible to identify some of the key features responsible for stabilization or destabilization. Clearly, the cases of predator evolution and prey evolution are limiting cases of coevolution, so it must be possible for the outcomes described above to occur under genuine coevolution, provided the species have very unequal magnitudes of evolutionary change in their capture-related parameters. However, there are additional mechanisms whereby stability can be affected when both species undergo significant evolutionary change. The interaction of predator and prey traits in determining a capture rate seems to be the most important factor determining the stability of the entire system. Most studies in which cycles are a commonly observed outcome assume that the predator maximizes its rate of capture of the prey by matching the prey's phenotype. For a measurable trait, this means that the prey have a "bidirectional" axis of vulnerability; they can reduce their risk by having a trait either larger than or smaller than an intermediate "most vulnerable" phenotype, where the latter is determined by the predator's phenotype. The presence of this sort of bidirectional axis of vulnerability seems to be the primary reason for the occurrence of cycles in the models by Marrow et al (71, 72), Marrow & Cannings (69), Dieckmann et al (37), Dieckmann & Law (36), Gavrilets (46), and one set of models in Abrams & Matsuda (14). The mechanism for cycling in these studies can be described roughly as follows. If the prey's phenotype is initially slightly larger than the "most vulnerable" value, it will continue to increase, and the predator's phenotype will increase in response. When the prey phenotype becomes sufficiently extreme, its rate of change is slowed by the greater costs of increasing an already extreme trait. This allows the predator to catch up and surpass the prey phenotype. Immediately thereafter, there is strong selection for the prey to reduce its character value, and the predator phenotype then chases the prey phenotype back to low values of the trait. This is again followed by a decreased evolutionary rate of the prey when extremely small values of the trait incur heavy costs, and the predator is able to overtake the prey in phenotypic space, restarting the cycle. These evolutionary cycles occur in models with fixed population sizes, as shown by Gavrilets (46), and stability conditions appear to be changed relatively little by population dynamics (13).

A bidirectional axis of vulnerability is not sufficient to guarantee that cycles will occur. If there is sufficiently strong stabilizing selection based on the costs of extreme trait values in both species, then stable equilibria are likely (13). Stabilizing selection on the predator's trait is especially likely to eliminate cycles (13). If it does so, there may be either one or two alternative stable equilibria (37). The possibility of alternative equilibria arises because the prey can reduce predation by being either larger or smaller than the phenotype that is most vulnerable to the predator. If there is stabilizing selection based on effects on other population growth parameters in each species, the number of equilibria will depend on whether the optimal traits based on stabilizing selection alone are similar or different. If similar, then alternative equilibria are likely because the prey can reduce risk by becoming larger or smaller in trait value, depending on initial conditions. If the optima in predator and prey are quite different, it is more likely that there is a single equilibrium with the difference in trait values being determined by the positions of these predation-independent optima. The efficiency of predators in reducing the prey population size is an important parameter determining the existence of cycling. If the efficiency is low, the predator will not be common enough to drive the prey population significantly away from the trait optimum attained in the absence of predators. If the predators are efficient, they reduce both their own and the prey's population sizes. In the model of Dieckmann et al (37), this reduces the strength of selection on the predator, since its prey intake near equilibrium is going to be low (based on the low prey density), regardless of its phenotype. In models with a nonlinear functional or numerical response (unlike the models of Dieckmann et al), high predator efficiency usually leads to population dynamic cycles, which entrain cycles in the traits. Both Dieckmann et al (37) and Abrams

& Matsuda (13) showed that the prey must evolve with sufficient rapidity relative to the predator that their trait value can temporarily change more rapidly than that of the predator.

Several other models assume bidirectional axes of prey vulnerability. Van der Laan & Hogeweg (96) made this assumption in their model of predator-prey coevolution, which also exhibits cycles. Their model makes the rather unusual additional assumption that the phenotypic axis is circular. This would be the case if both species were potentially active for 24 hours and the adaptations for escape and capture were shifts in activity time. This seems to be at best a rare scenario. Doebeli's (39) is one of the few studies that argues for a strongly stabilizing role for coevolution when there is a bidirectional axis of vulnerability. He adopted a discrete generation model that is identical to the Nicholson-Bailey host-parasitoid model except for the presence of genetic variation in capture-related traits and sexual reproduction. Stabilization in this case means a reduced amplitude of cycles and/or longer persistence times, rather than a locally stable equilibrium. The Nicholson-Bailey model itself predicts rapid extinction due to divergent oscillations of parasitoid and host. In Doebeli's coevolutionary version of this model, a large number of loci (all assumed to have small additive effects) is required before indefinite persistence occurs. It is not clear to what extent genetic variation stabilizes systems that would persist in the absence of variation, such as the Nicholson-Bailey model with added density dependence, although Doebeli (39) presented some simulations showing that increased stability is possible for this case.

Of the papers that assume a bidirectional axis of vulnerability, there appears to be only one that did not observe cycles in traits and population densities. This is the study of Brown & Vincent (27). However, there are several potential explanations for the apparent stability described in that study. First, their analysis does not include any explicit evolutionary dynamics. Instead, they simply calculate equilibria that represent local fitness maxima for mutant types and use these to determine the traits and population densities in the final community. Because cycling is usually associated with equilibria where mean fitness is minimized for one of the species, this outcome may have simply been missed because of the lack of any explicit evolutionary dynamics. In addition, they presented calculations for only a limited range of parameters; thus, it could be that they happened to choose parameters that were stable. Finally, Brown & Vincent's model makes assumptions that differ from those of most other studies of coevolution: 1) new predator and prey types can always enter the system if their traits allow them to increase; and 2) the predator's growth is a decreasing function of the ratio of predator to prey density. The first assumption often requires the occurrence of mutations of large effect that can breed true. The second assumption is difficult to reconcile with more mechanistic consumer-resource models (48). While not impossible, neither assumption seems to be very general. Thus, in general, it appears that a bidirectional axis is a potent mechanism for generating coevolutionary cycles.

A bidirectional axis of vulnerability is not a requirement for the existence of cycles in the traits of predator and prey. Both Abrams & Matsuda (13) and Sasaki &

Godfray (85) have published models in which cycling can be caused by coevolution in spite of a unidirectional axis of prey vulnerability. Here, for each species, the direction of change in the trait that reduces (or increases) predator capture rates is independent of the phenotype of the other species. Both of these studies assumed an S-shaped relationship between the difference in trait values and the capture-rate parameter, corresponding to C in Equations 1 and 2. Abrams & Matsuda (13) assumed a continuous time model similar to Equation 1; here cycles with a unidirectional axis occur for a relatively narrow range of parameters. The presence of cycles was associated with an evolutionary equilibrium at which the mean trait value of prey represented a relative fitness minimum. In Sasaki & Godfray (85), the basic model was discrete, with many similarities to Equation 2. Here, the most common form of cycling was one in which the mean abilities of both species increased over a number of generations, followed by a crash in host resistance, which then produced a more gradual decrease in parasitoid attack ability. When attack ability was low enough, the increasing phase of the trait cycle began again. These cycles occurred for parameters where population densities would cycle in the absence of any evolutionary change. However, the amplitude of population cycles was often less in a model with coevolution than in a comparable model with no evolutionary change. Sasaki & Godfray (85) assumed that evolution occurred via competition among asexual clones, and this assumption appears to be important for the existence of the cycles that they observed. These depend on the invasion and increase of prey (hosts) with minimal defense when the majority of the prey have high levels of defense and the predators have highly developed attack traits. If the prey trait was polygenically determined in a sexual population, mutants with traits much less than the mean would not breed true.

Frank (44) provided another example of a predator-prey model with a unidirectional axis of prey vulnerability, where cycles in traits and population densities occur. Frank called his model a host-parasite model, but the population dynamics are described by difference equations that are analogous in form to Equation 1. This Lotka-Volterra form makes the model more appropriate as a description of predator-prey systems, although his discrete generation form may yield unrealistically large fluctuations or extinction when prey growth is high. In any case, Frank's model exhibits some examples of fluctuating traits and populations that probably involve mechanisms similar to those described for the model of Sasaki & Godfray's (85). It is unclear to what extent the very complex and chaotic fluctuations observed with inherently unstable population dynamics may be affected by the unrealistic aspects of the assumed linear density dependence in this difference equation system.

Saloniemi (84) appears to be the only other study in which cycles were observed in a model that did not assume a bidirectional axis for prey defensive traits. However, the cyclic outcomes appear to be due to a questionable model structure, which did not include the demographic consequences implied by the assumed stabilizing selection. As Abrams & Matsuda (13) showed, when these demographic effects are included, cycles never occur. This represents another case with a unidirectional axis of prey defense, where coevolution is basically a force that promotes stability.

Most studies of predator-prey coevolution have considered single homogeneous populations of both species. However, many populations exist as arrays of semi-independent populations connected by dispersing individuals. Spatial clumping of parasitoids and hosts has been one of the main factors cited as stabilizing interactions between them (52). However, the clumping has generally been assumed to occur in models without any consideration of whether or not it is adaptive. Van Baalen & Sabelis (94) were the first to consider the movement strategies of predators and prey (parasitoids and hosts) as evolutionary variables. The main finding of their coevolutionary model was that very large differences in patch quality (measured by prey population size that could be supported in the absence of predators) were required for stability at a coevolutionary equilibrium; stability also required many patches of poor quality. Coevolution failed to stabilize the system when patches were similar in quality.

Extensions of the idea of patch selection as a component of predator-prey coevolution include Abrams (10) and van Baalen & Sabelis (95). Abrams (10) examined the stability of two-patch systems in which predators adaptively switched between two patches, each characterized by unstable population dynamics. Faster switching was always advantageous to the predator, but it often led to increased amplitude of the population cycles. Van Baalen & Sabelis (95) extended their earlier (94) work to examine the population dynamics that occurred when there was flexible and adaptive patch selection by both parasitoid and host in many-patch environments. They found that, in spite of chaotic population dynamics, conditions allowing coexistence of both species were broader than when both species had rigid patch selection behavior.

Conclusions Regarding Evolution and Stability

Several general conclusions can be drawn from studies of the impact of evolution on the stability of predator-prey systems. The first is that evolution in a predator's capture-related traits is most likely to be stabilizing. Even if it does not confer stability on a cycling system, by itself it is relatively unlikely to produce cycles in a system that would otherwise be stable. Prey evolution often increases the instability of a system that would cycle in the absence of evolutionary change; cycle amplitude may become larger, or stable cycles may become divergent, leading to extinction. However, prey evolution by itself seems relatively unlikely to destabilize otherwise stable systems and may stabilize some unstable systems. Coevolution of both species seems most likely to generate cycles when there is a bidirectional axis of vulnerability, meaning that prey can reduce their risk by increasing or decreasing trait values relative to a most-vulnerable form, which depends on the predator's phenotype. Even in this case, cycling requires additional conditions, including sufficiently rapid prey evolution relative to predator evolution, sufficiently weak stabilizing selection due to costs of the traits, and an intermediate level of predator efficiency in converting consumed prey into new predators. Seger (1992) drew attention to the fact that parasite-host models had often predicted cycles, while most previous predator-prey models had not. This was largely a consequence of the gene-for-gene model of virulence and resistance that had been adopted in most previous parasite-host models and the unidirectional axis of attack and defense assumed by the few previous predator-prey models based on continuous traits. In both bidirectional and gene-for-gene models, the selectively favored direction of change in the predator's trait depends on prey's trait and vice versa. This is not the case for unidirectional axes such as speed vs. speed, or strength vs. armor. Switches in the direction of selection on the trait are required for evolution to generate cycles in an otherwise stable system.

EFFECTS OF ENVIRONMENTAL PARAMETERS ON TRAITS

Effects of Traits of Other Species

This section begins by considering the effect of an evolutionary shift in a predation-related trait in one species on the evolutionarily-favored trait value of its partner. This response should reveal much about the long-term evolution of both species. The types of models discussed above assume that a certain spectrum of trait values exists in one or both populations, but costs or tradeoffs entailed by the traits are fixed. On a longer time scale, novel mutations will arise, or changes in the genetic architecture will allow one species to reduce the costs of a given predation-related trait, and rapidly evolve to a new evolutionary equilibrium. The long-term course of evolution is determined by how each partner in the interaction responds to such changes in the other species. There has been remarkably little work on this particular question. The most comprehensive analyses appeared 10 or more years ago (2, 4). These papers assume that there is a stable population-dynamical equilibrium, and that each species has trait values that maximize individual fitness. As is clear from the preceding discussion of stability, neither of these assumptions is always valid. However, if they are, some rather general results can be obtained.

First consider the question, how does a predatory trait change in response to an evolutionary improvement in prey defense? Some authors have assumed that the predator's attack-related traits should always increase in response to an increase in the prey's defense (33). However, this is often not an adaptive response for predators whose traits are maintained by a balance between the benefits of catching more prey and the costs of greater trait development. One significant factor, often neglected in the absence of mathematical analysis, is the change in prey density as the result of its greater escape ability. Prey population size is likely to increase as its mean defensive trait increases. There are some exceptions to this statement if the defensive trait affects the prey's exploitation rate of its own biological resources. However, if we neglect this possibility, then the prey's density response (increased population) will offset its decreased vulnerability. It thus becomes less clear whether there should be any change in selection on the predator's trait following the increase in prey defense. If the predator's per capita population growth rate is solely a function of its intake rate of prey, and if the capture rate can be factored into the product of a term dependent on the predator's trait and a term dependent on the prey's trait, then the prey trait has no effect on the evolutionary equilibrium of the predator's trait (2). Under these conditions, the increase in prey density exactly cancels the effect of the decreased capturability of individual prey, and there is no change in the selection pressure on the predator's trait. This simple result is unlikely to be literally true in any real system. The result is changed if the predator's density affects its per capita growth rate, if the predator and prey traits don't combine multiplicatively, or if the prey's trait affects other aspects of the interaction, such as handling time. Nevertheless, the existence of some compensatory response in prey density when the vulnerability of individual prey decreases is a relatively general phenomenon. Furthermore, there are many circumstances under which the predator's optimal capture ability decreases when the prey become better at evading capture. This outcome is likely if the ratio of the costs of increased capture ability to the benefits goes up as the capture trait increases (2).

The response of prey to an evolutionary improvement in the predator's ability to capture is more likely to be a decrease than an increase in its inherent vulnerability (2, 4). Increased predator capture ability may decrease predator population size if the predator is already overexploiting prey (i.e. the equilibrium prey density is below the most productive density). However, this decrease is generally smaller in magnitude than the original increase in predator capture ability that caused the decrease in density. This is more in accord with thinking based on the "arms race" analogy than is the normal predator response to increased prey defense. However, when the predator and prey traits combine additively to determine the per capita capture rate, then greater predator abilities will select for less defense in the prey whenever the prey are overexploited (4). In this additive case, the strength of selection for better escape/defense is proportional to predator population size, but independent of the predator's trait.

The analysis in Abrams (2, 4) suggests that the key factors determining the direction of response of predators to a change in prey, or vice versa, are: 1) the mathematical relationships between a measure of the predator's trait and the per capita capture rate, and the relationship between the prey's trait value and the capture rate; 2) the nonlinearity of the predators functional and numerical responses; and 3) whether the predator's population density affects its own per capita growth rate. Predictions of the directions of evolutionary response under different conditions are as yet untested. However, the asymmetry in predicted evolutionary responses resulting from prey density compensation supports the general finding that prey often evolve in response to the addition of predators, but predators are seldom observed to change (or inferred to have changed) in response to prey addition. Evidence is presented below.

The findings (2, 4) summarized above are based on an assumption that traits and populations reach a locally stable equilibrium point. This is clearly not generally valid because predator-prey cycles occur frequently in laboratory systems and at least occasionally in the field (43). There has been relatively limited exploration of the consequences of cycles for the nature of one species' response to a change in the other. However, it is clear that population dynamical cycles often result in evolutionary outcomes that differ significantly from those in stable systems. This is frequently true when individual fitness is nonlinearly related to the evolving character (9). For example, the scenario described above, in which the effect of prey density cancels the effect of a lower encounter rate, is usually not true when there are predator-prey cycles; in cyclic systems it is more common for predators to increase capture ability in response to decreased prey vulnerability (9).

Effects of Productivity

The final environmental parameter considered here is the productivity of the environment, reflected in carrying capacity, K, intrinsic (maximum) per capita population growth, r, or both r and K. Productivity is of interest because it is often easy to manipulate and has a significant effect on population dynamics. Frequently natural spatial gradients in productivity represent natural experiments. The so-called "paradox of enrichment" (predator-prey cycles caused by fertilization) (78) was the impetus for several early studies of predator-prey coevolution (79, 80, 89). Rosenzweig & Schaffer (80; p. 162) concluded that "Coevolution always opposes the destabilization induced by enrichment and, in case r and K vary proportionally, actually increases stability." However, this work assumes that population densities have their equilibrium values; this is not valid in the model Rosenzweig & Schaffer (80) considered because the equilibrium is unstable at high carrying capacities.

Several more recent studies have revisited the question of the effect of enrichment on both trait values and stability in an evolving predator-prey system. The results are both more varied and more interesting than the work of Rosenzweig & Schaffer (80) suggested. In the models of predator evolution in Abrams (7), greater carrying capacity always increases the probability that the system will cycle. In the event that it does cycle, the mean trait value is generally lower than the equilibrium value. When the system is stable, increasing the carrying capacity, K, can either increase or decrease the equilibrium value of the predator's trait. When there is an upper limit to the functional or numerical response, a decrease in the capture rate following increased K is more likely than an increased capture rate. In the case of prey evolution, the effects of increased productivity are discussed by Abrams (4) and Abrams & Matsuda (14). In stable systems, prey vulnerability usually decreases in response to increased K, but it often increases or has a unimodal response to an increased intrinsic growth rate, r (4). In unstable systems, it is likely that an increased carrying capacity will decrease prey investment in defense, since larger K implies cycles of greater amplitude, but generally causes a slight decline in average predator abundance (16). Both of these changes generally favor decreased defense. However, if increased productivity acts primarily by increasing r, prey defense is likely to increase. This is because a greater r has little effect on cycle amplitude while significantly increasing mean predator density (17).

The consequences of increased productivity for the coevolution of traits in both species have apparently been addressed in only a single study since the work of Rosenzweig & Schaffer (80). Hochberg & van Baalen (57) studied the effects of productivity gradients in a metapopulation model with coupled predator and prey evolution within patches. The dispersal between patches did little to change the patterns of attack and defense abilities predicted by considering evolution independently in each patch. The population dynamical assumptions ensured that each patch in isolation would reach a stable equilibrium, so the effect of evolution on the paradox of enrichment was not addressed. In the model of Hochberg & van Baalen (57), higher productivity (in this case, greater maximum growth and equilibrium density) led to higher levels of both attack and defense in predator and prey, respectively. This result stems from the accelerating cost functions they assumed for each trait, and the linear relationship between predator intake rate and birth rate. These ensure that, for each species, higher density of the other selects for greater investment in predatory functions. Greater productivity increases both densities, leading to greater trait development in both species. Other studies have shown that a nonlinear numerical response of the predator often leads to a decrease in the optimal values of predatory traits with increased prey density (5). Thus, the uniform increase in predatory traits with increased productivity is likely to depend on specific assumptions of the model as acknowledged by Hochberg & van Baalen (57). They assumed an additive combination of trait values, but alternative assumptions are known to give different results (4), at least when the predator's trait is relatively constant.

The traits and dynamics of predator-prey systems respond in a number of ways to increases in the productivity of the environment. Higher productivity may destabilize systems more rapidly in evolving systems than in the absence of evolution, but it is also possible that the evolution eliminates or reduces the paradox of enrichment. The former outcome seems most likely to be associated with a dominance of the prey's evolutionary response, while the latter is driven by a greater possible evolutionary shift in the predator. In stable systems, the change in trait values with increased productivity is influenced by how traits combine in determining a capture rate and by the shapes of the functional and numerical responses of the predator. Work is as yet insufficient for general conclusions about the conditions required for different effects of productivity.

GAPS IN PRESENT-DAY THEORY

In assessing the articles summarized above, it is important to realize that neither the range of models considered nor the comprehensiveness of the analysis of those models is very great. Most models assume that the traits affect only the per capita rate of successful attack of predator on prey (or parasitoid on host). Other traits that are important in most interactions include those determining the efficiency of conversion of prey into new predators, the amount of time required by the predator to handle captured prey, the amount of direct competition between predators, and the effect of prey availability on the strength of that competition. There has been limited consideration of some of these factors (e.g. see 83 for efficiency), but much remains to be explored. While there has been some work on patch selection by predator and prey in heterogeneous environments, more work is needed in this area. If the prey can escape by growing into larger, less vulnerable forms, then life history variables need to be considered, and the dynamics framework used in models (1) and (2) is not appropriate. Chase (28) has made a start toward considering this case by examining competition between a form that can grow to an invulnerable stage and one that cannot. One option for a stage-structured prey species is to have very high exposure to predation in early life-history stages in order to get through those stages rapidly, while another possible strategy is to reduce exposure and prolong the early stages (18, 77). Which strategy is favored by evolution will have a significant effect on population dynamics and the evolution of traits in the predator that influence their ability to capture later life-history stages of the prey.

Past theoretical work has also largely avoided consideration of evolution and coevolution when one or both species have two or more traits that influence predation rate. One exception to this is Hochberg's (55) analysis of the evolution of traits determining host exposure to attack by parasitoids and traits determining host ability to encapsulate (and thus kill) parasitoid eggs after they have been attacked. These two traits are obviously coupled, since concealment from parasitoids makes costly encapsulation abilities undesirable. Hochberg considers coevolution with a parasitoid having independent traits related to discovery of hosts and defenses of eggs against encapsulation. Unlike hosts, parasitoids generally have high values of both traits when hosts have significant defenses. There are cases similar to this one in other types of predator-prey systems. Endler (42) has broken the predation process down into the five sequential stages of detection, identification, approach, subjugation, and consumption. Clearly, traits influencing one stage will alter the selective value of traits at other stages. Traits to resist subjugation have no advantage to a prey individual that is not detected or is ignored because it is misidentified. It is possible that models with many traits will lead to different effects of coevolution on stability.

The influence of the genetics of trait determination on the course of coevolution is another area where little is known. Some early models with artificially simple one-locus-few-allele systems have been largely replaced by a quantitative genetic or simpler phenotypic description of character evolution. However, the consequences of asexual vs. sexual reproduction, assortative vs. random mating, and continuous vs. discrete traits have received relatively little attention. There are some cases where we know that the outcome was greatly affected by the genetic system of trait determination. For example, Doebeli (39) notes that, unlike multilocus sexual inheritance, asexual inheritance does not increase persistence in the simple host-parasitoid models he examined. There is reason to believe that the genetic system will often have a major effect on dynamics when one or more potential evolutionary equilibria are characterized by disruptive selection on one or more traits. Under some circumstances (e.g. asexual inheritance and some cases of sexual inheritance and assortative mating), disruptive selection is likely to result in splitting of the lineage undergoing disruptive selection into two separately evolving lines (35, 73). However, under randomly mating sexual systems, this circumstance will result in a stable equilibrium or cycles, but no increase in the number of independently evolving units (13, 15).

Theory has understandably concentrated on the simplest systems in which there is a single specialist predator consuming a single prey species. Even in this rather special case, the two species have typically been left out of their food web context. A more complete model would include the resource(s) of the prey species and any higher level predators or parasites that attack the focal predator. These omissions are not trivial. It is possible to represent some of the effects of the prey's resource by simply introducing density-independent growth of the prey. However, this automatically eliminates the time lag that occurs when the density dependence must be generated by changes in the abundance of a resource. Furthermore, the types of costs of defensive traits are likely to be represented differently in these artificially simplified models. When there is no explicit representation of the resource, the cost of greater defense is almost always assumed to be either a reduced maximum growth rate or an increased susceptibility to intraspecific competition. However, defense frequently reduces feeding (67), and this affects both maximum growth rate and competition in ways that differ from the way they are normally incorporated into models of coevolution (1, 6). Models of simultaneous behavioral optimization by both predator and prey embedded in the middle of a four-species food chain have quite different properties than do models in which the top and bottom populations are not represented (6). Most predators feed on more than a single prey species, and most prey are consumed by several predators (29). Thus, understanding the consequences of the coupled evolution of a given predator and a given prey will usually require consideration of other interacting species. There has been limited analysis of coevolution in the context of systems with three or more species (5, 8, 11); these studies have revealed a variety of potential evolutionary indirect effects and new mechanisms for cycles. However, multispecies systems still represent *terra incognita* in both empirical and theoretical realms.

The past decade has been characterized by an increasing focus by population and community ecologists on the spatial context in which population interactions occur (51). This focus has not spread to theoretical studies of predator-prey coevolution, with a few exceptions (e.g. 56). Thompson (92, 93) has emphasized the general importance of spatial heterogeneity for coevolution, but most well-studied examples involve parasites and hosts rather than predators and prey. Ironically, one of the earliest theoretical studies of predator evolution was a simulation study of the evolution of exploitation rates in a metapopulation in which predator and prey within subpopulations could either cycle or reach a stable equilibrium depending on the predator's capture rate (47). The main result of that study was that group selection on predators could reduce the values of otherwise cost-free traits that increased capture rates. This scenario for group selection has yet to receive any empirical confirmation, but this mechanism should operate to some extent in most metapopulations.

EVIDENCE

Connection between the above theory and the real world can be made in two ways: 1) demonstrations that key assumptions of the model are satisfied; and 2) demonstrations that predictions of the models actually occur in natural systems. Both issues are considered below.

The models reviewed above suggest the evolution of predation-related traits is largely determined by a small number of key assumptions. Thus, for example, if we knew whether most defensive traits could be characterized as bidirectional or unidirectional, we could reach some tentative conclusions about the probable stabilizing or destabilizing effects of coevolution. If we knew in more detail how trait values of the two species enter into an expression for the consumption rate, then we would have made a major step towards knowing how traits in each species should respond to evolutionary change in traits of the other species. The shapes of functional and numerical responses of predators are often key determinants of the stability of the coevolutionary system and of the direction of response of one trait to a change in a trait in the other species. Unfortunately, information about most of these key properties is generally lacking. For example, although more than ten theoretical papers have assumed a bidirectional axis of vulnerability, not one of these refers to an empirical study that has demonstrated such an axis. The most that has been done is to note that predators are generally size-selective and ignore both very large and very small prey items. This can be used as a justification for a bidirectional axis only if the range of size variation in prey species is known to be large enough so that safety via both large and small size within that species is possible. Such information should be obtainable, but it does not seem to have been noted in any published articles on predator-prey systems. It is known that functional responses in the laboratory are usually strongly saturating, but most theorists continue to adopt the simpler assumption of linear responses.

Unfortunately, population densities are usually impossible to assess from the fossil record, and experimental systems in which two species can be maintained for sufficiently many generations to observe evolution in both are generally limited to bacteria and phage, and other parasite-host systems among single-celled and smaller organisms. Few natural systems have been studied thoroughly enough and long enough to assess changes in both predation-related traits and population densities over periods of many generations. This has left a rather unfortunate gap between theory and experiment. This gap is most likely to be closed by studies that are able to deduce some of the properties of the evolutionary interaction from measurements over one or a few generations.

Empirical work on the coevolution of competitors has been a much more active area than the study of predator-prey coevolution (see e.g. 50, 86, 87). This is probably because different questions have been asked about the two types of systems. In the case of competitors, there exist essentially no data or theory about how one competitor changes in response to an evolutionary innovation in the other. There is no interest in stability, since competition is unlikely to generate population cycles. Thus, the relatively difficult questions asked about predators and prey are not asked about competitors. However, there are many natural experiments in which two or more competitors occur in isolation or together. Thus, the central question in the evolution of characters related to competition has been how they change when another species is introduced or removed. This question is rather uninteresting in the case of predator-prey systems. Specialist predators cannot exist in the absence of their prey, so a sympatric-allopatric comparison is not possible. Prey do exist without predators, but it is not particularly significant that prey increase defenses when predators are introduced and decrease defenses when predators are absent (41).

One of the major problems in empirical studies of predator-prey relationships is that it is often difficult to find species pairs with both members having easily measurable traits influencing the interaction. There are many cases where predators have selected for more cryptic color patterns in their prey species (41). However, the predator traits involved in detecting cryptic prey are much more difficult to measure. The fossil record provides examples of change in morphological traits that affect predation (97), but many of the potential counter-adaptations of the predators (e.g. increased perceptual acuity) would have left no trace in the record.

The question of stability is exceptionally difficult to examine empirically. We lack long-term records of both traits and population densities for natural systems undergoing cycles. Laboratory host-parasitoid systems have produced some suggestive evidence that host evolution may have decreased the amplitude of cycles (75). This would be consistent with the theory of Abrams & Matsuda (14) if it were shown that there were strong stabilizing selection on the prey's trait. Because many species exist as metapopulations, one possible method for examining stability in a short-term study is to examine the variation in traits of similar but isolated pairs of predator and prey populations. Lively (68) has done this for a parasite-host system, which appears to undergo cycles. One of the most consistent results from previous models is the association of bidirectional axes of vulnerability with population/evolutionary cycles. It is conceivable that such an association could be demonstrated if we knew more about the relationship between traits and capture-rate parameters. Cycles in the asymmetry of the jaws of scale eating fish is a possible example (58a).

In spite of the long list of difficulties, there are some empirical connections with theory. A topic that has received considerable attention over the years is the question of whether there is a trend for one or the other party of the predator-prey relationship to become more successful over evolutionary time. Success is difficult to define (3, 97), but it usually corresponds to the maximal rate at which a predator individual captures prey in some standardized circumstances. Evidence in some systems strongly suggests that one or the other party in a predator-prey interaction has increased its success in the interaction over time. Bakker's (19) analysis of locomotor traits in ungulates and their cursorial predators and of tooth wear in the ungulates suggests a decrease in the risk of mortality due to predators over the past 60 million years. Similarly, the fact that the shells of some species of snails became more heavily armored while their crab predators did not alter claw morphology suggests that those snail species have improved their defense over time (97). Endler (41; Table 5.1) reviewed several dozen examples of prey responding to selection by predators, but no cases of predators responding to prey (although there are some cases of herbivores responding to selection due to plant characteristics). These observations are consistent with the asymmetry in responses described above, in which prey density-compensation reduces any selective pressure on predators generated by better prey defense. However, there are alternative explanations (26), including the possibility that the predators may have responded with greater (unfossilized) perceptual or behavioral abilities. We still are uncertain whether predator lineages are on average less successful than prey lineages over evolutionary time. This asymmetry is certainly unlikely to characterize all interacting lineages, since predators are still very well represented among the earth's fauna.

The greatest progress in empirical studies of both predator-prey and parasite-host relationships seems likely to come from systems in which there is geographical variation in the interaction and in which key traits of both species can easily be measured. One of the more promising examples of such a system is the interaction between the garter snake, *Thamnophis sirtalis*, and the toxic newt, *Taricha granulosa*, studied by Brodie & Brodie (23–26). The newts produce a highly potent neurotoxin, tetrodotoxin, from skin glands, and *T. sirtalis* is the only predator able to survive consumption of the newt. Both species occur over a wide geographical range, and each species occurs in some areas where the other species is absent. There is genetically based geographic variation in the degree of resistance of the snake and geographic variation in the degree of toxicity of the newt. There is also within-population variation in the resistance of the snake. This is expected, since tetrodotoxin-resistance entails a cost in decreased locomotor performance by the snakes (25). The patterns of variation for the most part exhibit positive between-species correlations (more resistant snakes occur where newts are more toxic), but the correlation is not perfect, and good assays for newt toxicity have only recently been developed (26). Toxin-antitoxin traits are likely to combine additively in determining the ability of snakes to eat newts, since it is likely to be at least in part the difference in amounts of toxins and detoxifying substances that determines the prey's actual toxicity to the predator (2). This additive model suggests that predators should respond to higher levels of toxin by developing greater resistance to the toxin, resulting in a classic arms race scenario (2). This is supported by the observed geographic correlation of trait values.

The garter snake–newt system is unusual in that there is evidence of evolutionary change of each party brought about by the other. Another species pair in which there is a rapidly growing body of evidence for coevolution from between population comparisons is the interaction between *Drosophila* and its hymenopterous parasitoids in the genera *Asobara* and *Leptopilina* (63). Here again, there is a positive correlation between the virulence of the parasitoid (ability of eggs and larvae to survive within the host body) and resistance of the host (immune defenses against the early stages of parasitoids). Studies of geographical variation in predator-prey systems commonly find variability in prey traits based on presence or absence of predators, but they either do not find or are unable to study variation in the predators characteristics (e.g. 21, 40, 77).

The study of geographic variation is important for reasons beyond simply establishing or refuting the existence of coevolution. The work reviewed by Kraaijeveld & Godfray (63) has shown that resistance and virulence in the *Drosophila-Asobara tabida* interaction represents a unidirectional axis of ability. The study of geographic variation potentially allows measurement of the detailed shape of the relationship between predator and/or prey traits and the per capita consumption rate. Systems in which variation in both defense and attack abilities exists and can be measured are ideal for developing the quantitative descriptions required for population models. The studies of Henter (53) and Henter & Via (54) of variation in an aphid's vulnerability to a parasitoid, and of the parasitoid's ability to overcome those defenses, illustrate the type of measurements that are possible. The growing interest in studying geographically variable interactions (92, 93) should result in significant advances in our understanding of how evolution changes predator-prey systems.

ACKNOWLEDGMENTS

I thank the University of Toronto for financial support and Troy Day for discussions of coevolution.

Visit the Annual Reviews home page at www.AnnualReviews.org

LITERATURE CITED

1. Abrams PA. 1984. Foraging time optimization and interactions in food webs. *Am. Nat.* 124:80–96
2. Abrams PA. 1986. Adaptive responses of predators to prey and prey to predators: the failure of the arms race analogy. *Evolution* 40:1229–47
3. Abrams PA. 1989. The evolution of rates of successful and unsuccessful predation. *Evol. Ecol.* 3:157–71
4. Abrams PA. 1990. The evolution of antipredator traits in prey in response to evolutionary change in predators. *Oikos* 59:147–56
5. Abrams PA. 1991. The effects of interacting species on predator-prey coevolution. *Theor. Pop. Biol.* 39:241–62
6. Abrams PA. 1992. Predators that benefit prey and prey that harm predators: unusual effects of interacting foraging adaptations. *Am. Nat.* 140:573–600

7. Abrams PA. 1992. Adaptive foraging by predators as a cause of predator-prey cycles. *Evol. Ecol.* 6:56–72
8. Abrams PA. 1996. Evolution and the consequences of species introductions and deletions. *Ecology* 77:1321–28
9. Abrams PA. 1997. Evolutionary responses of foraging-related traits in unstable predator-prey systems. *Evol. Ecol.* 11:673–86
10. Abrams PA. 1999. The adaptive dynamics of consumer choice. *Am. Nat.* 153:83–97
11. Abrams PA. 2000. Character displacement of species that share predators. *Am. Nat.* In press
12. Abrams PA, Kawecki TJ. 1999. Adaptive host preference and the dynamics of host-parasitoid interactions *Theor. Pop. Biol.* 56:307–24
13. Abrams PA, Matsuda H. 1997. Fitness minimization and dynamic instability as a consequence of predator-prey coevolution. *Evol. Ecol.* 11:1–20, [reprinted with corrections from 1996, *Evol. Ecol.* 10:167–86]
14. Abrams PA, Matsuda H. 1997. Prey evolution as a cause of predator-prey cycles. *Evolution* 51:1740–48
15. Abrams PA, Matsuda H, Harada Y. 1993. Evolutionarily unstable fitness maxima and stable fitness minima in the evolution of continuous traits. *Evol. Ecol.* 7:465–87
16. Abrams PA, Namba T, Mimura M, Roth JD. 1997. Comment on Abrams and Roth: The relationship between productivity and population densities in cycling predator-prey systems. *Evol. Ecol.* 11:371–73
17. Abrams PA, Roth JD. 1994. The responses of unstable food chains to enrichment. *Evol. Ecol.* 8:150–71
18. Abrams PA, Rowe L. 1996. The effects of predation on the age and size of maturity of prey. *Evolution* 50:1052–61
19. Bakker RT. 1983. The deer flees, the wolf pursues: incongruencies in predator-prey coevolution. In *Coevolution*, ed. DJ Futuyma, M Slatkin, pp. 350–52. Sunderland: Sinauer
20. Beddington JR, Free CA, Lawton JH. 1978. Concepts of stability and resilience in predator-prey models. *J. Anim. Ecol.* 47:791–16
21. Benkman CW. 1999. The selection mosaic and diversifying coevolution between crossbills and lodgepole pine. *Am. Nat.* 153:S75–S91
22. Berry RJ, Crawford TJ, Hewitt GM, eds. 1992. *Genes in Ecology.* Oxford: Blackwell Sci.
23. Brodie ED III, Brodie ED Jr. 1990. Tetrodotoxin resistance in garter snakes: an evolutionary response of predators to dangerous prey. *Evolution* 44:651–59
24. Brodie ED III, Brodie ED Jr. 1991. Evolutionary response of predators to dangerous prey: reduction of toxicity of newts and resistance of garter snakes in island populations. *Evolution* 45:221–24
25. Brodie ED III, Brodie ED Jr. 1999. Costs of exploiting poisonous prey: evolutionary tradeoffs in a predator-prey arms race. *Evolution* 53:626–31
26. Brodie ED III, Brodie ED Jr. 1999. Predator-prey arms races. *Bioscience* 49:557–68
27. Brown JS, Vincent TL. 1992. Organization of predator-prey communities as an evolutionary game. *Evolution* 46:1269–83
28. Chase JM. 1999. To grow or reproduce? The role of life history plasticity in food web dynamics. *Am. Nat.* 154:571–86
29. Cohen JE, Briand R, Newman CM. 1990. *Community Food Webs: Data and Theory.* Berlin: Springer Verlag
30. Cott HB. 1940. *Adaptive Coloration in Animals.* London: Methuen
31. Darwin CR. 1859. *The Origin of Species.* London: John Murray
32. Dawkins R. 1982. *The Extended Phenotype.* Oxford: Oxford Univ. Press
33. Dawkins R, Krebs JR. 1979. Arms races between and within species. *Proc. R. Soc. Lond. B* 202:489–511

34. DeAngelis DL, Kitchell JA, Post WM. 1985. The influence of naticid predation on evolutionary strategies of bivalve prey: conclusions from a model. *Am. Nat.* 126:817–42
35. Dieckmann U, Doebeli M. 1999. On the origin of species by sympatric speciation. *Nature* 400:354–57
36. Dieckmann U, Law R. 1996. The dynamical theory of coevolution: a derivation from stochastic ecological processes. *J. Math. Biol.* 34:579–612
37. Dieckmann U, Marrow P, Law R. 1995. Evolutionary cycling in predator-prey interactions: population dynamics and the Red Queen. *J. Theor. Biol.* 176:91–102
38. Doebeli M. 1996. Quantitative genetics and population dynamics. *Evolution* 50:532–46
39. Doebeli M. 1997. Genetic variation and the persistence of predator-prey interactions in the Nicholson-Bailey model. *J. Theor. Biol.* 188:109–20
40. Downes S, Shine R. 1998. Sedentary snakes and gullible geckos: predator-prey coevolution in nocturnal rock-dwelling reptiles. *Anim. Behav.* 55:1373–85
41. Endler JA. 1986. *Natural Selection in the Wild.* Princeton, NJ: Princeton Univ. Press
42. Endler JA. 1986. Defense against predators. In *Predator-Prey Relationships*, ed. M Feder, G Lauder, pp. 109–34. Chicago: Univ. Chicago Press
43. Ellner S, Turchin P. 1995. Chaos in a noisy world: new methods and evidence from time-series analysis. *Am. Nat.* 145:343–75
44. Frank SA. 1994. Coevolutionary genetics of hosts and parasites with quantitative inheritance. *Evol. Ecol.* 8:74–94
45. Futuyma DJ. 1998. *Evolutionary Biology.* Sunderland, MA: Sinauer. 3rd ed.
46. Gavrilets S. 1997. Coevolutionary chase in exploiter-victim systems with polygenic characters. *J. Theor. Biol.* 186:527–34
47. Gilpin ML. 1975. *Group Selection in Predator-Prey Communities.* Princeton, NJ: Princeton Univ. Press
48. Ginzburg LR. 1998. Assuming reproduction to be a function of consumption raises doubts about some popular predator-prey models. *J. Anim. Ecol.* 67:325–27
49. Gould SJ. 1977. *Ever Since Darwin: Reflections in Natural History.* New York: Norton
50. Grant PR. 1986. *Ecology and Evolution of Darwin's Finches.* Princeton, NJ: Princeton Univ. Press
51. Hanski I, Gilpin M. ed. 1997. *Metapopulation Biology: Ecology, Genetics, and Evolution.* New York: Academic Press
52. Hassell MP, May RM. 1973. Stability in insect host-parasite models. *J. Anim. Ecol.* 42:693–736
53. Henter HJ. 1995. The potential for coevolution in a host-parasitoid system. II. Genetic variation within a population of wasps in the ability to parasitize an aphid host. *Evolution* 49:439–45
54. Henter HJ, Via S. 1995. The potential for coevolution in a host-parasitoid system. I. Genetic variation within an aphid population in susceptibility to a parasitic wasp. *Evolution* 49:427–38
55. Hochberg ME. 1997. Hide or fight? The competitive evolution of concealment and encapsulation in parasitoid-host associations. *Oikos* 80:342–52
56. Hochberg ME, Holt RD. 1995. Refuge evolution and the population dynamics of coupled host-parasitoid associations. *Evol. Ecol.* 9:633–61
57. Hochberg ME, van Baalen M. 1998. Antagonistic coevolution over productivity gradients. *Am. Nat.* 152:620–34
58. Holling CS. 1959. The components of predation as revealed by a study of small mammal predation of the European pine sawfly. *Can. Entomol.* 91:293–320
58a. Hori M. 1993. Frequency dependent natural selection in the handedness of scale-eating cichlid fish. *Science* 260:216–19
59. Ives AR, Dobson AP. 1987. Antipredator behavior and the population dynamics

of simple predator-prey systems. *Am. Nat.* 130:431–37
60. Janzen DH. 1980. When is it coevolution? *Evolution* 34:611–12
61. Jerison H. 1973. *Evolution of the Brain and Intelligence.* New York: Academic
62. Kaitala V, Ylikarjula J, Heino M. 1999. Dynamic complexities in host-parasitoid interaction. *J. Theor. Biol.* 197:331–41
63. Kraaijeveld AR, Godfray HCJ. 1999. Geographic patterns in the evolution of resistance and virulence in *Drosophila* and its parasitoids. *Am. Nat.* 153:S61–S74
64. Lande R. 1976. Natural selection and random genetic drift in phenotypic evolution. *Evolution* 30:314–34
65. Levin SA, Udovic, JD. 1977. A mathematical model of coevoling populations. *Am. Nat.* 111:657–75
66. Levy CK. 1999. *Evolutionary Wars.* New York: W. H. Freeman
67. Lima SL. 1998. Stress and decision-making under the risk of predation: recent developments from behavioral, reproductive and ecological perspectives. *Adv. Stud. Behav.* 27:215–90
68. Lively CM. 1999. Migration, virulence, and the geographic mosaic of adaptation by parasites. *Am. Nat.* 153 (suppl.):S34–S47
69. Marrow P, Cannings C. 1993. Evolutionary instability in predator-prey systems. *J. Theor. Biol.* 160:135–50
70. Marrow P, Dieckmann U, Law R. 1996. Evolutionary dynamics of predator-prey systems: an ecological perspective. *J. Math. Biol.* 34:556–78
71. Marrow P, Law R, Cannings C. 1992. The coevolution of predator-prey interactions: ESSs and Red Queen dynamics. *Proc. R. Soc. Lond. B* 250:133–41
72. Matsuda H, Abrams PA. 1994. Timid consumers: self-extinction due to adaptive change in foraging and anti-predator effort. *Theor. Pop. Biol.* 45:76–91
73. Metz JAJ, Geritz SAH, Meszèna G, Jacobs FJA, van Heerwaarden JS. 1996. Adaptive dynamics, a geometrical study of the consequences of nearly faithful reproduction. In *Stochastic and Spatial Structures of Dynamical Systems*, ed. SJ van Strien, SM Verduyn Lunel, pp. 183–231. Amsterdam: KNAW Verhandelingen
74. Pimentel D. 1961. Animal population regulation by the genetic feed-back mechanism. *Am. Nat.* 95:65–79
75. Pimentel D, Stone FA. 1968. Evolution and population ecology of parasite-host systems. *Can. Entomol.* 100:655–62
76. Radinsky L. 1978. Evolution of brain size in carnivores and ungulates. *Am. Nat.* 112:815–31
77. Reznick DN, Bryga H, Endler JA. 1990. Experimentally induced life-history evolution in a natural population. *Nature* 346:357–59
78. Rosenzweig ML. 1971. Paradox of enrichment: destabilization of exploitation ecosystems in ecological time. *Science* 171:385–87
79. Rosenzweig ML. 1973. Evolution of the predator isocline. *Evolution* 27:84–94
80. Rosenzweig ML, Schaffer WM. 1978. Homage to the Red Queen. II. Coevolutionary response to enrichment of exploitation ecosystems. *Theor. Pop. Biol.* 14:158–163
81. Rosenzweig ML, Brown JS, Vincent TL. 1987. Red Queens and ESS: the coevolution of evolutionary rates. *Evol. Ecol.* 1:59–94
82. Roughgarden J. 1979. *Theory of Population Genetics and Evolutionary Ecology*: An Introduction. New York: MacMillan
83. Roughgarden J. 1983. The theory of coevolution. In *Coevolution*, ed. DJ Futuyma, M Slatkin, pp. 33–64. Sunderland: Sinauer
84. Saloniemi I. 1993. A coevolutionary predator-prey model with quantitative characters. *Am. Nat.* 141:880–96
85. Sasaki A, Godfray HCJ. 1999. A model for the coevolution of resistance and virulence in coupled host-parasitoid interactions. *Proc. R. Soc. Lond. B* 266:455–63

86. Schluter D. 1994. Experimental evidence that competition promotes divergence in adaptive radiation. *Science* 266:798–801
87. Schluter D. 2000. The role of ecological character displacement in adaptive radiation. *Am. Nat.* In press
88. Seger J. 1992. Evolution of exploiter-victim relationships. In *Natural Enemies*, ed. M Crawley, pp. 3–26. Oxford: Blackwell
89. Shaffer WM, Rosenzweig ML. 1978. Homage to the Red Queen. I. Coevolution of predators and their victims. *Theor. Pop. Biol.* 14:135–57
90. Slobodkin LB. 1974. Prudent predation does not require group selection. *Am. Nat.* 108:665–78
91. Taper M, Case TJ. 1992. Models of character displacement and the theoretical robustness of taxon cycles. *Evolution* 46:317–34
92. Thompson JN. 1994. *The Coevolutionary Process*. Chicago: Univ. Chicago Press
93. Thompson JN. 1999. Specific hypotheses on the geographic mosaic of coevolution. *Am. Nat.* 153:S1–S14
94. Van Baalen M, Sabelis M. 1993. Coevolution of patch selection strategies of predators and prey and the consequences for ecological stability. *Am. Nat.* 142:646–70
95. Van Baalen M, Sabelis M. 1999. Nonequilibrium population dynamics of "ideal and free" prey and predators. *Am. Nat.* 154:69–88
96. Van der Laan JD, Hogeweg P. 1995. Predator-prey coevolution: interactions across different timescales. *Proc. R. Soc. Lond. B* 259:35–42
97. Vermeij GJ. 1987. *Escalation and Evolution*. Cambridge, MA: Harvard Univ. Press
98. Vermeij GJ. 1994. The evolutionary interaction among species: selection, escalation, and coevolution. *Annu. Rev. Ecol. Syst.* 25:219–36
99. Vincent TL, Cohen Y, Brown JS. 1993. Evolution via strategy dynamics. *Theor. Pop. Biol.* 44:149–76

Annu. Rev. Ecol. Syst. 2000. 31:107–38

THE ECOLOGY AND PHYSIOLOGY OF VIVIPAROUS AND RECALCITRANT SEEDS

Elizabeth Farnsworth
Smith College, Clark Science Center, Northampton, Massachusetts 01063; e-mail; efarnswo@mtholyoke.edu

Key Words desiccation intolerance, dormancy, plant reproduction, phytohormones

■ **Abstract** Understanding seed physiology is central to reconstructing how angiosperms have evolved, to characterizing dormancy and germination regimes shared by suites of species, and to devising sound strategies for seed bank conservation, agriculture, and forestry. While species with dormant seeds have received the lion's share of attention, hundreds of plant species exhibit no seed dormancy and germinate either viviparously on the parent plant or shortly after release. Embryos of these recalcitrant and viviparous species cannot tolerate the maturation drying that is usually prerequisite to dormancy; such desiccation intolerance creates challenges for storing and preserving such embryos. I review the physiology, morphology, and ecology of these desiccation-intolerant, nondormant lineages. Differences in the production and function of plant hormones are implicated in the occurrence of recalcitrance and vivipary in plant families. Plant hormones are key regulators of seed physiology and simultaneously coordinate responses of the seedling and mature plant to their environment. Desiccation-intolerant embryos occur most commonly among species of wet or flooded environments and have evolved multiple times in disparate lineages. Natural selection in wetland environments simply may not eliminate these seed types or may select for changes in hormone physiology that simultaneously affect both maternal and embryonic tissues. Integrative data from ecological, genetic, and physiological studies are needed to elucidate evolutionary origins and maintenance of reproductive strategies in organisms.

A COMPARATIVE STUDY OF SEED PHYSIOLOGY

Most species exhibit a critical capacity to control the timing of their reproduction and the establishment of a new generation of offspring. Seeds, cysts, planktonic larvae, and other mobile propagules—hardy and compact phases in the life history of many taxa—enable organisms to disperse in space and to persist in time. Many of these forms exhibit some type of dormancy or metabolic quiescence involving a delay between fertilization and subsequent establishment. Dormancy and other complex life history traits are under physiological control, which, in turn, is regulated through mechanisms of gene action that are now being elucidated through intensive study. In particular, comparative and experimental studies of the activity and evolution of the ubiquitous and powerful hormones that mediate responses of both embryos and mature organisms to their environment can help illuminate many aspects of life history.

I consider the case of angiosperm seed physiology as a model system for investigating the hormonal control of traits in general. I review the physiological, morphological, and ecological characteristics shared by 195 plant species that exhibit no dormancy (Table 1) and hypothesize—on the basis of evidence from these as well as other species that precociously germinate or exhibit hormonal mutations—that they share consistent features of hormonal physiology. These species are common to predominantly wet or flooded environments, indicating that these habitats select for (or do not select against) these types of hormonal pathways. I posit that integrative data from ecological, genetic, and physiological studies are needed to elucidate evolutionary origins and adaptive costs and benefits of reproductive strategies in organisms.

DEFINING STATES OF SEEDS: Dormancy, Vivipary, and Recalcitrance

This review focuses on two types of nondormant seeds: those that are viviparous and those that are recalcitrant. Plant species in which the embryo grows sufficiently to emerge visibly from within the seed tissues *before* dispersal are termed viviparous (42, 68, 173). Vivipary (cf "viviparity," a term used by zoologists) is a relatively rare form of plant reproduction among angiosperms, but has been remarked upon by botanists for centuries (Table 1). In some plant species, the viviparous embryo can attain prodigious sizes and can grow for several months prior to release. Truly viviparous plants should not be conflated with species that produce apomictic or asexual plantlets or bulbils instead of embryos derived from sexual fertilization, although many of these latter taxa bear the species name *viviparum* (42, 68, 181). A precise definition of vivipary implies that formation of a seed and growth of the sexually generated embryo are integral to the process. Thus, vivipary can be studied and compared with germination processes in the context of seed development, physiology, and dispersal.

In another set of species (Table 1), the embryo sustains metabolic activity throughout ontogeny but bursts the seed tissues shortly after dispersal. In natural populations, these seeds may germinate readily within the fruit or soon after dehiscence, and they do not persist in the soil seed bank. These types of embryos rapidly lose viability if they are dried or chilled; hence they are termed "recalcitrant" to storage (146). The term recalcitrant is generally applied to seeds that have been systematically tested to determine their ability to tolerate desiccation (40). The inability to store seeds of these species creates challenges

TABLE 1 Plant species with recalcitrant or viviparous seeds[a]

Family	Species	Life form	Latitude	Habitat	Seed status	Source
Aceraceae	*Acer* spp. (3)	T	T	Mesic forest	R	82
Alismataceae	*Sagittaria latifolia*	H	T	Wetland	R	124
Anacardiaceae	*Mangifera indica*	T	TR	Wet forest	R	28
Annonaceae	*Cymbopetalum baillonii*	ST	TR	Wet forest	R	142
Apocynaceae	*Hancornia speciosa*	T	TR	Wet forest	R	128
Apocynaceae	*Landolphia kirkii*	V	TR	Wet forest	R	12
Araceae	*Dieffenbachia longispatha*	S	TR	Wet forest	V	86
Araceae	*Xanthosoma sagittifolium*	S	TR	Swamp forest	R	42
Araliaceae	*Hedera helix*	V	T	Mesic cultivar	R	71
Araucariaceae	*Agathis robusta*	T	TR	Wet forest	R	201
Araucariaceae	*Araucaria* spp. (2)	T	TR	Coast	R	201
Arecaceae	*Areca catheca*	P	TR	Wet forest	R	28
Arecaceae	*Calamus* spp. (2)	P	TR	Swamp forest	R	115, 122
Arecaceae	*Chrysalidocarpus leutecens*	P	TR	Wet forest	R	11
Arecaceae	*Cocos nucifera*	P	TR	Coast	V	28
Arecaceae	*Elaeis guineensis*	P	TR	Wet forest (ag.)	R	28
Arecaceae	*Nypa fruticans*	P	TR	Coast	V	173
Arecaceae	*Sabal* spp. (2)	P	TR	Wet forest (ag.)	R	28
Asteraceae	*Abrotanella linearis*	S	ST	Wet forest	V	22
Asteraceae	*Acamptopappus* sp.	S	ST	Desert	R	201
Asteraceae	*Pachystegia insignis*	S	ST	Wet forest	V	22
Avicenniaceae	*Avicennia* spp. (8)	T	TR	Coast	V	53
Bombacaceae	*Durio zibethinus*	T	TR	Wet forest	R	28
Bombacaceae	*Montezuma speciosissima*	T	TR	Wet forest	V	118
Boraginaceae	*Cordia alliodora*	T	TR	Wet forest	R	185
Burseraceae	*Dacryodes excelsa*	T	TR	Wet forest	R	118
Caricaeae	*Jacaratia dolichaula*	ST	TR	Wet forest	R	86
Caryophyllaceae	*Scheidea diffusa*	S	TR	Montane	V	10
Celastraceae	*Salaciopsis ingifera*	V	TR	Wet forest	V	33
Ceratophyllaceae	*Ceratophyllum* sp.	H	TR	Riverine	R	190
Chenopodiaceae	*Chenopodium quinoa*	H	T	Agriculture	R	40
Chrysobalanaceae	*Coupeia polyandra*	ST	TR	Wet forest	R	142
Clusiaceae	*Garcinia mangostana*	ST	TR	Wet forest (ag.)	R	28
Clusiaceae	*Symphonia globulifera*	ST	TR	Wet forest	R	185
Combretaceae	*Conocarpus erectus*	ST	TR	Coast	R	173

(Continued)

TABLE 1 (*Continued*)

Family	Species	Life form	Latitude	Habitat	Seed status	Source
Combretaceae	*Laguncularia racemosa*	T	TR	Coast	R	173
Connaraceae	*Connarus grandis*	ST	TR	Wet forest	V	33
Cornaceae	*Corokia macrocarpa*	S	ST	Wet forest	V	22
Cornaceae	*Griselinia* spp. (2)	S	ST	Riverine	V	22
Corylaceae	*Corylus americana*	ST	T	Mesic forest	R	28
Cucurbitaceae	*Sechium edule*	H	T	Agriculture	V	28
Cucurbitaceae	*Telfaira occidentalis*	S	TR	Agriculture	R	40
Cupressaceae	*Cupressus macrocarpa*	ST	TR	Wet forest	R	2
Cymodoceaceae	*Amphibolus* spp. (2)	H	TR	Coast	V	38
Cymodoceaceae	*Thalassodendron* spp. (2)	H	TR	Coast	V	173
Dipterocarpaceae	*Anisoptera laevis*	T	TR	Wet forest	R	201
Dipterocarpaceae	*Dipterocarpus* spp. (8)	T	TR	Wet forest	R	126
Dipterocarpaceae	*Dryobalanops aromatica*	T	TR	Wet forest	R	126
Dipterocarpaceae	*Hopea* spp. (8)	T	TR	Wet forest	R	126
Dipterocarpaceae	*Parashorea densiflora*	T	TR	Wet forest	R	126
Dipterocarpaceae	*Shorea robusta*	T	TR	Wet forest	R	201
Dipterocarpaceae	*Stemonoporus oblongifolius*	T	TR	Wet forest	R	126
Ebenaceae	*Diospyros virginiana*	ST	ST	Mesic forest	R	201
Elaeocarpaceae	*Sloanea berteriana*	T	TR	Wet forest	R	126
Euphorbiaceae	*Dalechampia scandens*	V	TR	Mesic forest	R	127
Euphorbiaceae	*Hevea brasiliensis*	T	TR	Wet forest	R	28
Fabaceae	*Castanospermum australe*	T	TR	Coast	R	190
Fabaceae	*Inga* spp. (2)	ST	TR	Swamp forest	V	120
Fabaceae	*Mora oleifera*	T	TR	Coast	V	87
Fabaceae	*Pithecellobium racemosum*	T	TR	Wet forest	V	105
Fagaceae	*Castanea dentata*	T	T	Mesic forest	R	28
Fagaceae	*Lithocarpus densiflorus*	T	T	Mesic forest	R	201
Fagaceae	*Quercus* spp. (3)	T	T	Mesic forest	R	58
Flacourtiaceae	*Casearia corymbosa*	T	ST	Wet forest	R	86
Flacourtiaceae	*Dovyalis hebecarpa*	ST	T	Agriculture	R	28
Flacourtiaceae	*Flacourtia indica*	ST	TR	Wet forest	R	28
Flacourtiaceae	*Muntingia calabura*	ST	TR	Wet forest	R	66
Hippocastanaceae	*Aesculus hippocastanum*	ST	T	Mesic forest	R	175
Lauraceae	*Machilus thunbergii*	ST	TR	Swamp forest	R	112
Lauraceae	*Nectandra ambigens*	ST	TR	Wet forest	R	142
Lauraceae	*Persea americana*	S	ST	Wet forest	R	28
Lecythidaceae	*Barringtonia racemosa*	T	TR	Coast	V	126

(*Continued*)

TABLE 1 (*Continued*)

Family	Species	Life form	Latitude	Habitat	Seed status	Source
Lecythidaceae	*Bertholletia excelsa*	T	TR	Wet forest	R	86
Lecythidaceae	*Lecythis ampla*	T	TR	Wet forest	R	86
Liliaceae	*Crinum capense*	H	TR	Riverine	V	9, 71
Liliaceae	*Hymenocallis* spp. (2)	H	ST	Riverine	V	42
Liliaceae	*Nerine* sp.	H	TR	grassland	V	42
Liliaceae	*Ripogonum scandens*	H	ST	Wet forest	V	23
Lobeliaceae	*Lobelia* sp.	S	TR	Montane	R	190
Loganiaceae	*Fagraea fragrans*	S	TR	Wet forest	R	66
Magnoliaceae	*Magnolia portoricensis*	ST	TR	Wet forest	R	118
Magnoliaceae	*Michelia champaca*	T	TR	Wet forest	R	28
Melastomataceae	*Melastoma malabathricum*	S	TR	Swamp forest	R	66
Meliaceae	*Aglaia odorata*	S	TR	Wet forest	R	201
Meliaceae	*Carapa guianensis*	T	TR	Swamp forest	R	86
Meliaceae	*Guarea glabra*	T	TR	Wet forest	R	118
Meliaceae	*Turrianthus africana*	T	TR	Wet forest	R	185
Moraceae	*Artocarpus heterophyllus*	T	TR	Wet forest	R	28
Moraceae	*Morus latifolia*	ST	ST	Wet forest	V	40
Myristicaceae	*Myristica hollrungii*	T	TR	Swamp forest	V	190
Myrsinaceae	*Aegiceras* spp. (2)	ST	TR	Coast	V	173
Myrtaceae	*Amomyrtus lama*	T	ST	Wet forest	R	201
Myrtaceae	*Eugenia* spp. (2)	ST	TR	Wet forest	R	28
Nepenthaceae	*Nepenthes gracilis*	E	TR	Swamp forest	R	66
Nyctaginaceae	*Pisonia longirostris*	T	TR	Swamp forest	V	33
Nymphaceae	*Nymphaea* sp.	H	TR	Riverine	R	190
Nyssaceae	*Nyssa aquatica*	T	ST	Swamp forest	R	190
Oxalidaceae	*Averrhoa carambola*	ST	TR	Wet forest	R	28
Oxalidaceae	*Oxalis* sp.	S	T	Mesic forest	R	28
Pellicierieaceae	*Pelliciera rhizophorae*	T	TR	Coast	V	87
Piperaceae	*Piper hispidum*	S	TR	Wet forest	R	185
Plumbaginaceae	*Aegialitis* spp. (2)	S	TR	Coast	V	173
Poaceae	*Spartina anglica*	H	T	Coast	R	140
Poaceae	*Zizania aquatica*	H	T	Wetland	R	14, 140
Podocarpaceae	*Dacrycornus dacrydioides*	ST	ST	Mesic forest	R	37
Podocarpaceae	*Podocarpus henkelii*	S	ST	Mesic forest	R	37
Polygonaceae	*Fagopyrum esculentum*	H	T	Agriculture	V	40
Potamogentonaceae	*Potamogeton* sp.	H	T	Wetland	R	124
Proteaceae	*Macadamia ternifolia*	ST	TR	Agriculture	R	28

(*Continued*)

TABLE 1 (*Continued*)

Family	Species	Life form	Latitude	Habitat	Seed status	Source
Ranunculaceae	*Caltha palustris*	H	T	Wetland	R	190
Rhizophoraceae	*Bruguiera* spp. (6)	T	TR	Coast	V	173
Rhizophoraceae	*Carallia brachiata*	ST	TR	Swamp forest	R	201
Rhizophoraceae	*Ceriops* spp. (3)	ST	TR	Coast	V	173
Rhizophoraceae	*Kandelia candel*	ST	TR	Coast	V	173
Rhizophoraceae	*Rhizophora* spp. (8)	T	TR	Coast	V	173
Rosaceae	*Eriobotrya japonica*	ST	T	Mesic forest	R	28
Rubiaceae	*Coffea* spp. (2)	ST	TR	Montane	R	28
Rubiaceae	*Coprosma robusta*	S	ST	Wet forest	V	22
Rubiaceae	*Ixora* sp.	ST	TR	Wet forest	R	190
Rubiaceae	*Ophiorrhiza tomentosa*	S	TR	Wet forest	V	167
Rubiaceae	*Posoqueria latifolia*	ST	TR	Swamp forest	R	86
Rutaceae	*Citrus* spp. (2)	ST	TR	Agriculture	R	28
Rutaceae	*Clausena dentata*	ST	TR	Agriculture	R	190
Rutaceae	*Fortunella japonica*	ST	TR	Agriculture	R	190
Santalaceae	*Santalum album*	ST	TR	Wet forest	R	2
Sapindaceae	*Euphoria longan*	ST	TR	Wet forest	R	40
Sapindaceae	*Litchi chinensis*	ST	TR	Wet forest	R	201
Sapindaceae	*Magonia pubescens*	ST	TR	Wet forest	R	185
Sapindaceae	*Meliococcus bijugatus*	ST	TR	Wet forest	R	40
Sapindaceae	*Nephelium lappaceum*	T	TR	Wet forest	R	28
Sapotaceae	*Calocarpum sapota*	T	TR	Wet forest	R	28
Sapotaceae	*Chrysophyllum cainito*	ST	TR	Wet forest	R	28
Sapotaceae	*Manilkara zapota*	ST	TR	Wet forest	R	28
Sapotaceae	*Mimusops* sp.	T	TR	Agriculture	R	190
Sapotaceae	*Pouteria ramiflora*	T	TR	Wet forest	R	185
Sapotaceae	*Euphrasia disperma*	H	ST	Wet forest	V	28
Scrophulariaceae	*Quassia indica*	T	TR	Wet forest	R	190
Simaroubaceae	*Cola nitida*	ST	TR	Mesic forest	R	28
Sterculiaceae	*Theobroma cacao*	ST	TR	Wet forest	R	28
Surianaceae	*Guilfolylia monostylis*	T	TR	Wet forest	R	127
Theaceae	*Camellia sinensis*	ST	TR	Montane	R	28
Verbenaceae	*Vitex divaricata*	T	TR	Wet forest	R	118
Vochysiaceae	*Vochysia honurensis*	T	TR	Wet forest	R	185

[a]Listed are the taxonomic **family** to which they belong; the **species** (with a number indicating total number of species with trait, if more than one per genus); the **life form** of the adult plant (T, tree taller than 10 m; ST, small tree; S, shrub; V, vine/liana; P, palm; H, herbaceous); the **latitude** or native region of the species (T, temperate; TR, tropical; ST, sub-tropical); the native **habitat** of the species, including agricultural ("ag") of the species is cultivated; **seed status** (R, recalcitrant; V, viviparous; and **source**, the published paper(s) that documents seed status with a germination study).

for germ plasm conservationists, foresters concerned with tropical and temperate forest regeneration, and restoration ecologists. Thus, in devising seed storage schemes, substantial efforts have been devoted to systematically diagnosing types and causes of recalcitrant behaviors (28, 31, 139). Many of the recalcitrant species thus far identified are economically important tropical fruit crops (28) and timber species (184, 201). The degree of recalcitrance exhibited by seeds varies among maternal lines in some species, indicating a genetic component to its control (136).

Until recently, comprehensive accounts of seed dormancy regimes among the angiosperms have been widely scattered throughout the botanical and physiological literature (e.g. 5, 10, 33, 42, 74, 76, 104, 127). However, a few recent compendia provide systematic data on the specific phenomena of vivipary and recalcitrance and on anatomical features, germination characteristics, and habitats shared by predominantly economically important species (28, 42, 190, 191).

Compiling this literature, I enumerate here 78 families, including 195 species in 143 genera, containing members that exhibit some form of vivipary or recalcitrance (Table 1). This feature manifests itself in fully viviparous or cryptoviviparous (e.g. see 173) behavior in 65 species (cf ~50 species noted in 42) and recalcitrance in the remaining 130 species. This list omits the hundreds of species, primarily of tropical wet-forest and riverine habitats, whose seeds germinate within a few days of release but for which their desiccation intolerance has not been established (e.g. 67, 185), as well as taxa for which only one anecdotal account exists.

It is also illuminating to compare these types with other species that sprout viviparously under certain circumstances. Some of these species are recalcitrant, while others are normally dormant. For example, seeds of many crop plants, including wheat, rice, maize, sorghum, and barley, that are subjected to unusually high humidity or to flooding will exhibit precocious, preharvest sprouting (5, 145, 176). The external cue for such behavior in these species is an abundance of water, often enhanced by warm temperatures. Although the propensity to sprout prematurely is dependent on environmental stimuli, susceptibility to these cues varies heritably among cultivars and thus is genetically based (192). In addition, a small set of mutant genotypes also produce seeds that germinate prematurely on the parent plant. Mutants of tomato, wheat, corn, and *Arabidopsis* exhibit viviparous phenotypes that reflect altered production of phytohormones, reduced sensitivity to dormancy-inducing phytohormones, and modified embryonic and adult water relations (36, 69, 80, 98, 109, 119, 149, 183). Studies of both precocious sprouters and mutants suggest that these nondormant phenotypes may result from alterations in hormonal biosynthetic or signal transduction pathways. Many precocious sprouters and viviparous mutants exhibit nondormant behavior that is anomalous with respect to the normal seed behavior of the species. However, I make reference to them from time to time in this review because the biological phenomena behind these unusual phenotypes may, in some cases, be similar to those operating in viviparous and recalcitrant taxa.

EMBRYO DRYING DISTINGUISHES DORMANT SEEDS FROM VIVIPAROUS AND RECALCITRANT SEEDS

Viviparous and recalcitrant seeds differ in fundamental ways from seeds that undergo dormancy. Dormant seeds exhibit the following generalized chronology of development: (*a*) embryo growth and tissue differentiation; (*b*) seed expansion, reserve deposition, and vacuole filling; (*c*) internal desiccation, organellar de-differentiation, and membrane stabilization; (*d*) metabolic quiescence; (*e*) imbibition, reserve mobilization, and resumption of metabolism in response to environmental signals; and (*f*) germination, commonly by root protrusion through the seed coat (Figure 1). Germination commences when intensive metabolic activity is regained following dormancy. In contrast to dormant species, viviparous and recalcitrant taxa lack the third step of maturation drying and consequent metabolic quiescence and proceed directly to the germination phase.

Viviparous and recalcitrant embryos maintain high tissue moisture contents throughout ontogeny (146; Figure 1). Water is critical to embryo metabolism and development; indeed, its uncontrolled loss exerts deleterious impacts on cell structure, mitotic growth, and biochemistry in all plant tissues. Seed germination, growth, DNA integrity, protein synthesis, membrane structure, organellar formation, and normal embryo development are disrupted when internal hydration levels drop below critical thresholds, which themselves vary among species (3, 95, 113, 131). Loss of metabolic water during prolonged drying is accompanied frequently by fusion of vacuoles, vesiculation of the endoplasmic reticulum, and free-radical peroxidation of lipid and protein components of cell membranes, leading to eventual cellular collapse (77, 129, 132, 156). Much of this damage is manifested during rehydration. Both the rate of dehydration and the absolute percentage of water lost determine the extent of tissue damage sustained by desiccating embryos (40, 50, 51). All seeds are intolerant to premature drying early in their development, but dormant seeds acquire tolerance to slow desiccation within days after fertilization and increasing tolerance to rapid drying as the embryo ages, whereas recalcitrant seeds never do (Figure 1). Mechanisms conferring desiccation tolerance on the maturing seed have been inferred from numerous studies that use hormonal mutants, endogenous manipulation of water levels and hormone activities, and exogenous applications of hormones to excised embryos (97). The acquisition of desiccation tolerance during mid- to late embryogenesis has been correlated with increases in three substances with related activities: abscisic acid (ABA), dehydrin proteins, and oligosaccharides (Figure 2). Considerably less is known about the status of these substances in naturally desiccation-intolerant species.

ROLES OF PHYTOHORMONES IN EMBRYO DRYING

Abscisic Acid

The phytohormone ABA appears to play a central role in preparing embryos for maturation drying before dormancy. In many desiccation-tolerant seeds, levels of

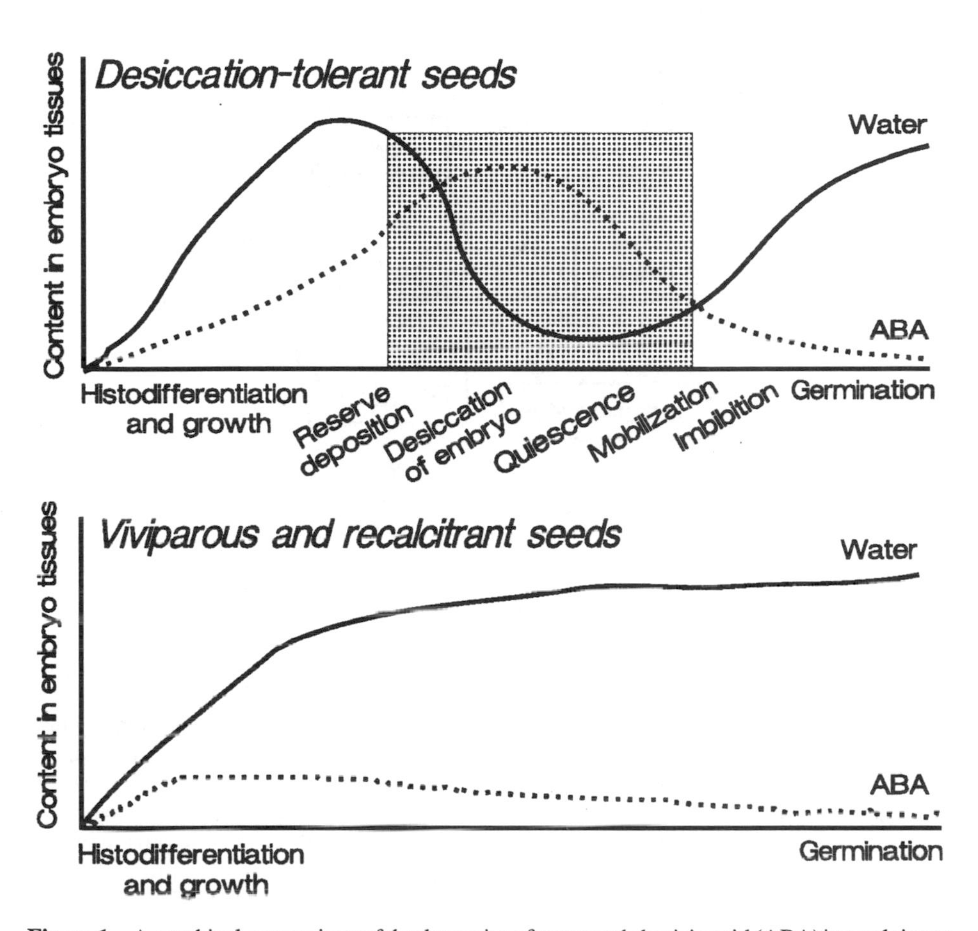

Figure 1 A graphical comparison of the dynamics of water and abscisic acid (ABA) in recalcitrant and viviparous (desiccation-intolerant) and dormant (desiccation-tolerant) seeds. Recalcitrant and viviparous embryos cannot dry, and they do not enter a dormant phase. Desiccation-tolerant embryos, by contrast, are capable of drying. While not all inherently desiccation-tolerant embryos enter dormancy in nature, embryos of all naturally dormant plant species show some form of desiccation tolerance. In dormant seeds (*upper diagram*, after 13), water content increases to a peak at the onset of reserve deposition, drops during maturation drying, remains low throughout dormancy, then rises again during imbibition. ABA content increases as the seed dries, but does not necessarily remain high during dormancy. *Lower diagram* (after data from 51, 95) shows the hypothesized accumulation of high levels of water during histodifferentiation in desiccation-intolerant seeds, levels that remain high throughout development. Metabolic quiescence does not occur. ABA levels may peak early during histodifferentiation, but generally remain low throughout maturation.

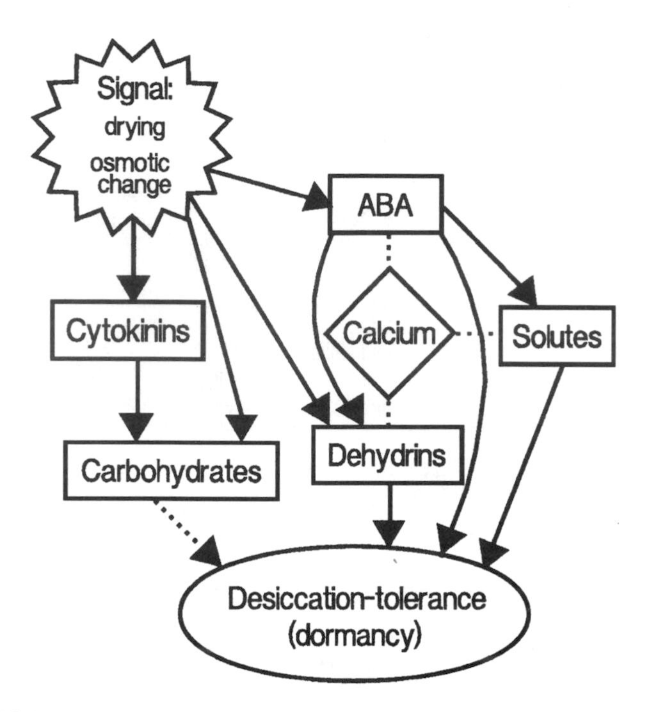

Figure 2 Linkages among multiple pathways for regulating desiccation tolerance in plant tissues, including seeds. An environmental signal that induces a change in osmotic status triggers production of the phytohormone abscisic acid (ABA) and its cofactor, calcium, which in turn transduces signals for the release of compatible solutes and transcription of dehydrin class proteins that protect membranes during desiccation stress. Independent pathways may lead to increased cytokinin production and creation of carbohydrate sinks within dehydrating tissues. Together, these linked mechanisms confer desiccation-tolerance on tissues, a prerequisite for dormancy in embryos.

ABA peak either once or twice during mid-embryogenesis (Figure 1). These peaks in ABA levels coincide with the onset of the maturation drying that is prerequisite to dormancy and subsequent germination. ABA may be supplied maternally at first, but later is produced endogenously in the embryo (81).

Several lines of evidence suggest that ABA is necessary (but not sufficient, in some cases) for the acquisition of desiccation tolerance and entry into dormancy. Seeds of ABA-deficient mutants of corn (*vp*), *Arabidopsis* (*abi*), *Nicotiana plumbaginifolia* (*iba*), wheat (*EH*-47), and tomato (*sitiens*) exhibit reduced protein accumulation and lack of dormancy frequently leading to viviparous germination (13, 60, 93, 98, 101, 107). Preharvest sprouting varieties of sorghum also exhibit abnormally low levels of endogenous ABA relative to sprouting-resistant cultivars (161). Embryos of naturally occurring recalcitrant species, including *Theobroma cacao* (135), *Quercus robur* (57), *Hopea odorata* (64), and *Machilus thunbergii* (112), as well as several species of mangroves (halophytic tropical trees) that are viviparous, exhibit low quantities of ABA throughout embryo development (49, 52, 53).

Artificial manipulation of ABA levels also modifies desiccation tolerance and consequent dormancy in seeds. For example, exogenous addition of ABA induces and prolongs desiccation tolerance in cultured embryos, ABA-deficient mutants (121), and recalcitrant species (58) and also inhibits germination (148, 200). Likewise, application of ABA-inhibitors at early stages of embryo maturation induces precocious germination in seeds (200).

Desiccation intolerance also may, in certain cases, be associated with a lack of sensitivity to ABA. Embryos of some recalcitrant species (65), preharvest sprouters (161), and one naturally viviparous species (163) exhibit reduced sensitivity to the normally inhibitory effects of ABA on metabolic activity and germination. Likewise, ABA-insensitive mutants are desiccation intolerant and germinate viviparously (36, 119, 183). More information on hormonal sensitivity is needed for a wider range of species (177). Together, these findings suggest the hypothesis that ABA is integrally involved in preparing the mature embryo for desiccation in anticipation of seed dormancy and, conversely, that ABA levels or sensitivity may be lacking in recalcitrant or viviparous species.

Both ABA concentrations (Figure 1) and embryo tissue sensitivity to ABA first peak and then diminish quickly following drying, however, indicating that ABA initiates, but does not necessarily enforce, long-term dormancy (95). Both ABA and external osmotic potentials may constitute analogous and complementary, but separate, signals to the embryo (95). Precise mechanisms by which ABA controls osmotic balance in embryo cells are largely unknown, as cellular ABA receptors have proven elusive (109, 194). ABA may regulate osmoticum in the seed directly (63). Alternatively, ABA may function primarily indirectly as a signaling molecule that binds to a membrane-bound ABA-response element (e.g. see 8), initiating a phosphorylation cascade that up-regulates gene expression for a constellation of stress-related proteins (19). Both free and membrane-bound calcium ions (Ca^{2+}), ubiquitous in plant cells, may help transduce ABA signals (155), a point to which I will return later in this review. ABA has been detected in all seed components, and its regulatory action in seed dormancy may not be restricted to embryonic tissues alone. Communication of ABA among endosperm and embryo to coordinate tissue dehydration, for example, may occur (78).

The pleiotropic (e.g. see 56) nature of ABA action in different plant compartments must be taken into account when studying changes in its production in isolated plant tissues. ABA, like other phytohormones, regulates suites of phenotypic traits in plants and coordinates integrated responses of plants to multiple, interacting environmental stresses (25; Table 2). In addition to its activities in seeds, abscisic acid regulates desiccation tolerance in vegetative tissues of mature plants (19, 129). Evidence for this role comes from studies of mature ABA-deficient mutants, which are prone to wilting, drought rhizogenesis, and other manifestations of impaired water balance. ABA levels transiently increase in roots of plants subjected to flooding, salt stress, and drought, and ABA may be transported to leaves as a signal to induce stomatal closure (36, 202) and to shoots to promote

TABLE 2 A summary of basic hormone actions in plants

Structure	Cytokinins	Auxin	Gibberellins	Abscisic acid	Ethylene
Organelle/ wall	Tonoplast integrity; chloroplast maturation; increases wall plasticity	Wall loosening through proton release	Increases cell wall elasticity	Regulates ion permeability of cell/tonoplast membranes	Increases tonoplast permeability; degrades cell walls
Whole cell	Reduces free radical production; promotes release of cellulases; promotes mitosis; increases turgor	Up-regulates proton pump; induces release of cellulases	Up-regulates sugar-digesting enzymes; Increases osmotic potential; promotes mitosis; promotes cell elongation	Down-regulates proton pump; promotes dehydrin protein production; promotes calcium import	Inhibits chlorophyll binding; up-regulates cellulase; promotes radial expansion; hastens cell senescence
Shoot meristem	Promotes lateral bud formation	Confers apical dominance; promotes epidermal growth	Releases bud dormancy; floral induction	Enforces bud dormancy	Promotes epinasty; slows stem growth; slows hook opening
Stem	May promote elongation	Promotes phototropic and geotropic bending	Promotes bolting and elongation	Unknown	Increases thickening; production of air spaces
Leaves	Delays senescence	Unknown	Hastens maturation; promotes growth	Wilting; stomatal closure; abscission	Promotes curling; hastens senescence
Flowers	Unknown	May inhibit flowering	Induces flowering; day-length response; vernalization; sex expression	Unknown	Flower curling; promotes female sex expression; may influence timing of flowers
Fruits/seeds	Hastens cotyledon maturation	Unknown	Promotes parthenocarpic fruit production; promotes seed germination	Enforces seed dormancy	Hastens fruit ripening
Roots	May promote meristem growth and root production	May promote meristem growth; promotes adventitious root production; regulates geotropism; up-regulates ethylene	Inhibits adventitious root production	Inhibits root growth; regulates membrane permeability to ions	Inhibits elongation; increases adventitious root number; aerates roots through production of aerenchyma

elongation during submergence (16). In roots and stomata, as in seeds, ABA alters cellular permeability to water and up-regulates production of versatile stress proteins (84). Together, ABA and osmotic stimuli can induce expression of salt-responsive genes (64).

Major evolutionary modifications in ABA levels and action in certain plant tissues may be necessitated by selection pressures experienced by plants as they colonize novel habitats via population differentiation and speciation (189). For example, high ABA concentrations in roots or leaves may be required to maintain whole-organism water balance under conditions of water stress or salinity. Because the plant is an integrated unit, however, high ABA levels in vegetative compartments could potentially inhibit or hinder metabolic processes in reproductive tissues. Plant species subjected to chronic stress may compartmentalize ABA production, activities, and tissue sensitivities, such that up-regulation of ABA in one sector does not impact another sector and cause a cascade of linked responses to ensue (24, 25). Evidence supporting this hypothesis comes from a study of four unrelated families of mangroves (49). These mangroves exhibit moderate to high levels of ABA in vegetative compartments and maternal tissues of the fruit, but very low ambient levels of ABA in the embryo throughout development (relative to nonviviparous related species). Vivipary and reductions in ABA have both arisen in mangrove lineages coincident with acquisition of the halophytic habit (92, 152, 173). A loss of desiccation tolerance and consequent seed dormancy in naturally occurring wetland or semi-aquatic species, precipitated by a significant reduction of ABA in the seed relative to that present in vegetative tissues, may be one notable consequence of evolutionary changes in ABA physiology that enable certain species to persist under conditions of osmotic stress (49, 52).

Other Phytohormones

ABA is one of five recognized phytohormones critical to plant growth and development (Table 2). Evidence for a role of other phytohormones in controlling desiccation tolerance or dormancy is scanty and somewhat inconsistent, however. For example, active gibberellic acids (GAs) are known to promote embryo germination and antagonize ABA activity (83, 165). Likewise, increased sensitivity to GAs may accompany or precede germination (106), and application of GA inhibitors can induce dormancy in formerly nondormant seeds (96). One might hypothesize that precociously germinating species of all types discussed here would show elevated levels or differing forms of gibberellins throughout embryogeny. Dormant and nondormant varieties of beech, for example, appear to possess different suites of GAs synthesized from divergent biochemical pathways (55). However, although active GAs have been detected in young embryonic axes of a viviparous mangrove species, *Rhizophora mangle* (133), they were not unusually high relative to nonviviparous species, nor is there consistent evidence that gibberellins are significantly elevated in recalcitrant seeds or perform functions different from those observed in dormant seed types (53, 125). Vivipary has not been reported among constitutive GA mutants (83, 98). Furthermore, GAs often cannot induce germination in embryos of some desiccation-tolerant species unless maturation drying, accompanied by peaks in ABA production, has first occurred (54, 95). Many facets of GA signal transduction and its role in seed physiology remain to be clarified (172).

Auxins, specifically indoleacetic acid, are found in all seed tissues, most abundantly in the cotyledons and the pericarp, which are both reserve-accumulating tissues. However, auxin concentrations in tissues generally decline as the embryo matures, and auxin does not appear to play a major role in either dormancy or germination. Its concentrations in recalcitrant seeds do not consistently differ from those of desiccation-tolerant seeds (53, 125).

Cytokinins are implicated both in the reserve accumulation phase of embryo development and in maintaining cell growth and division throughout embryogeny. Cytokinins promote cell division by accelerating rates of protein synthesis and decreasing the duration of cellular interphase (36). In seeds that undergo dormancy, cytokinins are most abundant during early histodifferentiation of the embryo—a period of rapid cell division and growth—and later in embryogeny, when they may influence seed germination through complementary signal transduction pathways sensitive to light (169). Cytokinins are present in relatively high concentrations in the recalcitrant seeds of *Citrus* spp. (43), *Avicennia marina* (53), and several other viviparous species of mangroves (46), indicating that they may be involved in maintaining continual cellular activity. Cytokinins may also help halophytes to overcome salt-induced inhibition of germination (70). Their role in seed metabolism and precocious germination deserves further study.

Ethylene exerts limited effects on embryo metabolism, and mechanisms of its action are largely unknown (94). When it attains high concentrations inside the testa, ethylene sometimes is credited with breaking dormancy (103). Ethylene production accelerates as fruits ripen, but this increase appears to be decoupled from the maturation of the embryo itself. Although ethylene production apparently rises during maturation drying of certain seeds, its dynamics vary between congeneric species, and its role in desiccation tolerance is uncertain (88). In fact, ethylene impedes germination of the recalcitrant seeds of *Quercus robur* (58). In flooded or aquatic habitats, ethylene concentrations in submerged organs (especially roots) can attain high levels due to its slowed diffusion from waterlogged tissues. Because recalcitrant and viviparous wetland taxa often inhabit seasonally or chronically flooded environments (Table 1), a possible role for ethylene in altered seed physiology warrants more investigation.

In determining the roles hormones play in the regulation and evolution of a complex life history trait such as seed dormancy, it is imperative to quantify the following: (*a*) hormone concentrations at the site of action, (*b*) sensitivity of the tissues involved in complex responses, and (*c*) relative significance of hormonal control assessed against a background of other controlling influences (177, 197). Manipulative tests of hormone action must be coupled with genetic analysis of loci involved in producing hormones and transducing their signals and performed in a broad array of desiccation-intolerant species.

ROLES OF PROTEINS IN EMBRYO DRYING

Several common proteins are implicated in the acquisition of desiccation tolerance (84). Many of these proteins are produced in a variety of vegetative and seed tissues during periods of drought stress, although similar responses occur during exposure to other stresses including salinization, heat-shock, and chilling. It is of interest that many of the recalcitrant and viviparous seeds that lack these proteins are also quite sensitive to chilling, suggesting that these proteins have multiple protective functions.

A specific class of dehydration proteins becomes prevalent in desiccation-tolerant seeds as maturation drying commences, ABA levels increase, and water content declines. These proteins [Em proteins (117), late-embryogenesis-abundant (LEA) proteins (39), and ABA-responsive (Rab) proteins (155)] are referred to as "dehydrins" throughout the literature (30). They have been identified in diverse species including cotton, several cereals, legumes, tomato, pine, *Arabidopsis*, "resurrection" plants in several families, and mosses (84, 95, 100). Their expression chronology, sequence motifs, biochemical characteristics, and hydrophilic nature are highly conserved within and among diverse taxa (30). Dehydrins bind water, sequester ions amassed during desiccation, and coat membrane components to preserve a stable configuration during water loss (30, 39). They also function as ABA-responsive promoters of gene expression (109, 117). ABA-deficient viviparous mutants show reduced levels of these proteins throughout embryogeny (116), indicating that ABA signaling may be prerequisite to the production of dehydrins. Dehydrins are absent from several recalcitrant and viviparous species (52, 53). However, their presence in some recalcitrant species, including *Quercus* species and wild rice, suggests that dehydrins are necessary, but not sufficient, to achieve desiccation tolerance in seeds (18, 59). Likewise, some ABA-insensitive mutants of *Arabidopsis* show little apparent reduction in dehydrins (60). Thus, ABA may not constitute the sole signal for dehydrin up-regulation in this complex pathway (44).

ROLES OF CARBOHYDRATES AND COMPATIBLE SOLUTES IN EMBRYO DRYING

Carbohydrates

Carbohydrates are a dominant component of all plant cells, providing the foundations of cellular structures, the fuel for cellular activities, and the solutes for maintaining osmotic equilibrium in the cytosol. The importance of various carbohydrates to the acquisition of desiccation tolerance by plant tissues, especially seeds, is continually debated, principally because the precise mechanisms by which sugars confer desiccation tolerance are still largely unknown. Certain soluble sugars increase in concentration within embryos of some species as dehydration commences (15, 108) and decrease in other species as desiccation tolerance is lost

(102). However, this phenomenon appears to be both species-specific and dependent upon the rate of drying applied to the seed (168). Sugars such as sucrose can promote desiccation tolerance by stabilizing cell membranes, either by replacing water with hydroxyl groups (35, 79, 95) or by vitrifying—forming highly viscous, aqueous glass (187). Seeds of desiccation-intolerant species, therefore, may be deficient in these sugars throughout development (46). However, evidence of water replacement and vitrification has been detected in seeds of recalcitrant species (187). Some recalcitrant seeds exceed desiccation-tolerant seeds in their oligosaccharide content (12, 51), for example, and consistent differences in sugar content between viviparous and nonviviparous mangroves have not been found (193). Farrant et al (51) proposed a link between high levels of cytokinins in the embryo and an accelerated rate of sugar import, as cytokinins are implicated in metabolic sink formation. Cells that are actively respiring, growing, and dividing also become sinks for carbohydrates, especially sucrose (which functions doubly to fuel metabolic activities of cells and, as a compatible osmolyte, to maintain turgor). Likewise, modifications in ABA production or action in viviparous, recalcitrant, early sprouting or mutant seeds may also affect sugar metabolism. For example, ABA inhibits acid invertase, sucrose synthase, and sucrose phosphate synthase activities in certain cells (29), acting antagonistically to cytokinins. Thus, a reduction in ABA, coupled with an increase in cytokinins in these embryos, may be associated with changes in rates of phloem unloading and sucrose processing. Mutants exhibiting altered carbohydrate biosynthetic pathways would be particularly useful study subjects in this regard, but their seeds have not yet been examined explicitly for desiccation intolerance (196). Because levels and types of sugars cannot be linked consistently with desiccation tolerance, it is unlikely that differences in oligosaccharide concentrations alone can distinguish recalcitrant and viviparous seeds from other types (129).

Solutes

Since viviparous and recalcitrant embryos do not normally develop tolerance to osmotic stress associated with maturation drying, it is logical to ask whether ion concentrations are consistently different in cells of these embryos and whether these differences contribute to the inability of these embryos to withstand drying. Simple inorganic ions, such as sodium, potassium, and calcium, can accumulate at various levels in drying cells as a function of water loss. They may be selectively released from vacuoles or selectively concentrated through preferential uptake by ion-specific membrane transporters (150). Nondormant barley varieties, for example, show higher cellular conductivity to potassium than dormant types (182). The preferential accumulation of potassium vs sodium in vegetative tissues has been well documented in salt- and drought-tolerant halophytes, including mangroves (6), in which active transport of potassium ameliorates the potential deleterious effects of sodium on photosynthetic processes. Joshi et al (90) observed that sodium:potassium ratios are lower in viviparous embryos than in nonviviparous mangrove embryos and proposed that this preferential uptake of potassium in competition with sodium reflects precocious development of salt tolerance. In contrast, observations that solutes decrease over time in viviparous embryos have led to the suggestion that viviparous reproduction is a desalinating process (203).

In addition to their ionic function in regulating osmotic potential, calcium ions are ubiquitous intracellular second messengers in plants, and they help to transduce ABA signals into stimuli for gene up-regulation (109, 195). Calcium inhibitors, as well as reductions in cytoplasmic calcium contents, inhibit production of ABA-responsive stress messenger RNAs (153). It is of interest, therefore, that calcium appears to occur at lower concentrations in the embryos of viviparous mangroves than those of nonviviparous nonmangroves (46). However, Joshi et al (90) found little difference in calcium levels in embryos of viviparous vs nonviviparous mangroves, indicating that reduced intracellular calcium may be more a characteristic of mangroves in general than a correlate of the viviparous habit. Likewise, abscisic acid-deficient mutants of *Arabidopsis* show similar responsiveness (compared with wild types) of internal calcium levels to applied salt stress, indicating that calcium responds to osmotic and ABA signals independently (34).

A small spectrum of compatible solutes (soluble compounds of low charge that do not harm cellular metabolism even at high concentrations) is produced in response to several stressors, including desiccation, salinity, and cold. In their evolutionary conservatism, they are reminiscent of the broad-response stress proteins that serve both as osmoregulatory solutes and as structural osmoprotectants (17, 84). Although compatible solutes are known to contribute to the development of desiccation tolerance in vegetative tissues, their roles during maturation drying in seeds only rarely have been investigated. Because proline can constitute a quarter or more of the amino acid profile of reproductive tissues, it has received the most attention. In desiccation-tolerant varieties of *Arabidopsis thaliana*, proline is found most abundantly in seed tissues with relatively low water content, and it is up-regulated in embryonic tissues subjected to artificial water stress or applications of exogenous ABA (27). Specifically focusing on recalcitrant seeds, Lin & Chen (112) found that low levels of proline characterized the desiccation-intolerant embryos of *Machilus thunbergii*. Viviparous mangroves do not appear to be impaired in their production of compatible solutes in vegetative tissues, but levels in embryonic tissues are unknown (138).

In summary, while considering physiological correlates with embryo drying in plant lineages, I have discussed certain regulators that are likely to be shared by several plant types that are intolerant to maturation drying. First and perhaps foremost, the recalcitrant and viviparous seeds studied to date most consistently exhibit either reduced levels of, or sensitivity to ABA, or both. Studies of the acquisition of desiccation tolerance in vegetative tissues and seeds suggest that changes in the production of stress proteins (particularly the late-embryogenesis-abundant dehydrins), cytokinins, sucrose, ions, and compatible solutes may coincide with either imposed water stress or ordinary maturation-related drying. A reasonable hypothesis, therefore, is that the loss of desiccation tolerance in recalcitrant and viviparous phenotypes is attributable to alterations in the regulation of one or many of these (physiologically linked) characters. Figure 2 illustrates how principal phytohormones, proteins, and solutes may interact during acquisition of desiccation tolerance and how reductions in their production can result in desiccation intolerance.

ECOLOGICAL AND STRUCTURAL COMMONALITIES AMONG RECALCITRANT AND VIVIPAROUS SPECIES

In addition to shared physiological characteristics, several broad ecological commonalities emerge from qualitative surveys of desiccation-intolerant taxa (Table 1). The taxa identified span a broad range of plant life forms including shrubs (12%), palms (5%), lianas/vines (2%), herbs (9%), and epiphytes (1%), with canopy trees (45%) and small understory trees (26%) constituting the majority. In terms of habitat, most recalcitrant and viviparous species (89%) occupy wet-forest, riverine, flooded, or coastal environments. Most species (79%) are native to the tropics [a recent review (10) posits that, in general, more than 60% of species of wet tropical zones possess minimal dormancy, but less is known specifically about their desiccation intolerance]. Few of the desiccation-intolerant taxa identified here occur in seasonally cold climes (15% of total), and most of these temperate-zone species inhabit riverine or swamp habitats (e.g. *Potamogeton*, *Caltha*, and *Sagittaria* species (10)]. A majority of the viviparous species occupy coastal tropical zones, especially mangrove forests, that are inundated daily or seasonally by tides. Many of the desiccation-intolerant taxa produce seeds that mature during tropical monsoons and rainy seasons and are unlikely to experience dry conditions.

These disparate species also share several fruit and seed characteristics, which contrast with seed traits of closely related dormant taxa. Over 70% of viviparous and recalcitrant species produce seeds that occur singly within the fruit. Their seeds are typically large (exceeding 4 cm in length). While the typically fleshy tissues surrounding the seed are sometimes soft, permeable to water, and high in moisture content, hard testae or exocarps occur among more than 40% of these species. Many (74%) of these species possess large embryos that occupy more than half the volume of the seed. Endosperm volume is correspondingly small or nonexistent in the majority of these species, and exalbuminous embryos that are independent of the endosperm are common. However, several viviparous mangroves possess a well-developed endosperm (32, 99), which in members of the Rhizophoraceae may even physically displace the embryo and hasten its bursting from the fruit (91). Copious cotyledonary starch reserves have also been noted among the recalcitrant taxa surveyed by von Teichman & van Wyk (191). Of interest is whether cotyledonary tissues function as storage tissue for the embryo and/or whether they provide nutrients to fuel continual metabolic activity (53).

We might expect large (especially viviparous) embryos to be supplied initially by maternal resources, but specific mechanisms of maternal transfer have not been studied. Certain viviparous species of mangroves appear to show discrete haustorial transfer tissues that do not occur in putatively ancestral, upland relatives (72, 164, 178, 199). Elaboration of maternal communication to the embryo may have proceeded during evolution once vivipary evolved in these lineages, as has been demonstrated in the Rhizophoraceae (92, 199). Alternatively, intensive maternal provisioning may preclude dormancy because water and nutrients are continually supplied to a metabolically active embryo. Is altered ABA production a consequence of modified modes of communication and water relations between the maternal plant and embryo among certain species of flooded or saline environments? Mature plants of these recalcitrant and viviparous species produce ABA in vegetative tissues, a likely mechanism for promoting tolerance to osmotic stress (16, 49), yet young embryos appear to postpone ABA up-regulation until after dispersal or establishment has occurred. As proposed earlier, physiological traits in the developing seed may evolve as a consequence or by-product of maternal adjustments to a range of external stimuli and selection pressures, including mechanisms of compartmentalizing hormonal production and function. However, mature plants and seeds of a wider range of recalcitrant and viviparous species must be studied to establish the generality of this hypothesis.

THE EVOLUTIONARY STATUS OF VIVIPARY AND RECALCITRANCE

To assess the evolutionary significance of correlations among desiccation intolerance, other seed characteristics, and habitat, it is necessary to establish whether these characteristics are ancient or recent evolutionary developments and whether they consistently appear simultaneously during evolution. Recalcitrance sensu lato has been viewed as pleisiomorphic based on its putative correlation with other primitive (e.g. see 190, 191) characteristics such as woodiness and a tropical habitat. It is difficult to assess the precise concordance of these characteristics in relation to desiccation intolerance at the species level, however, as previous authors have compiled their surveys at the level of families, providing only very coarse resolution (190, 191). Takhtajan (166) postulated trends in endosperm evolution, seed coat simplification, and dispersal modes that can indicate the evolutionary age of taxa and the relative status of seed traits, but simultaneously notes the occurrence of parallel character states in lineages of both ancient and recent origin. The seeds surveyed in the present review exhibit a variety of ancestral and derived traits according to Takhtajan's criteria. Phylogenetic evidence suggests that vivipary and recalcitrance are not always relict characteristics of ancient taxa;

rather, these traits have evolved repeatedly in descendants of desiccation-tolerant taxa. Recent phylogenies for the Araceae (62), Aceraceae (1), Araliaceae (130), and Rhizophoraceae (152), for example, place the recalcitrant or viviparous taxa in recent or terminal clades, while others addressing the Arecaceae conflict on the precise placement of the viviparous, monotypic genus *Nypa* (26, 179). Recalcitrance or vivipary occurs only in a single known species in each of 37 (47%) of the families, and 122 (85%) of the genera are monotypic for this trait. Thus, this characteristic has not proliferated among congeneric taxa within families, except among the mangrove members of the Rhizophoraceae, Avicenniaceae, Myrsinaceae, and Plumbaginaceae, as well as members of the species-rich Dipterocarpaceae native to tropical wet-forest habitats. Considering the fact that recalcitrance or vivipary occurs primarily in single taxa or genera within families, the most parsimonious explanation for their presence points to a few convergent losses of desiccation tolerance. A broader, quantitative examination of the evolutionary status and linkages among vivipary and recalcitrance awaits both the development of high-resolution phylogenies for more of the plant families listed in Table 1 and the systematic characterization of desiccation tolerance in their sister taxa.

EVALUATING COSTS AND BENEFITS OF VIVIPARY AND RECALCITRANCE

The benefits of seed dormancy as a means of maximizing seed output and optimizing dispersal distance and time have received most attention in the ecological and evolutionary literature (45, 75, 86, 144, 186, 188, 198). Similar models have been applied to animals, such as sponges and copepods, whose gemmules and eggs, respectively, exhibit long-term dormancy (73, 141). In contrast, the lack of dormancy has received comparatively little theoretical attention (42, 67, 85, 185).

When considering the putative advantages and disadvantages of particular modes of reproduction to explain their adaptive value, evolutionary biologists have emphasized three seed features as critically important to subsequent plant fitness: (*a*) maternal carbon costs of reproduction balanced against the early carbon needs of the offspring, (*b*) seed quantity (maternal fecundity) vs seed quality (namely, seed size and nutrient content), and (*c*) the value of dormancy or dispersal for ensuring establishment in spatially and temporally heterogeneous environments. In investigating recalcitrance and vivipary—the lack of dormancy—with reference to these features, authors have postulated that early germination proffers adaptive benefits and should also be associated with larger seed size, directed dispersal, long adult life spans, and less specialization to microhabitats (61, 144, 154, 162, 196, 188, 190). Several authors have postulated that dormancy and concomitant dispersal in time and space offer no selective advantage to seeds released into spatially or temporally homogeneous habitats or into coarsely mosaic environments where compatible patches are separated widely in space

(42, 110, 137, 171, 181). The seed phase can be short-lived, and selection pressures to promote the evolution of dormancy do not exist in these environments.

When one looks critically at the hypothetical benefits of vivipary and recalcitrance, particularly in the context of the ecology of specific species, one also notes the considerable costs of these modes. For example, large, viviparous seedlings of species such as mangroves may enjoy a considerable head start due to their early germination, assimilation of carbon from both maternal and (possibly) atmospheric sources (91), and prodigious growth before dispersal. Early germination is thought to expedite rapid rooting of mangroves following propagule release (114), for example, or to promote the salt tolerance of seedlings (89, 157). However, neither propagule size nor levels of nitrogen provisioning of viviparous mangrove propagules is correlated with subsequent establishment success or rates of growth (46, 111), and seedlings produced in a hypersaline maternal environment do not fare better in high-salinity growing conditions than do seedlings growing on trees of less saline areas (157). Other hypotheses suggest that, in nondormant species, establishment immediately follows reproduction, which itself is cued to environmental conditions that will optimally foster seedling growth (4, 159). However, this does not hold true consistently; some viviparous mangrove species reproduce copiously during certain seasons but undergo massive seedling mortality in most years (41, 47). Alternatively, others have proposed that large, nondormant seeds of tropical trees may be less inhibited by a lack of light (both a phytochrome germination signal and a critical resource for growth) in deep forest understories (185). However, studies of viviparous mangrove species show that early photosynthesis, growth, and survivorship are quite sensitive to light availability (41, 47). Another theory postulates that dormant seeds lingering in soil banks are quickly eaten or parasitized and that rapid germination enables accelerated establishment and escape from seed predators (126, 170). However, both predispersal and postdispersal predation of viviparous mangrove embryos, for example, can significantly reduce reproductive success (48, 147). Likewise, early emergence time is not strongly associated with resistance to herbivory in tropical rain forests (134).

Studies focusing on the adaptive value of these reproductive strategies yield evidence that precocious germination can also exact costs by entailing substantial maternal investment of carbon and nutrients in supplying a metabolically active embryo, reducing the quantity of seeds produced relative to the cost of provisioning each seed, and limiting dispersal latitude while hastening establishment of the vulnerable, metabolically demanding embryo. Comparative studies need to address mechanisms by which vivipary and recalcitrance have arisen convergently in so many unrelated taxa and, conversely, why such a seemingly advantageous strategy has not proliferated among temperate-zone halophytes and other angiosperms. Indeed, many (especially temperate) halophytes and freshwater wetland plant species exhibit desiccation tolerance and dormancy (180). A fruitful line of research would focus on specifically comparing the physiology and relative fitness of desiccation-tolerant and desiccation-intolerant taxa within these environments (82, 174).

INTEGRATING PHYSIOLOGY, ECOLOGY, AND EVOLUTION IN UNDERSTANDING SEEDS

To refine hypotheses about costs and benefits of recalcitrance and vivipary, we must critically examine whether desiccation tolerance actually confers higher survivorship, establishment success, or fitness across a range of species within particular habitat types and across the lifetimes of plants. To address the evolution of complex traits generally, we must (*a*) understand physiological pathways that control single traits and the linkages among them, (*b*) identify other traits with which the character in question is consistently associated across lineages, (*c*) determine that it has arisen independently in multiple lineages (and is not a by-product of phylogenetic relatedness), (*d*) investigate whether the evolutionary appearance of the trait within lineages consistently coincides with the evolutionary colonization of the habitat in question, and (*e*) compare the fitness of species sharing the same selective pressures that exhibit and do not exhibit the trait.

To develop coherent and credible optimality theories addressing the evolution of seed traits, we need to clarify genetic, physiological, and ecological similarities among species and learn from their differences (160). Recent promising comparative studies have begun to explore the physiological bases for convergent traits appearing in diverse lineages (46, 49, 52), identify genuine suites of correlated seed traits occurring within lineages (45, 134, 158), and investigate the transgenerational adaptive significance of seed traits and maternal effects (123, 162).

As Voesenek & Blom (189) and Blom (16) recently observed, a closer look at the status and evolution of plant hormones may help us to predict the appearance of traits under certain environmental conditions, to understand linkages among traits, and to manipulate traits to promote fitness. Within the seed, phytohormones figure importantly in controlling embryogenesis, dormancy, and germination. Hormonal pathways and sensitivities are pleiotropic and heritable, and hence they are subject to selection (109, 136). Little is as yet known, however, about how hormonal regulation of seed and whole-plant ecophysiology has evolved in angiosperms or has contributed to trends in dormancy regimes or contrasting seed bank dynamics in different environments (24, 25, 189). Insofar as hormones modulate source-sink relationships between the seed and parent plant, they may shape the evolution of life history tradeoffs in maternal investment during seed maturation (20, 143, 151). Indeed, the study of evolutionary physiology in general has lagged behind that of morphological evolution because (*a*) the precise nature of homology in physiological and biochemical traits is challenging to identify, (*b*) linkages between morphological homologies and their physiological correlates are difficult to delineate, (*c*) the plasticity and transience of physiological states demand that traits be characterized for a full range of conditions when comparing among taxa, and (*d*) physiological states generally do not leave a fossil record (21). Nevertheless, the comparative study of hormonal physiology in plant lineages is made possible by advances in phylogenetic methodology and by an exponentially increasing knowledge of hormonal mechanisms at both molecular and whole-organism levels.

ACKNOWLEDGMENTS

I thank E. A. Kellogg, N. M. Holbrook, F. A. Bazzaz, A. M. Ellison, D. G. Fautin, L. A. Meyerson, and D. Stein for comments that have improved various incarnations of this review. I am grateful to C. Baskin, F. S. Chapin III, P. Nel, P. B. Tomlinson, C. Vazquez-Yanes, M. Westoby, D. Haig, and especially J. M. Farrant for data, insights, and/or suggestions that have helped these thoughts gel over time. Staff of the Harvard University Herbaria helped immensely with researching seed traits. Research and writing were supported by National Science Foundation grants IBN-9623313 and DGE-9714522 to EJF, the DeLand Fund of the Arnold Arboretum, and Harvard University.

LITERATURE CITED

1. Ackerly DD, Donoghue MJ. 1998. Leaf size, sapling allometry, and Corner's rules: phylogeny and correlated evolution in maples (*Acer*). *Am. Nat.* 152:767–91
2. Akamine EK. 1951. Viability of Hawaiian forest tree seeds in storage at various temperatures and relative humidities. *Pac. Sci.* 5:36–46
3. Artlip TS, Madison JT, Settler TL. 1995. Water deficit in developing endosperm of maize: cell division and nuclear DNA endoreplication. *Plant Cell Environ.* 18:1034–40
4. Augspurger CK. 1979. Irregular rain cues and the germination and seedling survival of a Panamanian shrub (*Hybanthus prunifolius*). *Oecologia* 44:53–59
5. Auranen M. 1995. Pre-harvest sprouting and dormancy in malting barley in northern climatic conditions. *Acta Agric. Scand.* 45:89–95
6. Ball MC. 1996. Comparative ecophysiology of tropical lowland moist rainforest and mangrove forest. In *Tropical Forest Plant Ecophysiology*, ed. SS Mulkey, RL Chazdon, AP Smith, pp. 461–96. New York: Chapman & Hall
7. Baker HG. 1989. Some aspects of the natural history of seed banks. In *Ecology of Soil Seed Banks*, ed. MA Leck, VT Parker, RL Simpson, pp. 9–21. San Diego, CA: Academic
8. Baker SS, Wilhelm KS, Thomashow MF. 1994. The 5′-region of *Arabidopsis thaliana cor 15a* has *cis*-acting elements that confer cold-, drought- and ABA-regulated gene expression. *Plant Mol. Biol.* 24:701–13
9. Barton LV. 1961. *Seed Preservation and Longevity.* Plant Sci. Monogr. Ser. New York: InterScience
10. Baskin CC, Baskin JM. 1998. *Seeds: Ecology, Biogeography, and Evolution of Dormancy and Germination.* San Diego, CA: Academic
11. Becwar MR, Stanwood PC, Roos ER. 1982. Dehydration effects on imbibitional leakage from desiccation-sensitive seeds. *Plant Physiol.* 69:1132–35
12. Berjak P, Vertucci CW, Pammenter NW. 1992. Homoiohydrous (recalcitrant) seeds: developmental status, desiccation sensitivity and the state of water in axes of *Landolphia kirkii* Dyer. *Planta* 186:249–61

13. Bewley JD. 1997. Seed germination and dormancy. *Plant Cell* 9:1055–66
14. Bewley JD, Black M. 1994. *Seeds: Physiology of Development and Germination.* New York: Plenum
15. Black M, Corbineau F, Grzesik M, Guy P, Come D. 1996. Carbohydrate metabolism in the developing and maturing wheat embryo in relation to its desiccation tolerance. *J. Exp. Bot.* 47:161–69
16. Blom CWPM. 1999. Adaptations to flooding stress: from plant community to molecule. *Plant Biol.* 1:261–73
17. Bonhert HJ, Nelson DE, Jensen RG. 1995. Adaptations to environmental stresses. *Plant Cell* 7:1099–111
18. Bradford KJ, Chandler PM. 1992. Expression of "dehydrin-like" proteins in embryos and seedlings of *Zizania palustris* and *Oryza sativa* during dehydration. *Plant Physiol.* 99:488–94
19. Bray EA. 1997. Plant responses to water deficit. *Trends Plant Sci.* 2:48–54
20. Brenner ML, Cheikh N. 1995. The role of hormones in photosynthate partitioning and seed filling. In *Plant Hormones: Physiology, Biochemistry and Molecular Biology*, ed. PJ Davies, pp. 649–70. Dordrecht, The Netherlands: Kluwer Academic
21. Burggren WW, Bemis WE. 1990. Studying physiological evolution: paradigms and pitfalls. In *Evolutionary Innovations*, ed. MH Nitecki, pp. 191–228. Chicago, IL: Univ. Chicago Press
22. Burrows CJ. 1995. The seeds always know best. *NZ J. Bot.* 32:349–63
23. Burrows CJ. 1996. Germination behavior of the seeds of seven New Zealand vine species. *NZ J. Bot.* 34:93–102
24. Chapin FS III. 1991. Integrated responses of plants to stress. *BioScience* 41:29–36
25. Chapin FS III, Autumn K, Pugnaire F. 1993. Evolution of suites of traits in response to environmental stress. *Am. Nat.* 142 (Suppl.):S78–92
26. Chase MW, Duvall MR, Hills HG, Conran JG, Cox AV, et al. 1995. Molecular phylogenetics of Lilianae. In *Monocotyledons: Systematics and Evolution*, ed. PJ Rudall, PB Cribb, DF Cutler, CJ Humphries, pp. 109–37. London: R. Bot. Gard. Kew
27. Chiang HH, Dandekar AM. 1995. Regulation of proline accumulation in *Arabidopsis thaliana* (L.) Heynh. during development and in response to desiccation. *Plant Cell Environ.* 18:1280–90
28. Chin HF, Roberts EH. 1980. *Recalcitrant Crop Seeds.* Kuala Lumpur, Malaysia: Tropical
29. Chraibi A, Palms B, Druart N, Goupil P, Gojon A, et al. 1995. Influence of abscisic acid on nitrogen partitioning, sucrose metabolism and nitrate reductase activity of chicory suspension cells. *J. Exp. Bot.* 46:1525–33
30. Close TJ. 1996. Dehydrins: emergence of a biochemical role of a family of plant dehydration proteins. *Physiol. Plant.* 97:795–803
31. Cohn MA. 1996. Operational and philosophical decisions in seed dormancy research. *Seed Sci. Res.* 6:147–53
32. Cooke T. 1907. The embryology of *Rhizophora mangle. Bull. Torrey Bot. Club* 34:271–77
33. Corner EJH. 1976. *The Seeds of Dicotyledons.* Cambridge, UK: Cambridge Univ. Press
34. Cramer GR, Jones RL. 1996. Osmotic stress and abscisic acid reduce cytosolic calcium activities in roots of *Arabidopsis thaliana. Plant Cell Environ.* 19:1291–98
35. Crowe JH, Hoekstra FJ, Crowe LM. 1992. Anhydrobiosis. *Annu. Rev. Physiol.* 54:579–99
36. Davies PJ, ed. 1995. *Plant Hormones: Physiology, Biochemistry and Molecular Biology.* Dordrecht, The Netherlands: Kluwer Academic
37. Dodd MC, van Staden J. 1981. Germination and viability studies on the seeds of *Podocarpus henkelii* Stapf. *S. Afr. J. Sci.* 77:171–74
38. Ducker SC, Knox RB. 1976. Submarine

pollination in sea grasses. *Nature* 263:705–6
39. Dure LS III. 1993. The lea proteins of higher plants. In *Control of Plant Gene Expression*, ed. DPS Verma, pp. 325–35. Boca Raton, FL: CRC Press
40. Ellis RH, Hong TD, Roberts EH. 1985. *Handbook of Seed Technology for Genebanks*. Vol. II, *Compendium of Specific Germination Information and Test Recommendations*. Rome, Italy: Int. Board Plant Genet. Resour.
41. Ellison AM, Farnsworth EJ. 1993. Seedling survivorship, growth and response to disturbance in Belizean mangal. *Am. J. Bot.* 80:1137–45
42. Elmqvist T, Cox PA. 1996. The evolution of vivipary in flowering plants. *Oikos* 77:3–9
43. Elotmani M, Lovatt CJ, Coggins CW, Agusti M. 1995. Plant growth regulators in citriculture: factors regulating endogenous levels in citrus tissues. *Crit. Rev. Plant Sci.* 14:367–412
44. Espelund M, Debedout JA, Outlaw WH, Jakobsen KA. 1995. Environmental and hormonal regulation of barley late-embryogenesis-abundant (Lea) mRNAs is via different signal transduction pathways. *Plant Cell Environ.* 18:943–49
45. Evans AS, Cabin RJ. 1995. Can dormancy affect the evolution of post-germination traits? The case of *Lesquerella fendleri*. *Ecology* 76:344–56
46. Farnsworth EJ. 1997. Evolutionary and Ecological Physiology of Mangrove Seedlings: Correlates, Costs, and Consequences of Viviparous Reproduction. PhD thesis, Harvard Univ., Cambridge, MA. 308 pp.
47. Farnsworth EJ, Ellison AM. 1996. Sun-shade adaptability of the red mangrove, *Rhizophora mangle* (Rhizophoraceae): changes through ontogeny at several levels of biological organization. *Am. J. Bot.* 83:1131–43
48. Farnsworth EJ, Ellison AM. 1997. Global patterns of pre-dispersal seed predation in mangrove forests. *Biotropica* 29:318–30
49. Farnsworth EJ, Farrant JM. 1998. Reductions in abscisic acid are linked with viviparous reproduction in mangroves. *Am. J. Bot.* 85:760–69
50. Farrant JM, Pammenter NW, Berjak P. 1989. Germination-associated events and the desiccation sensitivity of recalcitrant seeds: a study on three unrelated species. *Planta* 178:189–98
51. Farrant JM, Pammenter NW, Berjak P. 1993. Seed development in relation to desiccation tolerance: a comparison between desiccation-sensitive (recalcitrant) seeds of Avicennia marina and desiccation-tolerant types. *Seed Sci. Res.* 3:1–13
52. Farrant JM, Pammenter NW, Berjak P, Farnsworth EJ, Vertucci CW. 1996. Presence of dehydrin-like proteins and levels of abscisic acid in recalcitrant (desiccation sensitive) seeds may be related to habitat. *Seed Sci. Res.* 6:175–82
53. Farrant JM, Pammenter NW, Cutting JGM, Berjak P. 1993. The role of plant growth regulators in the development and germination of the desiccation-sensitive (recalcitrant) seeds of *Avicennia marina*. *Seed Sci. Res.* 3:55–63
54. Fennimore SA, Foley ME. 1998. Genetic and phsyiological evidence for the role of gibberellic acid in the germination of dormant Avena fatua seeds. *J. Exp. Bot.* 49:89–94
55. Fernandez H, Doumas P, Bonnet-Masimbert M. 1997. Quantification of GA1, GA3, GA4, GA7, GA8, GA9, GA19, and GA20: and GA20 metabolism in dormant and non-dormant beechnuts. *Plant Growth Regul.* 22:29–35
56. Finch CE, Rose MR. 1995. Hormones and the physiological architecture of life history evolution. *Q. Rev. Biol.* 70:1–52
57. Finch-Savage WE. 1992. Seed development in the recalcitrant species *Quercus robur* L.: germinability and desiccation tolerance. *Seed Sci. Res.* 2:17–22
58. Finch-Savage WE, Clay HA. 1994.

Evidence that ethylene, light and abscisic acid interact to inhibit germination in the recalcitrant seeds of *Quercus robur* L. *J. Exp. Bot.* 45:1295–99
59. Finch-Savage WE, Pramanik SK, Bewley JD. 1994. The expression of dehydrin proteins in desiccation-sensitive (recalcitrant) seeds of temperate trees. *Planta* 193:478–85
60. Finkelstein RR. 1993. Abscisic acid-insensitive mutations provide evidence for stage-specific signal pathways regulating expression of an *Arabidopsis* late embryogenesis-abundant (*Lea*) gene. *Mol. Gen. Genet.* 238:401–8
61. Foster SA. 1986. On the adaptive value of large seeds for moist tropical forest trees: a review and synthesis. *Bot. Rev.* 52:260–99
62. French JC, Chung MG, Hur YK. 1995. Chloroplast DNA phylogeny of the Ariflorae. In *Monocotyledons: Systematics and Evolution*, ed. PJ Rudall, PB Cribb, DF Cutler, CJ Humphries, pp. 255–75. London: R. Bot. Gard. Kew
63. Galau GA, Jakobsen KS, Hughes DW. 1991. The controls of late dicot embryogenesis and early germination. *Physiol. Plant.* 81:280–88
64. Garcia AB, de Almeida Engler J, Claes B, Vuillarroel R, Montagu M, van Gerats T, Caplan A. 1998. The expression of the salt-response gene salt from rice is regulated by hormonal and developmental cues. *Planta* 207:172–80
65. Garello G, LePaige-Degivry MT. 1995. Desiccation-sensitive *Hopea odorata* seeds: sensitivity to abscisic acid, water potential, and inhibitors of gibberellin biosynthesis. *Physiol. Plant.* 95:45–50
66. Garrard A. 1955. The germination and longevity of seeds in an equatorial climate. *Gard. Bull. Singapore* 14:534–45
67. Garwood NC. 1983. Seed germination in a seasonal tropical forest in Panama: a community study. *Ecol. Monogr.* 53:159–81
68. Goebel KE. 1905. *Organography of Plants.* New York: Hafner
69. Grill E, Himmelbach A. 1998. ABA signal transduction. *Curr. Opin. Plant Biol.* 1:412–18
70. Gul B, Weber DJ. 1998. Effect of dormancy-relieving compounds on the seed germination of non-dormant *Allenrolfea occidentalis* under salinity stress. *Ann. Bot.* 82:555–60
71. Guppy HB. 1906. *Observations of a Naturalist in the Pacific Between 1896 and 1899.* Vol. 2, *Plant Dispersal.* London: Macmillan
72. Haberlandt G. 1895. Über die Ernährung der Keimlinge end die Bedeutung des Endosperms bei viviparen Mangrovepflanzen. *Ann. Jard. Bot. Buitenzorg* 12:102–5
73. Hairston NG Jr, Van Brunt RA, Kearns CM, Engstrom DR. 1995. Age and survivorship of diapausing eggs in a sediment egg bank. *Ecology* 76:1706–11
74. Hanelt P. 1977. Okologische und systematische Aspekte der Lebensdauer von Semen. *Biol. Rundsch.* 15:81–91
75. Harper JL, Lovell PH, Moore KG. 1970. The shapes and sizes of seeds. *Annu. Rev. Ecol. Syst.* 1:327–56
76. Harrington JF. 1972. Seed storage and longevity. In *Seed Biology*, Vol. III, ed. TT Kozlowski, pp. 145–250. New York: Academic
77. Hendry GAF. 1993. Oxygen free radical processes and seed longevity. *Seed Sci. Res.* 3:141–53
78. Hilhorst HWM. 1995. A critical update on seed dormancy. *Seed Sci. Res.* 5:61–73
79. Hoekstra FA, Haigh AM, Tetteroo FAA, van Roekel T. 1994. Changes in soluble sugars in relation to desiccation tolerance in cauliflower seeds. *Seed Sci. Res.* 4:143–47
80. Holdsworth M, Kurup S, McKibbin R. 1999. Molecular and genetic mechanisms regulating the transition from embryo development to germination. *Trends Plant Sci.* 4:275–80
81. Hole DJ, Smith JD, Cobb BG. 1989.

Regulation of embryo dormancy by manipulation of abscisic acid in kernels and associated cob tissue of *Zea mays* L. cultured in vitro. *Plant Physiol.* 91:101–5

82. Hong TD, Ellis RH. 1990. A comparison of maturation drying, germination, and desiccation tolerance between developing seeds of *Acer psuedoplatanus* L. and *Acer platanoides* L. *New Phytol.* 116:589–96
83. Huttly AK, Phillips AL. 1995. Gibberellin-regulated plant genes. *Physiol. Plant.* 95:310–17
84. Ingram J, Bartels D. 1996. The molecular basis of dehydration tolerance in plants. *Annu. Rev. Plant Physiol. Plant Mol. Biol.* 47:377–403
85. Janzen DH. 1978. Seeding patterns of tropical forest trees. In *Tropical Trees as Living Systems*, ed. PB Tomlinson, MH Zimmerman, pp. 83–128. Cambridge, UK: Cambridge Univ. Press
86. Janzen DH. 1983. Costa Rican Natural History. Chicago, IL: Univ. Chicago Press
87. Jiménez JA. 1994. *Los manglares del Pacífico de Centroamérica.* Heredia, Costa Rica: EFUNA
88. Johnson-Flanagan AM, Spencer MS. 1994. Ethylene production during development of mustard (*Brassica juncea*) and canola (*Brassica napus*) seed. *Plant Physiol.* 106:601–6
89. Joshi AC. 1933. A suggested explanation for the prevalence of vivipary on the seashore. *J. Ecol.* 21:209–12
90. Joshi GV, Jamale BB, Bhosale L. 1975. Ion regulation in mangroves. *Proc. Int. Symp. Biol. Manage. Mangroves,* Vol. 2, ed. GE Walsh, SC Snedaker, HJ Teas, pp. 595–607. Gainesville: Univ. Florida Press
91. Juncosa AM. 1982. *Embryo and Seedling Development in the Rhizophoraceae.* PhD thesis, Duke Univ., Durham, NC
92. Juncosa AM, Tomlinson PB. 1988. Systematic comparison and some biological characteristics of Rhizophoraceae and Anisophyllaceae. *Ann. Mo. Bot. Gard.* 75:1296–319
93. Kawakami K, Miyake Y, Noda K. 1997. ABA insensitivity and low ABA levels during seed development of non-dormant wheat mutants. *J. Exp. Bot.* 48:1415–21
94. Kepczynski J, Kepczynska E. 1997. Ethylene in seed dormancy and germination. *Physiol. Plant.* 101:720–26
95. Kermode AR. 1990. Regulatory mechanisms involved in the transition from seed development to germination. *Crit. Rev. Plant Sci.* 9:155–95
96. Khan AA. 1994. Induction of dormancy in non-dormant seeds. *J. Am. Hort. Soc.* 119:408–13
97. Kigel J, Galili G. 1995. *Seed Development and Germination.* New York: Marcel Dekker
98. King J. 1991. *The Genetic Basis of Plant Physiological Processes.* Oxford, UK: Oxford Univ. Press
99. Kipp-Goller A. 1940. Über Bau und Entwicklung der viviparen Mangrovekeimlinge. *Z. Bot.* 35:1–40
100. Knight CD, Sehgal A, Atwal K, Wallace JC, Cove DJ, et al. 1995. Molecular responses to abscisic acid and stress are conserved between moss and cereals. *Plant Cell* 7:499–506
101. Koornneef M, Rueling G, Karssen CM. 1984. The isolation and characterization of abscisic acid-insensitive mutants of *Arabidopsis thaliana. Physiol. Plant.* 61:377–83
102. Koster KL, Leopold AC. 1988. Sugars and desiccation tolerance in seeds. *Plant Physiol.* 88:829–32
103. Lalonde J, Saini HS. 1992. Comparative requirement for endogenous ethylene during seed germination. *Ann. Bot.* 69:423–28
104. Leck MA. Parker VT, Simpson RL, eds. 1989. *Ecology of Soil Seed Banks.* San Diego, CA: Academic
105. Leite AMC, Rankin JM. 1981. Ecologia de sementes de *Pithecelobium racemosum* Ducke. *Acta Amaz.* 11:309–18
106. Léon-Kloosterziel KM, Gil MA. Ruijs J,

Jacobsen SE, Olszewski NE, et al. 1996. Isolation and characterization of abscisic acid-deficient *Arabidopsis* mutants at two new loci. *Plant J.* 10:655–61
107. Léon-Kloosterziel KM, van de Bunt GA, Zeevaart JAD, Koornneef M. 1996. *Arabidopsis* mutants with a reduced seed dormancy. *Plant Physiol.* 110:233–40
108. LePrince O, Bronchart R, Deltour R. 1990. Changes in starch and soluble sugars in relation to the acquisition of desiccation tolerance during maturation of *Brassica campestris* seed. *Plant Cell Environ.* 13:539–46
109. Leung J, Giraudat J. 1998. Abscisic acid signal transduction. *Annu. Rev. Plant Physiol. Plant Mol. Biol.* 49:199–222
110. Levin SA, Cohen D, Hastings A. 1984. Dispersal strategies in patchy environments. *Theor. Popul. Biol.* 26:165–91
111. Lin G, da SL Sternberg L. 1995. Variation in propagule mass and its effect on carbon assimilation and seedling growth of red mangrove (*Rhizophora mangle*) in Florida, USA. *J. Trop. Ecol.* 11:109–19
112. Lin TP, Chen MH. 1995. Biochemical characteristics associated with the development of the desiccation-sensitive seeds of *Machilus thunbergii* Sieb and Zucc. *Ann. Bot.* 76:381–87
113. MacIntyre GI. 1987. The role of water in the regulation of plant development. *Can. J. Bot.* 65:1287–98
114. MacNae W. 1968. A general account of the flora and fauna of mangrove swamps and forests in the Indo-West Pacific region. *Adv. Mar. Biol.* 6:73–70
115. Manokoran N. 1978. Germination of fresh seeds of Malaysian rattans. *Malay. For.* 41:319–24
116. Mao ZY, Paiva R, Kriz AL, Juvik JA. 1995. Dehydrin gene expression in normal and viviparous embryos of *Zea mays* during seed development and germination. *Plant Physiol. Biochem.* 33:649–53
117. Marcotte WR Jr, Bayley CC, Quatrano RS. 1988. Regulation of a wheat promoter by abscisic acid in rice protoplasts. *Nature* 335:454–57
118. Marrero J. 1942. A seed storage study of Maga. *Caribb. For.* 3:173–84
119. McCarty DR. 1995. Genetic control and integration of maturation and germination pathways in seed development. *Annu. Rev. Plant Physiol. Plant Mol. Biol.* 46:71–93
120. McCormick JF. 1995. A review of the population dynamics of selected tree species in the Luquillo experimental forest, Puerto Rico. In *Tropical Forests: Management and Ecology*, ed. AE Lugo, C Lowe, pp. 224–57. New York: Springer-Verlag
121. Meurs C, Basra AS, Karssen CM, van Loon LC. 1992. Role of abscisic acid in the induction of desiccation tolerance in developing seeds of *Arabidopsis thaliana*. *Plant Physiol.* 98:1484–93
122. Mori T, Rahman ZBHA, Tan CH. 1980. Germination and storage of rattan (*Calamus manan*) seeds. *Malay. For.* 43:14–55
123. Mousseau TA, Fox CW. 1998. The adaptive significance of maternal effects. *TREE* 13:403–7
124. Muenscher WC. 1936. Storage and germination of seeds of aquatic plants. *Cornell Univ. Agric. Exp. Stn. Bull.* 642, Cornell Univ., Ithaca, NY
125. Musatenko LI, Berestetsky VA, Vedenicheva NP, Generalova VN, Martyn GI, Sytnik KM. 1995. Phytohormones and structure of cell of *Acer saccharinum* seed embryo. *Biol. Plant.* 37:553–59
126. Ng FSP. 1992. *Manual of Forest Fruits, Seeds and Seedlings*. Kuala Lumpur: For. Res. Inst. Malaysia
127. Nkang A, Chandler G. 1986. Changes during embryogenesis in rainforest seeds with orthodox and recalcitrant viability characteristics. *J. Plant Physiol.* 126:243–56
128. Oliveira LMQ, Valio IFM. 1992. Effects of moisture content on germination

of seeds of *Hancornia speciosa* Gom. (Apocynaceae). *Ann. Bot.* 69:1–5

129. Oliver MJ, Bewley JD 1997. Desiccation-tolerance of plant tissues: a mechanistic overview. *Hort. Rev.* 18:171–213
130. Olmstead RG, Bremer B., Scott KM, Pamer JD. 1993. A parsimony analysis of the Asteridae *sensu lato* based on *rbcL* sequences. *Ann. Mo. Bot. Gard.* 80:700–22
131. Osborne DJ, Boubriak II. 1994. DNA and desiccation tolerance. *Seed Sci. Res.* 4:175–85
132. Pammenter NW, Berjak P, Farrant JM, Smith MT, Ross G. 1994. Why do stored hydrated recalcitrant seeds die? *Seed Sci. Res.* 4:187–91
133. Pannier RF, Pannier F. 1973. Determinación de substancias de tipo gibberellina en tejidas de *Rhizophora mangle* en diferentes etapas de desorrollo. *Acta Cient. Venez.* 24(Suppl. 1):33–34
134. Paz H, Mazer SJ, Martínez-Ramos M. 1999. Seed mass, seedling emergence, and environmental factors in seven rain forest *Psychotria* (Rubiaceae). *Ecology* 80:1594–606
135. Pence VC. 1991. Abscisic acid in developing zygotic embryos of *Theobroma cacao*. *Plant Physiol.* 95:1291–93
136. Peroni PA. 1995. Field and laboratory investigations of seed dormancy in red maple (*Acer rubrum* L.) from the North Carolina piedmont. *For. Sci.* 41:378–86
137. Philbrick CT, Les DH. 1996. Evolution of aquatic angiosperm reproductive systems. *BioScience* 46:813–26
138. Popp M. 1995. Salt resistance in herbaceous halophytes and mangroves. *Prog. Bot.* 56:416–42
139. Pritchard HW, Tompsett PB, Manger KR. 1996. Development of a thermal time model for the quantification of dormancy loss in *Aesculus hippocastaneum* seeds. *Seed Sci. Res.* 6:127–35
140. Probert RJ, Longley PL. 1989. Recalcitrant seed storage physiology in three aquatic grasses (*Zizania palustris, Spartina anglicaa* and *Porteresia coarctica*). *Ann. Bot.* 63:53–63
141. Pronzato R, Manconi R. 1994. Life history of *Ephydatia fluviatilis*: a model for adaptive strategies in discontinuous habitats. In *Sponges in Time and Space, Proc. Int. Porifera Congr., 4th*, ed. RWM van Soest, TMG van Kempen, JC Braekman, pp. 327–34. Rotterdam, The Netherlands: Balkema
142. Puchet CE, Vazquez-Yanes C. 1987. Heteromorfismo criptico en las semillas recalcitrantes de tres especies de la selva tropical humeda de Veracruz, Mexico. *Phytologia* 62:100–6
143. Ravishankar KV, Shaanker RU, Ganeshaiah KN. 1995. War of hormones over resource allocation to seeds: strategies and counter-strategies of offspring and maternal parent. *J. Biosci.* 20:89–103
144. Rees M. 1994. Delayed germination of seeds: a look at the effects of adult longevity, the timing of reproduction, and population age/stage structure. *Am. Nat.* 144:43–64
145. Ringlund K, Mosleth E, Mares DJ, eds. 1990. *Fifth International Symposium on Pre-harvest Sprouting in Cereals*. Boulder, CO: Westview
146. Roberts EH. 1973. Predicting the storage life of seeds. *Seed Sci. Technol.* 1:499–514
147. Robertson, AI, Giddens R, Smith TJ III. 1990. Seed predation by insects in tropical mangrove forests: extent and effects on seed viability and the growth of seedlings. *Oecologia* 83:213–19
148. Rock CD, Quatrano RS. 1995. The role of hormones during seed development. In *Plant Hormones: Physiology, Biochemistry and Molecular Biology*, ed. PJ Davies, pp. 671–697. Dordrecht, The Netherlands: Kluwer Academic
149. Romagosa I, Han F, Clancy JA, Ulrich SE. 1999. Individual locus effects on dormancy during seed development and after ripening in barley. *Crop Sci.* 39:74–79

150. Schachtman D, Liu W. 1999. Molecular pieces to the puzzle of the interaction between potassium and sodium uptake in plants. *Trends Plant Sci.* 4:281–87
151. Schupp EW. 1995. Seed-seedling conflicts, habitat choice, and patterns of plant recruitment. *Am. J. Bot.* 82:399–409
152. Setoguchi H, Kosuge K, Tobe H. 1999. Molecular phylogeny of Rhizophoraceae based on *rbcL* gene sequences. *J. Plant Res.* 112:443–55
153. Sheen J. 1996. Ca^{2+}-dependent protein kinases and stress signal transduction in plants. *Science* 274:1900–2
154. Silvertown JW. 1981. Seed size, life span, and germination date as co-adapted features of plant life history. *Am. Nat.* 118:860–64
155. Skriver K, Mundy J. 1990. Gene expression in response to abscisic acid and osmotic stress. *Plant Cell* 2:503–12
156. Smirnoff N. 1993. The role of active oxygen in the response of plants to water deficit and desiccation. Tansley Review 52. *New Phytol.* 125:27–58
157. Smith SM, Snedaker SC. 1995. Salinity responses in two populations of viviparous *Rhizophora mangle* L. seedlings. *Biotropica* 27:435–40
158. Smith-Ramirez C, Armesto JJ, Figueroa J. 1998. Flowering, fruiting and seed germination in Chilean rain forest Myrtaceae: ecological and phylogenetic constraints. *Plant Ecol.* 136:119–31
159. Smythe M. 1970. Relationships between fruiting seasons and seed dispersal methods in a neotropical forest. *Am. Nat.* 104:25–35
160. Stearns SC, Schmid-Hempel P. 1987. Evolutionary insights should not be wasted. *Oikos* 49:118–25
161. Steinbach HS, Benecharnold RL, Kristof G, Sanchez RA, Marcuccipoltri S. 1995. Physiological basis of pre-harvest sprouting resistance in *Sorghum bicolor* (L.) Moench. ABA levels and sensitivity in developing embryos of sprouting-resistant and sprouting-susceptible varieties. *J. Exp. Bot.* 46:701–9
162. Stöcklin J, Fischer M. 1999. Plants with longer-lived seeds have lower local extinction rates in grassland remnants 1950–1985. *Oecologia* 120:539–43
163. Sussex I. 1975. Growth and metabolism of the embryo and attached seedling of the viviparous mangrove, *Rhizophora mangle*. *Am. J. Bot.* 62:948–53
164. Swamy BGL, Padmanabhan D. 1961. Notulae embryologieae. I. The functions of endosperm in *Avicennia officinalis*. *Curr. Sci.* 30:424–25
165. Takahashi N, Phinney BO, MacMillan J, eds. 1991. *Gibberellins*. Berlin: Springer-Verlag
166. Takhtajan A. 1991. *Evolutionary Trends in Flowering Plants*. New York: Columbia Univ. Press
167. Tan H, Rao AN. 1981. Vivipary in *Ophiorrhiza tomentosa* Jack (Rubiaceae). *Biotropica* 13:232–33
168. Tetteroo FAA, Bomal C, Hoekstra FA, Karssen CM. 1994. Effect of abscisic acid and slow drying on soluble carbohydrate content in developing embryoids of carrot (*Daucus carota* L.) and alfalfa (*Medicago sativa* L.). *Seed Sci. Res.* 4:203–10
169. Thomas TH, Hare PD, Van Staden J. 1997. Phytochrome and cytokinin responses. *Plant Growth Regul.* 23:105–22
170. Thompson K 1987. Seeds and seed banks. *New Phytol.* 106:23–34
171. Thompson K, Bakker JP, Bekker RM, Hodgson JG. 1998. Ecological correlates of seed persistence in soil in the north-west European flora. *J. Ecol.* 86:163–69
172. Thornton TM, Swain SM, Olszewski NE. 1999. Gibberellin signal transduction presents... the SPY who O-GlcNAc'd me. *Trends Plant Sci.* 4:424–28
173. Tomlinson PB. 1986. *The Botany of Mangroves*. Cambridge, UK: Cambridge Univ. Press
174. Tompsett PB. 1982. The effect of desiccation on the longevity of seeds of

Araucaria hunsteinii and *A. cunninghamii*. *Ann. Bot.* 50:693–704
175. Tompsett PB, Pritchard HW. 1998. The effect of chilling and moisture status on the germination, desiccation tolerance and longevity of *Aesculus hippocastanum* L. seed. *Ann. Bot.* 82:249–61
176. Trethowan RM, Rajaram S, Ellison FW. 1996. Pre-harvest sprouting tolerance of wheat in the field and under rain simulation. *Aust. J. Agric. Res.* 47:705–16
177. Trewavas A J. 1987. Sensitivity and sensory adaptation in growth substance responses. In *Hormone Action in Plant Development*, ed. GV Hoad, MB Jackson, JR Lenton, RK Atkin, pp. 19–38. London: Butterworth
178. Treub M. 1883. Notes dur l'embryon: le sac embryonnaire et l'ovule *Avicennia officinalis*. *Ann. Jard. Bot. Buitenzorg* 3:79–85
179. Uhl NW, Dransfield J, Davis JI, Luckow MA, Hansen KS, Doyle JJ. 1995. Phylogenetic relationships among palms: cladistic analyses of morphological and chloroplast DNA restriction site variation. In *Monocotyledons: Systematics and Evolution*, ed. PJ Rudall, PB Cribb, DF Cutler, CJ Humphries, pp. 623–61. London: R. Bot. Gard. Kew
180. Ungar IA. 1991. *Ecophysiology of Halophytes*. Boca Raton, FL: CRC Press. 209 pp.
181. van der Pijl L. 1983. *Principles of Dispersal in Higher Plants*. Berlin: Springer-Verlag
182. VanDuijn B, Flikweert MT, Heidekamp F, Wang M. 1996. Different properties of the inward rectifying potassium conductance of aleurone protoplasts from dormant and non-dormant barley grains. *Plant Growth Regul.* 18:107–13
183. Vartanian N. 1996. Mutants as tools to understand cellular and molecular drought tolerance mechanisms. *Plant Growth Regul.* 20:125–34
184. Vazquez-Yanes C, Arechiga MR. 1996. Ex situ conservation of tropical rain forest seed: problems and perspectives. *Interciencia* 21:293
185. Vazquez-Yanes C, Orozco Segovia A. 1984. Ecophysiology of seed germination in the tropical humid forests of the world: a review. In *Physiological Ecology of Plants in the Wet Tropics*, ed. E Medina, HA Mooney, C Vazquez-Yanes, pp. 37–50. The Hague: Junk
186. Venable DL, Brown JS. 1988. The selective interactions of dispersal, dormancy and seed size as adaptations for reducing risk in variable environments. *Am. Nat.* 131:360–84
187. Vertucci CW, Farrant JM. 1995. Acquisition and loss of desiccation tolerance. In *Seed Development and Germination*, ed. J Kigel, G Galili, pp. 237–71. New York: Marcel Dekker
188. Vleeshouwers LM, Bouwmeester HJ, Karssen CM. 1995. Redefining seed dormancy: an attempt to integrate physiology and ecology. *J. Ecol.* 83:1031–37
189. Voesenek LACJ, Blom CWPM. 1996. Plants and hormones: an ecophysiological view on timing and plasticity. *J. Ecol.* 84:111–19
190. von Teichman I, van Wyk AE. 1991. Trends in the evolution of dicotyledonous seeds based on character associations, with special reference to pachychalazy and recalcitrance. *Bot. J. Linn. Soc.* 105:211–37
191. von Teichman I, van Wyk AE. 1994. Structural aspects and trends in the evolution of recalcitrant seeds in dicotyledons. *Seed Sci. Res.* 4:225–39
192. Walker-Simmons M. 1988. ABA levels and sensitivity in developing wheat embryos of sprouting resistant and susceptible cultivars. *Plant Physiol.* 84:61–66
193. Walter H, Steiner M. 1936. Die Oekologie der Ostafrikanischen Mangroven. *Z. Bot.* 30:65–193
194. Ward JM, Pei Z-M, Schroeder JI. 1995. Roles of ion channels in initiation of

signal transduction in higher plants. *Plant Cell* 7:833–44

195. Webb AAR, McAinsh MR, Taylor JE, Hetherington AM. 1996. Calcium ions as intracellular second messengers in higher plants. *Adv. Bot. Res.* 22:45–96
196. Weber H, Borisjuk L, Wobus U. 1997. Sugar import and metabolism during seed development. *Trends Plant Sci.* 2:169–74
197. Weyers JDB, Peterson NW, Abrook R, Peng ZY. 1995. Quantitative analysis of the control of physiological phenomena by plant hormones. *Physiol. Plant.* 95:486–94
198. Willson MF. 1983. *Plant Reproductive Ecology.* New York: Wiley & Sons
199. Wise RR, Juncosa AM. 1989. Ultrastructure of the transfer tissues during viviparous seedling development in *Rhizophora mangle* (Rhizophoraceae). *Am. J. Bot.* 76:1286–98
200. Xu NF, Bewley JD. 1995. The role of abscisic acid in germination, storage protein synthesis and desiccation-tolerance in alfalfa (*Medicago sativa* L.) seeds, as shown by inhibition of its synthesis by fluridone during development. *J. Exp. Bot.* 46:687–94
201. Young JA, Young CG. 1992. *Seeds of Woody Plants of North America.* Portland, OR: Dioscorides
202. Zhang J, Davies WJ. 1989. Abscisic acid produced in dehydrating roots may enable the plant to measure the water status of the soil. *Plant Cell Environ.* 12:73–81
203. Zheng WJ, Wang WQ, Lin P. 1999. Dynamics of element contents during the development of hypocotyles and leaves of certain mangrove species. *J. Exp. Mar. Biol. Ecol.* 233:247–57

Annu. Rev. Ecol. Syst. 2000. 31:139–62

INBREEDING DEPRESSION IN CONSERVATION BIOLOGY

Philip W. Hedrick
Department of Biology, Arizona State University, Tempe, Arizona 85287; email: philip.hedrick@asu.edu

Steven T. Kalinowski
Conservation Biology Division, National Marine Fisheries Service, 2725 Montlake Blvd. East, Seattle, Washington 98112; email: steven.kalinowski@noaa.gov

Key Words endangered species, extinction, fitness, genetic restoration, purging

■ **Abstract** Inbreeding depression is of major concern in the management and conservation of endangered species. Inbreeding appears universally to reduce fitness, but its magnitude and specific effects are highly variable because they depend on the genetic constitution of the species or populations and on how these genotypes interact with the environment. Recent natural experiments are consistent with greater inbreeding depression in more stressful environments. In small populations of randomly mating individuals, such as are characteristic of many endangered species, all individuals may suffer from inbreeding depression because of the cumulative effects of genetic drift that decrease the fitness of all individuals in the population. In three recent cases, introductions into populations with low fitness appeared to restore fitness to levels similar to those before the effects of genetic drift. Inbreeding depression may potentially be reduced, or purged, by breeding related individuals. However, the Speke's gazelle example, often cited as a demonstration of reduction of inbreeding depression, appears to be the result of a temporal change in fitness in inbred individuals and not a reduction in inbreeding depression.

Down, July 17, 1870

My Dear Lubbock,

...In England and many parts of Europe the marriages of cousins are objected to from their supposed injurious consequences: but this belief rests on no direct evidence. It is therefore manifestly desirable that the belief should be either proved false, or should be confirmed, so that in this latter case the marriages of cousins might be discouraged...

It is moreover, much to be wished that the truth of the often repeated assertion that consanguineous marriages lead to deafness and dumbness, blindness, &c, should be ascertained: and all such assertions could be easily tested by the returns from a single census.

Believe me,
Yours very sincerely,
Charles Darwin

INTRODUCTION

The detrimental effects of close inbreeding on traits related to fitness have long been documented in humans and organisms commonly bred by humans (20, 21). However, the impact of inbreeding upon endangered species was not considered until 1979 (82), when it was documented that inbreeding lowered the juvenile survival in 41 of 44 populations of captive ungulates. In a follow-up survey, 36 of 40 captive populations exhibited decreased juvenile viability in inbred animals, although there was extensive variation among the estimates, and some populations had statistically nonsignificant inbreeding depression (81). Supporting the ubiquity of inbreeding depression, Lacy (51, p. 331) stated that he was "unable to find statistically defensible evidence showing that any mammalian species is unaffected by inbreeding. Moreover, endangered species seem no less impacted by inbreeding, on average, than are common taxa." Even the cheetah, argued to be already so inbred and low in fitness that further inbreeding would have no effect (75), has been shown to have inbreeding depression (32, 102). General recognition of the potential negative effect of inbreeding on fitness has made inbreeding depression a concern in small-population conservation and inbreeding avoidance a priority in captive breeding of endangered species. Therefore, we concentrate here on reviewing literature concerned with the relationship of inbreeding depression to conservation, primarily in endangered animal species.

Thorough reviews have covered various aspects of inbreeding depression such as the evolution of inbreeding (14, 97) and the purging of inbreeding depression in plant populations (12). Further, the effect of inbreeding on different categories of traits has also been considered as, for example, the magnitude and timing of inbreeding depression in plants (40), inbreeding depression of different fitness components or genetic abnormalities (56), and comparison of inbreeding depression for life-history and morphological traits (22). In addition, some recent reviews have discussed the amount of genetic variation for fitness components and the factors affecting the mutation-selection balance for these traits (15, 64). The genetic basis of inbreeding depression has been considered and debated for many years (14a). Recent experimental work is having some success in characterizing the genetic variation underlying inbreeding depression (23, 42, 83, 103, 104). The ability to map genes (QTLs) (68) affecting fitness-related traits portends imminent knowledge of the detailed architecture of genes affecting inbreeding depression, i.e. the number and location of the genes, the distribution of their effects and their dominance, and the interaction (epistasis) of different genes. There has also been discussion of other theoretical concepts relevant to our review (13, 30, 39, 72, 90, 100). We suggest that these papers, or references cited in them, be consulted for these and related topics.

Usually, inbreeding refers to the mating of closely related individuals, and inbreeding depression is defined as reduced fitness of the offspring of these matings compared to the offspring of randomly mated individuals. In addition, though, genetic drift in small populations causing fixation of detrimental alleles can result in all, or nearly all, individuals in some populations having a lower fitness than other populations. Because populations of endangered species may be small in size or may have gone through bottlenecks or founder events, this effect may be particularly important in conservation. Within a population of an endangered species, inbred offspring may not have a lower fitness than non-inbred ones (we use "non-inbred" synonymously with randomly mated or outbred individuals), but instead all individuals have a lower fitness than the ancestral individuals before the effect of genetic drift. This type of inbreeding depression generally may be documented only by crossing to individuals from another population and observing the fitness of their progeny. Below we discuss a theoretical context for this phenomenon and describe three recent examples in which populations recovered fitness as the result of the introduction of outside individuals.

Model organisms can be useful in understanding underlying phenomena in biology (however, see 31), and this approach may determine unifying patterns in conservation genetics so that each species does not have to be considered as a unique case study (26, 28). This goal has recently been advanced by a number of laboratory experiments related to inbreeding depression using insects, including *Drosophila* (5, 6, 25, 28, 60, 61, 71), houseflies (11), flour beetles (78, 79), crickets (86), and butterflies (88). Many of the insights into inbreeding depression have come from detailed studies in *Drosophila* (14, 16, 18, 93).

It appears that in *Drosophila* approximately half the genetic load is from nearly recessive lethals and half from detrimentals of small effect but with higher dominance (14, 100). However, it is important to ask whether such findings are completely generalizable to endangered species. Most genetic phenomena, such as segregation, independent assortment, linkage, mutation, etc., are virtually identical across species, but *Drosophila* and endangered species may differ in some evolutionary and ecological factors that are important in inbreeding depression. For example, *Drosophila* have a very large effective population size, whereas in many endangered species, genetic drift is quite important either because of a currently small population size or because of past severe bottlenecks or founder events.

As a result, a smaller population (with greater genetic drift) may translate into a different genetic architecture of detrimental variation than that for a larger population. In a small, finite population, relative to a large, infinite population, lethals (and other variants of large detrimental effect) have a lower expected frequency because selection will push them to a low frequency and then they will be lost from the population by genetic drift (17, 74). Similarly, the standing genetic variation in *Drosophila* may be larger than in endangered species and thereby prompt a pessimistic outlook on the amount of inbreeding depression and an optimistic perspective on adaptive potential. In addition, in a small, finite population, genetic drift becomes a stronger influence on allelic frequency than selection if $s < 1/2N$, where s is the selective disadvantage of homozygotes and N is the effective population size. In small populations, detrimental mutations with a selective disadvantage less than $1/2N$ become fixed much as if they were neutral (50, 58, 63, 66). As a result, fitness may decline over time, and the population may decrease in size so that detrimental mutants of larger effect become effectively neutral and subsequently are more likely to be incorporated. This feedback process has been named mutation meltdown and, in theory, may result in the extinction of small populations (65). It is not clear how significant mutation meltdown may actually be because extinction probability due to other factors may be high in such small populations (59); there are, however, examples of apparent fixation of deleterious variants in some endangered species with small population sizes.

IMPACT OF THE ENVIRONMENT ON LOWERED FITNESS IN INBRED INDIVIDUALS

The effects of inbreeding on endangered species have generally been examined in captive populations for which the environment may be less harsh than natural environments. For example, juvenile survival is generally much higher in captivity than it is in nature. Estimates of inbreeding depression from captivity or laboratory environments are thought to underestimate or at least to be different than the effects in a natural environment. Several studies have attempted to evaluate the effect of inbreeding on fitness in nature to determine the extent of this difference. It is difficult to have an appropriate experimental design in these instances, e.g., to replicate populations and have simultaneous controls (however, see 70a). Nevertheless, these experiments illustrate that inbreeding depression may have a significant effect on fitness in natural populations (see also 15a).

Jimenez et al (41) examined the survival of adult non-inbred and inbred white-footed mice (*Peromyscus leucopus noveboracensis*). Stock for the experiment was captured from the natural study site near Chicago, Illinois, brought into the laboratory, and bred to produce individuals with inbreeding coefficients of 0.00 or 0.25 (from full-sib matings). Almost 800 mice, nearly equally split between non-inbred and inbred, were released during three different periods. The area had a low number of mice during the release, suggesting that the environment was harsh because of some unknown cause. For the 10 weeks following release, non-inbred individuals had a higher weekly survivorship at all census times than the inbred individuals (Figure 1). Using capture-recapture data, the weekly survival of the

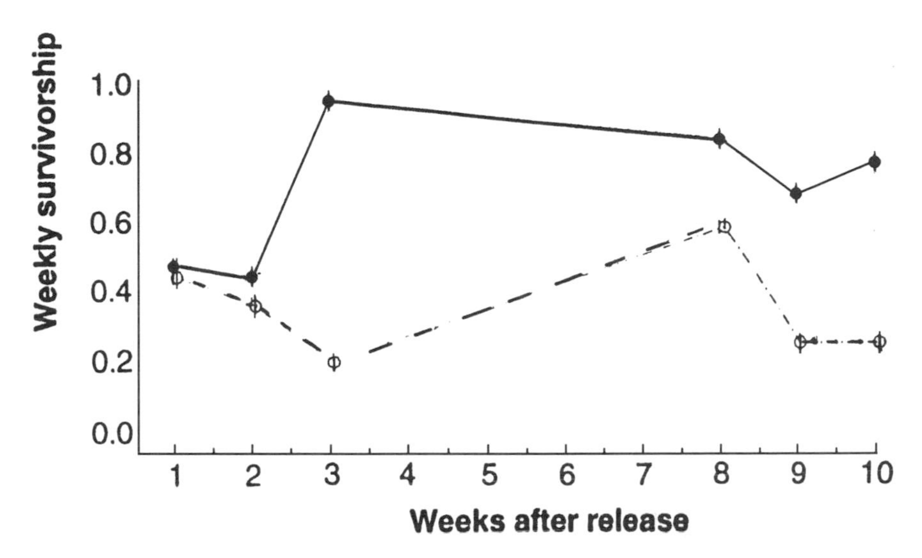

Figure 1 Weekly survivorship of non-inbred (solid line) and inbred (broken line) white footed mice over 10 weeks in a natural habitat (from 41).

inbred mice was estimated to be 56% that of the non-inbred mice. In addition, inbred male mice lost significant body mass throughout the experiment, while non-inbred male mice did not.

The population crash of the wild population of song sparrows (*Melospiza melodia*) on Mandarte Island, British Columbia, involves a documented case of inbreeding depression in the wild (49). This population appears to undergo periodic crashes, probably due to severe winter weather: The decline in 1989 killed 89% of the adult animals. Because there had been an extensive program to mark individuals in this population over several generations, the inbreeding coefficient for most individuals was known before the crash. After the crash, the inbreeding coefficient for the 10 survivors was 0.0065 (only three had known inbreeding), whereas the inbreeding coefficient for the 206 birds that died was significantly higher at 0.0312. All the birds with inbreeding coefficients of 0.0625 or higher (13% of the population) died during the crash. Later, high inbreeding depression was shown in the population (48), consistent with the differential survival between inbred and non-inbred individuals.

Saccheri et al (89) reported that the rate of extinction was negatively correlated with heterozygosity in 42 small populations of fritillary butterflies (*Melitaea cinzia*). Over the summer of 1996, seven of these populations went extinct. The heterozygosity, as determined by seven allozyme loci and one microsatellite locus, was significantly lower in the populations that went extinct than in the surviving populations. To avoid the confounding effect of population size on both heterozygosity and extinction, in their statistical analysis, Saccheri et al controlled for

TABLE 1 (*a*) A general way to predict the fitness of inbred individuals in a natural environment based on the relative fitness of inbreds in the captive environment and the relative fitness of non-inbreds in a natural environment and (*b*) an example using the observed relative survival in the white-footed field mouse (41). Here the observed survival of inbred mice in the natural environment is much less than that expected.

		Population	
	Environment	**Non-inbred**	**Inbred**
(a) General	Captive	1	$\overline{w}_I$
	Natural	$\overline{w}_N$	$\overline{w}_I\overline{w}_N$
(b) Mouse data	Captive	1	$\overline{w}_I = 0.935$
	Natural	$\overline{w}_N = 0.221$	$\overline{w}_I\overline{w}_N = 0.207$
			Observed = 0.046

the effect of ecological variables on population extinction. Laboratory studies indicate substantial inbreeding depression in this species from one generation of full-sib mating (74a, 89). This amount of inbreeding is probably similar to that in some isolated populations that may have been founded by a single female, a scenario confirmed by extinction risk differences in experimental inbred and outbred populations (74a).

To quantify the effects of inbreeding in nature, let us assume that the fitness, or a fitness component, in the captive or laboratory environment for non-inbred individuals is standardized to be unity. The fitness relative to this in the natural environment is $\overline{w}_N$(the effect of the environment), and the fitness relative to that in the inbred population is $\overline{w}_I$ (the effect of inbreeding). If we assume that the effects of inbreeding and the environment are multiplicative, then the expected fitness of the inbred group in the natural environment is $\overline{w}_I\overline{w}_N$ (part *a* of Table 1). In part *b* of Table 1, the data from the white-footed mouse example are given (41); these show that the survival of non-inbreds in the natural environment relative to the survival of non-inbreds in captivity is 0.221, and the survival of inbreds in captivity relative to non-inbreds in captivity is 0.935. Therefore, the predicted survival of inbreds in nature is 0.207, but the actual observed survival is only 0.046, about 22% of that predicted, suggesting that inbreds in nature survive much more poorly than expected.

To demonstrate that inbred individuals are more affected by a natural environment than expected, it is not enough just to show that inbred individuals have lower fitness in a natural environment than in a captive one. For example, the approach given in Table 1 provides a method to compare the predicted joint effects of a natural or more stressful environment and inbreeding on fitness and the reduction actually observed. This can also be measured by comparing estimates of the number of lethal equivalents in the captive and the natural environments (see below).

Another approach is to test for a significant environment-inbreeding interaction component in analysis of variance.

The Gila topminnow provides an unusual example of an endangered species that can be bred and evaluated in captive situations with replicates, simultaneous controls, and the other attributes that make model organisms useful. The Gila topminnow is a widely cited example of the importance of genetic variation to conservation because a sample from Sharp Spring, Arizona, was found to have both more allozyme variation and a higher value of traits potentially related to fitness than a sample from Monkey Spring, Arizona (80, 99). Recent studies with highly variable loci demonstrate that fish from Monkey Spring have substantial genetic variation (38, 76), but more relevant to our discussion is that measures of traits related to fitness in fish from different populations were, unlike in the previous study, quite similar (91). The differences between these studies were not trivial, and Sheffer et al (91) found that fish from Sharp Spring had neither higher survival, nor less bilateral asymmetry, nor larger size than did fish from other populations. In addition, the Sharp Spring sample of wild-caught fish had higher fecundity than that from the Monkey Spring, but a sample from another site had somewhat higher fecundity levels still. These fecundity differences disappeared, however, in the next laboratory-raised generation.

Sheffer et al (91) concluded that the differences were likely attributable to the laboratory environment used by Quattro & Vrijenhoek (80) in New Jersey being more stressful than theirs in Arizona. For example, Monkey Spring individuals experienced eight times as much mortality over the first 12 weeks and ten times as much bilateral asymmetry in New Jersey as in Arizona. Obviously, ability to cope with stressors is important for endangered species, but in this case, the stressors in the New Jersey environment are unknown, and they are probably unrelated to any stressors that would be encountered in natural populations in Arizona. Differences in fitness between populations in a stressful laboratory environment suggest that it would be useful to determine if these differences are also present in natural environments or if they cause the same effect between non-inbreds and inbreds in natural environments.

Sheffer et al (92) also examined populations from the four major watersheds containing Gila topminnows for inbreeding or outbreeding effects on several traits potentially related to fitness. These laboratory studies produced no evidence of either inbreeding or outbreeding depression: there was generally high survival, similar body size, and little bilateral asymmetry for all the inbred and outbred matings. Similarly, no evidence appeared of inbreeding or outbreeding effects for fecundity or sex ratio except for the sample from Monkey Spring, which had highly female-biased sex ratios and low fecundity after one generation of inbreeding. No evidence of an increase in fitness appeared in crosses between populations, suggesting there was no evidence of fixation of different detrimental alleles over the populations. However, as discussed above, in more extreme situations as often encountered in natural habitats, fitness components may be more influenced than in laboratory situations.

In captive breeding programs of endangered species, substantial effort is often focused on maintaining these species and, as a result, husbandry, diet, understanding of behavior, etc. improve over time. As a result, survival tends to increase over time even as the inbreeding coefficient increases if the population was started from a small number of founders. For example, in both the Mexican and red wolf captive breeding programs (46), a significant increase occurred in the survival of animals over time (Figure 2). Concurrently, the inbreeding coefficient increased so that any inbreeding depression might be cancelled out by, or confounded with, the temporal increase in viability. Similar temporal changes in survival have been observed in other species (45, 47).

GENETIC RESTORATION OF POPULATIONS WITH LOW FITNESS

Populations of some endangered species have become so small that they have lost genetic variation and appear to have become fixed for deleterious genetic variants. To avoid extinction from this genetic deterioration, some populations may benefit from the introduction of individuals from related populations or subspecies for genetic restoration, i.e., elimination of deleterious variants and recovery to normal levels of genetic variation. Hedrick (34) developed one way to assess the potential positive and negative effects of introducing individuals from genetically diverse but geographically isolated populations into apparently inbred populations, in evaluating the then-proposed genetic restoration of the Florida panther through the introduction of Texas cougars.

Some of these results are illustrated in Figure 3. First, consider the expected change in fitness after introduction of a gene causing lower fitness in the endangered population (bottom line), where the relative fitnesses of genotypes A_1A_1, A_1A_2, and A_2A_2, are 1, 1, and 0.5, respectively. In this case, the endangered population is fixed for detrimental allele A_2 and the outside population is fixed for allele A_1. If there is 20% gene flow from outside in the first generation and 2.5% every generation thereafter, the fitness quickly improves; before 10 generations, it has approached the maximum possible for this gene. One concern about this approach is that any locally adapted alleles may be swamped by gene flow from outside. To examine this scenario, it was assumed that the fitnesses of the genotypes are 1, 1.2, and 1.2, respectively, the endangered population is fixed for a dominant advantageous allele, and the outside population is fixed for an allele that is disadvantageous in the environment of the endangered population. In this case, the fitness is only slightly reduced as the result of gene flow (top line), so the advantageous allele is able to maintain itself in spite of gene flow from the outside. If these two effects are combined (broken line), then the expected fitness increases over time and approaches the maximum before 10 generations. These findings appear to be generally robust to the effects of finite population size, variation in the level of dominance, and other factors (34).

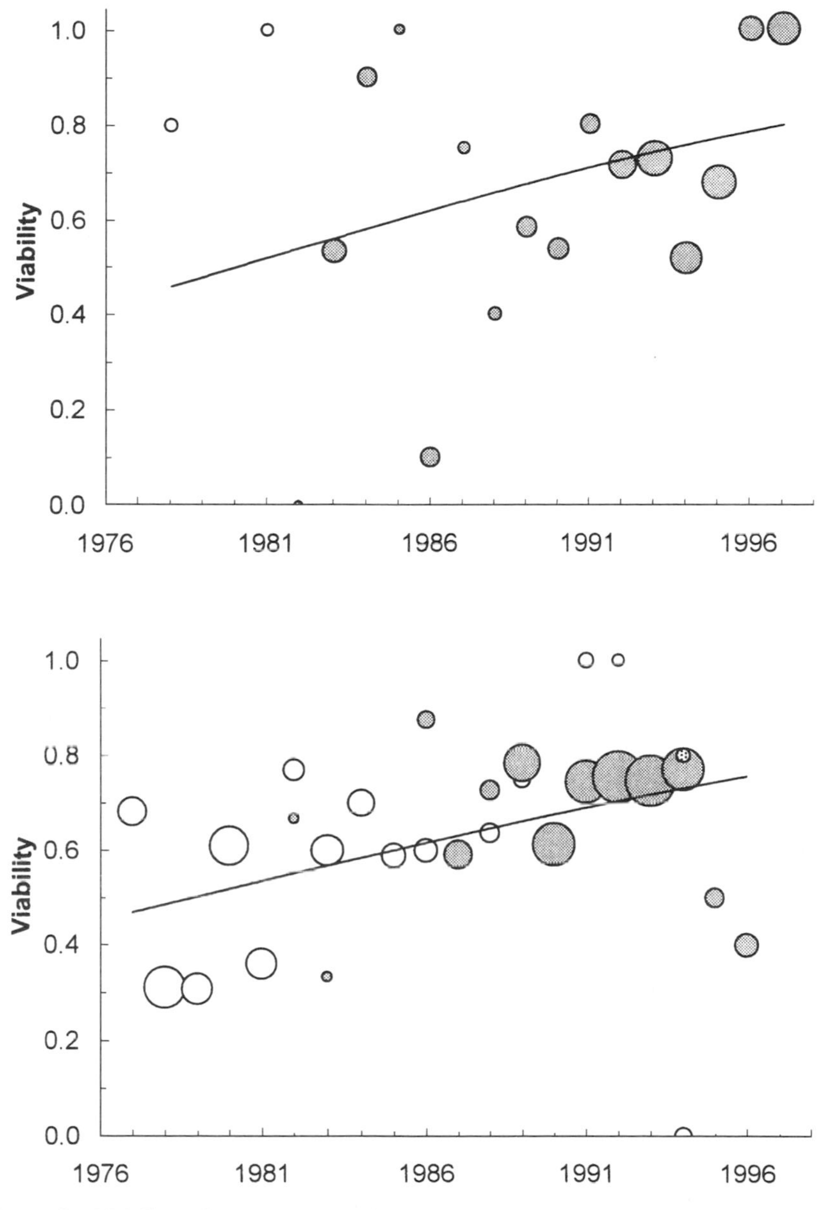

Figure 2 Viability of non-inbred individuals (open symbols) and inbred individuals (shaded symbols) for the Mexican (above) and red (below) wolves. The area of the circles is proportional to the number of individuals, and the fitted line shows the viability as a logistic function of year of birth (from 46).

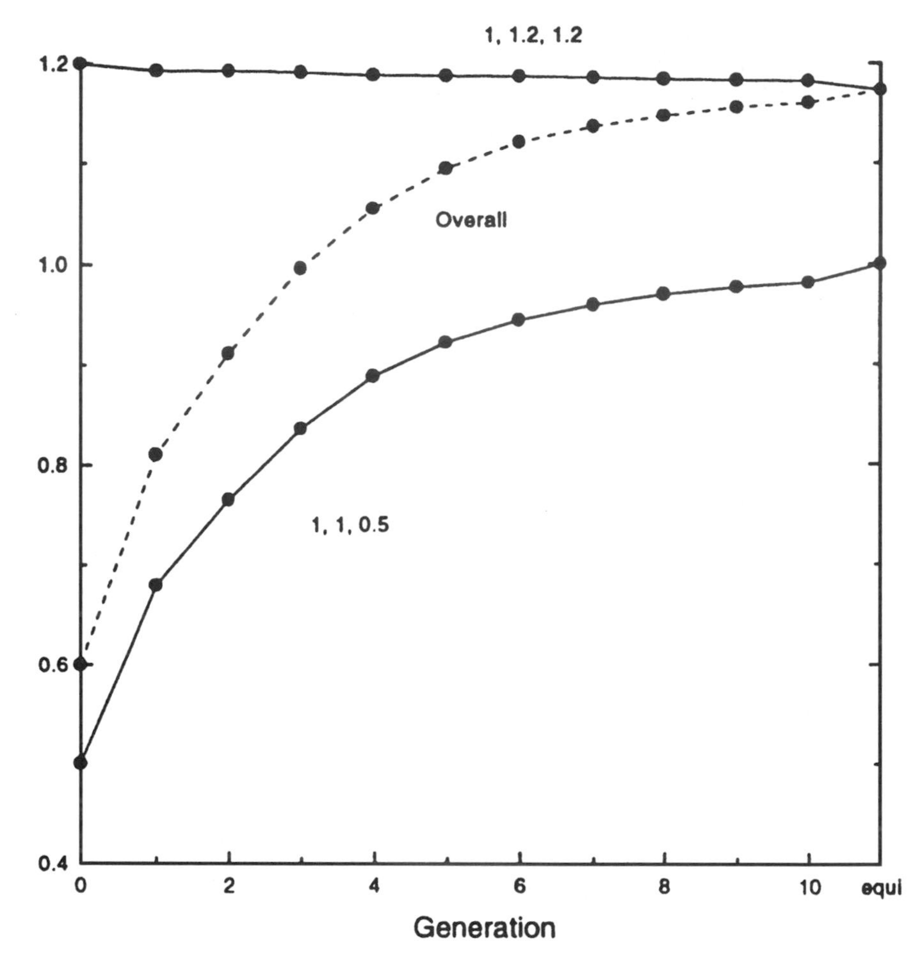

Figure 3 The change in relative fitness in a population over 10 generations with gene flow of 20% in the first generation and 2.5% per generation thereafter from an outside source, and with selection either causing lowered fitness (genotypes A_1A_1, A_1A_2, and A_2A_2 with fitnesses 1, 1, and 0.5) or an adaptive advantage (1, 1.2, 1.2), and the overall effect of these two loci (from 34).

Before we discuss several examples of apparently successful genetic restoration, several cautionary remarks are in order. If safeguards are not employed, introduction of animals into a population may have immediate detrimental effects on the population by the introduction of disease or other effects (94). If animals that have been in captivity are used for the introduction, they may be adapted to captivity and/or have lost adaptation to the natural environment. Further, introduction may potentially reduce the overall effective population size (36, 87), and outbreeding depression (in this case a lower fitness of the F_1s and future generations than that in the parental population) may further reduce the viability of the population.

Three experiments have suggested that gene flow from outside populations can restore the fitness of populations that seemed to be suffering from fixation of detrimental alleles. In these instances, the mean fitness of the F_1 individuals appears to be higher than that of individuals in the impacted population. Examination of inbreeding depression in these populations before infusion of outside individuals, using the traditional approach of comparing non-inbred and inbred offspring, may not reveal inbreeding depression because all individuals are similar genetically for detrimental alleles. Only by crossing to individuals from outside the population is the effect of inbreeding depression measurable and the fitness restored to a level found before the effect of genetic drift (14).

The last remaining population of the Florida panther (*Felis concolor coryi*) has a suite of traits that suggests genetic drift has fixed (or nearly fixed) the population for previously rare and potentially deleterious traits. These traits, which are found in high frequency only in the Florida panther and are unusual in other puma subspecies, include a high frequency of cryptochordism (unilateral undescended testicles), the poorest quality of semen recorded in any felid, kinked tail, and cowlick (85). In addition, a large survey of microsatellite loci have shown that Florida panthers have much lower molecular variation than other North American populations of mountain lions, and much lower variation today than found in samples from around 1900 (19). The Florida panther has been isolated in southern Florida since the early 1900s, and the effective population size in recent decades appears to be 25 or fewer. As a result, a program to release females from the closest natural population from Texas was initiated in 1995 to genetically restore fitness in this population (34). The introduced Texas females have bred with resident Florida panther males, and 14 F_1 offspring have been reproduced. Of these, none has a kinked tail and only one has a cowlick (57). Although the sample size is small, and cryptochordism and semen quality have not been evaluated, the frequency of detrimental traits appears to have been greatly reduced. This may become an important example of genetic restoration.

An isolated population of an adder (*Vipera berus*) in southern Sweden appears to have accumulated deleterious traits during a decline in population size to about 10 males in 1992 (70). During this decline, there was very low recruitment, a high proportion of deformed or stillborn offspring, and very low genetic variability. Twenty males from another population were captured and released for three years into the site. In 1996, the first year that new adult male adders were expected to be observed, there was increased recruitment of F_1 individuals (69). The number continued to increase and in 1999 32 male adders were observed, the most since 1981.

A remnant population of the greater prairie chickens (*Tympanuchus cupido pinnatus*) in Illinois provides another example of possible genetic restoration. The population decreased from around 2000 individuals in 1962 to fewer than 50 in 1994 (101). Fitness, as measured by fertility and hatching rates, declined over this period, and an estimate of genetic variation in the population was low compared both to populations from other states (7) and to historical specimens from Illinois (8). In 1992, 271 birds were translocated from large populations in other states; nests monitored after the translocation suggested restored fertility and hatching rates.

MEASURING INBREEDING DEPRESSION

Several experimental and observational methods have been used to examine the effect of inbreeding on fitness and its components, and several statistical approaches have been used to quantify the extent of inbreeding depression. The simplest method compares the mean value of non-inbred individuals with the mean value of inbred individuals. This difference is usually standardized by expressing it as a proportional change in fitness $\delta = (\overline{w}_N - \overline{w}_I)/\overline{w}_N$. δ has both been called the coefficient of inbreeding depression and the cost of inbreeding. This statistic is only meaningful when defined so that $\overline{w}_I$is expected to be less than $\overline{w}_N$ (e.g. measuring viability instead of mortality).

When data exist for multiple levels of inbreeding, the rate of decline in fitness with increased inbreeding is often of interest, which requires a model of how inbreeding affects fitness components. If loci causing inbreeding depression interact independently and additively, then fitness will decline linearly with increased inbreeding, as $\overline{w}_I = \overline{w}_N - bf$ where f is the inbreeding coefficient. The constant b can be used to compare the effects of inbreeding across variable levels of inbreeding but not across traits with different magnitudes. A standardized rate of decline of fitness, b', is obtained by dividing the rate at which fitness declines by the fitness of non-inbred individuals as $b' = b'/\overline{w}_N$ or, equivalently, by dividing the cost of inbreeding by how much inbreeding occurred, $b' = \delta/f$ (15). For example, if $\overline{w}_N = 80$ and $\overline{w}_I = 50$ for individuals with an inbreeding coefficient of 0.25 (full-sib or parent-offspring mating), then $\delta_{0.25} = 0.375$, $b = 120$, and $b' = 1.5$.

If loci determining fitness have independent, multiplicative effects, then fitness is expected to decline exponentially with inbreeding as $\overline{w}_I = \overline{w}_N e^{-Bf}$ so that where $B = -[\ln(\overline{w}_I/\overline{w}_N]/f$ where B is a constant characteristic of the population for the given trait (14). This model is most useful for examining the relationship between inbreeding and viability, in which case $2B$ is approximately equal to the number of lethal equivalents affecting viability in a diploid genome (73). The lethal equivalent is a unit (although commonly used as a statistic) that can be used to quantify the effects of genes upon survival. One lethal equivalent is defined as a set of alleles that, if dispersed in different individuals, would, on average, be lethal in one individual of the group. For example, two alleles, which each cause death 50% of the time, constitute one lethal equivalent. The number of lethal equivalents in a diploid genome provides a measure of the potential effects of deleterious, recessive alleles.

A good example of the calculation of $2B$ is from the white-footed mouse study (41) in which the survival from birth to weaning at day 20 of non-inbred mice in captivity was 0.879 and survival of inbred mice ($f = 0.25$) was 0.822. Thus, the estimated number of lethal equivalents in a diploid genome in juvenile mice in captivity is then 0.54. On the other hand, for the initial three-week release period in the natural environment, adult survival for non-inbreds was 0.194 and inbreds was 0.040. Using these data, the estimated number of lethal equivalents is 12.64 in the natural environment. This also illustrates the extreme environmental dependence of inbreeding depression in this case.

Although linear and exponential relationships between fitness and inbreeding are conceptually different, they predict similar response of fitness for low to moderate amounts of inbreeding. For example, if the cost of full-sib mating equals its median estimate among mammals of 0.32 (81), then both models predict a similar response of fitness to inbreeding from values of f less than 0.375.

In captive populations of endangered species, there are often various inbreeding categories, and generally least squares linear regression has been used to estimate B (81, 95). However, this approach does not work when there have been no survivors in a given inbreeding class because the logarithm of zero is undefined. As a result, a "small sample size correction" has been used to circumvent this problem (95), although this correction introduces a bias in the estimation procedure. To avoid this bias, Kalinowski & Hedrick (43) advocated using a maximum likelihood approach, which does not necessitate a small sample size correction.

An obstacle to measuring inbreeding depression in endangered species is the tendency for modern captive breeding programs to unintentionally reduce the statistical power to measure inbreeding depression (44). By preferentially pairing unrelated individuals, the distribution of inbreeding coefficients in a population that is managed to maximally preserve genetic variation will narrow until all individuals have approximately the same inbreeding coefficient. When this occurs, data from subsequent births will provide little additional information on the relationship between fitness and inbreeding.

Measuring the effect of inbreeding requires estimating inbreeding coefficients and a measure of fitness for a set of individuals. Estimating inbreeding coefficients for most populations is difficult because it generally requires knowing the pedigree of the individuals (1). Furthermore, calculating inbreeding coefficients from pedigrees requires specifying a degree of relatedness among founders of the pedigree. When no information is available, founders are usually assumed to be non-inbred and unrelated. If these assumptions are not true then inbreeding coefficients within the pedigree may be underestimated. Only captive populations generally have good enough pedigree information to calculate inbreeding coefficients. but in a few wild populations in which parents are known for several generations, good estimates of inbreeding coefficients have been obtained (49, 98).

Highly variable loci appear to provide an approach to estimating relatedness that does not require pedigree data (84). The heterozygosity at 29 microsatellite loci was determined in captive gray wolves with inbreeding coefficients known from pedigree information (24). The relationship of the average observed heterozygosity for these loci and f is given in Figure 4. The linear regression (solid line) explains 67% of the variation and is highly statistically significant. With this baseline,

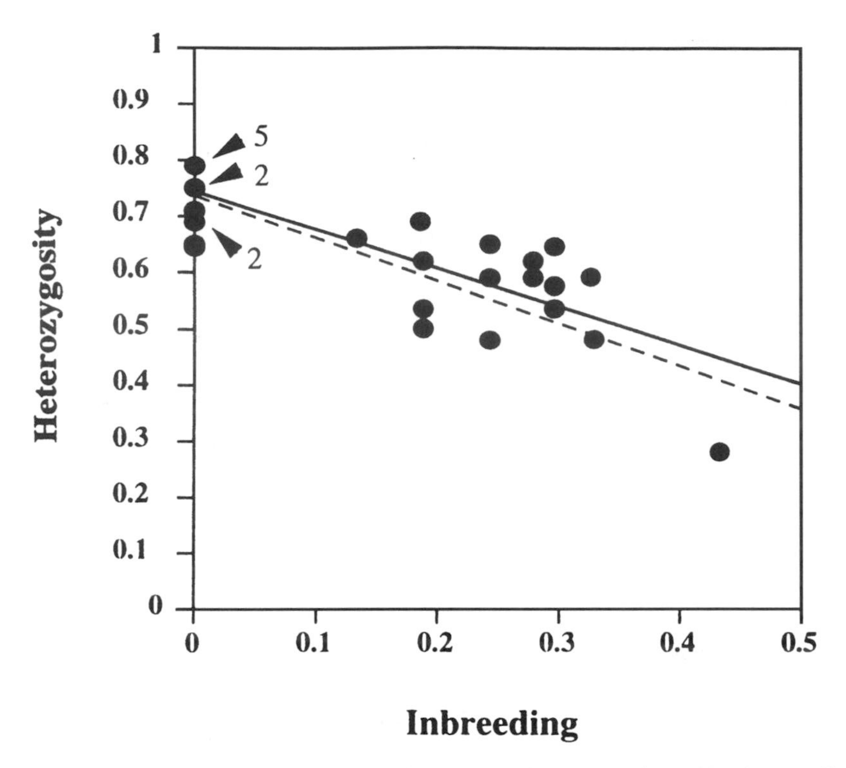

Figure 4 The relationship between individual average heterozygosity at 29 microsatellite loci in captive Scandinavian gray wolves. The straight line is the regression of individual heterozygosity on inbreeding coefficient, and the broken line indicates the expected relationship between H and f assuming that $H = 0.75$ between unrelated wolves (from 24).

Ellegren (24) evaluated the level of inbreeding in a sample of 13 of the 60 to 70 wild wolves left in Sweden. He found that the mean heterozygosity for the same loci in the wild wolves is 0.52, suggesting that their average inbreeding coefficient is around 0.2 to 0.3. In fact, one wild wolf had a heterozygosity of only 0.25, lower than the most inbred captive wolf examined. In addition, estimates of relatedness (67) of the wild wolves revealed pairs that appeared to be closely related, consistent with the suggestion that some of the wolves surveyed had high inbreeding coefficients.

Inbreeding depression may not be detected for several reasons, none of which indicates that inbreeding depression is absent from the population or species. First, the effects of inbreeding are generally examined in only one or a few components of fitness. Lack of inbreeding depression for juvenile survival does not mean there is no inbreeding depression for fecundity or mating success, or that different fitness components do not interact to reduce overall fitness. Second, often inbreeding depression is examined in a captive situation in which the environment may be more benign, or at least different, than in a natural environment. As a result, there may be no detectable inbreeding depression in the given environment, but there may be an effect in a different or a natural environment. Third, there may not be the statistical power to detect biologically important inbreeding depression. This can be the result of either small sample size or the structure of the pedigree in which many individuals have similar inbreeding coefficients.

Kalinowski et al (46) found no detectable evidence for inbreeding depression in captive populations of Mexican and red wolves, although there is substantial evidence for inbreeding depression in the highly inbred Scandinavian wolves (54, 55). However, they were able to obtain substantial data for only two fitness-related traits, juvenile survival and litter size, and all the animals were in captive environments. Further, although the sample sizes for juvenile viability were 251 and 688 animals in the Mexican and red wolf samples, respectively, there was low statistical power to demonstrate the absence of inbreeding depression because of the structure of the pedigrees(44). It is important to report such "negative" results (along with their limitations), or the overall perspective of the extent and pattern of inbreeding depression may become biased. The potential presence of undocumented inbreeding depression in the Mexican wolf captive breeding program was an important motivation for combining the three independent (and inbred) lineages of Mexican wolves into one population (37). The inclusion of the two other lineages into the captive breeding program resulted in increased genetic diversity (by increasing the number of founders from 3 to 7) and may overcome any reduction of fitness that may have resulted from fixation of detrimental alleles within the lineages.

Interpretation of experimental results requires an understanding of the many statistical pitfalls associated with phenotypic data in inbred populations. The issues that should be addressed in experimental design and data analysis are complex, and it is important to be cautious in the analysis and interpretation of most data (see 68 for a detailed review). For example, the phenotype of individuals in multigenerational studies can be influenced by environmental trends. Plant studies may avoid this problem by storing seeds from each generation, then raising them contemporaneously in a common environment (4). When this approach is not possible, an alternative is to infer environmental effects by contemporaneously observing a large, randomly mating control population. With these data, the mean phenotype of inbred lineages can be adjusted to compensate for the changing effects of the environment. Another possible way to minimize this potential effect is to examine multiple levels of inbreeding in one environment at one time. In addition, statistical examination of phenotypes must account for the dependence of the phenotype of sequential samples upon their predecessors, decreasing variance for phenotypic traits caused by loss of genetic variation, and variation in the genetic comparison of the individuals that founded lineages. Lynch (62) maintains that these issues require a reliable study to have a large number of replicate populations.

PURGING INBREEDING DEPRESSION—EVIDENCE AND SIGNIFICANCE

After Ralls et al (82) documented inbreeding depression, there was a general effort to avoid inbreeding in the management of captive populations. However, inbreeding is unavoidable in populations founded from a small number of individuals as were many captive populations of endangered species. Templeton & Read (95) proposed to eliminate (purge) inbreeding depression through carefully controlled breeding and selection, a program that they claimed was successful in a captive population of the endangered Speke's gazelle. The original captive Speke's gazelle population was descended from three females and one male, which were; therefore, soon faced with unavoidable half-sib or parent-offspring matings. Mating pairs for the Speke's gazelle population were selected for three years; the second and third generations of inbred births had significantly higher viability than did the first generation of inbred gazelles. It appeared that inbred gazelles with inbred parents had higher survival than inbred gazelles with non-inbred parents, suggesting reduction of inbreeding depression because of past inbreeding. Although purging has not become an accepted strategy for managing small populations, the Speke's gazelle captive breeding program has remained a prominent case study in the inbreeding depression and conservation biology literature and is widely cited as a successful example (77).

Several authors have questioned Templeton & Read's evaluation on a variety of grounds (26, 33, 51, 105, but see 96). Most recently, Kalinowski et al (47) argued that the evidence for selection reducing inbreeding depression in the Speke's gazelle breeding program is based on a mischaracterization of when viability increased. Previous analyses compared the viability of the first generation of inbred births (born to non-inbred parents) with the viability of the second and third generations of inbreeding (born to non-inbred parents). These analyses assumed that the observed increase in viability occurred after selection had operated on the first generation of inbred gazelles. In contrast, Kalinowski et al (47) showed that the viability of inbred births actually increased during the first generation of inbreeding, before selection could have been detected. More specifically, inbred gazelles with no ancestral inbreeding born prior to 1976 had low viability, whereas similar gazelles born in 1976 or later had higher viability. In addition, the second generation of inbred gazelles (inbred gazelles with inbred parents), which could have benefited from selection in the previous generation, had a viability similar to the first generation of inbred gazelles born after 1975.

One way to illustrate this temporal change is to categorize gazelles by year of birth and parental inbreeding coefficient (Figure 5). The offspring of non-inbred gazelles born before 1976 (curve Ia) have significant inbreeding depression, but the offspring of non-inbred gazelles born in or after 1976 have no significant inbreeding depression (curve Ib). As reported by Templeton & Read (95), the offspring of inbred gazelles (curve II), most of which were born in or after 1976, do not have

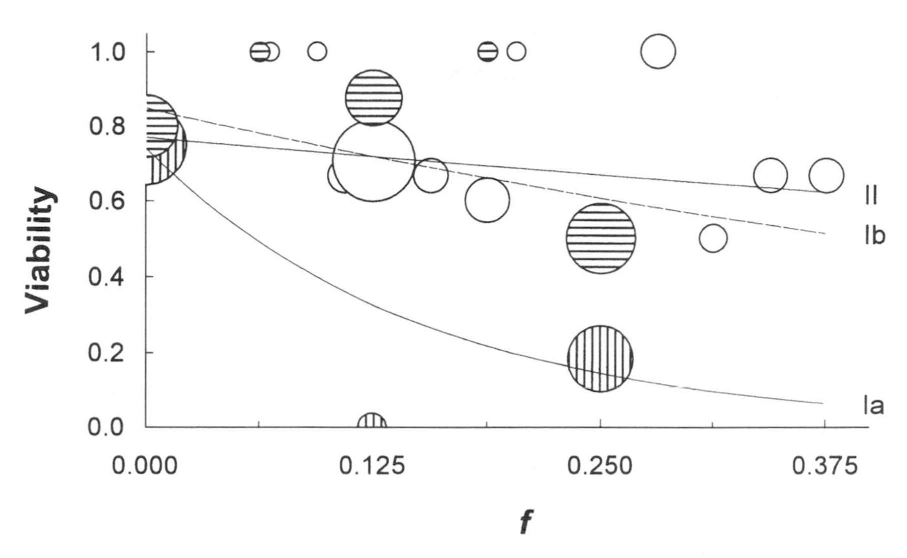

Figure 5 Observed (circles) and fitted (curved lines) viability in Speke's gazelles categorized by year of birth (before 1976, or in and after 1976) and parental inbreeding coefficient (non-inbred or inbred parents). Offspring of non-inbred gazelles born before 1976 (curve Ia, vertically lined circles), offspring of non-inbred gazelles born after 1976 (curve Ib, horizontally lined circles), and offspring of inbred gazelles (curve II, open circles) are given (from 47).

significant inbreeding depression. Thus, in the last half of the period examined, the offspring of neither non-inbred nor inbred gazelles exhibit inbreeding depression. In other words, inbred gazelles born in or after 1976 do not have reduced viability, whether or not their parents are inbred. In addition, simulation of the Speke's gazelle pedigree suggests that it is very unlikely that inbreeding depression could be eliminated in this short time period (47). In other words, it appears that the Speke's gazelle captive breeding program is a better example of the complexity of inbreeding depression, apparently related to unknown changes in the environment over time, than of purging. In either case, the Speke's gazelle captive breeding illustrates that populations with significant inbreeding depression can successfully be founded from a small number of individuals if their reproductive potential is large enough. Significantly, in a recent examination of the potential effects of purging in 17 mammalian species (2), a nonsignificant reduction in inbreeding depression in the updated Speke's gazelle captive population was found, and the inbreeding depression in the Speke's gazelle was the highest of any of species analyzed.

Since the Speke's gazelle captive breeding program fueled interest in purging as a conservation strategy, experimental work has examined the purging process in detail. Much of the work has centered on plants and has been thoroughly reviewed (12). Some experimental studies have closely examined the response of inbreeding depression to multiple generations of selection, and in several insect model systems, there appears to have been a rebound in fitness (10, 25, 88). In a 10-generation study of inbreeding depression in *Peromyscus* (53), three subspecies differed in response. In general, fitness measures declined in each subspecies during the first generation of inbreeding. Subsequent generations of inbreeding were accompanied by improvements for one of the subspecies, another subspecies showed consistent decreases in fitness with subsequent inbreeding, and in the third subspecies, inbreeding depression was exacerbated by further inbreeding.

In endangered species that have low reproductive potential, deliberate inbreeding to purge inbreeding depression is a risky strategy because the added reduction in fitness from inbreeding may result in extinction. Even if the population survives, detrimental alleles may become fixed permanently, lowering the population fitness; genetic variation for other loci may be lost (33, 100). Minimizing inbreeding is the only currently accepted method for minimizing inbreeding depression. This can be accomplished in several related ways. The effective size of a population can be increased by increasing its census size or by minimizing any of the ways a real population departs from the Wright-Fisher ideal population (35). In addition, the typical genetic goal in management of captive populations has been to minimize loss of genetic variation (and inbreeding). Most modern captive breeding programs with pedigreed populations have adopted maximization of genetic diversity (measured by expected heterozygosity) as their primary genetic goal (3), with the implicit assumption that this will restrict inbreeding to an acceptable level. This is accomplished by selecting individuals to mate whose offspring minimize the average relatedness of the population.

CONCLUSIONS

Influence of inbreeding on fitness-related traits in endangered species and other organisms appears to be variable over populations, traits, and environments. Because endangered species generally have small population sizes and may have gone through bottlenecks, the genetic characteristics of inbreeding depression may differ in endangered species from more cocies, and they may have lower numbers of lethals contributing to the genetic load. Or, the effects of genetic drift may be present in only some endangered species, resulting in a greater variation over populations or species in the characteristics of inbreeding depression than in more common species.

The effects of inbreeding on fitness vary over species. Some of this variation is due to chance, including variation in number of lethal equivalents in the founders (52), but much of it may be due to different mean levels of inbreeding depression over the species, traits, or environments examined. For example, a low inbreeding coefficient of 0.03125 appeared to have an effect in houseflies (9), whereas in some experiments with *Drosophila*, an inbreeding coefficient of 0.7 was necessary to have an impact (27). Therefore, it is problematic to predict the expected effect of inbreeding on fitness in an unexamined endangered species. However, the perspective provided here should give a context to evaluating inbreeding depression and suggest that its effects are generally likely to be more than first observed, rather than less.

Unless there is a high genetic load, detecting inbreeding depression in endangered species may be difficult in both natural and captive environments because of low statistical power, relatively benign conditions in captive environments, and inability to examine all aspects of fitness. Therefore, even if there is no statistical evidence for inbreeding depression, it is prudent to assume it is present. If it is assumed not to be present, then inbreeding may result in a loss of fitness that could have been avoided. On the other hand, just because there is inbreeding depression or low fitness because of past fixation from genetic drift, one should not give up on the population. For populations with low fitness, recent examples have shown that introduction of individuals from outside can genetically restore the fitness of the population. For populations with high inbreeding depression, sometimes alleles with large detrimental effects may be purged both in theory (33, 100) and in laboratory experiments with model insects. However, the oft-cited example of reduced inbreeding depression in the Speke's gazelle is most parsimoniously explained by a change in survival of inbred animals over time.

One of the early guidelines in captive breeding of endangered species to avoid inbreeding depression was based on the observation that "animal breeders accept inbreeding coefficients as high as a one percent increase per generation (i.e. an effective population size of 50) in domestic animals without great concern" (29). It was assumed that such slow inbreeding allows selection to remove deleterious alleles without endangering the population. However, a number of endangered species with successful breeding programs have had quite low founder numbers, e.g. Speke's gazelle with 4, Przewalski's horse with 13, black-footed ferret with 6, etc. Just because the numbers are low does not mean we should not make all attempts to save a species. All the surviving individuals have the unique characteristics of the species, and eventually the population number may be high enough to overcome the detrimental effects of the initial population size restriction. However, Wang et al (100), who have explored theoretically the effects of the nature of the mutational load, reproductive capacity, and the organization of the genome, demonstrate that when the effective size is 50, fitness generally declines because of genetic fixation of detrimental alleles.

ACKNOWLEDGMENTS

We appreciate support from the National Science Foundation (PWH), the US Fish and Wildlife Service (PWH, STK), National Marine Fisheries Service (STK), and the Ullman Distinguished Professorship (PWH) during the preparation of this review. Our thanks to the contributions of P. Miller and R. Sheffer on various aspects of this research.

Visit the Annual Reviews home page at www.AnnualReviews.org

LITERATURE CITED

1. Ballou JD. 1983. Calculating inbreeding coefficients from pedigrees. In *Genetics and Conservation*, ed. CM Schonewald-Cox, SM Chambers, B MacBryde, L Thomas, pp. 509–20. Menlo Park, CA: Benjamin/Cummings
2. Ballou JD 1997. Ancestral inbreeding only minimally affects inbreeding depression in mammalian populations. *J. Hered.* 88:169–78
3. Ballou JD, Gilpin M, Foose TJ, eds. 1995. *Population Management for Survival and Recovery.* New York: Columbia Univ. Press
4. Barrett SCH, Charlesworth D. 1991. Effect of a change in the level of inbreeding on the genetic load. *Nature* 352:522–24
5. Bijlsma R, Bundgaard J, Van Putten WF. 1999. Environmental dependence of inbreeding depression and purging in *Drosophila melanogaster. J. Evol. Biol.* 12:1125–37
6. Bijlsma R, Bundgaard J, Boerema AC. 2000. Does inbreeding affect the extinction risk of small populations? Predictions from *Drosophila. J. Evol. Biol.* 13:502–14
7. Bouzat JL, Cheng HH, Lewin HA, Westemeier RL, Brawn JD, Paige KN. 1998. Genetic evaluation of a demographic bottleneck in the greater prairie chicken. *Conserv. Biol.* 12:836–43
8. Bouzat JL, Lewin HA, Paige KN. 1998. The ghost of genetic diversity past: historical DNA analysis of the greater prairie chicken. *Am. Natur.* 152:1–6
9. Bryant EH, McCommas SA, Combs LM. 1986. The effect of an experimental bottleneck upon quantitative genetic variation in the housefly. *Genetics* 114:1191–1211
10. Bryant EH, Meffert LM, McCommas SA. 1990. Fitness rebound in serially bottlenecked populations of the house fly. *Am. Natur.* 136:542–49
11. Bryant EH, Backus VL, Clark ME, Reed DH. 1999. Experimental tests of captive breeding for endangered species. *Conserv. Biol.* 13:1487–96
12. Byers DL, Waller DM. 1999. Do plant populations purge their genetic load? Effects of population size and mating history on inbreeding depression. *Annu. Rev. Ecol. Syst.* 30:479–513
13. Charlesworth B. 1998. The effect of synergistic epistasis on the inbreeding load. *Genet. Res.* 71:85–89
14. Charlesworth D, Charlesworth B. 1987. Inbreeding depression and its evolutionary consequences. *Annu. Rev. Ecol. Syst.* 18:237–68

14a. Charlesworth D, Charlesworth B. 1999. The genetic basis of inbreeding depression. *Genet. Res.* 74:329–40

15. Charlesworth B, Hughes KA. 1999. The maintenance of genetic variation in life-history traits. In *Evolutionary Genetics from Molecules to Morphology*, ed. RS Singh, CB Krimbas CB, pp. 369–92. Cambridge, UK: Cambridge Univ. Press

15a. Coltman DW, Pilkington JG, Smith JA, Pemberton JM. 1999. Parasite-mediated selection against inbred Soay sheep in a free-living, island population. *Evolution* 53:1259–67

15b. Crnokrak P, Roff DA. 1999. Inbreeding depression in the wild. *Heredity* 83:260–70

16. Crow JF. 1993. Mutation, mean fitness and genetic load. *Oxford Rev. Evol. Biol.* 9:3–42
17. Crow JF, Kimura M. 1970. *An Introduction to Population Genetics Theory.* New York: Harper & Row
18. Crow JF, Simmons MJ. 1983. The mutation load in Drosophila. In *The Genetics and Biology of Drosophila.* Volume 3c,

ed. M Ashburner, HL Carson, JN Thompson, pp. 1–35. London: Academic Press
19. Culver M, Johnson WE, Pecon-Slattery J, O'Brien SJ. 2000. Genomic ancestry of the American puma (*Puma concolor*). *J. Hered.* 91:186–97
20. Darwin CR. 1868. *Variation of Animals and Plants under Domestication.* London: John Murray
21. Darwin CR. 1876. *The Effects of Cross and Self fertilization in the Vegetable Kingdom.* London: John Murray
22. DeRose, MA, Roff DA. 1999. A comparison of inbreeding depression in life-history and morphological traits in animals. *Evolution* 53:1288–92
23. Dudash MR, Carr DE. 1998. Genetics underlying inbreeding depression in *Mimulus* with contrasting mating systems. *Nature* 393:682–84
24. Ellegren H. 1999. Inbreeding and relatedness in Scandinavian grey wolves Canis lupus. *Hereditas* 130:239–44
25. Fowler K, Whitlock MC. 1999. The variance in inbreeding depression and the recovery of fitness in bottlenecked populations. *Proc. R. Soc. Lond B.* 266:2061–66
26. Frankham R. 1995. Conservation genetics. *Annu. Rev. Genet.* 29:305–27
27. Frankham R. 1995. Inbreeding and extinction: a threshold effect. *Conserv. Biol.* 9:792–99
28. Frankham R. 1999. Resolving conceptual issues in conservation genetics: the roles of laboratory species and meta-analyses. *Hereditas* 130:195–201
29. Franklin IR. 1980. Evolutionary change in small populations. In *Conservation Biology: An Evolutionary-Ecological Perspective*, ed. ME Soule, BA Wilcox, pp. 135–49. Sunderland, MA: Sinauer
30. Fu Y-B, Namkoong G, Carlson JE. 1998. Comparison of breeding strategies for purging inbreeding depression via simulation. *Conserv. Biol.* 12:856–64
31. Harshman LG, Hoffmann AA. 2000. Laboratory selection experiments using *Drosophila*: What do they really tell us? *Trends Ecol. Evol.* 15:32–36
32. Hedrick PW. 1987. Genetic bottlenecks. *Science* 237:963
33. Hedrick PW. 1994. Purging inbreeding depression and the probability of extinction: full-sib mating. *Heredity* 73:363–72
34. Hedrick PW. 1995. Gene flow and genetic restoration: the Florida panther as a case study. *Conserv. Biol.* 9:996–1007
35. Hedrick PW. 2000. *Genetics of Populations.* Boston: Jones & Bartlett. 2nd ed.
36. Hedrick PW, Hedgecock D, Hamelberg S. 1995. Effective population size in winter-run chinook salmon. *Conserv. Biol.* 9:615–24
37. Hedrick PW, Miller PS, Geffen E. Wayne R. 1997. Genetic evaluation of the three captive Mexican wolf lineages. *Zoo Biol.* 16:47–69
38. Hedrick PW, Parker KM. 1998. MHC variation in the endangered Gila topminnow. *Evolution* 52:194–99
39. Hedrick PW, Savolainen O, Karkkainen K. 1998. Factors influencing the extent of inbreeding depression: an example from Scots pine. *Heredity* 82:441–50
40. Husband BC, Schemske DW. 1996. Evolution of the magnitude and timing of inbreeding depression in plants. *Evolution* 50:54–70
41. Jimenez JA, Hughes KA, Alaks G, Graham L, Lacy RC. 1994. An experimental study of inbreeding depression in a natural habitat. *Science* 266:271–73
42. Johnston MO, Schoen DJ. 1995. Mutation rates and dominance levels of genes affecting total fitness in two angiosperm species. *Science* 27:226–29
43. Kalinowski ST, Hedrick PW. 1998. An improved method for estimating inbreeding depression in pedigrees. *Zoo Biol.* 17:481–97
44. Kalinowski ST, Hedrick PW. 1999. Detecting inbreeding depression is difficult in captive endangered species. *Anim. Conserv.* 2:131–36

45. Kalinowski ST, Hedrick PW. 2000. Inbreeding depression in captive bighorn sheep. *Zoo Biol.* Submitted
46. Kalinowski ST, Hedrick PW, Miller PS. 1999. No evidence of inbreeding depression in Mexican and red wolves. *Conserv. Biol.* 13:1371–77
47. Kalinowski ST, Hedrick PW, Miller PS. 2000. A close look at inbreeding depression in the Speke's gazelle captive breeding program. *Conserv. Biol.* In press
48. Keller, LF. 1998. Inbreeding and its fitness effects in an insular population of song sparrows (*Melospiza melodia*). *Evolution* 52:240–50
49. Keller LF, Arecese P, Smith JMN, Hochachka WM, Stearns SC. 1994. Selection against inbred song sparrows during a natural population bottleneck. *Nature* 372:356–57
50. Kimura M. *The Neutral Theory of Molecular Evolution.* Cambridge, UK: Cambridge Univ. Press
51. Lacy RC. 1997. Importance of genetic variation to the viability of mammalian populations. *J. Mammal.* 78:320–35
52. Lacy RC, Alak G, Walsh A. 1996. Hierarchial analysis of inbreeding depression in *Peromyscus polionotus*. *Evolution* 50:2187–2200
53. Lacy RC, Ballou JD. 1998. Effectiveness of selection in reducing the genetic load in populations of *Peromyscus polionotus*. *Evolution* 50:2187–2200
54. Laikre L, Ryman N. 1991. Inbreeding depression in a captive wolf (*Canis lupus*) population. *Conserv. Biol.* 5:33–41
55. Laikre L, Ryman N, Thompson ES. 1993. Hereditary blindness in a captive wolf population: frequency reduction of a deleterious allele in relation to gene conservation. *Conserv. Biol.* 7:592–601
56. Laikre L. 1999. Conservation genetics of Nordic carnivores: lessons from zoos. *Hereditas* 130:203–16
57. Land D, Lotz M, Shindle D, Taylor SK. 1999. Florida panther genetic restoration and management:Annual performance report 1998–1999. Florida Fish & Wildlife Conservation Commission, Naples, FL
58. Lande R. 1994. Risk of population extinction from fixation of new deleterious mutation. *Evolution* 48:1460–69
59. Lande R. 1995. Mutation and conservation. *Conserv. Biol.* 9:782–91
60. Latter BDH. 1998. Mutant alleles of small effect are primarily responsible for the loss of fitness with slow inbreeding in *Drosophila melanogaster*. *Genetics* 148:1143–58
61. Latter BDH, Mulley JC, Reid D, Pascoe L. 1995. Reduced genetic load revealed by slow inbreeding in *Drosophila melanogaster*. *Evolution* 139:287–97
62. Lynch M. 1988. Design and analysis of experiments on random drift and inbreeding depression. *Genetics* 120:791–807
63. Lynch M. 1996. A quantitative-genetic perspective on conservation issues. In *Conservation Genetics*, ed. J Avise, J Hamrick, pp. 471–501. New York: Chapman & Hall
64. Lynch M, Blanchard J, Houle D, Kibota T, Schultz S, Vassilieva L, Willis J. 1999. Perspective: spontaneous deleterious mutation. *Evolution* 53:645–63
65. Lynch, M, Conery J, Burger R. 1995. Mutation meltdowns in sexual populations. *Evolution* 49:1067–88
66. Lynch M, Gabriel W. 1990. Mutation load and the survival of small populations. *Evolution* 44:1725–37
67. Lynch M, Ritland K. 1999. Estimation of pairwise relatedness with molecular markers. *Genetics* 152:1753–66
68. Lynch M, Walsh B. 1998. *Genetics and Analysis of Quantitative Traits.* Sunderland, MA: Sinauer
69. Madsen T, Shine R, Olsson M, Wittsell H. 1999. Restoration of an inbred adder population. *Nature* 402:34–35
70. Madsen T, Stille B, Shine R. 1996. Inbreeding depression in an isolated population of adders *Vipera berus*. *Biol. Conserv.* 75:113–18

70a. Meagher S, Penn DJ, Potts WK. 2000. Male-male competition magnifies inbreeding depression in wild house mice. *Proc. Natl. Acad. Sci USA* 97:3324–29
71. Miller PS, Hedrick PW. 1993. Inbreeding and fitness in captive populations: lessons from *Drosophila*. *Zoo Biol.* 12:333–51
72. Mills LS, Smouse PE. 1994. Demographic consequences of inbreeding in remnant populations. *Am. Nat.* 144:412–31
73. Morton NE, Crow JF, Muller HJ. 1956. An estimate of the mutational damage in man from data on consanguineous marriages. *Proc. Natl. Acad. Sci. USA.* 42:855–63
74. Nei M. 1968. The frequency distribution of lethal chromosomes in finite populations. *Proc. Natl. Acad. Sci. USA* 60:517–24
74a. Nieminen M, Singer MC, Fortelius W, Schops K, Hanski I. 2000. Experimental confirmation of inbreeding depression increasing extinction risk in butterfly populations. *Am. Nat.* In revision
75. O'Brien SJ, Roelke ME, Marker L, Newman A, Winkler CA, et al. 1985. Genetic basis for species vulnerability in the cheetah. *Science* 227:1428–34
76. Parker KM, Sheffer RJ, Hedrick PW. 1999. Molecular variation and evolutionarily significant units in the endangered Gila topminnow. *Conserv. Biol.* 13:108–16
77. Pennisi E. 1999. The perils of genetic purging. *Science* 285:193
78. Pray LA, Goodnight CJ. 1995. Genetic variation in inbreeding depression in the red flour beetle *Tribolium castaneum*. *Evolution* 49:176–88
79. Pray LA, Schwartz, JM, Goodnight CJ, Stevens L. 1994. Environmental dependency of inbreeding depression: implications for conservation biology. *Conserv. Biol.* 8:562–68
80. Quattro JM, Vrijenhoek RC. 1989. Fitness differences among remnant populations of the endangered Sonoran topminnow. *Science* 245:976–78
81. Ralls K, Ballou JD, Templeton AR. 1988. Estimates of lethal equivalents and the cost of inbreeding in mammals. *Conserv. Biol.* 2:185–93
82. Ralls K, Brugger K, Ballou J. 1979. Inbreeding and juvenile mortality in small populations of ungulates. *Science* 206:1101–03
83. Ritland K. 1996. Inferring the genetic basis of inbreeding depression in plants. *Genome* 39:1–8
84. Ritland K. 1996. Estimators for pairwise relatedness and individual inbreeding coefficients. *Genet. Res.* 67:175–85
85. Roelke ME, Martenson JS, O'Brien SJ. 1993. The consequences of demographic reduction and genetic depletion in the endangered Florida panther. *Curr. Biol.* 3:340–50
86. Roff DA. 1998. Effects of inbreeding on morphological and life history traits of the sand cricket, *Gryllus firmus*. *Heredity* 81:28–37
87. Ryman N, Laikre L. 1991. Effects of supportive breeding on the genetically effective population size. *Conserv. Biol.* 5:325–29
88. Saccheri I, Brakefield PM, Nichols RA. 1996. Severe inbreeding depression and rapid fitness rebound in the butterfly *Bicyclus anynana* (Satyridae). *Evolution* 50:2000–13
89. Saccheri I, Kuussaari M, Kankare M, Vikman P, Fortelius W, Hanski I. 1998. Inbreeding and extinction in a butterfly metapopulation. *Nature* 392:491–94
90. Schultz ST, Willis JH. 1995. Individual variation in inbreeding depression: the roles of inbreeding history and mutation. *Genetics* 141:1209–23
91. Sheffer RJ, Hedrick PW, Minckley WL, Velasco AL. 1997. Fitness in the endangered Gila topminnow. *Conserv. Biol.* 11:162–71

92. Sheffer RJ, Hedrick PW, Velasco A. 1999. Testing for inbreeding and outbreeding depression in the endangered Gila topminnow. *Anim. Conserv.* 2:121–29
93. Simmons MJ, Crow JF. 1977. Mutation affecting fitness in Drosophila populations. *Annu. Rev. Genet.* 11:49–78
94. Snyder NH, Derrickson SR, Beissinger SR, Wiley JW, Smith TB, et al. 1996. Limitation of captive breeding in endangered species recovery. *Conserv. Biol.* 10:338–48
95. Templeton AR, Read B. 1983. The elimination of inbreeding depression in a captive herd of Speke's gazelle. In *Genetics and Conservation*, ed. CM Schonewald-Cox, SM Chambers, B MacBryde, L Thomas, pp. 41–61. Menlo Park, CA: Benjamin/Cummings
96. Templeton AR, Read B. 1998. Elimination of inbreeding depression from a captive population of Speke's gazelle: validity of the original statistical analysis and confirmation by permutation testing. *Zoo Biol.* 17:77–98
97. Uyenoyama MK, Holsinger KE, Waller DM. 1993. Ecological and genetic factors directing the evolution of self-fertilization. *Oxford Surv. Evol. Biol.* 9:327–81
98. Van Noordwijk AJ, Scharloo W. 1981. Inbreeding in an island population of the great tit. *Evolution* 35:674–88
99. Vrijenhoek RC, Douglas ME, Meffe GK. 1985. Conservation genetics of endangered fish populations in Arizona. *Science* 229:400–2
100. Wang J, Hill WG, Charlesworth D, Charlesworth B. 1999. Dynamics of inbreeding depression due to deleterious mutations in small populations: mutation parameters and inbreeding rate. *Genet. Res.* 74:165–78
101. Westemeier, RL, Brown JD, Simpson SA, Esker TL, Jansen RW, et al. 1998. Tracking the long-term decline and recovery of an isolated population. *Science* 282:1695–98
102. Wielebnowski N. 1996. Reassessing the relationship between juvenile mortality and genetic monomorphism in captive cheetahs. *Zoo Biol.* 15:353–69
103. Willis JH. 1999. Inbreeding load, average dominance and the mutation rate for mildly deleterious alleles in *Mimulus guttatus. Genetics* 153:1885–98
104. Willis JH 1999. The role of genes of large effect on inbreeding depression in *Mimulus guttatus. Evolution* 53:1678–91
105. Willis K, Wiese RJ. 1997. Elimination of inbreeding depression from captive populations: Speke's gazelle revisited. *Zoo Biol.* 16:9–16

Annu. Rev. Ecol. Syst. 2000. 31:163–96

AFRICAN CICHLID FISHES: Model Systems for Evolutionary Biology

Irv Kornfield[1] and Peter F. Smith[2]
[1]School of Marine Sciences and [2]Department of Biological Sciences, University of Maine, Orono, Maine 04469-5751;
e-mail: irvk@maine.edu; peter.smith@umit.maine.edu

Key Words Cichlidae, speciation, sexual selection, African lakes

■ **Abstract** Cichlid fishes (Perciformes: Teleostei) found in the lakes of Africa have served as model systems for the study of evolution. The enormous number of species (1000 in Lake Malawi alone), the great diversity of trophic adaptations and behaviors, and the extreme rapidity of their divergence (<50,000 y for some faunas) single out these organisms as examples of evolution in progress. Because these fishes are confined to discrete lacustrine environments and their origination is bounded by geological features, these groups provide models with which to study evolution. We review theoretical studies and empirical research on the cichlid faunas of Africa to provide a synthetic overview of current knowledge of the evolutionary processes at work in this group. This view provides the critical information needed to formulate and test hypotheses that may permit discrimination among the diverse theories and models that have been advanced to explain the evolution of these fishes.

INTRODUCTION

Periods of explosive speciation and adaptive radiation have punctuated the history of metazoan evolution. The cichlid fish faunas of lakes in Africa present singular opportunities to study such events. Lineages within the Cichlidae have diversified extensively in both the New World and the Old World (26, 67), but it is in Africa where the most spectacular examples of radiation and speciation are found. Because of this, African cichlids have been incorporated into many basic evolutionary texts (e.g. 172).

The cichlid fishes of Lakes Malawi, Tanganyika, Victoria (38, 66, 104), and the other, much smaller, African lakes (34, 155, 159) provide unique opportunities with which to investigate evolutionary processes (25, 67, 146). Times associated with diversification for many of these complexes are extremely short (83) (<50 ky for some faunas), the extent of trophic differentiation is extensive

(34) (ranging from plankton grazers to egg predators), and the number of endemic taxa is enormous [1000 species in Lake Malawi is a reasonable estimate (185)]. Because of these factors, the cichlids of African lakes provide geologically bounded systems within which to explore fundamental processes of speciation and diversification.

There is an extremely rich, speculative literature concerning the mechanisms underlying the processes of cichlid divergence (35, 118, 171, 184). The rapidity of radiations has suggested to many that some exceptional mechanism(s), including special water characteristics (91), habitat complexity (180), lake age (91), lake level fluctuation (32, 183), predation (42), mutation (34), hybridization (19), reproductive characteristics (22, 88), and trophic polymorphism (152) are necessary to explain the extent of cichlid radiation. Similarly, allopatric (45, 112), microallopatric (32), and sympatric (88, 155, 162, 185) modes of speciation have been suggested. There is general consensus that almost all of the endemic cichlid species and genera of Lakes Malawi and Victoria originated within the lakes proper (106); intralacustrine speciation is extensive in Lake Tanganyika, though it was clearly seeded several times (176). However, despite a wealth of recent theoretical and modeling efforts, limited empirical progress has been made in determining the critical factors that have accelerated or retarded diversification.

The possibility of sexual selection has raised enormous interest (184), but there have been few critical examinations of more conventional models of diversification. That is, alternative population processes such as subdivision, isolation, and drift may be significant driving forces for speciation at the geographic level. The extent of panmixia is not known, and the possibility of geographic structuring on local or regional scales has received limited attention (see below). Similarly, tests of many hypotheses have been hampered by the absence of reliable phylogenies for closely related species within the lakes. It may be beneficial to seek to reject null hypotheses of spatial variation prior to proposing more elaborate explanations for diversification.

In this contribution we assemble and review basic data on the geological and biological attributes that make African cichlid species unique and useful systems in which to study evolution. These faunas challenge simple application of species concepts; clarification of and consensus on this topic is of central importance to resolve persistent questions in cichlid evolution. Geographical models of speciation in African cichlids are summarized and empirical studies that have been offered as support for specific models are critiqued. The effects of fluctuating lake levels as a promoter of allopatric divergence are presented and examined. We offer a critical evaluation of the mechanisms potentially contributing to this spectacular diversification, discuss classical evolutionary forces as they apply to these cichlid fish systems, and offer speculation on alternative processes such as hybridization and the emerging mechanism of favor, sexual selection.

GEOLOGICAL FRAMEWORK AND CICHLID RADIATIONS

One of the most compelling aspects of the African cichlid fauna is that geology indicates that diversification of species and entire faunas has occurred at rates that many biologists have found difficult to accept. Further, geophysical processes have also played a central role in the formulation of hypotheses about mechanisms of diversification.

Rates and ages of cichlid diversification have been inferred from three elements: geological history, topography, and comparative genetic analysis. Early estimates for the ages of the African great lakes were based on gross geological features external to the lakes (34); more recent age estimates and, particularly, dates for lake level variations have been based on evidence from coring, acoustic profiling, climatic data, and modeling (reviewed in 17, 58). Estimates of maximum ages for Lake Tanganyika range from 9 to 12 my, 4 to 9 my for Lake Malawi, and 0.5 to 1 my for Lake Victoria (104). Extensive sediments (>4 km) in the basins of Lakes Malawi and Tanganyika testify to an aquatic environment of long duration, though such sedimentation need not be continuous.

Ages of the cichlid faunas estimated indirectly from molecular data assume that the markers being used behave as reasonably constant molecular clocks. This assumption may not hold when dealing with very divergent cichlid lineages (26) and is subject to modifications associated with geography, population structure, and behavior (53). Relative levels of sequence divergence suggest that the endemic cichlid fauna of Lake Victoria is of extremely recent origin (120, 121, but see 126). In Lake Malawi, divergence between the two major clades of endemic haplochromines may be on the order of 1 my (125), while divergence among taxa within these lineages is extremely recent (123, 138) and comparable to that observed in Victoria. In Tanganyika, divergence between cichlid tribes is extensive (76), and some predate the estimated age of the lake (176). Many lineages within tribes are well differentiated (76, 130), but some diverse genera possess numerous species which are also of extremely recent origin (173, 175).

Historical Lake Level Fluctuations and the Age of Cichlid Faunas

The time during which a lake has been able to support teleost faunas depends on lake characteristics, particularly topography, maximum depth, and number of basins. Evidence of desiccation and geological structures associated with shallow water sedimentation provide evidence for the boundaries of such events. However, sedimentary profiles are complex and rates of sedimentation are variable in time and space (59). Lake Tanganyika is now composed of three discrete basins separated by shallow sills; lake level declines in excess of 550 m serve to isolate these regions, and desiccation occurs with declines in excess of 1000 m (173). Approximately

1.1–0.67 mya the northern basin itself may have contained multiple lakes (17). Though originally separated, the three basins of Lake Tanganyika have remained continuously connected as a single lake for >21 ky. By contrast, Lake Malawi is composed of one basin that could not have been fragmented into isolated sub-basins. With a maximum depth of 800 m in the northern rift, Malawi has probably never been subject to complete desiccation, but given shoaling topography in the southern regions, lake level decreases of ~100 m would result in desiccation and could eliminate many narrow endemics. Similarly, the form of Victoria is a single, shallow basin (maximum depth of ~80 m). Such topography makes much of the fauna vulnerable to the effects of drying events.

New data suggest that earlier estimates of recent lake level fluctuations (157) may have been inflated. A review of data for lake level declines suggests that Lake Tanganyika was depressed 600 m approximately 200–75 kya and 300 m approximately 21–13 kya (58). Depressions of ~250 m for 190–170 kya and ~160 m for 40–35 kya have also been estimated (17). In Lake Malawi, lake level was 200–300 m lower than present 40–28 kya, and 100–150 m lower than present 10–6 kya (29). Desiccation of Lake Victoria has been estimated to have occurred 12 kya (60). Approximations of the timings of past lake levels depend upon rates of sedimentation. However, extrapolation of sedimentary rates is subject to considerable uncertainty, especially because the rates are based on relatively few radiocarbon dates that are themselves subject to considerable error (T Johnson, personal communication).

Numerous lake level fluctuations have been documented for all studied African lakes, but there is no consensus regarding continental coordination. Lake Tanganyika and Lake Victoria appear to vary in concert with glaciation events (13, 37), while Lake Malawi appears out of phase with these more northern lakes (29, 58). In consequence, it is difficult to establish a single geohydrological mechanism driving these potentially biologically significant events.

The hypothesis of complete desiccation of Lake Victoria (60) has had a singular impact on models of rapid cichlid diversification, particularly if, as claimed, refugia for endemics were absent during such events. In the absence of refugia, virtually all of the current diversity in the lake would have been generated since the most recent desiccation, 12 kya. However, paleolake Natron/Magadi in an adjacent basin experienced a significant increase in water level (60 m) at almost precisely the same time as Lake Victoria dried (158). Further, presumed endemicity of non-cichlid fishes and aquatic invertebrates challenges the idea that refugia were absent in this region (33). For example, at least 11 species of gastropods are endemic as well as two catfishes with deep water adaptations. Thus, the rates of diversification of the Victorian cichlid fauna are unclear. A similar situation exists for Lake Malawi where sediment analysis, hydrological modeling, and oral tradition suggest that the southern portion of the lake was dry approximately 200 ya (134). Because of the large number of endemic cichlids in this region (145), speciation rates would have to have been extremely rapid to generate the current diversity. However, a synthetic review of available information suggests only minor lake level fluctuations in the last millennium (128). In general, resolution of the timing involved in abiotic processes will permit a fuller understanding of the biological impact of hydrological events.

BASIC BIOLOGY

Given the extremely large number of cichlid species, there is immense variation in virtually all aspects of their ecology, behavior, and reproductive biology (34). Comprehensive field studies have focused on pelagic and sand communities of Lakes Victoria (39) and Tanganyika (66), and particularly on rocky littoral faunas of Lakes Malawi (145), Tanganyika (149), and Victoria (160).

Ecological Diversity

Taxa in all of the African great lakes provide extensive examples of adaptive radiation, spanning trophic guilds that include benthic algae grazers, sand-dwelling planktivores, pelagic piscivores, ambush predators, and specialized pursuit predators (for summaries, see 24, 34, 44, 56). This diversification may have been facilitated by a key innovation associated with pharyngeal jaws (56a). The extent of trophic diversification (and associated behavioral adaptations) is similar among lakes, but reaches its most extreme in Lake Tanganyika where cichlids are represented by multiple tribes with long and distinct evolutionary histories (129, 173). In each lake, faunas may be divided into several basic ecological guilds on the basis of habitat preference and, in several cases, also by genealogy (125). One group typically inhabits shallow waters surrounding rocky outcrops, a second occurs over sandy sediments at a variety of depths, and a third is principally pelagic. In all lakes, additional specialization occurs within habitats, exemplified by species inhabiting swamps or abandoned gastropod shells (154). In the large lakes, some taxa change habitats during the reproductive season. In the deep rift lakes of Tanganyika and Malawi, cichlids are absent from anoxic waters at depths >100–200 m.

In several regions within the African great lakes, habitats are discontinuous: rocky islands may be isolated by surrounding sandy substrate, and many coastal areas are characterized by alternating patches of rocks and sand on scales of 1–10 s of km. In other regions, habitats are continuous for 10–100 s of km.

Within these habitat-defined groups, there may be gross similarities in trophic adaptations among taxa, but virtually all sympatrically occurring species examined to date differ in obvious or subtle aspects of resource exploitation (10, 142, 149). For example, differential exploitation of invertebrates and associated microfloras in sandy habitats may be realized either via diversity of oral dentitions or via specialized grazing behaviors (24); mutualistic interactions are also common cichlid resource utilization strategies (65). Though trophic differentiation may be inferred through morphology and behavior, in situ observations have revealed food-switching with fluxes in prey abundance (115); this phenomenon may be particularly apparent during plankton blooms (11), suggesting that species-specific strategies and adaptations may be of greatest significance in the relatively leanest times. However, in a well-studied community in Victoria, niche differentiation was not obvious among some taxa (11). Comprehensive understanding of the importance of trophic specialization to generation and maintenance of species diversity may thus mandate long-term observation.

One ecological guild, the rockfishes, is of particular interest to speciation theorists; they include the mbuna in Lake Malawi, the mbipi in Lake Victoria, and multiple lineages in Lake Tanganyika. The mbuna, one of the most extensively studied groups (81, 145), is a useful example; species of *Tropheus* and other genera, such as *Simochromis* and *Petrochromis*, form an analogous group in Lake Tanganyika (82, 174). The mbuna constitutes a diverse monophyletic group of >300 species in 10 genera that feed and breed primarily on algae-covered rocks and cobbles in relatively shallow (<20 m) water. Some species are extremely abundant with $>10^6$ individuals along continuous habitats (145), while other taxa are rare. Most mbuna have extremely restricted dispersal abilities (50, 115) and are generally absent from areas of sand, sediment, and deep water, which act as barriers to dispersal (103). Preliminary genetic study suggests that within rocky outcrops, male dispersal may be greater than that of females (74), but both fine-scale and long-term analyses have yet to be conducted. In a similar manner, rockfish in Lake Tanganyika show extremely weak dispersal (173). In all of these faunas, many common species are widely distributed, but typically have discontinuous distributions.

Reproduction

Modes of reproduction in African lake cichlids are extremely varied (34, 66, 67). Because of intense predation, diverse strategies for protection of eggs and fry have evolved. Oral incubation has independently arisen many times in cichlids (40) and is practiced by all Lake Malawi and Lake Victoria endemics; in Lake Tanganyika, several lineages also utilize this form of parental care. Microsatellite-based paternity analysis for a number of mbuna species has demonstrated multiple paternity with up to six males fertilizing a single clutch (68, 137); polygyny has also been observed in sand-dwelling Lake Malawi cichlids (113). The evolutionary significance of multiple mating by females is problematic (193a).

In Malawi, males of *Metriaclima(=Pseudotropheus) zebra* (167) and related mbuna defend small ($\sim$1-m^2), permanent territories primarily against conspecific males. These areas serve as foci for breeding and male foraging. Territories are densely packed and often overlap those of other mbuna. In *M. zebra*, appropriate rocky substrate for such territories is limited; removal experiments demonstrate rapid species-specific colonization by non-territorial males (48; I Kornfield, unpublished data). In Lake Tanganyika, removal experiments for several species show similar results (36, 55, 78). Because of this space limitation, only a fraction of males in a population can hold territories and thus potentially breed. Although sneaker males may occur in some taxa (149), none have been observed in *M. zebra*. Further, males may maintain territories for several (>2.5) years (48; I Kornfield, unpublished data). Thus, for some taxa, space per se is a major reproductive resource (194). In these and related species, territory acquisition and defense may be the most critical factors influencing male reproductive success.

By contrast, females in *M. zebra* are non-territorial, although they are markedly philopatric (50). Females forage on algae-covered substrates across the territories of many species. When reproductively active females swim over areas of densely packed territories of conspecific males, they are solicited by territorial males using behavioral displays such as side presentation and leading (108). Comparative behavioral analyses and extensive in situ observation of the mbuna have not demonstrated species-specific variation in these behaviors (108). Spawning occurs when a female follows a male down to his territory, where he continues to display; the female deposits one or several eggs that she immediately puts in her mouth. In *M. zebra*, fertilization may be achieved while the eggs are still on the substrate or after they are in the mouth of the female. Females leave the male territory following the reproductive bout, although they may return for subsequent matings. Clutch size in rockfish from all lakes is small, ranging from ~10 to 50 eggs; with determinate growth, buccal cavity size may limit reproductive output (132). Following oral incubation for several weeks, females deposit free-swimming fry in rocky crevices (182). There is thus no opportunity for behavioral imprinting in these taxa. However, in other mouth brooders, females guard schools of foraging fry for extended periods and protect them from predation by gathering them back into the mouth (81). In these latter cases, post-hatching interactions between parent and offspring do represent an opportunity for cultural (i.e. extra-genetic) transmission of behavior to succeeding generations (153), although no such occurrence has been documented to date.

The reproductive behavior of many species of sand-dwelling fishes is similar to that of *M. zebra* (113, 149). In species that spawn on sandy substrates, males commonly build and defend nests or "bowers" (187, 193) in dense aggregations. Male-male interactions on such leks are extensive, and sneaking is a common strategy (67). Females mate with multiple males that, as for male mbuna, provide no resources other than fertilization.

In some of the substrate-guarding endemic Tanganyikan lamprologines, familial communities are established where young act as helpers at the nest (36, 154). Because of space limitations, some sub-adults cannot find a place to breed and help their parents instead. In these taxa, many of which are shell-brooders, mating systems are extremely polygynous, nest-brooding females adopt gregarious behaviors, and opportunities for social interactions are extensive. There may be great advantage in numbers to avoid predation.

In summary, there is a premium on available substrate for reproduction in all these systems. This is nearly universal for males, but extends to females in the case of substrate guarders. The intense predation on early life-history stages in these fishes has promoted the development of diverse protection strategies. In turn, these strategies have influenced the development of community-wide behaviors.

Sensory Modalities

In general, cichlids perceive chemical stimuli and differentiate species or reproductive condition on the basis of olfaction/taste (127). While little is known about specificity of odor production among closely related species, the dynamics of odor transmission in aquatic environments (28) suggests that this modality would be expected to be of limited value in rapid interactions among conspecifics. Similarly, acoustical communication has been demonstrated in a variety of cichlid species (127). Extensive ambient sound is associated with feeding activities and with some other behaviors. In situ measurement of sound production in Lake Malawi demonstrated the potential for conspecific recognition using auditory cues; significantly different pulse rates within calls were associated with courtship in two sand-dwelling species (99). Whether this sensory modality has played a role in evolutionary differentiation is unclear.

Males of most rockfishes in Lakes Malawi (81, 45), Tanganyika (79, 82), and Victoria (160) have dramatic breeding coloration in which there is tremendous variation. Female coloration is generally so drab that specific identification can be taxonomically challenging (e.g. 190). Variation in coloration is extensive within species complexes; the colors of males in Malawi mbuna have been analyzed qualitatively (21, 110), and certain hues are more common than others. In *M. zebra* and other taxa, quantitative reflective scanning demonstrated intense coloration ranging from UV through long wavelengths (84). Analysis of visual systems in a number of African species has demonstrated multiple absorption maxima across the visible spectrum (27, 188). Vision in the UV portion of the spectrum has been convincingly demonstrated for mbuna species based on microspectrophotometry, analysis of opsin genes (15), and electroretinograms of retinal bipolar cells (9). Thus UV colors, unperceived by human observers, could play diverse roles in cichlids, as they may in other fishes (100).

While the bulk of intraspecific phenotypic differentiation of males occurs among allopatric populations (79, 145), extensive variation may be present within populations. Variation in male coloration within populations has been documented in Lakes Malawi (81), Tanganyika (82), and Victoria (161). For example, in the *M. zebra* species complex, males in many populations exhibit continuous variation in colors and color patterns (Figure 1; see color insert). The relevance of color variation to taxonomic differentiation remains a central question in the evolution of the East African cichlid species flocks.

SPECIES CONCEPTS, DIVERSITY, AND TAXONOMY

Discussions about definitions of species have dominated the cichlid literature for many years. Although most research before 1980 assumed the Biological Species Concept (BSC) of Mayr, subsequent studies considered alternative definitions. While the BSC or "isolation concept" has been called the most useful species concept for the study of speciation, alternative approaches have been explicitly championed by some cichlid researchers. The recognition concept of Patterson (139), which defines species as groups that share common fertilization systems, and the cohesion concept (181) have been emphasized (46, 143, 144, 173, 185).

Understanding the speciation process may depend upon the species concept employed (181). For this reason, Ribbink (144), Sturmbauer (173), and Turner (185) have chosen non-relational species concepts in the study of cichlid evolution. However, the species concept debate can also be viewed as semantic. Indeed, regardless of the species concept used, the processes (genetic drift, natural selection, sexual selection, etc) involved in the early divergence of populations are the same, whether they are seen as divisive or cohesive. Confusing issues surrounding the use of the BSC may result from what has been observed as a tendency of biologists to create caricatures of one species concept in order to clarify the reasoning of another (47). Here we adhere to the reasoning of the BSC, but do not reject the incorporation of alternate views.

Species Definitions

How are species recognized? Because of their phenotypic similarity and patchy distribution, distinguishing species is problematic (173). When sympatric, biological species have been demonstrated by in situ observations of assortative mating by color morphs (e.g. 54, 166; P Reinthal, unpublished information; JA Stauffer, personal communication). More typically, species have been established (or their status confirmed) by studies that report significant differences in frequencies of molecular markers among sympatric morphs (83, 111, 116, 190) or by laboratory examination of mate discrimination (72, 73, 165).

In allopatry, differences in male breeding coloration have been uniformly employed to recognize species, while qualitative differences in dentition and osteology have been important in defining genera and species groups (96). It is nearly universally presumed that the existence of unique male breeding coloration is sufficient to delimit species (24, 45, 81, 96, 145, 185,). Though a few taxa have wide distributions, most species defined on the basis of color are "narrow endemics," restricted to specific islands or rocky outcrops (81, 145). Note that under an ecological species concept, closely related allopatric populations differing only in color may be the same species if they all occupy the same ecological niche. Indeed, many geographic variants, strikingly different in coloration, possess indistinguishable trophic adaptations and foraging ecologies (174; I Kornfield, unpublished information). Such geographical color variants may be of central importance to reproductive isolation and generation of diversity. Comparative studies of sympatric and allopatric populations in different stages of divergence may provide the most profound insights into geographic modes of speciation (30).

Taxonomic Diversity

Most species of African cichlids have yet to be formally described. The task of cataloging the diversity of the East African great lakes is particularly difficult both because of the large number of species (38) and because of the obscurity of species boundaries (173). In the case of allopatric populations showing geographic color variation, reproductive isolation may never be tested in nature. While allopatric populations are not necessarily biological species, the magnitude of novel coloration may merit taxonomic recognition. For example, newly discovered cichlids are given descriptive names in order to distinguish them from all other populations [e.g. "*Pseudotropheus* species tropheops mbenji blue" (see 81, p 53)]. With extremely few exceptions, allopatric variants have not been assigned subspecific epithets, although several such populations have recently been given specific status (167). However, because of the ubiquity of geographic variation, considerable care needs to be exercised in defining taxa and interpreting taxonomic definitions. In particular, we emphasize the need to examine adequate sample sizes of organisms and to make in situ observations of coloration; in the absence of such practices, artificial species may be created (Figure 1; cf 167). Additionally, it is important to recognize the subtle yet critical difference between population differentiation and speciation (102). The inability to distinguish between these two biological processes may confound evolutionary studies.

CENTRALITY OF PHYLOGENETIC RECONSTRUCTION

Differentiating alternative explanations for patterns of divergence and the processes that drive them requires robust hypotheses of relationships among species. Organismal phylogenies are essential for adjudication of the roles that special processes or characters may play in the speciation process. In the Cichlidae, there is extensive potential for convergence or parallelism in trophic or morphological domains (94). Among haplochromines, the most speciose lineage of African cichlids, morphological similarity and ecological equivalence are obvious among the three major lakes (34, 170). Within lakes, the magnitude of similarity is striking as well (24, 44). If convergence is real, the use of standard ichthyological characters for the construction of phylogenies, particularly those associated with dentition, can be misleading.

The study of lepidological variables (those derived from scale morphology and squamation pattern), pioneered by Lippitsch (97), has proved exceedingly valuable in establishing interrelationships among some groups of African cichlids. They have been used to demonstrate monophyly in the well-defined tribes in Lake

Tanganyika and to erect a phylogeny among some of these lineages (98); they also appear useful for defining lineages among some groups of more closely related haplochromines (160). While general application of this approach to some haplochromine taxa has been problematic (90), it shows promise for well-defined groups.

Given the potential for convergence, it is clear that phylogenies generated from characters independent of morphology are desirable. More specifically, findings from molecular studies are central to resolving debates involving such attributes as behavior and morphology. Molecular studies have clarified relationships for the Cichlidae in general (26) and among African great lake cichlids, and have begun to provide insights among taxa within lakes. The seminal paper by Meyer and coworkers (120) using mtDNA demonstrated monophyly of the faunas of Lakes Malawi and Victoria and established relationships of the cichlid faunas among lakes. The often cited convergence among the mbuna of Lake Malawi and the rockfishes of Lake Tanganyika was unambiguously confirmed by Kocher and colleagues (75) with comparative mtDNA sequence analysis; subsequent analyses established sister-group relationships between faunas of Lakes Malawi and Tanganyika (118). Relationships among tribes within Tanganyika and confirmation of multiple invasions of portions of this fauna were initially established with allozymes (129, 130) and have been extended and clarified using mtDNA data (76, 175, 176). SINES (Short Interspersed Nuclear Elements) have been exploited to support monophyly of several Tanganyikan tribes (179). The unique insertion events associated with SINES are compelling, but the sensitivity of this assay among very closely related cichlids may be limited (179).

Elucidating the relationships among more recently diverged taxa using molecular techniques has been problematic. While mtDNA proved useful to understand divergence within several lineages in Lake Tanganyika (174), it discriminates only major, divergent clades in the Lake Malawi fauna (125). The reason that mtDNA failed at lower levels is that many taxa retain ancestral polymorphisms (123, 124); comprehensive analysis of mtDNA revealed extensive retained polymorphism throughout the Lake Malawi fauna (138). The absence of lineage sorting of mtDNA compromises interpretation of molecular data in a phylogenetic context (12, 141) in which gene trees may masquerade as species trees (6). In Lake Victoria and surrounding waters, neutral nuclear polymorphisms have persisted in cichlids for extended periods of time (126). Extreme caution needs to be exercised in exploiting sequence data (70, 133, 178) for phylogenetic analyses in cichlid fishes, particularly where samples sizes are limited. Note, however, that molecular polymorphisms may be useful tools for examining other aspects of the evolutionary process (16, 71) such as effective population size.

Multilocus comparisons have demonstrated the utility of microsatellites to distinguish groups of closely related taxa (85). This class of markers appears to be sufficiently informative to erect statistically compelling phylogenies among some closely related species (PF Smith & I Kornfield, in preparation). In like manner, large numbers of AFLPs (Amplified Fragment Length Polymorphisms) have been used to examine phylogenetic relationships among several well-defined groups of mbuna (1). In distance analyses, significant phylogenetic affinities among groups of congeners were demonstrated, but statistical support for intergeneric relationships was more limited (1).

In general, the difficulty in establishing phylogenies reflects the short intervals of time that must have transpired during generic-level divergence. If diversification is concentrated in short periods associated with major hydrological events such as lake expansions, speciation may effectively occur over wide geographic areas. In such cases of near simultaneity, resultant star phylogenies may reflect hard polytomies (52) rather than an inability to resolve underlying signals. Distinguishing such bursts of speciation from soft polytomies may be possible if there is resolution above and below the event in question (95). Because of the desirability of a phylogenetic framework for evaluating hypotheses of origination and their associated mechanisms (14), additional research should be vigorously pursued.

GEOGRAPHICAL MODELS OF SPECIATION IN CICHLID FISH

Monophyly of Lake Malawi and of Lake Victoria cichlids suggests intralacustrine speciation, but strictly speaking, this is different from sympatric speciation. Several authors have developed scenarios whereby intralacustrine speciation occurs in an allopatric mode. Macro-allopatric speciation results from differentiation between separate basins, and micro-allopatric speciation is caused by differentiation between patches of fragmented habitat or via isolation by distance (121, 148). Sympatric speciation is used in reference to divergence in the absence of geographical barriers to gene flow (O Seehausen & JJA van Alphen, submitted for publication; 162). Here we consider speciation from allopatric and sympatric perspectives.

Allopatry

One general model for allopatric speciation, the "Nabugabo scenario," was articulated by Greenwood (42) for Lake Victoria. In this widely cited scheme, very minor fluctuations in lake level would serve to isolate fish in small, peripheral lakes over short periods of time (<4 ky); such lakes could generate endemics that might retain specific identity upon reunion with Victoria and thus contribute to cumulative diversity. Difficulties with the evaluation of this model based on the present condition of the lake (64) have included the inability to confirm that extant "peripheral endemics" are truly absent from Lake Victoria and the observation that some "peripheral endemics" have been found in multiple satellite lakes (31, 131). There has also been faunal interchange among other lakes in the region (63). Nevertheless, given that the extensive and exceedingly complex shoreline of Lake Victoria has not been systematically surveyed (63), the idea that novel assemblages of endemics might have persisted and reinvaded Lake Victoria remains appealing.

Fryer (32) presented a hypothesis of intralacustrine speciation that does not require fragmentation of a lake basin into separate bodies of water. In his scenario for Lake Malawi, allopatry is achieved by highly philopatric cichlids occupying isolated areas of habitat, allowing genetic divergence to take place over distance (see 83, 103). In Lake Tanganyika, secondary contact between divergent populations of *Tropheus* inferred from biogeography illustrate micro-allopatric divergence (34). Subsequent molecular studies of several Tanganyikan rockfish lineages (177) have revealed patterns ranging from localized differentiation to continuity. The extent of population structure is important to all allopatric models of cichlid speciation, regardless of the proposed mechanism, because greater population subdivision implies a greater number of discrete entities subject to genetic drift and selection. While divergence in allopatry (micro-allopatry) may provide some insight into the early stages of intralacustrine speciation, there is a great deal still unknown about why divergence of sexual characteristics in allopatry would be so rapid and how newly divergent populations could survive presumed repeated bouts of secondary contact to become new species.

Evolutionary Implications of Lake Level Fluctuations

Changes in habitat are associated with lake level fluctuations; these events could be central to understanding broader patterns of radiation and, perhaps, mechanisms controlling diversity and rates of speciation. Repetitive changes in lake level might continually create new opportunities for isolation in allopatry. Such potential effects were explicitly noted in early studies (32, 42) and have been considered by many others (112, 173). For example, two hypotheses have been presented to explain the presence of many species endemic to small islands in the southern shallow areas of Lake Malawi. First, it may be that these taxa predate the most recent rise in lake level; in this scenario, current island endemics arose through colonization of the newly flooded islands by species followed by their subsequent extinction in their former ranges. Alternately, when the southern portion of the lake was flooded, the newly formed islands were initially colonized at random by individuals of various mbuna species that are still extant elsewhere in the lake; these founding individuals subsequently diverged to produce new taxa. Repeated lake-level fluctuation events could be viewed as facilitating taxon-cycles (63) and, depending upon biogeography and lake basin topography, could act as "species pumps" (148) to increase endemic diversity.

An appreciation of the details of this process offers opportunities to understand the roles that natural selection and genetic interchanges may play in the overall process (Figure 2). In the most general model, discrete basins and their surrounding habitats would act as refugia during lake regressions. Extinction would be elevated during these periods; taxa that are resident in such areas would not be tested by intense competition to the same extent as species that were forced to move into them. Previously allopatric taxa that were trophically distinct, or those that exploited some novel aspect of reproduction, would be candidates for persistence.

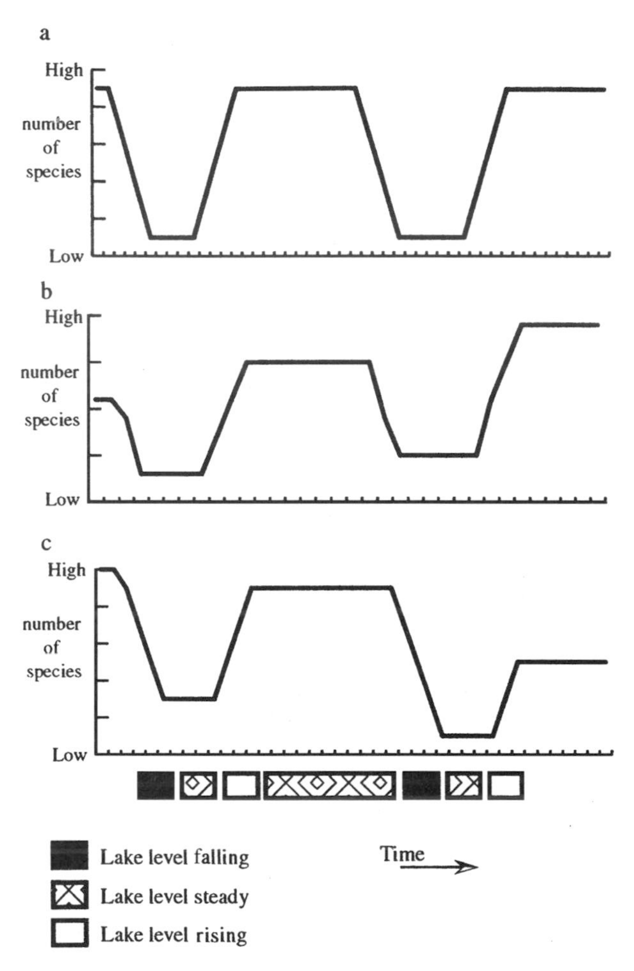

Figure 2 The effects of repeated lake fluctuations in cichlid species diversity. Extinction rates are high during periods of lake level drops (*black*), while speciation rates are high during lake level rises (*white*); when lake level is constant (*hatched*), speciation and extinction are in equilibrium. (*a*) Net number of species remains constant; (*b*) number of species increases with fluctuation cycles, the species pump (63, 148); (*c*) number of species decreases. See Sturmbauer (173) for alternative figure and discussion. See text for details.

During such ecological crunches, significant trophic differentiation might occur in some lineages, though overall species diversity should decrease. Reproductively isolated ecological cognates would be particularly susceptible to extinction. Narrow endemics that were uncommon or rare (145) would be expected to have high rates of extinction relative to taxa with large population sizes. In aggregate, diversity in refugia would increase as new taxa were added, but absolute space limitations for trophic and reproductive activities would suggest episodes of increased competition in such shrinking environments. Depending upon the extent of crowding, opportunities for hybridization may be extensive (see Mechanisms Driving Differentiation below). With lake level increases, new habitats would become available; during this phase, opportunities for speciation might increase due to rapid population growth and reduced competition. During periods of relatively stable lake levels, speciation and extinction might be at equilibrium or, depending upon the mechanisms driving speciation (below), there might be an increase in species diversity. Overall, species diversity might remain constant at extended high water stands if extinction during regressions balanced speciation during expansion (Figure 2*a*). However, net species diversity might expand with increases in lake level cycles if new trophically differentiated lineages regularly arose (Figure 2*b*). With a large number of cycles, if selection during low stands were sufficiently intense, net species diversity might decrease and eventually fluctuate around some equilibrium (Figure 2*c*). Lake level cycles over long periods of evolutionary time contributed to the differentiation of major lineages in Lake Tanganyika (173), but it is not clear whether such cycles continue to contribute to species diversity.

Sympatry

The origin of tilapias in the Cameroon lakes (155) has been extensively acknowledged as demonstrating sympatric speciation (e.g. 136), and has been a model for this mode of divergence in cichlids. Analysis of mtDNA sequences of eleven taxa endemic to the Barombi-Mbo crater lake suggested that these species formed a monophyletic group to the exclusion of geographically proximate riverine taxa, presumably the ancestors of this small species flock. Given the characteristics of the species involved, the study suggested that speciation in this endemic fauna was based on trophic differentiation and not sexual selection (155). In a wide diversity of fishes, radiation via trophic differentiation is a common pattern in recently founded populations (89). Extremely rapid specialization into planktonic and benthic feeding modes within multiple independent groups suggests the generality of this pattern (156).

Given its centrality, further consideration of the Barombi-Mbo radiation is warranted. Analyses of portions of the molecular data were ambiguous: species trees and gene trees may have been conflated, and sample size limitations may have influenced resolution of these data (158). Potential ancestors from the Niger River, an area that is periodically connected with Cameroon drainages (147), were not examined. As in most phylogenetic investigations, assessment of monophyly depends entirely on the taxa included in the analysis. mtDNA restriction assays of Barombi-Mbo cichlids supported the idea that at least two lineages of tilapias were present in this fauna (S Seyoum & I Kornfield, unpublished information). Regardless, despite the possibility that the species flock of Barombi-Mbo may have had more than one progenitor, sympatric speciation appears to have been the mode of origination of much of the current diversity.

A second example of rapid speciation of tilapias has emerged from careful study of the *Alcolapia* (*Oreochromis*) flock of Lakes Natron and Magadi in East Africa (158, 159). These two small soda lakes support four endemic taxa, three of them syntopic. Like the endemics of Barombi-Mbo, these cichlids are distinguished by morphological features associated with ecological specialization. While speciation in this system may have been sympatric, the complex hydrological history of the area suggests the possibility of repeated lake fragmentation and reunion; analysis of mtDNA sequences is equivocal (158).

Application of Sympatric Speciation Scenarios to the East African Great Lakes

The extreme youth of the species flocks of the African Great Lakes suggests that some mode of divergence different from the traditionally accepted models of allopatry has operated. It has been suggested that sympatric speciation may happen more rapidly than allopatric speciation (107). The inferences of sympatric speciation in the faunas of the Cameroons and in the soda lakes of Natron and Magadi have encouraged workers to consider sympatry as the dominant geographical mode of speciation in the Great Lakes of East Africa (35). Rapid sympatric divergence due to ecological adaptations could apply to initial diversification of lineages in the African great lakes (148). Under such scenarios, the deepest levels of divergence in any radiation would be based on marked trophic differences. Indeed, conventional (24) and molecular (173) phylogenies of cichlids from both Lakes Malawi and Tanganyika are consistent with this idea: major lineages are trophically divergent.

However, models of contemporary sympatric speciation in the African Great Lakes differ from other hypotheses of sympatric speciation. Because there is little evidence that ecological differences are associated with the most closely related species of African haplochromines, sympatric differentiation is presumed to occur by mechanisms not involving rapid ecological divergence, differences in habitat preference or performance mediated by disruptive natural selection. Hypotheses of sympatric speciation in African cichlids facilitated by runaway sexual selection are discussed below.

MECHANISMS DRIVING DIFFERENTIATION

Our review of geographical models of speciation in cichlid fishes (above) suggests the validity of various mechanisms by which reductions in gene flow may be realized and wherein evolutionary trajectories may diverge and give rise to new species. We now focus on the forces responsible for the divergence of hundreds of ecologically similar species in Lakes Malawi, Victoria, and Tanganyika. Many of the factors responsible for the great diversity of cichlid species in the African lakes remain unknown. While genetic drift and natural selection are thought to be the primary factors responsible for evolutionary divergence in general, the extent and rapidity of the African radiations have evaded conventional logic in many cases. The most important elements of any divergence scenario are those that lead to testable hypotheses. Table 1 summarizes several theoretical frameworks within which cichlid speciation has been studied and emphasizes testable predictions of each model. In this section, we review theoretical and empirical research on speciation by sexual selection following treatment of more conventional evolutionary forces.

How Important Is Local Adaptation to Species Diversity?

Some New World cichlids exhibit trophic polymorphisms (87, 152) or are phenotypically plastic (84, 117). In the Old World, however, aside from the mollusc crusher *Astatotilapia alluadi* (43), there are few examples of ecologically significant plasticity. Indeed, there is little evidence to date that these factors contribute to apparent species diversity in the African faunas.

Although cichlids are known for a vast diversity of trophic morphologies, it is generally considered unlikely that differentiation in feeding specializations leads to speciation (185). Even in the absence of the necessary phylogenetic information, it is clear that many sibling species differ only in coloration (145, 185), while significant trophic morphologies differ at the level of genera (34, 81). Unlike the Trinidadian guppy system studied by Endler, which demonstrated natural selection in the form of local adaptation to predators, many East African cichlids have few natural predators once they are mature. In Lake Tanganyika, predation influences reproductive behaviors in some sand-dwelling fishes (61). Geographical variation in predation pressures has not been documented in Lakes Victoria and Malawi. In general, local natural selection pressures on coloration and behavior will likely be subtle. Two studies that have attempted to correlate coloration with ecological and biological variables found no (21) or weak (110) associations. Interspecific competition (48) and changes in the biotic community may exert strong pressures in an episodic manner, but their presence and significance have yet to be documented. Knowledge of local variation in selective regimes would benefit from additional consideration.

Is Genetic Drift Critical to Diversification?

Large size is thought to buffer a population against genetic and phenotypic changes, making small population size requisite for rapid divergence by drift (105). The postulated effects of random genetic drift following colonization events have been central to several models of cichlid speciation (22, 173). Tests of this concept are beginning to appear with the recent advances in molecular genetics; objective protocols for evaluating the magnitude of genetic bottlenecks are available (101). Studies of mbuna population structure in presumably recently submerged habitat in southern portions of Lake Malawi have been informative. The very low level of mtDNA haplotype variation exhibited *by Melanochromis auratus* (12) is consistent with a historical bottleneck, although current population size is quite large.

TABLE 1 Models of cichlid speciation

Dominant geographical model	Mechanisms driving speciation	Model system	Testable assumptions	Author (citations)
Allopatry	None specified[a]	Malawi *mbuna*	Limited gene flow between populations	Fryer (32)
Allopatry	None specified[a]	Victorian haplochromines	Biogeography and phylogenetic relationships of extant taxa	Greenwood (46)
Allopatry	Sexual selection Founder events	Malawi *mbuna*	Bottlenecks Mate choice on male coloration[b]	Dominey (22)
Allopatry	None specified[a]	Tanganyikan haplochromines	Limited gene flow between populations	Rossiter (148)
Allopatry	Runaway sexual selection on bower size	Malawi sand dwellers	Mate choice on bower size/shape[b]	McKaye (113)
Sympatry	Runaway sexual selection on male coloration	Victorian haplochromines	Mate choice on male coloration[b] Genetics of color/preference	Seehausen & van Alphen (162, 164)
Sympatry	Runaway sexual selection	Malawi *mbuna*	Mate choice on male coloration Genetics of color/preference	Turner & Borrows (186)

[a]None specified. Mechanisms driving speciation are not emphasized. Classical allopatric divergence sensu Mayr (105) is implied.
[b]Proving mate choice is the minimal first step to inferring speciation by sexual selection. See text for additional consideration.

In contrast, an mtDNA study of Malawi zebras showed very little reduction in diversity when narrow endemics were compared with cosmopolitan species (124, see also 19a). In such cases, decreases in allelic diversity could also be artifacts of the non-uniform sampling of haplotypes due to small samples sizes and highly variable markers (151). Initial surveys of nine Malawi taxa using microsatellite loci showed high levels of allelic diversity (190).

In two other microsatellite-based studies in Lake Malawi, significant population structure was revealed over very small spatial scales, suggesting reduction of gene flow under a specific colonization scenarios (3, 103). These studies correlated loss of observed genetic diversity [either effective number of alleles (3) or heterozygosity (103)] with proxies for isolation [distance from nearest conspecific population (3) or depth of sand-rock interface (103)]. Although genetic diversity was consistently high and no evidence for severe bottlenecks was observed, the data suggested a slight loss of genetic diversity after the presumed colonization events. In each case, however, the significance of a bottleneck effect to speciation is speculative. In Lake Victoria, preliminary characterizations of several species using microsatellite markers detected substantial of genetic variation (192). Extensive mtDNA-based population studies on the fauna of Lake Tanganyika have revealed considerable variation within a number of narrow endemics (119, 173). In summary, there is no general evidence that founder events have contributed to cichlid radiation in the Africa Great Lakes.

What Role Might Hybridization Play in Faunal Evolution?

Hybridization of heterospecific cichlids in nature has been observed to occur as a result of anthropogenic insults or accidental translocations in Lake Malawi (168, 169), and by eutrophication of Lake Victoria (165, 191). These observations and the relative ease of making crosses of divergent lineages in the laboratory (109) indicate that diversity of sympatric forms in nature is maintained by pre-zygotic, rather than post-zygotic, factors. Although distinctiveness is maintained under normal conditions, rapid environmental changes in the past may have contributed to hybridization and the diversity of cichlids observed today. Sudden drops in lake level, for instance, would bring adjacent populations into secondary contact. Color-differentiated ecological cognates might be among the first wave of taxa to hybridize, because of their ecological identity. The results of such hybridizations are unknown and may differ depending on the genetic architecture of underlying reproductive traits. Interspecific hybridizations might be facilitated under conditions of extreme population density, perhaps exacerbated by rapid changes in community composition. Whether hybridization can generally cause speciation in animals (4, 23, 41) is a subject of speculation, as it is in cichlids (19, 173). Certain predictions of the effects of hybridization can be made. In general, levels of genetic variation in hybrids should increase, but distinctive characters such as unique reproductive coloration might be altered if they are polygenic in nature. Alternatively, if each reproductive character is encoded by a separate locus, mosaic phenotypes qualitatively different from parental strains might emerge (5). For example, hybrids between divergent mbuna geneva display novel morphology (109). Hybrids that do not suffer a decrease in fitness should thrive, making speciation by hybridization potentially a major contributor to overall species diversity. However, if hybrids suffer a decrease in fitness, the evolution of mate discrimination behaviors that facilitate pre-zygotic isolation in areas of secondary contact could be a result of reinforcement. Distinctive male colorations might evolve at the time of secondary contact, rather than during isolation. Reinforcement of female preferences has been modeled in *Drosophila* (69) and experimentally demonstrated in the three-spine stickleback (150). The role of reinforcement in cichlid speciation has not been appropriately addressed (21, 144, 185) and in part may have been lost in the semantic arguments of species concepts and adaptations (see 47 for a discussion). However, detection of reinforcement in cichlids might be extremely difficult. Variations in numbers of offspring normally observed in species with extremely low fecundities may make detection of post-zygotic fitness effects impossible.

Is Sexual Selection Driving Cichlid Speciation?

That sexual selection operates in cichlid fishes was suggested very early in the study of the group (88). Although critical empirical support is rare, many investigators have assumed that female mate choice on male coloration is a common phenomenon in cichlids (113, 118, 144, 173, 185). While interspecific discrimination has been extensively demonstrated (see above for molecular research and 72, 73, 164–166), strong evidence for intraspecific mate choice per se is limited.

Proposed Mechanisms of Divergence by Sexual Selection Models of cichlid speciation by sexual selection fall into two major categories: rapid divergence of sexual characters in allopatry (22, 113) and disruptive or divergent sexual selection in sympatry (cichlid references: 162, 164, 166, 186, 189; general references: 51, 80, 92, 93). Both scenarios employ the process of Fisherian runaway sexual selection. The distinction of the Fisherian process from other mate choice scenarios, such as "good genes" or "handicap" processes, is that the female preference for a sexually selected trait increases in frequency as a result of co-inheritance of trait and preference, rather than via direct selection owing to increased fitness of females that choose quality males (2). Assumptions of the runaway process include (*a*) heritable genetic variation in male trait(s), (*b*) heritable genetic variation in female preference, (*c*) increased mating success of males with exaggerated trait, and (*d*) genetic covariance of trait and preference (2). Allopatric models posit that geographical variation in sexual characteristics (e.g. coloration and bower form) is generated by founder effects and maintained by limited dispersal. Sympatric models have had to overcome the theoretical restriction imposed by the effects of recombination on reproductive traits and preferences, and the elimination of intermediate genotypes (57). This has been done by assuming genetic dominance

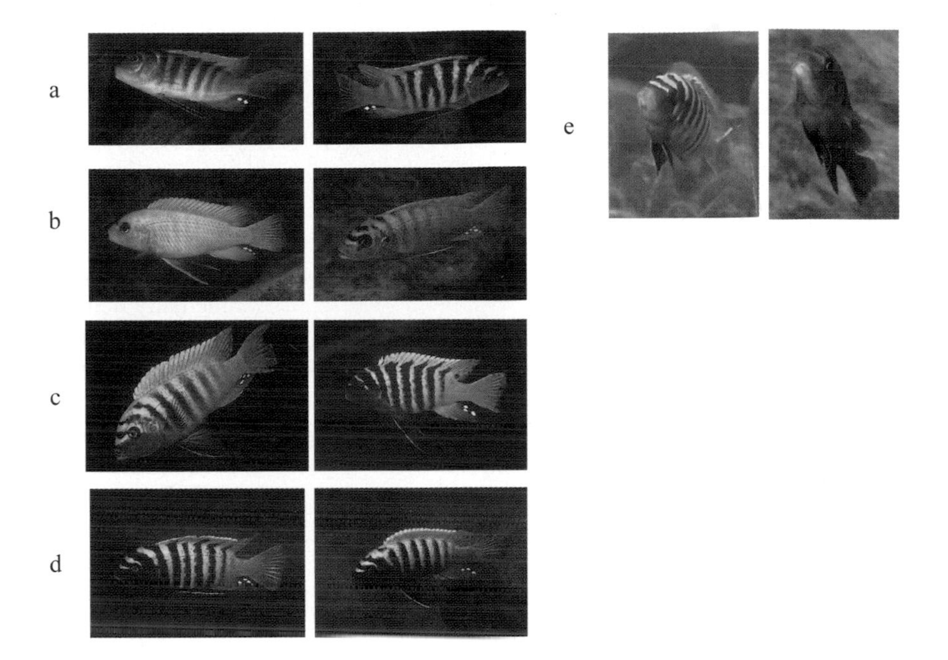

Figure 1 Intrapopulation color diversity within the Lake Malawi *Metriaclima zebra* species complex. Variation in the illustrated phenotypic traits is continuous; presented are examples of extreme phenotypes. (*a*) Pattern asymmetry (flanks of a single "red dorsal" individual—Chimwalani); (*b*) bar density [Mazinzi Reef—*Metriaclima benetos* (167)]; (*c*) dorsal fin bar penetration ("Blue-Black" Meponda Rocks); (*d*) dorsal fin bar penetration ["red dorsal" Nakantenga Island—*Metriaclima pyrsonotos* (167)]; (*e*) throat patch coloration (Eccles Reef—*Metriaclima thapsinogen* (167)]. (Photographs a—d by A Konings.)

in female preference traits (186), variation in habitat fidelity (189), and extreme mating preferences (51).

Appeal of Speciation by Sexual Selection The association of species diversity with exaggerated sexual characteristics is appealing. A recent body of literature has addressed this phenomenon among birds by identifying sexual dimorphism or ornamentation as a correlate of species richness (7, 122, 135, 140). However, this result begs the question of the causal relationship between these two factors. Despite the need for rigorous hypothesis testing and the data that are required for empirical support of sexual selection, the ability to infer mechanisms causing species diversity is limited by the time scales during which divergence events occur. Barraclough and his coworkers (8) suggested several methods for inferring past divergence events using species level phylogenies. The need for such reconstructions among closely related African cichlid species is clear.

Testing Divergence Hypotheses Female cichlids are able to discriminate males of their own species and closely related species by body color (54, 72, 73, 162, 164 166). This ability is important in the maintenance of diversity, especially where closely related species are sympatric, and not ecologically differentiated (165). Documentation of interspecific mate choice, though, does not facilitate assessment of competing hypotheses of divergence by sexual selection. Indeed, current data do not allow rejection of completely allopatric divergence in which sexual selection plays no role. For example, in evaluating evidence of assortative mating of closely related sympatric species (165), a hypothesis of secondary contact is equally parsimonious with that of divergence in situ.

Intraspecific mate choice has been recorded in field observations (62, 113) and field experimentation (49). Although these studies suffer from small sample size and limited replication, they present compelling evidence for the existence of mate choice within species. The male traits under selection in these studies were bower size (113), pelvic fin length and symmetry (62), and number of egg spots on the anal fin (49). In general, studies have not demonstrated heritability of male traits, female preferences, or the covariation of those traits through breeding experiments.

Alternative hypotheses for the evolution of exaggerated sexual dimorphism deserve attention. For instance, intrasexual selection via male contest competition is very likely to operate in cichlids and remains largely unstudied (61). Divergence of male coloration in allopatry may be as likely under intrasexual as it is under intersexual selection, and intrasexual selection does not require a change in female preference traits. For example, observations previously considered supportive of female choice (114) could be explained by the effects of male competition. McKaye's documentation of female mate choice on bower size and position in Lake Malawi cichlids depends on the use of bower morphology as an "extended phenotype" of males. Two later studies of mate choice in Lake Tanganyika (61) and Lake Malawi (86) bower-building cichlids failed to find evidence for choice in relation to bower morphology, and further found that males are often ousted from their bowers by males that defend territories in the absence of a bower (61). These additional data suggest that male competition is important in the acquisition and defense of breeding territory. Size affects competition between males (I Kornfield, unpublished data) but the use of conspicuous male coloration as an intermale signal is also an alternative.

In light of the available models (22, 113, 162, 164, 166, 186, 189) and limited data (49, 62, 113), several scientists have interpreted their experimental findings as de facto consistent with speciation by sexual selection (21, 159a). Such associative orientations are of limited value in evaluating alternative ideas, since no hypothesis-testing framework is formulated. Future studies will need to address alternative hypotheses more directly so that mechanisms of divergence or components of those mechanisms can be systematically supported or rejected. Phylogenetic hypotheses based on biogeographic patterns and competing methods of divergence are one example of such hypothesis testing (141; PF Smith & I Kornfield, in preparation).

EVOLUTIONARY GENERALIZATIONS AND FUTURE STUDIES

In this paper we have reviewed basic information as well as current research in the study of African cichlid evolution. Discussion of information on cichlids and their natural environments provides the basic data necessary to formulate testable hypotheses regarding the divergence of the many hundreds of cichlid species endemic to the lakes of Africa. For the last several decades, cichlid species flocks (25) have served as model systems in which to study the evolutionary process. As more empirical studies become available to critically evaluate the theories of cichlid evolution, resolution of some issues may be achieved. However, comprehensive understanding of the evolutionary history of these faunas and the processes currently at work remains elusive, and there is growing consensus that a single mechanism may be insufficient to explain the origins of these fishes (113, 173). We caution against premature adoption of particular divergence scenarios in the absence of sufficiently strong evidence.

Understanding the Genetic Basis of Relevant Traits

Cichlid diversification should be intensively studied at the genomic level. *What is the genetic basis of male phenotypes and of female choice? How is trophic differentiation controlled*? For example, preliminary breeding studies for some Malawi taxa do not support simplistic models for inheritance of coloration (20, 84; D McElroy, personal communication). The lack of post-zygotic isolation between divergent species lends itself to QTL (quantitative trait loci) analysis of color and behavioral characteristics. The genomic mapping studies pioneered by Kocher and colleagues (1a, 77) will provide a basis for such work.

Understanding Coloration in Cichlids

The role of color in cichlids is considered central to diversification, so knowledge about the physiological basis of color production and of color perception is required. *Are cichlids different than other fishes in their use of colors? What is the significance of UV coloration*? For example, in some cichlid systems, limitations to female color perception may be expected to differ because of habitat characteristics, yet sensory ranges may be great. Comparative differences between physiological abilities and abiotic constraints among taxa within and between lakes may reveal processes molding these attributes.

Studying Behavior

Male coloration varies extensively, but little is known about variation in female preference. *For what male traits is female preference displayed? Are some types of preferences common to diverse lineages within or between lakes*? Quantification of the magnitude and variation in female preference for coloration (as well as other attributes of male phenotypes) is necessary to understand the process of diversification. Information on search strategies is necessary to model and understand the evolution of male traits and female preferences (56b, CRB Boake, in preparation). Importantly, the implications of extensive polyandry and polygyny need be considered within this framework. Such research is demanding; in situ observations are opportunistic, although territoriality facilitates study of male characteristics subsequent to mating. Experimentation in aquaria must be undertaken with caution since mating is infrequent and artifacts may occur; common proxies for female behavior in choice experiments, such as differential time allocation or proximity to particular males, merit verification. Experimentation under seminatural conditions may be necessary, for example, in large ponds with adequate space for territories. In some aquatic systems, female choice could be determined by the use of molecular markers with subsequent assessment of male attributes.

Insights from Population Genetics

Preliminary population studies suggest very limited gene flow among taxa over short distances, but this is contradicted by early studies of migration (112). *How extensive is gene flow among discontinuously distributed populations? Is isolation by distance a universally demonstrable attribute in these systems*? A test for genetic differentiation at several scales (100 m–100 km) in putatively stable populations (e.g. those associated with deep basins) might provide information on the extent of isolation/gene flow. Observations of population differentiation over distance would suggest that the behavioral characteristics of the species studied are insufficient to produce population subdivision in stable populations in the absence of physical barriers to migration.

Criteria for Evaluating Models of Divergence

Construction of accurate phylogenies is central to evaluate the influence of geography on speciation. *Are molecular phylogenies consistent with hypotheses that can exclude allopatric differentiation*? For example, discrimination between recolonization scenarios, allopatric differentiation, and sympatric divergence can be achieved by examining relationships among clusters of closely related species over large distances (e.g. 141). Similarly, with comprehensive geographic coverage, secondary contact can be examined using sensitive molecular frameworks. Species-level phylogenies will also provide a basis on which to infer the evolution of important characteristics among sibling species. Such inferences may point to particular characters that are involved in early divergence.

ACKNOWLEDGMENTS

We thank A Parker, D McElroy, W Glanz, and C Campbell for commenting on the manuscript. A Konings generously shared his underwater photographs and K Carleton kindly provided a preprint of her research. Malawi fieldwork was facilitated by S. Grant. This work was supported by NSF DEB 97-07532.

LITERATURE CITED

1. Albertson RC, Markert JA, Danley PD, Kocher TD. 1999. Phylogeny of a rapidly evolving clade: the cichlid fishes of Lake Malawi, East Africa. *Proc. Natl. Acad. Sci. USA* 96:5107–10
1a. Albertson RC, Kocher TD. 2000. Shape differences in the trophic apparatus of two Lake Malawi cichlid species and their hybrid progeny—a landmark-based morphometric approach. Presentation, Annual Meetings of the Society for Integrative and Comparative Biology, Atlanta, Jan 4–8
2. Andersson M. 1994. *Sexual Selection.* Princeton, NJ: Princeton Univ. Press, 599 pp.
3. Arnegard ME, Markert JA, Danley PD, Stauffer JR, Ambali AJ, et al. 1999. Population structure and colour variation of the cichlid fish *Labeotropheus fuellerborni* Ahl along a recently formed archipelago of rocky habitat patches in southern Lake Malawi. *Proc. R. Soc. London Ser. B* 266:119–30
4. Arnold ML. 1997. *Natural Hybridization and Evolution.* New York: Oxford Univ. Press
5. Arnold ML, Bulger MR, Burke JM, Hempel AL, Williams JJ. 1999. Natural hybridization: how long can you go and still be important? *Ecology* 80:371–81
6. Avise JC. 1994. *Molecular Markers, Natural History and Evolution.* New York: Chapman & Hall. 511 pp.
7. Barraclough T, Harvey P, Nee S. 1995. Sexual selection and taxonomic diversity of passerine birds. *Proc. R. Soc. London Ser. B* 259:211–15
8. Barraclough T, Vogler AP, Harvey PH. 1998. Revealing the factors that promote speciation. *Philos. Trans. R. Soc. London Ser. B* 353:241–49
9. Bilotta J, Cassidy J, McElroy D, Kornfield I, Davis EC. 2000. Spectral sensitivity

of the ERG B- and D-waves of an African cichlid species. Annu. Meet. Assoc. Res. Vision. Ophthalmol. Fort Lauderdale presentation
10. Bootsma HA, Hecky RE, Hesslein RH, Turner GF. 1996. Food partitioning among Lake Malawi nearshore fishes as revealed by stable isotope analyses. *Ecology* 77:1286–90
11. Bouton N, Seehausen O, van Alphen JJM. 1997. Resource partitioning among rock-dwelling haplochromines (Pisces: Cichlidac) from Lake Victoria. *Ecol. Freshw. Fish* 6:225–40
12. Bowers N, Stauffer JR, Kocher TD. 1994. Intra- and interspecific mitochondrial DNA sequence variation within two species of rock dwelling cichlids. *Mol. Phylogenet. Evol.* 3:75–82
13. Broeker WS, Peteet D, Hajdas I, Lin JEC. 1998. Antiphasing between rainfall in Africa's rift valley and North America's Great Basin. *Quat. Res.* 50:12–20
14. Cannatella DC, Hillis DM, Chippindale PT, Weigt L, Rand AS, et al. 1998. Phylogeny of frogs of the *Physalaemus Pustulosus* species group, with an examination of data incongruence. *Syst. Biol.* 47:311–35
15. Carleton KL, Hárosi FI, Kocher TD. 2000. Visual pigments of African cichlid fishes: evidence for ultraviolet vision from microspectrophotometry and DNA sequences. *Vision Res.* 40:879–90
16. Clark AG. 1997. Neutral behavior of shared polymorphism. *Proc. Natl. Acad. Sci. USA* 94:7730–34
17. Cohen AS, Lezzar KE, Tiercelin JJ, Soreghan M. 1997. New paleographic and lake-level reconstructions of Lake Tanganyika: implications for tectonic, climatic and biological evolution in a rift lake. *Basin Res.* 9:107–132
18. Deleted in proof.
19. Crapon de Caprona M-D, Fritzsch B. 1984. Interspecific fertile hybrids of haplochromine Cichlidae (Teleostei) and their possible importance for speciation. *Neth. J. Zool.* 34:503–38
19a. Danley PD, Markert JA, Arnegard ME, Kocher T. 2000. Divergence with gene flow in the rock-dwelling cichlids of Lake Malawi. *Evolution.* In press
20. Danley PD, Kocher T. 1999. The inheritance of sexually selected male characteristics in a rock dwelling cichlid of Lake Malawi. Annu. Meet. Am. Soc. Ichthyol. Herpetol. College Park, PA: Penn State Univ. presentation
21. Deutsch JC. 1997. Colour diversification in Malawi cichlids: evidence for adaptation, reinforcement, or sexual selection? *Biol. J. Linn. Soc.* 62:1–14
22. Dominey W. 1984. Effects of sexual selection and life history on speciation: species flocks in African cichlids and Hawaiian *Drosophila*. See Ref. 25, pp. 231–49
23. Dowling TE, Secor CL. 1997. The role of hybridization and introgression in the diversification of animals. *Annu. Rev. Ecol. Syst.* 28:593–619
24. Eccles DE, Trewavas E. 1989. *Malawian Cichlid Fishes: The Classification of Some Haplochromine Genera.* Herten, Germany: Lake Fish Movies. 334 pp.
25. Echelle AA, Kornfield I, eds. 1984. *Evolution of Fish Species Flocks.* Orono, ME: Univ. Maine Orono Press. 257 pp.
26. Farias IP, Orti G, Sampaio I, Schneider H, Meyel A. 1998. Mitochondrial DNA phylogeny of the family Cichlidae: monophyly and fast molecular evolution of the neotropical assemblage. *J. Mol. Evol.* 48:703–11
27. Fernald RD. 1984. Vision and behavior in African cichlid fish. *Am. Sci.* 72:58–65
28. Fineli CM, Pentcheff ND, Zimmer-Faust RK, Wethey DS. 1999. Odor transport in turbulent flows: constraints on animal navigation. *Limnol. Oceanogr.* 44:1056–71
29. Finney BP, Scholz CA, Johnson TC, Trumbore S, Southon J. 1996. Late Quaternary lake-level changes of Lake

Malawi. In *The Limnology, Climatology, and Paleoclimatology of the East African Lakes*, ed. TC Johnson, EO Odada. pp. 495–508. Toronto: Gordon & Breach

30. Foster SA, Scott R, Cresko WA. 1998. Nested biological variation and speciation. *Philos. Trans. R. Soc. London Ser. B* 353:207–18
31. Fournieret Y, Plisnier PD, Micha JC. 1992. Feeding habits of four species belonging to the genus *Haplochromis* (Teleostei, Cichlidae) in Ihema Lake (Rwanda). *Ann. Limnologie* 28:58–69
32. Fryer G. 1959. The trophic interrelationships and ecology of some littoral communities of Lake Nyasa with especial reference to the fishes, and a discussion of the evolution of a group of rock-frequenting Cichlidae. *Proc. Zool. Soc. London* 132:153–281
33. Fryer G. 1997. Biological implications of a suggested Late Pleistocene desiccation of Lake Victoria. *Hydrobiologia* 354:177–82
34. Fryer G, Iles TD. 1972. *The Cichlid Fishes of the Great Lakes of Africa.* London: Oliver & Boyd. 641 pp.
35. Galis F, Metz J. 1998. Why are there so many cichlid species? *Trends Ecol. Evol.* 13:1–2
36. Gashagaza MM. 1991. Diversity of breeding habits in lamprologine cichlids in Lake Tanganyika. *Physiol. Ecol. Japan* 28:29–65
37. Gasse F, Ledee V, Massault M, Fontes J-C. 1989. Water-level fluctuations of Lake Tanganyika in phase with oceanic changes during the last glaciation and deglaciation. *Nature* 342:57–59
38. Goldschmidt T. 1996. *Darwin's Dreampond.* Cambridge, MA: MIT Press. 274 pp.
39. Goldschmidt T, Witte F, de Visser J. 1990. Ecological segregation in zooplanktivorous haplochromine species (Pisces: Cichlidae). *Oikos* 58:343–55
40. Goodwin NB, Balshine-Earn S, Reynolds JD. 1998. Evolutionary transitions in parental care in cichlid fish. *Proc. R. Soc. London Ser. B* 265:2265–72
41. Grant PR, Grant BR. 1997. Genetics and the origin of bird species. *Proc. Natl. Acad. Sci. USA* 94:7768–75
42. Greenwood PH. 1965. The cichlid fishes of Lake Nabugabo, Uganda. *Bull. Br. Mus. Nat. Hist. Zool* 12:315–57
43. Greenwood PH. 1965. Environmental effects on the pharyngeal mill of a cichlid fish. *Proc. Linn. Soc. London* 176:1–10
44. Greenwood PH. 1974. The cichlid fishes of Lake Victoria, East Africa: the biology and evolution of a species flock. *Bull. Br. Mus. Nat. Hist. Zool.* Suppl. 6:1–134
45. Greenwood PH. 1984. African cichlids and evolutionary theories. See Ref. 25, pp. 141–54
46. Greenwood PH. 1991. Speciation. See Ref. 67, pp. 84–102
47. Harrison R. 1998. Linking evolutionary pattern and process; the relevance of species concepts for the study of speciation. In *Endless Forms: Species and Speciation*, ed. DJ Howard, SH Berlocher, pp. 19–31. New York: Oxford Univ. Press
48. Hert E. 1990. Factors in habitat partitioning in *Pseudotropheus aurora* (Pisces: Cichlidae), and introduced species to a species-rich community of Lake Malawi. *J. Fish Biol.* 36:853–57
49. Hert E. 1991. Female choice based on egg-spots in *Pseudotropheus aurora* Burgess 1976, a rock-dwelling cichlid of Lake Malawi, Africa. *J. Fish Biol.* 38:951–53
50. Hert E. 1992. Homing and home-site fidelity in rock-dwelling cichlids (Pisces: Teleostei) of Lake Malawi, Africa. *Environ. Biol. Fishes* 33:229–37
51. Higashi M, Takimoto G, Yamamura N. 1999. Sympatric speciation by sexual selection. *Nature* 402:523–26
52. Hoelzer GA, Melnick DJ. 1994. Patterns of speciation and limits to phylogenetic resolution. *Trends Ecol. Evol.* 9:104–5
53. Hoelzer GA, Wallman J, Melnick DJ. 1998. The effects of social structure,

geographic structure, and population size on the evolution of mitochondrial DNA. II. Molecular clocks and the lineage sorting period. *J. Mol. Evol.* 47:21–31

54. Holzberg S. 1978. A field and laboratory study of the behaviour and ecology of *Pseudotropheus zebra* (Boulenger), and endemic cichlid of Lake Malawi (Pisces: Cichlidae). *Z. Zool. Syst. Evol. Forsch.* 16:171–87
55. Hori M. 1991. Feeding relationships among cichlid fishes in Lake Tanganyika: effects of intra- and interspecific variations of feeding behavior on their coexistence. *Ecol. Int. Bull.* 19:89–101
56. Hori M. 1993. Frequency-dependent natural selection in the handedness of scale-eating cichlid fish. *Science* 260:216–19

56a. Hunter JP. 1998. Key innovations and the ecology of macroevolution. *Trends Ecol. Evol.* 13:31–35

56b. Jennions M, Petrie M. 1997. Variation in mate choice and mating preferences: a review of causes and consequences. *Biol. Rev. Linn. Soc.* 72:283–327

57. Johnson P, Gulberg U. 1998. Theory and models of sympatric speciation. In *Endless Forms: Species and Speciation*, ed. D Howard, S Berlocher, pp. 79–89. New York: Oxford Univ. Press
58. Johnson TC. 1996. Sedimentary processes and signals of past climatic change in the large lakes of the East African rift valley. In *The Limnology, Climatology and Paleoclimatology of the East African Lakes*, ed. TC Johnson, EO Odada, pp. 367–412. Toronto: Gordon & Breach
59. Johnson TC, Halfman JD, Rosendahl BR, Lister GS. 1987. Climatic and tectonic effects on sedimentation in a rift valley lake: evidence from high-resolution seismic profiles, Lake Turkana, Kenya. *Geol. Soc. Am. Bull.* 98:439–47
60. Johnson TC, Scholz CA, Talbot MR, Kelts K, Ricketts RD, et al. 1996. Late Pleistocene desiccation of Lake Victoria and rapid evolution of cichlid fishes. *Science* 273:1091–93
61. Karino K. 1996. Tactic for bower acquisition by male cichlids, *Cyathopharynx furcifer,* in Lake Tanganyika. *Ichthyol. Res.* 43:125–32
62. Karino K. 1997. Female mate preference for males having long and symmetric fins in the bower-holding cichlid *Cyathopharynx furcifer. Ethology* 103:883–92
63. Kaufman L. 1997. Asynchronous taxon cycles in haplochromine fishes of the greater Lake Victoria Region. *S. Afr. J. Sci.* 93:601–6
64. Kaufman LS, Ochumba P. 1993. Evolutionary and conservation biology of cichlid fishes as revealed by faunal remnants in Lake Victoria. *Conserv. Biol.* 7:719–30
65. Kawanabe H, Cohen JE, Iwasaki K, eds. 1993. *Mutualism and Community Organization: Behavioural, Theoretical, and Food-web Approaches.* Oxford, UK: Oxford Univ. Press. 26 pp.
66. Kawanabe H, Hori M, Nagoshi M, eds. 1997. *Fish Communities in Lake Tanganyika.* Sakyo-ku, Japan: Kyoto Univ. Press. 298 pp.
67. Keenlyside M, ed. 1991. *Cichlid Fishes: Behavior, Ecology, and Evolution.* London: Chapman & Hall
68. Kellogg KA, Markert JA, Stauffer JRJ, Kocher TD. 1995. Microsatellite variation demonstrates multiple paternity in lekking cichlid fishes from Lake Malawi, Africa. *Proc. R. Soc. London Ser. B* 260:79–84
69. Kelly JK, Noor MAF. 1996. Speciation by reinforcement: a model derived from studies of *Drosophila. Genetics* 143:1485–97
70. Klein D, Ono H, O'hUigín C, Vincek V, Goldschmidt T, et al. 1993. Extensive MHC variability in cichlid fishes of Lake Malawi. *Nature* 364:330–34
71. Klein J, Sato A, Nagl S, O'hUigín C. 1998. Molecular trans-species polymorphism. *Annu. Rev. Ecol. Syst.* 29:1–21
72. Knight ME, Turner GF. 1999. Reproductive isolation among closely related Lake

Malawi cichlids: Can males recognize conspecific females by visual cues? *Anim. Behav.* 58:761–68
73. Knight ME, Turner GF, Rico C, vanOppen MJH, Hewitt GM. 1998. Microsatellite paternity analysis on captive Lake Malawi cichlids supports reproductive isolation by direct mate choice. *Mol. Ecol.* 7:1605–10
74. Knight ME, van Oppen MJH, Smith HL, Rico C, Hewitt GM, et al. 1999. Evidence for male-biased dispersal in Lake Malawi cichlids from microsatellites. *Mol. Ecol.* 8:1521–27
75. Kocher TD, Conroy JA, McKaye KR, Stauffer JR. 1993. Similar morphologies of cichlid fish in Lakes Tanganyika and Malawi are due to convergence. *Mol. Phylogenet. Evol.* 2:158–65
76. Kocher TD, Conroy JA, McKaye KR, Stauffer JR, Lockwood SF. 1995. Evolution of NADH dehydrogenase subunit 2 in East African cichlid fish. *Mol. Phylogenet. Evol.* 4:420–32
77. Kocher TD, Lee WJ, Sobolewska H, Penman D, McAndrew B. 1998. A genetic linkage map of a cichlid fish, the tilapia (*Oreochromis niloticus*). *Genetics* 148:1225–32
78. Kohda M. 1998. Coexistence of permanently territorial cichlids of the genus *Petrochromis* through male-mating attack. *Environ. Biol. Fishes* 52:231–42
79. Kohda M, Yanagisawa Y, Tetsa S, Nakaya K, Niimura Y, et al. 1996. Geographical colour variation in cichlid fishes at the southern end of Lake Tanganyika. *Environ. Biol. Fishes* 45:237–48
80. Kondrashov A, Shpak M. 1998. On the origin of species by means of assortative mating. *Proc. R. Soc. London Ser. B* 265:2273–78
81. Konings A. 1995. *Malawi Cichlids in Their Natural Habitat.* Zevenhuizen, Netherlands: Cichlid Press. 303 pp.
82. Konings A. 1998. *Tanganyika Cichlids.* Ahornweg, Germany: Cichlid Press. 272 pp.
83. Kornfield I. 1978. Evidence for rapid speciation in African cichlid fishes. *Experientia* 34:335–36
84. Kornfield I. 1991. Genetics. See Ref. 67, pp. 103–28
85. Kornfield I, Parker A. 1997. Molecular systematics and a rapidly evolving species flock: the *mbuna* of Lake Malawi and the search for phylogenetic signal. In *Molecular Systematics of Fishes*, ed. TD Kocher, CA Stepien, pp. 25–37. San Diego: Academic
86. Kornfield I, Reinthal PN, Lobel PS, Merrill S. 1991. Selection and constraints on bower morphology in a Malawi cichlid. 71st Meet. Am. Soc. Ichthyol. Herpetol. American Museum of Natural History presentation, New York
87. Kornfield I, Taylor JN. 1983. A new species of polymorphic fish, *Cichlasoma minckleyi* from Cuatro Cienegas, Mexico, (Teleostei: Cichlidae). *Proc. Biol. Soc. Wash.* 96:253–69
88. Kosswig C. 1947. Selective mating as a factor for speciation in cichlid fish of East African lakes. *Nature* 159:604–5
89. Kruuk L. 1999. Sticklers for sympatry. *Trends Ecol. Evol.* 14:465–66
90. Kuusipalo L. 1998. Scale morphology in Malawian cichlids. *J. Fish Biol.* 52:771–81
91. Ladiges W. 1968. Die bedutung okologischer faktoren fur die differenzierung der cichliden Tanganjika- and des Njassa-Sees. *Int. Rev. Gesamten. Hydrobiol.* 53:339–52
92. Lande R. 1981. Models of speciation by sexual selection on polygenic traits. *Proc. Natl. Acad. Sci. USA* 78:3721–25
93. Lande R. 1982. Rapid origin of sexual isolation and character divergence in a cline. *Evolution* 36:213–23
94. Lazzaro X. 1991. Feeding convergence in South American and African zooplanktivorous cichlids *Geophagus brasilensis* and *Tilapia rendalli*. *Environ. Biol. Fishes* 31:283–93
95. Lessa EP, Cook JA. 1998. The molecular phylogenetics of tuco-tucos (genus

Ctenomys, Rodentia: Octodontidae) suggests an early burst of speciation. *Mol. Phylogenet. Evol.* 9:88–99
96. Lewis DS. 1982. A revision of the genus *Labidochromis* (Teleostei: Cichlidae) from Lake Malawi. *Zool. J. Linn. Soc.* 75:189–265
97. Lippitsch E. 1993. A phyletic study on lacustrine haplochromine fishes (Perciformes, Cichlidae) of East Africa, based on scale and squamation characters. *J. Fish Biol.* 42:903–46
98. Lippitsch E. 1998. Phylogenic study of cichlid fishes in Lake Tanganyika: a lepidological approach. *J. Fish Biol.* 53:752–66
99. Lobel PS. 1998. Possible species specific courtship sounds by two sympatric cichlid fishes in Lake Malawi, Africa. *Environ. Biol. Fishes* 52:443–52
100. Losey GS, Cronin TW, Goldsmith TH, Hyde D, Marshall NJ, et al. 1998. The UV visual world of fishes: a review. *J. Fish Biol.* 54:921–43
101. Luikart G, Sherwin B, Steele B, Allendorf F. 1998. Usefulness of molecular markers for detecting population bottlenecks via monitoring genetic change. *Mol. Ecol.* 7:963–74
102. Magurran A. 1998. Population differentiation without speciation. *Philos. Trans. R. Soc. London Ser. B* 353:275–86
103. Markert JA, Arnegard ME, Danley PD, Kocher TD. 1999. Biogeography and population genetics of the Lake Malawi cichlid *Melanochromis auratus*: habitat transience, philopatry and speciation. *Mol. Ecol.* 8:1013–26
104. Martens K, Godderis B, Coulter G, eds. 1994. *Speciation in Ancient Lakes,* E. Schweizerbart'sche Verlagsbuchhandlund, Stuttgart: 44:508 pp.
105. Mayr E. 1963. *Animal Species and Evolution.* Cambridge, MA: Harvard Univ. Press
106. Mayr E. 1984. Evolution of fish species flocks: a commentary. In *Evolution of Fish Species Flocks*, ed. AA Echelle, I Kornfield, pp. 3–12. Orono, ME: Univ. Maine Press
107. McCune A, Lovejoy N. 1998. The relative rate of sympatric and allopatric speciation in fishes. In *Endless Forms: Species and Speciation*, ed. DJ Howard, SH Belocher. New York: Oxford Univ. Press, 172–85
108. McElroy D, Kornfield I. 1990. Sexual selection, reproductive behavior, and speciation in the *mbuna* species flock of Lake Malawi (Pisces: Cichlidae). *Environ. Biol. Fishes* 28:273–84
109. McElroy DM, Kornfield IL. 1993. Novel jaw morphology in hybrids between *Pseudotropheus zebra* and *Labeotropheus fullerborni* (Teleostei: Cichlidae) from Lake Malawi, Africa. *Copeia* 1993:933–45
110. McElroy DM, Kornfield IL, Everett J. 1991. Coloration in African cichlids: diversity and constraints in Lake Malawi endemics. *Neth. J. Zool.* 41:250–68
111. McKaye ER, Kocher T, Reinthal P, Kornfield I. 1982. A sympatric sibling species complex of *Petrotilapia* Trewavas from Lake Malawi analyzed by enzyme electrophoresis (Pisces: Cichlidae). *Zool. J. Linn. Soc.* 76:91–96
112. McKaye K, Gray N. 1984. Extrinsic barriers to gene flow in rock-dwelling cichlids of Lake Malawi: macrohabitat heterogeneity and reef colonization. See Ref. 25, pp. 169–83
113. McKaye KR. 1991. Sexual selection and the evolution of the cichlid fishes of Lake Malawi, Africa. See Ref. 67, pp. 241–57
114. McKaye KR, Howard JH, Stauffer JR, Morgan RP, Shonhiwa F. 1993. Sexual selection and genetic relationships of a sibling species complex of bower building cichlids in Lake Malawi, Africa. *Japan. J. Ichthyol.* 40:15–21
115. McKaye KR, Marsh AC. 1983. Food switching by two specialised algae-scraping cichlid fishes in Lake Malawi, Africa. *Oceologia* 56:245–48

116. McKaye MC, Kocher TD, Reinthal P, Harrison H, Kornfield I. 1984. Genetic evidence for allopatric and sympatric differentiation among color morphs of a Lake Malawi cichlid fish. *Evolution* 38:215–19
117. Meyer. 1987. Phenotypic plasticity and heterochrony in *Cichlasoma manguense* (Pisces: Cichlidae) and their implications for speciation in cichlid fish. *Evolution* 41:1357–69
118. Meyer A. 1993. Phylogenetic relationships and evolutionary processes in East African cichlid fishes. *Trends Ecol. Evol.* 8:279–85
119. Meyer A, Knowles LL, Verheyen E. 1996. Widespread geographical distribution of mitochondrial haplotypes in rock-dwelling cichlid fishes from Lake Tanganyika. *Mol. Ecol.* 5:341–50
120. Meyer A, Kocher T, Basasibwaki P, Wilson A. 1990. Monophyletic origin of Lake Victoria cichlid fishes suggested by mitochondrial DNA sequences. *Nature* 347:550–53
121. Meyer A, Ontero C, Spreinat A. 1994. Evolutionary history of the cichlid fish species flocks of the East African great lakes inferred from molecular phylogenetic data. *Arch. Hydrobiol. Beih. Limnol.* 44:407–23
122. Moller AP, Cuervo JJ. 1998. Speciation and feather ornamentation in birds. *Evolution* 52:859–69
123. Moran P, Kornfield I. 1993. Retention of an ancestral polymorphism in the mbuna species flock (Pisces: Cichlidae) of Lake Malawi. *Mol. Biol. Evol.* 10:1015–29
124. Moran P, Kornfield I. 1995. Were population bottlenecks associated with the radiation of the *mbuna* species flock of Lake Malawi? *Mol. Biol. Evol.* 12:1085–93
125. Moran P, Kornfield I, Reinthal P. 1994. Molecular systematics and radiation of the haplochromine cichlids (Teleostei: Perciformes) of Lake Malawi. *Copeia* 1994:274–88
126. Nagl S, Tichy H, Mayer WE, Takahata N, Klein J. 1998. Persistence of neutral polymorphisms in Lake Victoria cichlid fish. *Proc. Natl. Acad. Sci. USA* 95:14238–43
127. Nelissen MHJ. 1991. Communication. See Ref. 67, pp. 225–40
128. Nicholson SE. 1998. Fluctuations of rift valley lakes Malawi and Chilwa during historical times: a synthesis of geological, archaeological and historical information. In *Environmental Change and Response in East African Lakes*, ed. JT Lehman, pp. 207–31. Dordrecht, The Netherlands: Kluwer Academic
129. Nishida M. 1991. Lake Tanganyika as an evolutionary reservoir of old lineages of East African cichlid fishes: inferences from allozyme data. *Experientia* 47:974–79
130. Nishida M. 1997. Phylogenetic relationships and evolution of Tanganyikan cichlids: a molecular perspective. In *Fish Communities in Lake Tanganyika*, ed. H Kawanabe, M Hori, M Naagoshi, pp. 1–23. Sakyo-ku, Japan: Kyoto Univ. Press
131. Ogutu-Ohwayo R. 1993. The effects of predation by Nile Perch, *Lates niloticus* L., on the fish of Lake Nabugabo, with suggestions for conservation of endangered endemic cichlids. *Conserv. Biol.* 7:701–11
132. Okuda N, Tayasu I, Yanagisawa Y. 1998. Determinate growth in a paternal mouthbrooding fish whose reproductive success is limited by buccal capacity. *Evol. Ecol.* 12:681–99
133. Ono H, O'hUigín C, Tichy H, Klein J. 1993. Major-histocompatibility-complex variation in two species of cichlid fishes from Lake Malawi. *Mol. Biol. Evol.* 10:1060–72
134. Owen RB, Crossley R, Johnson TC, Tweddle D, Kornfield I, et al. 1990. Major low levels of Lake Malawi and their implications for speciation rates in cichlid fishes. *Proc. R. Soc. London Ser. B* 240:519–33

135. Owens I, Bennet P, Harvey P. 1999. Species richness among birds: body size, life history, sexual selection, or ecology? *Proc. R. Soc. London Ser. B* 266:933–39
136. Page RDM, Holmes EC. 1998. *Molecular Evolution.* Oxford, UK: Blackwell Sci. 346 pp.
137. Parker A, Kornfield I. 1996. Polygynandry in *Pseudotropheus zebra,* a cichlid fish from Lake Malawi. *Environ. Biol. Fishes* 47:345–52
138. Parker A, Kornfield I. 1997. Evolution of the mitochondrial DNA control region in the *mbuna* (Cichlidae) species flock of Lake Malawi, East Africa. *J. Mol. Evol.* 45:70–83
139. Patterson HEH. 1985. The recognition concept of species. In *Species and Speciation*, ed. ES Vrba, pp. 21–29. Pretoria: Transvaal Museum
140. Price T. 1998. Sexual selection and natural selection in bird speciation. *Philos. Trans. R. Soc. London Ser. B* 353:251–60
141. Reinthal P, Meyer A. 1997. Molecular phylogenetic tests of speciation models in African cichlid fishes. In *Molecular Phylogenetics of Adaptive Radiations*, ed. T Givinish, K Sytsma, pp. 375–90. Boston: Cambridge Univ. Press
142. Reinthal PN. 1990. The feeding habits of a group of herbivorous rock-dwelling cichlid fishes (Cichlidae: Perciformes) from Lake Malawi, Africa. *Environ. Biol. Fish.* 27:215–33
143. Ribbink AJ. 1986. The species concept, sibling species and speciation. *Ann. Mus. R. Afr. Centr.* 251:109–16
144. Ribbink AJ. 1994. Alternative perspectives on some controversial aspects of cichlid fish speciation. *Arch. Hydrobiol. Beih. Limnol.* 44:101–25
145. Ribbink AJ, Marsh BA, Marsh AC, Ribbink AC, Sharp BJ. 1983. A preliminary survey of the cichlid fishes of rocky habitats in Lake Malawi. *S. Afr. J. Zool.* 18:149–310
146. Ricklefs RE, Schluter D. 1993. Species diversity: regional and historical influences. In *Species Diversity in Ecological Communities*, ed. RE Ricklefs, D Schluter, pp. 350–63. Chicago: Univ. Chicago Press
147. Roberts TR. 1975. Geographical distribution of African freshwater fishes. *Zool. J. Linn. Soc.* 57:249–319
148. Rossiter A. 1995. The cichlid fish assemblages of Lake Tanganyika: ecology, behaviour and evolution of its species flocks. In *Advances in Ecological Research*, ed. M Begon, AH Fitter, pp. 401. Boston: Academic Press
149. Rossiter A, Yamagishi S. 1997. Intraspecific plasticity in the social system and mating behaviour of a lek-breeding cichlid fish. See Ref. 25, pp. 193–217
150. Rundle H, Schluter D. 1998. Reinforcement of stickleback mate preferences: sympatry breeds contempt. *Evolution* 52:200–8
151. Ruzzante D. 1998. A comparison of several measures of genetic distance and population structure with microsatellite data: bias and sampling variance. *Can. J. Fish Aquat. Sci.* 55:1–14
152. Sage D, Selander RK. 1975. Trophic radiation through polymorphism in cichlid fishes. *Proc. Natl. Acad. Sci. USA* 72:4669–73
153. Sage RD, Loiselle PV, Basasibwaki P, Wilson AC. 1984. Molecular versus morphological change among cichlid fishes of Lake Victoria. See Ref. 25, pp. 185–201
154. Sato T, Gashagaza MM. 1997. Shell-brooding cichlid fishes of Lake Tanganyika: their habits and mating systems. In *Fish Communities in Lake Tanganyika*, ed. H Kawanabe, M Hori, M Nagoshi, pp. 221–40. Sakyo-ku, Japan: Kyoto Univ. Press
155. Schliwewen U, Tautz D, Paabo S. 1994. Sympatric speciation suggested by monophyly of crater lake cichlids. *Nature* 368:629–32

156. Schluter D. 1998. Ecological causes of speciation. In *Endless Forms: Species and Speciation*, ed. DJ Howard, New York: Oxford Univ. Press, pp. 114–29
157. Scholz CA, Rosendahl BR. 1988. Low lake stands in Lakes Malawi and Tanganyika, East Africa, delineated with multifold seismic data. *Science* 240:1645–48
158. Seegers L, Sonnenberg R, Yamamoto R. 1999. Molecular analysis of the *Alcolapia* flock from lakes Natron and Magadi, Tanzania and Kenya (Teleostei: Cichlidae), and implications for their systematics and evolution. *Ichthyol. Explor. Fresh.* 10:175–99
159. Seegers L, Tichy H. 1999. The *Oreochromis alcalicus* flock (Teleostei: Cichlidae) from lakes Natron and Magadi, Tanzania and Kenya, with descriptions of two new species. *Ichthyol. Explor. Freshw.* 10:97–146
159a. Seehausen O. 2000. Explosive speciation rates and unusual species richness in haplochromine cichlid fishes: effects of sexual selection. *Adv. Ecol. Res.* 30:235–71
160. Seehausen O, Lippitsch E, Bouton N, Zwennes H. 1998. Mbipi, the rock-dwelling cichlids of Lake Victoria: description of three new genera and fifteen new species (Teleostei). *Ichthyol. Explor. Freshw.* 9:129–228
161. Seehausen O, Mayhew PJ, van Alphen JJM. 1999. Evolution of colour patterns in East African cichlid fish. *Evol. Biol.* 12:514–34
162. Seehausen O, van Alphen JJA. 1999. Can sympatric speciation by disruptive sexual selection explain rapid evolution of cichlid diversity in Lake Victoria? *Ecology Letters.* 2:262–71
163. Deleted in proof
164. Seehausen O, van Alphen JJM. 1998. The effect of male coloration on female mate choice in closely related Lake Victoria cichlids (*Haplochromis nyererei* complex). *Behav. Ecol. Sociobiol.* 42: 1–8
165. Seehausen O, van Alphen JJM, Witte F. 1997. Cichlid fish diversity threatened by eutrophication that curbs sexual selection. *Science* 277:1808–11
166. Seehausen O, Witte F, van Alphen JJM, Bouton N. 1998. Direct mate choice maintains diversity among sympatric cichlids in Lake Victoria. *J. Fish Biol.* 53:37–55
167. Stauffer JR, Bowers NJ, Kellogg KA, McKaye KR. 1997. A revision of the blue-black *Pseudotropheus zebra* (Teleostei: Cichlidae) complex from Lake Malawi, Africa, with a description of a new genus and ten new species. *Proc. Acad. Nat. Sci. Philos.* 148:189–230
168. Stauffer JR, Bowers NJ, Kocher TD, McKaye KR. 1996. Evidence of hybridization between *Cyanotilapia afra* and *Psuedotropheus zebra* (Teleostei: Cichlidae) following an intralacustrine translocation in Lake Malawi. *Copiea* 1996:203–8
169. Wtauffer JRJ, Hert E. 1992. *Pseudotropheus callainos* a new species of *mbuna* (Cichlidae), with analyses of changes associated with two intra-lucustrine transplantations in Lake Malawi, Africa. *Ichthyol. Explor. Freshw.* 3:253–64
170. Stiassny MLJ. 1981. Phylogenetic versus convergent relationship between piscivorous cichlid fishes from Lakes Malawi and Tanganyika. *Bull. Br. Mus. Nat. Hist. Zool.* 40:67–101
171. Stiassny MLJ, Meyer A. 1999. Cichlids of the rift lakes. *Sci. Am.* 280:44–49
172. Strickberger MW. 2000. *Evolution.* Sudbury, MA:Jones and Bartlett. 722 pp.
173. Sturmbauer C. 1998. Explosive speciation in cichlid fishes of the African Great Lakes: a dynamic model of adaptive radiation. *J. Fish Biol.* 53:18–36

174. Sturmbauer C, Meyer A. 1992. Genetic divergence, speciation and morphological stasis in a lineage of African cichlid fishes. *Nature* 358:578–81
175. Sturmbauer C, Meyer A. 1993. Mitochondrial phylogeny of the endemic mouthbrooding lineages of cichlid fishes from Lake Tanganyika, East Africa. *Mol. Biol. Evol.* 10:751–68
176. Sturmbauer C, Meyer A, Baric S, Verheyen E, Salzburger W. 2000. Submitted
177. Sturmbauer C, Verheyen E, Meyer A. 1994. Mitochondrial phylogeny of the Lamprologini, the major substrate spawning lineage of cichlid fishes from Lake Tanganyika in Eastern Africa. *Mol. Biol. Evol.* 11:691–703
178. Sturmbauer C, Verheyen E, Ruber L, Meyer A. 1997. Phylogeographic patterns in populations of cichlid fishes from rock habitats in Lake Tanganyika. In *Molecular Phylogeny of Fishes*, ed. TD Kocher, C Stepien, pp. 97–111. San Diego: Academic
178. Sültmann H, Mayer WE. 1997. Reconstruction of cichlid fish phylogeny using nuclear DNA markers. In *Molecular Systematics of Fishes*, ed. TD Kocher, CA Stepien, pp. 39–51. San Diego: Academic
179. Takahashi K, Terai Y, Nishida M, Okada N. 1998. A novel family of short interspersed elements (SINEs) from cichlids: the patterns of insertion of SINEs at orthologous loci support the proposed monophyly of four major groups of cichlid fishes in Lake Tanganyika. *Mol. Biol. Evol.* 15:391–407
180. Temple PH. 1969. Some biological implications of a revised geological history for Lake Victoria. *Biol. J. Linn. Soc.* 1:363–71
181. Templeton AR. 1989. The meaning of species and speciation: a genetic perspective. In *Speciation and Its Consequences*, ed. D Otte, JA Endler, pp. 3–27. Sunderland, MA: Sinauer
182. Trendall K. 1988. The distribution of dispersal of introduced fish at Thumbi West Island in Lake Malawi, Africa. *J. Fish Biol.* 33:357–69
183. Trewavas E. 1935. A synopsis of the cichlid fishes of Lake Nyasa. *Ann. Mag. Nat. Hist.* 10:65–118
184. Turner G. 1997. Small fry go big time. *New Sci.* 155:36–40
185. Turner GF. 1994. Speciation mechanisms in Lake Malawi cichlids: a critical review. *Arch. Hydrobiol. Beih. Limnol.* 44:139–60
186. Turner GF, Burrows MT. 1995. A model of sympatric speciation by sexual selection. *Proc. R. Soc. London Ser. B* 260:287–92
187. Tweddle D, Eccles DH, Frith CB, Fryer G, Jackson PBN, et al. 1998. Cichlid spawning structures-bowers or nests? *Environ. Biol. Fishes* 51:107–9
188. van der Meer HJ, Bowmaker JK. 1995. Interspecific variation of photoreceptors in four co-existing haplochromine cichlid fishes. *Brain Behav. Evol.* 5:232–40
189. van Doorn GS, Noest AJ, Hogeweg P. 1998. Sympatric speciation and extinction driven by environment dependent sexual selection. *Proc. R. Soc. London Ser. B* 265:1915–19
190. Van Oppen M, Turner G, Rico C, Robinson R, Deutch J, et al. 1998. Assortative mating among rock-dwelling cichlid fishes supports high estimates of species richness from Lake Malawi. *Mol. Ecol.* 7:991–1001
191. Witte F, Barel KDN, van Oijen MJP. 1997. Intraspecific variation of haplochromine cichlids from Lake Victoria and its taxonomic implications. *S. Afr. J. Sci.* 93:585–94
192. Wu L, Kaufman L, Fuerst P. 1999. Isolation of microsatellite markers in *Astatoreochromis alluaudi* and their cross-species amplifications in other African cichlids. *Mol. Ecol.* 8:895–906

193. Yanagisawa Y, Ochi H, Gashagaza MM. 1997. Habitat use in cichlid fishes for breeding. See Ref. 66, pp. 151–73
193a. Yasui Y. 1988. The 'genetic benefits' of female multiple mating reconsidered. *Trends Ecol. Evol.* 13:246–50
194. Yuma M, Kondo T. 1991. Interspecific relationships and habitat utilization among benthivorous cichlids. See Ref. 66, pp. 89–103

Annu. Rev. Ecol. Syst. 2000. 31:197–215

Shrub Invasions of North American Semiarid Grasslands

O. W. Van Auken
Division of Life Sciences, University of Texas at San Antonio, San Antonio, Texas 78249-0662; e-mail: ovanauken@utsa.edu

Key Words brush encroachment, desert grassland, desertification, encroachment, woody plant invasion

■ **Abstract** The composition and structure of the semiarid or desert grasslands of southwestern North America have changed over the past 150 y. Brushy or woody species in these communities have increased in density and cover. This increase in density of woody species is called brush encroachment because most of these species have been present in these communities at lower densities for thousands of years. The brushy or woody species were not introduced from other continents or from great distances. They are indigenous species that have increased in density or cover because of changes in local abiotic or biotic conditions. The brushy and woody plants are not the cause of these changes, but their increase is the result of other factors. The causes of changes that have led to the present woody-brushy composition of these semiarid grasslands has been difficult to determine. Warming of the climate seems to be a background condition, but the driving force seems to be chronic, high levels of herbivory by domestic animals. This herbivory has reduced the aboveground grass biomass, leading to the reduction of fine fuel and a concomitant reduction or complete elimination of grassland fires. This combination of factors favors the encroachment, establishment, survival and growth of woody plants. Less competition from grasses, dispersal of seeds of woody plants by domestic animals, and changes in rodent, lagomorph, and insect populations seem to modify the rate of change. Elevated levels of atmospheric CO_2 are not necessary to explain shrub encroachment in these semiarid grasslands. The direction of future change is difficult to predict. The density of brushy and woody plants will probably increase as will the stature and number of species. However, if soil nutrients increase, woody legumes may be replaced by other brushy or woody species. Reversing the changes that have been going on for 150 y will be a difficult, long-term, and perhaps impossible, task.

INTRODUCTION

Deglaciation has caused major fluctuations in plant populations throughout the world over the past 11,000–12,500 y. Woodlands with spruce, fir, and pine covered much of what is now mixed deciduous forest in the eastern United States (51–53). Subalpine woodlands were found in areas now covered with piñon-juniper

0066-4162/00/1120-0197$14.00

woodlands in the Great Basin (20, 120). Spruce forests or open spruce, pine, birch parkland occurred in areas of tall grass prairie in Kansas and in areas of juniper grassland or savanna in west central New Mexico. Pine parkland was found in what is now short-grass prairie in western Texas and in areas of desert grassland in southeastern Arizona (78, 79, 85, 147, 166).

In the past 150 y, changes and rates of change of plant populations and communities have been unparalleled (42, 56, 132). Most of the recent changes in woody plant populations associated with grasslands or savannas do not appear to have been caused directly by deglaciation nor are they to be considered invasions as suggested by many authors; rather, these phenomena are associated with adjacent communities that have existed for a considerable time (32, 120). In this chapter I examine, not the establishment and growth of newly introduced species of woody plants, but the increase in density or cover of local shrubs and woody species that have been present in the semiarid grasslands of southwestern North America for thousands of years (86, 93, 167, 169). I favor the word encroachment for this phenomenon.

At lower elevations, brushy or woody species usually associated with the Chihuahuan or Sonoran Deserts have increased in density in areas previously covered by semiarid grassland (29, 82). At higher elevations, various species of juniper, previously restricted to rocky outcrops, steep slopes, and shallow soils have spread downslope into grasslands throughout southwestern North America (21, 59, 90, 115, 120, 170, 174). Not all species have moved, however; for example, populations of *Quercus emoryi* that form woodlands above the semiarid grasslands in Arizona and Northern Mexico apparently have not migrated up or down in elevation for many years (173). However, seedling establishment is restricted to the area below the adult tree canopy (69, 173).

Alterations in density of woody plants have been attributed to climate change (29, 82), but in the semiarid grasslands of the southwestern United States, warming would presumably cause juniper to move northward and up in elevation (120). This did not occur, juniper populations seemed to move down in elevation and are affected by reduced grass biomass and a concomitant reduction in fire frequency (21, 32, 90, 118). Recent climatic or precipitation changes do not seem connected to these vegetation alterations in the semiarid grasslands (8). Anthropogenic forces have apparently caused most of the recent changes in semiarid grasslands either directly or indirectly. The process is known by a number of names, including desertification, shrub invasion, woody plant invasion, and bush or brush encroachment (2, 4, 5, 109, 115, 136, 141, 142, 144) and has occurred throughout the world (1, 2, 4, 7, 21, 29, 34, 77, 81, 82, 120, 141, 142, 144, 150, 164, 168).

AREA OF STUDY

Semiarid grasslands or desert grasslands are found in the basins and valleys of southwestern North America as well as on outwash plains (bajadas) and on the lower foothills (24, 25, 30, 55, 98, 109, 142). Similar vegetation regions occur throughout the world; among them are the Espinal and some parts of the Chaco in Argentina (151), the veld and some savanna types in southern Africa (30, 44), arid parts of northwestern India (44, 122), the acacia shrublands and semiarid low woodlands between the forests of the southeastern part of Australia, and the interior continental deserts of Australia (13, 63, 124, 130).

The semiarid grasslands of southwestern North America are extensive but discontinuous (24, 25, 30, 55, 98, 109, 142). In the north, they extended from northwest of Dallas in northern Texas, across southern New Mexico, to northwest of Phoenix and west of Tucson in southern Arizona (approximately 35° N 98° W to 31° N 111° W). In the south, they extended from west of Veracruz to northeast of Colima, Mexico (approximately 18° N 96° W to 19° N 104° W). These semiarid regions usually occur at elevations between 1100 and 2500 m. Rainfall may average 230 mm per year in the north to as much as 600 mm per year at higher elevations in the south. Approximately 50% of the rain falls between May and October in western Arizona, 75% during this time period in western Texas, and as much as 90% during this time period in central Mexico (109). Mean annual temperature ranges from 13°C to 16°C with fewer than 75 d with freezing temperatures (149); evaporation rates are high (142). Plant productivity in semiarid grasslands is the lowest of all North American grasslands, probably because of low and variable rainfall, high evapotranspiration, and shallow soil (109, 119, 142, 149). Aboveground net primary production values as low as 13 g m^{-2} yr^{-1} have been reported, but total plant production is thought to be about 250–350 g m^{-2} yr^{-1} (109).

Soils are mostly Aridisols, Mollisols, or Entisols (9, 105). Most are usually shallow, but they may be deep in places. They include sandy outwash material with little horizon development to very old, deep soils with well-developed profiles. Deeper soils may have clay-loam to clay subsoils and well-developed calcic horizons. Most of these semiarid grassland soils have less organic material than other grassland soils (142). The surface and subsurface properties of the soils, especially the distribution of water and the capacity to hold water, determine to a large extent the kind of plants found on the surface (107, 108).

Many different names have been used to describe the semiarid grasslands of the southwestern United States and central and northern Mexico, including, Desert grassland (55, 110), semidesert grassland (25), Chihuahuan desert grassland, Sonoran desert grassland, high desert bunchgrass, and high desert sod grass (142), and others (30). In Mexico, the names of the communities include pastizal, mesquital, matorral crasicaule, matorral desértico rosetófilo, and the matorral desértico micrófilo (138, 142).

Much of the area listed above as semiarid grassland has also been described as savanna, including various types of piñon-juniper savanna, southwestern oak savanna, and mesquite savanna (115), or as an ecotone (55, 142). Much of the biology and ecology of savanna species interactions have been considered recently (144). In the east, the semiarid grasslands merge with or interdigitate with the Chihuahuan Desert shrublands at elevations of about 1000 m and where precipitation falls to about 250 mm per year. In the northeast, the semiarid grassland blends into the short grass prairie, but the boundaries are vague. In the west, the semiarid grasslands merge with or interdigitate with the Sonoran Desert shrublands at low elevations and where precipitation is reduced; whereas at elevations of more than approximately 1750 m in the United States and more than 2500 m in Mexico, they merge with or interdigitate with evergreen-oak savanna, woodland, or chaparral (107, 142).

HISTORICAL BACKGROUND

Reports of increased density of brushy or woody plants in the semiarid grasslands of southwestern North America are not new. The timing of encroachment of woody plants in Arizona has been well documented and seems to be associated with large-scale cattle ranching in the 1870s and fire exclusion (6, 7). The documentation is also fairly good for New Mexico and Texas, although the process may have started sooner (2, 55, 86). However, all of these changes have occurred with a backdrop of climate change, which makes sorting out causal forces very difficult.

Some woody plants that were introduced into southwestern North America and are now apparently well established may be found in widespread areas whereas others are quite localized. *Euryops multifidus* (resin bush), a South African shrub that was introduced into southern Arizona in 1938 (107), is not eaten by domestic or native herbivores and continues to spread. It grows on hillsides and mesas in the semiarid grasslands, and native grasses and woody plants have disappeared from areas where it is found. Several species of *Tamarix* (salt cedar or Tamarisk) have been introduced into southwestern North America from the Mediterranean region, the Middle East or Africa (49). The two most widespread species, *Tamarix chinensis* and *T. gallica*, which seem to have the greatest densities (86, 152), occur in riparian zones and in salt flats and other salinized, wet soils associated with the semiarid grasslands but not in the grasslands themselves. *Elaeagnus angustifolia* (Russian olive) is also a widely distributed, introduced small tree that is found in riparian habitats associated with semiarid grasslands.

A number of herbaceous invaders from the Middle East and Africa have also been introduced into the southwestern semiarid grasslands (12, 23, 123). These heavily grazed, disturbed habitats are more likely to be invaded than similar non-disturbed habitats (10), and the invaders are usually introduced annuals from the families Poaceae, Asteraceae, and Brassicaceae (6, 30, 31, 100, 112, 131, 137, 152). In southwestern North America, approximately 10% of the local floras are established, non-native species, but it is unknown how many seeds of non-native species arrive, germinate, grow, die and never establish a viable population (60, 95, 103).

ENCROACHING SPECIES

It is difficult to estimate the area that has changed from semiarid grassland to shrubland in southwestern North America and the area occupied by the major encroaching species. Approximately 60 million ha are estimated (77, 86), but this area is larger than that estimated for the semiarid grasslands of southwestern North America (98). One cause of this apparent paradox is that more than one species of woody plant is found in the same area (77), and another is that some species of encroaching woody plants are found outside the semiarid grasslands but these areas have not been differentiated. The third cause is that some areas have a high density or cover of the woody plant, others are moderate, and some are light. Nonetheless, it seems that the composition and structure of most, if not all, semiarid southwestern grasslands has been changed by the encroachment of one or more shrubby or woody species.

Prosopis (mesquite, possibly *P. glandulosa*, *P. velutina*, *P. torreyana* or *P. juliflora*, depending on the systematics and location) is the dominant woody plant on more than 38 million ha of what has been considered semiarid southwestern grasslands. *Larrea tridentata* (creosotebush) is the dominant shrub on more than 19 million ha of similar grasslands. Other shrubs, small trees, or succulents that have increased in density and area can be locally important, but individually do not cover nearly the same area as do *Prosopis* and *Larrea*. They include various species of *Acacia, Yucca, Flourensia, Haplopappus, Opuntia, Gutierrezia, Juniperus,* and *Quercus* (55, 77, 86, 142).

CAUSES OF ENCROACHMENT

The causes of shrub or woody plant encroachment in semiarid grasslands throughout the world have been much debated. Most often cited as reasons are climate change, chronic high levels of herbivory, change in fire frequency, changes in grass competitive ability, spread of seed by livestock, small mammal populations, elevated levels of CO_2, and combinations of these factors (2–4, 6, 7, 18, 37, 64, 86, 121, 125, 134, 142, 158, 166, 171). Most of the changes in woody plant density have been associated with the introduction of cattle into these grasslands (3, 6, 7, 11, 127, 129).

Some small amount of herbivory is tolerated by plants without noticeable changes in productivity, biomass, growth, or reproduction, but higher levels result in depression or reduction of these factors (2, 14, 68, 80, 84, 99). The stimulation of plant growth directly by herbivory and benefits of herbivory to plants seems minimal and probably rarely occur, but these topics have received considerable attention in the literature (16, 17, 22, 58, 113, 128). Rather, chronic high levels of herbivory seem to negatively affect plants and are the dominant reasons for the encroachment of shrubs and other woody plants into the semiarid grasslands and for changes in woody plant density (3, 6, 7, 11, 127, 129). The introduction of cattle seems to be the primary factor in the conversion of semiarid grasslands into shrublands or woodlands, but the mechanisms involved are still not well understood, and the rates, dynamics, patterns, and successional processes are not well defined.

GLOBAL CHANGE

We are currently in an interglacial period that started about 12,500 y ago. The climate has been warming since the most recent glacial period ended (88, 111). Apparently plant communities migrate as the climate warms or cools. During the current warming trend, in what are now the semiarid grasslands or desert grasslands of southwestern North America, pine parkland and juniper woodland or savanna moved mostly in a northern direction or up in elevation to their approximate current location. Changes in the populations of plants and animals of the communities of the American southwest have been pieced together from pollen records and fossil packrat middens (20, 104, 120, 166). There does not seem to be conclusive evidence that changes in precipitation patterns or temperature in southwestern North America since the 1870s are linked to recent shrub or woody plant encroachment in the semiarid grasslands. The unevenness of this encroachment, especially for *Prosopis*, and dramatic differences in density in adjacent fenced, edaphically similar areas would seem to rule out large scale climatic influences as the major cause of woody plant increases (8). It has also been proposed that the rising level of atmospheric CO_2 is the cause of shrub encroachment (87, 92, 106, 133). This hypothesis could account for the synchronous, widespread encroachment of shrubs and other woody plants into semiarid grasslands and savannas throughout the world. It is based on observations that most woody plants have the C_3 photosynthetic pathway and in the semiarid grasslands, the grasses that are being replaced have the C_4 photosynthetic pathway.

Plants with the C_3 photosynthetic pathway have a growth advantage at higher levels of CO_2 compared to plants with the C_4 photosynthetic pathway. However, there are some difficulties with this hypothesis (3). Quantum yields, photosynthesis rates, and water-use efficiencies are comparable for a variety of C_3 and C_4 species at current levels of CO_2. Many C_4 grasses are more responsive to increased levels of CO_2 than previously supposed. The replacement of C_3 grasses by encroaching C_3 woody shrubs in the cold deserts is not explained by the hypothesis of elevated CO_2. Fences reduce the encroachment of C_3 woody shrubs in edaphically similar areas with C_4 grasses. Shifts in populations of C_4 grasses to C_3 grasses in these same areas have not occurred. There is a temporal disparity between the time of the greatest increase in CO_2 which occurred after approximately 1910 and the encroachment of woody plants that had by this time in many areas already started or occurred (3). Many shifts in dominance of woody plants and grasses during the Holocene do not appear to be related to elevated levels of CO_2. Finally, not all studies have shown a CO_2 fertilizer effect, suggesting other limitations or constraints.

COMPETITION

The importance of competition between grasses and woody species has been demonstrated in many arid and semiarid communities (41, 65). In addition, interactions between and among species are known to be important in determining community structure and function. Competition between species is considered one of the major factors determining community characteristics (75, 76, 80), but it is one of several factors (39, 40, 135, 173) and is continually debated (72–74, 155). Competition has been reported in many studies (143), but intraspecific competition was as strong as interspecific competition in 75% of the examples (48). However, the growth of plants in arid and semiarid communities is at least partially controlled by interference from neighbors (41, 65, 101, 102), and this growth can be modified by herbivory (37, 141, 158).

Changes in competition between grasses and woody plants are implicated in the encroachment of woody plants into semiarid grasslands (176), where competition is primarily belowground (36, 158, 176). Because of the low stature of the plants, relatively low plant density, high belowground biomass, and high root:shoot ratios (36, 158, 176) competitive ability of plants in these communities may depend on root biomass, root density, root branching, root radius, root hair characteristics, mycorrhizae, timing of growth, or interactions with other soil organisms (41, 163, 165). Competition in these semiarid grasslands seems to change depending on the species and environmental conditions. Grasses inhibit the woody species most during the germination, establishment, and early growth of the woody plants (37, 158, 163). However, the interaction seems to be reversed once the woody plant roots are below the root zone of the grasses and the woody plant shoot is above the shoot zone of the grasses (19, 28, 35, 97).

Probably the best-studied woody species in these semiarid grasslands are *Prosopis* (mesquite, *Prosopis glandulosa* and other species) and *Acacia* (*Acacia smallii* and other species). Aboveground, belowground, and total dry mass of *Prosopis glandulosa* was reduced or suppressed when it was grown in the greenhouse with some C_4 grasses (36, 157, 159, 160, 163) or when it was grown in the field with C_4 grasses (37, 62, 71, 158). However, several studies did not show growth inhibition (27, 28), probably because of lower levels of herbaceous biomass or site-specific factors. Similar trends have been shown for *Acacia smallii* in greenhouse competition studies with C_3 and C_4 grasses and in the field (43, 161, 162).

Although high density or biomass of grass reduces germination, survival, and growth of woody seedlings, some seedlings survive (27, 28, 158, 163; cf 37). Despite suppressed growth, some of these seedlings would finally escape the grass zone of suppression and ultimately convert the grassland into shrubland or woodland. If, for example, one woody plant ha^{-1} produced ten survivors ha^{-1} y^{-1}, in 100 y the density of that species of woody plant would be 1000 plants ha^{-1}, the grassland would have become a shrubland or woodland.

FIRE

Periodic burning is required to control or reduce the establishment and growth of woody plants in most if not all grasslands (47, 177). Fire interacts with other factors such as topography, soil, herbivores, and amount of herbaceous fuel to determine the nature, density, and location of woody plants in a landscape (77, 86, 117, 177). The occurrence and frequency of fires is linked to climate. In higher elevation forest communities of the southwestern United States, fire occurrence is determined by climatic patterns associated with Southern Oscillations (high phase-La Niña, low phase-El Niño) (154). Large fires usually occur after dry springs (La Niña) and smaller fires follow wet springs (El Niño). The same is probably true for the lower elevation semiarid grasslands in the American southwest.

Fire frequency in the semiarid grasslands has decreased in the past 150 y, as has fire size, while the size and density of the woody plants has increased and biomass of grass has decreased (6). There is historical evidence of fires in these semiarid grasslands from the earliest travelers and European settlers (6, 86, 89), although some authors do not think there was ever enough fuel in these semiarid grasslands to carry an extensive fire or do not agree with the evidence (29, 55, 82). Most changes in the composition of the semiarid grasslands in southeastern Arizona, and probably in New Mexico and western Texas as well, occurred after the beginning of large-scale cattle ranching and fire exclusion in the 1870s (6, 7, 86). Today, wildfires are rare. High densities of woody plants, low amounts of fuel, and extensive grassland fragmentation seem to be the cause.

The seedlings of many shrubs and other woody plants of the semiarid grasslands are sensitive to fire (117). Some will not resprout if their tops are killed (32), and others are susceptible to fire mortality until reaching an appreciable size (71). If these plants do not produce seeds before they are 10 y of age, then a fire return time of 10 y or less would keep these semiarid grasslands relatively free of woody plants (117). Fire-tolerant species would be suppressed by recurring fires and remain in the grassland at a small size (2). However, with a reduction of the fine fuel load by heavy and constant herbivory, fire frequencies would decrease. Further increases in woody plant cover and density would follow.

HERBIVORY

Herbivores may reduce the growth of individual plants by damaging the leaves, stems, or roots (14, 80, 99). Damage to plants by herbivores is determined by the timing of the encounter, location of the tissue eaten, amount of tissue eaten, and frequency of attack (50). By damaging plant parts, herbivores may alter a plant's ability to obtain resources or selectively eliminate a plant as a competitor, and thereby influence the outcome of species interactions (50, 65, 99). Furthermore, herbivory can increase the number of gaps present in the cover of these semiarid grassland communities, reduce the aboveground and belowground grass biomass, modify the pattern of resource availability, and alter biomass allocation (37, 38, 45, 46, 54, 146, 158). Therefore, at the population level, damage to individual plants by removal of biomass may lead to changes in plant abundance and distribution through alteration in fecundity or ability to regrow or through changes in mortality (50, 80). The ability of a plant to regrow after encounters with herbivores is usually reduced, and regrowth of grasses following removal of aboveground parts is usually associated with reductions in belowground growth and biomass (54, 145, 163; see however, 18, 22, 58, 113, 114, 128).

In the semiarid grasslands of the American southwest, brush encroachment has been coincident with or been preceded by development of the livestock industry (2, 3, 6, 7). Alterations in the grass species composition as well as reductions in herbaceous plant basal area, density, and aboveground and belowground biomass accompany chronic high levels of livestock grazing (84). Herbivory at low density and frequency may cause little change in a grassland community, but at high density and frequency, it can alter grassland composition, changing it to a shrubland or woodland (2).

If only grasses are consumed by herbivores, they may be at a disadvantage in their ability to interact with other plant species. If, on the other hand, browsers consume the shrubs or woody plants, the woody plants will be at a disadvantage (156). In systems in which woody plants are browsed, woody plant stature remains small and density remains static, but if the browsers are removed, woody plant size and density increase (66). Grazers and browsers in African grasslands and savannas exert a major influence on distribution and abundance of woody plants (15, 57, 153, 178). In the grasslands of central and southwestern North America, defoliation of woody seedlings by rodents, lagomorphs, and insects is an important source of mortality, particularly for *Prosopis glandulosa* (26, 70, 71, 83, 116, 126, 148). The black-tailed prairie dog, *Cynomys ludovicianus*, consumes seeds, pods, and seedlings of *Prosopis glandulosa* and maintains its colony's surface clear of seedlings and saplings (171, 172). Extensive anthropogenic eradication of prairie dogs and reduction of colonies by 98% in the early 1900s may have removed an important constraint to establishment of woody plants over a large area of the American southwest. However, this eradication apparently occurred after extensive encroachment of *Prosopis glandulosa* into many semiarid grasslands (6, 29, 55, 82).

SPREAD OF SEED

One of the theories concerning the maintenance of grasslands free of shrubs or woody plants during most of the Holocene was that dispersal of seeds of woody plants into grasslands was low because of a limited number seed dispersers. Some large and small mammals, including domestic livestock, feed on the fruit of woody plants and act as seed dispersal agents (27, 28, 96, 175). *Prosopis glandulosa*, a native legume, has a thick seed coat that requires scarification, which occurs during mastication; the seed survives passage through the gut of cattle and various native species (28, 96). The introduction of domestic herbivores may have increased the dispersal of *P. glandulosa* and other woody plant seeds, but many native herbivores could and probably still do the same thing (28, 96). In fact, *Prosopis* and other woody plants were present in many semiarid grasslands prior to increases in domestic herbivores, although plant density and probably stature have increased (8). These increases in woody plant density would not seem to require long-distance seed dispersal by domestic herbivores and could have been dispersed as far by native species.

MECHANISMS OF SHRUB OR WOODY PLANT ENCROACHMENT

The first and possibly the most critical factor causing woody plant encroachment is chronic high levels of herbivory by domestic animals. This causes a reduction of the aboveground grass biomass and of the light, fluffy fuel needed for grassland fires, thereby reducing fire frequency. With a lack of periodic fires, shrubs and other woody plants have a growth advantage over the grasses. Reducing aboveground and belowground biomass of grass through chronic high levels of herbivory by domestic ungulates would decrease the competitive ability of the grass (35, 37, 145, 158). At the same time, decreased grass cover would cause increased runoff and erosion leading to increases in temporal and spatial heterogeneity for soil resources, especially water, nitrogen, and probably phosphate (139–141). Low soil-nitrogen levels would favor establishment of species that have low soil nitrogen requirements, such as *Larrea tridentate* and many leguminous shrubs including acacias, paloverdes, and mesquite (61, 91). With the growth of woody plants, resources are partitioned differently. Clumps of shrubs or woody plants concentrate soil resources in many arid and semiarid communities; these areas are termed resource islands (4, 33, 34, 67, 94, 141). Soil resources are recycled in these resource islands, making them favored sites for germination and growth of woody plants that may establish readily, thereby making it more difficult for grass reestablishment. Cyclic droughts would seem to favor the deeper-rooted woody plants, but recent changes in the rainfall regime do not seem to be a cause of establishment. Past changes in temperature and atmospheric CO_2 do not appear to be the main cause of the rapid, recent shift in woody plant populations, but continued increases in temperature and atmospheric CO_2 concentration will probably play some future role. It is difficult to evaluate the past role of small native mammals and insects in reducing the survival and growth of brushy and woody plant seedlings thereby maintaining the grassland habitat. It would seem that their effects would not be uniform, but important in some areas. Shifts in plant populations and soil resources would reduce some populations of animals that feed on the seedlings of woody plants that could change woody seedling survival and further alter community structure.

CONCLUSIONS

The semiarid grasslands of southwestern North America have changed dramatically over the last 150 y. These grassland communities have not been invaded by non-native bushy, shrubby, or woody species. The process is encroachment and the species are native. Encroachment of native woody species has changed the appearance and structure of many of these former semiarid grasslands to shrublands, brushlands, or woodlands. *Prosopis* (mesquite) has probably increased in density and cover over a larger area than any other species. Although some authors have attributed the encroachment of shrubs or woody plants into the semiarid grasslands to one factor, often climate change, most recent studies have suggested an interaction of several factors. The major cause of the encroachment of these woody species seems to be the reduction of grass biomass (fine fuel) by chronic high levels of domestic herbivory coupled to a reduction of grassland fires, which would have killed or suppressed the woody plants to the advantage of the grasses. The role of plant competition and the spread of seeds by introduced domestic herbivores seem to be secondary and probably modified the rate of change. The role of many small native mammals and insects that consume woody plant seedlings seems to be secondary and possibly localized. Secondary factors probably modified the rate of change, rather than causing the change. Thus, the brushy and woody plants are not the cause of the changes in these semiarid grasslands as is so often presumed, but they are the result of the effect of changes of other factors on the species in these grassland communities.

The stature and composition of these communities has changed dramatically over the past 150 y and will continue to change in the future. Although chronic high levels of herbivory by domestic animals seems to have been the driving force behind the changes in stature and composition of these semiarid grasslands, the direction of future change or trends is difficult to predict. All of the above factors will probably continue to interact to regulate community composition and structure. The density of shrubby and woody plants will probably increase with some increase in stature and number of species. However, if soil nutrients increase, the woody legumes may be replaced by other shrubby or woody species. Reversing the trend or changes that have been going on for 150 y will be a difficult, probably long-term process, and possibly impossible task.

ACKNOWLEDGMENTS

I thank my colleagues AT Tsin, LE Gilbert, and WH Schlesinger for interesting discussions and considerable long-term encouragement. I also thank many of my students who have helped shape my thoughts about the semiarid grasslands and other desert communities. I especially thank my colleague and friend JK Bush for more help than I can ever repay.

Visit the Annual Reviews home page at www.AnnualReviews.org

LITERATURE CITED

1. Adamoli J, Sennhauser E, Acero JM, Rescia A. 1990. Stress and disturbance: vegetation in the dry Chaco region of Argentina. *J. Biogeogr.* 17:491–500
2. Archer S. 1994. Woody plant encroachment into southwestern grasslands and savannas: rates, patterns and proximate causes. In *Ecological Implications of Livestock Herbivory in the West*, ed. M Vavra, WA Laycock, RD Pieper, pp. 13–69. Denver: Soc. Range Manage.
3. Archer S, Schimel DS, Holland EA. 1995. Mechanisms of shrubland expansion: land use, climate or CO_2? *Clim. Change* 29:91–9
4. Archer S, Scifres C, Bassham CR, Maggio R. 1988. Autogenic succession in a subtropical savanna: conversion of grassland to thorn woodland. *Ecol. Monogr.* 52: 111–27
5. Archer SR. 1989. Have southern Texas savannas been converted to woodlands in recent history? *Am. Nat.* 134:545–61
6. Bahre CJ. 1995. Human impacts on the grasslands of southeastern Arizona. See Ref. 110, pp. 230–64
7. Bahre CJ. 1991. *A Legacy of Change: Historic Human Impact on Vegetation of the Arizona Borderlands.* Tucson: Univ. Ariz. Press
8. Bahre CJ, Shelton ML. 1993. Historic vegetation change, mesquite increases, and climate in southeastern Arizona. *J. Biogeogr.* 20:489–504
9. Bailey RG. 1978. *Description of the Ecoregions of the United States.* Ogden, UT: US Dep. Agric., For. Serv., Intermountain Region
10. Baker HG. 1986. Patterns of plant invasion in North America. See Ref. 20a, pp. 44–57
11. Bartolome JW. 1993. Application of herbivory optimization theory to rangelands of the western United States. *Ecol. Appl.* 3:27–9
12. Bazzaz FA. 1986. Life history of colonizing plants: some demographic, genetic, and physiological features. See Ref. 20a, pp. 96–110
13. Beadle NCW. 1981. *The Vegetation of Australia.* New York: Gustav Fischer Verlag
14. Belsky AJ. 1986. Does herbivory benefit plants? A review of the evidence. *Am. Nat.* 127:870–92
15. Belsky AJ. 1984. Role of small browsing mammals in preventing woodland regeneration in the Serengeti National Park, Tanzania. *Africa J. Ecol.* 22:271–9
16. Belsky AJ. 1996. Viewpoint: Western juniper expansion: Is it a threat to arid northwestern ecosystems? *J. Range Manage.* 49:53–9
17. Belsky AJ, Blumenthal DM. 1997. Effects of livestock grazing on stand dynamics and soils in upland forests of the interior west. *Conserv. Biol.* 11:315–27
18. Belsky AJ, Matzke A, Uselman S. 1999. Survey of livestock influences on stream and riparian ecosystems in the western United States. *J. Soil Water Conserv.* 54:419–31
19. Berendse FA. 1981. Competition between plant populations with different rooting depths. II. Pot experiments. *Oecologia* 48:334–41
20. Betancourt JL, Van Devender TR, Martin PS. 1990. Synthesis and prospectus. In *Packrat Middens: The Last 40,000 Years of Biotic Change*, ed. JL Betancourt, TR Van Devender, PS Martin, pp. 435–47. Tucson: Univ. Ariz. Press

20a. Billings WD, Golley F, Lange OL, Olson JS, Remmert H, ed. 1986. *Ecology of Biological Invasions of North America and Hawaii.* New York: Springer-Verlag

21. Blackburn WH, Tueller PT. 1970. Pinyon and juniper invasion in black sagebrush communities in east-central Nevada. *Ecology* 51:841–8
22. Briske D. 1993. Grazing optimization: a plea for a balanced perspective. *Ecol. Appl.* 3:24–6
23. Brock JH, Wade M, Pysek P, Green D, eds. 1997. *Plant Invasions: Studies from North America and Europe.* Leiden: Backhuys
24. Brown DE. 1982. Plains and Great Basin grassland. In *Biotic Communities of the American Southwest–United States and Mexico*, ed. DE Brown. *Desert Plants* 4:115–21
25. Brown DE. 1982. Semidesert grassland. In *Biotic Communities of the American Southwest–United States and Mexico*, ed. DE Brown. *Desert Plants* 4:123–31
26. Brown JH, Heske EJ. 1990. Control of a desert grassland transition by a keystone rodent guild. *Science* 250:1705–7
27. Brown JR, Archer S. 1999. Shrub invasion of grassland: recruitment is continuous and not regulated by herbaceous biomass or density. *Ecology* 80:2385–96
28. Brown JR, Archer S. 1989. Woody plant invasion of grasslands: establishment of honey mesquite (*Prosopis glandulosa* var. *glandulosa*) on sites differing in herbaceous biomass and grazing history. *Oecologia* 80:19–26
29. Buffington LC, Herbel CH. 1965. Vegetational changes on a semidesert grassland range from 1858 to 1963. *Ecol. Monogr.* 35:139–64
30. Burgess TL. 1995. Desert grassland, mixed shrub savanna, shrub steppe, or semidesert shrub? The dilemma of coexisting growth forms. See Ref. 110, pp. 31–67
31. Burgess TL, Bowers JE, Turner RM. 1991. Exotic plants at the Desert Laboratory, Tucson, Arizona. *Madrono* 38:96–114
32. Burkhardt JW, Tisdale EW. 1976. Causes of juniper invasion in southwestern Idaho. *Ecology* 57:472–84
33. Burquez A, de los Angeles-Quintana M. 1994. Islands of diversity: ironwood ecology and the richness of perennials in a Sonoran Desert biological reserve. In *Ironwood: An Ecological and Cultural Keystone on the Sonoran Desert*, ed. GP Nabhan, JL Carr, pp. 9–27. Tucson: Conserv. Int.
34. Bush JK, Van Auken OW. 1986. Changes in nitrogen, carbon, and other surface soil properties during secondary succession. *Soil Sci. Soc. Am. J.* 50:1597–1601
35. Bush JK, Van Auken OW. 1991. Importance of time of germination and soil depth on growth of *Prosopis glandulosa* (Leguminosae) seedlings in the presence of a C_4 grass. *Am. J. Bot.* 78:1732–39
36. Bush JK, Van Auken OW. 1989. Soil resource levels and competition between a woody and herbaceous species. *Bull. Torr. Bot. Club* 116:22–30
37. Bush JK, Van Auken OW. 1995. Woody plant growth related to planting time and clipping of a C_4 grass. *Ecology* 76: 1603–9
38. Caldwell MM, Richards JH, Johnson DA, Nowak RS, Dzurec RS. 1981. Coping with herbivory: photosynthetic capacity and resource allocation in two semiarid *Agropyron* bunchgrasses. *Oecologia* 50:14–24
39. Callaway FM, DeLucia EH, Moore D, Nowak R, Schlesinger WH. 1996. Competition and faciliation: contrasting effects of *Artemisa tridentata* on desert vs. montane pines. *Ecology* 77:2130–41
40. Callaway RM. 1995. Positive interactions among plants. *Bot. Rev.* 61:306–49
41. Casper BB, Jackson RB. 1997. Plant competition underground. *Annu. Rev. Ecol. Syst.* 28:545–70
42. Chapin FS III, Sala OE, Burke IC, Grime JP, Hooper DU, et al. 1998. Ecosystem consequences of changing biodiversity. *BioScience* 48:45–52
43. Cohn EJ, Van Auken OW, Bush JK. 1989. Competitive interactions between *Cynodon dactylon* and *Acacia smallii* seedlings

at different nutrient levels. *Am. Midl. Nat.* 121:265– 72
44. Cole MM. 1986. *The Savannas: Biogeography and Geobotany.* New York: Academic Press
45. Collins SL. 1987. Interaction of disturbances in tall grass prairie: a field experiment. *Ecology* 68:1243–50
46. Collins SL, Knapp AK, Briggs JM, Blair JM, Steinauer EM. 1998. Modulation of diversity by grazing and mowing in native tallgrass prairie. *Science* 280:745–47
47. Collins SL, Wallace LL. 1990. *Fire in North American Tallgrass Prairies.* Norman, OK: Univ. Okla. Press
48. Connell JH. 1983. On the prevalence and relative importance of interspecific competition: evidence from field experiments. *Am. Nat.* 122:661–96
49. Correll DS, Johnston MC. 1970. *Manual of the Vascular Plants of Texas.* Renner, TX: Texas Res. Foun.
50. Crawley MJ. 1997. Plant-herbivore dynamics. In *Plant Ecology*, ed. MJ Crawley, pp. 401–74. Oxford: Blackwell Sci.
51. Davis MB. 1981. Quaternary history and the stability of forest communities. In *Forest Succession: Concepts and Applications*, ed. DC West, HH Shugart, DB Botkin, pp. 132–53. New York: Springer–Verlag
52. Delcourt PA, Delcourt HR. 1981. Vegetation maps for the eastern United States: 40,000 yr B. P. to present. In *Geobotany II*, ed. RC Romans, pp. 123–65. New York: Plenum
53. Delcourt PA, Delcourt HR, Webb T. 1983. Dynamic plant ecology: the spectrum of vegetation change in space and time. *Q. Sci. Rev.* 1:153–75
54. Detling JK, Dyer MI, Winn DT. 1979. Net photosynthesis, root respiration, and regrowth of *Bouteloua gracilis* following simulated grazing. *Oecologia* 41:127–34
55. Dick–Peddie WA. 1993. *New Mexico Vegetation: Past, Present, and Future.* Albuquerque: Univ. New Mexico Press
56. Dobson AP, Bradshaw AD, Baker AJM. 1997. Hopes for the future: restoration ecology and conservation biology. *Science* 277:515–22
57. Dublin HT, Sinclair ARE, McGlade J. 1990. Elephants and fire as causes of multiple stable states in the Serengeti-Mara woodlands. *J. Animal Ecol.* 49:1147–64
58. Dyer MI, Turner CL, Seastedt TR. 1993. Herbivory and its consequences. *Ecol. Appl.* 3:10–16
59. Eddleman LE. 1987. Establishment and stand development of western juniper in central Oregon. In *Proc. Pinyon-Juniper Conf.*, ed. RL Everett, pp. 255–59. Ogden, UT: US For. Serv. General Tech. Rep. INT–215.
60. Enserink M. 1999. Biological invaders sweep in. *Science* 285:1834–36
61. Felker P, Clark PR. 1980. Nitrogen fixation (acetylene reduction) and cross inoculation in 12 *Prosopis* (mesquite) species. *Plant Soil* 57:114–26
62. Felker P, Smith D, Smith M, Bingham RL, Reyes I. 1984. Evaluation of herbicides for use in transplanting *Leucaena leucocephala* and *Prosopis alba* on semiarid lands without irrigation. *For. Sci.* 30: 747–55
63. Fitzpatrick EA, Nix HA. 1970. The climatic factor in Australian grassland ecology. In *Australian Grasslands,* ed. RM Moore, pp. 3–26. Canberra: Aust. Natl. Univ. Press
64. Fleischner TL. 1994. Ecological costs of livestock grazing in western North America. *Conserv. Biol.* 8:629–44
65. Fowler NL. 1986. The role of competition in plant communities in arid and semiarid regions. *Annu. Rev. Ecol. Syst.* 17:89–110
66. Friedel MH. 1985. The population structure and density of central Australian trees and shrubs, and relationships to range condition, rabbit abundance and soil. *Aust. Rang. J.* 7:130–39
67. Garcia-Moya E, McKell CM. 1970. Contribution of shrubs to the nitrogen economy

of a desert-wash plant community. *Ecology* 51:81–88
68. Gardener CJ, McIvor JG, Willians J. 1990. Dry tropical rangelands: solving one problem and creating another. *Proc. Ecol. Soc. Aust.* 16:279–86
69. Germain HL, McPherson GR. 1999. Effects of biotic factors on emergence and survival of *Quercus emoryi* at lower tree line. *Ecoscience* 6:92–99
70. Gibbens RP, Havstad KM, Billheimer DD, Herbel CH. 1993. Creosotebush vegetation after 50 years of lagomorph exclusion. *Oecologia* 94:210–17
71. Glendening GE, Paulsen HAJ. 1955. Reproduction and establishment of velvet mesquite as related to invasion of semidesert grasslands. *Tech. Bull., USDA* 1127:1–50
72. Goldberg DE. 1994. Influence of competition at the community level: an experimental version of the null models approach. *Ecology* 75:1503–6
73. Grace JB. 1993. The effects of habitat productivity on competition intensity. *Trends Ecol. Evol.* 8:229–30
74. Grace JB. 1995. On the measurement of plant competition intensity. *Ecology* 76:305–8
75. Grace JB, Tilman D. 1990. *Perspective on Plant Competition.* New York: Academic Press
76. Grime JP. 1979. *Plant Strategies and Vegetation Processes.* New York: Wiley
77. Grover HD, Musick HB. 1990. Shrubland encrochment in southern New Mexico, USA: an analysis of desertification processes in the American southwest. *Clim. Change* 17:305– 30
78. Gruger J. 1973. Studies on the vegetation of northeastern Kansas. *Geol. Soc. Am. Bull.* 84:239–50
79. Halfsten U. 1961. Pleistocene development of vegetation and climate in the southern high plains as evidenced by pollen. In *Paleoecology of the Llano Estacato. Fort Burgwin Research Center Report I*, ed. F Wencorf, pp. 59–91. Albuquerque: Univ. New Mexico Press
80. Harper JL. 1977. *Population Biology of Plants.* New York: Academic Press
81. Harrington GN, Wilson AD, Young MD. 1984. *Management of Australia's Rangelands.* Melbourne, Aust.: Commonwealth Sci. Industrial Res. Org.
82. Hastings JR, Turner RM. 1965. *The Changing Mile.* Tucson: Univ. Ariz. Press
83. Havstad KM, Gibbens RP, Knorr CA, Murray LW. 1999. Long-term influences of shrub removal and lagomorph exclusion on Chihuahuan Desert vegetation dynamics. *J. Arid Environ.* 42:155–66
84. Heitschmidt RK, Stuth JW. 1991. *Grazing Management: An Ecological Perspective.* Portland: Timberline
85. Hevly RH, Martin PS. 1961. Geochronology of pluvial Lake Cochise, southern Arizona. I. Pollen analysis of shore deposits. *J. Ariz. Acad. Sci.* 2:24–31
86. Humphrey RR. 1958. The desert grassland: a history of vegetational change and an analysis of causes. *Bot. Rev.* 24:193–252
87. Idso SB. 1992. Shrubland expansion in the American southwest. *Clim. Change* 22: 85–86
88. Imbrie J, Imbrie KP. 1979. *Ice Ages: Solving the Mystery.* Short Hills, NJ: Enslow
89. Inglis JM. 1964. *A History of Vegetation on the Rio Grande Plain.* Austin: Texas Parks Wildlife Dept.
90. Johnsen TN. 1962. One-seed juniper invasion of Northern Arizona grasslands. *Ecol. Monogr.* 32:187–207
91. Johnson HB, Meyeux HS Jr. 1990. *Prosopis glandulosa* and the nitrogen balance of rangeland: extent and occurrence of nodulation. *Oecologia* 84:176–85
92. Johnson HB, Polley HW, Mayeux HS. 1993. Increasing CO_2 and plant-plant interactions: effects on natural vegetation. *Vegetation* 104-105:157–70
93. Johnston MC. 1963. Past and present grasslands of southern Texas and northeastern Mexico. *Ecology* 44:456–66

94. Jurena PN, Van Auken OW. 1998. Woody plant recruitment under canopies of two Acacias in a southwestern Texas shrubland. *Southwest. Nat.* 43:195–203
95. Kaiser J. 1999. Stemming the tide of invading species. *Science* 285:1836–41
96. Kamp BA, Ansley RJ, Tunnell TR. 1998. Survival of mesquite seedlings emerging from cattle and wildlife feces in a semi-arid grassland. *Southwest. Nat.* 43: 300–12
97. Knoop WT, Walker BH. 1985. Interactions of woody and herbaceous vegetation in a southern African savanna. *J. Ecol.* 73:235–53
98. Lauenroth WK. 1979. Grassland primary production: North American grasslands in perspective. In *Perspectives in Grassland Ecology: Results and Applications of the US/IBP Grassland Biome Study*, ed. NR French, pp. 3–24. New York: Springer-Verlag
99. Louda SM, Keeler KH, Holt RD. 1990. Herbivory influences on plant performance and competative interactions. See Ref. 75, pp. 413–44
100. Mack RN. 1986. Alien plant invasion into the Intermountain West: a case history. See Ref. 20a, pp. 191–213
101. Mahall BE, Callaway RM. 1992. Root communication mechanisms and intracommunity distributions of two Mojave desert shrubs. *Ecology* 73:2145–51
102. Mahall BE, Callaway RM. 1991. Root communications among desert shrubs. *Proc. Natl. Acad. Sci.* 88:874–6
103. Malakoff D. 1999. Fighting fire with fire. *Science* 285:1841–43
104. Martin PS. 1999. Deep history and a wilder West. See Ref. 136a, pp. 255–90
105. Martin SC, Reynolds HG. 1973. The Santa Rita Experimental Range: your facility for research on semidesert ecosystems. *J. Ariz. Acad. Sci.* 8:56–67
106. Mayeux HS, Johnson HB, Polley HW. 1991. Global change and vegetation dynamics. In *Noxious Range Weeds*, ed. LF James, JO Evans, MH Ralphs, BJ Sigler, pp. 62–74. Boulder, CO: Westview
107. McAuliffe JR. 1995. Landscape evolution, soil formation, and Arizona's Desert Grassland. See Ref. 110, pp. 100–29
108. McAuliffe JR. 1999. The Sonoran Desert: landscape complexity and ecological diversity. See Ref. 136a, pp. 68–114
109. McClaran MP. 1995. Desert grasslands and grasses. See Ref. 110, pp. 1–30
110. McClaran MP, Van Devender TR, eds. 1995. *The Desert Grassland*. Tucson: Univ. Ariz. Press
111. McDowell PF, Webb T III, Bartlein PJ. 1995. Long-term environmental change. In *Ecological Time Series*, ed. TM Powell, JH Steele, pp. 327–70. New York: Chapman & Hall
112. McLaughlin SP, Bowers JE. 1999. Diversity and affinities of the flora of the Sonoran Floristic Province. See Ref. 136a, pp. 12–35
113. McNaughton SJ. 1993. Grasses and grazers, science and management. *Ecol. Appl.* 3:17–20
114. McNaughton SJ, Banyikwa FF, McNaughton MM. 1998. Root biomass and productivity in a grazing system: the Serengeti. *Ecology* 79:587–92
115. McPherson GR. 1997. *Ecology and Management of North American Savannas*. Tucson: Univ. Ariz. Press
116. McPherson GR. 1993. Effects of herbivory and herb interference on oak establishment in a semi-arid temperate savanna. *J. Veg. Sci.* 4:687–92
117. McPherson GR. 1995. The role of fire in the Desert Grassland. See Ref. 110, pp. 131–51
118. McPherson GR, Wright HW, Wester DB. 1988. Patterns of shrub invasion in semiarid Texas grasslands. *Am. Midl. Nat.* 120:391–97
119. Milchunas DG, Lauenroth WK. 1993. Quantitative effects of grazing on vegetation and soils over a global range of environments. *Ecol. Monogr.* 63:327–66

120. Miller RF, Wigand PE. 1994. Holocene changes in semiarid pinyon-juniper woodlands: response to climate, fire, and human activies in the Great Basin. *Bioscience* 44:465–74
121. Milton SJ, Dean WR, du Plessis MA, Siegfried WR. 1994. A conceptual model of arid rangeland degredation. *Bioscience* 44:70–6
122. Misra R. 1983. Indian savannas. In *Tropical Savannas,* ed. F Bourliere, pp. 151–66. New York: Elsevier
123. Mooney HA, Drake JA, eds. 1986. *Ecology of Biological Invasions of North America and Hawaii.* New York: Springer–Verlag
124. Moore RM, Condon RW, Leigh JH. 1970. Semi-arid woodlands. In *Australian Grasslands*, ed. RM Moore, pp. 228–45. Canberra: Aust. Natl. Univ. Press
125. Neilson RP, 1986. High-resolution climatic analysis and southwest biogeography. *Science* 232:27–34
126. Nilsen ET, Sharifi MR, Virginia RA, Rundel PW. 1987. Phenology of warm desert phreatophytes: seasonal growth and herbivory in *Prosopis glandulosa* var. *torreyana* (honey mesquite). *J. Arid Env.* 13:311–18
127. Noy-Meir I. 1993. Compensating growth of grazed plants and its relevance to the use of rangelands. *Ecol. Appl.* 3:32–4
128. Painter EL, Belsky AJ. 1993. Application of herbivore optimization theory to rangelands of the western United States. *Ecol. Appl.* 3:2–9
129. Patten DT. 1993. Herbivore optimization and overcompensation: Does native herbivory of western rangelands support these thories? *Ecol. Appl.* 3:35–6
130. Perry RA. 1970. Arid shrublands and grasslands. In *Australian Grasslands,* ed. RM Moore, pp. 246–59. Canberra: Aust. Natl. Univ. Press
131. Pimentel D. 1986. Biological invasions of plants and animals in agriculture and forestry. See Ref. 20a, pp. 149–62
132. Pimm SL, Russell GJ, Gittleman JL, Brooks TM. 1995. The future of biodiversity. *Science* 269:347–50
133. Polley HW, Johnson HB, Mayeux HS. 1992. Carbon dioxide and water fluxes of C_3 annuals and C_3 and C_4 perennials at subambient CO_2 concentrations. *Funct. Ecol.* 6:693–703
134. Polley HW, Johnson HB, Mayeux HS. 1994. Increasing CO_2: comparative responses of the C_4 grass *Schizachyrium scoparium* and grassland invader *Prosopis. Ecology* 75:976–88
135. Pugnaire FI, Haase P, Puigdefabregas J. 1996. Facilitation between higher plant species in a semiarid environment. *Ecology* 77:1420–6
136. Reynolds JF, Virginia RA, Kemp PR, de Soyza AG, Tremmel DC. 1999. Impact of drought on desert shrubs: effects of seasonality and degree of resource island development. *Ecol. Monogr.* 69:69–106
136a. Robichaux RH, ed. 1999. *Ecology of the Sonoran Desert Plants and Plant Communities*. Tucson: Univ. Ariz. Press
137. Roundy BA, Biedenbender SH. 1995. Revegetation in the Desert Grassland. See Ref. 110, pp. 265–303
138. Rzedowski J. 1978. *Vegetacion de Mexico.* Mexico: Limusa
139. Schlesinger WH, Abrahams AD, Parsons AJ, Wainwright J. 1999. Nutrient losses in runoff from grassland and shrubland habitats in southern New Mexico: rainfall simulation experiments. *Biogeochemistry* 45:21–34
140. Schlesinger WH, Raikes JA, Hartley AE, Cross AF. 1996. On the spatial pattern of soil nutrients in desert ecosystems. *Ecology* 77:364–74
141. Schlesinger WH, Reynolds JF, Cunningham GL, Huenneke LF, Jarrell WM, et al. 1990. Biological feedbacks in global desertification. *Science* 247:1043–48

142. Schmutz EM, Smith EL, Ogden PR, Cox ML, Klemmedson JO, et al. 1991. Desert grassland. In *Natural Grasslands: Introduction and Western Hemisphere*, ed. RT Coupland, pp. 337–62. Amsterdam: Elsevier
143. Schoener TW. 1983. Field experiments on interspecific competition. *Am. Nat.* 122:240–85
144. Scholes RJ, Archer SR. 1997. Tree-grass interactions in savannas. *Annu. Rev. Ecol. Syst.* 28:517–44
145. Schuster JL. 1964. Root development of native plants under three grazing intensities. *Ecology* 45:63–70
146. Seagle SJ, McNaughton J, Russ RW. 1992. Simulated effects of grazing on soil nitrogen and mineralization in contrasting Serengeti grassland. *Ecology* 73:1105–23
147. Sears PB, Clisby KH. 1956. San Agustin Plains-Pleistocene climate changes. *Science* 124:537–9
148. Simpson BB. 1977. *Mesquite: Its biology in two desert scrub ecosystems.* Stroudsberg, PA: Dowden, Hutchinson & Ross
149. Sims PL, Singh JS, Lauenroth WK. 1978. The structure and function of the ten western North American grasslands. I. Abiotic and vegetational characteristics. *J. Ecol.* 66:251–85
150. Smith TM, Goodman PS. 1987. Successional dynamics in an *Acacia nilotica–Euclea divinorum* savannah in southern Africa. *J. Ecol.* 75:603–10
151. Soriano A. 1979. Distribution of grasses and grasslands in South America. In *Ecology of Grasslands and Bamboolands in the World*, ed. M Numata, pp. 84–91. Boston: Dr. W. Junk
152. Stromberg JC, Gengarelly L, Rogers BF. 1997. Exotic herbaceous species in Arizona's riparian ecosystems. In *Plant Invasions: Studies from North America and Europe*, ed. JH Brock, M Wade, P Pysek, D Green, pp. 45–57. Leiden: Backhuys
153. Stuart-Hill GC. 1992. Effects of elephants and goats on the Kaffrarian succulent thicket on the eastern Cape, South Africa. *J. Appl. Ecol.* 29:699–710
154. Swetnam TW, Betancourt JL. 1990. Fire–southern oscillation relations in the southwestern United States. *Science* 249: 1017–20
155. Twolan–Strutt L, Keddy PA. 1996. Above- and belowground competitive intensity in two contrasting wetland plant communities. *Ecology* 77:256–70
156. Van Auken OW. 1994. Changes in competition between a C_4 grass and a woody legume with differential herbivory. *Southwest. Nat.* 39:114–21
157. Van Auken OW, Bush JK. 1988. Competition between *Schizachyrium scoparium* and *Prosopis glandulosa. Am. J. Bot.* 75:782–9
158. Van Auken OW, Bush JK. 1997. Growth of *Prosopis glandulosa* in response to changes in aboveground and belowground interference. *Ecology* 78:1222–9
159. Van Auken OW, Bush JK. 1990. Importance of grass density and time of planting on *Prosopis glandulosa* seedling growth. *Southwest. Nat.* 35:411–5
160. Van Auken OW, Bush JK. 1987. Influence of plant density on the growth of *Prosopis glandulosa* var. *glandulosa* and *Buchloe dactyloides. Bull. Torr. Bot. Club* 114:393–401
161. Van Auken OW, Bush JK. 1991. Influence of shade and herbaceous competition on the seedling growth of two woody species. *Madrono* 38:149–57
162. Van Auken OW, Bush JK. 1990. Interactions of two C_3 and C_4 grasses with seedlings of *Acacia smallii* and *Celtis laevigata. Southwest. Nat.* 35:316–21
163. Van Auken OW, Bush JK. 1989. *Prosopis glandulosa* growth: influence of nutrients and simulated grazing of *Bouteloua curtipendula. Ecology* 70:512–6
164. Van Auken OW, Bush JK. 1985. Secondary succession on terraces of the San Antonio River. *Bull. Torr. Bot. Club* 112:158–66

165. Van Auken OW, Manwaring JH, Caldwell MM. 1992. Effectiveness of phosphate acquisition by juvenile cold-desert perenials from different patterns of fertile–soil microsites. *Oecologia* 91:1–6
166. Van Devender TR. 1995. Desert grassland history: changing climates, evolution, biography, and community dynamics. See Ref. 110, pp. 68–99
167. Van Devender TR, Spaulding WG. 1979. Development of vegetation and climate in the southwestern United States. *Science* 204:701–10
168. Van Vegten JA. 1983. Thornbush invasion in a savanna ecosystem in eastern Botswana. *Vegetatio* 56:3–7
169. Vasek FC. 1980. Creosote bush: long-lived clones in the Mojave Desert. *Am. J. Bot.* 67:246–55
170. Wells PV. 1965. Scarp woodlands, transported grassland soils, and concept of grassland climate in the Great Plains Region. *Science* 148:246–9
171. Weltzin J, Archer S, Heitschmidt R. 1997. Small-mammal regulation of vegetation structure in a temperate savanna. *Ecology* 78:751–63
172. Weltzin JF, Archer S, Heitschmidt RK. 1998. Defoliation and woody plant (*Prosopis glandulosa*) seedling regeneration: potential vs realized herbivory tolerance. *Plant Ecol.* 138:127–35
173. Weltzin JF, McPherson GR. 1999. Facilitation of conspecific seedling recruitment and shifts in temperate savanna ecotones. *Ecol. Monogr.* 69:513–34
174. West NE. 1984. Successional patterns and productivity potentials of pinyon-juniper ecosystems. In *Developing Strategies for Range Management*, ed. NRCNAS, pp. 1301–32. Boulder, CO: Westview
175. Wilson M. 1993. Mammals as seed-dispersal mutualists in North America. *Oikos* 67:159–76
176. Wilson SD. 1998. Competition between grasses and woody plants. In *Population Biology of Grasses*, ed. GP Cheplick, pp. 231–54. Cambridge, UK: Cambridge Univ. Press
177. Wright HA, Bailey AW. 1980. *Fire Ecology and Prescribed Burning in the Great Plains–A Research Review.* Washington, DC: US Dept. Agric., For. Serv. General Tech. Rep. INT–77
178. Yeaton RI. 1988. Porcupines, fires and the dynamics of the tree layer of the *Burkea africana* savanna. *J. Ecol.* 76:1017–29

Annu. Rev. Ecol. Syst. 2000. 31:217–38

THE GRASSES: A Case Study in Macroevolution

Elizabeth A. Kellogg
Department of Biology, University of Missouri-St. Louis, St. Louis, Missouri 63121; e-mail: tkellogg@umsl.edu

Key Words Poaceae, genome, development, gene expression, unisexual flowers

■ **Abstract** Macroevolution in the grasses has often involved change in the position (heterotopy) of developmental programs, possibly via ectopic gene expression. Heterotopy apparently has been involved in the evolution of unique epidermal morphology in the grasses and their sister genus, *Joinvillea*; in the origin of the grass flower and possibly in the spikelet as well; in the formation of unisexual flowers in the panicoid grasses, and in the repeated origin of C_4 photosynthesis. Change in timing of development (heterochrony) may explain the novel morphology of the grass embryo. Changes in the structure and size of the nuclear genome correlate with phylogenetically informative cytogenetic characteristics. Most of the 10,000 species of grasses evolved tens of millions of years after the common ancestor of the family, indicating that the origin of novel morphologies did not lead to immediate radiation.

INTRODUCTION

Macroevolution has been defined as the evolution of higher taxa (51), large groups of organisms traditionally classified as tribes, subfamilies, and families, or higher ranks. Among multicellular organisms, taxa at such ranks often have been named because they have some distinctive morphological character. It follows that the study of macroevolution must involve, at least in part, the study of the origin and persistence of morphological novelties. In particular, a macroevolutionary study requires (*a*) a robust phylogeny, (*b*) a detailed developmental description of morphological characters, and (*c*) an understanding of the genetic basis of morphological characters. There are few groups that meet all three criteria for a truly integrated picture of the evolutionary process. Among plants, the best candidate is the grass family (75, 77, 78, 81).

Grasses occur on all continents and are ecologically dominant in ecosystems such as the African and South American savannas (108). The family includes some 10,000 species, among which are the cereals—wheat, rice, maize, sorghum, sugar cane, pearl millet, foxtail millet, common millet, finger millet, tef, barley, rye, oats—as well as pasture and forage grasses too numerous to list here (27, 122). Because of their ecological and economic importance, the grasses have been studied for centuries, and much is known about their morphology, physiology, ecology, population biology, developmental genetics, molecular biology, and cytogenetics.

Maize (*Zea mays*) is one of the major plant model systems, and has incomparable resources for the study of genetics (see http://www.agron.missouri.edu/). Genetic studies by Doebley and his colleagues (39,41–46) have identified five genetic loci that control morphological differences between cultivated maize and its wild ancestors, the teosintes. These studies directly address the question of the genes behind morphological characteristics, and offer some of the first clear hypotheses about how selection acts to produce major morphological change.

Extrapolation of the maize results to larger groups of grasses, however, assumes that events underlying major radiations are qualitatively the same as events underlying population differentiation and/or speciation. This assumption could be tested if we knew the genetic basis of the morphological characters distinguishing major clades. Such a test is difficult, because the common ancestor of major groups has been extinct for millions of years, and members of the groups long ago ceased to be interfertile. Thus the genetic approach used to dissect the maize/teosinte distinctions is inapplicable. Inferring the basis of morphological change instead rests heavily on circumstantial evidence and inference from comparative data. Nonetheless, considerable progress has been made in recent years.

In this review I summarize briefly our understanding of the phylogeny of the grasses. This is sufficiently robust that it can be taken as given. I will then describe, or re-describe, a handful of systematic characters, about which we have learned considerably more over the last several years. In general, many characteristics of major clades—some of which have been used for decades as taxonomic characters—can be explained as a change in position or timing of a developmental program. This implies that structures or pathways that were physically or temporally linked in ancestral taxa have become decoupled in the descendants.

PHYLOGENY

Sister Taxa

Phylogenetic studies have shown that the grass family (Poaceae or Gramineae) is most closely related to several small families of eastern Asia and the Southern Hemisphere (19, 21, 22, 33, 47, 48, 80, 94). These are Restionaceae (38 genera/approximately (ca.) 420 species), Centrolepidaceae (3/28), Joinvilleaceae (1/2), Ecdeiocoleaceae (1/1), Anarthriaceae (1/6), and Flagellariaceae (1/4), all much less speciose than the grasses. The families are of little or no economic importance, with the possible exception of members of the Restionaceae, some of which are used for thatch in southern Africa (93). In the most recent classification of the angiosperms, these are all members of the graminoid clade, a monophyletic subset of the order Poales (3a, 22).

The sister family of the grasses is Joinvilleaceae, forest-margin plants of Southeast Asia and the South Pacific. *Joinvillea* differs from the grasses in having conventional sepals and petals and a berry-like fruit. Although Cronquist (31) thought the grasses were closest to the sedges (Cyperaceae), there is little or no molecular evidence for this, nor is there evidence for Clifford's (28) suggestion that the grasses are close to the palms.

Like many members of the Poales, all families of the graminoid clade are apparently wind-pollinated, have the capacity to deposit silica somewhere in the plant, and have leaves borne in two vertical ranks (distichous phyllotaxy). These characteristics thus evolved long before the origin of the grass family itself.

Several members of the graminoid clade have unusual types of cells in their leaves or stems. Restionaceae and some grasses have photosynthetic cells with invaginated cell walls; these are known as peg cells in Restionaceae and arm cells in the grasses. Restionaceae lack leaves at maturity, and photosynthesis occurs in peg cells in the stems. Grass arm cells occur in the mesophyll of many species in the Bambusoideae, Ehrhartoideae (= Oryzoideae), and three other small subfamilies, and also are the site of photosynthesis. Ultrastructure of these cells has not been investigated, and their homology is only a matter of conjecture. It is possible, however, that a similar developmental program is used in the stems of Restionaceae and in the leaves of some grasses.

Species of *Joinvillea* and some grasses have large colorless cells in the central mesophyll (79, 112, 120). These cells appear empty at maturity, although in *Joinvillea* they may accumulate deposits of tannin or silica (120). In the grasses, these are called fusoid cells; they occur in all the basal grasses, the bamboos, and a handful of genera in the subfamily Panicoideae. Their development has not been studied and it is not known whether they are homologous wherever they occur.

Grass Phylogeny

Many phylogenetic studies have assessed relationships among the grasses themselves (6–8, 26, 34, 61–63, 76, 79, 82, 96, 97, 114). The greatest insights have come from studies that incorporate molecular data, and the largest molecular data sets have recently been combined into a single compilation of gene sequences (56). A robust and detailed phylogeny of the family is now available (Figure 1).

Whether the trees are combined to create a semistrict consensus (76), or the data are combined to produce a single large data set (56; http://www.ftg.fiu.edu/grass/gpwg/default.html), many results are the same. The earliest diverging lineages ["basal grasses" (73)] are the Anomochlooideae, Pharoideae, and Puelioideae (25), all herbaceous plants of tropical forest understories. The basal grasses were once included in the Bambusoideae (e.g. 113), but are clearly only distantly related to the bamboos. All molecular studies with sufficient sampling find that each of the subfamilies Bambusoideae, Ehrhartoideae (Oryzoideae), Pooideae, Aristidoideae, Chloridoideae, and Panicoideae is monophyletic. (The genus *Ehrharta* is clearly sister to the rest of the rice group and should be included with it, a result that is strongly supported by molecular data. Unfortunately, the subfamily name Ehrhartoideae is older than the more familiar Oryzoideae, so the

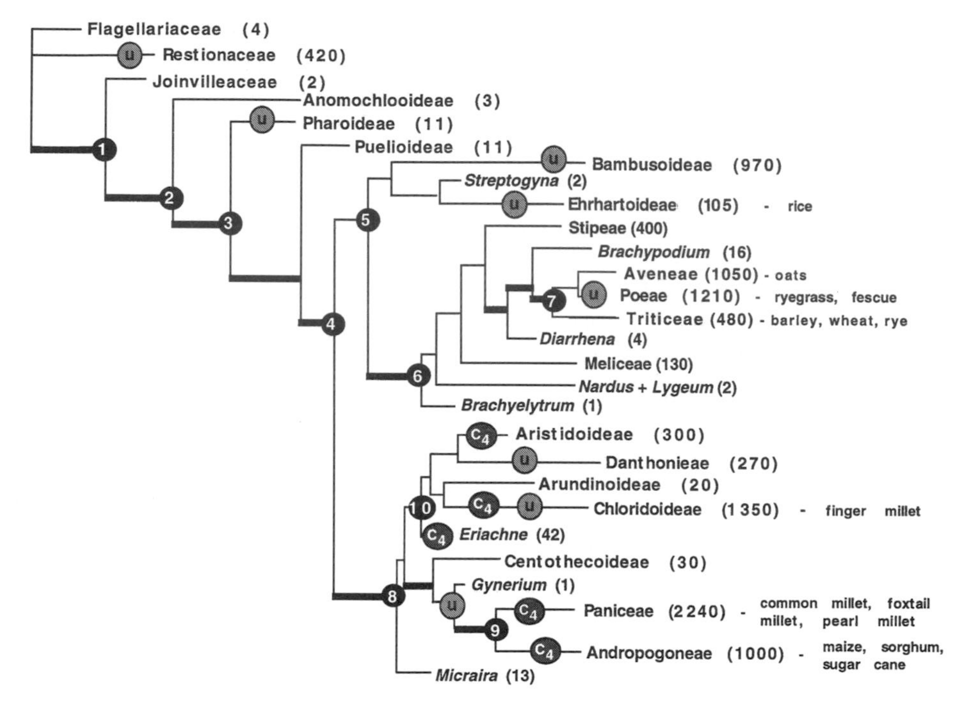

Figure 1 Phylogeny of the grass family and its sister families (based on data from 6–8, 26, 34, 56, 61–63, 76, 79, 82, 96, 97, 114). *Numbers in parentheses* indicate approximate numbers of species. Branch lengths are approximately proportional to numbers of mutations, as indicated by parsimony analyses. *Heavy branches* indicate groups that are found in all molecular phylogenies, and thus appear in the semistrict consensus (76) and/or are supported by 90–100% bootstrap values in analyses that include one or more molecular data sets. *Numbered nodes* refer to discussion in text. Lineages marked with C_4 are wholly or partly made up of taxa with the C_4 photosynthetic pathway. Lineages marked with U include taxa with unisexual flowers.

rules of nomenclature require that Ehrhartoideae be the name of the subfamily.) In all studies, the aristidoids, chloridoids, and panicoids are included with representatives of the Arundinoideae and Centothecoideae in a large, well-supported clade known by its acronym as the PACC clade (34; for Panicoideae, Arundinoideae, Centothecoideae, and Chloridoideae; node 8 on Figure 1).

The order of events after node 4 is uncertain. The phylogeny shown in Figure 1 indicates a speciation event that gave rise to the Bambusoideae plus Ehrhartoideae (Oryzoideae) and Pooideae ("the BEP clade," approximately 40% of the species in the family), and the PACC clade containing everything else (approximately 60% of the species). A number of characteristics are shared between the Pooideae and the PACC clade, however, which has led some authors (e.g. 114) to favor an interpretation that the two are sister groups.

The subfamily Pooideae (all descendants of the ancestor at node 6) is large and heterogeneous, but most species are derived from the common ancestor at node 7.

Descendants of this ancestor include such familiar north temperate species as wheat, barley, rye, oats, fescue, ryegrass, bluegrass, timothy, and other common forage grasses.

Relationships among the subfamilies in the PACC clade are unclear. This is partly because of poor sampling, which is biased toward taxa from the Northern Hemisphere even though most of the phylogenetically critical taxa are groups with a Gondwanan distribution. The clade may be the result of a rapid radiation, but the sample is too poor to determine the actual pattern.

Space does not permit discussion of the many morphological changes that occurred in the diversification of the grasses. Much detailed information on the morphological variation in the family can be obtained via a searchable database (122). Extensive discussion of character variation in a phylogenetic context can be found in Soreng & Davis (114), and in a forthcoming paper by the Grass Phylogeny Working Group (unpublished manuscript).

LONG-SHORT CELL ALTERNATION IN THE LEAF EPIDERMIS

Change in Position of Asymmetric Cell Divisions, Node 1

One of the central problems of developmental biology is the division of one cell into two cells whose fates differ (106). This can be accomplished by either an asymmetric division of the original cell or a symmetric division followed by differentiation of the two cells. Such divisions have been studied in many organisms and, among plants, in many cell types, particularly those of the epidermis (111). The epidermis of grass leaves is convenient for studying differentiation because the cells are arranged in longitudinal files. In some of these files, asymmetric divisions produce short cells that go on to make stomata, and long cells that do not. A gradient of differentiation extends along the file, so cells can be observed at different stages of development.

In monocots in general, alternation between stomata and long cells is common in the leaves (32, see figures). These have not been studied developmentally in most cases (120, see comments). Only in the grasses and *Joinvillea*, however, does long-short cell alternation appear in files of cells adjacent to the stomatal files, a characteristic that was postulated to be a synapomorphy for the two families (19). In maize leaves, the final, pre-differentiation divisions in all cells of the leaf blade epidermis—not just the cells that will produce the guard mother cells—are asymmetric (118).

The alternation of epidermal long and short cells apparently reflects a change in the place of activity of a developmental program, which must have occurred in the common ancestor of the Poaceae and Joinvilleaceae. The developmental program that controls asymmetric cell division, instead of being deployed only in stomatal cell files, appears to have changed its expression pattern in *Joinvillea* and the grasses to be expressed in all cell files in the leaf. With this change in position, the short cells became the sites of silica deposition. Evolutionary novelty may have been produced by use of an old developmental mechanism in a new position.

THE GRASS EMBRYO

Change in Timing of Differentiation, Node 2

In most monocots, embryos do not differentiate much while in the seed. This is particularly true of some families in the graminoid clade. Johri et al (69) described embryos in *Flagellaria* as "small," in Centrolepidaceae as "unorganized," and in Restionaceae as "not fully differentiated." The embryo of *Joinvillea* is similar in outline to the capitate type of Martin (95) and seems to be largely undifferentiated (19). [Martin's character states have been used in several recent phylogenetic studies (23, 94, 117), but are probably overly simplistic descriptors of the underlying variation.] Most differentiation of recognizable leaves and vasculature occurs after germination.

In the grasses, embryo development is accelerated relative to seed and fruit maturation, so that the embryo produces several leaves, a vascular system, and organized apical meristems at both the shoot and the root poles. This is clearly a heterochronic change, indicating that, in this case at least, development of the embryo has been decoupled from maturation of the seed. The genetic basis of this change is unknown.

The grasses exhibit a particularly dramatic example of heterochronic change, but similar changes in timing of embryo development appear in other families. For example, some Cyperaceae and some Commelinaceae produce a cotyledonary sheath and portion of the embryo representing the first leaf (69). In none of these does development proceed as far as in the grasses, however. The underlying mechanism of the change is not known, nor do we know if it is the same in each family.

FLORAL MORPHOLOGY

Conservation and Change in Position of Expression of Organ Identity Genes, Node 2 or 3

The grass flower is apparently unique among angiosperms. In the center of the flower is a single pistil with one anatropous ovule, and generally two styles and stigmas. Three or more stamens occur in one or two whorls outside the pistil. Outside the stamens are two or three lodicules, structures that become turgid at anthesis and force open the flower. Outside the lodicules is an adaxial, prophyll-like structure, the palea, which is commonly, but not invariably, two-keeled. Subtending the flower on the side away from the axis is another leaf-like bract, the lemma. The origins of lodicules, the palea, and the lemma are unknown, and have been the subject of much speculation (summarized in 29), but recent genetic studies are providing some clues to the sorts of changes that must have occurred.

Maize has two carpel identity genes, *zag1* and *zmm2* (98, 107). These are related to genes such as *agamous* (*ag*) that specify carpel and stamen identity in dicots, with *zag1* more similar to *ag* than *zmm2* is. Transposon-induced mutations in *zag1* lead to plants with extra carpels in the pistils of female flowers. Stamens in these mutants are functional, which shows that the function of *zag1* is not identical to that of *ag*. *zag1* is expressed preferentially in carpels, whereas *zmm2* appears more strongly in stamens. Assuming that the two genes were originally expressed equally in both carpels and stamens, during grass evolution their expression pattern apparently has been modified to become more organ-specific. Thus the duplication of the genes has been followed by a change in position of expression.

The genes involved in stamen and petal identity are known as B-class genes (14). When their place of expression overlaps with that of *agamous*-like (C-class) genes, floral organs become stamens: when B-class genes are expressed with A-class genes, floral organs become petal-like. When B-class genes are mutated, stamens are converted to carpels and petals to sepals. The maize mutant *silky* has a phenotype suggesting it belongs to the B-class, a hypothesis confirmed by recent cloning of the gene (3). In *silky* mutants, stamens are converted to carpels, and lodicules to organs that look like paleas or lemmas. In normal maize plants, *silky* RNA is found in stamens and lodicules. Thus the mutant phenotype and the gene expression pattern show that lodicules, which are in the position of petals, express petal identity genes and are converted to outer whorl structures when *silky* is nonfunctional. This implies that lodicules are either highly modified petals, or that bracts have become modified to become more petal-like. Previous authors have speculated that lodicules might be tepals, staminodes, stipules, ligules, reduced bracts, leaves, branch systems, or novel structures (29). The genetic data rule out the possibility that they are staminodes, because staminodes would be converted to carpels in *silky* mutants. The genetic data also suggest that lodicules are novel modifications of tepals. Nonetheless, although they are positionally similar to tepals, and share with them the expression pattern of one gene, they are morphologically and functionally unlike tepals.

CONVENTIONAL GRASS SPIKELETS

Change in Position of Expression of Organ Identity Genes, Node 2 or 3

Grass flowers are generally described as being subtended by two bracts, the lemma and palea. One or more of the bracteate flowers are then arranged in little spikes—spikelets—that are subtended by two more bracts known as glumes. All grasses originating after the common ancestor at node 3 have spikelets that fit this description. In the Anomochlooideae it is difficult to determine which of the many floral bracts constitute "lemmas" or "glumes." The conventional spikelet thus may have originated before node 2 and then been lost in Anomochlooideae, or may not have originated until after the origin of the earliest diverging lineage (between nodes 2 and 3).

The interpretation of the spikelet as a bracteate axis must be modified in light of recent studies of gene expression and mutagenesis. In the *silky* mutants described above, the similarity of mutant lodicules to paleas and lemmas suggests that the "bracts" are actually modifications of the outer perianth whorl. In maize, paleas and lemmas are not easily distinguished, so it is impossible to say whether the mutant lodicules are converted to paleas or lemmas, or to some hybrid organ.

A single gene mutant in barley converts lemmas to leaves (*leafy lemma*), suggesting that the distinction between lemma and leaf might be a simple one (12). The leaf that replaces the lemma in such mutants has a recognizable sheath and blade.

Mutant phenotypes in maize thus indicate that lemma and palea may be modified tepals: in barley, mutants indicate that the lemma at least might be a modified leaf. The two possibilities are not mutually exclusive, and it seems possible that two developmental programs that were spatially separate in grass ancestors have become co-expressed in the grasses. Alternatively, expression of A-class genes may be reduced in grasses relative to their ancestors, such that the outer perianth whorl reverts partially to the default identity of a leaf. This could be tested by cloning A-class genes from grasses and their relatives and examining their expression. This explanation would also predict that *leafy lemma*, which is not cloned, would be a regulator of A-class activity in lemmas.

UNISEXUAL FLOWERS

Change in Position of Expression of Cell Death Genes, Node 9

Unisexual flowers have appeared multiple times in the evolution of the grasses (marked by U in Figure 1). Recent evidence suggests that the mechanism of staminate flower production varies among the grasses, indicating that the multiple origins are not really parallelisms.

Staminate Flowers Produced by Cell Death Unisexual flowers evolved in or shortly before the common ancestor of the Panicoideae (node 9), the subfamily that includes the model system, *Zea mays*. The Panicoideae have been recognized as distinct for almost 200 years, since the work of Robert Brown (15, 16). In all members of the group the spikelets consist of two flowers, the lower one of which is reduced to be either male or sterile.

Development of male flowers is uniform throughout the subfamily (18, 24, 90, 91, 124). In all flowers of all species studied, both stamens and gynoecia initiate so that in early development all flowers are the same. After the gynoecium forms a ridge around the developing nucellus, however, the development of male flowers diverges from that of hermaphrodite or female ones. The subepidermal cells of the gynoecium of a male flower lose their cytoplasm and nuclei, whereas cells in the same position in a hermaphrodite or female flower remain densely cytoplasmic and nucleated. This cell death correlates with the cessation of growth, so that mature male flowers retain a gynoecium with gynoecial ridge and nucellus.

Research on maize has identified numerous mutations that affect sex determination (64–67). Among these are *tasselseed1* (49), *tasselseed2* (49, 50), and *silkless* (*sk1*; 70). Complete loss-of-function *ts2* mutants in maize cause all tassel flowers to become female, and also cause development of gynoecia in the proximal flowers of ear spikelets (49, 101); *sk1* mutants lack gynoecium development in both ear and tassel. Jones (71, 72) constructed the *tasselseed2*; *silkless1* double mutants and found that the phenotype was similar, but not identical, to that of *ts2* single mutants. In the ear of the double mutant, both flowers produced gynoecia, and in the tassel, some (but not all) flowers were female. This indicates that the *ts2* mutation is epistatic to *sk1*, and that the gene products both function in the same pathway, a result confirmed by Irish et al (66).

DeLong et al (35) cloned *tasselseed2*, which proved to be a short-chain alcohol dehydrogenase. In wild type maize plants, *ts2* is expressed in the sub-epidermal cell layers in the gynoecium of both tassel (male) and ear (female) flowers. The gynoecium in the tassel undergoes abortion through programmed cell death, as does the gynoecium of the proximal (sterile) floret in the ear; the gynoecium of the fertile floret in the ear does not abort. The timing and position of *ts2* expression suggest that TS2 is a candidate for direct regulation of programmed cell death in the gynoecium of aborting female organ primordia.

The sister genus to *Zea* is *Tripsacum* (82, 115). A feminizing mutation of *Tripsacum*, *gynomonoecious sex form1* (*gsf1*), is similar in phenotype to *ts2* mutations in maize. Li et al (91) found that *gsf1* and *ts2* failed to complement each other in a maize × *Tripsacum* cross. They found that the sequence of *gsf1* was similar to that of *ts2*, and that the expression pattern in *Tripsacum* was identical to that of *ts2* in maize.

Expression of *ts2* and *gsf1* in wild type maize and *Tripsacum* thus correlates with cell death and cessation of gynoecial development. Null mutations in each gene suggest that *ts2* is necessary for gynoecial abortion. From these data, and the histological investigations described, it seems likely that *ts2* is expressed in gynoecia of male flowers of all members of the subfamily Panicoideae.

Staminate Flowers Produced Without Cell Death Unisexual flowers evolved independently in *Zizania aquatica*, the North American wild rice (Ehrhartoideae, Figure 1). As in the panicoids, early floral development is identical for both sexes of flowers (124). Gynoecial development in male flowers, however, proceeds well beyond the stage of gynoecial ridge formation. Stigmatic arms begin to form and the ovule differentiates. Integuments are obvious, and the ovule becomes anatropous: early stages of female gametophyte development can be seen in some sections.

At this stage, the ovule ceases to develop, and dark-staining material is deposited through the ovule tissue, but not in the ovary wall. DNA does not break down, as indicated by DAPI staining, but cellular integrity is lost.

The histology of the arresting gynoecium is thus quite different from the precisely controlled cell death pattern seen in the panicoids. Either *ts2* is not involved in male flower formation in *Zizania*, or it is regulated differently so that it is expressed at a much later stage in development and is deployed in quite a different way.

Cell Death in Dicot Tapetum *ts2*-like genes have been cloned from two dicots (88). *Silene latifolia* (white campion) has unisexual flowers, and sex expression is controlled by sex chromosomes. Unlike male flowers in the grasses, male flowers in *Silene* never initiate a gynoecium at all, so it seems unlikely that cell death would be involved in sex determination. *sta1*, the *ts2*-like gene, is expressed only in tapetal cells in male flowers of *Silene*, and is not expressed at all in the center of the flower, as would be expected if it were involved in repressing gynoecial initiation or growth. A similar gene was cloned from *Arabidopsis thaliana*, with hermaphrodite flowers, and also found to be expressed in tapetal cells only. These data indicate that in most angiosperms, *ts2* is expressed in tissues, such as tapetal cells, that must die to be functional. This suggests that gynoecial expression of the gene is novel in the panicoids, and represents ectopic expression relative to the ancestors.

Indirect Evidence from Lack of Gynoecial Development in Other Taxa with Unisexual Flowers In male flowers of the dicot *Rumex acetosella*, as in *Silene*, the gynoecium never forms at all and cell death is apparently not involved (2). Also no gynoecium forms in male flowers of Araceae (5, 13) and Alismataceae (103). This constrains the hypothesis of when *ts2* changed its role from being a general cell death gene to being the controlling agent of maleness. Although less compelling than the expression data cited above, the general developmental morphology suggests that gynoecial expression of *ts2* originated in the panicoids.

All data are consistent with the hypothesis that *ts2* expression shifted in evolutionary time, from tapetal cells to the gynoecium. This change in position apparently occurred in the common ancestor of the panicoid grasses, thus preceding a radiation of over 3000 species.

C_4 PHOTOSYNTHESIS

Change in Position of Expression of Photosynthetic Genes, After Node 8

Striking variation in internal leaf anatomy was observed among the grasses as early as 1882 (57), and the variation was used as a character for refining the taxonomic classification (4, 17, 116). Many grasses, particularly those of warm regions of the world, have veins that are close together and bundle sheath cells that are dark green, in contrast to the more usual distant vein spacing and bundle sheaths that are clear. Only much later was this anatomical variation shown to reflect physiological differences (58).

The C_4 photosynthetic pathway is an addition to the more conventional C_3 pathway, in which the primary carbon acceptor ribulose 1,5 bisphosphate carboxylase/oxygenase (RuBisCO) is replaced by phosphoenol pyruvate carboxylase (PEPC) in the mesophyll. (For recent reviews of the biochemistry and regulation of the pathway see 74, 89, 109.) This is accomplished by changing the location of gene expression, such that RuBisCO is produced only in the cells surrounding the vascular bundle (the bundle sheath), and PEPC is expressed strongly in the mesophyll. Accompanying these changes in carboxylases are changes in decarboxylating malic enzymes, pyruvate orthophosphate dikinase (PPDK), malate dehydrogenase (MDH), and the light harvesting chlorophyll a,b binding proteins (LHCP). These are all enzymes that are present and active in C_3 leaves, but at a much lower level than in C_4. The pathway requires close contact between individual mesophyll and bundle sheath cells, and thus the veins of C_4 species are rarely separated by more than two mesophyll cells (60; Figure 2).

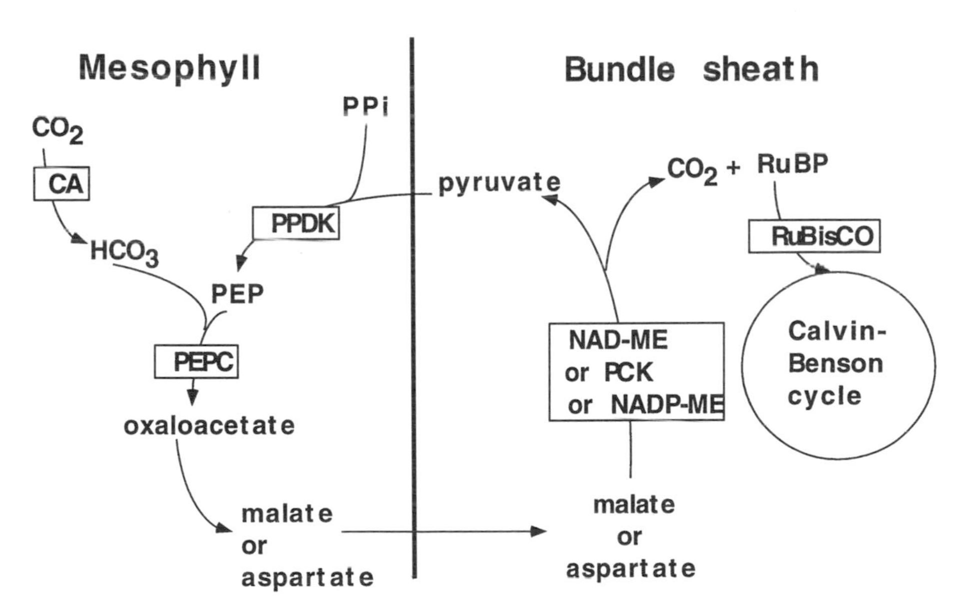

Figure 2 Diagram of the C_4 photosynthetic pathway. *CA*, carbonic anhydrase; *PEP*, phosphoenol pyruvate; *PEPC*, PEP carboxylase; *PPDK*, pyruvate orthophosphate dikinase; *RuBisCO*, ribulose 1,5-bisphosphate carboxylase/oxygenase; *NAD-ME*, NAD malic enzyme; *NADP-ME*, NADP malic enzyme; *PCK*, PEP carbosykinase.

The C_4 pathway occurs in only about 3% of the flowering plants, but in nearly half the grasses (104). All these species occur in the PACC clade (node 8). Mapping the character C_4 photosynthesis on the phylogeny in Figure 1, we can infer two to four origins, one at node 9 and one to three at node 10. Within the Paniceae is either one origin and multiple reversals, or multiple origins, or persistence of a polymorphism (LM Giussani, JH Cota, F Zuloaga, and EA Kellogg, unpublished data).

The C_4 "syndrome" is actually the co-occurrence of multiple characteristics, biochemical and histological. Sinha & Kellogg (110) dissected the pathway into several characters and assessed gene expression of multiple genes in the four C_4 lineages. They found that C_4 physiology is not a single phenomenon, with some genes being regulated quite differently in the different lineages (Table 1). Data in the literature indicate that histology also differs. What has occurred in parallel then, is down-regulation of RuBisCO in the mesophyll (by blocking transcription of *rbcS*; 109), up-regulation of PEPC, and production of an extra file of veins in the leaves. Regulation of PPDK, LHCP, NAD-ME, NADP-ME, and PCK all differ among the four lineages, as do the number of bundle sheaths, chloroplast ultrastructure, and chloroplast position in the cell (Table 1).

Variation among C_4 lineages indicates that many of the changes commonly associated with the pathway are not necessary for it to function. PEPC must be up-regulated in the mesophyll, and RuBisCO down-regulated. The veins must be close together, so that each bundle sheath cell is adjacent to a mesophyll cell. As indicated by Table 1 and Figure 2, however, there is variation in whether malate or aspartate is transported to the bundle sheath, the position of expression of PPDK, whether the LHC proteins are strongly expressed, and hence whether the plastids are granal, whether the outer bundle sheath is lost, and which of the three common decarboxylating enzymes is used. Although these may be important in any one instance, they represent different "machines" leading to the same output.

C_4 photosynthesis represents a change in tissue-specific gene expression. A histological distinction already exists between bundle sheath and mesophyll cells in C_3 grasses. C_4 species thus take advantage of this pre-pattern, and modify expression of the photosynthetic enzymes accordingly. In addition, an extra file of veins forms. The (unknown) developmental signals that cause formation of vasculature are expressed in cells that ancestrally produced chlorenchymatous mesophyll. Once again, ectopic gene expression appears to underlie evolutionary change.

GENOME EVOLUTION

Nuclear genomes of the grasses are largely colinear (10, 52, 99); synteny is not perfect, but large blocks of genes form similar linkage groups in species as disparate as rice and maize. Possibly as few as 20 rearrangements occurred in the time since rice and maize shared a common ancestor (123). This has led to the hope that genetic information on any grain crop can be applied with ease to any other (1, 10, 121).

TABLE 1 Comparison of enzyme localization and leaf anatomy among four C_4 lineages of grasses and their C_3 relatives (from 59, 110), showing differences in gene expression among the independent origins of the pathway listed across the top of the table (see also Figure 1). The photosynthetic enzymes listed are expressed in both bundle sheath cells (Bu) and mesophyll cells (Me) in C_3 taxa, whereas they are differentially localized in C_4 taxa (see also Figure 2). Decarboxylating enzymes may be malic enzyme using NAD (NAD-ME), malic enzyme using NADP (NADP-ME), or PEP carboxykinase (PCK)

		C_4 lineage			
	C_3	Panicoideae	Aristidoideae	Eriachneae	Chloridoideae
Photosynthetic enzymes					
RuBisCO	Bu + Me[a]	Bu	Bu	Bu	Bu
PEPC	Bu + Me	Me	Me	Me	Me
PPDK	Bu + Me	Bu + Me	Bu + Me	Bu + Me	Bu
LHCP	Bu + Me	Me; Bu − Me	Bu + Me	Bu + Me	Bu
Decarboxylating enzyme	Not applicable	NAD-ME NADP-ME PCK	NADP-ME	NADP-ME	NAD-ME PCK
Histology and anatomy					
Mesophyll cells between veins	7	2	2	2	2
Bundle sheaths	2	1 or 2	2	2	2
Bundle sheath plastids	Granal	Granal or agranal	Agranal	Granal	Granal

[a]Bu, expressed primarily or exclusively in bundle sheath; Me, expressed primarily or exclusively in mesophyll.

Rate of genome rearrangements varies. Too few nuclear genomes have been mapped to test if the rate is truly non-random, but superficially the data suggest a punctuated pattern of evolution, as though some sort of homeostatic mechanism operates most of the time, and then is occasionally disrupted. For example, bread wheat is an allohexaploid created by the combination of the genomes of *Triticum monococcum* (source of the A genome), *Aegilops tauschii* (source of the D genome), and *Aegilops speltoides* (similar to the source of the B genome; 84, 85). These three genomes differ from each other by only a few rearrangements (37, 121). When any of the three is compared with the closely related *Aegilops umbellulata*, however, at least 12 rearrangements separate them (52). This is approximately the same number of rearrangements as separate wheat from rye (the genus *Secale*, also in the tribe Triticeae; 36) and wheat from rice (in subfamily Ehrhartoideae; 52).

The two members of Ehrhartoideae that have been mapped show similar gene order. Comparison of the rice genome with the North American wild rice (*Zizanias*) shows that 12 of the 15 wild rice chromosomes correspond to the 12 chromosomes of rice (83). The remaining three *Zizania* chromosomes are apparently duplications of three *Oryza* chromosomes.

A major rearrangement marks the radiation of the Pooideae, and is the basis of a traditionally used taxonomic character. Avdulov (4) noticed that some grasses of temperate regions, all now placed in the core Pooideae, have only seven sets of unusually large chromosomes instead of the more usual 10–13 sets of small chromosomes. On the phylogeny, this represents a single change at node 7. This correlates with the formation of a novel linkage group in which the genes of rice chromosome 10 were inserted as a block into the middle of rice chromosome 5 (52). At the same time, parts of the ancestral chromosomes 7 and 4 were linked.

This point also marked the beginning of a trend to very large genomes, the largest diploid genome in the grasses now being that reported for *Secale* (rye) at 29–33 picograms of DNA per 4C nucleus (9). Whether the trend to larger genomes was gradual or abrupt depends heavily on assumptions about the mode of increase (11). The relatively large genome of maize was created by amplification of retrotransposons in intergenic regions (105), and a similar mechanism may have operated in the pooids.

Another reduction in chromosome number, correlating with genomic rearrangements, apparently occurred at node 9, at the origin of the subfamily Panicoideae. Rice linkage groups 1, 6, 8, and 11 correspond to separate chromosomes in *Sorghum* (102) and foxtail millet *Setaria* (38). In addition, rice group 10 is inserted into the middle of rice 3, and rice 9 into rice 7 (52, 76, 99). The latter two novel linkage groups can also be found in rearranged form in the genome of maize. Part of rice 12 is linked to rice 4 in (*Setaria*; 38); such a linkage is postulated for the progenitor of maize (123), but does not occur in *Sorghum* (102). Either the published map of *Sorghum* is not sufficiently resolved to detect the linkage, or the linkage has occurred in parallel in maize and foxtail millet.

The sister taxa of the Panicoideae all have chromosome numbers of 12 or 13, whereas the Panicoideae itself is divided into two major clades, one with a predominant base chromosome number of 10, and the other predominantly with $x = 9$ (54; LM Giussani, JH Cota, F Zuloaga, and EA Kellogg, unpublished data). The ancestral state is ambiguous. Within the $x = 10$ clade, there are multiple reductions to $x = 5$ (115); these were thought to be ancestral by some earlier workers (e.g. 20), but the phylogeny does not support that conclusion. There are also some taxa with $x = 9$, but this number appears to be derived within the clade rather than ancestral.

Data on nuclear genome structure among panicoid grasses are difficult to reconcile with the phylogeny. Maize is a member of the $x = 10$ clade, but it is an ancient allotetraploid (53). This implies that it should be derived from a species with $x = 5$. Phylogenetic data, however, place it as sister to *Tripsacum* ($x = 18$), and those two are in an ambiguous position among a number of genera with $x = 10$ (115). Wilson et al (123) argued that the progenitor of maize had $x = 8$. This contention is based on careful comparisons between maize and rice, and incorporation of previously published data on *Sorghum* (102). Very few of the panicoid grasses have $x = 8$, however, and none of those is closely related to maize (54, 122, LM Giussani, JH Cota, F Zuloaga, and EA Kellogg, unpublished data). Duplication of an $x = 9$ ancestor seems more consistent with phylogenetic data; it will be interesting to see if the $x = 8$ hypothesis continues to be supported as genome mapping becomes more precise and more species are sampled.

TEMPO AND MODE OF EVOLUTION

Diversification Follows a Punctuated Pattern

The earliest reliable records of grass pollen date from the Paleocene of South America and Africa, between 60 and 55 million years ago (mya) (68). Other pollen grains attributed to Poaceae appear in the upper Cretaceous in Maastrichtian deposits (ca. 70 mya), but it is not clear whether they represent grasses or one of the sister families, all of which have superficially similar pollen (92). The earliest grass macrofossil, a spikelet, appears in the early Eocene (ca. 55 mya) in North America (30, 68). The common ancestor of the grasses (node 2) thus must have lived before 55 mya, but probably not much earlier than 70 mya.

The major diversification of the family did not occur until after nodes 6 and 8 (Figure 1). Gaut & Doebley (53) have estimated the origin of the panicoid grasses, based on the divergence of maize and *Pennisetum* (node 9), to be 25–32 mya. The phylogeny shows that the ancestor appeared somewhat after the other C_4 lineages, so the 25-mya date should correlate approximately with the fossil record for C_4 species. The earliest isotopic record for C_4 photosynthesis is ca. 15 mya (86, 87), and the earliest C_4 macrofossil is dated 12.5 mya (100). Thus the approximate date for node 9 is no later than 15 mya, and possibly as old as 32 mya. Grass-dominated ecosystems did not appear until the Miocene in the Northern Hemisphere, although they may have arisen somewhat earlier in South America and Africa (68).

Simple arithmetic thus shows that diversification (as opposed to the origin) of the grasses occurred a minimum of 23 my (55–32 my) after the origin of their peculiar morphological characteristics. Morphological evolution, as indicated by speciation, followed a punctuated pattern, with long periods of morphological stasis followed by bursts of diversification. Wind pollination and silica deposition are shared with the sister taxa of the grasses and thus must have originated long before the major grass radiation. Other distinctive characters of the grasses—the unusual floral morphology, complex inflorescence structure, and differentiated embryo—did not lead immediately to speciation and diversification. They are either irrelevant to the current dominance of the grasses, or they represent exaptations (55), characteristics that were not built by natural selection for their current function. Of the characters discussed here, only C_4 photosynthesis is phylogenetically placed to be both adaptive (selected for its current function) and permitting an immediate adaptive radiation.

Evolution of Development

Heterotopy may play a major role in morphological evolution of higher taxa. Expressing existing genetic programs in different places in the plant is a common mechanism, as indicated by the change in asymmetric cell division in portions of the leaf epidermis, the differentiation of tissue specificity of floral organ identity genes, the cell death program that marks the panicoid grasses, and the shift in regulation of photosynthetic enzymes in C_4 lineages. Heterochrony, changes in the relative timing of events, is another possible mechanism, although the genetic basis of this is unknown. As we come to understand the basis for some of the changes described here, it may prove difficult to distinguish heterotopy from heterochrony. Because plants develop continuously, shifting gene expression earlier or later in development (heterochrony) may lead directly to a change in the position of developmental programs (heterotopy). Physical changes in the genome may correlate with differences in gene regulation: this is an unexplored area but one likely to see many advances over the coming years.

The study of macroevolution—morphological differences among large groups of organisms—is generally discussed separately from microevolution—the change in gene frequency in populations—even though we believe that the mechanisms underlying the two processes are basically the same. At least part of the reason for the distinction is our inability to describe morphological variation among higher taxa in terms of genes. Doebley & Lukens (40) suggested that most evolutionary change comes from modifications of gene regulation rather than modifications of structural genes themselves. From the evidence presented here, I suggest that modification of gene regulation has often led to a change in the position of gene expression, and that heterotopic change is frequent in macroevolution.

ACKNOWLEDGMENTS

The work cited here was supported by grants from the National Science Foundation, and by the E. Desmond Lee Endowment at the University of Missouri-St. Louis. I thank H. J. Stevens for helpful comments.

Visit the Annual Reviews home page at www.AnnualReviews.org

LITERATURE CITED

1. Ahn SN, Tanksley SD. 1993. Comparative linkage maps of the rice and maize genomes. *Proc. Natl. Acad. Sci. USA* 90:7980–84
2. Ainsworth C, Crossley S, Buchanan-Wollaston V, Thangavelu M, Parker J. 1995. Male and female flowers of the dioecious plant sorrel show different patterns of MADS box gene expression. *Plant Cell* 7:1583–98
3. Ambrose BA, Lerner DR, Ciceri P, Padilla CM, Yanofsky MF, Schmidt RJ. 2000. Molecular and genetic analyses of the *Silky1* gene reveal conservation in floral organ specification between eudicots and monocots. *Mol. Cell.* 5:569–79

3a. Angiosperm Phylogeny Group. 1998. An ordinal classification for the families of flowering plants. *Ann. Mo. Bot. Gard.* 85:531–53

4. Avdulov NP. 1931. Kario-sistematicheskoye issledovaniye semeystva zlakov. *Bull. Appl. Bot. Gen. Plant Breed.* Suppl. 44:1–428
5. Barabé D, Bertrand C. 1996. Organogénie florale des genres *Culcasia* and *Cercestis* (Araceae). *Can. J. Bot.* 74:898–908
6. Barker NP. 1997. The relationships of *Amphipogon, Elytrophorus* and *Cyperochloa* (Poaceae) as suggested by *rbcL* sequence data. *Telopea* 7:205–13
7. Barker NP, Linder HP, Harley E. 1995. Phylogeny of Poaceae based on *rbcL* sequences. *Syst. Bot.* 20:423–35
8. Barker NP, Linder HP, Harley EH. 1998. Sequences of the grass-specific insert in the chloroplast *rpoC2* gene elucidate generic relationships of the Arundinoideae (Poaceae). *Syst. Bot.* 23:327–50
9. Bennett MD, Cox AV, Leitch IJ. 1998. *Angiosperm DNA C-Values Database.* http://www.rbgkew.org.uk/cval/database1.html
10. Bennetzen JL, Freeling M. 1993. Grasses as a single genetic system: genome composition, collinearity and compatibility. *Trends Genet.* 9:259–61
11. Bennetzen JL, Kellogg EA. 1997. Do plants have a one-way ticket to genomic obesity? *Plant Cell* 9:1509–14
12. Bossinger G. 1990. *Klassifizierung von Entwicklungsmutanten der Gerste anhand einer Interpretation des Pflanzenaufbaus der Poaceae aus Phytomeren.* Inaug.-Diss., Rheinischen Friedrich-Wilhelms-Universität, Bonn
13. Boubes C, Barabé D. 1996. Développement de l'inflorescence et des fleurs du *Philodendron acutatum* Schott (Araceae). *Can. J. Bot.* 74:909–18
14. Bowman JL, Smyth DR, Meyerowitz EM. 1991. Genetic interactions among floral homeotic genes of *Arabidopsis*. *Development* 112:1–20
15. Brown R. 1810. *Prodromus Florae Novae Hollandiae.* London: J Johnson
16. Brown R. 1814. *General Remarks, Geographical and Systematical, on the Botany of Terra Australis.* London: G & W Nicol
17. Brown WV. 1958. Leaf anatomy in grass systematics. *Bot. Gaz.* 119:170–78
18. Calderon-Urrea A, Dellaporta SL. 1999. Cell death and cell protection genes

determine the fate of pistils in maize. *Development* 126:435–41

19. Campbell CS, Kellogg EA. 1987. Sister group relationships of the Poaceae. In *Grass Systematics and Evolution*, ed. TR Soderstrom, KW Hilu, CS Campbell, ME Barkworth, pp. 217–24. Washington, DC: Smithsonian Inst. Press
20. Celarier RP. 1957. Cytotaxonomy of the Andropogoneae. 2. Sub-tribes Ischaeminae, Rottboelliinae, and the Maydeae. *Cytologia* 22:160–83
21. Chase MW, Soltis DE, Olmstead RG, Morgan D, Les DH, et al. 1993. Phylogenetics of seed plants: an analysis of nucleotide sequences from the plastid gene *rbcL*. *Ann. Mo. Bot. Gard.* 80:528–80
22. Chase MW, Soltis DE, Soltis PS, Rudall PJ, Fay MF, et al. 2000. Higher-level systematics of the monocotyledons: an assessment of current knowledge and a new classification. *Monocots: systematics and evolution*, ed. KL Wilson, DA Morrison. Sydney, Australia: CSIRO. pp. 3–16
23. Chase MW, Stevenson DW, Wilkin P, Rudall PJ. 1995. Monocot systematics: a combined analysis. In *Monocotyledons: Systematics and Evolution*, ed. PJ Rudall, PJ Cribb, DF Cutler, CJ Humphries, pp. 685–730. Kew, UK: R. Bot. Gard.
24. Cheng PC, Greyson RI, Walden DB. 1983. Organ initiation and the development of unisexual flowers in the tassel and ear of *Zea mays*. *Am. J. Bot.* 70:450–62
25. Clark LG, Kobayashi M, Mathews S, Spangler RE, Kellogg EA. 2000. The Puelioideae, a new subfamily of Poaceae. *Syst. Bot.* 25:181–87
26. Clark LG, Zhang W, Wendel JF. 1995. A phylogeny of the grass family (Poaceae) based on *ndhF* sequence data. *Syst. Bot.* 20:436–60
27. Clayton WD, Renvoize SA. 1986. *Genera Graminum.* London: Her Majesty's Stationery Office
28. Clifford HT. 1970. Monocot classification with special reference to the origin of the grasses (Poaceae). In *New Research in Plant Anatomy*, ed. NKB Robson, DF Cutler, M Gregory, pp. 25–34. London: Academic
29. Clifford HT. 1987. Spikelet and floral morphology. In *Grass Systematics and Evolution*, ed. TR Soderstrom, KW Hilu, CS Campbell, ME Barkworth, pp. 21–30. Washington, DC: Smithsonian Inst. Press
30. Crepet WL, Feldman GD. 1991. The earliest remains of grasses in the fossil record. *Am. J. Bot.* 78:1010–14
31. Cronquist A. 1981. *An Integrated System of Classification of Flowering Plants.* New York: Columbia Univ. Press
32. Dahlgren RMT, Clifford HT, Yeo PF. 1985. *The Families of the Monocotyledons.* New York: Springer-Verlag
33. Davis JI. 1995. A phylogenetic structure of the monocotyledons, as inferred from chloroplast DNA restriction site variation, and a comparison of measures of clade support. *Syst. Bot.* 20:503–27
34. Davis JI, Soreng RJ. 1993. Phylogenetic structure in the grass family (Poaceae), as determined from chloroplast DNA restriction site variation. *Am. J. Bot.* 80:1444–54
35. DeLong A, Calderon-Urrea A, Dellaporta SL. 1993. Sex determination gene *TASSELSEED2* of maize encodes a short-chain alcohol dehydrogenase required for stage-specific floral organ abortion. *Cell* 74:757–68
36. Devos KM, Atkinson MD, Chinoy CN, Francis HA, Harcourt RL, et al. 1993. Chromosomal rearrangements in the rye genome relative to that of wheat. *Theor. Appl. Genet.* 85:673–80
37. Devos KM, Dubcovsky J, Dvorák J, Chinoy CN, Gale MD. 1995. Structural evolution of wheat chromosomes 4A, 5A, and 7B and its impact on recombination. *Theor. Appl. Genet.* 91:282–88
38. Devos KM, Wang ZM, Beales J, Sasaki T, Gale MD. 1998. Comparative genetic maps of foxtail millet (*Setaria italica*) and

rice (*Oryza sativa*). *Theor. Appl. Genet.* 96: 63–68
39. Doebley J. 1995. Genetics, development, and the morphological evolution of maize. In *Experimental and Molecular Approaches to Plant Biosystematics*, ed. PC Hoch, AG Stephenson, pp. 57–70. St. Louis, MO: Missouri Bot. Garden
40. Doebley J, Lukens L. 1998. Transcriptional regulators and the evolution of plant form. *Plant Cell* 10:1075–82
41. Doebley J, Stec A. 1991. Genetic analysis of the morphological differences between maize and teosinte. *Genetics* 129:285–95
42. Doebley J, Stec A. 1993. Inheritance of the morphological differences between maize and teosinte: comparison of results for two F_2 populations. *Genetics* 134:559–70
43. Doebley J, Stec A, Gustus C. 1995. *Teosinte branched1* and the origin of maize: evidence for epistasis and the evolution of dominance. *Genetics* 141: 333–46
44. Doebley J, Stec A, Hubbard L. 1997. The evolution of apical dominance in maize. *Nature* 386:485–88
45. Dorweiler J, Stec A, Kermicle J, Doebley J, 1993. *Teosinte glume architecture 1*: a genetic locus controlling a key step in maize evolution. *Science* 262:233–35
46. Dorweiler JE, Doebley J. 1997. Developmental analysis of *Teosinte glume architecture1*: a key locus in the evolution of maize (Poaceae). *Am. J. Bot.* 84:1313–22
47. Doyle JJ, Davis JI, Soreng RJ, Garvin D, Anderson MJ. 1992. Chloroplast DNA inversions and the origin of the grass family (Poaceae). *Proc. Natl. Acad. Sci. USA* 89:7722–26
48. Duvall MR, Clegg MT, Chase MW, Clark WD, Kress WJ, et al. 1993. Phylogenetic hypotheses for the monocotyledons constructed from *rbcL* sequence data. *Ann. Mo. Bot. Gard.* 80:607–19
49. Emerson RA. 1920. Heritable characters in maize. II. Pistillate flowered maize plants. *J. Hered.* 11:65–76
50. Emerson RA, Beadle GW, Fraser AC. 1935. A summary of linkage studies in maize. *Cornell Univ. Agric. Exp. Stn. Memoir* 180:1–83
51. Futuyma DJ. 1998. *Evolutionary Biology*. Sunderland, MA: Sinauer. 3rd ed.
52. Gale MD, Devos KM. 1998. Comparative genetics in the grasses. *Proc. Nat. Acad. Sci. USA* 95:1971–74
53. Gaut BS, Doebley JF. 1997. DNA sequence evidence for the segmental allotetraploid origin of maize. *Proc. Natl. Acad. Sci. USA* 94:68090–94
54. Gómez-Martinez R, Culham A. 2000. Phylogeny of the subfamily Panicoideae with emphasis on the tribe Paniceae: evidence from the *trnL-F* cpDNA region. *Grasses: systematics and evolution*, ed. SWL Jacobs, JE Everett. Melbourne, Australia: CSIRO. pp. 136–140
55. Gould SJ, Vrba ES. 1982. Exaptation—a missing term in the science of form. *Paleobiology* 8:4–15
56. Grass Phylogeny Working Group. 2000. A phylogeny of the grass family (Poaceae), as inferred from eight character sets. *Grasses: systematics and evolution*, ed. SWL Jacobs, JE Everett. Melbourne, Australia: CSIRO. pp. 3–7
57. Haberlandt G. 1882. Vergleichende Anatomie des Assimilatorischen Gewebesystems der Pflanzen. *Jahrb. Wiss. Bot.* 13: 74–188
58. Hatch MD, Slack CD, Johnson HS. 1967. Further studies on a new pathway of photosynthetic carbon dioxide fixation in sugar cane and its occurrence in other plant species. *Biochem. J.* 102:417–22
59. Hattersley PW, Watson L. 1992. Diversification of photosynthesis. In *Grass Evolution and Domestication*, ed. G Chapman, pp. 38–116. Cambridge, UK: Cambridge Univ. Press
60. Hattersley PW, Watson L. 1975. Anatomical parameters for predicting

photosynthetic pathways of grass leaves: the 'maximum lateral cell count' and the 'maximum cells distant count'. *Phytomorphology* 25:325–33
61. Hsiao C, Chatterton NJ, Asay KH, Jensen KB. 1994. Phylogenetic relationships of 10 grass species: an assessment of phylogenetic utility of the internal transcribed spacer region in nuclear ribosomal DNA in monocots. *Genome* 37:112–20
62. Hsiao C, Jacobs SWL, Barker NP, Chatterton NJ. 1998. A molecular phylogeny of the subfamily Arundinoideae (Poaceae) based on sequences of rDNA. *Aust. Syst. Bot.* 11:41–52
63. Hsiao C, Jacobs SWL, Chatterton NJ, Asay KH. 1999. A molecular phylogeny of the grass family (Poaceae) based on the sequences of nuclear ribosomal DNA (ITS). *Aust. Syst. Bot.* 11:667–88
64. Irish EE. 1997. Class II tassel seed mutations provide evidence for multiple types of inflorescence meristems in maize (Poaceae). *Am. J. Bot.* 84:1502–15
65. Irish EE. 1997. Experimental analysis of tassel development in the maize mutant *Tassel Seed 6*. *Plant Physiol.* 114: 817–25
66. Irish EE, Langdale JA, Nelson TM. 1994. Interactions between *Tassel Seed* genes and other sex determining genes in maize. *Dev. Genet.* 15:155–71
67. Irish EE, Nelson TM. 1989. Sex determination in monoecious and dioecious plants. *Plant Cell* 1:737–44
68. Jacobs BF, Kingston JD, Jacobs LL. 1999. The origin of grass-dominated ecosystems. *Ann. Mo. Bot. Gard.* 86:590–643
69. Johri BM, Ambegaokar KB, Srivastava PS. 1992. *Comparative Embryology of Angiosperms*, Vol. 2. Berlin: Springer-Verlag
70. Jones DF. 1925. Heritable characters in maize. XXIII. Silkless. *J. Hered.* 16: 339–41
71. Jones DF. 1932. The interaction of specific genes determining sex in dioecious maize. *Proc. Int. Cong. Genet, 6th*, 2:104–7
72. Jones DF. 1934. Unisexual maize plants and their bearing on sex differentiation in other plants and animals. *Genetics* 19:552–67
73. Judziewicz EJ, Clark LG, Londoño X, Stern MJ. 1999. *American Bamboos.* Washington, DC: Smithsonian Inst. Press
74. Kanai R, Edwards GE. 1999. The biochemistry of C_4 photosynthesis. In *C_4 Plant Biology*, ed. RF Sage, RK Monson, pp. 49– 87. San Diego, CA: Academic
75. Kellogg EA. 1996. Integrating genetics, phylogenetics, and developmental biology. In *The Impact of Plant Molecular Genetics*, ed. BWS Sobral, pp. 159–72. Boston, MA: Birkhauser
76. Kellogg EA. 1998. Relationships of cereal crops and other grasses. *Proc. Nat. Acad. Sci. USA* 95:2005–10
77. Kellogg EA. 1998. Who's related to whom? Recent results from molecular systematic studies. *Curr. Opin. Plant Biol.* 1:149–58
78. Kellogg EA, Birchler JA. 1993. Linking phylogeny and genetics: *Zea mays* as a tool for phylogenetic studies. *Syst. Biol.* 42:415–39
79. Kellogg EA, Campbell CS. 1987. Phylogenetic analyses of the Gramineae. In *Grass Systematics and Evolution*, ed. TR Soderstrom, KW Hilu, CS Campbell, ME Barkworth, pp. 310–22. Washington, DC: Smithsonian Inst. Press
80. Kellogg EA, Linder HP. 1995. Phylogeny of the Poales. In *Monocotyledons: Systematics and Evolution*, ed. PJ Rudall, PJ Cribb, DF Cutler, CJ Humphries, pp. 511–42. Kew, UK: R. Bot. Gard.
81. Kellogg EA, Shaffer HB. 1993. Model organisms for evolutionary studies. *Syst. Biol.* 42:409–14
82. Kellogg EA, Watson L. 1993. Phylogenetic studies of a large data set. I. Bambusoideae, Andropogonodae, and Pooideae (Gramineae). *Bot. Rev.* 59:273–343
83. Kennard W, Phillips R, Porter R, Grombacher A. 1999. A comparative map of

wild rice (*Zizania palustris* L. 2n = 2x = 30). *Theor. Appl. Genet.* 99:793–99
84. Kihara H. 1954. Consideration on the evolution and distribution of *Aegilops* species based on the analyser method. *Cytologia* 19:336–57
85. Kimber G, Feldman M. 1987. *Wild wheat: An Introduction.* Columbia, MO: Univ. Mo. Coll. Agric.
86. Kingston JD, Marino BD, Hill A. 1994. Isotopic evidence for Neogene hominid paleoenvironments in the Kenya Rift Valley. *Science* 264:955–59
87. Latorre C, Quade J, McIntosh WC. 1997. The expansion of C_4 grasses and global change in the late Miocene: stable isotope evidence from the Americas. *Earth Planet. Sci. Lett.* 146:83–96
88. Lebel-Hardenack S, Ye D, Koutnikova H, Saedler H, Grant SR. 1997. Conserved expression of a *TASSELSEED2* homolog in the tapetum of the dioecious *Silene latifolia* and *Arabidopsis thaliana*. *Plant J.* 12:515–26
89. Leegood RC, Walker RP. 1999. Regulation of the C_4 pathway. In C_4 *Plant Biology*, ed. RF Sage, RK Monson, pp. 89–131. San Diego, CA: Academic
90. LeRoux LG, Kellogg EA. 1999. Floral development and the formation of unisexual spikelets in the Andropogoneae (Poaceae). *Am. J. Bot.* 86:354–66
91. Li D, Blakey CA, Dewald CL, Dellaporta SL. 1997. Evidence for a common sex determination mechanism for pistil abortion in maize and its wild relative *Tripsacum*. *Proc. Natl. Acad. Sci. USA* 94:7217–22
92. Linder HP. 1987. The evolutionary history of the Poales/Restionales: a hypothesis. *Kew Bull.* 42:297–318
93. Linder HP. 1991. A review of the Southern African Restionaceae. *Contrib. Bolus Herb.* 13:209–64
94. Linder HP, Kellogg EA. 1995. Phylogenetic patterns in the commelinid clade. In *Monocotyledons: Systematics and Evolution*, ed. PJ Rudall, PJ Cribb, DF Cutler, CJ Humphries, pp. 473–96. Kew, UK: R. Bot. Gard.
95. Martin AC. 1946. The comparative internal morphology of seeds. *Am. Midl. Nat.* 36:513–660
96. Mason-Gamer RJ, Weil CF, Kellogg EA. 1998. Granule-bound starch synthase: structure, function, and phylogenetic utility. *Mol. Biol. Evol.* 15:1658–73
97. Mathews S, Tsai RC, Kellogg E. 2000. Phylogenetic structure in the grass family (Poaceae): evidence from the nuclear gene phytochrome B. *Am. J. Bot.* 87: 96–107
98. Mena M, Ambrose B, Meeley RB, Briggs SP, Yanofsky MF, Schmidt RJ. 1996. Diversification of C-function activity in maize flower development. *Science* 274:1537–40
99. Moore G, Devos KM, Wang Z, Gale MD. 1995. Grasses, line up and form a circle. *Curr. Biol.* 5:737–39
100. Nambudiri EMV, Tidwell WD, Smith BN, Hebbert NP. 1978. A C_4 plant from the Pliocene. *Nature* 276:816–17
101. Nickerson NH, Dale EE. 1955. Tassel modifications in *Zea mays*. *Ann. Mo. Bot. Gard.* 42:195–212
102. Pereira MG, Lee M, Bramel-Cox P, Woodman W, Doebley J, Whitkus R. 1994. Construction of an RFLP map in sorghum and comparative mapping in maize. *Genome* 37:236–43
103. Ronse Decraene LP, Smets EF. 1995. The androecium of monocotyledons. In *Monocotyledons: Systematics and Evolution*, ed. PJ Rudall, PJ Cribb, DF Cutler, CJ Humphries, pp. 243–54. Kew, UK: R. Bot. Gard.
104. Sage RF, Li M, Monson RK. 1999. The taxonomic distribution of C_4 photosynthesis. In C_4 *Plant Biology*, ed. RF Sage, RK Monson, pp. 551–84. San Diego, CA: Academic
105. SanMiguel P, Tikhonov A, Jin YK, Motochoulskaia N, Zakharov D, et al. 1996. Nested retrotransposons in the

intergenic regions of the maize genome. *Science* 274:765–68

106. Scheres B, Benfey PN. 1999. Asymmetric cell division in plants. *Annu. Rev. Plant Physiol. Plant Mol. Biol.* 50:505–37
107. Schmidt RJ, Veit B, Mandel MA, Mena M, Hake S, Yanofsky MF. 1993. Identification and molecular characterization of *ZAG1*, the maize homolog of the *Arabidopsis* floral homeotic gene *AGAMOUS*. *Plant Cell* 5:729–37
108. Shantz HL. 1954. The place of grasslands in the earth's cover of vegetation. *Ecology* 35:143–45
109. Sheen J. 1999. C_4 gene expression. *Annu. Rev. Plant Physiol. Plant Mol. Biol.* 50:187–217
110. Sinha NR, Kellogg EA. 1996. Parallelism and diversity in multiple origins of C_4 photosynthesis in grasses. *Am. J. Bot.* 83:1458–70
111. Smith LG. 1996. What is the role of cell division in leaf development? *Semin. Cell Dev. Biol.* 7:839–48
112. Smithson E. 1957. The comparative anatomy of the Flagellariaceae. *Kew Bull.* 11:491–501
113. Soderstrom TR, Ellis RP. 1987. Position of bamboo genera and allies in a system of grass classification. In *Grass Systematics and Evolution*, ed. TR Soderstrom, KW Hilu, CS Campbell, ME Barkworth, pp. 225–38. Washington, DC: Smithsonian Inst. Press
114. Soreng RJ, Davis JI. 1998. Phylogenetics and character evolution in the grass family (Poaceae): simultaneous analysis of morphological and chloroplast DNA restriction site character sets. *Bot. Rev.* 64:1–85
115. Spangler R, Zaitchik B, Russo E, Kellogg E. 1999. Andropogoneae evolution and generic limits in *Sorghum* (Poaceae) using *ndhF* sequences. *Syst. Bot.* 24: 267–81
116. Stebbins GL. 1956. Cytogenetics and evolution of the grass family. *Am. J. Bot.* 43:890–905
117. Stevenson DW, Loconte H. 1995. Cladistic analysis of monocot families. In *Monocotyledons: Systematics and Evolution*, ed. PJ Rudall, PJ Cribb, DF Cutler, CJ Humphries, pp. 543–78. Kew, UK: R. Bot. Gard.
118. Sylvester AW, Cande WZ, Freeling M. 1990. Division and differentiation during normal and *liguleless-1* maize leaf development. *Development* 110:985–1000
119. Deleted in proof
120. Tomlinson PB. 1969. Commelinales-Zingiberales. In *Anatomy of the Monocotyledons*, ed. CR Metcalfe, 3:1–446. Oxford, UK: Clarendon
121. VanDeynze AE, Dubcovsky J, Gill KS, Nelson JC, Sorrells ME, et al. 1995. Molecular-genetic maps for group 1 chromosomes of Triticeae species and their relation to chromosomes in rice and oat. *Genome* 38:45–59
122. Watson L, Dallwitz MJ. 1999. *Grass Genera of the World: Descriptions, Illustrations, Identification, and Information Retrieval; including Synonyms, Morphology, Anatomy, Physiology, Phytochemistry, Cytology, Classification, Pathogens, World and Local Distribution, and References*, Version 18 August 1999. *http://biodiversity.uno.edu/delta/*
123. Wilson WA, Harrington SE, Woodman WL, Lee M, Sorrells ME, McCouch SR. 1999. Inferences on the genome structure of progenitor maize through comparative analysis of rice, maize and the domesticated panicoids. *Genetics* 153:453–73
124. Zaitchik BF, LeRoux LG, Kellogg EA. 2000. Development of male flowers in *Zizania aquatica* (North American wildrice; Gramineae). *Int. J. Plant Sci.* 161: In press

Annu. Rev. Ecol. Syst. 2000. 31:239–63

THE ECOLOGY OF TROPICAL ASIAN RIVERS AND STREAMS IN RELATION TO BIODIVERSITY CONSERVATION

David Dudgeon
Department of Ecology & Biodiversity, The University of Hong Kong, Hong Kong SAR, China; e-mail: dddudgeon@hkucc.hku.hk

Key Words fishes, benthos, dams, pollution, freshwater

■ **Abstract** Tropical Asian rivers support a rich but incompletely known biota, including a host of fishes, a diverse array of benthic invertebrates, and an assemblage of mammals adapted to riverine wetlands. River ecology is dominated by flow seasonality imposed by monsoonal rains with profound consequences for fishes and zoobenthos. Information on life histories, feeding, and the trophic base of production of these animals is summarized. Widespread use of allochthonous foods by fishes and zoobenthos is apparent. Migration by fishes is often associated with breeding and results in seasonal occupation of different habitats. Riverine biodiversity is threatened by habitat degradation (pollution, deforestation of drainage basins), dams and flow regulation, as well as over-harvesting. Conservation efforts in tropical Asia are constrained by a variety of factors, including lack of ecological information, but the extent of public awareness and political commitment to environmental protection are likely determinants of the future of riverine biodiversity.

INTRODUCTION

Tropical Asia—the Oriental Region—has a rich flora and fauna yet, compared to the Neotropics and Africa, does not seem to invoke the same concern over biodiversity conservation. In addition, comparatively little ecological research undertaken in tropical Asia appears in the mainstream scientific literature. This is surprising. Asia is home to elephants, orangutan, gibbons and other primates (monkeys, tarsiers, and lorises), three of the world's five species of rhinos, buffalo and other wild cattle, bears, and tigers, lions, three species of leopards, and an assortment of smaller cats. Indonesia alone supports around 15% of the world's species, including more birds and flowering plants than the whole of Africa. This is megadiversity on a global scale! Of particular concern is the extent to which freshwater biodiversity in Asia has been neglected, especially that associated with rivers and streams. Hynes' (50) landmark treatise on the ecology of running waters devotes hardly any space to tropical streams (and virtually none to Asian waters), indicating the state of knowledge of these habitats in the late 1960s. Subsequent reviews of limnology (i.e. the study of inland waters) in the tropics either focus mainly upon lakes (e.g. 5), or give scant coverage to Asia (57, 82), or both (71). The incompleteness of such reviews is not surprising when we consider the extent of the subject area encompassed by "tropical limnology." The freshwater fisheries literature likewise highlights Africa and the Neotropics (88a), and the major treatise on tropical fish ecology (60) contains relatively little about Asian rivers. This lacuna in our knowledge is due, in part, to a lack of primary research but is also a reflection of a scattered, highly fragmented literature, some of which is inaccessible. Important information may be published in obscure local or regional journals that frequently have limited distribution, or data appear only in reports (for government, private environmental consultants, and the like) and are never formally published. It is thus not surprising that Allan's (1) recent text on stream ecology matches Hynes (50) in the space devoted to the tropics.

Does this matter? Given time and effort, data accumulate and information gaps are filled. Gradually our knowledge of Asian rivers will be enriched. Unfortunately, we do not have the luxury of unlimited time for further studies. Asia is the most populous region of the planet, in terms of both absolute abundance (over 50% of the global human total) and density (in 13% of the world's land area). More people live in poverty in Asia than in Africa and Latin America combined. These facts contribute to a situation in which economic growth takes precedence over other considerations. "Development now, clean up later" has become the credo of most Asian governments, leading to environmental degradation and increasing demands on natural resources.

Anthropogenic modification of riverine habitats is not new. Several of the great rivers of Asia (e.g. the Ganges, Indus, and Yangtze) have been cradles of ancient civilizations. The Chinese hero-emperor Yu the Great, who reigned from 2205 to 2198 BC, was instrumental in building dams and dikes and channelizing rivers. Sustained and pervasive human impacts are thus typical of many Asian rivers and their drainage basins. None is pristine, and the extent of past and ongoing human impacts on riverine biodiversity cannot be assessed accurately because we know little of pre-impact conditions. Pollution from agricultural areas and non-point sources is largely uncontrolled, and domestic wastewater treatment is limited (41) such that (for example) almost all sewage entering the Ganges and the Yangtze (= Chang Jiang) is untreated. Many rivers in Southeast Asia are in such poor condition that fisheries have collapsed. Legislation concerning discharge of untreated industrial effluents has been enacted in several countries (e.g. Thailand, Indonesia, Malaysia) but is weakly enforced (41). Deforestation is another major conservation issue in Asia. It results in changed runoff patterns (typically making hydrographs more "flashy") and increased siltation of rivers. Around 2% of the remaining forest area is lost each year throughout tropical Asia, but local variation is considerable (13, 43a). In the Philippines, for example, only 5% of the land area remains under natural forest compared to about 35% for Southeast Asia as a whole (13). In addition to deforestation, river flows are altered by dams, water extraction, and other river engineering works (36). Nevertheless, despite centuries of human impact in parts of Asia, large rivers such as the Mahakam, Kapuas, and Baram in Borneo retain a significant degree of ecological integrity. But even in these rivers, damaging fishing practices and over-harvest of fishes and turtles are causes of concern (16–18).

One conclusion to be drawn from these facts is that our opportunity to study the ecology of tropical Asian rivers is finite and, if we wish to preserve what remains of the biodiversity of these habitats, we must apply the ecological information we have now, notwithstanding its incompleteness. We should also decide what additional research is needed, and establish priorities so effort is put into studies that provide the information we need to underpin conservation and management strategies. This requires that we take stock. What do we know? What are the main threats to rivers and their biota? What do we need to know? How can we direct our research in order to generate the information we require? In this article I will address these questions, drawing upon and synthesizing information from a recent monograph on tropical Asian streams and rivers (33). The geographic scope is monsoonal Asia south of latitude 30°N (i.e. China north of the Yangtze is excluded), which corresponds to the Oriental Biogeographic Region, but many of my examples are from east Asia. For the sake of brevity, I have not cited primary sources for data which have been thoroughly reviewed elsewhere (33; see also 24, 25, 28, 29, 36–38, 41), although new research or noteworthy case studies and examples are referenced in full.

WHAT DO WE KNOW?

Riverine Biodiversity

Aquatic Biodiversity Asian freshwater habitats support an exceptionally rich biota. Indonesia, for instance, has 900 amphibian species, at least 1200 fishes (the final total may exceed 1700 species), and more dragonflies (>660 species) than any other country (13). The global proportion of riverine biodiversity in Asia is high. The region is, for example, home to three species of "true" river dolphins (i.e. dolphins that never enter the sea) of a global total of only five species: *Platanista gangetica* (in the Ganges and Brahmaputra), *Platanista minor* (in the Indus only), and *Lipotes vexillifer* (confined to the Yangtze, in which fewer than 200 individuals remain). All three are endangered by pollution, hunting, accidents with boat traffic, fisheries by catch, and population fragmentation by dams and barrages. Crocodilians (crocodiles, gharials, and one alligator) are well represented in tropical Asia, with eight of the global total of 23 species occurring in the region. All are endangered. Other herpetofauna are diverse also (6), and tropical Asia supports the world's richest assemblage of freshwater turtles and terrapins (83).

Tropical Asia is home to the world's largest and one of the smallest lotic fishes (the Mekong giant catfish *Pangasias* (= *Pangasianodon*) *gigas* and the tiny cyprinid *Danioella translucida*). The regional fauna is diverse, and the Indochinese Peninsula has over 930 fish species in 87 families (56). Individual countries have diverse faunas: Thailand, for example, has over 500 species belonging to 49 families. China's rivers are home to 717 freshwater fish species in 33 families (58); a further 66 species spend part of their lives in rivers. Most of these Chinese fishes are Oriental in distribution: 586 of the 717 primary freshwater species occur in the Yangtze or farther south, and the tropical Zhujiang (or Pearl River, China's second largest river) supports 262 of them (59). Species totals for other Asian rivers are likewise impressive: 290 in the Kapuas River (Borneo), 245 in the Yangtze, 222 in the Chao Phraya (Thailand), 150 in the Salween River (Burma), 147 in the Mahakam River (Borneo), 141 in the Ganges (India), and 115 in the Baram River (Borneo) (56, 58, 66, 93). Despite evident richness, existing inventories of Asian fish biodiversity are far from complete, and the general picture is one of a rather poorly known fauna (93).

How does fish biodiversity in Asian rivers measure up in global terms? The Mekong is the largest river in Asia. Its basin (802,900 km^2) contains more than 500 species (56), and perhaps as many as 1200 (73). That places it among the top three rivers in the world (after the Amazon and the Zaïre) in terms of fish species richness, although the Mekong ranks only 16th in the world in terms of length, and 15th in terms of discharge (28). If we consider richness of higher taxa, tropical Asia has more than 105 families of freshwater fishes compared to 74 in Africa and only 60 in South America. Moreover, the composition of these faunae is very different. Among 316 freshwater fish genera recorded from tropical Asia (56), the top three families are Cyprinidae (147 genera), Balitoridae (38 genera), and Sisoridae (19 genera). The Balitoridae (previously Homalopteridae) is an exclusively Oriental group of small benthic fishes, widespread in fast-flowing waters throughout the region. We know almost nothing about their ecology (21). Whereas cyprinids dominate in Asia, they are lacking in the Neotropics; Characidae is important in Africa and the Neotropics but does not occur in Asia, while Cichlidae—a major component of the fish fauna of Africa and America—is represented by only two species of *Etroplus* in Sri Lanka (although introduced African *Oreochromis* and *Tilapia* are now ubiquitous).

Invertebrate biodiversity in Asian rivers has not been studied thoroughly. The general composition of the benthos of large Asian rivers appears similar to that of such habitats the world over, and includes Tubificidae, Chironomidae (Chironominae), Gastropoda (Prosobranchia), and Bivalvia (50). However, a distinctive feature of rivers such as the Ganges, Zhujiang, and Chang Jiang is the regular occurrence of freshwater polychaetes. To this list can be added freshwater prawns (especially *Macrobrachium* spp.), which are circumtropical. Among meso- and macrocrustaceans, amphipods and isopods are scarce compared to their abundance in north-temperate habitats. Instead, Decapoda, comprising freshwater crabs, shrimps (Atyidae), and prawns (Palaemonidae), occupy a pre-eminent position in tropical Asian rivers, and have penetrated fast-flowing upland streams.

Benthic community composition in smaller rivers matches Hynes' (50) assertion that "...one of the most striking features of the faunas of stony streams is their remarkable similarity the world over." That said, the lotic mollusc fauna of tropical Asia has some distinctive elements, such as the importance of thiarid (and sometimes neritid) gastropods and corbiculid bivalves. Among aquatic insects, the diversity of plecopteran families is low in tropical Asian streams. The order is typical of cooler, more northern latitudes, and is represented mainly by Perlidae (especially Neoperlinae) plus Nemouridae, Leuctridae, and Peltoperlidae. Odonata and Naucoridae (Hemiptera: Heteroptera) are conspicuous elements of the zoobenthos in many tropical Asian streams, and Lepidoptera (Pyralidae) may be diverse and abundant.

Although most invertebrate species are still undescribed, high diversity is displayed by certain taxa. For example, the Mekong contains endemic species-flocks of stenothyrid and pomatiopsid gastropods (more than 110 species), and a similar radiation has occurred in the Yangtze. Likewise, freshwater crabs are astonishingly diverse, comprising six families with more than 80 genera in Asia (67); at least 185 species occur in China, but many more can be expected (see 33). Schmid (78) estimated that India alone may support 4000 species of caddisfly (Trichoptera), with perhaps 50,000 species in the Oriental Region as a whole, and the single genus *Chimarra* (Philopotamidae) may contain 500 Southeast Asian species (61).

Mammalian Diversity in Riverine Wetlands A striking fact about tropical Asia is that many nominally terrestrial species of mammals are associated with riverine wetlands for part or all of the year (6, 36, 37). Many of these animals (some of them "charismatic megafauna") are listed as "vulnerable" or "endangered" by the IUCN (90a). Dudgeon (37) gave a full list of species, among them otters (*Lutra*, *Lutragale*, and *Aonyx*), otter civets (*Cynogale* spp.), and fishing cats (e.g. *Prionailurus planiceps* and *P. viverrinus*). The proboscis monkey (*Nasalis larvatus*) depends upon forested riverine wetlands and sleeps in tall trees along riverbanks. These monkeys have webbed fingers and toes to aid swimming. Other primates such as leaf monkeys (*Presbytis* spp.) and crab-eating macaques (*Macaca fascicularis*) are abundant in riparian forest, while swamp forest is key habitat for orangutans (*Pongo pygmaeus*) in central Kalimantan (Borneo). With around 100 individuals remaining in the wild, this is probably the rarest large mammal in the world (43). Malayan tapirs (*Tapirus indicus*) are riverine animals par excellence. They inhabit dense vegetation and swamp forest by day, but venture onto marshy grasslands or floodplains to feed at night. A tapir can remain submerged for long periods, breathing through its long snout. Other species of megafauna, including Asian elephants (*Elephas maximus*) and Javan rhinoceros (*Rhinoceros sondaicus*), are more wide-ranging, but use riverine wetlands during the dry season because of the availability of water and green forage.

One habitat of particular importance to some elements of the Asian megafauna are the seasonally inundated, grassy floodplains of large rivers. The Indian rhino (*Rhinoceros unicornis*)—which is rarer than either African species of rhino—is confined to such habitats, and formerly ranged along the Indus, Ganges, and Brahmaputra Rivers (43). The historical distribution of the Indian rhino indicates that, in the recent past, significant expanses of swampy grassland mixed with forest, covered floodplains in parts of Asia, providing habitat for a complex of large grazing animals including rhinos, deer, and water buffalo and other wild cattle. For example, Asian water deer (*Hydropotes inermis*) graze vegetation on periodically inundated alluvial soils; Père David's deer (*Elaphurus davidianus*) was restricted to wetlands along the Yangtze; Hog deer (*Axis porcinus*) occur on floodplains and marshy areas with tall grass; and Sambar deer (*Cervus unicolor*), which make opportunistic use of floodplains and riparian forest, often feed while wading. Species or subspecies of marshland deer [e.g. *Cervus* (*Ruvicervus*) *duvauceli, C.* (*R.*) *eldi, C.* (*R.*) *schomburgki,* and *E. davidianus*] are—or were—confined to particular river systems. These riverine species of *Cervus* are larger than congeneric species from drylands, a fact that may be related to the high productivity of floodplain grazing lands (44), Marshland deer (e.g. *C.* (*R.*) *eldi eldi* and *E. davidianus*) have splayed or unusually large hooves which are adapted to floodplain grasslands. Marshland deer are restricted to open floodplains. Their movement in forests and overhanging vegetation is limited due to their expansive antlers, which bear acutely angled tines or have radial branching.

Patterns and Processes

An Asian scientist—Sunder Lal Hora—contributed to the study of lotic ecology during the second quarter of the twentieth century. He highlighted the role of substratum and current as determinants of the distribution and morphology of fauna in Indian torrent streams (45–47). He was aware of the importance of human impacts on Indian rivers and drew attention to declining fish stocks (48). In 1973, 50 y after Hora's first publication, research in Malaysia established that invertebrates are abundant deep in the streambed, highlighting the importance of the hyporheic zone (10). Yet, despite an early start in tropical Asia, lotic ecology remains in its infancy. We have detailed information from rather few localities (e.g. Sungai Gombak in Peninsular Malaysia; Tai Po Kau Forest Stream in Hong Kong); quantitative data on the magnitude of primary and secondary production (benthos or fishes) are in very short supply. Apart from some in Hong Kong (e.g. 26, 30, 39, 41a), manipulative experiments have been undertaken rarely. The work of Benzie (8) on colonization mechanisms of zoobenthos is a notable exception, while Wikramanayake & Moyle (89) performed a novel translocation experiment to investigate resource partitioning among Sri Lankan stream fishes.

Seasonal and Spatial Dynamics of Aquatic Communities What is the ecological backdrop for biodiversity in tropical Asian rivers? Some of the greatest rivers in the world (ranked by magnitude of discharge) are located in the region: the Yangtze,

Brahmaputra, Ganges, Irrawaddy, Zhujiang, Mekong, and Indus. Natural lakes, apart from those associated with river floodplains, are of relatively minor importance, and are far exceeded in area by riverine swamps. The ecology of Asian rivers is profoundly influenced by monsoons, which create a characteristic pattern of seasonality: predictable periods of drought and water scarcity during the dry season alternate with intervals of increased discharge, when flood plains are inundated during the wet season. This has important implications for aquatic productivity, although our understanding of the effects of monsoons on all elements of the biota is far from complete. There are also implications for the ionic composition of river waters plus the transported organic and inorganic loads. In general, concentrations of major ions are inversely proportional to flow rate, and hence decline during the monsoon season. Elevations in nitrogen may occur during floods or in wet-season discharge, especially from densely settled catchments, and there is often a direct correlation between river discharge and the transport of inorganic suspended loads (for details, see 33).

Seasonal patterns in abundance of phytoplankton and zooplankton in large rivers are related to the monsoon, with a wet season low (due to washout, dilution, and turbidity) and a dry season high. Likewise, scouring of periphytic algae during wet-season spates reduces standing stocks, as in subtropical Australian streams (65). Declines in zoobenthos abundance might be expected during the wet season, at least in small streams, where spates associated with monsoonal rains and tropical storms could initiate catastrophic drift and washout. Data from Hong Kong suggest that the picture is far from simple; for example, hydropsychid caddisflies and heptageniid mayflies in one forest stream showed dry-season peaks in density and wet-season lows (31, 32), while polyvoltine calamoceratid caddisflies showed a boom and bust cycle as populations were depleted by spates but recovered when flow conditions stabilized (34). By contrast, in a second, smaller stream, the abundance and species richness of the benthos increased during the wet season (26). In the second stream, base-flow conditions during the dry season imposed a greater disturbance than periodic spates and high discharge during the wet season (see also 65). Evidently, there is inter-stream variation with respect to seasonal fluctuations in zoobenthos densities such that either the wet or the dry season may be the period of greater numbers. Arunachalam et al (2) recorded such variation at a smaller scale, noting that seasonal trends in invertebrate abundance differed among pools in a South Indian river.

While the seasonal dynamics of Asian rivers are driven by monsoons, we know little about the details of spatial or temporal dynamics of almost all elements of the aquatic biota. Studies that document changes in composition of the lotic flora and fauna from the headwaters to mouth are lacking entirely, although some components of the biota have been investigated. Where a sufficiently wide altitudinal range is considered, there are longitudinal changes in the composition of zoobenthos assemblages, diatoms, bryophytes, and fishes (e.g. 70). An important corollary of such longitudinal patterns is the concentration of diversity at lower altitudes. This has important conservation implications as human impacts on the landscape are greatest in the lowlands. Understanding of longitudinal zonation in (especially) large Asian rivers is confounded by the influence of pollution, so it can be difficult to distinguish "natural" communities from those under anthropogenic influences. Local land use also affects the composition of the stream fauna and alters longitudinal zonation patterns. In his landmark monograph on Sungai Gombak (Peninsular Malaysia), Bishop (11) noted that clearance of riparian vegetation resulted in a marked decline in benthic invertebrate richness. More recent studies have also demonstrated the influence of riparian vegetation on community composition (22).

Zoobenthos: Life Histories, Trophic Relations, and Production While the seasonality of zoobenthos reflects, in part, washout and scouring during spates caused by monsoonal rains, the timing of life-cycle events influences population densities and community composition. The strategy of small polyvoltine insects, certain atyid shrimps, and some molluscs involves flexible, poorly synchronized life histories of the type that probably represents an adaptive response to streams with variable or unpredictable discharge. Periods of extended recruitment and multiple overlapping cohorts seem typical of many aquatic insects in which water temperatures remain above 15°C for most of the year (e.g. 92). Adult emergence prior to or at the start of the monsoon is common to a number of Hong Kong Odonata, Trichoptera, and Ephemeroptera. Such seasonality involves relatively large species. Compared to larger conspecifics, smaller larvae or eggs are less likely to be crushed or injured by the substratum movement during spates, and may be able to seek shelter deep in the hyporheic zone. Concordance of life histories among unrelated species from three insect orders lends support to the suggestion that timing of adult emergence (and subsequent oviposition) is an adaptive response to avoid spate-induced mortality of mature larvae during the peak of the summer monsoon, a degree of synchrony arising from physiological responses to increasing day length. Coincidence of reproduction with the monsoon—seen in some freshwater crabs and shrimps—could also reflect increased habitat availability (e.g inundated floodplain) for hatchlings in large rivers swollen by monsoonal rains. The breeding habitats of these decapods in large rivers are similar to those of fishes (see below), and may involve migrations. Decapods in smaller streams tend to reproduce before the peak monsoon floods, and hatchlings are benthic forms resembling miniature adults. This life-cycle strategy avoids spate-induced washout of planktonic larvae. Molluscs also seem to have evolved strategies to cope with discharge seasonality: some pomatiopsid snails (e.g. *Neotricula*) in the Mekong River mate during the high-water period but oviposition does not occur until water levels have fallen. Other Mekong genera (e.g. *Paraprososthenia*) mate before the floods, but egg development is delayed so that recruitment takes place when low-flow conditions prevail (4).

Whether seasonal or longitudinal changes in zoobenthic community composition have implications for productivity is unclear. The River Continuum Concept (RCC; 85) makes predictions (albeit rather broad or approximate ones) about downstream changes in representation of functional feeding groups in response to changes in the food base (especially allochthonous inputs) in lotic habitats, and can serve as an a template for comparing Asian running waters with those elsewhere. The functional organization of benthic communities in shaded streams in tropical Asia—which have high standing stocks of allochthonous detritus—does differ in a rather general way from that of communities in unshaded streams (23, 24; see also 91). However, the paucity of shredders (which comminute coarse detritus) in the upper course of some tropical Asian streams is a significant deviation from the RCC prediction that shredders and collectors should codominate in headwaters. Shredders comprise between 0.1 and 8.8% of the zoobenthos in Hong Kong streams (23, 38) and are scarce in leaf packs (41a). Even in primary rainforest streams in New Guinea, shredders do not exceed 2% of benthic populations (27, 91), and shredders were no more abundant in forested than unshaded streams in Nepal (70). If this under-representation of shredders is a general feature of tropical Asian rivers, it may reflect trophic flexibility and hence functional feeding-group misclassification, that is, the same taxon acting as a shredder or a collector of fine organic material under different circumstances. Alternatively (or in addition), a lack of shredders could reflect limited stream retentiveness for leaf litter (e.g. 69), an increased importance of microbes in litter breakdown in tropical streams (51), or higher investment in chemical defense by tropical leaves making them unpalatable to shredders (79). Leaf palatability (as influenced by toughness or tannin content) does affect the composition of invertebrate assemblages colonizing litter in Hong Kong streams, but most taxa are collector-gathers, not shredders (41a). Breakdown rates for the leaf species that have been studied are rapid, with complete disappearance of litter in less than three months (and within a month for some species). The paucity of shredders does not prevent litter breakdown, but the relative importance to this process of microbes, physical fragmentation, and high water temperatures has yet to be determined (51).

Our understanding of the functional organization of stream communities is based upon data on relative abundance of various feeding groups or, more rarely, absolute densities. Information on the biomass of benthos is limited, and data on periphyton primary production and invertebrate secondary production are fragmentary. Values seem to lie within the same order of magnitude as they do for temperate streams, but there can be considerable inter-annual variation in production by individual invertebrate taxa. For instance, rheophilic species were more abundant when stream discharge was above average in a Hong Kong stream (35). Despite considerable inter-year fluctuation (8 to 537%) in the production of individual species of eight species of filter-feeding caddisflies, total production by filter-feeders varied by only 1.8% between years. Thus, changes in the production of one or more filter-feeding species may be offset by alterations in others so that energy flow through the filter-feeder functional group remains rather constant. Likewise inter-annual variation in total production of heptageniid mayflies (8.6% for five species) was much less than that of individual species (28 to 674%). Generalizations about zoobenthos production in Asian streams must be qualified. Studies to date have focused mainly on larger species—most univoltine insects with rather synchronous growth and clear life-history patterns. However, most invertebrate taxa exhibit asynchronous growth, with year-round emergence of adults, short life cycles, and polyvoltine life histories. Production estimates for these animals require in situ measurement of the growth of individuals maintained in cages (e.g. 7). Preliminary studies of this type in Asia indicate generation times of a month or less for some mayflies (M Salas, D Dudgeon, submitted for publication), which is similar to values from subtropical and tropical streams elsewhere (7, 53, 62). The turnover of zoobenthos in tropical Asian streams is thus higher than that of the univoltine insects for which we have production estimates, so the present data we have are derived from slow-growing, atypical species with rather low production.

The relatively low proportion of shredders in tropical Asian streams suggests that consumption of allochthonous organic matter is unlikely to make a major contribution to zoobenthos production, but other data imply that this supposition is incorrect. Food webs in streams and small rivers (11, 19, 20) involve widespread use of allochthonous foods by primary consumers, and a relatively high percentage of nominally predatory forms. Nevertheless, the contribution of autochthonous algal food should not be underestimated. A small-scale manipulative study in a Hong Kong stream showed that reductions in periphyton standing stocks led to declines in zoobenthos abundance—especially mayflies (39). Low standing stocks of algae in forest streams may be attributed to shading and nutrient limitation, but an alternative explanation is that periphyton biomass is limited by intense grazing pressure, in which case standing stocks will not give a reliable indication of the importance of algal food to primary consumers. Direct measurements of community metabolism (R) and primary production (P) would help resolve this matter. The only figures we have are from a single day in a Hong Kong stream: ($P/R = 0.17$), indicating heterotrophy in a shaded reach and 1.02 in an unshaded section (24). Data are needed from additional streams during different seasons.

Fishes: Life Histories, Trophic Relations, and Production Information on the ecology of fishes in tropical Asian rivers is considerably more comprehensive than that for invertebrates, but is nonetheless surprisingly meager. Standing stocks in small rivers seem to be somewhat lower than those reported for north-temperate streams with comparable water chemistry, but this generalization is based on few studies. Fish biomass is reduced markedly in acidic ($pH \leqq 5.5$) blackwater streams draining peat swamp forest, although such waters may contain a diverse array of stenotopic species (54, 68). Fisheries yields from river floodplains may be high but are subject to substantial inter-annual variation according to the intensity of the monsoon, which drives flood pulses (i.e. the predictable advance and retraction of water over the floodplain). Generally, the result is higher yields during stronger monsoons (88a).

Many fishes show highly seasonal feeding activity, with alternating periods of resource scarcity during the dry season and resource glut during the wet season. A wide range of foods is exploited; allochthonous dietary items, such as terrestrial insects and fruits, are important and their use seems much greater than has been reported for lotic fishes in temperate latitudes (33, 88a). Many Asian river fishes migrate within river systems. The migrations involve a significant upstream or lateral component, in addition to return or downstream movements. They are frequently, but not invariably, associated with breeding, and are timed to exploit allochthonous foods from inundated floodplains or riparian forests. Thus these are not only breeding migrations but represent migrations to superior living space. Most species that depend upon the floodplain for food and breeding sites show marked inter-annual variation in production (88a).

The Mekong River provides an example of the interaction between fish ecology and discharge. The timing of fish migrations differs somewhat according to species and among different parts of the Mekong Basin, but this complex situation is amenable to generalization (88a). Breeding migrations are made upstream during the wet season as water levels rise; downstream migrations occur when levels fall during the dry season. Planktivorous or piscivorous pelagic fishes (whitefishes) migrate a greater distance than do bottom-dwelling species (blackfishes). The latter feed on benthic organisms, and make shorter lateral movements between the main channel and floodplain or inundated forest than do the former. The fishes differ with respect to breeding habits also: many benthic species are multibrooded and practice parental care (mouthbrooding, bubble-nest building). In addition, these blackfishes can tolerate deoxygenation of stagnant floodplain waters because they have accessory breathing organs, whereas most pelagic whitefishes spawn only once per season and scatter their eggs. An intermediate, relatively species-poor assemblage of generalists (greyfishes) have both migratory and static/territorial behavioral components, enabling them to respond facultatively to changing flow conditions; they may thrive where the natural flood regime has been modified.

In smaller Asian rivers, most fishes synchronize breeding activity with the monsoon or flood season and, where there are two monsoonal floods each year (e.g. in south India, Sumatra and Sri Lanka), they may reproduce twice. There are a few exceptions to the general pattern of wet season spawning: some species of small *Barbus* breed throughout the year or during the dry inter-monsoon period, while spawning by the Taiwanese cyprinid *Zacco pachycephalus* takes place prior to the onset of monsoonal spates which reduce recruitment success (86). All generalizations must be qualified, because our knowledge of fish ecology in Asian rivers is extremely limited. Indeed, writing of the Mekong (a relatively well-known system), Roberts (74) states "...the entire field of fish reproductive biology, including timing and stimulus of reproductive migrations, time, place and requirements for spawning, physiological adaptations of eggs and larvae, and comparative reproductive biology of carps, catfishes, and other groups under natural conditions, is largely untouched..." p. 58.

What Are the Threats to Biodiversity?

Anthropogenic influences imperil the biodiversity of rivers and their associated wetlands at a variety of scales. The main threats are:

1. Deforestation and drainage-basin alteration that destroy or degrade instream and riparian habitat;
2. River regulation, including flow modification and impoundment by dams, water extraction for irrigation, etc;
3. Pollution;
4. Over-harvesting (mainly of fishes and reptiles).

The combined effects of these threats may be more damaging than the sum of each. A further threat—global climate change—will affect river discharge patterns and interact with threats 1 and 2 above to further impact riverine biota.

Deforestation and Drainage Basin Degradation While the fauna of tropical Asian rivers is adapted to seasonal flow fluctuations, unregulated rivers do not necessarily have natural discharge regimes. The consequences of drainage-basin misuse and deforestation are evident throughout the region, and include increased runoff, sedimentation, and flash floods (33). In particular, the effects of deforestation interact with seasonally variable river flows, increasing the frequency and intensity of flood flows with direct and profound consequences for human inhabitants of floodplains. In China, for example, devastating floods along the Yangtze in 1998 focused attention on the link between catchment conditions and runoff. Government blamed the floods on years of uncontrolled logging in upland catchments. Similar consequences of drainage-basin degradation led to a logging ban in Thailand after devastating floods in 1988 claimed several hundred lives.

The loss of forest cover in tropical Asia (estimates range from 0.9% to 2.1% per annum, but local variation is substantial) has important implications for river conservation since a significant proportion of the rich fish fauna of the region exploits allochthonous foods in inundated forests. For example, forest clearance around Le Grand Lac of Cambodia—a floodplain lake connected to the Mekong by the Tonlé Sap River—has led to declining fish catches through a combination of reduced food, erosion, and siltation (88a). Increased turbidity limits primary production, and results in lower fish populations. Forest clearance also threatens fishes confined to soft, acidic, blackwater streams that drain peat swamps. Many such fishes will become extinct (as has occurred already in parts of Malaysia and Indonesia) if logging and other degradation of their habitats continue (68).

Changes in habitat characteristics of drainage basins do not affect forests only; floodplains are also modified or degraded. Because marshland deer (see above) have a narrow reliance on grass as food and highly specific habitat preferences that preclude the use of drylands, they declined in numbers as floodplains were drained, settled, and converted to rice cultivation. During the 19th century, herds of them thrived along the major rivers of India and Thailand; ecologically equivalent species occurred in southern and central China until around 200 y ago (44). Père David's deer was exterminated on the Yangtze floodplain through a combination of habitat modification and hunting (although captive-bred animals have re-established a population in a small part of the original range). Extinction of *Cervus* (*R.*) *schomburgki* in Thailand reflected, in part, possession of the most elaborate antlers of any marshland deer, restricting them to open floodplains and magnifying the risk from humans. Elimination of the Indian rhino over much of its range, the near extinction of wild water buffalo (*Bubalus arnee* [= *B. bubalis*]), and threats to other large grazers likewise reflect conversion of floodplains for agriculture (6, 37).

River Regulation and Dams Fishes and other elements of the lotic fauna are adapted to the floods that are typical of rivers in monsoonal Asia. Although there is some inter-annual variation in flood peaks, the flood-pulse is not a disturbance in the ecosystem. Instead, significant departures from the typical pattern of seasonal flow fluctuation can be regarded as disturbance. This sets the scene for conflict between the needs of humans and of riverine biota. The engineering response to flow seasonality is to capture and store water during flood times for use during the dry period, the benefits being amelioration of peak discharge (i.e. flood prevention) and a predictable water supply during the rest of the year. As many Asian rivers drain degraded and deforested catchments, the desire to limit peak flows by trapping flood waters behind huge dams is understandable. The conflict between human desires to limit or control flows and the requirements of the ecosystem for natural or semi-natural flows is heightened in Asia by high population densities that place severe constraints on the preservation of riverine biodiversity.

The recent trend in Asia has been toward more and bigger dams, culminating in the Three Gorges Scheme on the Yangtze—the largest dam in the world, which will impound Asia's largest river. Approximately 65% of large dams ($\geq$15 m tall) are in Asia (63); China has almost half of these, including 10 major dams (>150 m tall). India (second in Asia and fifth in world ranking) has more than 1100 large dams and 7 major dams. The Mekong has vast potential hydroelectricity-generating capacity and, in 1994, the Mekong Commission (an autonomous coordinating committee set up by three riparian states) identified 12 sites for dams (most with planned outputs >10,000 MW) along the river mainstream in Laos, Thailand, and Cambodia. A predictable result of these dams is that natural flow variability—to which the fauna are adapted and upon which they depend—will be smoothed out. This averaging of flows is significant because ecological responses to environmental change may be non-linear: fish breeding migrations, for example, may not begin until flows have passed a critical threshold.

The consequences of the Mekong Dam array for individual taxa cannot be predicted with certainty. We know little about the zoobenthos of the Mekong but, for most species, the effects are unlikely to be positive. A possible exception is the pomatiopsid snail *Neotricula aperta*, which is a host of the human parasite *Schistosoma mekongi* (Trematoda: Schistosomatidae) and may be favored by modified flow regimes (4). Mainstream dams change the timing and extent of floodplain inundation and, in combination with altered flow and temperature regimes, may suppress fish breeding. They will obstruct the passage of long-distance whitefish migrants, some of which cover 500–1000 km, including species forming the basis of Mekong wild-capture fisheries (36, 74–76). Dams will have their greatest impacts on species confined to the Mekong. Among them are *Pangasius gigas* and *P. krempfi* (Pangasiidae), the giant gouramy *Osphronemus exodon*, the carps *Probarbus labeamajor* and *Aaptosyax grypus* (a 1.3 m-long, 30 kg predator), and the endangered freshwater herring *Tenualosa thibaudeaui*. Some of them could be driven to extinction by a single mainstream dam. The life history of *Pangasius* (= *Pangasianodon*) *gigas*—the 3 m-long Mekong giant catfish—and the routes taken during breeding migrations have yet to be elucidated. Other significant Mekong species are poorly known also: *Aaptosyax grypus*, *Probarbus labeamajor*, and *Osphronemus exodon* were described by scientists within the last decade, yet all are fishes ≥10 kg in weight, with *P. labeamajor* attaining 70 kg!

Likely effects of dams on Mekong River fishes can be extrapolated also from changes in other large rivers. The more than 3000 dams that have been built along the course of the Zhujiang during the last 40 years have obstructed longitudinal migrations by *Clupanodon thrissa* (Clupeidae), and stocks of major carp species (including *Cirrhinus molitorella*) have fallen to levels that no longer sustain a viable fishery (59). Abundance of the Chinese sturgeon (*Acipenser sinensis*) has been greatly reduced, which has conservation implications because this fish is restricted to the Zhujiang and Yangtze. The anadromous clupeid *Tenualosa reevesii* has been eliminated from parts of the Zhujiang. Landings of *Tenualosa* (= *Hilsa*) *ilisha* likewise dwindled to almost nothing after completion of the Farakka Barrage on the Ganges (66).

The impacts of river regulation involve not only barriers to longitudinal migration but also limitations on lateral movements. Fisheries in the Jamuna River (Bangladesh) declined by over 50% when flood-control embankments reduced fish access to floodplains, and the major carp fishery of the Ganges practically disappeared after seasonal inundation of the floodplain was blocked (66). Other consequences of dams relate to the effects of thermal stratification (9). Water released from the hypolimnion is cool and oxygen-poor, and may even be anoxic and contain toxic hydrogen sulfide. This has predictable consequences for the downstream biota.

The detrimental effects of dams on river fish are often ignored because of the view that negative impacts are more than compensated for by fisheries in newly created impoundments (e.g. 12, 42). Most fishes adapted to the fluctuating discharge of a river will not thrive in a stagnant impoundment. High fisheries yields reported from Asian reservoirs depend on cage culture, stocking (usually of major carp), or introduction of exotic species (9). Such practices are accompanied by a loss of native biodiversity.

Pollution Water quality varies considerably through tropical Asia. However, most major rivers, as well as a host of minor streams, are grossly polluted, and some are among the most degraded in the world. Anti-pollution legislation is in place in some countries, but is weak and/or poorly enforced (41). Pollution from agriculture and non-point sources is largely uncontrolled and, as cities and industries expand without adequate waste-treatment facilities, rivers receive increasing quantities of effluent. Many rivers are oxygen-poor for much of the year.

In parts of Peninsular Malaysia and Java, some rivers are so severely degraded that fisheries have collapsed. The situation is especially bad in China where, by 1995, some 80% of the 50,000 km of major rivers were too polluted to sustain fisheries (84a). Fish have been entirely eliminated from more than 5% of total river length in China. Fish kills (due to industrial waste water, pesticides, etc) occur frequently during the dry season when water flows are insufficient to dilute pollutants. Tributaries are more heavily contaminated than the main channels where dilution of pollutants is greater. Even in the Yangtze, which once produced 70% of China's freshwater catch, landings fell by more than half between 1954 and 1981, and they have declined ever since.

Things are no better in India. Fish kills are common in the Ganges, which ranks third in the region by flow volume and is probably the most polluted large Asian river. With approximately 400 million people in the catchment, human population density is high, and 600 km of the river course is grossly polluted by sewage, animal wastes, and industrial effluents. Most industries discharging into the river lack operational waste treatment plants, and some cities (e.g. the holy city of Varanasi) lack sewage treatment facilities entirely. Environmental degradation has continued despite the fact that the Ganges is viewed by Hindus as sacred, and pilgrimages to bathe in the cleansing river waters are an important religious practice. This contradiction epitomizes the challenges facing those wishing to protect Asian rivers.

Over-Harvesting Freshwater fishes and herpetofauna are widely harvested across the region for local and overseas consumption. The effects on wild populations of some reptiles have been significant. For example, all eight species of Asian crocodiles are endangered mainly because of hunting pressure. Huge numbers of river turtles are exported from Thailand and Indonesia to China, and there is concern for the long-term viability of some species in the wild (18). Amphibians (especially edible frogs) may also be threatened by over-harvesting, but causes of population decline in this group are complex and reliable data are wanting (6).

Information on fish harvest is more plentiful, albeit frustratingly incomplete. Asia accounts for 64% of the global total of inland fisheries landings (84a). China, India, Bangladesh, and Indonesia are the four most important inland fishery countries in the world, with Thailand ranking seventh (84a). Available figures do not separate yields of aquaculture-based fisheries and enhanced fisheries (in which stock is added to water bodies) from natural capture fisheries, nor do they distinguish reservoir from river fisheries. The dispersed and informal nature of many fisheries leads to under-reporting of landings, and subsistence fisheries are rarely reported. One thing is clear: high catches in Asia reflect heavy exploitation of virtually all freshwaters, and capture fisheries based on wild stocks appear to be at or exceeding the limits of sustainable yield.

The effects of overfishing first become manifest among large species (17). Thus, in east Sumatran rivers, the freshwater "shark" (*Wallago attu*: Siluridae), which can reach a length of 2 m, has been exterminated over part of its range. Populations of the 1-m long cyprinid *Probarbus jullieni* in Sungai Perak, Peninsular Malaysia, were devastated by capture of egg-bearing individuals during upstream migrations, and stocks collapsed when passage was blocked by the Chenderoh Dam (55). Reductions in *Probarbus jullieni*, *P. labeamajor* (which reaches 80 kg), *Catlocarpio siamensis* (120 kg, formerly the basis of an important fishery), and *Pangasius gigas* (300 kg) have occurred in Laos, Cambodia, and Thailand, apparently due to overfishing. *Pangasius gigas* was the basis of an important fishery in Cambodia because oil could be extracted from its fatty flesh. Population declines began decades ago, and stocks have collapsed in recent years so that the fish is now endangered (90a).

Dramatic declines in capture fisheries—such as a 40% reduction in landings by Laotian artisanal fishers between 1984 and 1992—may reflect the use of unsustainable fishing techniques such as explosives. Increased availability and use of synthetic chemicals, rather than natural poisons (such as rotenone extracted from plants), may be a contributing factor also (16). Motorboats and refrigeration technology allow the exploitation and marketing of previously inaccessible stocks. They contribute to overharvesting by reducing the extent of refuges in which stocks are not fished and from which recruits are derived. Unsustainable fishing practices, such as the use of ever-finer meshed nets and overfishing of spawning grounds, combined with the application of poisons, explosives, and electric shocking, have led to marked declines in major carp recruitment in the Zhujiang. In this instance, the damage of such fishing practices cannot be separated from the effects of dams and fish kills caused by pollution (59).

Combined Impacts The synergism of anthropogenic impacts on the aquatic biota is exemplified by the Acipenseriformes (sturgeons and paddlefish) of the Yangtze (88). Acipenseriformes and other large fishes are especially vulnerable to overfishing and other environmental hazards because the long maturation time limits rates of population recovery. Spawning migrations of the anadromous Chinese sturgeon (*Acipenser sinensis*) were blocked by the Gezhouba Dam (or Yangtze Low Dam); fish passes were not provided. Alteration in flow and sediment characteristics have reduced the spawning success downstream of the dam. The Gezhouba Dam fragmented populations of the endemic Yangtze sturgeon (*Acipenser dabryanus*). This potamodromous species undertakes breeding migrations along (but within) the Yangtze. It is now virtually extinct in downstream reaches because populations stranded below the dam cannot travel upstream to breed. Sedimentation (due largely to erosion from deforested uplands) has impacted the populations upstream of the dam. Completion of the Three Gorges High Dam on the Yangtze will bring additional environmental alterations that are likely to further endanger remaining sturgeons. Pollution from industrial and domestic sources has degraded habitat quality for sturgeons throughout the river, and overfishing has contributed to declines of both species. The same combination of factors has led to the decline of the anadromous Chinese paddlefish (*Psephurus gladius*: Polydontidae) in the Yangtze. The Gezhouba Dam prevents access to upstream spawning sites, and this paddlefish (one of only two polydontids on the planet) is likely to become extinct.

Global climate change will affect river discharge patterns and may have other effects that are likely to impact riverine biota in tropical Asia (and elsewhere). The influence of water quantity on the life history of tropical fishes and other aquatic taxa seems to parallel the importance of temperature to fishes in temperate latitudes (36, 64). Possible scenarios of climate change include wetter wet seasons and drier dry seasons, with an increased frequency of extreme flow events. Because fish catches (and secondary production) from rivers are positively correlated with the extent of inundated floodplain, the increased frequency of extreme flow events will increase the variability of fisheries yields. In addition, lower flows during the dry season will concentrate pollution loads in many rivers. The combination of extreme flow events and the effects of deforestation of drainage basins on run-off can be anticipated to spur construction of even more dams and flood-control projects to the detriment of aquatic biodiversity.

What Do We Need To Know?

Existing conservation efforts are piecemeal and reactive—a result, at least in part, of a lack of awareness or interest in mitigation of anthropogenic impacts on rivers. Conservation must be proactive. Our ability to ameliorate or mitigate the effects of human activities is predicated upon an adequate understanding of river ecology. But information on the prevailing situation with respect to habitat integrity and biodiversity is rarely available. Even for fishes, which have direct relevance to human welfare, data on faunal composition are inadequate. Few countries have made systematic assessments of biodiversity. Some published catalogs are based on dated literature or old museum collections instead of recent surveys. Information on fish conservation status is thus unreliable, and extinctions may not be evident until long after they have occurred (72, 93).

The basic task of compiling species inventories is one priority for biodiversity conservation, but it should be accompanied by assessments of population size and long-term viability—especially for populations of wide-ranging species that are fragmented by dams. These tasks will require monitoring strategies, such as those developed by Humphrey et al (49) in tropical Australia, and formulation of indices of river health that can be applied to individual habitats. Study design and replication should unambiguously address the putative environmental impacts to the exclusion of confounding variables. Unfortunately, studies aimed at investigating the effects of pollutants in tropical Asian streams and rivers have failed to adopt appropriate, statistically rigorous designs (33) as advocated by Underwood (84), among others.

Detecting the onset of environmental degradation is problematic and may result in measures to reverse the situation being put in place only after population declines and habitat degradation are irreversible. Subtle but important trends in species loss are generally not discernible early enough to permit appropriate remedial action. In the case of Asian river fishes, for example, we lack data on the extent of inter-annual variation in community composition although, like fish production, it may reflect the extent of floodplain inundation in large rivers (88a). The use of zoobenthos abundance data to assess anthropogenic impacts is unlikely to be feasible if natural variation is high. For example, in a Hong Kong stream in which inter-annual variation in the densities of 20 benthic insects was 238% (range 2–1001%), an annual population decrease of 30% might be undetectable!

Assessment of environmental impacts is an important part of the activities of many ecologists, but empirical studies of impacts need to be buttressed by process-oriented studies on the mechanisms underlying the changes caused by the impact. This requires identification of priorities about where knowledge of underlying mechanisms (or even identification of fundamental patterns) is needed. A model of river ecosystems that might allow us to make predictions about natural conditions (and hence recognize significant deviations from such states) would be of great value to environmental managers. Moreover, such a model might indicate the relative importance of contributions of allochthonous energy to aquatic consumers. Does the RCC have any utility in Asia? We cannot—and should not—ignore prevailing paradigms such as the RCC, but this model was derived from research in temperate North America where catchments are dominated by deciduous forest species and climatic factors impose fundamentally different regimes on aquatic biota and ecological processes. Not surprisingly, the functional organization of zoobenthic communities in Asian streams do not agree closely with the predictions of the RCC (38). However, it is not clear what kind of field data at what sampling scale would be required to determine which predictions of the RCC apply to Asian rivers. Quantification of longitudinal changes in the functional organization of benthic communities along several pristine Asian rivers is needed to develop a generally applicable model; a large-scale, multi-national study devoted to this objective would be valuable. Its usefulness could be increased if fish communities were included with measurements of the biomass (or better still, production) of various feeding groups, and if investigations were carried out so as to take account of extremes of flow in the dry and wet seasons. It will be essential to include the consequences of floodplain inundation in the model, since this was not incorporated in the original RCC. Among the obvious practical difficulties is finding several unimpacted rivers of a suitable size since no large, pristine rivers remain in Asia. Nevertheless, without such research, at least on sections of those rivers that have escaped the worst effects of human activities, we have no means of determining whether the RCC serves as anything other than a caricature of lotic ecosystems in Asia.

Two techniques used by ecologists in recent years that have yet to be applied in Asia would aid understanding of the trophic basis of production in rivers. Examination of the stable isotopic signature of consumers allows an assessment of the relative importance of allochthonous and autochthonous energy sources in supporting secondary production. The ratios of stable carbon isotopes differ between terrestrial and aquatic plants; the stable isotope ratios of nitrogen also yield information about the trophic level of consumers. These methods have been used in tropical Australian rivers (14, 15). Preliminary results from Hong Kong streams have been surprising because they show that, despite low periphyton biomass and high detrital standing stocks, isotopic signatures indicate that mayfly larvae derive most of their assimilated energy from algae (M Salas, D Dudgeon, submitted for publication). Another technique that yields information about the importance of allochthonous and autochthonous energy sources involves the direct measurement of primary production and community respiration (and hence calculation of P/R ratios). When repeated several times over a year, annual production can be estimated. Workers in tropical Australia have deployed microcosms in situ to obtain habitat-specific measures of stream community metabolism (14, 15). Such measurements from tropical Asia could yield information that would be laborious to collect by other means and, because the RCC makes predictions about P/R ratios and the relative importance of various energy sources, could also be used to investigate the applicability of this model to Asian rivers.

Constraints and Challenges: Applying What We Know

Habitat destruction or degradation in and along Asian rivers is epidemic, with predictable consequences for resident and migratory species. How can the results of ecological research (including tests of the RCC) be applied to biodiversity preservation in such circumstances? Conservation of individual species will depend more on knowing the details of the natural history of the target organisms (although such data are generally lacking; 74) and less on a general understanding of ecosystem functioning. Conservation strategies must take account of the facts such as that Asian river fishes have complex life cycles and different habitat requirements at various stages of their life history, which means there are many routes of exposure to toxins, direct harvesting, and other detrimental influences. Data on habitat use must be collected on a case-by-case basis and used to develop species-based plans for in situ conservation. It can be hoped that strategies that are applicable to species of particular interest (such as the Mekong giant catfish) will protect other taxa that share the same habitat and have similar habits. However, there are few signs that relevant studies are being undertaken or even that there is much awareness of the need for such research (74, 75). For that reason, I (36, 37) have stressed the need to identify flagship species and keystone habitats that may help to increase public and government awareness of the existence of aquatic biodiversity. Effective habitat conservation will require forceful demonstrations of the benefits to be gained from the integrated use of floodplains, rivers, and riparian habitats as wetlands if we are to prevent their conversion to other uses. Promulgation of information on the value of these environments and their biota is the first step in this process.

Species that use distinct environments during different stages of their life histories (migratory fish for example) are at risk from elimination of crucial habitats or erection of structures that impede movement. Fragmentation of resident populations also results from barriers, and may ultimately lead to species loss. If these statements are correct, why does maintenance of the free passage of fishes along rivers receive so little attention during the planning stages of many large dam projects? One reason may be that scientists are not good at applying what is already known about the consequences of human impacts. Is this because we need more data to underpin conservation strategies? Calls for time to do additional research such as those made in the previous section will come at the cost of further biodiversity loss.

We should apply what we already know (e.g. fragmentation of river dolphin populations increases their probability of extinction), and place less emphasis on what we do not know (e.g. how long will the fragmented populations persist before extinction? Does the RCC apply in Asia?). We must try to get the science right but must do so within the constraint of limited time and money for research versus rapid habitat degradation. The factors that limit research in tropical countries include a shortage of funds, which has consequences for laboratory facilities, equipment, computers, library resources, and so on. Access to the international scientific literature is problematic over much of Asia. Where English is not a native language, there are constraints on reading, reporting, and writing. In addition, there appears to be a failure of tropical ecologists to recognize the importance of disseminating their findings by publishing them in refereed, internationally reputable journals (90). This failure is often one of implementation, also. The integral value of tropical ecology is not always recognized in its own right, or as a contribution to global knowledge, and this makes it difficult for inexperienced researchers writing in a second language to get their work accepted for publication. Manuscripts dealing with tropical rivers seem generally to be perceived as interesting regional studies first and contributions to limnology second; studies of north-temperate waters are viewed in the opposite way (90). One solution to this "temperate hegemony" is for tropical scientists to recognize the significance of their work, relate it to a wider international perspective, and draw attention to it in their own publications. This is easier for me to write than for others to achieve; journal editors and reviewers have a role to play too.

All research efforts are in vain if the resulting knowledge is not translated into social and/or political action. This may be the greatest challenge facing conservation ecologists in Asia. Even if we provide policy-makers with the best ecological information, we cannot assume they will act upon it. For instance, we know that river fish migrate. Most dams in Asia lack fish passes, which can mitigate the impact of dams, or include inappropriate fish ladders, or do not allow for downstream post-breeding migrations of adults (except through the dam turbines). Conventional fish ladders (designed for salmonids) seem likely to have limited success because most Asian fishes do not jump, but fish passes have been effective on some impounded rivers in southern China (Jiangsu Province; 59). Other conservation strategies for Asian river fishes could include application of controlled flooding and drainage permitting adults to access spawning sites and allowing the passage of water containing spawn and fry in a 'fish friendly' manner (80). Water allocation strategies that will maintain aquatic communities in river reaches downstream of dams (81) have yet to receive attention from water resource managers or funding agencies.

Non-scientific concerns are much greater obstacles to conservation in Asia than a paucity of data or uncertainty over the functioning of ecological systems. Institutional commitment to conservation over the long term is needed. Unfortunately, there is little pressure from the public to protect aquatic biodiversity, and scant evidence that biodiversity conservation or environmental protection are priorities of legislators. Non-government organizations promulgating conservation projects typically work within the political constraints set by governments; where these constraints include corruption and weak or inconsistent enforcement of legislation, conservation goals are compromised or unattainable. Aside from apathy and ignorance about aquatic biodiversity conservation, the major obstacle to conservation of remaining wildlands and rivers Asia is growing human populations. Tropical Asia is overpopulated, and many people are poor, landless, and crowded in burgeoning cities. All hope to improve their lives. The result will be per capita increases in resource use that will be accompanied by greater water consumption and further pollution, flow regulation, and habitat degradation. At the beginning of the third millennium, the prognosis for Asian rivers is grim.

Visit the Annual Reviews home page at www.AnnualReviews.org

LITERATURE CITED

1. Allan JD. 1995. *Stream Ecology.* London: Chapman & Hall. 388 pp.
2. Arunachalam M, Nair KCM, Vijverberg J, Kortmulder K, Suriyanaraynan H. 1991. Substrate selection and seasonal variation in densities of invertebrates in stream pools of a tropical river. *Hydrobiologia* 213:141–48
3. Deleted in proof
4. Attwood SW. 1995. A demographic analysis of *y-Neotricula aperta* (Gastropoda: Pomatiopsidae) populations in Thailand and southern Laos, in relation to the transmission of schistosomiasis. *J. Mollusc Stud.* 61:29–42
5. Beadle LC. 1981. *The Inland Waters of Tropical Africa: An Introduction to Tropical Limnology.* London: Longman. 475 pp.
6. Belsare DK. 1994. Inventory and status of vanishing wetland wildlife of Southeast Asia and an operational management plan for their conservation. In *Global Wetlands Old World and New*, ed. WJ Mitsch, pp. 841–56. Amsterdam: Elsevier
7. Benke AC, Jacobi DI. 1986. Growth rates of mayflies in a subtropical river and their implications for secondary production. *J. N. Am. BenthologicalSoc.* 5:107–14
8. Benzie JAH. 1984. The colonization mechanisms of stream benthos in a tropical river (Menik Ganga: Sri Lanka). *Hydrobiologia* 111:171–79
9. Bernacsek GM. 1997. *Large Dam Fisheries of the Lower Mekong Countries: Review and Assessment (MKG/R 97023, Vol. 1).* Bangkok: Mekong River Comm. 118 pp.
10. Bishop JE. 1973. Observations on the vertical distribution of the benthos in a Malaysian stream. *Freshw. Biol.* 3:147–56
11. Bishop JE. 1973. *Limnology of a Small*

Malayan River Sungai Gombak. The Hague: Dr W. Junk. 485 pp.
12. Biswas AK, El-Habr HN. 1993. Environment and water resources management: the need for a holistic approach. *Water Resour. Dev.* 9:117–25
13. Braatz S, Davis G, Shen S, Rees C. 1992. Conserving biological diversity: a strategy for protected areas in the Asia-Pacific Region. *World Bank Tech. Pap.* 193:1–66
14. Bunn SE, Davies PM, Kellaway DM. 1997. Contributions of sugar cane and invasive pasture grass to the aquatic food web of a tropical lowland stream. *Mar. Freshw. Resourc.* 48:173–79
15. Bunn SE, Davies PM, Mosisch TD. 1999. Ecosystem measures of river health and their response to riparian and catchment degradation. *Freshw. Biol.* 41:333–45
16. Caldecott J. 1996. *Designing Conservation Projects*. Cambridge, UK: Cambridge Univ. Press. 312 pp.
17. Christensen MS. 1993. The artisanal fishery of the Mahakam River floodplain in East Kalimantan, Indonesia. III. Actual and estimated yields, and their relationship to water levels and management options. *J. Appl. Ichthyol.* 9:202–9
18. Collins DE. 1999. Turtles in peril: the China crisis. *Vivarium* 10(4):6–9
19. Costa HH. 1974. Limnology and fishery biology of the streams at Horton Plains, Sri Lanka (*Ceylon*). *Bull. Fish. Res. Stn, Sri Lanka (Ceylon)* 25:15–26
20. Costa HH, Fernando ECM. 1967. The food and feeding relationships of the common meso- and macrofauna in the Maha Oya, a small mountainous stream at Peradeniya, Ceylon. *Ceylon J. Sci.* (*Biol. Sci.*) 7:74–90
21. Dudgeon D. 1987. Niche specificities of four fish species (Homalopteridae, Cobitidae, Gobiidae) from a Hong Kong forest stream. *Arch. Hydrobiol.* 108:349–64
22. Dudgeon D. 1988. The influence of riparian vegetation on macroinvertebrate community structure in four Hong Kong streams. *J. Zool. London* 216:609–27
23. Dudgeon D. 1989. The influence of riparian vegetation on the functional organization of four Hong Kong stream communities. *Hydrobiologia* 179:183–94
24. Dudgeon D. 1992. *Patterns and Processes in Stream Ecology: A Synoptic Review of Hong Kong Running Waters*. Stuttgart: Schweiz. Verlagsbuchhand. 147 pp.
25. Dudgeon D. 1992. Endangered ecosystems: a review of the conservation status of tropical Asian rivers. *Hydrobiologia* 248:167–91
26. Dudgeon D. 1993. The effects of spate-induced disturbance, predation and environmental complexity on macroinvertebrates in a tropical stream. *Freshw. Biol.* 30:189–97
27. Dudgeon D. 1994. The influence of riparian vegetation on macroinvertebrate community structure and functional organization in six New Guinea streams. *Hydrobiologia* 294:65–85
28. Dudgeon D. 1995. The ecology of rivers and streams in tropical Asia. In *Ecosystems of the World 22: River and Stream Ecosystems*, ed. CE Cushing, KW Cummins, GE Minshall, pp. 615–57. Amsterdam: Elsevier
29. Dudgeon D. 1995. River regulation in southern China: ecological implications, conservation and environmental management. *Reg. Riv.: Res. Managage.* 11:35–54
30. Dudgeon D. 1996. The influence of refugia on predation impacts in a Hong Kong stream. *Arch. Hydrobiol.* 138:145–59
31. Dudgeon D. 1996. Life histories, secondary production and microdistribution of heptageniid mayflies (Ephemeroptera) in a tropical forest stream. *J. Zool. London* 240:341–61
32. Dudgeon D. 1997. Life histories, secondary production and microdistribution of hydropsychid caddisflies (Trichoptera) in a tropical forest stream. *J. Zool. London* 243:191–210
33. Dudgeon D. 1999. *Tropical Asian Streams: Zoobenthos, Ecology and Conservation.*

Hong Kong: Hong Kong Univ. Press. 830 pp.

34. Dudgeon D. 1999. The population dynamics of three species of Calamoceratidae (Trichoptera) in a tropical forest stream. In *Proc. Int. Symp. Trichoptera, 9th*, ed. H Malicky, P Chantaramongkol, pp. 83–91. Chiang Mai, Thailand: Fac. Sci., Univ. Chiang Mai
35. Dudgeon D. 1999. Patterns of variation in secondary production in a tropical stream. *Arch. Hydrobiol.* 144:271–81
36. Dudgeon D. 2000. Going with the flow: large-scale hydrological alterations and the fate of riverine biodiversity in tropical Asia. *BioScience.* In press
37. Dudgeon D. 2000. Riverine wetlands and biodiversity conservation in tropical Asia. In *Biodiversity in Wetlands: Assessment, Function and Conservation*, ed. B Gopal, WJ Junk, JA Davis. pp. 1–26. The Hague: Backhuys
38. Dudgeon D, Bretschko G. 1995. Allochthonous inputs and land-water interactions in seasonal streams: tropical Asia and temperate Europe. In *Tropical Limnology, Past and Present*, ed. F Schiemer, pp. 161–79. The Hague: SPB Academic
39. Dudgeon D, Chan IKK. 1993. An experimental study of the influence of periphytic algae on invertebrate abundance in a Hong Kong stream. *Freshw. Biol.* 27:53–63
40. Deleted in proof
41. Dudgeon D, Choowaew S, Ho SC. 2000. River conservation in Southeast Asia. In *Global Perspectives on River Conservation: Science, Policy and Practice*, ed. PJ Boon, GE Petts, BR Davies. pp. 279–308. Chichester, UK: Wiley-Intersci.

41a. Dudgeon D, Wu KKY. 1999. Leaf litter in a tropical stream: food or substrate for macroinvertebrates? *Arch. Hydrobiol.* 146:65–82

42. Economic and Social Commission for Asia and the Pacific. 1992. *Towards Environmentally Sound and Sustainable Development of Water Resources in Asia and the Pacific* (*Water Res. Ser. No. 71*). Bangkok: Econ. Soc. Commiss. Asia Pac. 214 pp.
43. Foose TJ, van Strien N. 1997. *Asian Rhinos—Status Survey and Conservation Action Plan.* Cambridge, UK: World Conserv. Monit. Cent. 112 pp.

43a. Fu CB, Kim JW, Zhao ZC. 1998. Preliminary assessment of impacts on global change in Asia. In *Asian Change in the Context of Global Climate Change*, ed. Galloway JN, Melillo JM, pp. 308–41. Cambridge, UK:Cambridge Univ. Press

44. Giest V. 1998. *Deer of the World: Their Evolution, Behaviour and Ecology*. Mechanicsburg, PA: Stackpole. 421 pp.
45. Hora SL. 1923. Observations of the fauna of certain torrential streams in the Khasi Hills. *Rec. Indian Mus.* 25:579–600
46. Hora SL. 1930. Ecology, bionomics and evolution of the torrential fauna, with special reference to the organs of attachment. *Philos. Trans. R. Soc. London Ser. B* 218:171–282
47. Hora SL. 1936. Nature of substratum as an important factor in the ecology of torrential fauna. *Proc. Nat. Inst. Sci. India* 2:45–47
48. Hora SL. 1952. Major problems of the fisheries of India with suggestions for their solution. *J. Asiatic Soc.* (*Sci.*) 18:83–101
49. Humphrey CL, Faith DP, Dostine PL. 1995. Baseline requirements for assessment of mining impact using biological monitoring. *Aust. J. Ecol.* 20:150–66
50. Hynes HBN. 1970. *The Ecology of Running Waters*. Liverpool: Liverpool Univ. Press. 555 pp.
51. Irons JG, Oswood MW, Stout RJ, Pringle CM. 1994. Latitudinal patterns in leaf litter breakdown: is temperature really important? *Freshw. Biol.* 32:401–11
52. Deleted in proof
53. Jackson JK, Sweeney BW. 1995. Egg and larval development times of 35 species of tropical stream insects from Costa Rica.

J. N. Am. Benthological Soc. 14:115–30
54. Johnson DS. 1967. Distributional patterns of Malayan freshwater fish. *Ecology* 48:722–30
55. Khoo KH, Leong TS, Soon FL, Tan SP, Wong SY. 1987. Riverine fisheries in Malaysia. *Arch. Hydrobiol. Beih., Ergeb. Limnol.* 28:261–68
56. Kottelat M. 1989. Zoogeography of the fishes from Indochinese inland waters with an annotated checklist. *Bull. Zool. Mus, Univ. Amst.* 12:1–56
57. Lewis WM. 1987. Tropical limnology. *Annu. Rev. Ecol. Syst.* 18:159–84
58. Li S. 1981. *Studies on Zoogeographical Divisions for Freshwater Fishes of China*. Beijing: Sci. Press. 292 pp. (In Chinese)
59. Liao GZ, Lu KX, Xiao XZ. 1989. Fisheries resources of the Pearl River and their exploitation. *Can. Spec. Publ. Fish. Aquat. Sci.* 106:561–68
60. Lowe-McConnell RH. 1987. *Ecological Studies of Tropical Fish Communities*. Cambridge, UK: Cambridge Univ. Press. 382 pp.
61. Malicky H. 1989. Köcherfliegen (Trichoptera) von Sumatra und Nias: Die Gattungen *Chimarra* (Philopotamidae) und *Marilia* (Odontoceridae), mit Nachtrögen zu *Rhyacophila* (Rhyacophilidae). *Mitt. Schweiz. Entomol. Ges.* 62:131–43
62. Marchant R, Yule CM. 1996. A method for estimating larval life spans of aseasonal aquatic insects from streams on Bougainville Island, Papua New Guinea. *Freshw. Biol.* 35:101–7
63. McCully P. 1996. *Silenced Rivers: The Ecology and Politics of Large Dams*. London: Zed Books. 350 pp.
64. Meisner JD, Shuter BJ. 1992. Assessing potential effects of global climate change on tropical freshwater fishes. *GeoJournal* 28:21–27
65. Mosisch TD, Bunn SE. 1997. Temporal patterns of rainforest stream epilithic algae in relation to flow-related disturbance. *Aquat. Bot.* 58:181–93
66. Natarajan AV. 1989. Environmental impact of Ganga Basin development on genepool and fisheries of the Ganga River system. *Can. Spec. Publ. Fish. Aquat. Sci.* 106:545–60
67. Ng PKL. 1988. *The Freshwater Crabs of Peninsular Malaysia and Singapore*. Singapore: Dep. Zool., Natl. Univ. Singap. 156 pp.
68. Ng PKL, Tay JB, Lim KKP. 1994. Diversity and conservation of blackwater fishes in Peninsular Malaysia, particularly in the north Selangor peat swamp forest. *Hydrobiologia* 285:203–18
69. Nolen JA, Pearson RG. 1993. Factors affecting litter processing by *Anisocentropus kirramus* (Trichoptera: Calamoceratidae) from an Australian tropical rainforest stream. *Freshw. Biol.* 29:469–79
70. Ormerod SJ, Rundle SD, Wilkinson SM, Daly GP, Dale KM, Juttner I. 1994. Altitudinal trends in the diatoms, bryophytes, macroinvertebrates and fish of a Nepalese river system. *Freshw. Biol.* 32:309–22
71. Payne AI. 1986. *The Ecology of Tropical Lakes and Rivers*. Chichester, UK: Wiley-Intersci. 301 pp.
72. Pethiyagoda R. 1994. Threats to the indigenous freshwater fishes of Sri Lanka and remarks on their conservation. *Hydrobiologia* 285:189–201
73. Rainboth WJ. 1996. *Fishes of the Cambodian Mekong*. Rome: Food Agric. Organ. U. N. 265 pp.
74. Roberts TR. 1993. Artisanal fisheries and fish ecology below the great waterfalls in the Mekong River in southern Laos. *Nat. Hist. Bull. Siam Soc.* 41:31–62
75. Roberts TR. 1995. Mekong mainstream hydropower dams: run-of-the-river or ruin-of-the-river? *Nat. Hist. Bull. Siam Soc.* 43:9–19
76. Roberts TR, Baird IG. 1995. Traditional fisheries and fish ecology on the Mekong River at Khone Waterfalls in southern Laos. *Nat. Hist. Bull. Siam Soc.* 43:219–62

77. Deleted in proof
78. Schmid F. 1984. Essai d'évaluation de la faune mondiale des Trichoptéres. In *Proc. Int. Symp. Trichoptera, 4th*, ed. JC Morse, pp. 337. Clemson University, South Carolina, USA. The Hague: Dr W Junk (Abstr.)
79. Stout RJ. 1989. Effects of condensed tannins on leaf processing in mid-latitude and tropical streams: a theoretical approach. *Can. J. Fish. Aquat. Sci.* 46:1097–106
80. Sultana P, Thompson PM. 1997. Effects of flood control and drainage of fisheries in Bangladesh and the design of mitigating measures. *Reg. Riv.: Res. Manage.* 13:43–55
81. Sutton RJ, Miller WJ, Patti SJ. 1997. Application of the Instream Flow Incremental Methodology to a tropical river in Puerto Rico. *Rivers* 6:1–9
82. Talling JF, Lemoalle J. 1998. *Ecological Dynamics of Tropical Inland Waters*. Cambridge, UK: Cambridge Univ. Press. 451 pp.
83. Thirakhupt K, Van Dijk PP. 1994. Species diversity and conservation of turtles in western Thailand. *Nat. Hist. Bull. Siam Soc.* 42:207–59
84. Underwood AJ. 1994. On beyond BACI: sampling designs that might reliably detect environmental disturbances. *Ecol. Appl.* 4:3–15
84a. United Nations Food and Agricultural Organization. 1999. *Review of the State of World Fishery Resources: Inland Fisheries FAO Fish. Circ. No. 942*. Food and Agriculture Organization of the United Nations, Rome. 53 pp.
85. Vannote RL, Minshall GW, Cummins KW, Sedell JR, Cushing CE. 980. The River Continuum Concept. *Can J. Fish. Aquat. Sci.* 37:130–37
86. Wang JT, Liu MC, Fang LS. 1995. The reproductive biology of an endemic cyprinid, *Zacco pachycephalus*, in Taiwan. *Environ. Biol. Fish.* 43:135–43
87. Deleted in proof
88. Wei Q, Ke F, Zhang J, Zhaung P, Luo J, et al. 1997. Biology, fisheries, and conservation of sturgeons and paddlefish in China. *Environ. Biol. Fish.* 48:241–55
88a. Welcomme RL. 1979. *Fisheries Ecology of Floodplain Rivers*. London: Longman. 317 pp.
89. Wikramanayake ED, Moyle PB. 1989. Ecological structure of tropical fish assemblages in wet-zone streams of Sri Lanka. *J. Zool. London* 218:503–26
90. Williams WD. 1994. Constraints to the conservation and management of tropical inland waters. *Mitt. Int. Ver. Limnol.* 24:357–63
90a. World Conservation Monitoring Centre. 1996. *1996 IUCN Red List of Threatened Animals*. Cambridge, UK: World Conserv. Monit. Cent. 228 pp.
91. Yule CM. 1996. Trophic relationships and food webs of the benthic invertebrate fauna of two aseasonal tropical steams on Bougainville Island, Papua New Guinea. *J. Trop. Ecol.* 12:517–34
92. Yule CM, Pearson RG. 1996. Aseasonality of benthic invertebrates in a tropical stream on Bougainville Island, Papua New Guinea. *Arch. Hydrobiol.* 137:95–117
93. Zakaria-Ismail M. 1994. Zoogeography and biodiversity of the freshwater fishes of Southeast Asia. *Hydrobiologia* 285: 41–8

Annu. Rev. Ecol. Syst. 2000. 31:265–91

HARVESTER ANTS (*POGONOMYRMEX* SPP.): Their Community and Ecosystem Influences

James A. MacMahon,[1] John F. Mull,[2] and Thomas O. Crist[3]

[1]*Department of Biology and Ecology Center, Utah State University, Logan, Utah 84322; e-mail: jam@cc.usu.edu*

[2]*Department of Zoology, Weber State University, Ogden, Utah 84408; e-mail: jmull@weber.edu*

[3]*Department of Zoology, Miami University, Oxford, OH 45056; e-mail: cristto@muohio.edu*

Key Words granivores, community structure, ecosystem functioning, deserts, seeds

■ **Abstract** We summarize the influences of harvester ants of the genus *Pogonomyrmex* on communities and ecosystems. Because of nest densities, the longevity of nests, and the amount of seed harvested and soil handled, harvester ants have significant direct and indirect effects on community structure and ecosystem functioning. Harvester ants change plant species composition and diversity near their nests. These changes result from differential seed predation by the ants, their actions as seed dispersers and competitors with other granivores, and the favorable soil conditions they create through their digging. Their nest building creates islands of increased nutrient density. In some areas, the effects of their activities may be so pervasive that plant community structure is strongly influenced. Ant removal studies, which would reveal their total impact, have generally not been done. Granivore removals have been conducted in North America where ants are of lesser importance than small mammals, in contrast to other areas (except Israel) where ants are dominant granivores. We review the influence of harvester ants on their competitors, predators, and nest associates, and catalog the factors that influence their foraging patterns and consequently their local distribution.

> The harvesting habit in ants since it was first scientifically confirmed by Moggridge has excited an exceptional degree of interest and surprise. But in truth, when one considers all the conditions, the wonder is that it is not more widely distributed.
>
> Henry Christopher McCook (138, p. 116)

INTRODUCTION

Significance of Ants

Ants (Formicidae) have numerous effects in a variety of ecosystems worldwide (62). They form relationships that range from parasitism to mutualism with other ants and other organisms. They have strong symbiotic relationships with plants as seed dispersers, pollinators, removers of leaves, and guardians (19). Some ants are specialized predators that feed on organisms as varied as small arthropods and crustaceans and small vertebrates. Ants can grow fungi, change the physical and chemical properties of soil, and, sometimes, dominate entire landscapes with their workings (95). One activity—seed predation—has been well studied. Ants that collect and store seeds for later consumption are referred to as harvester ants, a name that encompasses >150 species in ≥ 18 genera and 3 subfamilies. Harvester ants are especially common in areas of semiarid to arid vegetation, although some occur in habitats as moist as the lowland rain forests of New Guinea and Southeast Asia (95).

Harvester ants can be important for a number of reasons. Because of their seed-eating habits, they may be pests in areas that produce grain for human consumption (127). In some places, nearly 20% of the land surface is covered by the mound areas that these ants have cleared of vegetation (218). This site preemption has economic consequences for rangelands and some crops (e.g. corn and alfalfa) (192). Nest-related activities have both direct (e.g. plant clipping) and indirect (e.g. soil modifications) effects on plant survival, growth, and reproduction. Even the biomass of ants can be significant. In one case, the biomass of harvester ants was estimated to equal or exceed that of rodents (200). An extreme example of their importance is the damage they cause to airport runways (222) and highway edges (44), where nest construction causes cavitation. Harvester ants occur in most places in the world except those of extreme cold, so their influence is widespread.

Pogonomyrmex spp.

Half of the harvester ant species in the world belong to *Pogonomyrmex*, a genus of 60 species found in North, Central, and South America (33, 191, 192). *Pogonomyrmex* spp., while favoring arid areas like other harvesting ants, also occur in forested environments. One species, *P. longibarbis*, occurs in the South American Andes up to elevations approaching 14,250 m (114) and another species, *P. mayri*, is found in dry tropical forests (113).

The potential impact of *Pogonomyrmex* spp. in natural systems is apparent from the building activities of a few species that clear a roughly circular area (disk) around their nesting mound. Both the mound and disk are kept free of vegetation by workers. In extreme cases, this disk may be ≤ 5.5 m in diameter, and the nesting mound may be ≤ 1.1 m high (184). Maximum reported density of nests ranges from 20 to 150 colonies per hectare (ha) (13, 42, 53, 87, 209). Estimated colony longevity ranges from 15 to 50 years (31, 80, 105, 164).

We limit our discussion to *Pogonomyrmex* spp., even though they co-occur with harvester ants in other genera— *Pheidole*, *Aphaenogaster*, *Messor*, and *Solenopsis* —in North America, because more data are available for *Pogonomyrmex* spp. and, except for the Old World genus *Messor*, they are the only harvester ants to build large nests, so their influence extends beyond that associated with seed gathering.

DIRECT INFLUENCES OF HARVESTER ANTS ON COMMUNITY STRUCTURE

Effects on Plants

Removal and Consumption of Seeds and Other Materials Harvester ants are single-load, central-place foragers (18). Most *Pogonomyrmex* spp. rely heavily upon seeds (95, 122, 192, 203), but some species rely on dead arthropods (113). Ants collect a variety of other food and nonfood materials. Among the nonseed items returned to the nest are leaves, twigs, pollen, flowers, vertebrate feces, and parts of various arthropods (90, 104, 112, 122, 129, 192, 194, 218). Foragers of some species are opportunistic predators of termites and other arthropods, but in most species such prey do not account for a significant fraction of total colony food intake (27, 48, 93, 95, 192, 203).

Pogonomyrmex spp. collect seeds from a wide range of plants (40, 46, 79, 143, 167, 192, 203, 212), but show obvious preferences for some. Seed attributes that affect seed selection under both natural and experimental conditions include relative abundance, nutritional quality (e.g. caloric content and percent soluble carbohydrate), morphology, viability, and size (40, 47, 85, 89, 106, 146, 158, 167, 204). Seeds infected with endophytic and saprophytic fungi are differentially rejected by workers (36, 111). Harvester ants specialize in abundant, small-seeded species (50). Large colony sizes and high colony densities enable ants to remove enormous numbers of seeds from a site (e.g. see 195). Still, most estimates of the fraction of total annual seed production removed by harvester ants have been 10% or less (128, 167, 203, 204; but see 128). Ants may have a much greater impact on the seeds of preferred species (89, 168, 204, 218), from which foragers may remove up to 100% of available seeds (40). Selective seed predation can cause qualitative changes in the community structure of plants by altering the relative abundance of plant species (18, 22, 99, 100, 183, 204). Based on long-term experimental studies conducted in the Sonoran and Chihuahuan Deserts (51, 52, 183), Davidson (50) concluded that removal of harvester ants (in several genera) has not dramatically altered the structure of plant communities. Comparable studies are lacking from the Great Basin, shrub-steppe, and short-grass steppe regions of North America, where *Pogonomyrmex occidentalis* and *Pogonomyrmex salinus* are abundant in many vegetation types.

The spatial effects of seed predation depend on several factors: the colony foraging strategy (individual or group), the relationship between colony foraging effort and distance from the nest, the dispersion pattern of seed resources, and seed density. Several common species of ants are trunk-trail foragers (38, 45, 76, 92, 95, 175). Consequently, rates of seed discovery and removal may be highest near trails (149, 150). As central-place foragers, harvester ants should deplete seed resources near their nests differentially (22, 40, 212). Harvester ants preferentially collect seeds from high-density patches, thereby reducing the spatial heterogeneity of seeds (168). In contrast to rodents, which collect seeds from various depths, ants collect seeds only at the surface (51, 168), a behavior that should increase the vertical patchiness of seeds. Differences between the structures of the seed rain and seed bank, including larger seeds and more temporal and less spatial variance in seed number in rains vs banks, may be caused by granivores, especially *Pogonomyrmex* spp., which remove large numbers of seeds from the surface (166). Harmon & Stamp (86) showed that ants reduce seed densities of *Erodium cicutarium*, a desert annual, which affects plant competition and, ultimately, reproductive output.

Storage and Rejection of Seeds Harvester ant colonies disperse some fraction of all seeds collected and returned to the nest by foragers (95). "Accidental" seed dispersal (see 19), or dyszoochory (220), contrasts with the directed dispersal of species known as myrmecochores (10). The seeds of these plants bear an appendage, the elaiosome, that attracts ants and encourages dispersal. Harvester ant-dispersed myrmecochores are known from arid regions of North America (20, 157; cf 72) but they represent a tiny fraction of the flora. *Pogonomyrmex* foragers may exert strong selective pressures on seed morphology that reduce levels of seed predation (66, 167) rather than facilitating dispersal.

Pogonomyrmex spp. disperse seeds largely by accident. Harvester ants occasionally remove seeds from the parent plant (128), but most dispersal occurs once seeds have fallen to the ground. This phase-II dispersal (i.e. movement of a seed once it has reached the ground) (24) involves three possible fates. First, a seed may be discarded along the path to the nest (83, 106). Second, it may be returned to the nest and subsequently discarded into a midden by another worker (111, 155, 170, 218). Finally, a seed returned to the nest may germinate and grow after nest abandonment, especially if it was stored in superficial granaries (32).

Seeds may be discarded from nest stores when foragers mistakenly collect seeds that do not correspond to current colony needs and preferences or when seeds are infected with fungal endophytes (111, 173). Possibly, seeds are discarded as part of a colony's bet-hedging strategy. Seeds of a nonpreferred species may be collected early in the season when few other seeds are available and be discarded from colony stores when seeds of preferred species become available. Colony foraging distances may extend for several meters (see below) and represent significant dispersal distances for seeds. The long-tailed curves generated by ant dispersal can be critical in reducing parent-offspring competition among plants (4) and the risk of predation by other granivores (157). Discarded seeds may have increased fitness when placed in microsites favorable to germination and growth (173).

Construction and Maintenance of the Nest Colonies allocate exterior workers to other tasks besides foraging. These include patrolling, midden work, and nest maintenance (73, 74, 163). Nest maintenance workers help create and maintain the disk surrounding the nest by removing debris and clipping vegetation. Herbaceous species, grasses, and shrub seedlings that establish themselves on the disk are removed by workers (30, 33, 123, 164, 176). *Pogonomyrmex rugosus* (174) and *Pogonomyrmex barbatus* (221) will also defoliate mature shrubs.

The disk is clearly an important extension of the nest structure, given the colony's investment in developing and maintaining it. Speculations on the adaptive value of the disk include reduced transit time for foragers, decreased risk of exposure to fire or predation, and increased nest exposure to solar radiation (reviewed in 123, 221). Cole (34) provided support for the value of increased solar radiation in *P. occidentalis* colonies. He showed that the shape and orientation of the nest cones, which act as solar collectors, increase the time available for foraging and brood development.

Effects on Other Taxa

Pogonomyrmex spp. may compete with granivorous ants in at least three genera—*Aphaenogaster* (*Novomessor*), *Messor* (*Veromessor*), and *Pheidole*—as well as with granivorous birds and mammals. The abundance of ants makes them an important food source for a variety of predators. Moreover, many species of soil organisms and various arthropods are either obligately associated with harvester ant nests or greatly facilitated by them.

Other Granivores Competitive interactions are important in structuring ant communities (95). In warm deserts, *Pogonomyrmex* spp. compete with each other and with other granivorous ants (18). Local species diversity, as well as ant biomass and density, is strongly correlated with mean annual precipitation and thus plant productivity, suggesting that ants are food-limited. Further evidence of food limitation comes from observations that fluctuations in the size and activity level of established colonies are related to seed availability (18, 45). Most of the evidence for interspecific competition among harvesters comes from studies documenting the overdispersion of biological traits (see 95).

In particular, competition has been invoked to account for the distribution of certain behavioral and morphological traits in sympatric species of granivorous ants. *Pogonomyrmex* spp. can be broadly categorized as group foragers that specialize in high-density seed resources or as individual foragers that specialize in low-density seed resources (18, 46, 92, 95, 193). Density specialization allows species of similar body size to coexist by utilizing distinct portions of available seed resources. Behavioral differences in activity times indicate temporal partitioning of foraging. These may be diurnal or seasonal differences (26, 85, 122, 205, 211, 212). Because desert seed resources are renewed once or twice per year, seasonal differences in activity patterns can promote coexistence (18); diurnal differences, on the other hand, serve only to reduce aggressive encounters between species. Body size and the size of seeds selected are positively correlated; thus, interspecific divergence in body size promotes coexistence. Body size differences that reduce overlap in the use of seed resources are common in sympatric, desert harvesters (18, 25–27, 46, 85, 95, 128, 143). Where three or more competing species are present, diffuse competition may be important. *P. rugosus* indirectly facilitates *Aphaenogaster cockerelli* by its interference with and local exclusion of *Pogonomyrmex desertorum*, a species that competes exploitatively with *A. cockerelli* (49).

Other well-documented cases of interspecific interference (26, 92, 94, 112) indicate that *Pogonomyrmex* spp. do not fare well in interactions with aggressive species in other genera. Hölldobler (93) and Gordon (75) provide two striking examples. In the first, workers of at least five *Pogonomyrmex* spp. are regularly robbed of insect prey by *Myrmecocystus mimicus*; in the second, *A. cockerelli* plugs the nest entrances of *P. barbatus*, an interference strategy that reduces the effectiveness of its competitor. This competitive disadvantage is clearly evident in interactions between harvester ants and exotic species. Like much of the North American ant fauna (8), *Pogonomyrmex* spp. are vulnerable to the invasions of both the red imported fire ant *Solenopsis invicta* (55, 97, 165) and the Argentine ant *Linepithema humile* (56, 190). Invasive species displace the native ant fauna through exploitative and interference competition (96, 98). Given the influence of harvester ants on community and ecosystem processes, this has important consequences for affected communities.

Seed-eating rodents and ants show significant overlap in seed use and are thus potential competitors (16). Long-term experiments in the Sonoran and Chihuahuan Deserts provide insight into the interactions among granivorous ants and rodents (15, 18). Following the removal of granivorous rodents, populations of Sonoran Desert harvester ants initially increased but then declined owing to vegetation changes in the absence of rodent granivory. Rodent populations showed a small, positive response to ant removal (15, 51; but see 14, 67). Because *Pheidole* spp. accounted for most of the granivorous colonies in these experiments (51), this study provides limited insight into *Pogonomyrmex*-rodent interactions. Comparable and longer-term experiments conducted in the Chihuahuan Desert detected no response by rodents to ant removal (17). Valone et al (196) argued that slight increases in rodent numbers on plots where ants have been removed may be biologically relevant but are too small to detect with the limited replication that was done. Ants (including *P. desertorum* and *Pheidole* spp.) showed a numerically stronger, positive response to rodent removal (196). The Chihuahuan Desert research shows that interactions between these taxa are indirect and mediated through a complex set of responses that involves the plants whose seeds are consumed by these granivores (52, 183, 196).

Predators Because they are long-lived, are sessile, contain large numbers of workers, and occur at locally high densities, harvester ant colonies represent a concentrated and highly apparent food source for predators (151, 163, 185). Vertebrate predators include blind snakes (*Leptotyphlops*), >25 species of birds, and lizards, especially horned lizards (*Phrynosoma* spp). Invertebrate predators include other ants, solpugids, beetles, asilids, reduviids, wasps and spiders, with the last group of invertebrates most frequently observed to eat ants (18, 21, 29, 68, 91, 123, 162, 163, 185, 219). With the exception of sphecid wasps in two genera (*Clypeadon* and *Listropygia*) that specialize in preying on *Pogonomyrmex* spp. workers (3, 57), the degree of prey specificity on *Pogonomyrmex* spp. by most invertebrate predators is unknown.

Horned lizards are specialized and widely studied vertebrate predators of harvester ants; several apparently coevolved traits occur in predator and prey. These include the potent venom and autotomous sting of many *Pogonomyrmex* spp. and the specialized digestive tracts and venom resistance of *Phrynosoma* spp. (88, 185). Each group has important ecological effects on the other. Some species of horned lizards rely heavily on *Pogonomyrmex* spp. ants (151, 152, 171, 185, 214), consuming up to 72% of the standing crop of harvester ants (209). Such high levels of predation and the colony-level changes in foraging behavior that occur when workers are lost to predation (discussed below) may reduce reproductive output (153). Predation by *Phrynosoma* spp. can be species-specific and may mediate interactions between species and thus structure harvester ant communities (171).

Predation influences harvester ant-plant interactions by reducing foraging activity. Despite expectations that predation is less of a foraging constraint in social insects than in other animals (18), significant changes in the behavior of *Pogonomyrmex* colonies occur in response to predation by horned lizards and spiders. These range from changes in the location of nest entrances, increased intercolony aggression, and predator mobbing (68, 91, 171) to the reduction or cessation of the foraging activity of individuals and colonies (41, 129, 152, 171, 209). As few as 15 spiders (*Latrodectus hesperus*) can halt all colony foraging behavior in *P. rugosus* for several days (123). MacKay & MacKay (129) suggest that the hallmark harvester ant activity of seed storage evolved as a protection against predation. The temporal progression of worker duties from interior to exterior tasks has been similarly interpreted. Predation risk is lowest in young, energy-rich workers that remain in the nest, and it is highest in the older, exterior workers. With fewer energy reserves, the exterior ants, particularly foragers, are characterized as a disposable caste; their susceptibility to predation represents a tolerable risk to the colony (124, 126, 163, 185).

Myrmecophiles The nests and associated soils of harvester ants influence the distribution and abundance of a variety of arthropods that live on or in them. Beetles from at least eight families, orthopterans, termites, homopterans, collembollans, thysanurans, diplurans, millipedes, spiders, and mites, are associated with the nests of *Pogonomyrmex* spp. (23, 28, 37, 122, 125, 154, 161, 198, 201). Such arthropods are known as myrmecophiles (95). This term encompasses a broad spectrum of symbiotic interactions between ants and other invertebrates that range from facultative to obligate and from species-general to species-specific (110). Myrmecophilous arthropod species associated with harvester ants span this range of symbioses. Little is known about the biology or specificity of *Pogonomyrmex* myrmecophiles. In contrast, two species of *Pogonomyrmex*, *P. colei* and *P. anergismus*, are obligate social parasites of either *P. rugosus* or *P. barbatus* (33, 103, 126, 172).

DIRECT INFLUENCES OF HARVESTER ANTS ON ECOSYSTEM FUNCTIONING

Harvester ants influence ecosystem processes through their effects on food-web structure, energy flows, nutrient transport, and soil modification. Some species also have important roles as disturbance agents. On a fine scale, the flows of material and energy by ants create resource patches that are distinct from surrounding areas. On a broader scale, the population densities of ant colonies determine the amount of energy flow, degree of soil modification, and frequency of disturbance within ecosystems.

Energy Flows and Food Webs

The amount of energy flowing through *Pogonomyrmex* spp. varies considerably among species and ecosystems. The estimated total energy flow (production plus respiration) through *P. badius* was 59–200 kJ m^{-2} y^{-1} in an old field, exceeding values estimated for populations of old-field mice (*Peromyscus polionotus*) and savannah sparrows (*Passerculus sandwichensis*) (71). These values were comparable to those estimated for total seed production. Thus, *P. badius* not only has an important role as a seed consumer but also relies on insects as an energy source (see also 58) because most of the energy assimilated by ants is used in respiration (71). In a short-grass steppe, the amount of energy flowing through *P. occidentalis* was estimated at only 1–6 kJ m^{-2} y^{-1} (176), considerably less than energy use by *P. badius* even though colony densities were similar. Parmenter et al (158) compared energy use by three *Pogonomyrmex* spp. (*P. occidentalis*, *P. rugosus*, and *P. maricopa*) with that of rodents and birds in four desert ecosystems. Sonoran and Mojave Desert rodents had far greater energy demands (75–112 kJ m^{-2} y^{-1}) than harvester ants or birds (1–9 kJ m^{-2} y^{-1}). The three taxa had similar energy demands at Chihuahuan Desert and shrub-steppe sites (1–30 kJ m^{-2} y^{-1}) (158). MacKay (126) provided detailed energy budgets for *P. rugosus*, *P. subnitidus*, and *P. montanus* that occupied different habitats along an elevational gradient from coastal scrub to montane pine forest. Energy flow ranged from <1 kJ m^{-2} y^{-1} to 32 kJ m^{-2} y^{-1} for *P. montanus* and *P. rugosus*, respectively. About 65% of the energy returned to nests by *P. rugosus* was not assimilated, but was discarded as seed hulls or uneaten seeds; *P. subnitidus* and *P. montanus* assimilated most of their food intake because of a lower proportion of seeds in their diets (126).

Thus, *Pogonomyrmex* spp. that are seed specialists may have a greater role in the redistribution of energy than omnivorous species such as *P. montanus*.

Harvester ants redistribute energy and materials between food webs both above and below-ground (see 160). Lauenroth & Milchunas (115) described above-ground energetics as primarily driven by biophagic feeding (consumption of live biomass) and below-ground energetics as mostly saprophagic (consumption of dead biomass). Harvester ants transfer live biomass from the soil surface (seeds, insects, and plant material) to below-ground nests where it is consumed or incorporated into the detrital food web. This pathway differs from that of other important consumers in deserts and grasslands, in which above-ground biophagic feeders, such as grasshoppers, transfer uneaten plant clippings and feces to the soil surface, and root feeders, such as larval June beetles (*Phyllophaga* spp.) accelerate detritus formation below ground (35, 145).

Soil Modification and Nutrient Transport

Ants redistribute soil and organic matter in several ways: excavation by nest workers may bring subsoil to the surface; coarse particles are retrieved by workers near nests; and foragers retrieve seeds, insects, and litter from foraging areas. Nest excavation by *P. occidentalis* results in soil movement rates of 80–280 kg ha^{-1} y^{-1} (22, 159). Soils in ant nests may have a coarser texture and a lower bulk density than those in surrounding areas (22, 133, 178).

Higher amounts of soil water occur in nests of some *Pogonomyrmex* spp. than in surrounding soils. Rogers & Lavigne (178) found no differences in soil water between *P. occidentalis* nests and surrounding areas in the upper 10–20 cm of soil, but nests and cleared disks had a greater water content at greater depths. Laundré (116) recorded seasonal patterns of soil moisture below *P. salinus* nest mounds and in off-mound areas in a sagebrush steppe. Soil water in the upper 20 cm was lower in the mounds than in surrounding areas during the spring recharge, but higher at greater depths within nests throughout the year. A greater soil pore volume (and lower field capacity) in the nest mound probably increases water infiltration to greater depths (116). Variation in sampling depth might partially explain decreases in soil water in nests near the surface (*P. occidentalis*, 22) or similar amounts compared to adjacent areas (*P. rugosus*, 210; *P. barbatus*, 198). Water infiltration and retention probably vary among *Pogonomyrmex* spp. because nests differ in the size and vertical distribution of galleries (69, 117, 122, 212).

Nest enrichment of organic matter and nutrients is widespread among harvester ants. Higher levels of organic matter were recorded in nests of *P. occidentalis* in piñon-juniper and sagebrush-steppe ecosystems (22, 64, 65). At the latter site, conspicuous mats of organic matter occurred at the base of nest mounds. These were densely packed roots that proliferated from nearby sagebrush (*Artemisia tridentata*) and were clipped by nest workers as they grew into the mound (64). Thus, both lateral expansion of plant roots and material transport by workers contribute to increased levels of organic matter in ant nests. Enriched levels of N, P, or K were recorded in nests of *P. badius* (69), *P. barbatus* (198), *P. rugosus* (210), and *P. occidentalis* (22, 65, 133). Forms of N that are available to plants—NO_3^- and NH_4^+—are 5- to 10-fold higher in nests than in adjacent soils (22, 65, 178, 198, 210). Potential sources of available N include metabolic wastes from ants and N mineralization from decomposition of organic matter within mounds. The latter may be especially important because functional groups of microbes such as decomposers of cellulose, chitin, and protein are higher in nests than in surrounding soils (65).

Spatial variation in soil nutrient enrichment among ant nests in different soils or vegetation is likely to be important but poorly studied. Soil in nests of *P. rugosus* is more nutrient-enriched at mid-slope than at base-slope locations along a Chihuahuan Desert bajada (210). Differences in the degree of enrichment were found in nests of *P. occidentalis* located in piñon-juniper and ponderosa pine vegetation (22). Finally, colony longevity and nest-site duration are important predictors of the degree of soil modification among ant species (119, 207).

Ants as Disturbance Agents

P. occidentalis and *P. salinus* are disturbance agents on rangelands because they clear large areas of vegetation around nests (109, 177, 188, 217, 218). The total land surface area cleared by ants is typically 1%–3% for *P. occidentalis* (189); values of 8% were reported on favorable soils in a saltbush (*Atriplex nuttallii*) community in Wyoming (109). *P. salinus* frequently clears 3%–6%, and values of 18% were estimated in sagebrush steppe of Oregon (189, 218). Apart from the area affected by ants, little is known about disturbance frequency and turnover rates. *P. occidentalis* disturbs an average of 2.5 m^{-2} ha^{-1} y^{-1} in short-grass steppe (31). Using spatial simulations of ant colony dynamics, Crist & Wiens (42) estimated that turnover rates at the same short-grass steppe site could reach 6–8 m^{-2} ha^{-1} y^{-1} on coarse-textured upland soils where colony densities are 30–40 ha^{-1}.

Functional Roles of Pogonomyrmex

Pogonomyrmex spp. have functional roles in granivory, food-web dynamics, and soil modification. In some arid environments, numerous species of seed-harvesting ants (8, 46, 148, 204) share functional roles as granivores with rodents and birds. Considerable resources are partitioned among granivore taxa, even among *Pogonomyrmex* spp. (see above), so the degree of functional redundancy among diverse granivore taxa is unclear. In many ecosystems, one *Pogonomyrmex* spp. (e.g. *P. salinus*, *P. badius*, *P. occidentalis*) accounts for virtually all of the seed predation by ants (41). The role of *Pogonomyrmex* spp. in food webs is important because of the dependence of specialist predators such as *Phrynosoma* lizards on them. Both ants and termites have important roles in soil modification within arid and semiarid ecosystems. *Pogonomyrmex* spp. have few functional equivalents for their roles in granivory, food-web structure, and soil modification.

INDIRECT EFFECTS

Nest construction and maintenance by some harvester ants create biotic and abiotic conditions that indirectly affect associated species. Vegetation clipping increases ground surface temperature (211) and percolation of water into the denuded zone (206, 211). An increase in nutrient enrichment and lowered bulk density may also influence both plants and animals to various degrees.

The removal of adjacent plants changes the competitive balance in plant communities. Nest disturbances may form a mosaic of patch types in the landscape as colonies die. These patches enhance community diversity and alter patterns of vegetation recovery (22). Species that invade abandoned nest sites are often different from those that invade disturbances created by other agents. In a Mojave Desert site, the richness, density, and frequency of plant species changed with distance from mounds of *P. rugosus*. Of 17 plant species, 4 were less dense on disks than in the surrounding vegetation, and 13 species were absent from disks (118). More annuals occurred on *P. occidentalis* mounds than on the disks, and a greater number of annual seeds were stored in the mounds than occurred on the disks in a semiarid grassland (32). *Bouteloua gracilis*, a dominant plant in short-grass communities, was absent from abandoned nest sites, and its seeds were absent from mounds.

The increased concentration of nutrients and lack of competition at the edges of disks cause woody plants and annuals to grow larger than in surrounding areas beyond the influence of ant nests (206). Similarly, grasses (e.g. *Aristida*) are more luxuriant on the edges of disks of *P. badius* nests than they are in the adjacent vegetation in South Carolina (71). Roots of plants surrounding the disk invade the soil beneath the disk in impressively dense mats that have 2,000–5,000 times the mycorrhizal fungal spore density found in the surrounding vegetation (64). However, mycorrhizae may provide no advantage to plants growing in the nutrient-rich soils of harvester ant nests (140).

The increased nutrient concentration and possibly greater availability of water can increase the reproductive productivity of individual plants. Mounds of *P. rugosus* and *Messor* spp. in Arizona had a 15.6-fold and 6.5-fold increase in the number of fruits or seeds produced per plant of *Schismus arabicus* and *Plantago insularis*, respectively (173). Nowak et al (156) attempted to pinpoint the reasons for the differences in plants on ant mounds. The bunchgrass *Oryzopsis hymenoides* is often the only plant on the cleared disk and mounds of *P. salinus*. The proportional number of seeds of *O. hymenoides* in the seed bank of mounds, cleared disks, and the native plant community did not differ, but size and density of these plants were greater near the mounds than away from them. Nowak et al concluded that competition among plants on ant mounds was mediated by the differential removal of plants by the ants.

The effects of ant activities on rangelands have been debated. *Pogonomyrmex* spp. denuded an estimated 36,450 ha of rangeland in Wyoming (187). In the same area, increased productivity of *Atriplex nuttallii* at the edges of disks more than offset the loss of productivity on the cleared disks (217). Additional studies suggested there was no need for widespread control programs (188).

Although seldom studied, ant mounds influence the community of soil animals. *P. barbatus* mounds and disks had a 30-fold higher density of microarthropods and five-fold higher density of protozoans than surrounding soils (198). Amoebae and ciliates occurred in higher numbers, and flagellates occurred in lower numbers than in the control areas. Densities of bacteria and fungi were similar on mound and control soils, even though NO_3^-, NH_4^+, P, and K were found in higher concentrations on the mounds. Crist & Friese (37) frequently found termites in nests of *P*. occidentalis when ants were less active.

Compared with the surrounding landscape, nest mounds and disks (and adjacent vegetation) of *P. occidentalis* may be used disproportionately for oviposition and thermoregulation by tenebrionid beetles (141), for mating displays by sage grouse (*Centrocercus urophasianus*) (70), and for foraging by some shrub-steppe passerines (215).

Harvester ant disturbances and the adjacent zone of increased plant growth create heterogeneity in vegetation cover and nutrient distribution. This can influence landscapes by increasing overall productivity, enhancing diversity (2), and affecting recovery time after disturbance (31, 42).

FACTORS AFFECTING *POGONOMYRMEX* FORAGING AND DISTRIBUTION

Considerable literature deals with *Pogonomyrmex* spp. foraging behavior (reviewed in 193) as it relates to granivore-seed interactions (see above) or predictions of foraging theory (47, 59, 61, 90, 146, 169, 199). Here we consider the factors affecting spatial and temporal patterns of foraging activity, the interactions among neighboring colonies, and the broad-scale distributions of ant colonies across landscapes and regions.

Daily and Seasonal Foraging Activity

Seasonal variation in foraging activity of *Pogonomyrmex* spp. is coupled with surface temperatures, seed availability, or the distribution of rainfall (12, 38, 126, 167, 186, 204, 211, 218). Warm-desert *Pogonomyrmex* spp. (*P. rugosus, P. californicus, P. desertorum, P. barbatus*, and *P. maricopa*) forage between March and November, with one or two peaks in foraging activity that differ among species (11, 126, 204, 211). Ant activity is closely coupled to seasonal patterns of rainfall (205), which are distinctly unimodal or bimodal in the warm deserts of North America (131). *Pogonomyrmex* spp. that are characteristic of higher elevations or latitudes (*P. subnitidus*, *P. montanus*, *P. occidentalis*, and *P. salinus*) forage from April to October, with an activity peak in summer foraging (38, 126, 218). Seasonal timing of foraging activity was correlated with seed availability in different environments (38, 167, 204, 211).

Daily variation in foraging activity is influenced by soil surface temperature and potential water loss. *Pogonomyrmex* spp. forage when surface temperatures range from 5° to 55°C (39, 43, 71, 130, 211), which correspond to laboratory determinations of critical thermal minima and maxima (211). Most foraging occurs at surface temperatures of 20° to 50°C, and peak activity occurs from 30° to 45°C in seven species found in a wide range of environments (38, 43, 71, 130, 211). Most species are diurnal, so daily foraging activity on hot days occurs mostly in the morning and afternoon, and on cooler days there is one midday peak in activity (130, 212). Foraging patterns in *P. mayri*, a tropical scavenger, are similarly determined by surface temperatures (113). Some species, such as *P. rugosus*, show significant periods of nocturnal activity during midsummer (205, 211). Potential water loss is also important: foraging activity increases with lower vapor pressure deficits or experimental watering (211).

Biotic factors influence foraging activity. Foraging decreases when rates of forage return are experimentally decreased (76) or when colonies become satiated (211). Predation has a strong effect on levels of foraging activity (see above). Foraging also depends on other worker behaviors not directly related to foraging, for example, daily changes in nest maintenance and patrolling (73, 74).

Hypotheses about the ecological significance of daily and seasonal variation in foraging include interspecific partitioning of foraging among seed-harvesting ants (see above), seasonal changes in seed availability (39, 40, 46, 167, 204, 205, 211), and matching of the abiotic tolerance responses of foragers to the seasonal patterns of seed availability (11, 211). These hypotheses are not mutually exclusive. In warm deserts, seed production and availability are linked to rainfall events, which also produce favorable abiotic conditions for ant foraging (204, 205, 211).

Foraging is largely unaffected by above-ground disturbances such as fire. In sandhill oak-pine habitat, *P. badius* colonies had similar pre- and post-fire levels of foraging activity, but increased the size of foraging areas following burns (139). In desert grassland, fire had no effect on activity patterns, foraging rates, or total food intake; foraging distances and the number of insect parts returned to nests increased slightly (223). Foraging is more likely influenced by soil disturbances, such as high surface erosion or livestock trampling, that result in increased nest maintenance costs (9, 42, 43, 74, 78).

Spatial Patterns of Foraging

Foraging decreases exponentially with distance from ant nests (38, 53, 202). Ants venture 15–25 m from nests (39, 170), but most foraging occurs within 10 m of nests: mean foraging distances were 1.5–3.0 m for *P. californicus* in the Mojave Desert (53), 2.6–5.1 m for *P. occidentalis* in short-grass steppe (37), 9.7 m for *P. occidentalis* in sagebrush steppe (37), and 8.0 m for *P. salinus* in a sagebrush-greasewood community (104). In *P. occidentalis*, foragers move farther in areas with lower vegetation cover (37, 59).

Foraging may be concentrated near trunk trails in *P. rugosus, P. barbatus, P. occidentalis*, and *P. salinus* (38, 61, 76, 87, 92, 104). Location and relative activity of trails vary in response to seed availability (38, 61, 76, 79), vegetation structure (38, 60, 92), and encounter rates with neighboring colonies (77, 80). Patterns of trunk-trail branching and seed capture by the ant *Messor barbarus* are analogous to the modular growth of stoloniferous plants (6, 120). Similarly, *P. barbatus* colonies expand foraging trails to fill available space and retract or redirect trails when there are frequent encounters with neighboring colonies (77, 80). Thus, foraging areas interdigitate among neighboring colonies (80), a pattern also observed in *P. rugosus* (202) and *P. salinus* (104).

Colony Spacing and Mortality

Colony location in *Pogonomyrmex* spp. is strongly influenced by the position of intraspecific or interspecific nests. Intraspecific colony spacing was regular in *P. badius* (87) and in three *P. occidentalis* populations located in different environments (22, 42, 216). These studies were conducted where other specialized seed-harvesting ants were absent. In contrast, both *P. rugosus* and *P. barbatus* had random dispersion patterns in the Chihuahuan Desert, where intra- and interspecific encounters were infrequent (212). In the Mojave Desert, *P. californicus* colonies were uniformly spaced and intraspecific aggression was common among colonies; interspecific encounters with *Messor pergandei* were rare (53). At another Mojave Desert site, intraspecific spacing of nests in *P. californicus* and *M. pergandei* was regular; interspecific spacing of nests was clumped (179, 180). Intraspecific competition is stronger than interspecific competition in territory use by foragers and in the establishment of founding queens (181, 182). Colonies can become regularly spaced through frequent nest relocation in *P. badius* (87) and *P. californicus* (53). Intraspecific competition does not always produce regular spacing: colonies of *P. barbatus* are clumped but encounters between ants of neighboring colonies partition foraging ranges into irregularly shaped areas that reduce overlap (80).

Colony establishment and mortality are strongly influenced by spacing patterns and colony age. Founding queens have high mortality during colony establishment, with <1% surviving the first year (87, 182, 216). Colony establishment in *P. occidentalis* and *P. badius* is highest near existing colonies that are small or <2 years old (82, 216); young colonies are less likely than older ones to overlap in foraging areas (81). The number of newly established colonies is also greater near mating leks than away from them (216) and varies considerably from year to year (81, 105). Mortality of 1-year-old colonies is higher than that of older colonies (81, 105, 216). Beyond 2 years of age, colonies have relatively low mortality (81, 216) and are likely to persist for several years during which mortality increases gradually (82). Thus, *Pogonomyrmex* colonies are similar to long-lived vertebrates or woody plants that have high juvenile mortality, high adult survivorship, and increased mortality with senescence (type III or IV survivorship; see also 8, 82, 101).

Broad-Scale Patterns of Ant Density and Distribution

Neighborhood interactions affect establishment, survivorship, and colony spacing within habitats where colony densities are high. On broader scales, variability in soils, vegetation, or land-use practices affects the density and distribution of *Pogonomyrmex* spp. In short-grass steppe, the spatial distribution of *P. occidentalis* is strongly influenced by topography: colony densities are $>20\ ha^{-1}$ on upland plains, $10–20\ ha^{-1}$ on slopes, and $0–10\ ha^{-1}$ in lowland swales (42). South-facing slopes have greater colony densities than north-facing slopes. High colony densities on uplands occur in deep (>1.5 m), well-drained, sandy loams; unfavorable sites include clay loams and shallow or poorly drained soils. Crist & Wiens (42) found that spacing of 1450 ant colonies is regular at the scale of the nearest neighbor (16 m), random between the first and second nearest neighbors (16 –23 m), and clumped at the scale of third nearest neighbor and beyond (>28 m). Thus, colonies are regularly spaced at fine scales, and variation in topography and soils produces patchy distributions of *P. occidentalis* colonies at broader scales. These scale-dependent patterns of density disappear in areas heavily grazed by cattle (42).

Johnson (102) measured soil characteristics along topographic gradients where *P. rugosus* and *M. pergandei* occurred alone or had overlapping distributions. *P. rugosus* occurred on upper bajadas in coarse-textured soils, and *M. pergandei* occurred largely in fine-textured soils on lower bajadas. Their distribution patterns were reversed in ridge-swale topography. In both types of topographic gradients, *P. rugosus* was associated with soils that had higher clay content and water retention; the two species co-occurred in areas with intermediate soil characteristics where they overlapped in tolerances to soil conditions (102). Colony densities of *P. salinus* varied within and among seven plant communities (13): highest densities ($\leq 164\ ha^{-1}$) were in deep sandy soils, whereas few colonies occurred in shallow or poorly drained soils. Vegetation differed in response to soils, and variation in colony density within sites was related to plant community types or past vegetation disturbance (13). In North Dakota, regional range expansion of *P. occidentalis* was facilitated by coarse-textured soils in roadside ditches. Movement along roadside corridors was greater on south-facing than on north-facing slopes, suggesting that both soil characteristics and topographic exposure are important to distribution of *P. occidentalis* across landscapes and regions (54).

COMPARISONS WITH OTHER HARVESTER ANTS

Birds, mammals, and harvester ants are the most important seed predators in arid areas worldwide. In North America, the majority of seeds taken from experimental feeding stations are removed by heteromyids, followed by ants and then birds (18, 135, 158). Birds may have their greatest influence on large-seeded winter annuals (84). In semiarid grasslands, the relative importance of ants may increase (167). Rodents remove more seeds than ants do in Israel (1), but small mammals are not the dominant seed-removers in other arid systems. In the Karoo of South Africa, ants take more seeds than do small mammals, presumably because there are no specialist granivore mammals (107). In the Monte Desert of Argentina, seeds do not occur at the high densities characteristic of North America (121, 135).

Nonetheless, ants take a higher proportion of seeds than do rodents or birds, although fewer of the ant species are specialist granivores; birds remove more seeds than do ants during some seasons (121, 137), a situation similar to that in the matorral of Chile (197). The Monte lacks granivorous mammals (134), in contrast with the Atacama, where granivorous rodents occur (144). However, Marone et al (136) recently suggested that, in the Monte, rodents may be more important and *Pogonomyrmex* spp. more strictly granivorous than previously thought.

In many places, seed harvesting by ants predominates in arid lands but is uncommon in other habitats. Australia differs in that there are few specialized seed-harvesting ants, but seeds constitute a portion of the diet of many species in virtually every community type except rain forests (7). Australian harvester ants are increasingly granivorous in arid areas, where diets may consist of 90% seed. Ants, which remove more seeds than small mammals or birds in most Australian communities, may have significant impacts, at least on ephemeral species and on some species of economic significance (5).

Few data are available to indicate the dominant granivores or the level of granivory in the Sahara Desert or Eurasian deserts (108, 208).

Weights of seeds removed by the granivore guild suggest that intensity of predation ranks highest in North America, followed by Australia, followed in turn by South America (108, 147), although some South American sites may experience significant seed loss seasonally (197). This ranking probably is not the same as that of biomass, although North America, where seed predation is greatest, is dominated by the activities of small mammals. Folgarait et al (63) inferred that the equable environments of the Southern Hemisphere do not select for granivory (i.e. reliance on high-energy food such as seeds). There is little indication that these granivorous taxa have converged from continent to continent in abundance, foraging patterns, and body size (108, 142, 208). This suggests that historical influences are more important than convergent evolution in determining some community attributes.

SUMMARY AND FURTHER STUDIES

Many aspects of the biology of *Pogonomyrmex* spp. are unknown. Details of ant impact on community and ecosystem structure and functioning are scant and often anecdotal. For example, we know that harvester ants may consume large portions of the seeds available on the soil surface, but few experiments have demonstrated that this alters the composition of the plant community. Despite the large percentage of the annual seed rain that may be taken by *Pogonomyrmex* spp., the most important community and ecosystem influences of the ants may stem from their vegetation clearing and nest construction. Little is known about the extent to which harvester ants disperse seeds. Finally, much information indicates soil is enriched around nests and disks, but details concerning nutrient and water dynamics are lacking. Conducting radioisotopic experiments in the field might provide relevant mechanistic data.

Ant community structure, including that of harvesters, may indicate environmental change or degradation (5, 132). Given the wide distribution of ants, their ecological amplitude, and the ease with which they are collected, it is reasonable to assume that ants are good indicators of environmental quality, and there is some evidence that this has been a useful approach in Australia. In contrast, Whitford et al (213) found no differences in ant communities among a variety of disturbed rangeland sites in New Mexico and Arizona, except that *Pogonomyrmex* spp. were less abundant in rangelands with introduced grasses.

Comparisons among continents suggest that we understand few details of the community and ecosystem consequences of harvester ants, yet the literature suggests that they are important worldwide. If studies of ants are to provide general insights into community and ecosystem ecology or if ants are to be used as predictors of change in human-altered environments, it will be necessary to analyze a greater series of habitats to elucidate generalizations, if they exist. Despite our limited knowledge, it is clear that *Pogonomyrmex* spp. influence communities and ecosystems in North America but perhaps less so in South America. Other genera of harvesters, particularly *Messor*, may have equally important influences on community and ecosystem composition where they occur.

Pogonomyrmex ants alter community and ecosystem characteristics. Given the amount of literature, it is surprising there is still so little certainty about causal mechanisms, especially the relative importance of direct and indirect effects.

ACKNOWLEDGMENTS

Research support to T.O.C. was provided by funding from the USDA NRI Competitive Grants Program (No. 95031420). We thank Linda Finchum for her assistance and patience during the preparation of the manuscript. We also thank Annie Mull for help in preparing the manuscript.

LITERATURE CITED

1. Abramsky Z. 1983. Experiments on seed predation by rodents and ants in the Israeli desert. *Oecologia* 57:328–32
2. Aguiar MR, Sala EO. 1999. Patch structure, dynamics and implications for the functioning of arid ecosystems. *Trends Ecol. Evol.* 14:273–77
3. Alexander B. 1985. Predator-prey interactions between the digger wasp *Clypeadon laticinctus* and the harvester ant *Pogonomyrmex occidentalis*. *J. Nat. Hist.* 19:1139–54
4. Andersen AN. 1988. Dispersal distance as a benefit of myremcochory. *Oecologia* 75:507–11
5. Andersen AN. 1990. The use of ant communities to evaluate change in Australian terrestrial ecosystems: a review and a recipe. *Proc. Ecol. Soc. Aust.* 16:347–57
6. Andersen AN. 1991. Parallels between ants and plants: implications for community ecology. In *Ant-Plant Interactions*, ed. CR Huxley, DF Culver, pp. 539–53. Oxford, UK: Oxford Univ. Press

7. Andersen AN. 1991. Seed harvesting by ants in Australia. In *Ant-Plant Interactions*, ed. CR Huxley, DF Cutler, pp. 493–503. Oxford, UK: Oxford Univ. Press
8. Andersen AN. 1997. Functional groups and patterns of organization in North American ant communities: a comparison with Australia. *J. Biogeogr.* 24:433–60
9. Anderson CJ, Mull JF. 1992. Emigration of a colony of *Pogonomyrmex occidentalis* Cresson (Hymenoptera: Formicidae) in southwestern Wyoming. *J. Kans. Entomol. Soc.* 65:456–58
10. Beattie AJ. 1985. *The Evolutionary Ecology of Ant-Plant Mutualisms.* New York: Cambridge Univ. Press
11. Bernstein RA. 1974. Seasonal food abundance and foraging activity in some desert ants. *Am. Nat.* 108:490–98
12. Bernstein RA, Gobbel M. 1979. Partitioning of space in communities of ants. *J. Anim. Ecol.* 48:931–42
13. Blom PE, Clark WH, Johnson JB. 1991. Colony densities of the seed harvesting ant *Pogonomyrmex salinus* (Hymenoptera: Formicidae) in seven plant communities on the Idaho National Engineering Laboratory. *J. Ida. Acad. Sci.* 27:28–36
14. Brown JH, Davidson DW. 1986. Reply to Galindo. *Ecology* 67:1423–25
15. Brown JH, Davidson DW, Munger JC, Inouye RS. 1986. Experimental community ecology: the desert granivore system. In *Community Ecology*, ed. J Diamond, TJ Case, pp. 41–61. New York: Harper & Row
16. Brown JH, Grover JJ, Davidson DW, Lieberman GA. 1975. A preliminary study of seed predation in desert and montane habitats. *Ecology* 56:987–92
17. Brown JH, Munger JC. 1985. Experimental manipulation of a desert rodent community: food addition and species removal. *Ecology* 66:1545–63
18. Brown JH, Reichman OJ, Davidson DW. 1979. Granivory in desert ecosystems. *Annu. Rev. Ecol. Syst.* 10:201–27
19. Buckley RC. 1982. Ant-plant interactions: a world review. In *Ant-Plant Interactions in Australia*, ed. RC Buckley, pp. 111–41. The Hague: Dr W Junk
20. Bullock SH. 1974. Seed dispersal of *Dendromecon* by seed predator *Pogonomyrmex. Madroño* 22:378–79
21. Cade WH, Simpson PH, Breland OP. 1978. *Apiomerus spissipes* (Hemiptera: Reduviidae): a predator of harvester ants in Texas. *Southwest. Nat.* 3:195–97
22. Carlson SR, Whitford WG. 1991. Ant mound influence on vegetation and soils in a semiarid mountain ecosystem. *Am. Midl. Nat.* 126:125–39
23. Cazier MA Mortenson MA. 1965. Bionomical observations on myrmecophilous beetles of the genus *Cremastocheilus* (Coleoptera: Scarabaeidae). *J. Kans. Entomol. Soc.* 38:19–44
24. Chambers JC, MacMahon JA. 1994. A day in the life of a seed: movements and fates of seeds and their implication for natural and managed systems. *Annu. Rev. Ecol. Syst.* 25:263–92
25. Chew AE, Chew RM. 1980. Body size as a determinant of small-scale distributions of ants in evergreen woodland southeastern Arizona. *Ins. Soc.* 27:189–202
26. Chew RM. 1977. Some ecological characteristics of the ants of a desert-shrub community in southeastern Arizona. *Am. Midl. Nat.* 98:33–49
27. Chew RW, De Vita J. 1980. Foraging characteristics of a desert ant assemblage: functional morphology and species separation. *J. Arid Environ.* 3:75–83
28. Clark WH. 1983. Cicada nymphs, *Okanagana* sp. (Homoptera: Cicadidae), in nests of the Owyhee harvester ant, *Pogonomyrmex owyheei* (Hymenoptera: Formicidae) in Idaho with a note on mound rebuilding by *Pogonomyrmex. J. Ida. Acad. Sci.* 19:29–32
29. Clark WH, Blom PE. 1992. Notes on spider (Theridiidae, Salticidae) predation of the harvester ant, *Pogonomyrmex salinus* Olsen (Hymenoptera: Formicidae:

Myrmicinae), and a possible parasitoid fly (Chloropidae). *Gt. Basin Nat.* 55:385–86
30. Clark WH, Comanor PL. 1975. Removal of annual plants from the desert ecosystem by western harvester ants, *Pogonomyrmex occidentalis*. *Environ. Entomol.* 4:52–56
31. Coffin DP Lauenroth WK. 1989. Small scale disturbances and successional dynamics in a shortgrass plant community: interactions of disturbance characteristics. *Phytologia* 67:258–86
32. Coffin DP, Lauenroth WK. 1990. Vegetation associated with nest sites of western harvester ants (*Pogonomyrmex occidentalis* Cresson) in a semiarid grassland. *Am. Midl. Nat.* 123:226–35
33. Cole AC Jr. 1968. *Pogonomyrmex Harvester Ants: A Study of the Genus in North America.* Knoxville, TN: Univ. Tenn. Press
34. Cole BJ. 1994. Nest architecture in the western harvester ant, *Pogonomyrmex occidentalis* (Cresson). *Ins. Soc.* 41:401–10
35. Crist TO. 2000. Insect populations, community interactions, and ecosystem processes in shortgrass-steppe. In *Ecology of the Shortgrass-Steppe: Perspectives from Long-Term Research*, ed. IE Burke, WK Lauenroth. Oxford, UK: Oxford Univ. Press. In press
36. Crist TO, Friese CF. 1993. The impact of fungi on soil seeds: implication for plants and granivores in a semiarid shrub-steppe. *Ecology* 74:2231–39
37. Crist TO, Friese CF. 1994. The use of ant nests by subterranean termites in two semiarid ecosystems. *Am. Midl. Nat.* 131:370–73
38. Crist TO, MacMahon JA. 1991. Foraging patterns of *Pogonomyrmex occidentalis* (Hymenoptera: Formicidae) in a shrub-steppe ecosystem: the roles of temperature, trunk trails, and seed resources. *Environ. Entomol.* 20:265–75
39. Crist TO, MacMahon JA. 1991. Individual foraging components of harvester ants: movement patterns and seed patch fidelity. *Ins. Soc.* 38:379–96
40. Crist TO, MacMahon JA. 1992. Harvester ant foraging and shrub-steppe seeds: interactions of seed resources and seed use. *Ecology* 73:1768–79
41. Crist TO, Wiens JA. 1994. Scale effects of vegetation on forager movement and seed harvesting by ants. *Oikos* 69:37–46
42. Crist TO, Wiens JA. 1996. The distribution of ant colonies in a semiarid landscape: implications for community and ecosystem processes. *Oikos* 76:301–11
43. Crist TO, Williams JA. 1999. Simulation of topographic and daily variation in colony acitvity of *Pogonomyrmex occidentalis* (Hymenoptera: Formicidae) using a soil temperature model. *Environ. Entomol.* 28:659–68
44. Crowell HH. 1963. Control of the western harvester ant *Pogonomyrmex occidentalis*, with poisoned baits. *J. Econ. Entomol.* 56:295–98
45. Davidson DW. 1977. Foraging ecology and community organization in desert seed-eating ants. *Ecology* 58:725–37
46. Davidson DW. 1977. Species diversity and community organization in desert seed-eating ants. *Ecology* 58:711–24
47. Davidson DW. 1978. Experimental tests of the optimal diet in two social insects. *Behav. Ecol. Sociobiol.* 4:35–41
48. Davidson DW. 1980. Some consequences of diffuse competition in a desert ant community. *Am. Nat.* 116:92–105
49. Davidson DW. 1985. An experimental study of diffuse competition in harvester ants. *Am. Nat.* 125:500–6
50. Davidson DW. 1993. The effects of herbivory and granivory on terrestrial plant succession. *Oikos* 68:25–35
51. Davidson DW, Inouye RS, Brown JH. 1984. Granivory in a desert ecosystem: experimental evidence for indirect facilitation of ants by rodents. *Ecology* 65:1780–86
52. Davidson DW, Samson DA, Inouye RS. 1985. Granivory in the Chihuahuan Desert:

interactions within and between trophic levels. *Ecology* 66:486–502

53. De Vita J. 1979. Mechanisms of interference and foraging among colonies of the harvester ant *Pogonomyrmex californicus* in the Mojave Desert. *Ecology* 60:729–37
54. DeMers MN. 1993. Roadside ditches as corridors for range expansion of the western harvester ant (*Pogonomyrmex occidentalis* Cresson). *Landsc. Ecol.* 8:93–102
55. Donaldson W, Price AH, Morse J. 1994. The current status and future prospects of the Texas horned lizard (*Phrynosoma cornutum*) in Texas. *Tex. J. Sci.* 46:97–113
56. Erickson JM. 1971. The displacement of native ant species by the introduced Argentine ant, *Iridomyrmex humilis* Mayr. *Psyche* 78:257–66
57. Evans HE, O'Neill KM. 1988. *The Natural History and Behavior of North American Beewolves*. Ithaca, NY: Comstock
58. Ferster B, Traniello JFA. 1995. Polymorphism and foraging behavior *Pogonomyrmex badius* (Hymenoptera: Formicidae): worker size, foraging distance, and load in size associations. *Environ. Entomol.* 24:673–78
59. Fewell JH. 1988. Energetic and time costs of foraging in harvester ants, *Pogonomyrmex occidentalis*. *Behav. Ecol. Sociobiol.* 22:401–8
60. Fewell JH. 1988. Variation in foraging patterns of the western harvester ant, *Pogonomyrmex occidentalis*, in relation to habitat structure. In *Interindividual Behavioral Variability in Social Insects*, ed. RL Jeanne, pp. 257–82. Boulder, CO: Westview
61. Fewell JH. 1990. Directional fidelity as a foraging constraint in the western harvester ant, *Pogonomyrmex occidentalis*. *Oecologia* 82:45–51
62. Folgarait PJ. 1998. Ant biodiversity and its relationship to ecosystem functioning: a review. *Biodivers. Conserv.* 7:1221–44
63. Folgarait PJ, Monjeau JA, Kittlein M. 1998. Solving the enigma of granivory rates in Patagonia and throughout other deserts of the world: Is thermal range the explanation? *Ecol. Austral* 8:251–63
64. Friese CF, Allen MF. 1993. The interaction of harvester ants and vesicular-arbuscular mycorrhizal fungi in a patchy semi-arid environment: the effects of mound structure on fungal dispersion and establishment. *Funct. Ecol.* 7:13–20
65. Friese CF, Morris SJ, Allen MF. 1997. Disturbance in natural ecosystems: scaling from fungal diversity to ecosystem functioning. In *The Mycota IV. Environmental and Microbial Relationships*, ed. DT Wicklow, B Söderström, pp. 47–65. Berlin: Springer-Verlag
66. Fuller PJ, Hay ME. 1983. Is glue production by seeds of *Salvia columbariae* a deterrent to desert granivores? *Ecology* 64:960–63
67. Galindo C. 1986. Do desert rodent populations increase when ants are removed? *Ecology* 67:1422–23
68. Gentry JB. 1974. Response to predation by colonies of the Florida harvester ant, *Pogonomyrmex badius*. *Ecology* 55:1328–38
69. Gentry JB, Stiritz KL. 1972. The role of the Florida harvester ant, *Pogonomyrmex badius*, in old field mineral nutrient relationships. *Environ. Entomol.* 1:39–41
70. Giezentanner KI, Clark WH. 1974. The use of western harvester ant mounds as strutting locations by sage grouse. *Condor* 76:218–19
71. Golley FB, Gentry JB. 1964. Bioenergetics of the southern harvester ant, *Pogonomyrmex badius*. *Ecology* 45:217–25
72. Gonzalez-Espinosa M, Quintana-Ascencio PF. 1986. Seed predation and dispersal in a dominant desert plant: Opuntia, ants, birds, and mammals. In *Frugivores and Seed Dispersal*, ed. TH Fleming, pp. 273–84. Dordrecht, The Netherlands: Dr W Junk
73. Gordon DM. 1984. Species-specific patterns in the social activities of harvester

ant colonies (*Pogonomyrmex*). *Ins. Soc.* 31:74–86

74. Gordon DM. 1986. The dynamics of the daily round of the harvester ant colony (*Pogonomyrmex barbatus*). *Anim. Behav.* 34:1402–19
75. Gordon DM. 1988. Nest-plugging: interference competition in desert ants (*Novomessor cockerelli* and *Pogonomyrmex barbatus*). *Oecologia* 75:114–18
76. Gordon DM. 1991. Behavioral flexibility and the foraging ecology of seed-eating ants. *Am. Nat.* 138:379–411
77. Gordon DM. 1992. How colony growth affects forager intrusion in neighboring harvester ant colonies. *Behav. Ecol. Sociobiol.* 31:417–27
78. Gordon DM. 1992. Nest relocation in harvester ants. *Ann. Entomol. Soc. Am.* 85:44–47
79. Gordon DM. 1993. The spatial scale of seed collection by harvester ants. *Oecologia* 95:479–87
80. Gordon DM. 1995. The development of an ant colony's foraging range. *Anim. Behav.* 49:649–59
81. Gordon DM, Kulig AW. 1996. Founding, foraging and fighting: relationships between colony size and the spatial distribution of harvester ant nests. *Ecology* 77:2393–409
82. Gordon DM, Kulig AW. 1998. The effect of neighbours on the mortality of harvester ant colonies. *J. Anim. Ecol.* 67:141–48
83. Gordon SA. 1980. Analysis of twelve Sonoran Desert seed species preferred by the desert harvester ant. *Madroño* 27:68–78
84. Guo Q, Thompson DB, Valone TJ, Brown JH. 1995. The effects of vertebrate granivores and folivores on plant community structure in the Chihuahuan Desert. *Oikos* 73:251–59
85. Hansen SR. 1978. Resource utlization and coexistence of three species of *Pogonomyremx* ants in an upper Sonoran grassland community. *Oecologia* 35:109–17
86. Harmon GD, Stamp NE. 1992. Effects of postdispersal seed predation on spatial inequality and size variabililty in an annual plant, *Erodium cicutarium* (Geraniaceae). *Am. J. Bot.* 79:300–5
87. Harrison JS, Gentry JB. 1981. Foraging pattern, colony distribution, and foraging range of the Florida harvester ant, *Pogonomyrmex badius*. *Ecology* 62:1467–73
88. Hermann HH, Blum MS. 1981. Defensive mechanisms in the social hymenoptera. In *Social Insects*, Vol. II, ed. HR Hermann, pp. 77–197. New York: Academic
89. Hobbs RJ. 1985. Harvester ant foraging and plant species distribution in annual grassland. *Oecologia* 67:519–23
90. Holder Bailey K, Polis GA. 1987. Optimal and central-place foraging theory applied to a desert harvester ant, *Pogonomyrmex californicus*. *Oecologia* 72:440–48
91. Hölldobler B. 1970. *Steatoda fulva* (Theridiidae), a spider that feeds on harvester ants. *Psyche* 77:202–8
92. Hölldobler B. 1976. Recruitment behavior, home range orientation and territoriality in harvester ants, *Pogonomyrmex*. *Behav. Ecol. Sociobiol.* 1:3–44
93. Hölldobler B. 1986. Food robbing in ants, a form of interference competition. *Oecologia* 69:12–15
94. Hölldobler B Stanton RC, Markl H. 1978. Recruitment and food retrieving behavior in *Novomessor* (Hymenoptera: Formicidae). I. Chemical signals. *Behav. Ecol. Sociobiol.* 4:163–81
95. Hölldobler B, Wilson EO. 1990. *The Ants*. Cambridge, MA: Harvard Univ. Press
96. Holway DA. 1999. Competitive mechanisms underlying the displacement of native ants by the invasive Argentine ant. *Ecology* 80:238–51
97. Hook AW, Porter SD. 1990. Destruction of harvester ant colonies by invading fire ants in south-central Texas (Hymenoptera: Formicidae). *Southwest. Nat.* 35:477–78
98. Human KG, Gordon DM. 1996. Exploitation and interference competition between

the invasive Argentine ant, *Linepithema humile*, and native ant species. *Oecologia* 105:405–12

99. Inouye RS. 1991. Population biology of desert annual plants. In *The Ecology of Desert Communities*, ed. GA Polis, pp. 27–54. Tucson, AZ: Univ. Ariz. Press
100. Inouye RS, Byers GS, Brown JH. 1980. Effects of predation and competition on survivorship, fecundity, and community structure of desert annuals. *Ecology* 61:1344–51
101. Jeanne RL, Davidson DW. 1984. Population regulation in social insects. In *Ecological Entomology*, ed. CB Huffaker, RL Rabb, pp. 559–87. New York: Wiley & Sons
102. Johnson RA. 1992. Soil texture as an influence on the distribution of the desert seed-harvester ants *Pogonomyrmex rugosus* and *Messor pergandei*. *Oecologia* 89:118–24
103. Johnson RA. 1995. Distribution and natural history of the workerless inquiline ant *Pogonomyrmex anergismus* Cole (Hymenoptera: Formicidae). *Psyche* 101:257–62
104. Jorgensen CD, Porter SD. 1982. Foraging behavior of *Pogonomyrmex owyheei* in southeast Idaho. *Environ. Entomol.* 11:381–84
105. Keeler KH. 1993. Fifteen years of colony dynamics in *Pogonomyrmex occidentalis*, the western harvester ant, in western Nebraska. *Southwest. Nat.* 38:286–89
106. Kelrick MI, MacMahon JA, Parmenter RR, Sisson DV. 1986. Native seed preferences of shrub-steppe rodents, birds and ants: the relationships of seed attributes and seed use. *Oecologia* 68:327–37
107. Kerley GIH. 1992. Small mammal seed consumption in the Karoo, South Africa: further evidence for divergence in desert biotic processes. *Oecologia* 89:471–75
108. Kerley GIH, Whitford WG. 1994. Desert-dwelling small mammals as granivores: intercontinental variations. *Aust. J. Zool.* 42:543–55
109. Kirkham DR, Fisser HG. 1972. Rangeland relations and harvester ants in north central Wyoming. *J. Range Manage.* 25:55–60
110. Kistner DH. 1982. The social insects' bestiary. In *Social Insects*, Vol. III, ed. HR Hermann, pp. 1–244. New York: Academic
111. Knoch TR, Faeth SH, Arnott DL. 1993. Endophytic fungi alter foraging and dispersal by desert seed-harvesting ants. *Oecologia* 95:470–73
112. Kugler C. 1984. Ecology of the ant *Pogonomyrmex mayri*: foraging and competition. *Biotropica* 16:227–34
113. Kugler C, Hincapie MdC. 1983. Ecology of the ant *Pogonomyrmex mayri*: distribution, abundance, nest structure, and diet. *Biotropica* 15:190–98
114. Kusnezov N. 1951. El genero *Pogonomyrmex* Mayr (Hym. Formicidae). *Acta Zool. Lilloana* 11:227–333
115. Lauenroth WL, Milchunas DG. 1992. Short-grass steppe. In *Ecosystems of the World*, Vol. 8A, ed. RT Coupland, pp. 183–226. Amsterdam: Elsevier
116. Laundré JW. 1990. Soil moisture patterns below mounds of harvester ants. *J. Range Manage.* 43:10–12
117. Lavigne RJ. 1969. Bionomics and nest structure of *Pogonomyrmex occidentalis* (Hymenoptera: Formicidae). *Ann. Entomol. Soc. Am.* 62:1166–75
118. Lei SA. 1999. Ecological impacts of *Pogonomyrmex* on woody vegetation of a *Larrea-Ambrosia* shrubland. *Gt. Basin Nat.* 59:281–84
119. Lobry de Bruyn LA, Conacher AJ. 1990. The role of termites and ants in soil modification: a review. *Aust. J. Soil Res.* 28:55–93
120. Lopez F, Serrano JM, Acosta FJ. 1994. Parallels between the foraging strategies of ants and plants. *Trends Ecol. Evol.* 9:150–53

121. Lopez de Casenave J, Cueto VR, Marone L. 1998. Granivory in the Monte desert, Argentina: Is it less intense than in other arid zones of the world? *Glob. Ecol. Biogeogr. Lett.* 7:197–204
122. MacKay WP. 1981. A comparison of the nest phenologies of three species of Pogonomyrmex harvester ants (Hymenoptera: Formicidae). *Psyche* 88:25–74
123. MacKay WP. 1982. The effect of predation of western widow spiders (Araneae: Theridiidae) on harvester ants (Hymenoptera: Formicidae). *Oecologia* 53:406–11
124. MacKay WP. 1983. Stratification of workers in harvester ant nests (Hymenoptera: Formicidae). *J. Kans. Entomol. Soc.* 56:538–42
125. MacKay WP. 1983. Beetles associated with the harvester ants, *Pogonomyrmex montanus, P. subnitidus* and *P. rugosus* (Hymenoptera: Formicidae). *Coleopt. Bull.* 37:239–46
126. MacKay WP. 1985. A comparison of the energy budgets of three species of *Pogonomyrmex* harvester ants (Hymenoptera: Formicidae). *Oecologia* 66:484–94
127. MacKay WP. 1990. The biology and economic impact of *Pogonomyrmex* harvester ants. In *Applied Myrmecology: A World Perspective*, ed. RK Vander Meer, K Jaffe and A Cedeno, pp. 533–42. Boulder, CO: Westview
128. MacKay WP. 1991. The role of ants and termites in desert communities. In *The Ecology of Desert Communities*, ed. GA Polis, pp. 113–50. Tucson, AZ: Univ. Ariz. Press
129. MacKay WP, MacKay EE. 1984. Why do harvester ants store seeds in their nests? *Sociobiology* 9:31–47
130. MacKay WP, MacKay EE. 1989. Diurnal foraging patterns of *Pogonomyrmex* harvester ants (Hymenoptera: Formicidae). *Southwest. Nat.* 34:213–18
131. MacMahon JA, Wagner FH. 1985. The Mojave, Sonoran, and Chihuahuan Deserts of North America. In *Hot Deserts and Arid Shrublands*, Vol. 12A, ed. M Evenari, I Noy-Meir, pp. 105–202. Amsterdam: Elsevier
132. Majer JD. 1983. Ants: bio-indicators of mine site rehabilitation, land-use, and land conservation. *Environ. Manage.* 7:375–83
133. Mandel RD, Sorenson CJ. 1982. The role of the western harvester ant (*Pogonomyrmex occidentalis*) in soil formation. *Soil Sci. Soc. Am. J.* 46:785–88
134. Mares MA, Blair WF, Enders FA, Greegor D, Hulse AC, et al. 1977. The strategies and community patterns of desert animals. In *Convergent Evolution in Warm Deserts*, ed. GH Orians, OT Solbrig, pp. 107–63. Stroudsburg PA: Dowden, Hutchinson & Ross
135. Mares MA, Rosenzweig ML. 1978. Granivory in North and South American deserts: rodents, birds, and ants. *Ecology* 59:235–41
136. Marone L, Lopez de Casenave J, Cueto VR. 2000. Granivory in southern South American deserts: conceptual issues and current evidence. *BioScience* 50:123–32
137. Marone L, Rossi BE, Lopez de Casenave J. 1998. Granivore impact on soil-seed reserves in the central Monte desert, Argentina. *Funct. Ecol.* 12:640–45
138. McCook HC. 1909. *Ant Communities and How They Are Governed. A Study in Natural Civics*. New York: Harper & Brothers
139. McCoy ED, Kaiser BW. 1990. Changes in foraging activity of the southern harvester ant *Pogonomyrmex badius* (Latreille) in response to fire. *Am. Midl. Nat.* 123:112–23
140. McGinley MA, Dhillion SS, Neumann JC. 1994. Environmental heterogeneity and seedling establishment: ant-plant-microbe interactions. *Funct. Ecol.* 8:607–15
141. McIntyre NE. 1999. Use of *Pogonomyrmex occidentalis* (Hymenoptera:

Formicidae) nest-sites by tenebrionid beetles (Coleoptera: Tenebrionidae) for oviposition and thermoregulation in a temperate grassland. *Southwest. Nat.* 44:379–84

142. Medel RG. 1995. Convergence and historical effects in harvester ant assemblages of Australia, North America, and South America. *Biol. J. Linn. Soc.* 55:29–44
143. Mehlhop P, Scott NJ. 1983. Temporal patterns of seed use and availability in a guild of desert ants. *Ecol. Entomol.* 8:69–85
144. Meserve PL. 1981. Trophic relations among small mammals in a Chilean semiarid thorn scrub community. *J. Mammal.* 62:304–14
145. Moore JC, Walter DE, Hunt HW. 1988. Arthropod regulation of micro- and mesobiota in below-ground detrital food webs. *Annu. Rev. Entomol.* 33:419–39
146. Morehead SA, Feener Jr. DH. 1998. Foraging behavior and morphology: seed selection in the harvester ant genus, *Pogonomyrmex*. *Oecologia* 114:548–55
147. Morton SR. 1985. Granivory in arid regions: comparison of Australia with North and South America. *Ecology* 66:1859–66
148. Morton SR, Davidson DW. 1988. Comparative structure of harvester ant communities in arid Australia and North America. *Ecol. Monogr.* 58:19–38
149. Mull JF, MacMahon JA. 1996. Factors determining the spatial variability of seed densities in a shrub-steppe ecosystem: the role of harvester ants. *J. Arid. Environ.* 32:181–92
150. Mull JF, MacMahon JA. 1997. Spatial variation in rates of seed removal by harvester ants (*Pogonomyrmex occidentalis*) in a shrub-steppe ecosystem. *Am. Midl. Nat.* 138:1–13
151. Munger JC. 1984. Optimal foraging? Patch use by horned lizards (Iguanidae: *Phrynosoma*). *Am. Nat.* 123:654–80
152. Munger JC. 1984. Long-term yield from harvester ant colonies: implications for horned lizard foraging strategy. *Ecology* 65:1077–86
153. Munger JC. 1992. Reproductive potential of colonies of desert harvester ants (*Pogonomyrmex desertorum*): effects of predation and food. *Oecologia* 90:276–82
154. Neece KC, Bartell DP. 1982. A faunistic survey of the organisms associated with ants of western Texas. Grad. Stud. No. 25 MS thesis. Tex. Tech Univ., Lubbock
155. Nickle DA Neal TM. 1972. Observations on the foraging behavior of the southern harvester ant, *Pogonomyrmex badius*. *Fla. Entomol.* 55:65–66
156. Nowak RS, Nowak CL, DeRocher T, Cole N, Jones MA. 1990. Prevalence of *Oryzopsis hymenoides* near harvester ant mounds: indirect facilitation by ants. *Oikos* 58:190–98
157. O'Dowd DJ, Hay ME. 1980. Mutualism between harvester ants and a desert ephemeral: seed escape from rodents. *Ecology* 61:531–40
158. Parmenter RR, MacMahon JA, Vander Wall SB. 1984. The measurement of granivory by desert rodents, birds and ants: a comparison of an energetics approach and a seed-dish technique. *J. Arid. Environ.* 7:75–92
159. Petal J. 1978. The role of ants in ecosystems. In *Production Ecology of Ants and Termites*, ed. MV Brian, pp. 327–40. Cambridge, UK: Cambridge Univ. Press
160. Polis GA, Strong DR. 1996. Food web complexity and community dynamics. *Am. Nat.* 147:813–46
161. Porter SD. 1985. *Masoncus* spider: a miniature predator of Collembola in harvester ant colonies. *Psyche* 92:145–50
162. Porter SD, Eastmond DA. 1982. *Euryopis coki* (Theridiidae), a spider that preys on *Pogonomyrmex* ants. *J. Arachnol.* 10:275–77
163. Porter SD, Jorgensen CD. 1981. Foragers of the harvester ant, *Pogonomyrmex owyheei*: A disposable caste? *Behav. Ecol. Sociobiol.* 9:247–56

164. Porter SD, Jorgensen CD. 1988. Longevity of harvester ant colonies in southern Idaho. *J. Range Manage.* 41: 104–7
165. Porter SD, Savignano DA. 1990. Invasion of polygyne fire ants decimates native ants and disrupts arthropod community. *Ecology* 71:2095–106
166. Price MV, Joyner JW. 1997. What resources are available to desert granivores: Seed rain or soil seed bank? *Ecology* 78:764–73
167. Pulliam HR, Brand MR. 1975. The production and utilization of seeds in plains grassland of southeastern Arizona. *Ecology* 56:1158–67
168. Reichman OJ. 1979. Desert granivore foraging and its impact on seed densities and distributions. *Ecology* 60:1085–92
169. Reyes Lopez JL. 1987. Optimal foraging in seed harvester ants: computer-aided simulation. *Ecology* 68:1630–33
170. Rissing SW. 1981. Foraging specializations of individual seed harvester ants. *Behav. Ecol. Sociobiol.* 9:149–52
171. Rissing SW. 1981. Prey preferences in the desert horned lizard: influence of prey foraging method and aggressive behavior. *Ecology* 62:1031–40
172. Rissing SW. 1983. Natural history of the workerless inquiline ant *Pogonomyrmex colei* (Hymenoptera: Formicidae). *Psyche* 90:321–32
173. Rissing SW. 1986. Indirect effects of granivory by harvester ants: plant species composition and reproductive increase near ant nests. *Oecologia* 68:231–34
174. Rissing SW. 1988. Dietary similarity and foraging range of two seed-harvester ants during resource fluctuations. *Oecologia* 75:362–66
175. Rissing SW, Pollock GB. 1989. Behavioral ecology and community organization of desert seed-harvester ants. *J. Arid. Environ.* 17:167–73
176. Rogers L, Lavigne R, Miller JL. 1972. Bioenergetics of the western harvester ant in the shortgrass plains ecosystem. *Environ. Entomol.* 1:763–68
177. Rogers LE. 1987. Ecology and management of harvester ants in the shortgrass plains. In *Integrated Pest Management on Rangeland*, ed. JL Capinera, pp. 261–70. Boulder, CO: Westview
178. Rogers LE, Lavigne RJ. 1974. Environmental effects of western harvester ants on the shortgrass plains ecosystem. *Environ. Entomol.* 3:994–97
179. Ryti RT, Case TJ. 1984. Spatial arrangement and diet overlap between colonies of desert ants. *Oecologia* 62:401–4
180. Ryti RT, Case TJ. 1986. Overdispersion of ant colonies: a test of hypotheses. *Oecologia* 69:446–53
181. Ryti RT, Case TJ. 1988. The regeneration niche of desert ants: effects of established colonies. *Oecologia* 75:303–6
182. Ryti RT, Case TJ. 1988. Field experiments on desert ants: testing for competition between colonies. *Ecology* 69:1993–2003
183. Samson DA, Philippi TE, Davidson DW. 1992. Granivory and competition as determinants of annual plant diversity in the Chihuahuan desert. *Oikos* 65:61–80
184. Schmidt JO, Schmidt PJ. 1986. *Pogonomyrmex occidentalis*, an addition to the ant fauna of Mexico, with notes on other species of harvester ants from Mexico (Hymenoptera: Formicidae). *Southwest. Nat.* 31:395–420
185. Schmidt PJ, Schmidt JO. 1989. Harvester ants and horned lizards predator-prey interactions. In *Special Biotic Relationships in the Arid Southwest*, ed. JO Schmidt, pp. 25–51. Albuquerque, NM: Univ. N. M. Press
186. Schumacher A, Whitford WG. 1976. Spatial and temporal variation in Chihuahuan Desert ant faunas. *Southwest. Nat.* 21:1–8
187. Sharp LA, Barr WF. 1960. Preliminary investigations of harvester ants on southern Idaho rangelands. *J. Range Manage.* 13:131–34
188. Sneva FA. 1979. The western harvester

ants: their density and hill size in relation to herbaceous productivity and big sagebrush cover. *J. Range Manage.* 32:46–47

189. Soulé PT, Knapp PA. 1996. *Pogonomyrmex owyheei* nest site density and size on a minimally impacted site in central Oregon. *Gt. Basin Nat.* 56:152–66
190. Suarez AV, Bolger DT, Case TJ. 1998. Effects of fragmentation and invasion on native ant communities in coastal southern California. *Ecology* 79:2041–56
191. Taber SW. 1990. Cladistic phylogeny of the North American species complexes of *Pogonomyrmex* (Hymenoptera: Formicidae). *Ann. Entomol. Soc. Am.* 83:307–16
192. Taber SW. 1998. *The World of the Harvester Ant*. College Station, TX: Tex. A & M Univ. Press
193. Traniello JFA. 1989. Foraging strategies of ants. *Annu. Rev. Entomol.* 34:191–210
194. Traniello JFA, Beshers SN. 1991. Polymorphism and size-pairing in the harvester ant *Pogonomyrmex badius*: a test of the ecological release hypothesis. *Ins. Soc.* 38:121–27
195. Tschinkel WR. 1999. Sociometry and sociogenesis of colony-level attributes of the Florida harvester ant (Hymenoptera: Formicidae). *Ann. Entomol. Soc. Am.* 92:80–89
196. Valone TJ, Brown JH, Heske EJ. 1994. Interactions between rodents and ants in the Chihuahuan desert: an update. *Ecology* 75:252–55
197. Vasquez RA, Bustamante RO, Simonetti JA. 1995. Granivory in the Chilean matorral: extending the information on arid zones of South America. *Ecography* 18:403–9
198. Wagner D, Brown MJF, Gordon DM. 1997. Harvester ant nests, soil biota and soil chemistry. *Oecologia* 112:232–36
199. Weier JA, Feener Jr. DH. 1995. Foraging in the seed-harvester ant genus *Pogonomyrmex*: are energy costs important? *Behav. Ecol. Sociobiol.* 36:291–300
200. Went FW, Wheeler J, Wheeler GC. 1972. Feeding and digestion in some ants (*Veromessor* and *Manica*). *BioScience* 22:82–88
201. Wheeler GC, Wheeler JN. 1986. *The Ants of Nevada*. Los Angeles, CA: Nat. Hist. Mus. Los Angeles Cty
202. Whitford WG. 1976. Foraging behavior of Chihuahuan Desert harvester ants. *Am. Midl. Nat.* 95:455–458
203. Whitford WG. 1978. Foraging in seed-harvester ants *Pogonomyrmex* spp. *Ecology* 59:185–89
204. Whitford WG. 1978. Structure and seasonal activity of Chihuahua Desert ant communities. *Ins. Soc.* 25:79–88
205. Whitford WG. 1978. Foraging by seed-harvesting ants. In *The Production Ecology of Ants and Termites*, ed. MV Brian, pp. 107–10. Cambridge, UK: Cambridge Univ. Press
206. Whitford WG. 1988. Effects of harvester ant (*Pogonomyrmex rugosus*) nests on soils and a spring annual, *Erodium texanum*. *Southwest. Nat.* 33:482–85
207. Whitford WG. 1997. The importance of the biodiversity of soil biota in arid ecosystems. *Biodivers. Conserv.* 5:185–95
208. Whitford WG. 1999. Comparison of ecosystem processes in the Nama-karoo and other deserts. In *The Karoo: Ecological Patterns and Processes*, ed. WRJ Dean, SJ Milton, pp. 291–302. Cambridge, UK: Cambridge Univ. Press
209. Whitford WG, Bryant M. 1979. Behavior of a predator and its prey: the horned lizard (*Phrynosoma cornutum*) and harvester ants (*Pogonomyrmex* spp.). *Ecology* 60:686–94
210. Whitford WG, DiMarco R. 1995. Variability in soils and vegetation associated with harvester ant (*Pogonomyrmex rugosus*) nests on a Chihuahuan Desert watershed. *Biol. Fertil. Soils* 20:169–73
211. Whitford WG, Ettershank G. 1975. Factors affecting foraging activity in

Chihuahuan Desert harvester ants. *Environ. Entomol.* 4:689–96
212. Whitford WG, Johnson P, Ramirez J. 1976. Comparative ecology of the harvester ants *Pogonomyrmex barbatus* (F. Smith) and *Pogonomyrmex rugosus* (Emery). *Ins. Soc.* 23:117–132
213. Whitford WG, Van Zee J, Nash MS, Smith WE, Herrick JE. 1999. Ants as indicators of exposure to environmental stressors in North American desert grasslands. *Environ. Monit. Assess.* 54:143–71
214. Whiting MJ, Dixon JR, Murray RC. 1993. Spatial distribution of a population of Texas horned lizards (*Phrynosoma cornutum*: Phrynosomatidae) relative to habitat and prey. *Southwest. Nat.* 38:150–54
215. Wiens JA. 1976. Population responses to patchy environments. *Annu. Rev. Ecol. Syst.* 7:81–120
216. Wiernasz DC, Cole BJ. 1995. Spatial distribution of *Pogonomyrmex occidentalis*: recruitment, mortality and overdispersion. *J. Anim. Ecol.* 64:519–27
217. Wight JR, Nichols JT. 1966. Effects of harvester ants on production of a saltbush community. *J. Range Manage.* 19:68–71
218. Willard JR, Crowell HH. 1965. Biological activities of the harvester ant, *Pogonomyrmex owyheei*, in central Oregon. *J. Econ. Entomol.* 58:484–89
219. Wing K. 1983. *Tutelina similis* (Araneae: Salticidae): An ant mimic that feeds on ants. *J. Kans. Entomol. Soc.* 56:55–58
220. Wolff A, Debussche M. 1999. Ants as seed dispersers in a Mediterranean old-field succession. *Oikos* 84:443–52
221. Wu H. 1990. Disk clearing behavior of the red harvester ant, *Pogonomyrmex barbatus*. *Smithson. Bull. Inst. Zool. Acad. Sin.* 29:153–64
222. Zelade R. 1986. Feasting on asphalt. *Tex. Mon.* August:92–94
223. Zimmer K, Parmenter RR. 1998. Harvester ants and fire in a desert grassland: Ecological responses of *Pogonomyrmex rugosus* (Hymenoptera: Formicidae) to experimental wildfires in central New Mexico. *Environ. Entomol.* 27:282–87

Annu. Rev. Ecol. Syst. 2000. 31:293–313

ORIGINS, EVOLUTION, AND DIVERSIFICATION OF ZOOPLANKTON

Susan Rigby[1] and Clare V. Milsom[2]

[1]Department of Geology and Geophysics, University of Edinburgh, Edinburgh EH9 3JW, United Kingdom; e-mail: suerigby@glg.ed.ac.uk; [2]School of Biological and Earth Sciences, Liverpool John Moores University, Liverpool L3 3AF, United Kingdom; e-mail: BECSMILS@livjm.ac.uk

Key Words plankton, migration, benthos, geology, invertebrate, palaeontology

■ **Abstract** Fossil plankton are difficult to identify but have formed a major component of most marine ecosystems throughout geological time. The earliest fossil heterotrophs include planktic forms, and subsequent adaptive innovations quickly appeared in the plankton; these include metazoans and animals with hard parts. Movement into the plankton occurred sporadically throughout geological history and seems to have been independent of any biological or environmental forcing mechanism. Subsequent radiations and extinctions in the cohort of plankton closely reflect events in the benthos. The diversity of zooplankton rose quickly during the early Paleozoic era, but low plankton diversity characterized the late Paleozoic. Significant radiations during early Mesozoic times led to an overall increase in diversity through the Phanerozoic eon. As the composition of the zooplankton has changed, so has their effect on biogeochemical cycles.

INTRODUCTION

In this paper, we review the origins of the major modern groups of zooplankton and of extinct groups that have been geologically significant. To accomplish this, we review the ways in which fossil plankton have been identified and defined. We review the likely controls on the zooplankton over geological time and discuss the pattern of origination and evolution of planktic organisms. Finally, we consider the biogeochemical implications of this evolutionary history. The approach is chronological, moving from the Proterozoic era to recent time. Previous reviews have partially covered the Holocene and include discussions of the Paleozoic plankton of North America (92), early Paleozoic plankton in general (13), the origin and early evolution of the planktic system (102), and the origins and evolution of microplankton (62, 115).

Members of the zooplankton have the widest geographical spread and greatest numerical abundance of any animals. Modern zooplankton are important

0066-4162/00/1120-0293$14.00

contributors to global biomass and to the chemistry of the oceans, a dominant means of flux to the seabed (76), and a source of food for many large animals. The microzooplankton are dominated by flagellate protists, including some dinoflagellates and zooflagellates, and by amoebae such as foraminifera and radiolarians. Planktic ciliates are common, although the major group of these, the tintinnids, have proteinaceous tests and leave little record in the sediment (59). The macrozooplankton include a wide range of solitary and colonial cnidarians, chaetognath and polychaete worms, and holoplanktic gastropods. Crustaceans are among the most common macrozooplankton, with copepods, euphausiids, amphipods, ostracodes, and decapods all abundant and diverse (83). Urochordates are widespread with two planktic groups, appendicularians and salps. Larval stages of invertebrates and fish make up a significant proportion of the heterotrophic plankton in the modern ocean, remaining as part of the plankton for periods ranging from minutes to years.

Preservation of plankton in the fossil record is generally poor because of postmortem transport and because morphological adaptations, particularly the reduction of skeletal material, reduce the fossilization potential of plankton. The characterization of fossil plankton is also rather different from that of modern zooplankton, and this has led to an historical bias against including organisms with a limited swimming capacity in this classification.

The zooplankton form a cohort of organisms that are unusually dependent on the physical environment surrounding them. They are supported, transported, and fed by the surrounding water. The degree to which intra- or interspecific competition occurs in this environment is partly dictated by chance encounters, and, as a result, it appears that very similar species are able to coexist indefinitely (43). This long-term coexistence of species that might be expected to compete is known as the paradox of the plankton (43). Because of the likely dominance of abiotic over biotic controls and the weak interspecific competition that is characteristic of the plankton, it has been possible for organisms to migrate into this environment over a prolonged period (88). A phylogenetic analysis of the zooplankton shows that high-level clades of organisms have colonized this realm from the benthos rather than evolving in situ (88). Such colonization has occurred throughout geological time, contrasting strongly with the patterns of colonization seen, for example, on land (29, 87).

In an analysis of ecosystem evolution, Vermeij suggested (121, p. xii) that organisms, including predators, have become better adapted through time, making many environments "riskier places in which to live." The long geological history of migration into the planktic realm suggests that this theory is not applicable to the zooplankton. Kinetic models of evolutionary change and diversification based on analogy with the logistic equation suggest that an ecosystem will experience early and rapid diversification of organisms, followed by a slower increase as organisms begin to compete actively (100). The planktic ecosystem likewise seems to be an exception to this model.

IDENTIFYING FOSSIL PLANKTON

Analysis of open-ocean fossil ecosystems has been impeded by a lack of consensus regarding the actual organisms that inhabited this regime. Two problems of definition exist. The first is that some authors have failed to distinguish between a pelagic position and a planktic life habit. Thus a suggested criterion for the definition of planktic organisms was their "...wide distribution, their lack of ecological connection with any exclusive sedimentary environment, and their predominance in sediments representing de-oxygenated bottom conditions" (13, p. 456). This definition is actually of pelagic rather than planktic organisms, because it defines a spatial position in the water column rather than a mode of life.

The second problem is that the definition of fossil plankton, as used by palaeontologists, has been much more rigid and prescriptive than that used by modern plankton specialists. The modern definition of planktic habit is that the organisms concerned must be unable to move against prevailing currents (83). However, the majority of modern planktic organisms are active swimmers, moving vertically by hundreds of meters each day. They can also be large, with siphonophores reaching lengths of 10 m, giant salps exceeding 1 m, and bathypelagic euphausiids attaining lengths of 10 cm. Members of the plankton, such as many planktic gastropods, can be active predators (58). In the modern oceans, all pelagic invertebrates are considered planktic, with the exception of some species of squid (83). However, in case studies of fossils, organisms that could move are generally excluded from the plankton, and only those with no apparent means of locomotion, such as graptolites, have been defined as planktic in habit. Organisms that appear to have been adapted for active swimming are defined as pelagic (35) or even as nektonic (RJ Aldridge, personal communication). A sensible solution to this problem would be to regard all fossilized pelagic invertebrates, apart from cephalopods, as part of the plankton in a broad sense, because this would bring fossil analyses into parity with the definitions applied to modern systems.

Recent studies of planktic life habit have tended to be organism specific, and a definitive list of criteria used to define such an ecology has not been established. Ruedemann (92) proposed that faunal association was fundamental in determining a planktic mode of life, and Fortey (35) established three lines of evidence that, he argued, should be applied in any attempt to deduce mode of life: arguments from analogy, arguments based on suitability of design, and the use of independent geological evidence. Discussed below are morphological and sedimentological factors that are considered indicative of a pelagic and/or planktic mode of life. In fact, most of these factors apply to the determination of location rather than life habit.

Defining a Pelagic Position

Rudemann (92) stated that any Paleozoic fauna associated with graptolites must be planktic, unless it also occurred with obviously benthic forms. Two problems with a more general application of this method are that plankton do occur with obviously benthic faunas and that a determinant group, such as graptolites, may have its life habit misdiagnosed. A potential circularity is inherent in any argument that identifies one species' mode of life from that inferred for another. Above all, the result of such an analysis is a determinant for living position and not for life habit, and it can therefore contribute to the inference that an organism was pelagic, but can say nothing meaningful about its planktic or nektonic lifestyle.

Organisms with a widespread and transoceanic distribution are usually assumed to have been pelagic (37, 103), with the obvious exception of individuals that drift for long distances postmortem. Facies has been cited as a critical indicator of a planktic mode of life (92). The occurrence of Ordovician inarticulate brachiopods in black shales has led to the conclusion that they were pelagic (92, 96). However, arguments based on facies distribution have also been used to suggest that most inarticulate brachiopods did not have holopelagic lifestyles (46). As with faunal relationships, fossil distribution is informative with regard to life position rather than habit.

Defining a Planktic Habit

A single feature of a fossil that seems to have a modern analog in a planktic organism allows the design of the rest of the fossil to be assessed for suitability for a planktic habit (37). Ruedemann (92) listed the morphological characteristics of plankton as small size and thin, chitinous shells. These characteristics have been used to determine a planktic life habit for the bivalves *Bositra radiata parva* and *Bositra radiata magna* (77), but they have also been used to support suggestions of a benthic mode of life for these organisms (52, 53).

Modern plankton are characterized by reduced skeletons, gas or fat bodies that reduce their overall density, flattened forms, and the frequent presence of spines or other features that increase drag. These adaptations might be used as good markers to determine a planktic habit in fossils.

Bulman (13) demonstrated, from an assessment of the functional morphology of the nema, that graptolites were probably holoplanktic rather than epiplanktic on seaweed. Analysis of the gross morphology of graptolites and of their "design" (see 37) shows that they were hydrodynamically stable and possessed modifications of form that would have enhanced their efficiency in a truly planktic mode of life (85, 89).

A strong argument for a planktic habit can be made if close living relatives of the group share this habit. It can surely be accepted that most single-celled organisms with a broad geographical distribution were planktic if this is the case for their living relatives. This argument applies, for example, to radiolaria and coccolithophorids. It also applies to foraminifera, despite the multiple evolution of planktic from benthic forms (26). Clearly, this is the strongest single line of evidence for mode of life. However, its applicability declines when older fossils, which are increasingly less likely to have living relatives, are considered.

TEMPORARY MEMBERS OF THE ZOOPLANKTON: Pseudoplankton and Planktotrophic Larvae

Within the zooplankton there is a long evolutionary history of facultative or obligate pseudoplankton and of organisms with a planktic larval stage of long duration. The constraints on these groups are substantially different from those of the true plankton, but their persistence in the planktic realm makes their contribution important.

Pseudoplanktic organisms are difficult to distinguish from deep-water-benthos and from transported elements of fossilized faunal assemblages. This is particularly true of facultative elements of the pseudoplankton present coevally in benthic settings. One criterion to distinguish a floating life from a benthic one is rarity—the argument is that floating substrates are rare and so should their colonizers be (126). Pseudoplankton are also identified by their distribution across a wide variety of substrates and with a wide geographical distribution. Uncertainty remains in the identification of possible in vivo colonization while the substrate was floating, as opposed to postmortem colonization on the seabed. Some specimens of epiplankton have been found overgrown by their host (71) or orientated toward a possible source of food at the aperture of a cephalopod (98), but in many cases convincing arguments cannot be made for either the locale of colonization (cf the arguments of 53 and 98) or the health of the substrate organism.

Pseudoplankton diversity appears to have been variable through the Phanerozoic eon, controlled largely by the availability of appropriate floating substrates. The first pseudoplankton appear to have been Ordovician inarticulate brachiopods and bryozoans, some of which are host specific, implying an obligate pseudoplanktic lifestyle (4). High pseudoplankton diversity in the Ordovician and Devonian periods is attributed to the abundance of large orthoconic nautiloids and driftwood, respectively, but no simple correlation explains diversity peaks in the Permian and Cretaceous periods (126).

Few members of the obligate pseudoplankton have a good fossil record. When a fossil record is known and evolutionary lineages can be traced, it appears that pseudoplankton evolve repeatedly from benthic ancestors rather than from one another (126).

The other common, temporary constituents of the zooplankton are the larval stages of benthic invertebrates and nektonic vertebrates. It has been suggested that possession of a planktic larval stage is the primitive metazoan condition (49), although this is disputed (33, 110). It is likely that the majority of benthic invertebrate species have planktic larvae (118). Certain groups may tend to evolve away from this strategy (80), possibly because of the likelihood of dispersal into unfavorable habitats and the high energetic cost of metamorphosis. Jablonski (48) suggested that there is a link between oceanic conditions and the evolution of planktotrophic, planktic larvae, as exemplified by the evolutionary history of neogastropods. This group originated during the Cretaceous period, at a time of high sea levels and broad geographical habitats, when planktic larvae had the advantage of suitable settlement sites at a distance from their point of origin. As climatic conditions became more heterogeneous, nonplanktotrophic or nonplanktic strategies were favored, because suitable sites would most likely have been encountered close to the site of spawning (48). If this explanation is generally valid, there should be a broad correlation between periods of climatic homogeneity and times of common planktotrophic planktic larvae. However, the fossil record of larvae is inadequate to quantify this relationship at present.

MULTIPLE ORIGINS OF ZOOPLANKTON

Nineteen major groups of zooplankton can be demonstrated to have had a benthic origin, either through their fossil record or from their phylogenetic trees (88; Figure 1). Plankton recruitment is achieved by one of three methods: (*a*) adopting a pseudoplanktic strategy, (*b*) paedomorphosis, or (*c*) the migration of mature individuals. The pseudoplanktic route requires the development of a sustainable method of attachment and, in common with paedomorphosis, the possession of a planktic larval stage. Because eggs and larvae have been present in the plankton since the origin of metazoans, neoteny may be considered as a long established route into the plankton.

Absence of a planktic larval stage does not preclude recruitment into the zooplankton. The benthic-to-planktic transition may be possible at a later stage of ontogeny by mature individuals with planktic preadaptations. Thus, Silurian ostracodes adopted a planktic lifestyle by virtue of their enhanced swimming capabilities (103) and thecosome gastropods through the finlike adaptation of their muscular foot (49, 58).

THE EVOLUTIONARY HISTORY OF HOLOZOOPLANKTON

Precambrian Era

The first metazoan zooplankton were probably jellyfish, but their age and detailed taxonomic affinities are disputed. The occurrence of pre-Ediacaran metazoans, said to include ancestral planktic cnidarians, is recorded from India (97). These faunas contain fossils interpreted as large medusae and are suggested to have formed part of an extensive and diverse zooplankton community. They are found

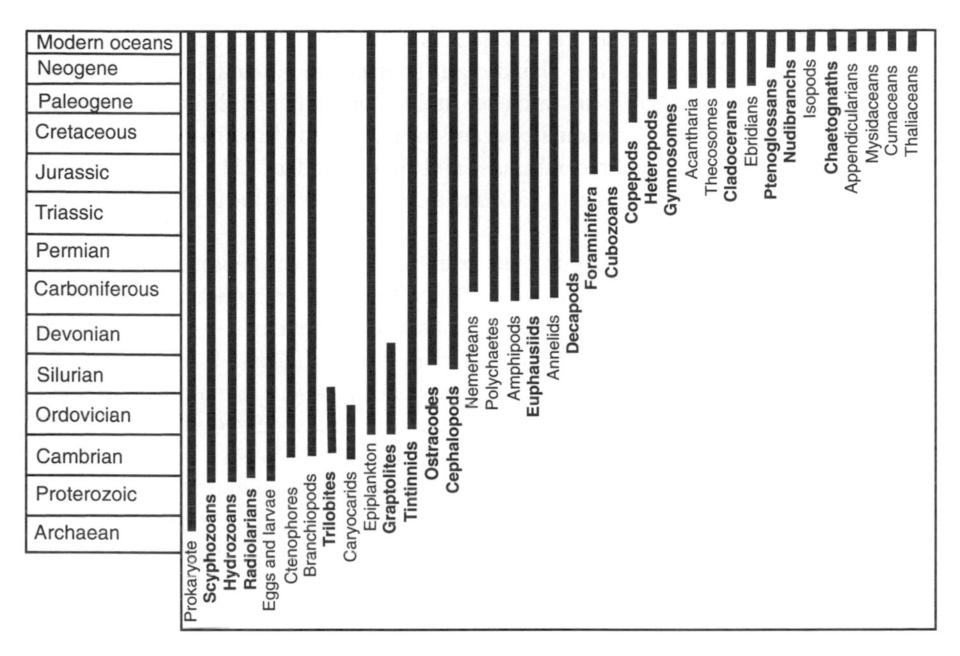

Figure 1 The pattern of origination of zooplankton through the Phanerozoic eon. *Bold type* indicates groups with a known ancestry in the benthos.

in rocks dated at between 900 and 1000 MYA and arguably represent the oldest metazoan fossils to date.

It is more generally agreed that the first medusoid cnidarians belonged to the Ediacaran fauna (44, 123, 124), which appeared around 565 MYA. The lifestyle and taxonomy of Ediacaran fauna have received much attention. If they were metazoans rather than vendobionts (99) or lichens (84), it seems likely that at least some were planktic; some medusiforms have been reinterpreted as benthic organisms (50), but others, notably *Ovatoscutum* and *Rugoconites*, remain as probable plankters. More precise taxonomy is problematic, although Runnegar & Fedonkin (93) were confident of the presence of three genera of Cyclomedusidae and four hydrozoans within the Ediacaran biota.

Experiments conducted on the preservation potential of modern jellyfish suggest it is unsafe to attempt a detailed taxonomic comparison of these Precambrian sand molds and modern taxa (76). However, Norris (76) considered it likely that broad similarities in form between Ediacaran radial fossils and extant jellyfish are a product of evolutionary convergence (76). Evolutionary convergence tends to imply functional convergence, making it likely that, even if their phylogeny is distinct from that of modern jellyfish, the Ediacaran radial fossils were probably planktic.

Second candidates for the earliest metazoan members of the zooplankton are the organisms that appear to have deposited fecal pellets in anoxic shales up to

1900 MYA (90). If these are real fossils rather than artifacts, as the authors convincingly argue (90), then planktic gut-bearing animals might be much older than previously thought. Research on molecular sequence divergence shows a likely point of origin for coelomate organisms in the early Proterozoic era (12, 129) rather than close to the Precambrian-Cambrian boundary (31). However, these molecular results are disputed (3, 24), and a consensus as to the timing of metazoan origins is still lacking.

The other zooplankton group that appears to have been represented in the late Precambrian is the chitinozoans, vase-shaped fossils of ambiguous affinity (47) of possibly the remains of heterotrophic protists or of metazoan eggs. Precambrian fossils of this form have been recovered from Arizona (8) and from Sweden (56). Planktic and planktotrophic larvae must have been present as temporary elements of the zooplankton from the time when metazoans first appeared, sometime during the Proterozoic era, within the envelope of 1500–600 MYA (12, 129). Finally, in considering early growth stages of the Precambrian zooplankton, it is possible that some of the fossils identified as acritarchs and interpreted as autotrophic eukaryotes were actually egg cases (119), although, until more analysis has been done, it is impossible to quantify this.

Cambrian Period

During the early Cambrian period, phytoplankton diversity rose rapidly (122). This may have coincided with the beginning of regular migration of metazoan benthos into the plankton, perhaps as a response to increased grazing pressures on the seabed (14). It is possible that the build-up of zooplankton during the early Cambrian reduced net carbon export from the water column and dampened the magnitude of boom and bust cycles seen in Precambrian phytoplankton (16). The composition of the Cambrian zooplankton was a mixture of Precambrian survivors, such as medusoids and chitinozoans, and new migrants from the benthos, predominantly representatives of newly skeletonized clades of coelomate metazoans. Single-celled animals also migrated into the plankton during the Cambrian period, most importantly radiolarians (62, 74, 81).

Various Cambrian medusoids have been identified (27). Most of these are chondrophores, and several, including *Gelenoptron tentaculum* and *Kullingia delicata*, seem to be Ediacaran survivors (23, 73). Early Cambrian rocks of Canada yield the filtering apparatus of a branchiopod-like crustacean likely to have been planktic for at least part of its life cycle (15). As reconstructed, these crustaceans would have been $\leq$10 mm long, and the regularity of their filtering nets suggests a planktic source of food. Agnostid trilobites, which reached a diversity peak in the mid- to late Cambrian period, may have been planktic, although some species were probably nektobenthic (72). Other trilobites may have also been pelagic: the early Cambrian genus *Kootenia* is cosmopolitan in distribution, whereas morphological considerations have led to the suggestion of a pelagic habit for *Irvingella* (late Cambrian) and *Centropleura* (mid-Cambrian)

(37). Some of these arthropods were filter feeders, but others were probably predatory (36).

The Cambrian plankton are best recorded in fossil *konservat-lägerstatten*, such as the Chengjiang fauna and the Burgess Shale, which provide a window onto the soft-bodied and lightly skeletonized elements of the Cambrian fauna. Medusiform metazoans are common on certain bedding planes within the lower Cambrian Chengjiang fauna (20) and are assumed to have had a planktic mode of life. However, putative gut traces within the body cavity mean that these may not have been true cnidarians. Two species of chondrophore are also known from this locality, as are two possible scyphozoans whose taxonomic affinity is less certain (111).

Although undisputed planktic forms compose <0.17% of the Burgess Shale in terms of numbers of individuals (excluding the problematic organism *Eldonia*), they account for ≥7% of the genera present (based on data from 11, 22). This assemblage begins to approach the degree of diversity seen in modern plankton communities and may contain five of the eight major phyla of modern macrozooplankton (although the taxonomic affinities of these organisms are often disputed). This high diversity indicates that, at least by middle Cambrian times, the Cambrian radiation had reached the pelagic realm.

The rise in benthic metazoan diversity must also have led to an increase in the abundance and diversity of planktotrophic larvae. If planktotrophy is a primitive larval state (49), it might be expected that the diversity of larvae in the zooplankton would have been higher than it is today, although this inference is disputed (102). Fossil evidence of planktic larval stages is sparse, but such an early life stage is predicted from fossil evidence for some Cambrian hyolithids (28) and for actrotretide brachiopods (7, 91).

In summary, it appears that zooplankton diversity rose slightly later than benthic metazoan diversity, possibly in the late Cambrian (102) but more likely in the earlier Cambrian period (16). However, the evolution of the two systems was connected by a variety of feedback mechanisms associated with primary production and transport of material to the seafloor (16, 63, 102). It is even possible that a top-down effect precipitated by the origin of mesozooplankton facilitated the Cambrian radiation itself (16). At present the fossil record and our capacity to model the effects of changes in biota are too poor to clarify what was happening in detail over this critical period in Earth history.

Ordovician and Silurian Periods

Zooplankton continued to migrate into the pelagic realm throughout the late Cambrian and early Ordovician periods, with radiolarians becoming dominant components of the Ordovician microzooplankton (115). Their distribution and changes in their evolutionary rate correlate well with similar changes in macrozooplankton (74). Tintinnids appeared in the Ordovician but were rare (21). Chitinozoa are diverse in rocks of this age (78). The most significant new zooplankton were probably the graptoloids, which originated by neoteny from benthic graptolites at the beginning of the Ordovician period (37). These colonial hemichordates radiated quickly into a wide variety of shapes, and presumably of niches (86). Trilobites with a likely pelagic habit include *Opipeuter*, *Carolinites*, *Girvanopyge*, *Selenopeltis*, the Bohemillidae, and the cyclopygids (35). These organisms probably radiated into a range of trophic niches and into different parts of the water column (39). Other arthropods also appeared in the plankton. Caryocarids were common elements of the Ordovician and Silurian plankton (92, 109), and ostracodes migrated into the planktic habitat in the Silurian period (103). The full complement of Lower Paleozoic plankton probably included at least one species of inarticulate brachiopod (82). This may have been an epipelagic form, and epiplankton thrived during the Ordovician and Silurian periods, due to the abundance of suitable floating substrates, including the shells of large orthoconic nautiloids (42).

Signor & Vermeij (102) observed that the rise in planktic diversity among metazoans parallels the rise of the Paleozoic fauna documented by Sepkoski (101). They suggested that the cause of this increase in planktic diversity may have been the use of the plankton as a refuge from bioturbation and predation on the seafloor, but this is disputed (120). Martin (68) suggested that the Lower Paleozoic era was a time of superoligotrophy, which limited planktic diversity. However, this assessment was based primarily on microfossil data, and we feel that this apparently low diversity is an artifact caused by the dominance of macrozooplankton over microzooplankton at this time. Although the Lower Paleozoic was generally a time of at least moderate plankton diversity, there is evidence for a major extinction within this realm at the end of the Ordovician period. Sequences of this age in central Nevada show that this extinction affected inshore plankton and graptolites first, extending later to offshore graptolites and to chitinozoans and radiolarians (34).

Devonian to Permian Periods

In the early Devonian period, a noticeable change in faunal composition occurred that correlates approximately with the time at which planktic graptoloids became extinct and with the appearance of ammonoids (57, 74). These changes in the planktic cohort at a relatively high trophic level may have contributed to changes in the system as a whole. In contrast to the previous periods, the Upper Paleozoic was a time of low plankton diversity and muted primary productivity. Late Devonian extinctions particularly affected the plankton and the system was unable to recover until the Jurassic period. This appears to be an unusual and rather protracted extinction event, precipitated by a catastrophic fall in origination rate rather than by a rise in rates of extinction (69).

The long period of impeded plankton development in the late Paleozoic era is a puzzle. The major radiation among vascular land plants and the formation of the first forests in the Devonian period have led various authors to suggest a causal link between vascular plant radiation and the crisis within the plankton. It has been suggested that the spread of vascular plants caused carbon, nitrogen, and phosphorus to be progressively retained on land (113). As existing marine nutrients were gradually lost into deep-sea sediments, a major biotic crisis was precipitated. An alternative view is that the development of land ecosystems over this period led to increased nutrient runoff and with it increased primary productivity (5). Another hypothesis is that the evolution of increased root masses altered patterns of soil formation, which led to transient pulses of nutrients entering the marine realm and stimulating eutrophication of epicontinental seaways (1). It is possible that prolonged disequilibrium in terrestrial systems may have contributed to the problem, or even that an unpreserved but diverse form of primary plankton producer was present (5), although the latter suggestion seems improbable.

However, even during this crisis in the plankton, old groups of zooplankton persisted. Jellyfish are found throughout the Upper Paleozoic, sometimes in great abundance (107). Over 6000 specimens of soft-bodied cnidarians are reported (40) from the Essex fauna, the marine portion of the Mazon Creek lagerstätte, and these include representatives of all the modern planktic scyphozoan orders. It is likely that new recruits also appeared in the zooplankton during this time of low productivity (88). The oldest pelagic nemertean worms are found in the Mazon Creek fauna (51), and the Carboniferous Granton Shrimp Bed contains the first planktic tomopterid polychaete, *Eotomopteris aldridgei* (9, 11). Carboniferous aeschronectid crustaceans were pelagic (95) as was *Waterstonella*, the most common Granton crustacean (94).

Mesozoic Era

The Mesozoic era saw a massive radiation of microzooplankton and probably also of macrozooplankton. Innovation of skeletal types, particularly the origination of calcareous microplankton, had far-reaching implications throughout the water column. Because the radiation included taxa with siliceous and calcareous skeletons and unmineralized groups, a change in oceanic chemistry is unlikely to have driven the diversification (55). It is postulated that strengthened circulation and increased upwelling may have been generated by continental break-up (55) or that the rise of bioturbating benthic animals may have resulted in more nutrients being recycled back into the water from earlier benthic systems.

Planktic foraminifera evolved first in the Middle Jurassic and were initially meroplanktic, evolving a holoplanktic mode of life in the Hauterivian (25). Radiation patterns within planktic foraminifera can be related to changes in sea level. High sea levels were associated with warmer climates and resulted in more sluggish circulation and increased water column stratification. This resulted in increased species numbers, the development of complex communities, and intense depth specificity of species. Lower sea levels were associated with cooler periods when ocean waters were well mixed. This led to decreased diversity and to blooms of opportunistic genera such as *Calpionella* in the late Jurassic and early Cretaceous and *Pithonella* in the Albian-Turonian (18).

The Mesozoic microzooplankton also included diverse species of radiolaria, especially nassellarians, which experienced an evolutionary burst in the Jurassic and Cretaceous periods (2, 115). Tintinnids underwent a similar radiation in the late Jurassic-early Cretaceous period, becoming rare once more in the late Cretaceous (115). The peak of microzooplankton diversity for the Phanerozoic eon seems to have been reached in the Cretaceous period (115).

The Mesozoic record of macrozooplankton is sparse, although chondrophores persist (106). Seven genera of scyphozoans are recorded from the Jurassic Solnhöfen Limestone, including *Rhizostomites* and *Leptobrachites*, and there are also three genera of hydrozoans, at least one of which is preserved in its medusoid stage (*Hydrocraspedota*) (6). Crustaceans were clearly abundant in the Mesozoic plankton although their fossil occurrence is limited to lagerstätten. Enigmatic fossils, found in the stomachs of fish from the Santana formation, have been identified as crab larvae, to which a planktic life habit is attributed (65). Juvenile shrimps, which may also have been planktic, are also preserved in this way, as well as in mass-mortality concretions.

With the exception of radiolarians, all elements of the fossilized calcareous and siliceous zooplankton show high levels of extinction across the Cretaceous-Tertiary boundary. Extinctions were highly selective—more severe in open-water surface dwellers and less severe for inhabitants of deep ocean waters (117). Paul & Mitchell (79) noted that, at both the K-T and Cenomanian-Turonian boundaries, phytoplankton were badly affected. This caused a rapid fall in primary productivity and mass starvation higher up the food chain. Smit & Romein (104) also proposed that the zooplankton were killed by starvation. Survivors among the zooplankton were predominantly small species, and individuals tended to be small even for their species (79).

Cenozoic Era

The enormous wealth of data available for reconstructions of Cenozoic oceanography allows the impact of oceanographic changes on the planktic system to be assessed. It seems likely that the origination of the circum-Antarctic current toward the end of the Miocene set the scene for evolutionary diversification in several microplankton groups through the rest of the Neogene (54). Among microzooplankton, the dominant forms of the Cenozoic era have been foraminifera and radiolaria (115). Tertiary planktic foraminifera are considered to have descended from two small planktic species that survived the Cretaceous-Tertiary decimation (25) and from benthic ancestors (26). Although diversity in Oligiocene faunas was limited, the Miocene epoch saw renewed radiation, probably as a consequence of increasing water mass heterogeneity and the development of vertical and lateral gradients (25). Ebridians were common for most of the Cenozoic era (30). Radiolarians show a conspicuous faunal change during the Cenozoic (19).

These changes seem to be linked to radiations in diatoms particularly in the Miocene epoch (112). The two groups compete for silica, and it seems that shallow-, warm-water radiolarians have responded to this competition by becoming lighter (45). The skeletal construction of the deep, cold-water radiolarians has remained unaffected. Tintinnids have a sparse Cenozoic record; examples of well-preserved calcareous forms are known from the Eocene and more rarely the Oligocene epochs, where they occur in association with diverse foraminiferal assemblages (114).

Holoplanktic gastropods are probably Cenozoic in origin. Heteropod and thecosome gastropods became planktic by migration from benthic habits prior to the Eocene. Gymnosomes, ptenoglossans, and nudibranchs are probably Cenozoic in origin as well. All of these groups evolved from benthic ancestors, probably by adult migration (41, 58). Most of the extant groups of planktic crustaceans lack a fossil record but probably migrated into the plankton in the Cenozoic era (88). Salps and thaliaceans are a major soft-bodied component of modern planktic ecosystems, but they have no fossil record and their time of origin is unknown.

During the Cenozoic era, plankton diversity in most groups recovered partially from the end Cretaceous event, but did not return to its former levels (61). Most microplankton seem to have been highly diverse in the Paleocene-Eocene and in the Miocene epochs, with low diversity between. Extinctions were survived preferentially by plankton with simple morphologies, and subsequent radiations produced iterative evolution, with complicated forms arising from these simple stocks (61).

TIMING OF MOVEMENT INTO THE PLANKTON

The previous section demonstrates that migration from the benthos to the plankton has been ongoing for at least the last 650 million years; hence, the planktic realm has never been closed to migration by increased competition for resources or through biotic innovation (Figure 1). Biotic crisis does not precipitate movement from the benthos into the plankton (88), although major oceanographic events may inspire radiations within the planktic cohort (60). For example, diversification in planktic foraminifers in the Mesozoic epoch was coincident with the development of more extensive seas and reduced oceanic circulation (116), but their multiple origins from benthic ancestors appear to have been independent of such a forcing mechanism (26). The zooplankton appear to be unique in the facility with which their composition changes through time. This ability is related to their intense dependence on physical factors for survival, such as water mass stability, temperature, salinity, and upwelling, which have precluded the development of increasingly "better" adaptations [in the sense of Vermeij (121)]. The same phenomenon is reflected today in the paradox of the plankton (43) and identifies the plankton as unique in their macroevolutionary constraints.

BIOGEOCHEMICAL EFFECTS OF THE ZOOPLANKTON THROUGH TIME

The planktic ecosystem contributes major amounts of matter to the seafloor. This matter is derived from the biomass generated in the plankton and from the controls that zooplankton exert on the sinking of small inorganic particles. Of the total flux of particulate matter to the seabed, $\leq$90% is mediated by zooplankton (76). The main materials derived from the biological activities of the realm are organic carbon compounds and the inorganic skeletal elements of planktic organisms.

Most of the organic matter in the oceans is derived from phytoplankton, and >90% of the plant matter produced is eaten (70), which means that the residue entering various biochemical cycles is the product of metabolism of animals, mainly zooplankton (108). It may now be possible to assess the point in the history of zooplankton when it assumed its modern importance in moving organic material to the seabed and into the sedimentary record. This could represent one of the most important reorganizations of biogeochemical cycles to have occurred in the oceans. It is suggested (63) that the appearance, in the late Precambrian era, of metazoan zooplankton with muscular guts able to produce large fecal pellets radically altered the rate at which organic matter left the water column for the seafloor and, hence, the chemistry of that material when it arrived on the seabed. Late Proterozoic hydrocarbons show isotopic signatures indicative of extensive reworking of organic matter in the water column, whereas later hydrocarbons show signatures indicative of faster rates of migration to the seafloor. This primitive, extensive reworking of material in the water column may have maintained the seafloor in an anoxic condition, promoting the dissolution of carbonate minerals in seawater and raising the carbonate lysocline. Phosphorus at this time would have been recycled in the photic zone and rarely would have reached the seafloor. Faster sinking rates of fecal pellets allowed the seafloor to become oxygenated and facilitated phosphate and carbonate precipitation into increasingly deeper water. Walter (125) proposed that the transition in biogeochemical state occurred in the interval between the deposition of two rock formations, one dated at 590–570 MYA and the other at about 540 MYA, although it is possible that fecal pellets appeared in the rock record much earlier than this (90). It should be noted, however, that Butterfield (16) proposed that elongating food chains within the water column might have increased residence times for organic matter, the opposite effect to that suggested by Logan et al (63).

Many organisms use mineralized hard parts for aggression or defense, and the origin of skeletons precipitated the Cambrian "explosion" of recorded animal diversity. Of 21 types of biologically precipitated inorganic substances, only 4 are commonly found in the holozooplankton: opaline silica, aragonite, calcite, and celestite ($SrSO_4$). Gypsum ($CaSO_4 \cdot H_2O$) may be precipitated by one group of planktic cnidarians (64). Planktotrophic larvae can have phosphatic skeletons, as can pseudoplanktic inarticulate brachiopods (64). Of these six inorganic compounds, by far the most volumetrically important are carbonates and silica.

Pelagic oozes currently make up 7% of total carbonate mass and 60% of carbonate flux. Since the Mesozoic era, there has been a transfer of mass between shallow and deep marine settings. Southam & Hays proposed (105) that this trend correlates with the rise of planktic calcifers, and modeling suggests that changes in accumulation rate are best explained by a gradual rise in the diversity and abundance of planktic calcifers, including foraminiferids and pteropods in the zooplankton. The model suggests that planktic carbonate was first delivered to the major oceanic basins about 145 MYA and that the total oceanic accumulation rate has been increasing since that time at a rate of $\sim 0.04 \times 10^{20}$ g of calcium per million years (127). It may be that high atmospheric pCO_2 and low $CaCO_3$ saturation in the ocean made surface waters corrosive to potential calcifers during the Paleozoic era (67). However, it is also apparent that deep-water accumulations of carbonate ooze were common in the Paleozoic oceans, based on limestone-dolostone sequence masses (127).

Organisms precipitate calcium carbonate in one of two crystal forms, calcite or aragonite. Aragonite is the less stable of the two in the physical and chemical conditions encountered at the Earth's surface. Heteropod and pteropod molluscs are the only source of aragonite in the open sea, and, in 98% of oceanic regions, this aragonite dissolves before reaching the seabed (32). It is suggested that unexpected alkalinity maxima in midwater may be attributable to this source (32), and these can only have been present since the Cenozoic era.

In modern oceans, most dissolved silica is removed from the water by organisms. It has been carried into the deep sea since the Ordovician radiation of radiolarians, most of which produce an opaline silica skeleton. A major change in distribution pattern is correlated with the radiation of diatoms from the Cretaceous period to the Eocene epoch (66). There is a concomitant change in radiolarian skeletons, which appears to support the suggestion that diatoms are better at removing silica and have reduced its availability to the zooplankton (45). In the modern day, the input of 10^{14} g of dissolved Si per year into the oceans from rivers and mid-ocean ridges is largely balanced by diatom sedimentation (128), and radiolarians contribute only a small amount to silica flux.

DISCUSSION

Members of the zooplankton have migrated into place throughout the late Precambrian era and Phanerozoic eon. The clades of organisms composing the zooplankton have changed significantly over this time, but no major biological innovation appears to have made organisms "better" at being planktic. This facilitates comparison between modern and fossil systems. However, it is important that the chemistry of the oceans has changed significantly with the changing groups of plankton. Although planktic organisms can be difficult to identify within fossil communities, their study has implications for a wide range of research areas, from macroevolution to biogeochemistry.

Visit the Annual Reviews home page at www.AnnualReviews.org

LITERATURE CITED

1. Algeo TJ, Maynard JB, Berner RA, Scheckler SE. 1995. Late Devonian oceanic anoxic events and biotic crises: 'rooted' in the evolution of vascular land plants? *GSA Today* 5:64–66
2. Anderson OR. 1983. *Radiolaria*. New York: Springer-Verlag. 273 pp.
3. Ayala FJ, Rzhetsky A, Ayala FJ. 1998. Origin of the metazoan phyla: molecular clocks confirm palaeontological estimates. *Proc. Natl. Acad. Sci. USA* 95:606–11
4. Baird GC, Brett CE, Frey RC. 1989. Hitchiking epizoans on orthoconic cephalopods: preliminary review of the evidence and its implications. *Senckenb. Lethaea* 69:439–65
5. Bambach RK. 1983. Seafood through time: changes in biomass, energetics and productivity in the marine ecosystem. *Paleobiology* 19:372–97
6. Barthel KW, Swinburne NHM, Conway Morris S. 1994. *Solnhofen: A Study in Mesozoic Palaeontology*. Cambridge, UK: Cambridge Univ. Press. 236 pp.
7. Biernat G, Williams A. 1970. Ultrastructure of the protegulum of some acrotretide brachiopods. *Palaeontology* 13:491–502
8. Bloeser B, Schopf JW, Horodyski RJ, Breed WJ. 1977. Chitinozoans from the late Precambrian Chuar group of the Grand Canyon, Arizona. *Science* 195:676–79
9. Briggs DEG, Clarkson ENK. 1987. The first tomopteris, a polychaete from the Carboniferous of Scotland. *Lethaia* 20:257–62
10. Deleted in proof
11. Briggs DEG, Erwin DH, Collier FJ. 1994. *The Fossils of the Burgess Shale*. Washington, DC: Smithsonian Inst. Press. 238 pp.
12. Bromham L, Rambaut A, Fortey RA, Cooper A, Penny D. 1998. Testing the Cambrian explosion hypothesis by using a molecular dating technique. *Proc. Natl. Acad. Sci. USA* 95:12386–89
13. Bulman OMB. 1964. Lower Paleozoic plankton. *Geol. Soc. London Q. J.* 120: 455–76
14. Butterfield NJ. 1994a. The Precambrian/Cambrian plankton record: ecological and evolutionary implications. *Terr. Nova* 3(Suppl.):1 (Abstr.)
15. Butterfield NJ. 1994b. Burgess Shale-type fossils from a lower Cambrian shallow-shelf sequence in northwestern Canada. *Nature* 369:477–79
16. Butterfield NJ. 1997. Plankton ecology and the Proterozoic-Phanerozoic transition. *Paleobiology* 23:247–62
17. Deleted in proof
18. Caron M, Homewood P. 1983. Evolution of early planktic foraminifers. *Mar. Micropaleontol.* 7:453–62
19. Casey RE. 1993. Radiolaria. In *Fossil Prokaryotes and Protists*, ed. JH Lipps, pp. 249–85. Cambridge, MA: Blackwell Sci.
20. Chen J-Y, Erdtmann B-D. 1992. Lower Cambrian fossil Lagerstatte from Chengjiang, Yunnan, China: insights for reconstructing early metazoan life. In *The Early Evolution of Metazoa and the Significance of Problematic Taxa*, ed. AM Simonetta, S Conway Morris, pp. 57–76. Cambridge, UK: Cambridge Univ. Press
21. Chennaux G. 1968. Presence de tintinnoidiens dans l'Ordovician du Sahara. *C. R. Acad. Sci. (Paris)*266(D):86–87
22. Conway Morris S. 1986. The community structure of the middle Cambrian

phyllopod bed (burgess shale). *Palaeontology* 29:423–67
23. Conway Morris S. 1993. Ediacaran-like fossils in Cambrian burgess shale-type faunas of North America. *Palaeontology* 36:593–635
24. Conway Morris S. 1997. Molecular clocks: defusing the Cambrian 'explosion'? *Curr. Biol.* 7:R71–R74
25. Culver SJ. 1993. Foraminifera. In *Fossil Prokaryotes and Protists*, ed. JH Lipps, pp. 203–48. Cambridge, MA: Blackwell Sci.
26. Darling KF, Wade CM, Kroon D, Brown AJL. 1997. Planktic foraminiferal molecular evolution and their polyphyletic origins from benthic taxa. *Mar. Micropalaeontol.* 30:251–66
27. Debrenne F. 1994. Cambrian ecological radiation of sponges and coelenterates. *Terr. Nova Abstr* 3(Suppl.):2 (Abstr.)
28. Dzik J. 1980. Ontogeny of bactrotheca and related hyoliths. *Geologiska Foereningers. Stockholm: Foerhandlingar.* 102: 223–33
29. Edwards D, Selden PA. 1993. The development of early terrestrial ecosystems. *Bot. J. Scotl.* 46:337–66
30. Ernissee JJ, McCartney K. 1993. Ebridians. In *Fossil Prokaryotes and Protists*, ed. JH Lipps, pp. 131–41. Cambridge, MA: Blackwell Sci.
31. Erwin DH. 1993. The origin of metazoan development: a paleobiological perspective. *Biol. J. Linn. Soc.* 50:255–74
32. Fabry VJ. 1990. Shell growth rates of pteropod and heteropod molluscs and aragonite production in the open ocean: complications for the marine carbonate system. *J. Mar. Res.* 48:209–22
33. Farmer JD. 1977. An adaptive model for the evolution of the ectoproct life cycle. In *Biology of Bryozoans*, ed. RM Woolacott, RL Zimmer, pp. 487–517. New York: Academic
34. Finney SC, Berry WBN, Cooper JD, Ripperdan RL, Sweet WC, et al. 1999. Late Ordovician mass extinction: a new perspective from stratigraphic sections in central Nevada. *Geology* 27:215–18
35. Fortey RA. 1985. Pelagic trilobites as an example of deducing the life habits of extinct arthropods. *R. Soc. Edinb. Trans.* 76:219–30
36. Fortey RA. 1994. Adaptive deployment in feeding habits in Cambrian trilobites. Ecological aspects of the Cambrian radiation (IGCP 366). *Terr. Nova* 63(Suppl.):3 (Abstr.)
37. Fortey RA, Landing E, Skevington D. 1982. Cambrian-Ordovician boundary sections in the cow head group, western Newfoundland. In *The Cambrian-Ordovician Boundary; Sections, Fossil Distributions and Correlations: Geol. Ser. 3*, ed. M Bassett, pp. 95–130. Cardiff, Wales: Natl. Mus. Wales
38. Deleted in proof
39. Fortey RA, Owens RM. 1999. Feeding habits in trilobites. *Palaeontology* 42:429–65
40. Foster MW. 1979. Soft-bodied coelenterates in the Pennsylvanian of Illinois. In *Mazon Creek Fossils*, ed. MH Nitecki, pp. 191–267. New York: Academic
41. Fretter V, Graham A. 1994. British Prosobranch molluscs. London: Ray Soc. 820 pp.
42. Gabbott SE. 1999. Orthoconic cephalopods and associated fauna from the late Ordovician Soom Shale lagerstatte, South Africa. *Palaeontology* 42:123–49
43. Ghilarov M. 1984. The paradox of the plankton; or, why do species coexist? *Oikos* 43:46–52
44. Glaessner MF. 1984. *The Dawn of Animal Life, A Biohistorical Study*. Cambridge, UK: Cambridge Univ. Press. 244 pp.
45. Harper HE, Knoll AH. 1975. Silica, diatoms and Cenozoic radiolarian evolution. *Geology* 3:175–77
46. Holmer L. 1989. Middle Ordovician phosphatic inarticulate brachiopods from

Vastergotland and Dalarna, Sweden. *Foss. Strata* 26:1–172
47. Horodyski RJ. 1993. Paleontology of Proterozoic shales and mudstones—examples from the Belt supergroup, Chuar group ad Pahrump group, Western USA. *Precambrian Res.* 61:241–78
48. Jablonski D. 1979. *Paleoecology, paleobiogeography and evolutionary patterns of late Cretaceous Gulf and Atlantic coastal plain mollusks*. PhD thesis. Yale Univ., New Haven, Conn. 604 pp.
49. Jagersten G. 1972. *Evolution of the Metazoan Life Cycle (a Comprehensive Theory)*. London/New York: Academic. 234 pp.
50. Jenkins GH. 1992. Function of ediacaran faunas. In *Origin and Early Evolution of the Metazoa*, ed. JH Lipps, PW Signor, pp. 131–76. New York: Plenum
51. Johnson RG, Richardson ES. 1966. A remarkable Pennsylvanian fauna from the Mazon Creek area, Illinois. *J. Geol.* 74:626–31
52. Kauffman EG. 1978. Benthic environments and paleoecology of the Posidonienschiefer (Toarcian). *Neues Jahrb Geol. Palaeontol. Abh.* 157:18–36
53. Kauffman EG. 1981. Ecological reappraisal of the German Posidoniensheifer (Toarcian) and the stagnant basin model. In *Communities of the Past*, ed. J Gray, AJ Boucot, WBN Berry, pp. 311–81. Stroudsberg, Pennsylvania: Hutchinson Ross
54. Kennett JP. 1986. Miocene paleoceanography and plankotn evolution. In *Mesozoic and Cenozoic Oceans*, ed. KJ Hsü, pp. 119–22, Washington, DC: Am. Geophys. Union
55. Knoll AH, Lipps JH. 1993. Evolutionary history of prokaryotes and protists. In *Fossil Prokaryotes and Protists*, ed. JH Lipps, pp. 19–31. Oxford, UK: Blackwell Sci.
56. Knoll AH, Vidal G. 1980. Late Proterozoic vase-shaped microfossils from the Vinsingo Beds, Sweden. *Geol. Foeren. Stockh. Foerh.* 102:207–11
57. Koren TN, Rickards RB. 1979. Extinction of graptolites. In *The Caledonides of the British Isles, Spec. Publ.*, ed. M. Bassett, pp. 457–66. Geol. Soc. London
58. Lalli CM, Gilmer RW. 1989. *Pelagic Snails: The Biology of Holopelagic Gastropod Mollusks*. Palo Alto, CA: Stanford Univ. Press. 184 pp.
59. Lalli CM, Parsons TR. 1993. *Biological Oceanography: An Introduction*. Oxford, UK: Pergamon. 310 pp.
60. Leckie RM. 1989. A palaeoceanographic model for the early evolutionary history of planktic foraminifera. *Palaeogeogr. Palaeoclimatol. Palaeoecol.* 73:107–38
61. Lipps JH. 1970. Plankton evolution. *Evolution* 24:1–22
62. Lipps JH. 1993. Proterozoic and Cambrian skeletonised protists. In *The Proterozoic Biosphere*, ed. JW Schopf, C Klein, pp. 237–40. Cambridge, UK: Cambridge Univ. Press
63. Logan GA, Hayes JM, Hleshima GB, Summons RE. 1995. Terminal Proterozoic reorganisation of biogeochemical cycles. *Nature* 376:53–56
64. Lowenstam HA. 1974. Impact of life on chemical and physical processes. In *The Sea*, Vol. 5: *Marine Chemistry*, ed. ED Goldberg, pp. 715–97. New York: Wiley & Sons
65. Maisey JG, de Carvalho MGP. 1995. Fossil decapods from the Santana Formation. *Atas XIV Congr. Bras. Paleontol. 90.*
66. Maliva RG, Knoll AH, Siever R. 1990. Secular change in chert distribution: a reflection of evolving biological participation in the silica cycle. *Palaios* 4:519–32
67. Martin RE. 1995. Cyclic and secular variation in microfossil biomineralization—clues to the biogeochemical evolution of phanerozoic oceans. *Glob. Planet. Chang.* 11:1–23
68. Martin RE. 1996. Secular increase in nutrient levels through the phanerozoic: implications for productivity, biomass and diversity of the marine biosphere. *Palaios* 11:209–19

69. McGhee GR. 1988. The Late Devonian extinction event: evidence for abrupt ecosystem collapse. *Paleobiology* 14:250–57
70. Menzel DW. 1974. Primary productivity, dissolved and particulate organic matter and the sites of oxidation of organic matter. In *The Sea*, Vol. 5: *Marine Chemistry*, ed. ED Goldberg, pp. 659–79, New York: Wiley &Sons
71. Merkt J. 1966. Uber Austen und Serpeln als Epoken auf Ammonitengehausen. *Neues Jahrb Geol. Palaontol. Abh.* 125:467–79
72. Muller KJ, Walossek D. 1987. Morphology, ontogeny, and life habit of *Agnostus pisiformis* from the upper Cambrian of Sweden. *Foss. Strata* 19:1–123
73. Narbonne GM, Myrow P, Landing E, Anderson MM. 1991. A chondrophorine (medusoid hydrozoan) from the basal Cambrian (Placentian) of Newfoundland. *J. Paleontol.* 65:186–91
74. Nazarov B, Ormiston AR. 1985. Evolution of radiolaria in the Paleozoic and its correlation with the development of other marine fossil groups. *Lethaia* 66:203–35
75. Noji TT. 1991. The influence of macrozooplankton on vertical particulate flux. *Sarsia* 76:1–9
76. Norris RD. 1989. Cnidarian taphonomy and affinities of the Ediacara biota. *Lethaia* 22:381–93
77. Oschmann W. 1994. Adaptive pathways of benthic organisms in marine oxygen-controlled environments. *Neues Jahrb Geol. Palaontol. Abh.* 191:393–444
78. Paris F, Nolvak J. 1999. Biological interpretation and palaeobiodiversity of a cryptic fossil group: the 'chitinozoan animal.' *Geobios* 32:315–24
79. Paul CRC, Mitchell SF. 1994. Is famine a common factor in mass extinctions? *Geology* 22:679–82
80. Pechenik JA. 1999. On the advantages and disadvantages of larval stages in benthic marine invertebrate life cycles. *Mar. Ecol. Prog. Ser.* 177:269–97
81. Petruschevskaya MG. 1977. O proiskhozhdenii radioyariy. *Zool. Zur.* 56:1448–58
82. Popov LE, Zesina ON, Nolvak J. 1982. Mikrostruktura apikal'noj chasti rakoviny bezzamkovykh brakhiopod I ee ekologicheskoe znachenie. *Byull MoskoObshch Ispyt Prir* 87:94–104
83. Raymont JEG. 1983. *Plankton and Productivity in the Oceans*, Vol. 2: *Zooplankton*. New York: Pergamon. 824 pp. 2nd ed.
84. Retallack GJ. 1994. Were the ediacaran fossils lichens? *Paleobiology* 20:523–44
85. Rickards RB, Rigby S, Rickards J, Swales C. 1998. The hydrodynamics of graptolites assessed by laser Doppler anemometry. *Palaeontology* 41:737–52
86. Rigby S. 1991. Feeding strategies in graptoloids. *Palaeontology* 34:797–813
87. Rigby S. 1997. A comparison of the colonization of the planktic realm and land. *Lethaia* 30:11–18
88. Rigby S, Milsom CV. 1996. Benthic origins of zooplankton: an environmentally mediated macroevolutionary effect. *Geology* 24:52–54
89. Rigby S, Rickards B. 1989. New evidence for the life habit of graptoloids from physical modelling. *Paleobiology* 15:402–13
90. Robbins EI, Porter KG, Haberyan KA. 1985. Pellet microfossils: possible evidence for metazoan life in early Proterozoic time. *Proc. Natl. Acad. Sci. USA* 82:5809–13
91. Rowell AJ. 1986. The distribution and inferred larval dispersion of *Rhondellina dorei*: a new Cambrian brachiopod (Acrotretida). *J. Paleontol.* 60:1056–65
92. Ruedemann R. 1934. *Paleozoic Plankton of North America*. Boulder, CO: Geol. Soc. Am. Mem. 2. 141 pp.
93. Runnegar BN, Fedonkin MA. 1992. Proterozic metazoan body fossils. In *The Proterozoic Biosphere*, ed. JW Schopf, C Klein, pp. 369–88. Cambridge, UK: Cambridge Univ. Press
94. Schram FR. 1979. British carboniferous malacostraca. *Fieldana Geol.* 40:1–129

95. Schram FR. 1982. The fossil record and the evolution of crustacea. In *The Biology of Crustacea*, Vol. 1: *Systematics, The Fossil Record and Biogeography*, ed. LG Abele, pp. 93–147. New York: Academic
96. Schuchert C. 1911. Paleogeographic and geologic significance of recent brachiopoda. *Geol. Soc. Am. Bull. Mem.* 22:258–75
97. Schukla M, Sharma M, Bansal R, Venkatachala BS. 1991. Pre-Ediacaran fossil assemblages from India and their evolutionary significance. *Mem. Geol. Soc. India* 20:169–79
98. Seilacher A. 1982. Ammonites as habitats in the Posidonia Shales of Holzmaden—floats or benthic islands? *Neues Jahrbuch Geol. Palaontol. Monatsh* V:98–114
99. Seilacher A. 1989. Vendozoa: organismic construction in the Proterozoic biosphere. *Lethaia* 22:229–39
100. Sepkoski JJ. 1978. A kinetic model of taxonomic diversity. I. Analysis of marine orders. *Paleobiology* 4:223–51
101. Sepkoski JJ. 1981. A factor analytic description of the Phanerozoic marine fossil record. *Paleobiology* 7:36–53
102. Signor PW, Vermeij GJ. 1994. The plankton and the benthos: origins and early evolution of an evolving relationship. *Paleobiology* 20:297–320
103. Siveter DJ, Vannier JMC, Palmer D. 1991. Silurian myodocopes: pioneer pelagic ostracodes and the chronology of an ecological shift. *J. Micropaleontol.* 10:151–73
104. Smit J, Romein AJT. 1985. A sequence of events across the K-T boundary. *Earth Planet. Sci. Lett.* 74:155–70
105. Southam JR, Hays WW. 1977. Time scales and dynamic models of deep sea sedimentation. *J. Geophys. Res.* 82:3825–42
106. Stanley GD, Kanie Y. 1985. The first Mesozoic chondrophorine (medusoid hydrozoan) from the lower cretaceous of Japan. *Paleontology* 28:101–9
107. Stanley GD, Yancey TE. 1986. A new late Paleozoic chondrophorine (Hydrozoa, Velellidae)—by the wind sailor from Malaysia. *J. Paleontol.* 60:76–83
108. Steele JH. 1972. Factors controlling marine ecosystems. In *Changing Chemistry of the Oceans*, ed. D Dryssen, D Jagner, pp. 209–21. London/New York: Wiley & Sons
109. Stormer L. 1936. Planktic crustaceans from the lower didymograptus shale (3b) of Oslo. *Nor. Geol. Tidsskr.* 16:267–78
110. Strathmann RR. 1993. Hypotheses on the origins of marine larvae. *Annu. Rev. Ecol. Syst.* 24:89–117
111. Sun W. 1992. Early medusiform fossils from Chengjiang, Yunnan, China. In *The Early Evolution of Metazoa and the Significance of Problematic Taxa*, ed. AM Simonetta, S Conway Morris, pp. 131. Cambridge, UK: Cambridge Univ. Press
112. Tappan H. 1980. *The Paleobiology of Plant Protists*. San Francisco: Freeman
113. Tappan H. 1982. Extinction or survival: selectivity and causes of Phanerozoic crises. *Geol. Soc. Am. Spec. Pap.* 190:265–76
114. Tappan H. 1993. Tintinnids. In *Fossil Prokaryotes and Protists*, ed. JH Lipps, pp. 285–304, Oxford, UK: Blackwell Sci.
115. Tappan H, Loeblich AR. 1973. Evolution of the oceanic plankton. *Earth Sci. Rev.* 9:207–40
116. Tappan H, Loeblich AR. 1988. Foraminiferal evolution, diversification and extinction. *J. Paleontol.* 62:695–714
117. Thierstein HR. 1982. Terminal cretaceous plankton extinctions: a critical assessment. *Geol. Soc. Am. Spec. Pap.* 190:385–99
118. Thorson G. 1960. Reproductive and larval ecology of marine bottom invertebrates. *Biol. Rev.* 25:1–45
119. Van Waveren IM, Marcus NH. 1993. Morphology of recent copepod egg envelopes and their implication for acritarch affinity. *Spec. Pap. Palaeontol.* 48:111–25

120. Verity PG, Smetacek V. 1996. Organism life cycles, predation and the structure of marine pelagic ecosystems. *Mar. Ecol. Prog. Ser.* 130:277–93
121. Vermeij GJ. 1987. *Evolution and Escalation, an Ecological History of Life*. Princeton, NJ: Princeton Univ. Press. 527 pp.
122. Vidal G, Knoll AH. 1983. Proterozoic plankton. *Geol. Soc. Am. Mem.* 161:265–77
123. Wade M. 1971. Bilateral Precambrian chondrophores from the Ediacara fauna, South Australia. *Proc. R. Soc. Vic.* 84:183–88
124. Wade M. 1972. Hydrozoa and scyphozoa and other medusoids from the Precambrian ediacara fauna, South Australia. *Palaeontology* 15:197–225
125. Walter M. 1993. Fecal pellets in world events. *Nature* 376:16–17
126. Wignall PB, Simms MJ. 1990. Pseudoplankton. *Palaeontology* 33:359–78
127. Wilkinson BH, Walker JCG. 1989. Phanerozoic cycling of sedimentary carbonate. *Am. J. Sci.* 289:525–48
128. Wollast R, McKenzie FT. 1983. The global cycle of silica. In *Silicon Geochemistry and Biochemistry*, ed. RS Astor, pp. 39–76. New York: Academic
129. Wray GA, Levinton JS, Shapiro LH. 1996. Molecular evidence for deep Precambrian divergences among metazoan phyla. *Science* 274:568–73

Annu. Rev. Ecol. Syst. 2000. 31:315–41

EVOLUTIONARY PHYSIOLOGY[1]

Martin E. Feder,[1] Albert F. Bennett,[2] and Raymond B. Huey[3]
[1]*Department of Organismal Biology & Anatomy and Committee on Evolutionary Biology, The University of Chicago, 1027 East 57th Street, Chicago, Illinois 60637; e-mail: m-feder@uchicago.edu*
[2]*Department of Ecology & Evolutionary Biology, University of California, Irvine, California 92697; e-mail: abennett@uci.edu*
[3]*Department of Zoology, University of Washington, Seattle, Washington 98195-1800; e-mail: hueyrb@u.washington.edu*

Key Words diversity, fitness, mechanism, phylogenetic approaches, variation

■ **Abstract** Evolutionary physiology represents an explicit fusion of two complementary approaches: evolution and physiology. Stimulated by four major intellectual and methodological developments (explicit consideration of diverse evolutionary mechanisms, phylogenetic approaches, incorporation of the perspectives and tools of evolutionary genetics and selection studies, and generalization of molecular techniques to exotic organisms), this field achieved prominence during the past decade. It addresses three major questions regarding physiological evolution: (*a*) What are the historical, ecological, and phylogenetic patterns of physiological evolution? (*b*) How important are and were each of the known evolutionary processes (natural selection, sexual selection, drift, constraint, genetic coupling/hitchhiking, and others) in engendering or limiting physiological evolution? and (*c*) How do the genotype, phenotype, physiological performance, and fitness interact in influencing one another's future values? To answer these questions, evolutionary physiology examines extant and historical variation and diversity, standing genetic and phenotypic variability in populations, and past and ongoing natural selection in the wild. Also, it manipulates genotypes, phenotypes, and environments of evolving populations in the laboratory and field. Thus, evolutionary physiology represents the infusion of paradigms, techniques, and approaches of evolutionary biology, genetics, and systematics into physiology. The reciprocal infusion of physiological approaches into evolutionary biology and systematics can likewise have great value and is a future goal.

> ...each level [of biological integration] offers unique problems and insights, and .... each level finds its explanations of mechanism in the levels below, and its significance in the levels above.
>
> George A. Bartholomew (7, p. 8)

[1]Dedicated to George Bartholomew on the occasion of his 80th birthday.

0066-4162/00/1120-0315$14.00

INTRODUCTION

Evolutionary physiology represents an explicit fusion of two complementary approaches: evolution and physiology. This field has been the subject of several recent reviews and symposia (18, 25, 53, 72, and papers following 72). Rather than recapitulating those here, we focus on how perspectives and approaches infused from evolutionary biology, genetics, and systematics are changing the scope and nature of physiological studies and how, in turn, physiological perspectives and approaches may contribute to evolutionary biology.

HOW EVOLUTIONARY BIOLOGY, EVOLUTIONARY AND POPULATION GENETICS, AND SYSTEMATICS HAVE INFORMED PHYSIOLOGY

Physiology has often incorporated both ecological and evolutionary perspectives. Since the field's inception, many physiologists have sought to understand how the environment affects function and how function has undergone evolutionary modification. The field's principal focus, however, has been on the mechanisms of function and description of their variation in cells, species, and environments. Beginning in the 1980s, a complementary focus, "evolutionary physiology," achieved prominence. Evolutionary physiology investigates (*a*) the evolutionary mechanisms underlying or constraining diversification of physiological mechanisms and (*b*) the discrete historical patterns of physiological evolution (104). While physiology has always readily borrowed from other disciplines, evolutionary physiology represents a novel importation of theory, paradigms, techniques, and questions from genetics, population biology, evolutionary biology, and systematics. The variables examined, such as metabolic rate, locomotor speed, thermoregulatory performance, and the physiological mechanisms that underlie them, are those that comparative physiology and physiological ecology have measured for decades. What is new is the analytical context of these studies.

Attributing evolutionary thinking in physiology (and vice versa) exclusively to evolutionary physiology is clearly erroneous. Evolutionary biologists such as Sewall Wright, Theodosius Dobzhansky, and Richard Goldschmidt had major research foci on "physiological genetics" (115, 144, 168). Large numbers of Russian physiologists contributed to a field that they had entitled, in 1914, "evolutionary physiology" (3, 155, 186) and that continues to flourish largely independent of the evolutionary physiology reviewed here. Moreover, many classical studies of comparative and environmental physiology (e.g. 7, 10, 167, 181, 182,) interpreted patterns as the outcome of adaptive evolution (138). Also, physiologists have long exploited the results of evolution in choosing the most appropriate species for investigation of physiological problems (116, 123, 208). Still other investigations, either in advance of or independent of evolutionary physiology, included an explicit evolutionary analysis of the physiological impact of specific gene alleles (28, 29, 79, 114, 124, 135–137, 211). Nonetheless, physiology and evolutionary biology often remained isolated from one another in the past. For example, standard textbooks on evolutionary biology (67) had little or no discussion of the evolution of physiological traits (more recent editions of this textbook discuss physiological evolution).

But in the late 1970s and early 1980s, several developments (both conceptual and methodological) elicited a substantially increased infusion of evolutionary thinking into the physiological sciences. Undoubtedly the most influential of these was an unwillingness to assume that all patterns of biological traits result from adaptation (174)—a development that achieved its greatest notoriety in a paper by Gould & Lewontin (82). This polemic not only criticized prevailing standards of evidence of adaptation, but also challenged comparative biologists to scrutinize their assumptions about the operation of evolution. One outcome was physiologists' explicit consideration of evolutionary hypotheses alternative to adaptation (e.g. nonadaptive forces such as drift or constraint) (15, 36), which in turn, necessitated explicit examinations of evolution in physiological studies.

A second, contemporaneous intellectual development was the recognition of the nonindependence of species as analytical units for comparative studies. Because of their phylogenetic relatedness, species share common ancestry and common genes to greater or lesser degrees. Consequently, conclusions based on traditional statistical methodologies became suspect. This recognition came first from behavioral ecologists undertaking comparative studies of behavior (e.g. 35). This awareness began to spread to other fields of comparative biology through the primary stimulus of Felsenstein's 1985 paper (62), which not only clearly described the problem but also provided a robust analytical solution (phylogenetically independent contrasts) for the analysis of comparative data. Indeed, immediately studies in evolutionary physiology began to incorporate a phylogenetic perspective. This new perspective not only influenced how investigators compare species, but also motivated comparative biologists to analyze evolutionary patterns from an ancestor-descendent perspective (103, 104). Thus it was a conceptual—and not merely a statistical—advance.

A third factor was the incorporation of the perspectives and tools of evolutionary genetics and selection studies (both field and laboratory (1–2, 125)). The impact here was fundamental. Evolutionary genetics contributed explicit expectations of the patterns of genetic and phenotypic variation that were necessary and/or sufficient conditions for physiological evolution by natural selection. Physiologists could then sample populations to ascertain whether these conditions were met. Also, whereas most previous evolutionary studies (at least those in physiology and morphology) investigated the results of past evolution, evolutionary genetics and selection studies enabled the monitoring of evolution in contemporary populations (i.e. in real time) and prediction of future evolutionary trajectories. In addition, these approaches permitted the design and execution of rigorous evolutionary experiments in which the experimentalist could manipulate putative selective forces, replicate treatments, and observe outcomes. Thus many assertions and hypotheses concerning physiological evolution, which had previously been only speculative, became falsifiable.

Finally, the techniques of molecular biology and genetic engineering escaped the constraints of standard laboratory model organisms and became broadly applicable to many of the diverse species of interest to evolutionary physiologists (55). While this development is only now yielding information of genomic and proteomic scope, even at the beginning it provided information with either deeper insights or far greater ease than had previously been possible.

The field of evolutionary physiology was greatly influenced by all of these developments and quickly exploited them. The first formative steps in the emergence of the field came from a workshop sponsored by the U.S. National Science Foundation, held in Washington, DC, in 1986, which resulted in an edited volume (53). Pough (163) first used the term "evolutionary physiology" to entitle a review of that volume, Diamond (39, 40) rechristened the field, and Garland & Carter (72) soon codified the term to designate the entire emerging area. In 1994, the U.S. National Science Foundation established a formal Program in Ecological and Evolutionary Physiology. The growth since that time, in both number and breadth of the studies encompassed, has been impressive. We provide only a few examples of relevant studies, and refer readers to successive reviews of this growing field (18, 25, 53, 72).

Major Questions in Evolutionary Physiology

By "physiological evolution," we mean change (or stasis) through time in traits and characters that are typically the subject of physiological studies. These traits may be at diverse levels of biological organization (molecular through organismal, as well as colonial and symbiotic) and may be biochemical, morphological, and/or behavioral as well as strictly physiological (1, 7–10, 50, 72, 74).

Pattern: What Transformations Has Physiology Undergone as Organisms Have Evolved and Diversified and as Their Environments Have Changed? Organisms inhabit a great range of environments, some seemingly inimical to life, and vary extensively in their physiological processes and capacities, morphology, and behavior. Explaining *how* an organism's phenotype enables it to exploit its environment was a central heuristic of pre-evolutionary physiology (10) and remains important. The resultant explanations are typically environment- or taxon-specific and post hoc (e.g. seasonal changes in insulation in arctic mammals, increasing cutaneous Na^+ influx and expression of Na^+ channels by amphibians in extremely hyposmotic media, and facultative anaerobiosis in animals undergoing temporary hypoxia or anoxia). [Vogel (205) likened such research programs to "shooting at a wall and drawing targets around the bullet holes."] Evolutionary physiology, by contrast, more often focuses on the discrete transformations occurring during physiological evolution; for example, how and why did endothermic vertebrates arise from ectothermic ancestors (19, 23, 95, 176, 177)? Alternatively, evolutionary physiology often proceeds from a priori hypotheses or predictions about the distribution of phenotypes in relationship to specific ecological, evolutionary, or genetic regimes and uses taxon-independence (i.e. convergent and/or parallel evolution) and meta-analyses to test the robustness of these predictions. For example, a large body of theory predicts that phenotypic plasticity should be greater in variable than in constant environments (180). This theory would be supported if, in all three of the foregoing examples (arctic mammals, amphibians, and facultative anaerobes), the magnitude of physiological plasticity were correlated with the magnitude of environmental variation. If the theory were supported, then subsidiary predictions would arise concerning the cost of phenotypic plasticity and the magnitude of genetic variability underlying it. Similar issues concern the rate of physiological evolution and how this rate differs in stressful vs benign environments and central vs peripheral populations of a species (99), physiological niche breadth in specialized vs generalized species (78), and the closeness of the match between organismal phenotypes and environment.

Evolutionary physiology and other aspects of physiology are clearly not separate endeavors, but may examine identical phenomena with similar techniques from their different starting points. Mechanistic physiology often makes predictions from principles of physics and/or engineering regarding distributions of phenotypes. These might concern, for example, the nature and diversity of respiratory gas and ion exchangers in air vs water (38), the general features of the design of gas exchangers (162), morphologies of organisms living in high vs low flow regimes (205), and the maximum body temperatures of animals (91, 96, 195). Evolutionary and other physiological approaches perhaps most closely coalesce in studies of physiological optimality. Evolutionary optimality models have a long heritage (184), but within the past 20 years, physiologists have undertaken explicit examinations of whether physiological supply and demand are in fact closely matched ("symmorphosis") or whether overdesign and safety margins are commonplace (39, 40, 219). This issue is still contentious (43, 219).

Process: How Important Have the Known Evolutionary Processes (Natural Selection, Sexual Selection, Drift, Constraint, Genetic Coupling/Hitchhiking, and Others) Been in Engendering or Limiting Physiological Evolution? The Modern Evolutionary Synthesis has long recognized the multiplicity of processes that result in or constrain evolution. Nonetheless, the footprint of one of these (adaptation as the outcome of natural selection) upon physiological diversity has been so manifest that many physiological investigators have understandably focused on it to the near or total exclusion of the others. Stepping back from this focus, evolutionary physiology ideally asks, How much of physiological diversity (or its lack) is due to each of the known evolutionary processes (15)? At a more basic level, can we rigorously deduce the evolutionary processes that led to and/or maintain the extant array of physiological phenotypes? One approach to these questions has been to take from evolutionary theory the conditions that are necessary and/or sufficient for each evolutionary process to occur (effective population size, genotypic and phenotypic variability, heritability, differential survival/reproduction, and so on) and to survey these conditions in natural or experimental populations with reference to physiological traits. The rigor of this approach will only increase with time as the genetic basis of complex physiological traits becomes better understood with the advent of functional genomics and proteomics.

Another approach is to survey the rate of physiological diversification in taxa separated naturally or experimentally in different environments for known lengths of time. Both approaches, however, are tempered with the realization that the outcomes of evolution are heavily contingent on the genetic and demographic conditions prevailing at its outset.

As explained below, laboratory and experimental evolution studies (77, 173, 175) are especially promising in that the experimentalist can manipulate these starting conditions, run multiple replicated evolutionary trials, and determine the probability of specific evolutionary outcomes directly. Moreover, modern techniques of genetic engineering allow this manipulation to occur at the level of the single gene or even nucleotide, with all other factors controlled. Obviously such work is still not feasible for every species of interest to physiologists. Nonetheless, it permits evolutionary physiologists to address still more significant questions.

First, within the range of feasible evolutionary outcomes, does physiological evolution generally follow only one, a few, or perhaps a multitude of these? If the evolution of a biological lineage were to occur again, would it result in the same, similar, or entirely different results (81)? In other words, are evolutionary trajectories predictable, given similar starting conditions?

Second, what is the importance of neutral evolution, specifically the neutral fixation of traits or genes? Such work collectively may reveal the relative importance of history, adaptation, and chance in the formation of physiological diversity (36, 199–202).

Components: How Do the Genotype, Phenotype, Physiological Performance, and Fitness Interact in Influencing One Another's Future Values? Physiologists have long studied the detailed chain of events that ensue between the reception of a physiological stimulus and the manifestation of its corresponding physiological response, the molecular and cellular components of these events, and the impact of each component on each subsequent element of the chain. The elucidations of signal transduction, homeostatic mechanisms, and neurotransmission, for example, are only a few of the success stories of mechanistic physiology. By contrast, we know much less about the detailed events that ensue between the reception of an ecological or evolutionary stimulus (e.g. stress) by a natural population and the manifestation of the corresponding response (selection, response to selection, extinction, and so on) in terms of physiological traits. Evolutionary physiology strives to discover these connections. The general paradigm is that genes encode the phenotype, the phenotype determines the performance of organisms in natural environments in response to ecological or evolutionary stimuli, the performance determines the evolutionary fitness of alternative genotypes, and the fitness determines the frequency of genotypes in the next generation, in recursive fashion (1, 7, 61, 72, 74, 102). Determining the details of each step is an ongoing challenge for evolutionary physiologists (61). For genotype to phenotype, for example, what is the genetic basis of complex physiological traits, what is the importance of epigenetic and nongenetic factors in determining physiological phenotypes, and how is the genotypic specification of the phenotype manifested during ontogeny? For phenotype to performance, what exactly is the impact of phenotypic variation on the performance of unrestrained organisms in nature? Mechanistic physiologists are routinely successful in explaining the impact of phenotypic variation on proxies of performance in the laboratory; are these explanations extensible to nature (16)? Similar questions apply to the other links as well. That evolutionary physiologists have one foot in mechanistic physiology and another in evolutionary biology prospectively poises them to make major contributions in this area, where multidisciplinary work is clearly needed.

Analytical Approaches in Evolutionary Physiology

Evolutionary physiology currently exploits two major approaches. The first is to analyze the outcome of "natural experiments"; that is, the extant and paleontological genotypes and phenotypes of evolved organisms, with environment and/or phylogeny (i.e. the genes and traits present at the start of evolution) as variables that have differed in the past. This approach, then, is one of historically based comparison and correlation, and usually best suits (by necessity) studies of physiological evolution among species or higher-level taxa. The second approach is to manipulate genotype, phenotype, and/or environment directly and to observe subsequent evolutionary outcomes.

Phylogenetically Based Comparisons: Examining Extant and Historical Variation and Diversity to Test Hypotheses The statistical analysis of physiological evolution within an explicit phylogenetic framework began only in the mid-1980s but has already become a central paradigm in evolutionary physiology and other comparative fields (76, 94, 104, 133, 141, 206). Phylogenetically based comparative studies analyze physiological and/or ecological variation and covariation against an independently derived phylogeny of the taxa involved. The first of two primary motivations for developing this approach was the realization that the best way to choose species for comparison was with respect to phylogenetic relatedness (104, 206). Many early studies compared very distantly related species (e.g. a hibernating marsupial and a white rat), undoubtedly to increase signal-to-noise ratio, but the results of such studies were inherently ambiguous in the sense that one was comparing apples with oranges (103). The second motivation was a growing awareness that species data are nonindependent, such that conclusions based on standard statistical models (e.g. regression) were at least suspect and potentially misleading (35, 62). Because of these realizations, comparative biologists and systematists have developed and are developing a variety of new analytical procedures (26, 62, 73, 76, 94, 141) that are leading both to more robust answers to pre-existing research questions and to entirely new questions for physiologists.

In evolutionary physiology, a comparative approach can reveal whether a particular character state (e.g. stenothermy or eurythermy) is ancestral or derived (103, 104), the most likely ancestral condition of a discrete or quantitative trait (75, 104), rates of evolutionary change (103), whether prior evolution of a trait has been necessary and/or sufficient for an ecological or evolutionary outcome, and whether evolution of one trait has evolved in advance of, simultaneously with, or after another (see 26, 62, 76, 84, 94, 133, for a discussion of methods and interpretations). Analyses may concern populations, species, genera, or even higher levels of biological organization or may encompass a combination of organizational levels. Any such analysis requires a minimum of three taxa, in part because of the necessity of incorporating a more distantly related "outgroup" into any comparison (69, 103). Examples of the use of phylogenetic approaches to study the evolution of physiological characters are now very diverse and include thermoregulatory patterns in fish (23, 24), evolution of locomotion in lizards (11, 30), locomotor performance in lizards (132), salinity tolerance in mosquitoes (85), diving physiology of pinnipeds (98, 153), metabolic rate in amphibians (207), dietary modulation in omnivorous birds (140), development in Antarctic birds (48), expression of glycolytic enzymes in fish (161), plasma osmotic concentration in amniotes (75), anaerobic metabolic end products in chordates (178), and nocturnality in geckos (4). These have revealed novel insights. For example, Mottishaw et al (153) showed that "diving bradycardia," long assumed to have evolved in mammals to facilitate diving, likely arose long before the evolution of diving habits. Block et al (24) showed that endothermy has evolved multiple times in fishes (as opposed to evolving only once in a common ancestor of endothermic fishes).

Rigorous phylogenetic approaches to physiological comparisons are not without drawbacks. They require physiological data on multiple species, which can be a formidable barrier for sophisticated mechanistic studies (138), as well as nonphysiological data (i.e. a phylogenetic topology with branch lengths and large numbers of taxa) that may be difficult or even impossible to obtain (e.g. too few taxa may exist for adequate statistical power). In addition, the robustness of any phylogenetic interpretation depends on the hypothesized phylogenetic relationships as well as on the model of evolution underlying the formal statistics (62). Moreover, phylogenetic approaches can establish only correlation and not causation (69, 104, 129); unfortunately, historical patterns can seldom be tested by direct and replicated experimental manipulation (104). A different kind of drawback associated with phylogenetic approaches has been their unfortunate tendency to impede, if not to stifle, comparative studies that employ traditional, nonphylogenetic approaches ["phylogenetic correctness" (73, p. 279)]. We advocate both greater tolerance when investigators eschew phylogenetic approaches with good reason, and explicit discussion of the rationale for "phylogenetic incorrectness" when it is warranted.

Despite difficulties of implementation, the incorporation of phylogenetic approaches appears positive for evolutionary physiology. Such approaches have improved the choice of species, even when evolutionary considerations are not of primary concern. Moreover, phylogentic considerations enhance the reliability of statistical inference, the kinds of evolutionary questions asked of physiological data, and the evolutionary relevance and robustness of comparative physiological studies.

Although comparative methods often infer ancestral physiological states from those of extant organisms, paleobiological studies may infer the physiological states of long-dead organisms from fossil anatomy (176, 177). Sometimes the essential features of such analyses (e.g. shared characters and parsimony) are implicit rather than explicit, and a phylogeny with appropriate character mapping may or may not be available. The form of the argument, however, is essentially parsimonious and phylogenetic (19). An excellent example of this approach is Hillenius' study of endothermy in the mammalian lineage (97), in which evidence of nasal turbinates in fossil skulls suggests endothermy in therocephalian therapsids. This result indicates that endothermy probably evolved before the emergence of mammals as a group and was therefore an inheritance rather than a novel evolutionary development in Mammalia.

Dormant stages of organisms can sometimes be resuscitated so that physiological states of recent "ancestors" can be determined directly. Many organisms in nature have dispausing or dormant eggs (87, 88) or seeds (130, 146) that persist in the environment for long periods (90). When resurrected, such time travelers can be compared with contemporary individuals in a common garden. Thus, ancestral and derived stocks from nature are compared directly, much the same way as with laboratory stocks with certain species [e.g. *Escherichia coli, Caenorhabditis elegans, and Drosophila melanogaster* (18)]. A fascinating example comes from a time-series analysis of *Daphnia* spp. from a lake that experienced eutrophication (and associated increases in cyanobacteria) in the 1960s and 1970s. After hatching dormant eggs of *Daphnia* spp. from sequential time periods, Hairston et al (89) found that *Daphnia* spp. rapidly evolved increased resistance to cyanobacteria in their diets during eutrophication. Future physiological studies can potentially explore the evolved mechanisms underlying such increased resistance.

Standing Genetic and Phenotypic Variability in Populations Evolutionary biology has established that the modes, rates, and outcomes of physiological evolution will depend critically on the pre-existing variation within the evolving population (or higher taxon), its heritability, and its relation to fitness. Because comparative studies are not sufficient to address these issues (129), evolutionary physiologists have increasingly attempted to characterize variation, heritability, and fitness consequences directly. These studies have used two types of approach. In the physiology-to-genetics approach (2), evolutionary physiologists assume that physiological traits of interest have a genetic basis and seek to establish the genetic properties of this basis without ascertaining the identities and natures of the actual underlying genes. At the simplest level, an evolutionary physiologist might ask whether sprint speed varies or is invariant within a lizard population, with the outcome having implications for the evolutionary malleability of sprint speed in the population. This approach differs from that of typical physiology, in which an investigator might regard exceptionally fast or slow individuals as atypical and therefore exclude them from analysis (14). At a more complex level, evolutionary physiologists have applied the techniques and insights of quantitative genetics (see 49, 137) to estimate the heritability of diverse functional characters (both organismal and suborgansimal), characterize phenotypic vs genotypic variation in traits, establish the relationship between traits and fitness, and quantify evolution. Locomotor performance (e.g. 42, 197, 203) and its mechanistic basis (e.g. 70, 71) have received particular attention.

A second approach, that of gene to physiology (2), examines the performance and fitness implications of discrete genes (or the products they encode) on organismal characters (114, 145, 212, 213): for example, lactate dehydrogenase on locomotor performance (164–166, 183) and temperature adaptation (63, 101, 196); hemoglobin on oxygen transport capacity (194); aminopeptidase on osmoregulatory ability (113, 114); alcohol dehydrogenase on ethanol tolerance (64–66); and phosphoglucose isomerase (31, 215–218), glycerol-3-phosphate dehydrogenase (6, 37, 126, 127), juvenile hormone esterase (86, 222–226), and troponin expression (139) on flight capacity.

Perhaps the initial motivation of these studies was from the discovery of unexpectedly large amounts of genetic variation in natural populations and attempts to test subsequent theory that most such variation was selectively neutral (114, 212, 213). This motivation continues as a debate on how allelic variation in genes of large effect can persist in populations without selection eliminating them (212, 213) and has its counterpart in the functional genomics of human disease. A repeated finding of gene-to-physiology studies, that the genes under investigation are often nonneutral, continues to provoke much interest. An additional goal, coincident with the emergence of evolutionary physiology, has been to rigorously explore the recursive relationships of genes to traits to organismal performance to fitness to genes against the background that these relationships can be nonstraightforward and nonobvious (61).

From its inception, a criticism of the gene-to-physiology approach has been that it is not sufficient to explain variation in complex physiological traits, which must be the product of numerous interacting genes. One response has been the exploitation of metabolic network theory to explain how single-gene variation can affect complex metabolic pathways and entire physiologies (212). Another is that advances in developmental biology, cell biology, and molecular physiology of model systems are revealing precisely how single gene changes can be manifested in whole-organism variation and how such variations have evolved (204).

The physiology-to-gene and gene-to-physiology approaches obviously converge with one another and with scientific progress in general. Already, the increasing technical sophistication of DNA arrays permits simultaneous screening of all genes of an organism or tissue for changes in expression in response to physiological change (33, 142, 198). Another development is increasing ease of precise genetic mapping, facilitating the description of quantitative trait loci. Quantitative trait loci, in turn, can establish or reject the polygeny of traits, permit the formal genetic analysis of quantitative traits, and facilitate the direct sequencing of the genes at the quantitative trait loci. On a more theoretical level, the two approaches bear on global genetic issues concerning organismal performance, such as pleiotropic effects on the evolution of physiological characters (41, 83, 185), the role of overall heterozygosity in influencing performance and fitness (112, 150), the effects of genetic correlation on rates of evolutionary change (2), and the relationship between numbers of deleterious mutations and overall viability and fitness (46, 154).

Past and Ongoing Natural Selection in the Wild Evolutionary biologists have developed a variety of methods to study the presence, intensity, and directionality of natural selection on character traits in wild populations (reviewed in 47, 151). Such studies of selection on physiological traits would be enormously valuable to our understanding of their evolution and current ecological importance. To date, however, relatively few such studies have been attempted (47), partly because of the difficulty of measuring physiological variables on very large numbers of animals. Available studies have concentrated on locomotor capacities (e.g. 16, 106, 148, 149, 209) and have sometimes demonstrated, through differential survival, that traits such as maximal speed and endurance contribute to fitness in natural populations. Such work forms an important bridge between the many laboratory studies on activity capacity and its mechanistic bases and its actual ecological and evolutionary significance. Future field selection studies may involve manipulation of such putative selective factors as predator density to test evolutionary hypotheses experimentally.

Although the above studies document patterns of selection on traits, they must be repeated over time to document actual evolutionary responses to selection. An alternative way to document short-term evolutionary responses involves monitoring shifts in species recently introduced into novel environments (159). An example is *Drosophila subobscura*, which was introduced from the Old World into North and South America in the late 1970s and spread rapidly on both continents (5). In only one to two decades, the North American populations have evolved a latitudinal wing-length cline, which parallels that in native Old World populations (105).

Experimental Approaches: Manipulating Variation and Diversity to Test Hypotheses Although natural experiments most clearly reflect the actual past and ongoing processes of physiological evolution as they occur in the wild, they have limitations. They can be poorly controlled, their sampling of organisms and environmental factors can be biased, they are often nonreplicated and unrepeatable, and the signal-to-noise ratio of the trait under investigation can be insubstantial. Most importantly, genetic, epigenetic, and phenotypic linkages among traits mean that seldom will a gene, trait, or suite of genes/traits of interest vary in isolation without covariation of interacting elements, thereby confounding the interpretation of natural experiments. Therefore, as in most other areas of the life sciences, evolutionary physiology includes a strong component of intentional manipulation or engineering of genes or traits of interest, but with outcomes typically characterized in relation to performance and/or fitness. A long heritage of experimental physiology has provided a wealth of surgical and pharmacological techniques to manipulate the traits themselves as well as diverse means of manipulating specific internal or external environmental variables. Joining these are laboratory and field techniques from experimental ecology for manipulating the number and variety of interacting organisms and their environments, experimental evolution approaches, and genetic engineering of allelic series, knock-outs, knock-ins, complementation, rescue, etc. Most of these techniques themselves have the drawback of manifesting processes that seldom, if ever, occur in nature and thus, by themselves, can reveal little about the likelihood or potential for physiological evolution in nature. For this reason, the complementary analysis of natural and deliberate experiments may yield the greatest insights (59). Thus, for example, whereas the analysis of *Ldh-B* genotype in natural populations of the fish *Fundulus* can implicate *Ldh-B* genotype as a likely component of fitness, it can neither establish that *Ldh-B* genotype is sufficient for variation in fitness nor exclude the influence of linked genes (165). Such demonstrations require manipulations such as the replacement of a genotype's allozyme with an alternative allozyme via microinjection (165). Whether evolution is likely to achieve comparable manipulations, however, can emerge only from study of the natural *Fundulus* populations.

Phenotypic Engineering A powerful approach to studying the mechanistic and adaptive significance of phenotypic characters is to manipulate them directly and subsequently study the performance and/or fitness consequences in the laboratory or the field (190). Such approaches have been termed "allometric engineering" when involving manipulations of body size (192) or as "phenotypic engineering" for more general manipulations (107, 108). An example of the latter involves manipulating butterfly wing color (by altering developmental temperature or by painting) and then monitoring field-released individuals for heat balance, mating success, and survival (e.g. 109–111). Other investigators have engineered intestinal transport capacity (27, 92), milk production capacity (93), hormone status (107, 108), and total body size (187, 191, 221). Performance consequences of such changes can be examined either in staged encounters in the laboratory or in animals released into natural populations. These latter studies then become a type of natural selection study, as discussed below.

Phenotypic engineering permits direct experimental tests of the significance of a character and is thus a valuable tool to expand upon comparative studies. It can not only expand the magnitude of variation in traits beyond that occurring in nature (188, 193), but also verify causal, mechanistic links between traits [e.g. between body size and performance (192)].

A complementary approach involves manipulating the environment rather than the phenotype. The classical methods involve reciprocal transplants (34, 157) and common gardens (189). The latter are widely used to factor out environmental effects in studies of geographic or interspecies variation (68, 69).

Genetic Engineering An evolutionary physiologist may choose to manipulate the gene(s) encoding a trait rather than manipulating the trait directly (51, 169). Such manipulations have long been possible, if not simple, in genetically tractable organisms such as yeast, *Drosophila*, and many bacteria. A few examples for bacteria concern the consequences of excess tryptophan synthesis for growth (44), the effects of lactose permease and beta-galactosidase expression on metabolic flux (45), and interactions between structural and regulatory genes that control expression of an efflux protein and their effects on growth in the presence and absence of an antibiotic (128, 156). Advances in molecular biology already (or will soon) make such manipulations feasible at any level down to the individual nucleotide, and in an expanding diversity of experimental species.

One example concerns the heat-shock genes, whose expression was correlated with inducible stress tolerance and thus were implicated as a mechanism of stress tolerance (52, 57). In yeast, this implication was first confirmed for the single-copy gene *HSP104* when deleting the gene abolished a significant component of inducible thermotolerance, and reintroducing the gene restored inducible thermotolerance to control levels (179). Moreover, site-directed mutagenesis of a single nucleotide in a critical region of this gene was sufficient to abolish inducible thermotolerance, whereas control mutagenesis had no effect (158). For other heat-shock genes and in more complex multicellular eukaryotes (where multiple copies of some heat-shock genes have evolved), more complex techniques are necessary (e.g. 220). Increasing the haploid copy number of the *hsp70* gene from 5 to 11 in *Drosophila*, for example, dramatically increases the resistance of whole larvae and pupae to natural thermal stress (56, 171, 172), and gut-specific expression of the *hsp70* gene off a heterologous promoter protects the gut against heat-induced damage during feeding (60). Many additional transgenic manipulations of heat-shock genes are now available and confirm the suspected consequences of these genes for fitness (52, 57). Similar genetic manipulations will undoubtedly be one of the most exciting and productive areas in future studies of evolutionary physiology.

Selection Studies Selection experiments on populations of organisms in the laboratory, long a mainstay of geneticists and evolutionary biologists, have also been incorporated into evolutionary physiology (18, 77). These permit the direct observation of evolutionary change resulting from an alteration in the selective environment and allow physiologists to experiment on physiological evolution itself. Laboratory experimentation facilitates control of the environment and selective factors, maintenance of nonselected populations, and replication of experimental groups, permitting a rigorous statistical evaluation of the evolutionary response.

Experimental selection studies follow three designs: natural selection in the laboratory, artificial truncation selection, and laboratory culling (173). The first manipulates an environmental variable (e.g. temperature or water or nutrient availability) and monitors the consequent changes in replicated experimental populations for many generations. In laboratory natural selection, the experimenter does not directly choose which organisms possessing which characters will be permitted to breed: those that are fittest in the new experimental environment will produce more offspring, irrespective of which characters result in higher fitness. In contrast, artificial truncation selection permits only organisms possessing certain traits to breed. This type of selection is familiar from plant and animal breeding. It requires the a priori establishment of the traits to be selected and the screening of individual organisms prior to reproduction. Laboratory culling creates a selective environment that permits only a small portion of each population to survive to reproduce. Choice of the type of selection experiment depends on the principal experimental goal. Testing of hypotheses concerning environmental adaptation would probably employ either natural selection in the laboratory or laboratory culling, while the production of organisms with certain combinations of physiological traits might be done through artificial truncation selection.

Perhaps the greatest utility of selection experiments in evolutionary physiology is their ability to test general predictions concerning physiological evolution, specifically in regard to patterns and consequences of evolutionary adaptation (12, 13, 18). Many formal and informal models of environmental adaptation begin with assumptions concerning evolutionary constraints and patterns. For example, in regard to adaptation to the thermal environment, most models (e.g. 78, 131) assume that adaptation to one thermal environment requires loss of fitness in other environments (trade-off) and consequent changes in the range of temperatures tolerated (niche shift). The ability to do direct experiments changes these assumptions into testable hypotheses, subject to falsification. In regard to these particular assumptions, experimental studies on evolutionary adaptation of bacterial populations to different temperatures in general fail to support them (17, 152). While the expected pattern of fitness trade-off and niche shift occured in one thermal environment, it was completely absent in four others. Some studies of *Drosophila*, however, are consistent with these assumptions (32, 160). Such results question facile assertions and assumptions concerning the course that evolution will or must take and require revision of evolutionary models that incorporate such assumptions.

Conclusion: On Finding the Right Organism for Study Although we present the foregoing analytical approaches individually, they clearly have the greatest power when used in complementary fashion (59). Each approach is best suited to reveal certain aspects of physiological evolution but may ignore or obscure others. To return to the example of heat-shock proteins in *Drosophila*, genetic engineering can unambiguously establish the phenotypic and fitness consequences of variation in Hsp70 protein expression (56, 58, 60, 117, 119–121, 171–172), but it can establish neither the likelihood that such variation will evolve, persist, or even occur in natural populations nor the ecological relevance of such variation (59). That goal requires direct studies of variation (20, 118, 122) and its ecology (54, 171, 172) in natural populations, but these, in turn, cannot unambiguously establish the physiological phenotypes of the genes under study. Similarly, comparative studies of past physiological evolution (e.g. the evolution of endothermy or diving bradycardia; see above), descriptive studies of contemporary physiological or morphological evolution (e.g. 105, 134), and experimental or laboratory evolution (see above) each provides special insights. The latter suite of approaches, however, may be most powerful when each approach is applied to a common species or population of interest and/or when new techniques are deployed. For the moment, the joint requirements of these approaches and techniques may be so restrictive as to exclude problems (e.g. the evolution of endothermy or diving bradycardia) and species (e.g. endothermic tuna) of traditional interest to physiologists. Thus, an increasingly common practice in evolutionary physiology has been to study nontraditional (at least for physiologists) models such as *Drosophila*, *E. coli*, and *C. elegans* either alongside or in place of the more traditional exotic subjects of ecological and comparative physiologists. "Choosing the right organism" for study is a hallowed tradition in physiology (116, 123, 208) and is yielding truly surprising outcomes. For example, genomic screens of *E. coli* undergoing experimental evolution at high temperatures suggest that the same genes are evolving in independent evolutionary trials (170). In pursuing such models, approaches, and techniques and in searching for insights from allied fields, evolutionary physiologists are continuing a long tradition of multidisciplinary physiology but in new directions.

HOW EVOLUTIONARY PHYSIOLOGY CAN INFORM EVOLUTIONARY BIOLOGY, EVOLUTIONARY AND POPULATION GENETICS, AND SYSTEMATICS

Much current research in evolutionary biology proceeds without explicit or detailed reference to the mechanisms encoded by the genes. Amechanistic (214) evolutionary biology has several bases, some of which are pragmatic: Statistical analyses of the genotype alone can yield considerable insights into evolution, as can the amechanistic scoring of traits to phylogenetic relationships. Also, incorporating functional analyses into evolutionary studies can be both logistically and intellectually challenging. But other bases of amechanistic evolutionary biology are ideological, stemming from a belief that detailed understanding of the phenotype is irrelevant or insufficient for understanding of evolution, or it can contribute little. Thus Mayr (143, p. 115) has written: "...the mechanistic approach, although quite indispensable in the study of proximate causations, is usually quite meaningless in the analysis of evolutionary causations," and Gould (80, p. 101) has written "...the flowering of [functional studies of evolution] has yielded a panoply of elegant individual examples and few principles beyond the unenlightening conclusion that animals work well." Watt (214) has analyzed these ideological bases in detail. Clearly, amechanistic approaches will continue to reward evolutionary biologists for years to come. We disagree, however, with the premise that functional and/or integrative studies have little to offer evolutionary biology. Our purpose here (as also stated in 72, 104, 138, 214) is to emphasize the value that mechanistic approaches can add to the research programs of evolutionary biologists.

The inclusion of mechanistic perspectives is becoming increasingly important and, indeed, necessary in the following areas of evolutionary biology:

1. Understanding the implications of genetic diversity. As evolutionary biologists increasingly examine the actual nucleotide sequences of genes under study and their variation, their principal challenge will be to explain the origin and consequences of such variation. Foreseeably, demonstrating that a given nucleotide is/isn't under selection or comparing a sequence to a null model may no longer represent an acceptable level of proof. Instead, meeting this challenge may require tests of hypotheses of the functional significance of variants, which in turn will require detailed understanding of the function these genes' products perform in intact organisms in natural environments (i.e. an evolutionary and ecological functional genomics).
2. Practical implications of evolutionary theory. While evolutionary biology has historically been the most curiosity-driven of the biological sciences, its bearing is increasing on applied issues of great significance to the national research agenda (147). These include the origin and spread of disease, conservation of biodiversity, global climate change, impact of genetically modified and exotic organisms, and evolutionary paradigms in engineering of drugs, biomaterials, and organisms. The devising of meaningful solutions to such problems will require detailed understanding of mechanisms underlying organismal function (e.g. 22).
3. Environmental influences on evolutionary diversification. Much evolutionary research and interpretation occur in an environmental context. Environmental stress, for example, is a recurrent motif in evolutionary studies (21, 99, 100). Theoretical models and laboratory studies of the role of the environment in evolution, however, are now far in advance of rigorous characterizations of natural environments and their impact on organisms. These disparate aspects must be brought into register.

Clearly, interaction with evolutionary (and nonevolutionary) physiologists could provide the mechanistic expertise for which the foregoing three examples call and could therefore substantially enhance the research programs of evolutionary biologists. In the past, the different goals and research foci of evolutionary biologists and physiologists have tended to isolate these two communities. The growth of evolutionary physiology, in which these communities focus on common problems and speak a common language, therefore represents a novel opportunity for evolutionary biologists to partner with mechanistic biologists and for physiologists to reciprocate to further the influx of evolutionary thought into their discipline.

ACKNOWLEDGMENTS

We gratefully acknowledge the National Science Foundation for its support of our individual research programs (MEF, IBN97-23298; AFB, IBN99-05980; RBH, IBN95-14203 and DEB96-29822) and for its ongoing support of the field of evolutionary physiology since its inception. RBH was also supported by a fellowship from the JS Guggenheim Foundation. We thank T Garland Jr. for helpful comments on the manuscript and acknowledge an intellectual debt to numerous mentors, colleagues, and students for helping our own thinking to evolve.

LITERATURE CITED

1. Arnold SJ. 1983. Morphology, performance, and fitness. *Am. Zool.* 23:347–61
2. Arnold SJ. 1987. Genetic correlation and the evolution of physiology. See Ref. 53, pp. 189–212
3. Arshavskii IA. 1985. Foundation of the principles and tasks of evolutionary physiology in light of the data of comparative-ontogenetic research. *J. Evol. Biochem. Physiol.* 21:105–10
4. Autumn K, Jindrich D, DeNardo D, Mueller R. 1999. Locomotor performance at low temperature and the evolution of nocturnality in geckos. *Evolution* 53:580–99
5. Ayala FJ, Serra L, Provosti A. 1989. A grand experiment in evolution: the *Drosophila subobscura* colonization of the Americas. *Genome* 31:246–55
6. Barnes PT, Laurie-Ahlberg CC. 1986. Genetic variability of flight metabolism in *Drosophila melanogaster*. 3. Effects of GPDH isoenzymes and environmental temperature on power output. *Genetics* 112:267–94
7. Bartholomew GA. 1964. The roles of physiology and behaviour in the maintenance of homeostasis in the desert environment. In *Homeostasis and Feedback Mechanisms*, ed. GM Hughes, pp. 7–29. Cambridge, UK: Cambridge Univ. Press
8. Bartholomew GA. 1982. Scientific innovation and creativity: a zoologist's point of view. *Am. Zool.* 22:227–35
9. Bartholomew GA. 1986. The role of natural history in contemporary biology. *BioScience* 36:324–29
10. Bartholomew GA. 1987. Interspecific comparison as a tool for ecological physiologists. See Ref. 53, pp. 11–35
11. Bauwens D, Garland T Jr, Castilla AM, Vandamme R. 1995. Evolution of sprint speed in lacertid lizards: morphological, physiological, and behavioral covariation. *Evolution* 49:848–63
12. Bell G. 1997. *The Basics of Selection*. New York: Chapman & Hall
13. Bell G. 1997. *Selection: The Mechanism of Evolution*. New York: Chapman & Hall

14. Bennett AF. 1987. Interindividual variability: an underutilized resource. See Ref. 53, pp. 147–69
15. Bennett AF. 1997. Adaptation and the evolution of physiological characters. In *Handbook of Physiology; Section 13*, ed. WH Dantzler, pp. 3–16. New York: Oxford Univ. Press
16. Bennett AF, Huey RB. 1990. Studying the evolution of physiological performance. *Oxford Surv. Evol. Biol.* 7:251–84
17. Bennett AF, Lenski RE. 1993. Evolutionary adaptation to temperature. 2. Thermal niches of experimental lines of *Escherichia coli*. *Evolution* 47:1–12
18. Bennett AF, Lenski RE. 1999. Experimental evolution and its role in evolutionary physiology. *Am. Zool.* 39:346–62
19. Bennett AF, Ruben JA. 1986. The metabolic and thermoregulatory status of therapsids. In *The Ecology and Biology of Mammal-Like Reptiles*, ed. N Hotton, PD MacLean, JJ Roth, EC Roth, pp. 207–18. Washington, DC: Smithsonian Inst. Press
20. Bettencourt BR, Feder ME, Cavicchi S. 1999. Experimental evolution of Hsp70 expression and thermotolerance in *Drosophila melanogaster*. *Evolution* 53:484–92
21. Bijlsma R, Loeschcke V, eds. 1997. *Environmental Stress, Adaptation, and Evolution*. Basel: Birkhäuser Verlag
22. Block BA, Dewar H, Farwell C, Prince ED. 1998. A new satellite technology for tracking the movements of Atlantic bluefin tuna. *Proc. Natl. Acad. Sci. USA* 95:9384–89
23. Block BA, Finnerty JR. 1994. Endothermy in fishes: a phylogenetic analysis of constraints, predispositions, and selection pressures. *Environ. Biol. Fishes* 40:283–302
24. Block BA, Finnerty JR, Stewart AFR, Kidd J. 1993. Evolution of endothermy in fish: mapping physiological traits on a molecular phylogeny. *Science* 260:210–14
25. Bradley TJ, Zamer WE. 1999. Introduction to the symposium: What is evolutionary physiology? *Am. Zool.* 39:321–22
26. Brooks DR, McLennan DA. 1991. *Phylogeny, Ecology, and Behavior: A Research Program in Comparative Biology*. Chicago: Univ. Chicago Press
27. Buchmiller TL, Shaw KS, Chopourian HL, Lloyd KCK, Gregg JP, et al. 1993. Effect of transamniotic administration of epidermal growth factor on fetal rabbit small-intestinal nutrient transport and disaccharidase development. *J. Ped. Surg.* 28:1239–44
28. Bult A, Lynch CB. 1996. Multiple selection responses in house mice bidirectionally selected for thermoregulatory nest-building behavior: crosses of replicate lines. *Behav. Genet.* 26:439–46
29. Bult A, Lynch CB. 1997. Nesting and fitness: lifetime reproductive success in house mice bidirectionally selected for thermoregulatory nest-building behavior. *Behav. Genet.* 27:231–40
30. Carrascal LM, Moreon E, Mozetich I. 2000. Locomotion mode as a link between leg morphology and habitat preferences. A phylogenetic and ecologomorphological study with Passeroidea (Aves: Passeriformes). *Evol. Ecol.* In press
31. Carter PA, Watt WB. 1988. Adaptation at specific loci. 5. Metabolically adjacent enzyme loci may have very distinct experiences of selective pressures. *Genetics* 119:913–24
32. Cavicchi S, Guerra V, Natali V, Pezzoli C, Giorgi G. 1989. Temperature-related divergence in experimental populations of *Drosophila melanogaster*. II. Correlation between fitness and body dimensions. *J. Evol. Biol.* 2:235–51
33. Chu S, DeRisi J, Eisen M, Mulholland J, Botstein D, et al. 1998. The transcriptional program of sporulation in budding yeast. *Science* 282:699–705
34. Clausen J, Keck DD, Heisey WM. 1948. Experimental studies on the nature of species. III. Environmental responses of

climatic races of *Achillea. Carnegie Inst. Wash. Publ.* 581:1–129
35. Clutton-Brock TH, Harvey PH. 1977. Primate ecology and social organization. *J. Zool. London* 183:1–39
36. Cohan FM, Hoffmann AA. 1986. Genetic divergence under uniform selection. 2. Different responses to selection for knockdown resistance to ethanol among *Drosophila melanogaster* populations and their replicate lines. *Genetics* 114:145–64
37. Connors EM, Curtsinger JW. 1986. Relationship between alpha-glycerophosphate dehydrogenase activity and metabolic rate during flight in *Drosophila melanogaster. Biochem. Genet.* 24:245–57
38. Dejours P. 1988. *Respiration in Water and Air: Adaptations-Regulation-Evolution.* Amsterdam: Elsevier
39. Diamond JM. 1992. Evolutionary physiology: the red flag of optimality. *Nature* 355:204–6
40. Diamond JM. 1993. Evolutionary physiology: quantitative design of life. *Nature* 366:405–6
41. Djawdan M, Sugiyama TT, Schlaeger LK, Bradley TJ, Rose MR. 1996. Metabolic aspects of the trade-off between fecundity and longevity in *Drosophila melanogaster. Physiol. Zool.* 69:1176–95
42. Dohm MR, Hayes JP, Garland T Jr. 1996. Quantitative genetics of sprint running speed and swimming endurance in laboratory house mice (*Mus domesticus*). *Evolution* 50:1688–701
43. Dudley R, Gans C. 1991. A critique of symmorphosis and optimality models in physiology. *Physiol. Zool.* 64:627–37
44. Dykhuizen D. 1978. Selection for tryptophan auxotrophs of *Escherichia coli* in glucose-limited chemostats as a test of the energy conservation hypothesis of evolution. *Evolution* 32:125–50
45. Dykhuizen DE, Dean AM. 1990. Enzyme activity and fitness: evolution in solution. *Trends Ecol. Evol.* 5:257–62
46. Elena SF, Lenski RE. 1997. Test of synergistic interactions among deleterious mutations in bacteria. *Nature* 390:395–98
47. Endler JA. 1986. *Natural Selection in the Wild.* Princeton, NJ: Princeton Univ. Press
48. Eppley ZA. 1996. Charadriiform birds in Antarctica: behavioral, morphological, and physiological adjustments conserving reproductive success. *Physiol. Zool.* 69:1502–54
49. Falconer DS, Mackay TFC. 1996. *Introduction to Quantitative Genetics.* Burnt Mill, Harlow, Essex, UK: Longman. 4th ed.
50. Feder ME. 1987. The analysis of physiological diversity: the future of pattern documentation and general questions in ecological physiology. See Ref. 53, pp. 38–75
51. Feder ME. 1999. Engineering candidate genes in studies of adaptation: the heat-shock protein Hsp70 in *Drosophila melanogaster. Am. Nat.* 154:S55–66
52. Feder ME. 1999. Organismal, ecological, and evolutionary aspects of heat-shock proteins and the stress response: established conclusions and unresolved issues. *Am. Zool.* 39:857–64
53. Feder ME, Bennett AF, Burggren WW, Huey RB. 1987. *New Directions in Ecological Physiology.* Cambridge, UK: Cambridge Univ. Press
54. Feder ME, Blair N, Figueras H. 1997. Natural thermal stress and heat-shock protein expression in *Drosophila* larvae and pupae. *Funct. Ecol.* 11:90–100
55. Feder ME, Block BA. 1991. On the future of physiological ecology. *Funct. Ecol.* 5:136–44
56. Feder ME, Cartaño NV, Milos L, Krebs RA, Lindquist SL. 1996. Effect of engineering *hsp70* copy number on Hsp70 expression and tolerance of ecologically relevant heat shock in larvae and pupae of *Drosophila melanogaster. J. Exp. Biol.* 199:1837–44
57. Feder ME, Hofmann GE. 1999. Heat-shock proteins, molecular chaperones, and

the stress response: evolutionary and ecological physiology. *Annu. Rev. Physiol.* 61:243–82
58. Feder ME, Karr TL, Yang W, Hoekstra JM, James AC. 1999. Interaction of *Drosophila* and its endosymbiont *Wolbachia*: natural heat shock and the overcoming of sexual incompatibility. *Am. Zool.* 39:363–73
59. Feder ME, Krebs RA. 1997. Ecological and evolutionary physiology of heat-shock proteins and the stress response in *Drosophila*: complementary insights from genetic engineering and natural variation. In *Environmental Stress, Adaptation, and Evolution*, ed. R Bijlsma, V Loeschcke, pp. 155–73. Basel: Birkhäuser
60. Feder ME, Krebs RA. 1998. Natural and genetic engineering of thermotolerance in *Drosophila melanogaster. Am. Zool.* 38:503–17
61. Feder ME, Watt WB. 1993. Functional biology of adaptation. In *Genes in Ecology*, ed. RJ Berry, TJ Crawford, GM Hewitt, pp. 365–91. Oxford, UK: Blackwell Sci.
62. Felsenstein J. 1985. Phylogenies and the comparative method. *Am. Nat.* 125:1–15
63. Fields PA, Somero GN. 1997. Amino acid sequence differences cannot fully explain interspecific variation in thermal sensitivities of gobiid fish A(4)-lactate dehydrogenases (A(4)-LDHs). *J. Exp. Biol.* 200:1839–50
64. Freriksen A, Deruiter BLA, Groenenberg HJ, Scharloo W, Heinstra PWH. 1994. A multilevel approach to the significance of genetic variation in alcohol dehydrogenase of *Drosophila. Evolution* 48:781–90
65. Freriksen A, Deruiter BLA, Scharloo W, Heinstra PWH. 1994. *Drosophila* alcohol dehydrogenase polymorphism and C-13 fluxes: opportunities for epistasis and natural selection. *Genetics* 137:1071–78
66. Freriksen A, Seykens D, Scharloo W, Heinstra PWH. 1991. Alcohol dehydrogenase controls the flux from ethanol into lipids in *Drosophila* larvae: a C-13 NMR study. *J. Biol. Chem.* 266:21399–403
67. Futuyma DJ. 1979. *Evolutionary Biology*. Sunderland, MA: Sinauer. 1st ed.
68. Garland T Jr, Adolph SC. 1991. Physiological differentiation of vertebrate populations. *Annu. Rev. Ecol. Syst.* 22:193–228
69. Garland T Jr, Adolph SC. 1994. Why not to do two-species comparative studies: limitations on inferring adaptation. *Physiol. Zool.* 67:797–828
70. Garland T Jr, Bennett AF. 1990. Quantitative genetics of maximal oxygen consumption in a garter snake. *Am. J. Physiol.* 259:R986–92
71. Garland T Jr, Bennett AF, Daniels CB. 1990. Heritability of locomotor performance and its correlates in a natural population. *Experientia* 46:530–33
72. Garland T Jr, Carter PA. 1994. Evolutionary physiology. *Annu. Rev. Physiol.* 56:579–621
73. Garland T Jr, Dickerman AW, Janis CM, Jones JA. 1993. Phylogenetic analysis of covariance by computer simulation. *Syst. Biol.* 42:265–92
74. Garland T Jr, Losos JB. 1994. Ecological morphology of locomotor performance in squamate reptiles. See Ref. 205a, pp. 240–302
75. Garland T Jr, Martin KLM, Diaz-Uriarte R. 1997. Reconstructing ancestral trait values using squared-change parsimony: plasma osmolarity at the origin of amniotes. In *Amniote Origins: Completing the Transition to Land*, ed. SS Sumida, KLM Martin, pp. 425–501. San Diego: Academic
76. Garland T Jr, Midford PE, Ives AR. 1999. An introduction to phylogenetically based statistical methods, with a new method for confidence intervals on ancestral values. *Am. Zool.* 39:374–88
77. Gibbs AG. 1999. Laboratory selection for the comparative physiologist. *J. Exp. Biol.* 202:2709–18
78. Gilchrist GW. 1995. Specialists and generalists in changing environments. I. Fitness

landscapes of thermal sensitivity. *Am. Nat.* 146:252–70

79. Gillespie JH. 1991. *The Causes of Molecular Evolution*. New York: Oxford Univ. Press
80. Gould SJ. 1980. The promise of paleobiology as a nomothetic, evolutionary discipline. *Paleobiology* 6:96–118
81. Gould SJ. 1989. *Wonderful Life: The Burgess Shale and the Nature of History*. New York: Norton
82. Gould SJ, Lewontin RC. 1979. The spandrels of San Marco and the Panglossian paradigm. A critique of the adaptationist program. *Proc. R. Soc. London Ser. B* 205:581–98
83. Graves JL, Toolson EC, Jeong C, Vu LN, Rose MR. 1992. Desiccation, flight, glycogen, and postponed senescence in *Drosophila melanogaster*. *Physiol. Zool.* 65:268–86
84. Greene HW. 1986. Diet and arboreality in the emerald monitor, *Varanus prasinus*, with comments on the study of adaptation. *Fieldiana Zool.* 31:1–12
85. Grueber WB, Bradley TJ. 1994. The evolution of increased salinity tolerance in larvae of *Aedes* mosquitos: a phylogenetic analysis. *Physiol. Zool.* 67:566–79
86. Gu X, Zera AJ. 1996. Quantitative genetics of juvenile hormone esterase, juvenile hormone binding and general esterase activity in the cricket *Gryllus assimilis*. *Heredity* 76:136–42
87. Hairston NG. 1996. Zooplankton egg banks as biotic reservoirs in changing environments. *Limnol. Oceanogr.* 41:1087–92
88. Hairston NG Jr, Ellner SP, Kearns CM. 1996. Overlapping generations: the storage effect and the maintenance of biotic diversity. In *Population Dynamics in Ecological Space and Time*, ed. OE Rhodes, RK Chesser, MH Smith, pp. 109–45. Chicago: Univ. Chicago Press
89. Hairston NG, Lampert W, Caceres CE, Holtmeier CL, Weider LJ, et al. 1999. Lake ecosystems: rapid evolution revealed by dormant eggs. *Nature* 401:446–446
90. Hairston NG Jr, Van Brunt RA, Kearns CM, Engstrom DR. 1997. Age and survivorship of diapausing eggs in a sediment egg bank. *Ecology* 76:1706–11
91. Hamilton WJ. 1973. *Life's Color Code*. New York: McGraw-Hill
92. Hammond KA, Lam M, Lloyd KCK, Diamond J. 1996. Simultaneous manipulation of intestinal capacities and nutrient loads in mice. *Am. J. Physiol.* 34:G969–79
93. Hammond KA, Lloyd KCK, Diamond J. 1996. Is mammary output capacity limiting to lactational performance in mice? *J. Exp. Biol.* 199:337–49
94. Harvey PH, Pagel MD. 1991. *The Comparative Method in Evolutionary Biology*. Oxford, UK: Oxford Univ. Press
95. Hayes JP, Garland T Jr. 1995. The evolution of endothermy: testing the aerobic capacity model. *Evolution* 49:836–47
96. Heinrich B. 1977. Why have some animals evolved to regulate a high body temperature? *Am. Nat.* 111:623–40
97. Hillenius WJ. 1994. Turbinates in therapsids: evidence for Late Permian origins of mammalian endothermy. *Evolution* 48:207–29
98. Hochachka PW. 1997. Is evolutionary physiology useful to mechanistic physiology? The diving response in pinnipeds as a test case. *Zool. Anal. Complex Syst.* 100:328–35
99. Hoffmann AA, Parsons PA. 1991. *Evolutionary Genetics and Environmental Stress*. Oxford, UK: Oxford Univ. Press
100. Hoffmann AA, Parsons PA. 1997. *Extreme Environmental Change and Evolution*. Cambridge, UK: Cambridge Univ. Press
101. Holland LZ, McFall-Ngai M, Somero GN. 1997. Evolution of lactate dehydrogenase-A homologs of barracuda fishes genus (*Sphyraena*) from different

thermal environments: differences in kinetic properties and thermal stability are due to amino acid substitutions outside the active site. *Biochemistry* 36:3207–15
102. Huey RB. 1982. Temperature, physiology, and the ecology of reptiles. In *Biology of the Reptilia*, ed. C Gans, FH Pough, pp. 25–91. London: Academic
103. Huey RB. 1987. Phylogeny, history and the comparative method. See Ref. 53, pp. 76–101
104. Huey RB, Bennett AF. 1986. A comparative approach to field and laboratory studies in evolutionary biology. In *Predator-Prey Relationships: Perspectives and Approaches from the Study of Lower Vertebrates*, ed. ME Feder, GV Lauder, pp. 82–98. Chicago: Univ. Chicago Press
105. Huey RB, Gilchrist GW, Carlson ML, Berrigan D, Serra L. 2000. Rapid evolution of a geographic cline in size in an introduced fly. *Science*. 287:308–9
106. Jayne BC, Bennett AF. 1990. Selection on locomotor performance capacity in a natural population of garter snakes. *Evolution* 44:1204–29
107. Ketterson ED, Nolan V. 1999. Adaptation, exaptation, and constraint: a hormonal perspective. *Am. Nat.* 154:S4–25
108. Ketterson ED, Nolan V, Cawthorn MJ, Parker PG, Ziegenfus C. 1996. Phenotypic engineering: using hormones to explore the mechanistic and functional bases of phenotypic variation in nature. *Ibis* 138:70–86
109. Kingsolver JG. 1995. Fitness consequences of seasonal polyphenism in western white butterflies. *Evolution* 49:942–54
110. Kingsolver JG. 1995. Viability selection on seasonally polyphenic traits: wing melanin pattern in western white butterflies. *Evolution* 49:932–41
111. Kingsolver JG. 1996. Experimental manipulation of wing pigment pattern and survival in western white butterflies. *Am. Nat.* 147:296–306
112. Koehn RK. 1987. The importance of genetics to physiological ecology. See Ref. 53, pp. 170–85
113. Koehn RK, Newell RI, Immermann F. 1980. Maintenance of an aminopeptidase allele frequency cline by natural selection. *Proc. Natl. Acad. Sci. USA* 77:5385–89
114. Koehn RK, Zera AJ, Hall JG. 1983. Enzyme polymorphism and natural selection. In *Evolution of Genes and Proteins*, ed. M Nei, RK Koehn, pp. 115–36. Sunderland, MA: Sinauer
115. Kohler RE. 1994. *Lords of the Fly: Drosophila Genetics and the Experimental Life*. Chicago: Univ. Chicago Press
116. Krebs HA. 1975. The August Krogh principle: "For many problems there is an animal on which it can be most conveniently studied." *J. Exp. Zool.* 194:221–26
117. Krebs RA, Feder ME. 1997. Deleterious consequences of Hsp70 overexpression in *Drosophila melanogaster* larvae. *Cell Stress Chaperones* 2:60–71
118. Krebs RA, Feder ME. 1997. Natural variation in the expression of the heat-shock protein Hsp70 in a population of *Drosophila melanogaster*, and its correlation with tolerance of ecologically relevant thermal stress. *Evolution* 51:173–79
119. Krebs RA, Feder ME. 1997. Tissue-specific variation in Hsp70 expression and thermal damage in *Drosophila melanogaster* larvae. *J. Exp. Biol.* 200:2007–15
120. Krebs RA, Feder ME. 1998. Experimental manipulation of the cost of thermal acclimation in *Drosophila melanogaster*. *Biol. J. Linn. Soc.* 63:593–601
121. Krebs RA, Feder ME. 1998. Hsp70 and larval thermotolerance in *Drosophila melanogaster:* How much is enough and when is more too much? *J. Insect Physiol.* 44:1091–101
122. Krebs RA, Feder ME, Lee J. 1998.

Heritability of expression of the 70-kD heat-shock protein in *Drosophila melanogaster* and its relevance to the evolution of thermotolerance. *Evolution* 52:841–47
123. Krogh A. 1929. Progress of physiology. *Am. J. Physiol.* 90:243–51
124. Lacy RC, Lynch CB. 1979. Quantitative genetic analysis of temperature regulation in *Mus musculus*. I. Partitioning of variance. *Genetics* 91:743–53
125. Lande R, Arnold SJ. 1983. The measurement of selection on correlated characters. *Evolution* 37:1210–26
126. Laurie-Ahlberg CC, Barnes PT, Curtsinger JW, Emigh TH, Karlin B, et al. 1985. Genetic variability of flight metabolism in *Drosophila melanogaster*. 2. Relationship between power output and enzyme activity levels. *Genetics* 111:845–68
127. Laurie-Ahlberg CC, Bewley GC. 1983. Naturally occurring genetic variation affecting the expression of alpha-glycerol-3-phosphate dehydrogenase in *Drosophila melanogaster*. *Biochem. Genet.* 21:943–61
128. Lenski RE, Souza V, Duong LP, Phan QG, Nguyen TNM, Bertrand KP. 1994. Epistatic effects of promoter and repressor functions of the Tn10 tetracycline-resistance operon on the fitness of *Escherichia coli*. *Mol. Ecol.* 3:127–35
129. Leroi AM, Rose MR, Lauder GV. 1994. What does the comparative method reveal about adaptation. *Am. Nat.* 143:381–402
130. Levin DA. 1990. The seed bank as a source of genetic novelty in plants. *Am. Nat.* 135:563–72
131. Levins R. 1968. *Evolution in Changing Environments*. Princeton, NJ: Princeton Univ. Press
132. Losos JB. 1990. Ecomorphology, performance capability, and scaling of West Indian *Anolis* lizards: an evolutionary analysis. *Ecol. Monogr.* 60:369–88
133. Losos JB, Miles D. 1994. Adaptation, constraint, and the comparative method: phylogenetic issues and methods. See Ref. 205a, pp. 60–98
134. Losos JB, Warheit KI, Schoener TW. 1997. Adaptive differentiation following experimental island colonization in *Anolis* lizards. *Nature* 387:70–73
135. Lynch CB. 1980. Response to divergent selection for nesting behavior in *Mus musculus*. *Genetics* 96:757–65
136. Lynch CB. 1994. Evolutionary inferences from genetic analyses of cold adaptation in laboratory and wild populations of the house mouse. In *Quantitative Genetic Studies of Behavioral Evolution*, ed. CRB Boake, pp. 278–301. Chicago: Univ. Chicago Press
137. Lynch M, Walsh B. 1998. *Genetics and Analysis of Quantitative Traits*. Sunderland, MA: Sinauer
138. Mangum CM, Hochachka PW. 1998. New directions in comparative physiology and biochemistry: mechanisms, adaptations and evolution. *Physiol. Zool.* 71:471–84
139. Marden JH, Fitzhugh GH, Wolf MR. 1998. From molecules to mating success: integrative biology of muscle maturation in a dragonfly. *Am. Zool.* 38:528–44
140. Martinez Del Rio C, Brugger KE, Rios JL, Vergara ME, Witmer M. 1995. An experimental and comparative study of dietary modulation of intestinal enzymes in European starlings (*Sturnus vulgaris*). *Physiol. Zool.* 68:490–511
141. Martins EP, ed. 1996. *Phylogenies and the Comparative Method in Animal Behavior*. Oxford, UK: Oxford Univ. Press
142. Marton MJ, DeRisi JL, Bennett HA, Iyer VR, Meyer MR, et al. 1998. Drug target validation and identification of secondary drug target effects using DNA microarrays. *Nature Med.* 4:1293–301
143. Mayr E. 1982. *The Growth of Biological Thought: Diversity, Evolution, and Inheritance*. Cambridge, MA: Belknap

144. Mayr E, Provine WB, eds. 1980. *The Evolutionary Synthesis: Perspectives on the Unification of Biology*. Cambridge, MA: Harvard Univ. Press
145. McCarrey JR, VandeBerg JL. 1998. Proceedings from the 9th International Congress on Isozymes, Genes, and Gene Families. *J. Exp. Zool.* 282:1–283
146. McGraw JB, Vavrek MC, Bennington CC. 1991. Ecological genetic variation in seed banks. 1. Establishment of a time transect. *J. Ecol.* 79:617–25
147. Meagher TR. 2000. Evolution, science, and society: evolutionary biology and the national research agenda. *Am. Nat.* 156: In press
148. Miles DB. 1987. Habitat related differences in locomotion and morphology in two populations of *Urosaurus ornatus*. *Am. Zool.* 27:44A
149. Miles DB. 1994. Population differentiation in locomotor performance and the potential response of a terrestrial organism to global environmental change. *Am. Zool.* 34:422–36
150. Mitton JB. 1994. Molecular approaches to population biology. *Annu. Rev. Ecol. Syst.* 25:45–69
151. Mitton JB. 1997. *Selection in Natural Populations*. Oxford, UK: Oxford Univ. Press
152. Mongold JA, Bennett AF, Lenski RE. 1996. Evolutionary adaptation to temperature. IV. Adaptation of *Escherichia coli* at a niche boundary. *Evolution* 50:35–43
153. Mottishaw PD, Thornton SJ, Hochachka PW. 1999. The diving response mechanism and its surprising evolutionary path in seals and sea lions. *Am. Zool.* 39:434–50
154. Mukai T. 1969. The genetic structure of natural populations of *Drosophila melanogaster*. VII. Synergistic interaction of spontaneous mutant polygenes controlling viability. *Genetics* 61:749–61
155. Natochin YV, Chernigovskaya TV. 1997. Evolutionary physiology: history, principles. *Comp. Biochem. Physiol.* 118A:63–79
156. Nguyen TNM, Phan QG, Duong LP, Bertrand KP, Lenski RE. 1989. Effects of carriage and expression of the Tn10 tetracycline-resistance operon on the fitness of *Escherichia coli* K12. *Mol. Biol. Evol.* 6:213–25
157. Niewiarowski PH, Roosenburg W. 1993. Reciprocal transplant reveals sources of variation in growth rates of the lizard *Sceloporus undulatus*. *Ecology* 74:1992–2002
158. Parsell DA, Sanchez Y, Stitzel JD, Lindquist S. 1991. Hsp104 is a highly conserved protein with two essential nucleotide-binding sites. *Nature* 353:270–73
159. Parsons PA. 1983. *The Evolutionary Biology of Colonizing Species*. Cambridge, UK: Cambridge Univ. Press
160. Partridge L, Barrie B, Barton NH, Fowler K, French V. 1995. Rapid laboratory evolution of adult life history traits in *Drosophila melanogaster* in response to temperature. *Evolution* 49:538–44
161. Pierce VA, Crawford DL. 1997. Phylogenetic analysis of glycolytic enzyme expression. *Science* 276:256–59
162. Piiper J, Scheid P. 1982. Models for a comparative functional analysis of gas exchange organs in vertebrates. *J. Appl. Physiol.* 53:1321–29
163. Pough FH. 1988. Evolutionary physiology. *Science* 240:1349–51
164. Powers DA, Lauerman T, Crawford D, Dimichele L. 1991. Genetic mechanisms for adapting to a changing environment. *Annu. Rev. Genet.* 25:629–59
165. Powers DA, Schulte PM. 1998. Evolutionary adaptations of gene structure and expression in natural populations in relation to a changing environment: a multidisciplinary approach to address the million-year saga of a small fish. *J. Exp. Zool.* 282:71–94

166. Powers DA, Smith M, Gonzalez-Villasenor I, DiMichele L, Crawford DL, et al. 1993. A multidisciplinary approach to the selectionist/neutralist controversy using the model teleost *Fundulus heteroclitus*. In *Oxford Surveys in Evolutionary Biology*, ed. D Futuyma, J Antonovics, pp. 43–107
167. Prosser CL. 1986. *Adaptational Biology: Molecules to Organisms*. New York: Wiley & Sons
168. Provine WB, ed. 1986. *Sewall Wright and Evolutionary Biology*. Chicago: Univ. Chicago Press
169. Purrington CB, Bergelson J. 1997. Fitness consequences of genetically engineered herbicide and antibiotic resistance in Arabidopsis thaliana. *Genetics* 145:807–14
170. Riehle MM, Bennett AF, Long AD. 1999. Genetic analysis of adaptation to temperature stress. *Am. Zool.* 39:58A
171. Roberts SP, Feder ME. 2000. Changing fitness consequences of *hsp70* copy number in transgenic *Drosophila* larvae undergoing natural thermal stress. *Funct. Ecol.* 14:In press
172. Roberts SP, Feder ME. 1999. Natural hyperthermia and expression of the heat-shock protein Hsp70 affect developmental abnormalities in *Drosophila melanogaster*. *Oecologia* 121:323–29
173. Rose MR, Graves JL, Hutchison EW. 1990. The use of selection to probe patterns of pleiotropy in fitness characters. In *Insect Life Cycles: Genetics, Evolution and Co-Ordination*, ed. FS Gilbert, pp. 29–42. London: Springer-Verlag
174. Rose MR, Lauder GV, eds. 1996. *Adaptation*. New York: Academic
175. Rose MR, Nusbaum TJ, Chippindale AK. 1996. Laboratory evolution: the experimental wonderland and the Cheshire cat syndrome. See Ref. 174, pp. 221–41
176. Ruben J. 1995. The evolution of endothermy in mammals and birds: from physiology to fossils. *Annu. Rev. Physiol.* 57:69–95
177. Ruben J. 1996. Evolution of endothermy in mammals, birds and their ancestors. In *Animals and Temperature: Phenotypic and Evolutionary Adaptation*, ed. IA Johnston, AF Bennett, pp. 347–76. Cambridge, UK: Cambridge Univ. Press
178. Ruben JA, Bennett AF. 1980. Antiquity of the vertebrate pattern of activity metabolism and its possible relation to vertebrate origins. *Nature* 286:886–88
179. Sanchez Y, Lindquist SL. 1990. HSP104 required for induced thermotolerance. *Science* 248:1112–15
180. Schlicting CD, Pigliucci M. 1998. *Phenotypic Evolution: A Reaction Norm Perspective*. Sunderland, MA: Sinauer
181. Schmidt-Nielsen K. 1998. *The Camel's Nose and Other Lessons: Memoirs of a Curious Scientist*. Washington, DC: Island Press
182. Scholander PF, Hock R, Walters V, Irving L. 1950. Adaptation to cold in arctic and tropical mammals and birds in relation to body temperature, insulation, and basal metabolic rate. *Biol. Bull.* 99:259–71
183. Schulte PM, Gomez Chiarri M, Powers DA. 1997. Structural and functional differences in the promoter and 5′ flanking region of Ldh-B within and between populations of the teleost *Fundulus heteroclitus*. *Genetics* 145:759–69
184. Seger J, Stubblefield JW. 1996. Optimization and adaptation. See Ref. 174, pp. 93–123
185. Service PM, Hutchinson EW, MacInley MD, Rose MR. 1985. Resistance to environmental stress in *Drosophila melanogaster* selected for postponed senescence. *Physiol. Zool.* 58:380–89
186. Severtsov AN. 1914. *Current Problems in Evolutionary Theory*. Moscow: Bios
187. Sinervo B. 1990. The evolution of maternal investment in lizards: an experimental and comparative analysis of egg size and its effects on offspring performance. *Evolution* 44:279–94
188. Sinervo B. 1994. Experimental tests of

reproductive allocation paradigms. In *Lizard Ecology: Historical and Experimental Perspectives*, ed. LJ Vitt, ER Pianka, pp. 73–90. Princeton, NJ: Princeton Univ. Press

189. Sinervo B, Adolph SC. 1989. Thermal sensitivity of growth rate in hatchling *Sceloporus* lizards: environmental, behavioral and genetic aspects. *Oecologia* 78:411–19
190. Sinervo B, Basolo AL. 1996. Testing adaptation using phenotypic manipulations. See Ref. 174, pp. 149–85
191. Sinervo B, Doughty P, Huey RB, Zamudio K. 1992. Allometric engineering: a causal analysis of natural selection on offspring size. *Science* 258:1927–30
192. Sinervo B, Huey RB. 1990. Allometric engineering: an experimental test of the causes of interpopulational differences in performance. *Science* 248:1106–9
193. Sinervo B, Licht P. 1991. Hormonal and physiological control of clutch size, egg size, and egg shape in side-blotched lizards (*Uta stansburiana*): constraints on the evolution of lizard life histories. *J. Exp. Zool.* 257:252–64
194. Snyder LRG. 1981. Deer mouse hemoglobins: Is there genetic adaptation to high altitude? *BioScience* 31:299–304
195. Somero GN. 1975. Temperature as a selective factor in protein evolution: the adaptational strategy of "compromise." *J. Exp. Zool.* 194:175–88
196. Somero GN. 1995. Proteins and temperature. *Annu. Rev. Physiol.* 57:43–68
197. Sorci G, Swallow JG, Garland T Jr, Clobert J. 1995. Quantitative genetics of locomotor speed and endurance in the lizard *Lacerta vivipara*. *Physiol. Zool.* 68:698–720
198. Spellman PT, Sherlock G, Zhang MQ, Iyer VR, Anders K, et al. 1998. Comprehensive identification of cell cycle-regulated genes of the yeast *Saccharomyces cerevisiae* by microarray hybridization. *Mol. Biol. Cell* 9:3273–97
199. Travisano M. 1997. Long-term experimental evolution in *Escherichia coli*. 5. Environmental constraints on adaptation and divergence. *Genetics* 146:471–79
200. Travisano M, Lenski RE. 1996. Long-term experimental evolution in *Escherichia coli*. 4. Targets of selection and the specificity of adaptation. *Genetics* 143:15–26
201. Travisano M, Mongold JA, Bennett AF, Lenski RE. 1995. Experimental tests of the roles of adaptation, chance, and history in evolution. *Science* 267:87–90
202. Travisano M, Vasi F, Lenski RE. 1995. Long-term experimental evolution in *Escherichia coli*. 3. Variation among replicate populations in correlated responses to novel environments. *Evolution* 49:189–200
203. Tsuji JS, Huey RB, VanBerkum FH, Garland T Jr, Shaw RG. 1989. Locomotor performance of hatchling fence lizards (*Sceloporus occidentalis*)—quantitative genetics and morphometric correlates. *Evol. Ecol.* 3:240–52
204. Valentine JW, Jablonski D, Erwin DH. 1999. Fossils, molecules and embryos: new perspectives on the Cambrian explosion. *Development* 126:851–59
205. Vogel S. 1981. *Life in Moving Fluids: The Physical Biology of Flow*. Boston: Grant

205a. Wainwright PC, Reilly SM. 1994. *Ecological Morphology: Integrative Organismal Biology*. Chicago: Univ. Chicago Press

206. Wake MH. 1990. The evolution of integration of biological systems: an evolutionary perspective through studies on cells, tissues, and organs. *Am. Zool.* 30:897–906
207. Walton BM. 1993. Physiology and phylogeny: the evolution of locomotor energetics in hylid frogs. *Am. Nat.* 141:26–50
208. Waterman TH. 1975. Expectation and achievement in comparative physiology. *J. Exp. Zool.* 194:309–43

209. Watkins TB. 1996. Predator-mediated selection on burst swimming performance in tadpoles of the Pacific tree frog, *Pseudacris regilla*. *Physiol. Zool.* 69:154–67
210. Watt WB. 1977. Adaptation at specific loci. 1. Natural selection on phosphoglucose isomerase of *Colias* butterflies: biochemical and population aspects. *Genetics* 87:177–94
211. Watt WB. 1985. Isozymes: allelic isozymes and the mechanistic study of evolution. *Curr. Top. Biol. Med. Res.* 12:89–132
212. Watt WB. 1994. Allozymes in evolutionary genetics: self-imposed burden or extraordinary tool. *Genetics* 136:11–16
213. Watt WB. 1995. Allozymes in evolutionary genetics: beyond the twin pitfalls of neutralism and selectionism. *Rev. Suisse Zool.* 102:869–82
214. Watt WB. 2000. Avoiding paradigm-based limits to knowledge of evolution *Evol. Biol.* 32:73–96
215. Watt WB, Carter PA, Blower SM. 1985. Adaptation at specific loci. 4. Differential mating success among glycolytic allozyme genotypes of *Colias* butterflies. *Genetics* 109:157–75
216. Watt WB, Carter PA, Donohue K. 1986. Females choice of good genotypes as mates is promoted by an insect mating system. *Science* 233:1187–90
217. Watt WB, Cassin RC, Swan MS. 1983. Adaptation at specific loci. 3. Field behavior and survivorship differences among *Colias* PGI genotypes are predictable from in vitro biochemistry. *Genetics* 103:725–39
218. Watt WB, Donohue K, Carter PA. 1996. Adaptation at specific loci. 6. Divergence vs parallelism of polymorphic allozymes in molecular function and fitness: component effects among *Colias* species (Lepidoptera, Pieridae). *Mol. Biol. Evol.* 13:699–709
219. Weibel ER, Taylor CR, Bolis L. 1998. *Principles of Animal Design: The Optimization and Symmorphosis Debate*. Cambridge, UK: Cambridge Univ. Press
220. Welte MA, Tetrault JM, Dellavalle RP, Lindquist SL. 1993. A new method for manipulating transgenes: engineering heat tolerance in a complex, multicellular organism. *Curr. Biol.* 3:842–53
221. Zamudio KR, Huey RB, Crill WD. 1995. Bigger isn't always better: body size, developmental and parental temperature, and territorial success in *Drosophila melanogaster*. *Anim. Behav.* 49:671–77
222. Zera AJ, Huang Y. 1999. Evolutionary endocrinology of juvenile hormone esterase: functional relationship with wing polymorphism in the cricket, *Gryllus firmus*. *Evolution* 53:837–47
223. Zera AJ, Potts J, Kobus K. 1998. The physiology of life-history trade-offs: experimental analysis of a hormonally induced life-history trade-off in *Gryllus assimilis*. *Am. Nat.* 152:7–23
224. Zera AJ, Sall J, Schwartz R. 1996. Artificial selection on JHE activity in Gryllus assimilis: nature of activity differences between lines and effect on JH binding and metabolism. *Arch. Insect Biochem. Physiol.* 32:421–28
225. Zera AJ, Sanger T, Cisper GL. 1998. Direct and correlated responses to selection on JHE activity in adult and juvenile *Gryllus assimilis*: implications for stage-specific evolution of insect endocrine traits. *Heredity* 80:300–9
226. Zera AJ, Zeisset M. 1996. Biochemical characterization of juvenile hormone esterases from lines selected for high or low enzyme activity in *Gryllus assimilis*. *Biochem. Genet.* 34:421–35

Annu. Rev. Ecol. Syst. 2000. 31:343–66

MECHANISMS OF MAINTENANCE OF SPECIES DIVERSITY

Peter Chesson
Section of Evolution and Ecology University of California, Davis, California, 95616; e-mail: PLChesson@UCDavis.edu

Key Words coexistence, competition, predation, niche, spatial and temporal variation.

■ **Abstract** The focus of most ideas on diversity maintenance is species coexistence, which may be stable or unstable. Stable coexistence can be quantified by the long-term rates at which community members recover from low density. Quantification shows that coexistence mechanisms function in two major ways: They may be (a) *equalizing* because they tend to minimize average fitness differences between species, or (b) *stabilizing* because they tend to increase negative intraspecific interactions relative to negative interspecific interactions. Stabilizing mechanisms are essential for species coexistence and include traditional mechanisms such as resource partitioning and frequency-dependent predation, as well as mechanisms that depend on fluctuations in population densities and environmental factors in space and time. Equalizing mechanisms contribute to stable coexistence because they reduce large average fitness inequalities which might negate the effects of stabilizing mechanisms. Models of unstable coexitence, in which species diversity slowly decays over time, have focused almost exclusively on equalizing mechanisms. These models would be more robust if they also included stabilizing mechanisms, which arise in many and varied ways but need not be adequate for full stability of a system. Models of unstable coexistence invite a broader view of diversity maintenance incorporating species turnover.

INTRODUCTION

The literature is replete with models and ideas about the maintenance of species diversity. This review is about making sense of them. There are many commonalities in these models and ideas. The ones that could work, that is, the ones that stand up to rigorous logical examination, reveal important principles. The bewildering array of ideas can be tamed. Being tamed, they are better placed to be used.

The most common meaning of diversity maintenance is coexistence in the same spatial region of species having similar ecology. These species are termed here "the community," but are in the same trophic level and may be described as belonging to the same guild, a term that commonly means species having overlapping

0066-4162/00/1120-0343$14.00

resource requirements (124). Another meaning of diversity maintenance refers not to coexistence of fixed sets of species but to the maintenance of species richness and evenness over long timescales, necessitating consideration of speciation and extinction rates, and infrequent colonizations (35, 74). The primary concern of this review is with diversity maintenance as species coexistence.

Many models of species coexistence are thought of as models of coexistence in some defined local area. However, to make any sense, the area addressed must be large enough that population dynamics within the area are not too greatly affected by migration across its boundary (103). At some spatial scale, this condition will be achieved, but it may be much larger than is considered in most models and field studies. There is a temptation to consider diversity maintenance on small areas and to treat immigration into the area as part of the explanation for coexistence (81), but that procedure becomes circular if immigration rates are fixed and are not themselves explained. Species that continue to migrate into an area are found there even if the habitat is a sink. Continued migration of a suite of species into a local area depends on diversity maintenance in the areas that are the source of the immigrants. Thus, nothing is learned about diversity maintenance beginning with the assumption that migration rates into local areas are constant.

STABLE COEXISTENCE

Species coexistence may be considered as stable or unstable. Stable coexistence means that the densities of the species in the system do not show long-term trends. If densities get low, they tend to recover. Unstable coexistence means that there is no tendency for recovery and species are not maintained in the system on long timescales. In early approaches to species coexistence, stable coexistence meant stability at an equilibrium point. These days, it is commonly operationalized using the invasibility criterion or related ideas (7, 13, 36, 51, 83, 94, 133). The invasibility criterion for species coexistence requires each species to be able to increase from low density in the presence of the rest of the community. A species at low density is termed the invader, with the rest of the community termed residents. In calculations, the residents are assumed unaffected by the invader because the invader's density is low. The important quantity is the long-term per capita growth rate of the invader, $\bar{r}_i$, which is referred to here as the "long-term low-density growth rate." If this quantity is positive, the invader increases from low density. This criterion has been justified for a variety of deterministic and stochastic models (36, 51, 94), and has been shown in some cases to be equivalent to more ideal definitions of coexistence such as stochastic boundedness (36). Most important, the long-term low-density growth rate can be used to quantify species coexistence (27).

For a species to have a positive, $\bar{r}_i$, it must be distinguished from other species in ecologically significant ways (34). Indeed, the question of stable species coexistence is largely the question of the right sorts of ecological distinctions for the given circumstances. The Lotka-Volterra competition model is a useful beginning for discussing the basic principles. For this reason, it continues to have an important place in text books and also in the primary literature where it has roles approximating, interpreting, or describing the outcome of more sophisticated models (37, 104, 127, 130), as submodels (16), with variations on it increasingly fitted to field data (15, 42, 92, 115, 139). Unfortunately, textbooks muddy the water by parameterizing Lotka-Volterra competition in terms of carrying capacities and relative coefficients of competition. In terms of absolute competition coefficients, the two-species Lotka-Volterra competition equations can be written as follows (30):

$$\frac{1}{N_i} \cdot \frac{dN_i}{dt} = r_i\left(1 - \alpha_{ii}N_i - \alpha_{ij}N_j\right), \quad i = 1, 2, j \neq i. \qquad 1.$$

The quantities α_{ii} and α_{ij} are, respectively, absolute intraspecific and interspecific competition coefficients, and the defining feature of Lotka-Volterra competition is that per capita growth rates are linear decreasing functions of the densities of each of the species. Parameterized in this way, species i can increase from low density in the presence of its competitor resident in the community if $\alpha_{jj} > \alpha_{ij}$, which may be read biologically as "species j cannot competitively exclude species i if the effect that species j has on itself is more than the effect that species j has on species i." The criteria for species coexistence in a two-species system are therefore $\alpha_{11} > \alpha_{21}$ and $\alpha_{22} > \alpha_{12}$, which can be read very simply as "intraspecific competition must be greater than interspecific competition." This criterion is equivalent to requiring the relative coefficients β of Bolker & Pacala (16), which equal α_{ij}/α_{jj}, to be less than 1. Unfortunately, the usual relative competition coefficients of textbooks (1), which equal α_{ij}/α_{ii}, are not instructive because they compare how the growth of a species is depressed by the other species with how much it depresses its own growth.

Per capita growth rates are linear functions of density in Lotka-Volterra models, which makes Lotka-Volterra models special and may bias their predictions (1, 30, 68, 121). However, models with nonlinear per capita growth rates can be written in the form of Equation 1 by making the competition coefficients functions of density [$\alpha_{ij} = f_{ij}(N_i, N_j)$], and the results above remain true, provided the competition coefficients are evaluated for the resident at equilibrium and the invader at zero.

Lotka-Volterra models and their nonlinear extensions may be thought of as models of direct competition (56): Individual organisms have immediate direct negative effects on other individuals. However, they may also be derived from models with explicit resource dynamics (104, 120). Such multiple interpretations mean that these models are phenomenological: They are not defined by a mechanism of competition. A mechanistic understanding of competition usefully begins with Tilman's resource competition theory in which species jointly limited by a single resource are expected to obey the R^* rule (127). For any given species, R^* is the resource level at which the species is just able to persist. The winner in competition is the species with the lowest R^* value. A species' R^*, however, reflects not the ability of members of the species to extract resources when they are in low concentration, but their ability to grow and reproduce rapidly enough, at low resource levels, to compensate for tissue death and mortality, which are affected by such factors as grazing and predation. A species experiencing low grazing and mortality rates will have a lower R^* than other species, all else being equal (30). In essence, it is the overall fitness of a species that leads to its R^* value.

Tradeoffs play a major role in species coexistence: Advantages that one species may have over others are offset by compensating disadvantages (129, 136). However, examined in relation to the R^* rule there is clearly more to stable coexistence. With just a single resource as a limiting factor, tradeoffs may make the R^* values of different species more nearly equal, but that would not lead to stable coexistence. This conclusion is amply illustrated in the case of a special linear form of this resource limitation model (13, 30, 141) where the long-term low-density growth rate of an invader i competing with a resident, s, is

$$\bar{r}_i = b_i \left(\frac{\mu_i}{b_i} - \frac{\mu_s}{b_s} \right), \qquad 2.$$

with the μs representing mean per capita growth rates of the species in the absence of resource limitation, and the bs representing the rates at which the per capita growth rates decline as resources decline in abundance (30). In this system, the ratios μ/b measure the average fitnesses of the species in this environment, and they have the appropriate property of predicting the winner in competition (13, 30): The species with the larger μ/b is the winner (has the smaller R^*). Tradeoffs may lead to similar values of μ/b for different species. However, such similarity in average fitness does not lead to stable coexistence as Formula 2 necessarily has opposite sign for any pair of species, meaning that only one of them can increase from low density in the presence of the other.

Coexistence with resource partitioning contrasts with this. Using MacArthur's mechanistic derivation of Lotka-Volterra competition (6, 26, 104), the per capita growth rate of an invader can be written in the form

$$\bar{r}_i = \frac{1}{\mathrm{N}_i} \cdot \frac{dN_i}{dt} = b_i(k_i - k_s) + b_i(1 - \rho)k_s, \qquad 3.$$

where the ks correspond to the μ/bs and ρ is a measure of resource-use overlap of the two species. [The ks are the $h' - m'$ of (30), and b_i is $b_i\sqrt{a_{ii}}$ of (26, 30).] The first term on the right in Equation 3 is the average fitness comparison of Equation 2 and therefore has opposite sign for the two species. Whenever, $\rho < 1$ (resource overlap is less than 100%) the last term in Equation 3 is positive for both species. This last term is therefore a stabilizing term that offsets inequalities in fitness expressed by the first term. The two species coexist if the stabilizing term is greater in magnitude than the fitness difference term because then both have positive growth as invaders. Alternatively, note that $\alpha_{is}/\alpha_{ss} = (k_s/k_i)\rho$, which means that to satisfy the coexistence requirement that intraspecific competition exceed interspecific competition, ρ must be less than 1, and the smaller ρ is, the easier it is to satisfy the condition, i.e. the larger the difference in k values compatible with coexistence (26, 30).

The stabilizing term in Equation 3 arises through a tradeoff in resource use. The assumptions of the model entail that doing well on some resources means doing less well on others. Each species has density-dependent feedback loops with its resources that limit it intraspecifically and limit other species interspecifically. However, limited resource overlap and tradeoffs in resource benefits mean that intraspecific limitation is enhanced relative to interspecific limitation. This concentration of intraspecific effects relative to interspecific effects is the essence of stabilization. Tradeoffs associated not with resource use, but for example, with mortality rates (30, 68), may minimize the fitness difference term, making it easier for domination by the stabilizing term, but such tradeoffs cannot create stability alone.

Equation 3 generalizes to multispecies communities involved in diffuse competition, that is, where competition between species involves comparable interaction strengths for all pairs of species. In a variety of different models of diffuse competition (including stochastic models), the following common approximation to the long-term low-density growth rate is found:

$$\bar{r}_i \approx b_i(k_i - \bar{k}) + \frac{b_i(1-\rho)D}{n-1}, \qquad 4.$$

where n is the number of species in the system, the ks are again measures of fitness of individual species, $\bar{k}$ is the average fitness of residents (the competitors of species i), ρ is again niche overlap, but not necessarily strictly defined in terms of resource use (27), and D is a positive constant. Like Equation 3, the first term is an average fitness comparison and the second term is a stabilizing term. Without this term, the first term of necessity leads to loss of all species but the most fit on average, which in the context of Tilman's R^* rule would be the species with the lowest R^* value. However, if the stabilizing term is larger in magnitude than the relative average fitness term for the worst species, then all species coexist. These two general terms may involve different mechanisms. Those reducing the magnitude of the fitness difference term will be referred to as *equalizing mechanisms*, while those increasing the magnitude of the stabilizing term will be referred to as *stabilizing mechanisms*. In the absence of the stabilizing term, equalizing mechanisms can, at best, slow competitive exclusion; but in the presence of stabilizing mechanisms, equalizing mechanisms may enable coexistence. Stabilizing and equalizing mechanisms are concepts applicable beyond diffuse competition, but their implementation and their sufficiency may differ for different competitive arrangements. For example, invasion of species occupying a one-dimensional niche axis means that neighboring species on the niche axis would be most important to the invader (105, 112), whereas the diffuse competition formula (Equation 4) implies that only the number of species and their average fitnesses matter to the invader. In addition, some mechanisms, as we shall see below, have both stabilizing and equalizing properties.

The key question to be addressed below is how mechanisms with stabilizing properties arise in various situations. The theoretical literature supports the concept that stable coexistence necessarily requires important ecological differences between species that we may think of as distinguishing their niches (34, 95) and that often involve tradeoffs, as discussed above. For the purpose of this review, niche space is conceived as having four axes: resources, predators (and other natural enemies), time, and space. In reality, each axis itself is multidimensional, a feature that does not intrude on the discussion here. A species' niche is not a Hutchinsonian hypervolume (95), but instead is defined by the *effect* that a species has at each point in niche space, and by the *response* that a species has to each point. For example, consider the resource axis. A species consumes resources, and therefore has an effect on resource density (54, 95). Individuals of a species may also reproduce, grow, or survive in response to resources (54, 95).

The essential way in which stabilization occurs is most clearly seen with resource competition. If a species depends most on a particular resource (strong response), and also reduces that resource (strong effect), then it has a density-dependent feedback loop with the resource and is limited by it. If a second species has a similar relationship with a different resource, then even though the species each consume some of the resource on which the other depends most strongly (limited resource overlap), each species depresses its own growth more than it depresses the growth of other species (60, 127). The result is stable coexistence. This conclusion, however, depends on explicit and implicit assumptions, which if varied, alter the conclusion. A symmetric situation in which each resource is equally rich and each species is equally productive, would lead to identical average fitnesses, and therefore Equation 3 would have only the stabilizing term. However, asymmetries, in which one resource is much richer (127), one species produces more per unit resource, or has a lower mortality rate (30, 68), would mean that $k_i - k_s$ would not be zero, and if sufficiently large, this fitness difference would counteract the stabilizing effect of low resource overlap, causing competitive exclusion. This result would occur because the advantaged species would be at such high density that it would consume too much of the resource on which the other species depends. An equalizing mechanism would then be necessary to reduce $k_i - k_s$ before coexistence were possible.

Another implicit assumption in the argument above is that the resources have independent dynamics apart from consumption by common species. However, that is not true if, for example, the resources are simply two stages of the same food species (17) or light intercepted at different heights in a forest canopy (31, 89), requiring special conditions to allow each species to be partially independently limited by the resource supply on which they depend (17, 89). The idea that species must be somewhat independently limited is critical to their depressing their own growth rates rate more than they depress the growth rates of other species. And it is critical also that this phenomenon involves a density-dependent feedback loop from the species to itself, either directly through some form of interference or indirectly, as discussed here, through a resource. Predators and other natural enemies may also provide density-dependent feedback loops for their prey (8, 70, 77). Space and time may modify such feedback loops applicable to the community as a whole in ways that intensify intraspecific density dependence relative to interspecific density dependence.

FLUCTUATION-DEPENDENT AND FLUCTUATION-INDEPENDENT MECHANISMS

Stable coexistence mechanisms may be fluctuation dependent or fluctuation independent (27). Examples of fluctuation-independent mechanisms are resource partitioning (6, 60, 118, 127) and frequency-dependent predation (53, 77). Community dynamics in both of these cases are commonly modeled by deterministic equations that have stable equilibrium points and are sometimes termed "equilibrium mechanisms." However, these mechanisms can function in the presence of environmental fluctuations. Indeed, incorporating environmental variability in Lotka-Volterra models by making the per capita rate of increase fluctuate with time need not change the conditions for species coexistence in any important way (27, 133, 134). Thus, we can think of the operation of the mechanism as independent of the fluctuations in the system. The tendency to dismiss such mechanisms because populations in nature fluctuate is not supported by the results of models.

Some stable coexistence mechanisms critically involve the fluctuations in the system, that is, without the fluctuations, the mechanism does not function. Thus, they are termed *fluctuation dependent* (27). In the case of temporal fluctuations, these mechanisms can be divided into two broad classes: *relative nonlinearity of competition* and *the storage effect*.

Relative Nonlinearity of Competition or Apparent Competition

The per capita growth rate of a population is commonly a nonlinear function of limiting factors, such as limiting resources [e.g. the light saturation curve of plant productivity (137)], or predators [e.g. if predators interfere with one another (45)]. Stable coexistence may result from different nonlinear responses to common fluctuating limiting factors (9, 13, 18, 27, 50, 76, 101, 125). As first thoroughly investigated by Armstrong & McGehee (13), for the case of a single limiting factor, the per capita growth rate of a species takes the form

$$r_i(t) = E_i(t) - \phi_i(F), \qquad 5.$$

where $E_i(t)$ is the maximum per capita growth rate as a function of possibly fluctuating environmental conditions (an environmental response), F is the limiting factor, for example the amount by which resources are depressed below their optimal values, and $\varphi_i\,(F)$ is the response defining the dependence of the per capita growth rate on F (27). The departure of the function φ_i from a linear function is its nonlinearity, which is measured by a quantity τ (27). For example, a type II

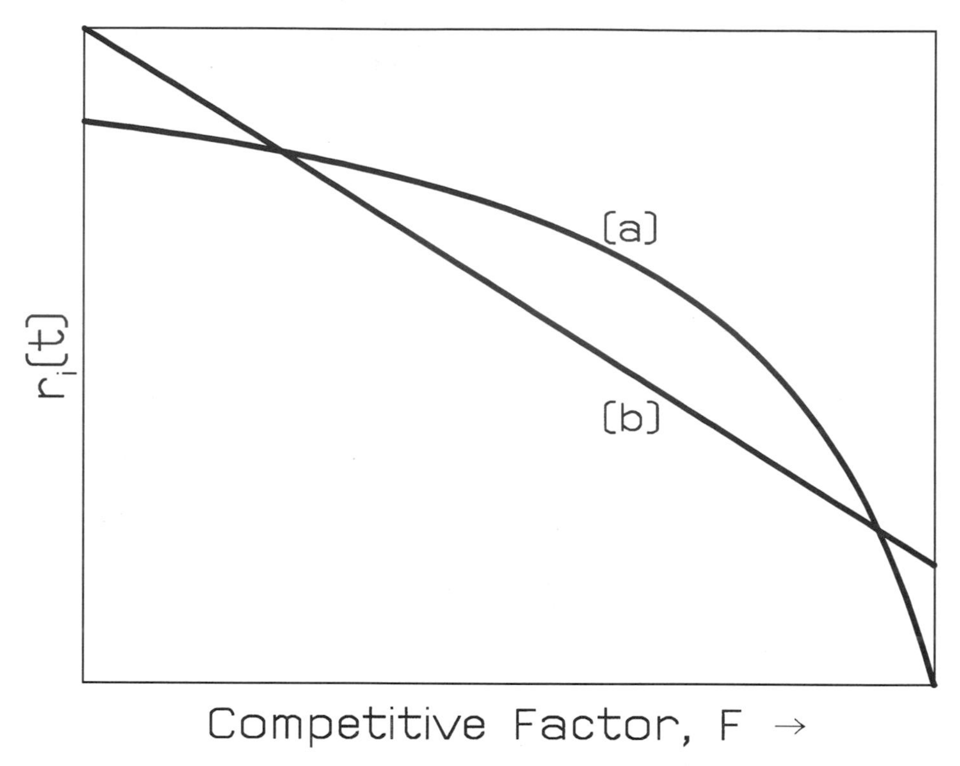

Figure 1 Per capita rates of growth of two species (a and b) as functions of a common limiting factor. In the case where the limiting factor is shortage of a resource, larger F means less resource, and curve a with positive τ would be generated by a type II functional response, while curve b with zero τ would occur with a linear response.

functional response to a limiting resource gives a positive τ, while a linear functional response gives a zero τ (Figure 1). Two species with different values of τ may coexist stably provided the species with the larger value of τ (a) has a mean fitness advantage in the absence of fluctuations in the limiting factor and (b) experiences lower fluctuations in the limiting factor when it is an invader than when it is a resident. Armstrong & McGehee showed that condition (b) may arise naturally with resource competition because a large value of τ may cause resources to have large-amplitude cycles over time.

These conditions for stable coexistence can be understood from the following general approximation to the long-term low-density growth rate of a species, i, in the presence of its competitor, s:

$$\bar{r}_i \approx b_i(k_i - k_s) - b_i(\tau_i - \tau_s)V(F^{-i}), \qquad 6.$$

where the first term is the fitness comparison in the absence of fluctuations in the limiting factor, $\tau_i - \tau_s$ measures relative nonlinearity, and $V(F^{-i})$ is the variance of the limiting factor calculated for invader i and resident s (27). In contrast to Equation 3 for the Lotka-Volterra model, the second term of this equation is not positive for both species but changes sign when resident and invader are exchanged. The species with more positive nonlinearity (larger value of τ) is disadvantaged by fluctuations in the limiting factor. However, if it is also the species with higher average fitness as measured by the first term, then this mechanism can have an equalizing role decreasing the advantage that a species gains from the first term. The fact that $V(F^{-i})$ may vary depending on which species is the invader means that it is possible for this mechanism to have a stabilizing role too. An algebraic rearrangement of Equation (6) clarifies these stabilizing versus equalizing roles. Defining B to be the average of $V(F^{-i})$ for each species as invader and A to be half the absolute value of the difference between these values, and assuming that the species with the smaller τ experiences larger $V(F)$ as an invader, then

$$\bar{r}_i \approx \{b_i(k_i - k_s) - b_i(\tau_i - \tau_s)B\} + b_i|\tau_i - \tau_s|A. \qquad 7.$$

The term in braces becomes the equalizing term—it is the relative fitness taking into account fluctuations in the limiting factor, and has opposite sign for the two species. The final term has the same sign for both species and is therefore a stabilizing term. Thus, relative nonlinearity (nonzero $\tau_i - \tau_s$ combined with variance in F) can have both stabilizing and equalizing roles. However, the stabilizing role might be negative, viz destabilizing if the species with the larger value of τ also has the larger value of $V(F^{-i})$, as then A in Equation 7 must be replaced by $-A$, which moves both species, long-term low density growth rates closer to zero promoting competitive exclusion (or apparent competitive exclusion if F relates to common predation on the two species).

Although there is a generalization of Equation 6 to multiple limiting factors (27), relative nonlinearity in the presence of multiple limiting factors has been poorly investigated. Huisman & Weissing (76), however, have demonstrated enormous potential for this mechanism with multiple competitive factors. For the dynamics of phytoplankton in lakes they use a model of competition for essential resources (98, 127) that has strong nonlinearities in a multidimensional sense. Fluctuations in these essential resources driven by their interactions with phytoplankton species appear to stabilize coexistence very strongly. This result contrasts with the feeble effects of relative nonlinearity on coexistence of phytoplankton previously demonstrated for the case of a single limiting factor (57–59).

The Storage Effect

Models imply that temporal environmental fluctuations can have major effects on population and community dynamics. It has been widely noted that negative environmental effects may reduce population densities and therefore reduce the magnitude of competition (55, 78, 79), but this possibility does not necessarily translate into less competitive exclusion (30). Instead, models in which stable coexistence results from environmental fluctuations are models of temporal niches: Species are not distinguished by the resources they use but by when they are most actively using them (2, 12, 30, 102). For stabilization to result, intraspecific competition must be concentrated relative to interspecific competition, and this requires three important ingredients, whose collective outcome is referred to as *the storage effect* (27). The most obvious of these is *differential responses to the environment*. Species from the same community may have different responses to their common varying environment. Environmental variation may be regular and deterministic, for example seasonal (12, 37, 47, 60, 86, 91, 102, 109, 140) or stochastic, attributable, for example, to weather on a variety of timescales (2, 27, 30, 32, 33, 36, 37, 51, 52, 62, 112, 123).

How can differential responses to the environment concentrate intraspecific density dependence relative to interspecific density dependence, and hence act as a stabilizing mechanism, when the physical environment is not altered by population densities? The answer is that population responses to the physical environment modify competition, as measured by the second essential feature of the storage effect, *covariance between environment and competition* (30, 33). Just as negative environmental effects may decrease competition, positive environmental effects may increase it. Covariance between environment and competition is measured by calculating the standard statistical covariance between the effects of the environment on the per capita growth rate of a population (*the environmental response*) and of competition (both intraspecific and interspecific) on the growth rate (*the competitive response*).

With covariance between environment and competition, and differential responses to the environment, intraspecific competition is strongest when a species is favored by the environment, and interspecific competition is strongest when the species' competitors are favored. The final ingredient, *buffered population growth*, limits the impact of competition when a species is not favored by the environment and is therefore experiencing mostly interspecific competition. Buffered population growth may result from a variety of life-history traits: seed banks in annual plants (29, 48, 49, 113, 114), resting eggs in freshwater zooplankton (22, 62, 63), and long-lived adults in perennial organisms (38, 86, 91, 96, 97, 112, 119). Organisms not having specific life-history stages with a buffering effect may be buffered in other ways. For example, desert rodents and herbaceous plants may have times of the year when they are dormant and therefore relatively immune to unfavorable environmental and competitive conditions (19, 37). Alternatively, subdivision of a population into different phenotypes, or local populations in space, with different exposure to environmental effects and competition, may also have a buffering role (33).

By diminishing the effects of interspecific competition when a species is not favored by the environment, buffered population growth, combined with the other two ingredients of the storage effect, leads to concentration of intraspecific effects on population growth relative to interspecific effects, which is stabilizing. In a variety of models of diffuse interactions involving the storage effect, the long-term low-density growth rate can be given approximately in the form

$$\bar{r}_i \approx b_i(k_i - \bar{k}) + \frac{b_i(1-\rho)(-\gamma)\sigma^2}{n-1}, \qquad 8.$$

where the symbols follow those in the general diffuse competition formula (Equation 4), with $D = (-\gamma)\sigma^2$ (27). Note that $(-\gamma)$ measures buffered population growth and is positive for buffering. The variance in the environmental response is σ^2, and ρ is the correlation between the environmental responses of different species. This formula has a fitness comparison (here representing an average over all environmental states), which is opposed by a stabilizing term (here due to the storage effect) promoting stable coexistence.

Mechanisms in Combination Integrated over Temporal Scales

Two general factors, competition and the physical environment, drive the mechanisms above. Two-factor analysis of variance shows how the effects of two factors can be divided into their separate or "main" effects and their interaction (126). Applying this technique to population growth (27) expresses the long-term low-density growth rate as

$$\bar{r}_i \approx \bar{r}'_i - \Delta N + \Delta I, \qquad 9.$$

where $\bar{r}'_i$ is the effect of fluctuation-independent mechanisms, plus mean fitness differences (i.e. everything that is independent of fluctuations), ΔN is relative nonlinearity of competition, and ΔI is the storage effect. Formulae for $\bar{r}'_i$, ΔN, and ΔI can be found in Chesson (27), but examples of them are respectively Equation 4 and the second terms of Equations 6 and 8. The storage effect, ΔI, is the interaction between fluctuating environmental and competitive responses, which might serve as its formal definition. The other two terms arise from the main effects with adjustments to remove all fluctuation-independent effects from ΔN and ΔI, and place them in $\bar{r}'_i$.

Although the derivation of Equation 9 in (27) involves approximations (hence "$\approx$"), these approximations occur in the formulae for the components, not the division into three terms. The important assumption limiting the generality of the results is that there should be fewer limiting factors than community members so that competition experienced by a low density invader can be expressed as a function of competition experienced by the resident species. With this one caveat, Equation 9 shows that in models of temporally fluctuating environmental and competitive effects, three broad classes of mechanisms (fluctuation-independent mechanisms, relative nonlinearity, and the storage effect) are exhaustive: Any mechanism of stable coexistence must be one of these or a combination of them.

Although the storage effect and relative nonlinearity may appear to be rather involved, there are no simpler fluctuation-dependent stable coexistence mechanisms waiting to be discovered. In particular, no credence can be given to the idea that disturbance promotes stable coexistence by simply reducing population densities to levels where competition is weak (30, 141). More sophisticated views of disturbance (23–25, 39, 40, 64), however, have yielded important hypotheses, as discussed below under the spatial dimension.

It is now commonly believed that spatial and temporal scale have major effects on the perception and functioning of diversity maintenance mechanisms (10, 14 43, 44, 74, 78, 80–82, 100, 116, 143). An important finding is that fluctuation-dependent mechanisms can give community dynamics indistinguishable from those of fluctuation-independent mechanisms when viewed on a longer timescale than the period of the fluctuations (37, 91), as is illustrated in Figure 2. Similar effects are often apparent in spatial models (16, 46, 64, 84, 122), although not always explicitly noted by the authors. In this regard, the term $\overline{r}_i'$ in Equation 9 can be regarded as containing not just fluctuation-independent mechanisms, but also fluctuation-dependent mechanisms on timescales shorter than the fluctuations considered in the ΔN and ΔI terms. When this is done, Equation 9 can be viewed as an iteration allowing the effects of fluctuation-dependent mechanisms operating on different timescales to be combined to give their total effect on very long timescales (37).

THE SPATIAL DIMENSION

Nature is strikingly patchy in space. Naturally, if species live in different habitats and have no direct or indirect interactions with each other, they should have no difficulty coexisting in a region combining these separate habitats. However, species do not have to be strictly segregated in space for regional coexistence (16, 20, 38, 90, 93, 99, 112, 122, 123). Spatial variation is similar to temporal variation in its effects on species coexistence with some important differences (32, 38, 123). In particular, there is a spatial analog of Equation 9 expressing growth from low density in terms of fluctuation-independent mechanisms, spatial relative nonlinearity, and the spatial storage effect (28a). In addition, there is a fourth term in the spatial analogue of Equation 9 involving the spatial covariance of local population growth with local population density, which behaves like a spatial storage effect in some circumstances, but like spatial relative nonlinearity in others (28a).

Spatial storage effects commonly occur in spatial models when spatial environmental variation is included. There are two common ways in which the requirement of differential responses to the environment is met in such models. First, relative fitnesses of different species may vary in space (21, 32, 38, 85, 107, 112, 122, 131). For example, for plant species, the identity of the species with the lowest R^* value may vary spatially due to differential dependence of R^* on spatially varying physical factors, such as temperature and pH (131). Second, there may be relative variation in dispersal into different habitats, for example, in marine habitats due to the complexities of spawning, currents, and developmental interactions (21, 32, 42, 85) and in insects potentially due to spatially varying habitat preferences (34, 84, 88). Such relative variation satisfies the requirement for differential responses to the environment.

Buffered population growth automatically occurs when populations are subdivided in space over a spatially varying environment (32, 38), but the strength of the buffering is affected by other aspects of the biology of the organisms, including their life-history attributes (28a). Covariance between environment and competition naturally arises when dispersal varies in space but not necessarily when relative fitnesses vary in space. The presence of covariance is easy to determine in a model, and is a powerful tool for determining if coexistence can be promoted by spatial environmental variation. For example, in models of sessile marine invertebrates with competition between adults and new settlers, but not among adults, spatial variation in adult death rates leads to strong covariance between environment and competition if the environmental differences between localities remain constant over time. Then adult densities build up in low-mortality locations, leading to strong competition from adults in such localities. Thus, covariance between environment and competition occurs, and regional coexistence by the spatial storage effect is possible (32, 107). However, if environmental differences between localities constantly change with time [i.e. environmental variation is spatio-temporal rather than purely spatial, sensu Chesson (32)], local population buildup occurs at best weakly before the environment changes. Covariance between environment and competition is weak or nonexistent, and so the storage effect is weak or nonexistent (32, 107).

There has been little explicit attention to relative nonlinearity of competition in spatial models, but Durrett & Levin (46) demonstrate coexistence in a spatial competition model dependent on spatial variation and different nonlinear competition functions that correspond biologically to different competitive strategies. The spatial Lotka-Volterra model of plant competition of Bolker & Pacala (16) does not have relatively nonlinear competition, as the per capita rates are linear. However, a similar nonlinear effect arises from covariance between local population density and competition. This covariance is due to chance variation in local population densities, which covaries with local competition in ways that give advantage, under some circumstances, to species with short-distance dispersal strategies. Similar phenomena have been found in other spatial competition models (93).

There are many other spatial models of competitive coexistence, few of which have been investigated within the framework presented here or a related framework (66, 89, 106, 117). Nevertheless, most do have some of the key elements of this framework, including nonlinearities, spatial covariances, and buffered growth rates, which appear actively involved in model behavior. From these other models, two ideas emerge as especially important: Tilman's resource-ratio hypothesis (127, 128), and colonization-competition tradeoffs.

The resource ratio hypothesis postulates that the ratio of the rates of supply of two limiting resources at a particular locality determines a unique pair of plant species from a given regional pool able to coexist at that locality. Given variation in space in the supply rate ratio, different pairs of species would be able to coexist at different localities in the absence of dispersal between localities. However, a full spatial model exploring this idea in the presence of dispersal has never been developed. Moreover, Pacala & Tilman (112) pointed out that under certain conditions, a given resource ratio may determine a particular best competitor that would dominate that site. With site-to-site variation in the resource ratio, this scenario leads to coexistence as a simple example of a spatial storage effect (112). Although there would be only one best species at any given locality, as determined by the resource ratio there, many species could be present at a locality due to dispersal from other localities.

Competition-colonization tradeoffs have been discussed in two distinct sorts of models. In one, a patch in space supports a local community that is destroyed at random by disturbance. It is then recolonized from other patches (24, 64). In other interpretations, a patch supports a single individual organism (87, 110, 130), whose death by some means (not necessarily disturbance) opens that locality for recolonization. Coexistence in these models requires the colonizing ability of different species to be ranked inversely to competitive ability, which therefore tends to drive a successional process in each locality. Localities becoming vacant randomly in space and time ensures a landscape in a mosaic of successional states. When vacant sites are interpreted as resulting from disturbance, this model provides perhaps the most satisfactory expression of the intermediate disturbance hypothesis (30), as diversity tends to be maximized at intermediate values of disturbance frequency (64).

PREDATORS, HERBIVORES, AND PATHOGENS

Predation is a common hypothesis for high biological diversity. However, the argument that predators maintain diversity by keeping populations below levels at which they compete is now known to be highly simplistic (4, 30). Predators may add density dependence to their prey populations through functional, numerical, and developmental responses to prey (108). In the absence of frequency-dependent functional responses or similar complications (65), a common predator of several prey is a single limiting factor with analogous effects to limitation by a single resource (67). One density-dependent limiting factor (the resource) is simply replaced by another (the predator). Indeed, there is a P^* rule exactly analogous to the R^* rule that says that the species with the highest tolerance of predation will drive other species extinct (69). Thus, if predation is so severe that it does eliminate competition, then competitive exclusion may be replaced by apparent competitive exclusion.

There are many ways in which predators may help species coexist, but it is far from an easy solution (7, 30, 65, 68, 70, 77, 135). Predators may promote species coexistence when each species has its own specialist predators, or more generally, specialist natural enemies that hold down the density of each species independently. This idea, known as the Janzen-Connell hypothesis, is an important hypothesis for the coexistence of trees in tropical forests (11) where tree seeds and seedlings may be subject to density- or distance-dependent seed predation (41, 138). By providing or attracting species-specific seed and seedling predators, an individual tree would have a greater negative effect on a conspecific growing nearby than on a heterospecific growing nearby (all else equal), thus providing the critical requirement of a stabilizing mechanism.

Many systems support generalist predators or herbivores, which prey or graze on a range of species (65, 77, 108). Murdoch has emphasized that predators may have frequency-dependent functional responses (switching; 108), which are stabilizing mechanisms. More complex intergenerational learning responses of parasitoids may have similar effects (65). Alternatively, a predator may not be frequency-dependent but have unequal effects on prey species that are also limited by resources (135). For example, two competing prey species may coexist if one of them is more strongly limited by the resource and the other is more strongly limited by the predator (61). In all of the above instances, the role of predation is to generate feedback loops in which an individual prey species depresses its own per capita growth rate more than it depresses the per capita growth rates of other species, thus meeting the requirements of stabilizing mechanisms. However, when these density-dependent and frequency-dependent requirements are not met, a generalist predator may instead have an equalizing role by inflicting greater predation on the competitive dominant. If a stablizing mechanism such as resource partitioning is present, then although the predator is not the stabilizing agent, stable coexistence may occur in its presence (30).

Predators are often invoked as biological disturbance agents inflicting mortality patchily in space and time. When mortality is not species specific, predators may be the agent of local extinction in the competition-colonization tradeoff models (23). If mortality is species specific, but density independent, then predators may have a role analogous to a spatio-temporally variable environment, which may lead to coexistence by the spatial storage effect (111). Thinking of predators as organisms, and the environment as the physical environment, invites generalizations of the storage effect to stable species coexistence in the presence of apparent competition. However, at present there is no theory of covariance between environment and apparent competition. As discussed above, predators can be fluctuating nonlinear limiting factors, but the potential of this mechanism has not been explored in any detail.

UNSTABLE COEXISTENCE

Hubbell (73, 74) has championed and steadily refined a model of community dynamics in which coexistence is not stable: The species compete for space but are ecologically identical and therefore have equal fitnesses under all conditions.

Thus, in Equation 4, both the average relative fitness term and the stabilizing term are close to zero, and the species in the system undergo a very slow random walk to extinction. To many people, a very slow loss of species is equivalent to indefinite coexistence, and it is certainly one model of how nature is (78, 79, 81, 123). Several objections have been raised against it. First, equal average fitnesses seem highly unlikely (28), and significant violation of this assumption destroys the conclusion of slow extinction (142) by creating nonzero average fitness difference terms for the species. Minor realistic variations on the model also lead to stabilizing components (28). Indeed, such features seem difficult to avoid in most systems (28, 142).

Other approaches to unstable coexistence have not assumed the extreme neutrality of Hubbell's model, and have sought means by which fitness differences may be minimized (78, 79, 81, 123). But these approaches have not recognized that stabilizing components are difficult to avoid (28), and may have overestimated the effectiveness of purported equalizing mechanisms (28). Nevertheless, there is undeniable merit in the question of unstable coexistence because it must be that in many systems at least some species are only weakly persistent because their fitness disadvantages are comparable in magnitude to the stabilizing component of their long-term low-density growth rate. At this point, the study of diversity maintenance needs to take account of macroevolutionary issues such as speciation and extinction, processes (35, 74), biogeographic processes of migration of species between communities on large spatial scales, and climate change on large temporal scales (37). Hubbell (74) argued that on such large scales, speciation, extinction, and migration processes are dominant, rendering the admitted oversimplifications of his neutral model unimportant. Independent data on the rates of these critical processes are needed to test this perspective.

NONEQUILIBRIUM COEXISTENCE

Stable and unstable coexistence is one view of the distinction between equilibrium and nonequilibrium coexistence (75). Another view is fluctuation-dependent versus fluctuation-independent coexistence (35). But equilibria are everywhere. For example, Hubbell's (74) model of unstable coexistence nevertheless has an equilibrium for species diversity on the large spatial scale. Thus, it seems best to ask, Is species coexistence stabilized or not? And if it is, Is that stabilization dependent on fluctuations or is it independent of fluctuations? The term nonequilibrium coexistence is better avoided.

LIMITING SIMILARITY

Unstable coexistence requires species to have very similar average fitnesses for long-term coexistence. Stable coexistence benefits from similar average fitnesses, but requires niche differences between species that intensify negative intraspecific

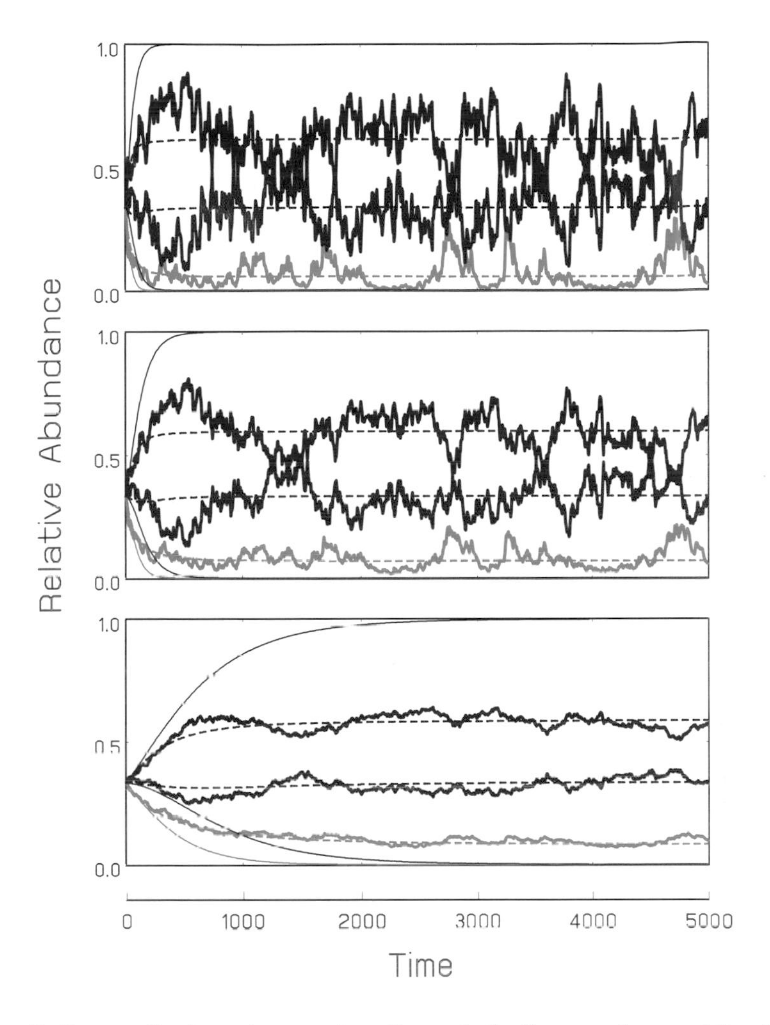

Figure 2 Community dynamics on a long timescale for three species coexisting by the storage effect (fluctuating lines) in the lottery model for intense competition for space (no space is left vacant, 37) compared with an approximating Lotka-Volterra model (37) modified for intense space competition (dashed lines) and the lottery model with no environmental fluctuations (solid smooth curves). The different panels are for mean adult longevities of 10, 20 and 100 respectively from top to bottom. Comparing the lottery model with and without fluctuations demonstrates the necessity of environmental fluctuations for species coexistence. The approximating Lotka-Volterra model from (37) shows that this mechanism can be mimicked by fluctuation-independent mechanisms with increasing precision as the longevity of the organism is increased. Lotka-Volterra models with noise added (133) give dynamics indistinguishable from the lottery model. The intraspecific and interspecific interaction coefficients in the Lotka-Volterra model are equal respectively to the variances and covariances of the environmental responses in the lottery model (37).

effects of density on population growth relative to negative interspecific density effects. This requirement of niche differences has often been referred to as limiting similarity (3, 112). It is clear that not any kind of niche differences between species will do. For example, the discussion of the storage effect above showed that species-specific responses to the environment, which are one sort of niche difference, lead to stability only if they are linked appropriately to density-dependent feedback loops. This condition applies generally, as has been emphasized here, but has not been studied in as much detail for most other mechanisms. The right sorts of niche differences might be called stabilizing niche differences. A precise limit to similarity implies a particular minimum value for stabilizing niche differences. It is clear there can be no such value (3, 5). For example, in the various equations above, the magnitude of the stabilizing term is naturally an increasing function of stabilizing niche differences (although this feature cannot be expected under all circumstances 6). The smaller the average fitness differences, the smaller the stabilizing niche differences can be. One may be tempted to suggest that average fitness differences can be zero, but there are plenty of reasons to expect them to differ from zero in most situations in nature, and as yet we have no theory that predicts average fitness differences. Thus, we cannot predict a particular limit to similarity, although the concept that niche dissimilarities of the right sort promote coexistence is generally supported (3).

Models of species on one-dimensional niche axes predict particular limits to similarity when one asks whether a species can invade between two particular resident's with a given niche spacing and given average fitness differences (105, 112). Such an invader is automatically at a disadvantage to the residents, however, because it has more competitors close to it than the residents do. Thus, this sort of analysis does not seem truly to answer the question of how close the niches of coexisting species can be, but it does emphasize that variation in the spacing of the niches of various species in a community, not just the average spacing, is an important factor in species coexistence (5). This feature is not captured by the simple equations for diffuse competition given here where there is no such variation in spacing.

Recently, particular limits to similarity were also claimed for coexistence by colonization-competition tradeoffs. However, the limit in question is the difference in colonizing ability of successively ranked competitors (87, 130). This difference has an equalizing effect because it compensates for a species' inferior competitive rank. It is not a stabilizing niche difference, but instead reduces average fitness inequalities that stabilizing niche differences must overcome. There should be no paradox in the idea that dissimilarities in niche differences of the stabilizing sort, which may be characterized as keeping species out of each other's way, promote coexistence, while differences in average fitness, which determine how much "better" one species is than another overall, favor competitive exclusion (14).

An objection that can be raised about all of the above analyses is that they take no consideration of the increasing sparseness of populations, and the decreasing absolute numbers of individuals in those populations, as more species are packed into a community. Allee effects in sparse (low density) populations (71, 72) and stochastic extinction in small populations (132) both potentially limit how similar the niches of coexisting species can be when similar niches mean sparser or smaller populations. These possibilities deserve further study as they have the unique property that they would still work when species are equal in average fitness, that is, they potentially lead to the requirement that stabilizing mechanisms of a certain minimum strength (depending on population sizes supportable in the system) must exist for coexistence regardless of the strength of equalizing mechanisms.

ACKNOWLEDGMENTS

I am grateful for many very helpful comments on the manuscript from P Abrams and more than a dozen colleagues and students.

LITERATURE CITED

1. Abrams P. 1980. Are competition coefficients constant? Inductive versus deductive approaches. *Am. Nat.* 116:730–35
2. Abrams P. 1984. Variability in resource consumption rates and the coexistence of competing species. *Theor. Popul. Biol.* 25:106–24
3. Abrams PA. 1983. The theory of limiting similarity. *Annu. Rev. Ecol. Syst.* 14:359–76
4. Abrams PA. 1987. The competitive exclusion principle: other views and a reply. *Trends Ecol. Evol.* 1:131–32
5. Abrams PA. 1996. Limits to the similarity of competitors under hierarchical lottery competition. *Am. Nat.* 148:211–19
6. Abrams PA. 1998. High competition with low similarity and low competition with high similarity: exploitative and apparent competition in consumer-resource systems. *Am. Nat.* 152:114–28
7. Abrams PA. 1999. Is predator-mediated coexistence possible in unstable systems? *Ecology* 80:608–21
8. Abrams PA, Holt RD, Roth RD. 1998. Apparent competition or apparent mutualism? Shared predation when populations cycle. *Ecology* 79:202–12
9. Adler FR. 1990. Coexistence of two types on a single resource in discrete time. *J. Math. Biol.* 28:695–713
10. Allen T, Hoekstra T. 1990. The confusion between scale-defined levels and conventional levels of organization in ecology. *J. Veg. Sci.* 1:1–8
11. Armstrong RA. 1989. Competition, seed predation, and species coexistence. *J. Theor. Biol.* 141:191–95
12. Armstrong RA, McGehee R. 1976. Coexistence of species competing for shared resources. *Theor. Popul. Biol.* 9:317–28
13. Armstrong RA, McGehee R. 1980. Competitive exclusion. *Am. Nat.* 115:151–70
14. Bengtsson J, Fagerstrom T, Rydin H. 1994. Competition and coexistence in plant communities. *Trends Ecol. Evol.* 9:246–50
15. Berlow EL, Navarrette SA, Briggs CJ, Power ME, Menge BA. 1999. Quantifying variation in the strengths of species interactions. *Ecology* 80:2206–24
16. Bolker B, Pacala S. 1999. Spatial moment equations for plant competition: understanding spatial strategies and the advantages of short dispersal. *Am. Nat.* 153:575–602
17. Briggs CJ, Nisbet RM, Murdoch WW.

1993. Coexistence of competing parasitoids on a host with a variable life cycle. *Theor. Popul. Biol.* 44:341–73
18. Brown JS. 1989. Coexistence on a seasonal resource. *Am. Nat.* 133:168–82
19. Brown JS. 1989. Desert rodent community structure: a test of four mechanisms of coexistence. *Ecol. Monogr.* 59:1–20
20. Brown JS, Rosenzweig ML. 1986. Habitat selection in slowly regenerating environments. *J. Theor. Biol.* 123:151–71
21. Butler AJ, Chesson P. 1990. Ecology of sessile animals on sublittoral hard substrata: the need to measure variation. *Aust. J. Ecol.* 15:520–30
22. Caceres CE. 1997. Temporal variation, dormancy, and coexistence: a field test of the storage effect. *Proc. Natl. Acad. Sci. USA* 94:9171–75
23. Caswell H. 1978. Predator-mediated coexistence: a nonequilibrium model. *Am. Nat.* 112:127–54
24. Caswell H, Cohen JE. 1991. Communities in patchy environments: a model of disturbance, competition, and heterogeneity. In *Ecological Heterogeneity*, ed. J Kolasa, STA Pickett, pp. 98–122. New York: Springer Verlag
25. Caswell H, Cohen JE. 1991. Disturbance, interspecific interaction and diversity in metapopulations. In *Metapopulation Dynamics: Empirical and Theoretical Investigations*, ed. M Gilpin, I Hanski, pp. 193–218. London: Academic
26. Chesson P. 1990. MacArthur's consumer-resource model. *Theor. Popul. Biol.* 37:26–38
27. Chesson P. 1994. Multispecies competition in variable environments. *Theor. Popul. Biol.* 45:227–76
28. Chesson P. 1997. Diversity maintenance by integration of mechanisms over various scales. In *Proc. Eighth Int. Coral Reef Symp.* 1:405–10. Panama City: Smithsonian Trop. Res. Inst., Balboa, Republic of Panama.
28a. Chesson P. 2000. General theory of competitive coexistence in spatially varying environments. *Theor. Popul. Biol.* In press
29. Chesson P, Huntly N. 1989. Short-term instabilities and long-term community dynamics. *Trends Ecol. Evol.* 4:293–98
30. Chesson P, Huntly N. 1997. The roles of harsh and fluctuating conditions in the dynamics of ecological communities. *Am. Nat.* 150:519–53
31. Chesson P, Pantastico-Caldas M. 1994. The forest architecture hypothesis for diversity maintenance. *Trends Ecol. Evol.* 9:79–80
32. Chesson PL. 1985. Coexistence of competitors in spatially and temporally varying environments: a look at the combined effects of different sorts of variability. *Theor. Popul. Biol.* 28:263–87
33. Chesson PL. 1990. Geometry, heterogeneity and competition in variable environments. *Philos. Trans. R. Soc. London Ser. B* 330:165–73
34. Chesson P. 1991. A need for niches? *Trends Ecol. Evol.* 6:26–28
35. Chesson PL, Case TJ. 1986. Overview: nonequilibrium community theories: chance, variability, history, and coexistence. In *Community Ecology*, ed. J Diamond, TJ Case, pp. 229–39. New York: Harper & Row
36. Chesson P, Ellner S. 1989. Invasibility and stochastic boundedness in monotonic competition models. *J. Math. Biol.* 27:117–38
37. Chesson P, Huntly N. 1993. Temporal hierarchies of variation and the maintenance of diversity. *Plant Species Biol.* 8:195–206
38. Comins HN, Noble IR. 1985. Dispersal, variability, and transient niches: species coexistence in a uniformly variable environment. *Am. Nat.* 126:706–23
39. Connell JH. 1978. Diversity in tropical rain forests and coral reefs. *Science* 199:1302–10
40. Connell JH. 1979. Tropical rainforests

and coral reefs as open non-equilibrium systems. In *Population Dynamics*, ed. RM Anderson, BD Turner, LR Taylor, pp. 141–63. Oxford: Blackwell Scientific
41. Connell JH. 1983. On the prevalence and relative importance of interspecific competition: evidence from field experiments. *Am. Nat.* 122:661–96
42. Connolly SR, Roughgarden J. 1999. Theory of marine communities: competition, predation, and recruitment-dependent interaction strength. *Ecol. Monogr.* 69:277–96
43. Cornell HV. 1993. Unsaturated patterns in species assemblages: the role of regional processes in setting local species richness. In *Species Diversity in Ecological Communities: Historical and Geographical Perspectives*, ed. RE Ricklefs, D Schluter, pp. 243–52. Chicago: Univ. Chicago Press
44. Cornell HV, Karlson RH. 1997. Local and regional processes as controls of species richness. In *Spatial Ecology: the Role of Space in Population Dynamics and Interspecific Interactions*, ed. D Tilman, P Kareiva, pp. 250–68. Princeton, NJ: Princeton Univ. Press
45. Cosner C, DeAngelis DL, Ault JS, Olson DB. 1999. Effects of spatial grouping on the functional response or predators. *Theor. Popul. Biol.* 56:65–75
46. Durrett R, Levin S. 1994. The importance of being discrete (and spatial). *Theor. Popul. Biol.* 46:363–94
47. Ebenhoh W. 1992. Temporal organization in a multispecies model. *Theor. Popul. Biol.* 42:152–71
48. Ellner S. 1985. ESS germination strategies in randomly varying environments. I. Logistic-type models. *Theor. Popul. Biol.* 28:50–79
49. Ellner S. 1985. ESS germination strategies in randomly varying environments. II. Reciprocal yield-law models. *Theor. Popul. Biol.* 28:80–116
50. Ellner S. 1986. Alternate plant life history strategies and coexistence in randomly varying environments. *Vegetatio* 69:199–208
51. Ellner S. 1989. Convergence to stationary distributions in two-species stochastic competition models. *J. Math. Biol.* 27:451–62
52. Ellner SP. 1984. Asymptotic behavior of some stochastic difference equation population models. *J. Math. Biol.* 19:169–200
53. Gendron RP. 1987. Models and mechanisms of frequency-dependent predation. *Am. Nat.* 130:603–23
54. Goldberg D. 1990. Components of resource competition in plant communities. In *Perspectives on Plant Competition*, ed. JB Grace, D Tilman, pp. 27–49. San Diego, CA: Academic
55. Goldberg D, Novoplansky A. 1997. On the relative importance of competition in unproductive environments. *J. Ecol.* 85:409–18
56. Greenman JV, Hudson PJ. 1999. Host exclusion and coexistence in apparent and direct competition: an application of bifurcation theory. *Theor. Popul. Biol.* 56:48–64
57. Grover JP. 1988. Dynamics of competition in a variable environment: experiments with two diatom species. *Ecology* 69:408–17
58. Grover JP. 1990. Resource competition in a variable environment: phytoplankton growing according to Monod's model. *Am. Nat.* 136:771–89
59. Grover JP. 1991. Resource competition in a variable environment: phytoplankton growing according to the variable-internal-stores model. *Am. Nat.* 138:811–35
60. Grover JP. 1997. *Resource Competition.* London: Chapman & Hall. 342 pp.
61. Grover JP, Holt RD. 1998. Disentangling resource and apparent competition: realistic models for plant-herbivore communities. *J. Theor. Biol.* 191:353–76
62. Hairston NG Jr, Ellner S, Kearns CM. 1996. Overlapping generations: the storage effect and the maintenance of biotic

diversity. In *Population Dynamics in Ecological Space and Time*, ed. OE Rhodes Jr, RK Chesser, MH Smith, pp. 109–45. Chicago: Chicago Univ. Press
63. Hairston NG Jr, Van Brunt RA, Kearns CM. 1995. Age and survivorship of diapausing eggs in a sediment egg bank. *Ecology* 76:1706–11
64. Hastings A. 1980. Disturbance, coexistence, history, and competition for space. *Theor. Popul. Biol.* 18:363–73
65. Hastings A, Godfray HCJ. 1999. Learning, host fidelity, and the stability of host-parasitoid communities. *Am. Nat.* 153:295–301
66. Heard SB, Remer LC. 1997. Clutch-size behavior and coexistence in ephemeral-patch competition models. *Am. Nat.* 150:744–70
67. Holt RD. 1984. Spatial heterogeneity, indirect interactions, and the coexistence of prey species. *Am. Nat.* 124:377–406
68. Holt RD. 1985. Density-independent morality, nonlinear competitive interactions and species coexistence. *J. Theor. Biol.* 116:479–93
69. Holt RD, Grover J, Tilman D. 1994. Simple rules for interspecific dominance in systems with exploitative and apparent competition. *Am. Nat.* 144:741–71
70. Holt RD, Lawton JH. 1994. The ecological consequences of shared natural enemies. *Ann. Rev. Ecol. Syst.* 25:495–520
71. Hopf FA, Hopf FW. 1985. The role of the Allee effect in species packing. *Theor. Popul. Biol.* 27:27–50
72. Hopf FA, Valone TJ, Brown JH. 1993. Competition theory and the structure of ecological communities. *Evolutionary Ecology* 7:142–54
73. Hubbell SP. 1979. Tree dispersion, abundance, and diversity in a tropical dry forest. *Science* 203:1299–309
74. Hubbell SP. 1997. A unified theory of biogeography and relative species abundance and its application to tropical rain forests and coral reefs. *Coral Reefs* 16:S9–21
75. Hubbell SP, Foster RB. 1986. Biology, chance, history and the structure of tropical rainforest communities. In *Community Ecology*, ed. J Diamond, TJ Case, pp. 314–29. New York: Harper & Row
76. Huisman J, Weissing FJ. 1999. Biodiversity of plankton by species oscillations and chaos. *Nature* 402:407–10
77. Huntly N. 1991. Herbivores and the dynamics of communities and ecosystems. *Annu. Rev. Ecol. Syst.* 22:477–503
78. Huston M. 1979. A general hypothesis of species diversity. *Am. Nat.* 113:81–101
79. Huston M. 1994. *Biological Diversity.* Cambridge: Cambridge Univ. Press. 681 pp.
80. Huston MA. 1997. Hidden treatments in ecological experiments: re-evaluating the ecosystem function of biodiversity. *Oecologia* 110:449–60
81. Huston MA, DeAngelis DL. 1994. Competition and coexistence: the effects of resource transport and supply rates. *Am. Nat.* 144:954–77
82. Hutchinson GE. 1961. The paradox of the plankton. *Am. Nat.* 95:137–45
83. Hutson V, Law R. 1985. Permanent coexsistence in general models of three interacting species. *J. Math. Biol.* 21:285–98
84. Ives AR. 1988. Covariance, coexistence and the population dynamics of two competitors using a patchy resource. *J. Theor. Biol.* 133:345–61
85. Iwasa Y, Roughgarden J. 1986. Interspecific competition among metapopulations with space-limited subpopulations. *Theor. Popul. Biol.* 30:194–214
86. Iwasa Y, Sato K, Kakita M, Kubo T. 1993. Modelling biodiversity: latitudinal gradient of forest species diversity. In *Biodiversity and Ecosystem Function*, ed. E-D Schulze, HA Mooney, pp. 433–51. Berlin: Springer-Verlag
87. Kinzig AP, Levin SA, Dushoff J, Pacala S. 1999. Limiting similarity, species packing, and system stability for hierarchical

competition-colonization models. *Am. Nat.* 153:371–83
88. Klopfer ED, Ives AR. 1997. Aggregation and the coexistence of competing parasitoid species. *Theor. Popul. Biol.* 52:167–78
89. Kohyama T. 1993. Size-structured tree populations in gap-dynamic forest—the forest architecture hypothesis for the stable coexistence of species. *J. Ecol.* 81:131–43
90. Kotler BP, Brown JS. 1988. Environmental hetergeneity and the coexistence of desert rodents. *Annu. Rev. Ecol. Syst.* 19:281–307
91. Kubo T, Iwasa Y. 1996. Phenological pattern of tree regeneration in a model for forest species diversity. *Theor. Popul. Biol.* 49:90–117
92. Laska MS, Wootton JT. 1998. Theoretical concepts and empirical approaches to measuring interaction strength. *Ecology* 79:461–76
93. Lavorel S, Chesson P. 1995. How species with different regeneration niches coexist in patchy habitats with local disturbances. *Oikos* 74:103–14
94. Law R, Morton RD. 1996. Permanence and the assembly of ecological communities. *Ecology* 77:762–75
95. Leibold MA. 1995. The niche concept revisited: mechanistic models and community context. *Ecology* 76:1371–82
96. Leigh EG Jr. 1982. Introduction: Why are there so many kinds of tropical trees. In *The Ecology of a Tropical Forest: Seasonal Rhythms and Long-Term Changes*, ed. EG Leigh Jr, AR Rand, DW Windser, pp. 64–66. Washington, DC: Smithsonian Inst. Press
97. Leigh EG Jr. 1999. *Tropical Forest Ecology: A View from Barro Colorado Island.* New York: Oxford Univ. Press. 245 pp.
98. Leon JA, Tumpson DB. 1975. Competition between two species for two complimentary or substitutable resources. *J. Theor. Biol.* 50:185–201
99. Levin SA. 1974. Dispersion and population interactions. *Am. Nat.* 108:207–28
100. Levin SA. 1992. The problem of pattern and scale in ecology. *Ecology* 73:1943–67
101. Levins R. 1979. Coexistence in a variable environment. *Am. Nat.* 114:765–83
102. Loreau M. 1992. Time scale of resource dynamics, and coexistence through time partitioning. *Theor. Popul. Biol.* 41:401–12
103. Loreau M, Mouquet N. 1999. Immigration and the maintenance of local species diversity. *Am. Nat.* 154:427–40
104. MacArthur R. 1970. Species packing and competitive equilibrium for many species. *Theor. Popul. Biol.* 1:1–11
105. MacArthur RH, Levins R. 1967. The limiting similarity, convergence and divergence of coexisting species. *Am. Nat.* 101:377–85
106. Molofsky J, Durrett R, Dushoff J, Griffeath D, Levin S. 1998. Local frequency dependence and global coexistence. *Theor. Popul. Biol.* 55:270–82
107. Muko S, Iwasa Y. 2000. Species coexistence by permanent spatial heterogeneity in a lottery model. *Theor. Popul. Biol.* 57:273–84
108. Murdoch WW, Bence J. 1987. General predators and unstable prey populations. In *Predation: Direct and Indirect Impacts on Aquatic Communities*, ed. WC Kerfoot, A Sih, pp. 17–30. Hanover/London: University Press of New England
109. Namba T, Takahashi S. 1993. Competitive coexistence in a seasonally fluctuating environment II. *Theor. Popul. Biol.* 44:374–402
110. Neuhauser C. 1998. Habitat destruction and competitive coexistence in spatially explicit models with local interactions. *J. Theor. Biol.* 193:445–63
111. Pacala SW, Crawley MJ. 1992. Herbivores and plant diversity. *Am. Nat.* 140:243–60
112. Pacala SW, Tilman D. 1994. Limiting similarity in mechanistic and spatial models

of plant competition in heterogeneous environments. *Am. Nat.* 143:222–57

113. Pake CE, Venable DL. 1995. Is coexistence of Sonoran desert annuals mediated by temporal variability in reproductive success? *Ecology* 76:246–61
114. Pake CE, Venable DL. 1996. Seed banks in desert annuals: implications for persistence and coexistence in variable environments. *Ecology* 77:1427–35
115. Pfister CA. 1995. Estimating competition coefficients from census data: a test with field manipulations of tidepool fishes. *Am. Nat.* 146:270–91
116. Ricklefs RE, Schulter D. 1993. Species diversity: regional and historical influences. In *Species Diversity in Ecological Communities: Historical and Geographical Perspectives*, ed. RE Ricklefs, D Schluter, pp. 350–63. Chicago: Univ. Chicago Press
117. Rosenzweig. 1995. *Species Diversity in Space and Time*. Cambridge: Cambridge Univ. Press. 436 pp.
118. Roughgarden J. 1995. *Anolis lizards of the Caribbean: Ecology, Evolution, and Plate Tectonics*. New York: Oxford Univ. Press. 200 pp.
119. Runkle JR. 1989. Synchrony of regeneration, gaps, and latitudinal differences in tree species diversity. *Ecology* 79:546–47
120. Schoener TW. 1978. Effects of density-restricted food encounter on some single-level competition models. *Theor. Popul. Biol.* 13:365–81
121. Schoener TW. 1982. The controversy over interspecific competition. *Am. Sci.* 70:586–95
122. Shigesada N, Roughgarden J. 1982. The role of rapid dispersal in the population dynamics of competition. *Theor. Popul. Biol.* 21:253–373
123. Shmida A, Ellner S. 1984. Coexistence of plant species with similar niches. *Vegetatio* 58:29–55
124. Simberloff D, Dayan T. 1991. The guild concept and the structure of ecological communities. *Annu. Rev. Ecol. Syst.* 22:115–43
125. Smith HL. 1981. Competitive coexistence in an oscillating chemostat. *SIAM J. Appl. Math.* 40:498–522
126. Sokal RR, Rohlf FJ. 1995. *Biometry*. New York: Freeman. 885 pp. 3 ed.
127. Tilman D. 1982. *Resource Competition and Community Structure*. Princeton, NJ: Princeton Univ. Press. 296 pp.
128. Tilman D. 1988. *Plant Strategies and the Dynamics and Structure of Plant Communities*. Princeton, NJ: Princeton Univ. Press. 359 pp.
129. Tilman D. 1990. Constraints and tradeoffs: toward a predictive theory of competition and succession. *Oikos* 58:3–15
130. Tilman D. 1994. Competition and biodiversity in spatially structured habitats. *Ecology* 75:2–16
131. Tilman D, Pacala S. 1993. The maintenance of species richness in plant communities. In *Species Diversity in Ecological Communities: Historical and Geographical Perspectives*, ed. RE Ricklefs, D Schluter, pp. 13–25. Chicago: Univ. Chicago Press
132. Turelli M. 1980. Niche overlap and invasion of competitors in random environments. II. The effects of demographic stochasticity. In *Biological Growth and Spread, Mathematical Theories and Applications*, ed. W Jager, H Rost, P. Tautu, pp. 119–29. Berlin: Springer-Verlag
133. Turelli M. 1981. Niche overlap and invasion of competitors in random environments. I. Models without demographic stochasticity. *Theor. Popul. Biol.* 20:1–56
134. Turelli M, Gillespie JH. 1980. Conditions for the existence of stationary densities for some two dimensional diffusion processes with applications in population biology. *Theor. Popul. Biol.* 17:167–89
135. Vandermeer J, Maruca S. 1998. Indirect effects with a keystone predator:

coexistence and chaos. *Theor. Popul. Biol.* 54:38–43

136. Vincent TLS, Scheel D, Brown JS, Vincent TL. 1996. Trade offs and coexistence in consumer resource models: It all depends on what and where you eat. *Am. Nat.* 148:1039–58
137. Weissing FJ, Huisman J. 1993. Growth and competition in a light gradient. *J. Theor. Biol.* 168:323–36
138. Wills C, Condit R, Foster RB, Hubbell SP. 1997. Strong density- and diversity-related effects help to maintain tree species diversity in a neotropical forest. *Proc. Natl. Acad. Sci. USA* 94:1252–57
139. Wilson JB, Roxburgh SH. 1992. Application of community matrix theory to plant competition data. *Oikos* 65:343–48
140. Woodin SA, Yorke DA. 1975. Disturbance, fluctuating rates of resource recruitment and increased diversity. In *Ecosystem Analysis and Prediction*, ed. SA Levin, pp. 38–41. Philadelphia: Proc. SIAM-SIMS Conf., Alta, Utah, 1974
141. Wootton JT. 1998. Effects of disturbance on species diversity: a multitrophic perspective. *Am. Nat.* 152:801–25
142. Zhang DY, Lin K. 1997. The effects of competitive asymmetry on the rate of competitive displacement: How robust is Hubbell's community drift model? *J. Theor. Biol.* 188:361–67
143. Zobel M. 1992. Plant species coexistence: the role of historical, evolutionary and ecological factors. *Oikos* 65: 314–20

Annu. Rev. Ecol. Syst. 2000. 31:367–93

Temporal Variation in Fitness Components and Population Dynamics of Large Herbivores

J.-M. Gaillard,[1] M. Festa-Bianchet,[2] N. G. Yoccoz,[3] A. Loison,[1] and C. Toïgo[4]

[1]Unité Mixte de Recherche No. 5558 "Biométrie et Biologie Evolutive," Université Claude Bernard Lyon 1, 69622 Villeurbanne Cedex, France; e-mail: gaillard@biomserv.univ-lyon1.fr; loison@biomserv.univ-lyon1.fr
[2]Département de Biologie, Université de Sherbrooke, Sherbrooke, Québec, Canada J1K 2R1; e-mail: mbianche@courrier.usherb.ca
[3]Department of Arctic Ecology, Norwegian Institute for Nature Research, Polar Environmental Centre, N-9296 Tromsø, Norway; e-mail: nigel.yoccoz@ninatos.ninaniku.no
[4]Office National de la Chasse, Division Recherche et Développement, 75017 Paris, France; e-mail: cnerafm@mail.sky.fr

Key Words environmental variation, density dependence, ungulates, demography, critical life-history stage, long-term studies

■ **Abstract** In large-herbivore populations, environmental variation and density dependence co-occur and have similar effects on various fitness components. Our review aims to quantify the temporal variability of fitness components and examine how that variability affects changes in population growth rates. Regardless of the source of variation, adult female survival shows little year-to-year variation [coefficient of variation (CV <10%)], fecundity of prime-aged females and yearling survival rates show moderate year-to-year variation (CV <20%), and juvenile survival and fecundity of young females show strong variation (CV >30%). Old females show senescence in both survival and reproduction. These patterns of variation are independent of differences in body mass, taxonomic group, and ecological conditions. Differences in levels of maternal care may fine-tune the temporal variation of early survival. The immature stage, despite a low relative impact on population growth rate compared with the adult stage, may be the critical component of population dynamics of large herbivores. Observed differences in temporal variation may be more important than estimated relative sensitivity or elasticity in determining the relative demographic impact of various fitness components.

0066-4162/00/1120-0367$14.00

INTRODUCTION: Widespread Large Herbivores in Variable Environments

Large terrestrial mammalian herbivores (with an adult mass of $\geq$10 kg) are found in most ecosystems, from arctic tundra to tropical forest (134). They face not only very different climates, but also great variation in predation pressure, risk of disease, and human interference. Despite these potential sources of temporal variability in survival and recruitment, populations of large herbivores are often considered to be only weakly affected by temporal variation and are often described with deterministic age-structured models (46). In this review we first examine and quantify temporal variation in some fitness components for populations of large herbivores, including survival and reproduction at different stages of an individual's life cycle (Figure 1). Temporal variation in at least one fitness component has been measured for >30 species of large herbivores. We then assess the effects of taxonomic position, ecosystem, and body size on patterns of

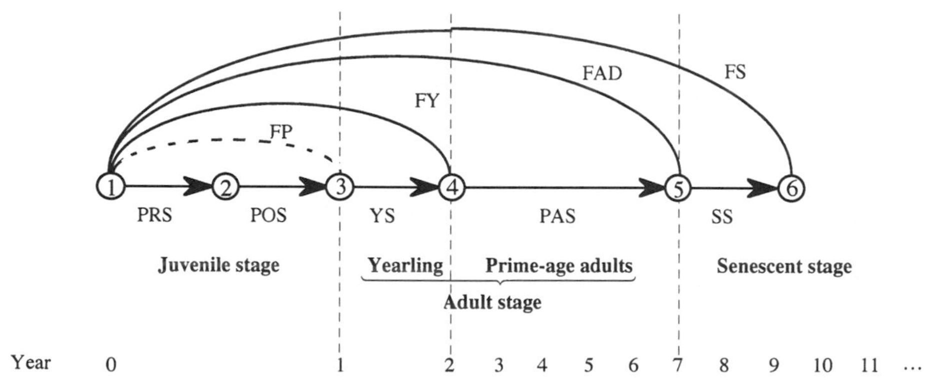

Figure 1 Life cycle graph of a large herbivore female: *Circled numbers*: 1, newborn [most ungulates are birth-pulse species (155)]; 2, weaned young (around 6 months for most species); 3, yearling; 4, 2-year-old (often the minimum age of primiparity); 5, prime-aged; and 6, senescent (older than a threshold age). *Straight lines* indicate transitions from one age group to the next, and *curved lines* indicate reproduction (and therefore production of newborns). These fitness components describe the development of individuals through the life cycle: PRS, pre-weaning survival [summer survival of young in temperate species (e.g. 29)]; POS, postweaning survival [winter survival in temperate species (e.g. 29)]; YS, yearling survival (survival probability between 1 and 2 years); PAS, prime-age adult survival [yearly survival between 2 and 7 years in small- and medium-sized species like Soay sheep (22), roe deer (64), or bighorn sheep (87); between 2 and 12 years in red deer (SD Albon, personal communication)]; SS, senescent survival [yearly survival of females older than a threshold age (7 or 10 years)]; FP, fecundity of yearling females [product of average litter size and proportion of females pregnant; differs from 0 in only a few medium-sized species like white-tailed deer or mule deer (117)]; FY, fecundity of young females [2-year-olds in most cases; in large species, primiparity is at 3 or 4 years or even older (e.g. see 56)]; FA, fecundity of prime-aged females; FS, fecundity of senescent females.

temporal variation. Finally, we propose that there may be a trade-off between the potential importance of a fitness component for changing population growth rate and its observed temporal variation: Those fitness components with the greatest potential impact on population growth rate tend to have the least temporal variability.

Populations of large herbivores display four major types of temporal variation. First, in seasonal habitats, there is predictable environmental variation over each year. Second, year-to-year fluctuations in climate lead to unpredictable, sometimes marked environmental variation. Third, density-dependent responses occur when populations overshoot a threshold density (58). Fourth, changes in abundance or in behavior of predators, prevalence of diseases, or human activities may produce environmental variation. Consequently, temporal variation should play a prominent role in the population dynamics of large herbivores (67, 156), contrary to the simplistic approach that downplays environmental stochasticity for long-lived species. Considerable progress has been made during the last 10 years to better understand demography in stochastic environments (178, 179) and to account for environmental variation, which has been shown to strongly affect estimates of population growth (178) and fitness of various life history strategies (135). It has also been suggested that, for large populations, the extinction risk from demographic stochasticity may be less important than the risk from environmental stochasticity (97), although reliable empirical data are lacking for large herbivores. Therefore, the effects of temporal variation on population dynamics are likely to have strong fundamental and applied implications.

OVERVIEW OF POPULATION DYNAMICS OF LARGE HERBIVORES

Covariation of Body Size, Lifespan, and Iteroparity

Large herbivores are among the heaviest mammals, ranging from <10 to >1000 kg [we excluded species exceeding 800 kg, because their population dynamics may differ from those of smaller species (136) and there is very little information on temporal variation of their fitness components]. Thus, the strong allometric relationships commonly found for most life history traits (144) lead large herbivores to show low fecundity and high adult survivorship (190), with only one or two offspring produced once per year (81) over a potential female lifespan exceeding 15 years (108). Large herbivores have generation times of >4 years (125) and low adult turnover (125). They are strongly iteroparous (69): Females generally reproduce >5 times (12, 13, 20, 29), and a few individuals may reproduce 15 times during a lifetime (63). Only suids deviate from this general model by having large litters (43) and short generation times, at least in heavily hunted populations (71). But because they are omnivores, we exclude suids from our review.

A Life Cycle Graph for Large Herbivores

Populations of large herbivores are strongly age- and sex-structured. In most species, a polygynous mating system leads to pronounced sexual size dimorphism (109), which correlates with marked sexual differences in life history traits (29). In particular, male survival is typically lower than female survival at all ages (29). Age has very strong effects on both reproduction and survival (Figure 1). Large herbivores fit Caughley's model of a dome-shaped age-dependent survival rate (23), with clearly identifiable juvenile (pre- and postweaning), prime-age (adults), and senescent (old adults) stages (64, 87). Here we refer to prime-aged females as those in age classes before the onset of survival senescence [often from 2 to 7–8 years (108)]. Juvenile survival can be subdivided into a preweaning component, during which mortality is mostly dependent on maternal care, and a postweaning component, when most mortality is care independent (66, 111). Yearling survival (from 1 to 2 years of age) is often lower and more variable than survival of prime-aged adults (29, 87) and therefore must be considered separately. Age-related variation in fecundity and litter size is also common, although it is often less pronounced than variation in survival (45). Primiparity is generally at 2 or 3 years, but in some small- or medium-sized species, females can breed during their first year (117). After first reproduction in most species, females attempt to reproduce every year, but in some populations females will not conceive for 1–2 years after weaning an offspring (92, 130, 176). In most ecosystems, births are highly synchronous (155), timed to maximize offspring survival by reducing predation risk (61) and synchronized with seasonal differences in vegetation quality or availability (52, 153).

Demographic Patterns of Populations of Large Herbivores

Populations of large herbivores have low growth rates (16), but compared with other vertebrates, they have high maximum population growth rates relative to their body size. Thus, monotocous species (those with a fixed litter size of one) like horses, red deer, or muskox may reach finite rates of increase (λ) of 1.25 to 1.35 (113, 148, 198), whereas polytocous species (with variable litter size and generally between 1 and 3 offspring per litter) like white-tailed deer may have $\lambda > 1.5$ (181). High potential population growth allows large herbivores to rapidly exploit areas where they may be introduced (24).

Demographic analyses reveal that the elasticity of adult survival is at least threefold higher than that of juvenile survival or of fecundity rates (48, 83, 132, 187). Elasticity measures relative sensitivity, which can be defined as the effect on population growth rate of a proportional change in a given fitness component (39). Therefore, a proportional change in a fitness component with high elasticity will have a greater effect on population growth rate than the same change in a fitness component with low elasticity. Thus, the population growth rate of large herbivores is much more sensitive to a given relative variation in adult survival than to the same relative variation in any other fitness component.

Temporal variation in abundance of large herbivores can have widely different sources, including density- and climate-dependent food limitation or control by humans, predation, and disease. Density-independent limitation (165) and density-dependent regulation (165) co-occur in most populations (115, 165), so that the impact on population growth of density-independent factors such as bad weather typically increases with population density (127, 146).

HOW AND WHY FITNESS COMPONENTS OF POPULATIONS OF LARGE HERBIVORES VARY OVER TIME

Heterogeneity of Data Type and Statistical Analyses: A Methodological Caveat

Ideally, our review should have included only studies based on long-term monitoring of individually recognizable animals, analyzed with methods that account for differences in recapture probability, because those studies minimize errors in estimates of fitness components and reduce the risk of sampling bias. Currently, however, only a handful of studies fit those criteria.

We included studies lasting ≥3 years and providing yearly estimates of at least one fitness component (see supplemental appendix at http://www.annurev.org). We thus faced considerable heterogeneity of data quality and statistical analyses. As a result, it was not always possible to transform the results of different studies into the fitness components defined in Figure 1. Standardized information on juvenile survival and fecundity rates was particularly difficult to extract from the literature. Several studies reported age ratios such as young:female or young:adult female. These ratios are of limited usefulness (116) because they combine juvenile survival and fecundity rate and ignore changes in female age structure. We thus analyzed age ratios separately. Studies also used widely different techniques to estimate fitness components, especially for survival. The quality of the data was highest for studies that monitored individually marked animals of known age and estimated survival by accounting for differences in detection probability of marked animals (49 of 141 studies). When juvenile and yearling survival rates were analyzed separately by sex, we used female survival. Studies based on comparing age ratios in successive years (59 of 141 studies) were of the lowest quality. Survival rates extracted from count ratios are affected by large sampling errors, partly because ratios can change after changes in either the numerator or the denominator. For example, a change in the juvenile:adult ratio could be due to changes in survival (or in sightability) of adults or of juveniles (116).

Temporal Variation in Survival

Preweaning Survival (Immature Stage 1) Most estimates of preweaning survival were based on individually marked newborns. Therefore, data quality was good [28 estimates were of high quality vs 9 of low quality (see supplemental appendix at http://www.annurev.org)]. Preweaning survival is generally low (mean of 0.638, $N = 46$) and varies markedly over time (CV of 0.265, $N = 39$) in most populations (Figure 2) in response to a great diversity of proximal factors. Most preweaning mortality occurs within 1 month of birth (2, 20, 76, 96, 129). It is likely that several studies overestimated preweaning survival, because unless all juveniles are caught and marked immediately after birth, some preweaning deaths, especially of neonates, will not be detected by researchers.

Survival to weaning is generally dependent on maternal care, especially when predation on neonates is not a major source of mortality. Thus, maternal attributes

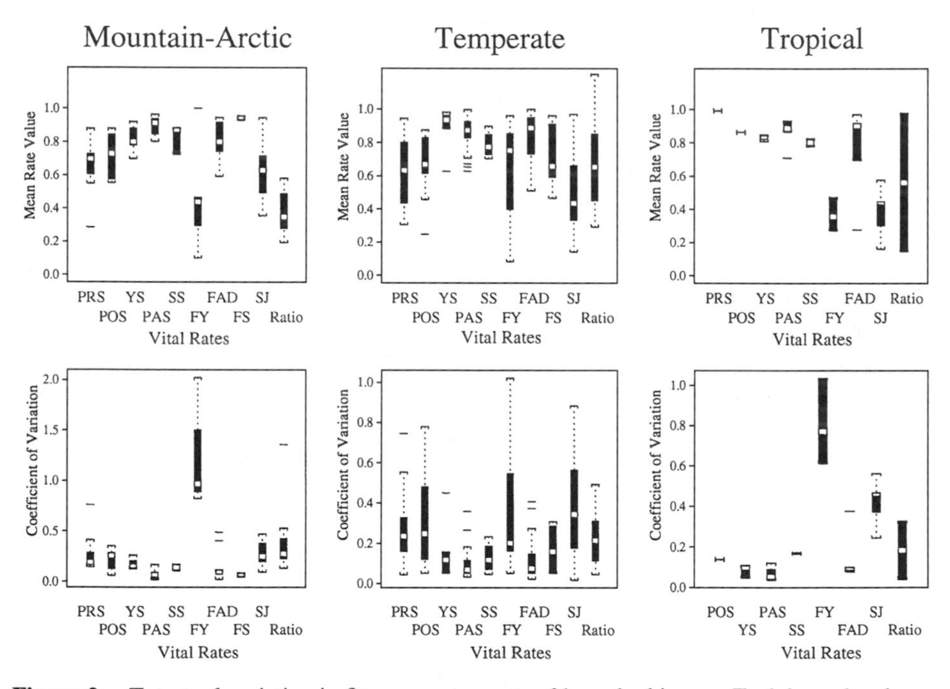

Figure 2 Temporal variation in fitness components of large herbivores. Each box-plot shows, for the mean yearly estimates and their coefficients of variation in a given population for a given ecosystem type (Mountain-Arctic, Temperate, and Tropical), the interquartile (25%–75%) range (given by the *filled box*), 1.5 times this range (*brackets*), the median (*white bar*) and the "outliers" (*horizontal bars*). Note that the scale for CVs for mountain ungulates is different from the others. PRS, preweaning survival; POS, postweaning survival; YS, yearling survival; PAS, yearly survival of prime-age adults; SS, yearly survival of old adults; FY, fecundity of young females; FAD, fecundity of prime-age adults; FS, fecundity of old females; SJ, juvenile survival; Ratio, calf:cow ratio.

such as age (3, 55, 76, 90, 96), size (21, 96, 168), reproductive experience (95, 139, 149), dominance status (31, 106), previous reproductive status (31, 53), or the size of female kin groups (138) can have strong effects on preweaning survival. Generally, the reproductive success of large multiparous prime-aged females of high social rank is much less affected by environmental conditions than is that of small, primiparous, and young females of low social rank.

Weather during gestation can affect preweaning survival. In temperate species, high snowfall and long duration of snow cover during the previous winter often reduce preweaning survival, likely by affecting maternal nutrition during late gestation (2, 114). Likewise, the amounts of precipitation during parturition and lactation cause preweaning survival to vary over years, especially for polytocous species (19, 65, 91) and at high density for monotocous species (146). Weather generally affects early survival by changing the availability of high-quality forage (19); direct effects of adverse weather on survival are exceptional (126). The strong positive correlation between preweaning survival, maternal condition, and weather during gestation and lactation is likely due to the importance of birth weight for early survival, an effect that tends to become stronger with adverse weather or high population density (3, 7, 33, 50, 65, 76, 100, 157, 171, 177). High forage quality and quantity often increase preweaning survival of large herbivores (131, 152, 197), possibly by improving milk quality (197).

Malnutrition appears to be a major cause of early mortality during some years (151) and, in extreme cases, may lead females to abandon their offspring (99). Malnutrition can also predispose juveniles to other sources of mortality. In particular, the transfer of passive immunity to newborns may be compromised at high population density (157).

Density dependence in preweaning survival has been reported in many large herbivores (26, 33, 65, 114, 120, 166), but preweaning survival of red deer (31), reindeer (171), and bighorn sheep (146) did not decrease with increasing density. Preweaning survival may therefore be more sensitive to population density in polytocous species that have a high energy expenditure per breeding attempt than in monotocous species (65).

Where predators are present, predation is often the main source of early mortality (105), and preweaning survival is generally <50% (47) and can be as low as 1% (20). Predation risk, however, is not always independent of maternal care. Maternal experience can decrease fawn vulnerability to predators (20, 139). The timing of birth may also account for temporal variation in preweaning survival. Late birth dates often lower survival in temperate ungulates (31, 33, 52) by shortening the period for access to growing forage. In the presence of predators, caribou calves born during the birth peak may enjoy higher survival than those born earlier or later (2). Finally, parasites (107), disease (126), and high levels of inbreeding (102, 162) may reduce preweaning survival to 10% in some years. Because all of these factors may vary spatially within a population in a given year, spatial heterogeneities in preweaning survival are expected to occur and have been found in pronghorn (50) and caribou (195), but not in red deer (76).

Postweaning Survival (Immature Stage 2) The data available to assess temporal variation in survival from weaning to 1 year include a narrower range of species than those available for preweaning survival. Similar to preweaning survival, data quality was good [20 estimates of high quality vs 7 of low quality (see supplemental appendix at http://www.annurev.org)]. Postweaning survival varied widely from year to year in most populations [mean of 0.697, $N = 30$; CV of 0.279, $N = 26$ (Figure 2)], but fewer factors were reported to affect postweaning survival than preweaning survival. Most reported causes of postweaning mortality, such as winter severity and density dependence, were care independent, and the proximate cause of mortality was usually starvation. Negative effects of severe winters have been reported for several temperate species (62, 78, 101, 166), but muskoxen seem to be highly resistant to deep snow and ice, at least in colonizing populations (86). Density dependence in postweaning survival was reported in several species (11, 31, 33, 85, 146, 166, 171) and appears to be more common in postweaning than in preweaning survival for monotocous species. Bartmann et al (11) provided the best experimental demonstration of density dependence in juvenile survival: By allowing density of mule deer to increase from 44 to 133 per km^2, they caused postweaning survival to decrease from 0.456 to 0.176. Predation (11) and late birth (166) can decrease postweaning survival in some years. Other factors, such as spring weather (161), birth weight (33), and maternal dominance status (31) may affect postweaning survival, but appear to be less important than for preweaning survival.

Because many studies did not distinguish pre- and postweaning survival, we examined overall variation in first-year survival. Data quality was reasonable [21 estimates of high quality vs 19 of low quality (see supplemental appendix at http://www.annurev.org)]. Juvenile survival displayed wide yearly variations in most populations [mean of 0.518, $N = 51$; CV of 0.346, $N = 43$ (Figure 2)]. Predation (41, 49, 73, 159, 180, 192), population density (89, 137, 160, 167, 174, 175, 189, 191), winter severity (9, 110, 146, 148, 159, 167, 192), weather affecting food supply during lactation (49, 137, 172, 189), genotype (143), and care-dependent factors such as birth weight (192), nursing time (160), or mother aggressiveness (159) affect juvenile survival. Therefore, juvenile survival is highly variable within and among populations of large herbivores (67).

Yearling Survival (Immature Stage 3) Because most studies did not estimate yearling survival separately from adult survival, there is a limited amount of reliable information [10 estimates of high quality vs 3 of low quality (see supplemental appendix at http://www.annurev.org)] on temporal variation for this fitness component. Yearling survival showed the same patterns of variation as postweaning survival, but the mean was higher and the variability was lower [mean of 0.872, $N = 16$; CV of 0.124, $N = 14$ (Figure 2)]. Population density (31, 33, 87, 164), adverse weather (137, 148, 175, but see 86), predation (73), and disease (164, 175) accounted for most yearly variation in yearling survival in the absence of hunting.

Adult Female Survival (Prime-Age Stage) Estimates of adult female survival are generally based on high-quality data involving long-term monitoring of recognizable individuals [28 estimates of high quality vs 19 of low quality (see supplemental appendix at http://www.annurev.org)]. The data reveal a striking and consistent pattern of high survivorship and very low yearly variation regardless of the sources of mortality [mean of 0.874, $N = 57$; CV of 0.087, $N = 48$ (Figure 2)].

We found very limited evidence of density dependence in adult survival of large herbivores [in buffalo (120), caused by undernutrition and affecting mainly old animals, and in island populations of Soay sheep (34) and red deer (30), although in both cases senescent animals were included in the estimate of female survival]. Stable adult survival despite wide changes in density has been reported in many species (35, 42, 68, 87, 191).

Adult survival also appears to be partly buffered against environmental sources of variation. Although winter severity (124, 128, 140, 161), adverse spring weather (161), severity of dry season (60), or severe overgrazing of the summer range (57) can decrease adult survival in some species, the survival of adult females was not correlated with any variable in roe deer (68) and was not affected by forage availability in bison (184), by severe drought in greater kudu (137), by food limitation in reindeer (167), by climate in red deer (110), moose (9), mule deer (94), and pronghorn (10), or by pneumonia in chamois (27). Disease may affect survival of adult ungulates, but there is limited evidence from long-term studies. For example, pneumonia epizootics had moderate and short-lived effects on survival of adult female bighorn sheep (87) and mouflon (36). Where large predators occur, they often account for much of the variation in adult survival, typically causing >50% of yearly mortality (73, 126, 164, 188, 194).

It appears that moderate or high levels of temporal variation in adult-female survival are mostly associated with rare events, such as epizootics of exotic diseases and high predation risk due to individual specialist predators or "predator-pit" situations, in which high levels of predation on a preferred but declining prey species can be sustained because of the availability of alternative prey species (73). In large herbivores, the stability of adult female survival relative to other fitness components may reflect a strategy of risk minimization involving a reduction of the maternal expenditure before any serious deterioration of female condition (5, 54, 82, 154).

Old Adult Female Survival (Senescent Stage) In most species of large herbivores, the ages of live females can be estimated reliably only in animals < 3 years old. Consequently, temporal variation in survival of old females can be measured only by very long-term monitoring: Typically, ≥10 years are required before known-age "old" females can be monitored. Very few studies have considered separately the survival of old and prime-age females, but available data were of very high quality [six estimates of high quality vs one of low quality (see supplemental appendix at http://www.annurev.org)]. Compared with prime-age females, old females have lower survival and are more sensitive to environmental variation [mean of 0.811, $N = 9$; CV of 0.164, $N = 8$ (Figure 2)], possibly because of tooth wear (64, 169). Old females are more affected than prime-age females by die-offs in nyala (6), rainfall variation in greater kudu (137), and variation in food availability in reindeer (169). It has recently become evident that individual heterogeneity plays a large role in survival to old age; life expectancy is greater for larger than for smaller prime-aged females (15, 66). Failure to distinguish age classes and the widespread occurrence of survival senescence may also bias the interpretation of reported density dependence in adult survival, as unmanaged high-density populations typically include a high proportion of older females.

Temporal Variation in Reproductive Traits

Fecundity of Young Females Reproductive patterns of large mammals are easier to measure than are patterns of survivorship. Therefore, we found abundant data of high quality on temporal variation in fecundity of young females, defined as those of the youngest age at which $\geq$10% of females were primiparous in a given population. Fecundity of young females is highly variable both within and among populations (8, 120, 163) and is more sensitive to adverse environmental conditions than adult fecundity in both temperate (25) and tropical (196) ungulates [mean of 0.519, $N = 32$; CV of 0.612, $N = 28$ (Figure 2)]. In populations of medium-sized species with abundant nutrition, however, the fecundity of young females can be as high and as stable from year to year as that of prime-age females (20, 63, 84). The main sources of variation in fecundity of young females are population density, weather, and food supply, especially in medium- to large-sized species. Density-dependent responses in age at primiparity have been reported in many large herbivores (25, 55, 59, 70, 85, 91, 120, 168). Female mass during the rut is often the proximate factor of variation in age at first breeding (77, 158, 170). A threshold mass must be reached before young females can reproduce (40, 51, 70, 85, 98), but in bighorn sheep, mass during the rut may play a limited role (88). Finally, adverse weather, such as drought (91) or severe winters (112), can lead to low fecundity of young females. Interpopulation variability in age of primiparity is often caused by differences in nutrition of young females, which can be independent of population density or weather. Thus, some populations in poor habitats may be characterized by late primiparity [e.g. 4–5 years instead of 2 for mountain goats (56)].

Fecundity of Prime-Age Females Many studies provide measurements of temporal variation in fecundity of prime-age females. Unfortunately, however, most studies pool prime-age and old adult females, leading to underestimation of mean values and overestimation of the magnitude of variation. The importance of these biases should depend on the proportion of old females included in the sample, and most studies did not provide that information. As previously mentioned, for most species it is impossible to know the exact age of females first marked as adults.

Fertility of prime-aged females is generally high and varies little from year to year [mean of 0.818, $N = 59$; CV of 0.125, $N = 51$ (Figure 2)]. Density dependence in adult fecundity has been reported in several species, although density effects are generally less evident than for age of primiparity (28, 85, 91, 93, 163, 184). Other species, however, show either a weak (53) or no decline of adult fecundity despite very high population densities (59, 70, 141, 147, 168). Skogland (168) suggested that the fecundity of migratory populations should be less sensitive to environmental variation than that of sedentary populations. Migratory populations of wildebeest, caribou, and elk do show stable fecundity of prime-aged females, but constant adult fecundity has also been reported in sedentary populations of roe deer (70) and fallow deer (147), indicating that adult fecundity of most large herbivores is resilient to a wide variety of environmental conditions and may be a species-specific life history trait. Studies of moose (18), pronghorn (20), and gazelles (8) confirm the high resilience of adult fecundity in ungulates. The limited density-independent, year-to-year variation in adult fecundity usually originates from yearly variation in weather such as March temperature (90), winter and spring precipitation (184), winter severity causing high fetal mortality (10), rainfall (14, 91), or snow depth and summer temperature (37). Body mass may affect adult fecundity (21, 84, 193), and poor nutrition may depress it (1, 150, 163). Body mass and population density can have an interactive effect, so that females of a given mass are less likely to conceive at high than at low population density (5), suggesting a reproductive strategy that minimizes risks to the mother. In some species, individual females may not reproduce in some years, particularly after having weaned an offspring (32, 92, 154).

Fecundity of Old Females Very few studies have investigated variation in fecundity of old females. Similar to what we found for survival, fecundity of old females is lower and more variable than that of prime-age females [mean of 0.783, $N = 7$; CV of 0.134, $N = 6$ (Figure 2)], suggesting reproductive senescence. However, reproductive senescence appears to be less precipitous and to have a later onset than survival senescence (72). Successful reproduction by all but the very oldest females has been reported in medium-sized species (15, 63, 133).

Litter Size In polytocous species, year-to-year variation in litter size is moderate for primiparous females (mean of 1.267, $N = 12$; CV of 0.164, $N = 10$), low for multiparous females (mean of 1.569, $N = 27$; CV of 0.092, $N = 26$), and mostly associated with female nutrition. For captive white-tailed deer, litter size increased from 1.11 for does on a low nutritional plane to 1.96 for those on a high nutritional plane (185). Both winter severity (121) and population density (84, 186) shape yearly variation in litter size by affecting female body mass. In moose, the largest polytocous species, twinning rates are the most variable component of fecundity and may be a sensitive indicator of habitat quality (18). Conversely, in populations of medium-sized species with abundant food, litter size may be fixed and independent of female age (20, 63).

DO TAXONOMY, PHYLOGENY, ECOSYSTEM, AND BODY SIZE AFFECT VARIATION IN POPULATION DYNAMICS OF LARGE HERBIVORES?

Temporal Variation in Fitness Components and Taxonomy: Cervids vs Bovids

Large herbivores include two major families within the order Artiodactyla: cervids and bovids. The mean duration of studies included in our survey did not differ significantly between these groups (7.5 years for cervids and 8.6 years for bovids, $P = 0.38$). Although both yearling survival and litter size of primiparous females are greater for cervids than for bovids, mean estimates of fitness components are generally close (Table 1). In both families, survivorship of yearlings, adults, and old individuals is higher than that of juveniles, while fecundity and litter size of prime-aged females are higher than those of young females (Table 1). Bovids and cervids also display the same patterns of temporal variation in fitness components, with no significant difference between groups (Table 1).

Allometric Component of Temporal Variation in Fitness Components

Allometric relationships are widespread among vertebrates (144), and we expected that body size would have a marked effect on population dynamics of large herbivores. Surprisingly, however, variation in adult mass explained little of the variation in either mean estimates ($r = 0.14$, $P = 0.53$ for prime-age survival; $r = 0.21$, $P = 0.31$ for juvenile survival; $r = -0.26$, $P = 0.24$ for fecundity of adult females; $r = -0.55$, $P = 0.05$ for fecundity of young females; Figures 3*A* and 3*B*) or temporal variation ($r = -0.04$, $P = 0.87$ for prime-age survival; $r = -0.31$, $P = 0.14$ for juvenile survival; $r = 0.40$, $P = 0.06$ for fecundity of adult females; $r = 0.48$, $P = 0.10$ for fecundity of young females; Figures 3*C* and 3*D*) of fitness components. We obtained similar results when we accounted for nonindependence of traits of related species due to phylogenetic inertia (75), using the independent contrasts method (79; Figure 3), by a taxonomy-based phylogeny (see 142 for further details). In particular, prime age ($r = -0.12$, $P = 0.60$) and juvenile survival ($r = 0.08$, $P = 0.70$), as well as their temporal variation ($r = 0.07$, $P = 0.74$ for prime-age survival; $r = -0.07$, $P = 0.74$ for juvenile survival) appeared to be independent of body mass (Figure 3*A*–*C*). On the other hand, there may be an allometric constraint on fecundity, especially for primiparous females, as longer development with increasing size prolongs the period before maturation. The progressive decrease in mean value ($r = -0.59$, $P = 0.04$; Figure 3*B*) and increase in variation ($r = 0.63$, $P = 0.02$; Figure 3*D*) of fecundity of young females as body mass increases support this allometric interpretation. A similar pattern occurs for prime-age fecundity ($r = 0.23$, $P = 0.30$; Figure 3*D*), suggesting that temporal variation in recruitment rates increases with adult body mass.

TABLE 1 Mean estimates (SE) and coefficients of variation (SE) of fitness components for cervids and bovids[a]

Fitness Component	Mean Estimate for Bovids (SE)	Mean Estimate for Cervids (SE)	P-Value	Mean CV for Bovids (SE)	Mean CV for Cervids (SE)	P-Value
Preweaning survival	0.749 (0.084)	0.622 (0.031)	0.110	0.274 (0.098)	0.254 (0.030)	0.797
Postweaning survival	0.676 (0.048)	0.708 (0.032)	0.640	0.296 (0.066)	0.282 (0.048)	0.881
Yearling survival	**0.829 (0.021)**	**0.929 (0.022)**	**0.009**	0.143 (0.027)	0.095 (0.019)	0.209
Prime-aged female survival	*0.895 (0.013)*	*0.855 (0.015)*	*0.076*	0.073 (0.008)	0.094 (0.013)	0.298
Senescent-female survival	0.825 (0.025)	0.794 (0.039)	0.498	0.178 (0.026)	0.140 (0.054)	0.504
Fecundity of young females	0.495 (0.077)	0.539 (0.076)	0.712	0.693 (0.186)	0.578 (0.084)	0.526
Fecundity of adult females	0.834 (0.048)	0.851 (0.021)	0.696	0.160 (0.043)	0.101 (0.014)	0.108
Fecundity of old females	0.944 (0.013)	0.781 (0.092)	0.307	0.065 (0.021)	0.123 (0.073)	0.584
Litter size of primiparous females	**1.119 (0.041)**	**1.372 (0.085)**	**0.04**	0.135 (0.044)	0.183 (0.055)	0.555
Litter size of multiparous females	1.624 (0.137)	1.559 (0.057)	0.633	0.098 (0.044)	0.091 (0.012)	0.834
Calf:cow ratio	*0.411 (0.077)*	*0.596 (0.060)*	*0.069*	0.400 (0.114)	0.245 (0.027)	0.109
Juvenile survival	0.547 (0.041)	0.492 (0.039)	0.338	0.304 (0.033)	0.393 (0.052)	0.283

[a] *P* values were obtained from one-way ANOVAs for each of the 12 fitness components surveyed in this review. Fitness components with significant between-group differences are shown in bold, and those with marginally significant between-group differences are in italics. CV, Coefficient of variation.

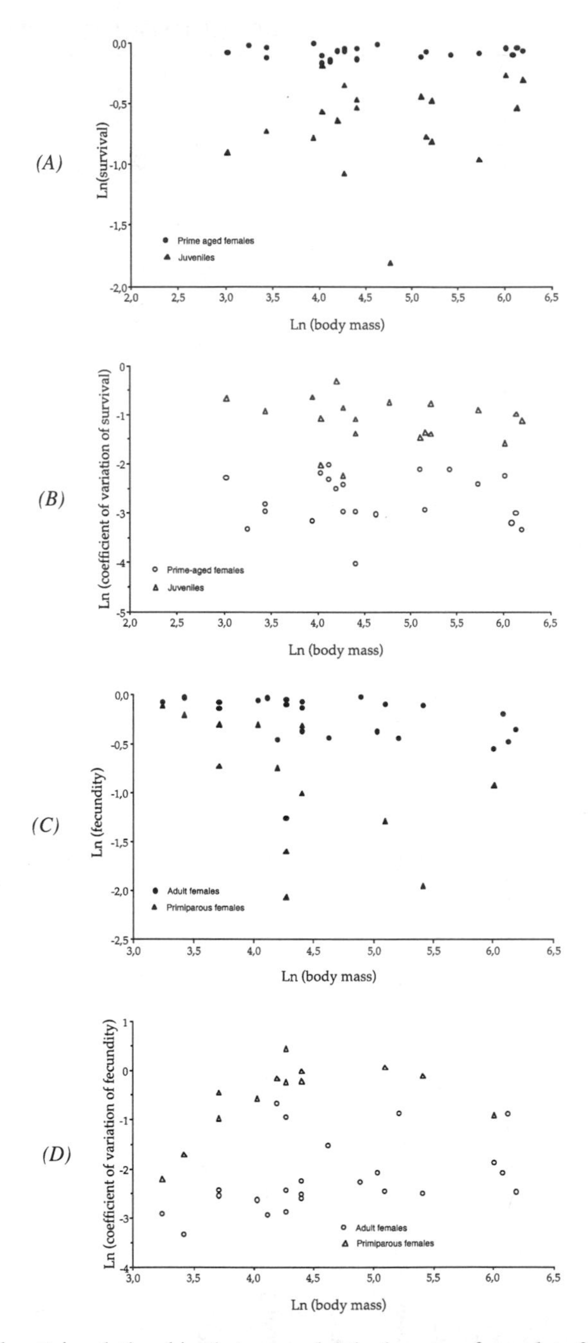

Figure 3 Allometric relationships between adult body mass of ungulate females and (*A*) mean log-transformed estimates of prime-age and juvenile survival, (*B*) log-transformed coefficients of variation of prime-age and juvenile survival, (*C*) mean log-transformed estimates of fecundity for adult and young females, and (*D*) log-transformed coefficients of variation of fecundity for adult and young females. For species with data from more than one population, the points indicate the median for each fitness component.

Temporal Variation in Fitness Components in Different Ecosystems: the Role of Seasonality and of Cover

Large herbivores in our survey have been studied in three major ecosystem types: mountains and the arctic, temperate lowlands, and tropical areas. Despite marked differences in climate and primary production, populations of large herbivores show roughly similar demographic profiles and the same patterns of temporal variation in fitness components in all ecosystems. In most populations, survival and fecundity of prime-age adults are high and constant over time, whereas juvenile survival and fecundity of young females are low and highly variable over time (Figure 2). In each ecosystem type, populations were studied in a wide range of ecological conditions, including considerable variation in food availability, population density, and predation pressure. However, there were no obvious between-ecosystem differences in temporal variation in fitness components. On the other hand, the amount of maternal energy allocated to reproduction seems to affect both magnitude and temporal variation of early survival. In temperate ecosystems, where most polytocous species occurred (81), they tended to show both lower (mean of 0.594, SE $= 0.038$, $N = 22$ after excluding moose and Soay sheep, whose twinning frequencies vary widely from year to year) and more variable (CV of 0.288, SE $= 0.040$, $N = 20$) preweaning survival than monotocous species (mean of 0.728, SE $= 0.038$, $N = 8$; and CV of 0.155, SE $= 0.025$, $N = 7$; $P = 0.10$ and 0.07 for mean and CV, respectively).

IDENTIFYING THE CRITICAL COMPONENTS OF POPULATION DYNAMICS OF LARGE HERBIVORES

Elasticity and Temporal Variability in Fitness Components: Is There a Trade-off?

The demography and population dynamics of large herbivores can be summarized as follows: Recruitment parameters (juvenile survival and some measures of fecundity) combine low elasticity with high temporal variability, whereas adult survival has the highest elasticity and the lowest temporal variability. It is important that these age-related differences in temporal variability occur regardless of whether the source of variation is stochastic (67) or density dependent (44); as we have previously reported (67), the fitness components that are more susceptible to stochastic effects are also more likely to show density dependence. These observations suggest that, in ungulates, there may be a trade-off between the potential importance of a fitness component for changing population growth rate (its elasticity) and the degree of observed temporal variation of that fitness component (its coefficient of variation). It has been suggested that a similar trade-off may also occur in other organisms (145). The resilience of adult survival to environmental variation may be an example of canalization (see 173) of a trait with a very strong influence on fitness.

The relative importance of a fitness component for changes in population growth rate depends on both its elasticity and its temporal variation; a trait such as adult survival with high elasticity but little variability may not have a greater importance in determining changes in population growth rate than a trait such as juvenile survival with low elasticity but high variability. If there was a trade-off between elasticity and temporal variation, then juvenile and adult survival could explain similar amounts of the observed variation in population growth rate. To assess this possibility, we considered three age classes of large herbivores: immature (before the minimum age of primiparity), prime age (from the age of primiparity to the onset of survival senescence), and old, and compared their contribution to temporal variation in population growth rates. We then developed a simple model based on recent developments of demography in stochastic environments (178, 179) and applied it to populations of large herbivores for which temporal variation of all fitness components had been measured.

Temporal Variation Makes the Difference

From five populations, the immature stage accounted for more of the observed variation in growth rate [from 51% to 94% (Table 2)] than either the prime-age or the old stage. For bighorn sheep and roe deer, long-term data were available for two populations. Those studies suggest that the importance of the immature stage for population growth rate may increase with environmental harshness; the

TABLE 2 Proportion of the total variation in population growth rate accounted for by the main life history stages (immature, prime-age, and old) in five populations of large herbivores monitored for >16 years[a]

		Proportion (%) of Population Growth Rate (Accounted for) by Life Stage		
Population	**Species**	**Immature**	**Prime-age**	**Old**
Ram Mountain	Bighorn sheep	69	13	18
Sheep River	Bighorn sheep	55	22	23
Chizé	Roe deer	94	5	1
Trois Fontaines	Roe deer	73	19	7
Rum	Red deer	51	44	4

[a]Temporal variation in the population growth rate can be expressed as the summation of the products of squared elasticities and squared coefficient of variation of fitness components (179). Total variation in population growth rate was thus calculated as $[e^2(PRS) \times CV^2(PRS)] + [e^2(POS) \times CV^2(POS)] + [e^2(YS) \times CV^2(YS)] + [e^2(PAS) \times CV^2(PAS)] + [e^2(SS) \times CV^2(SS)] + [e^2(FY) \times CV^2(FY)] + [e^2(FAD) \times CV^2(FAD)] + [e^2(FS) \times CV^2(FS)] + [e^2(LSP) \times CV^2(LSP)] + [e^2(LSM) \times CV^2(LSM)]$, where *e* is the elasticity, *CV* is the coefficient of variation, *PRS* is the preweaning survival, *POS* is the postweaning survival, *YS* is the yearling survival, *PAS* is the yearly survival of prime-aged females, *SS* is the yearly survival of old adults, *FY* is the fecundity of young females, *FAD* is the fecundity of prime-aged females, *FS* is the fecundity of old females, *LSP* is the litter size of primiparous females, and *LSM* is the litter size of multiparous females. The contribution of each life history stage was calculated as the proportion of total variation in growth rate of a given population that was explained by that stage. (Note: the numbers represent % that sum to 100% for a given population).

contribution of this stage was higher for populations at Ram Mountain (bighorn sheep) and at Chizé (roe deer), both of which showed strong density dependence, than in the more productive populations at Sheep River (bighorn sheep) and Trois Fontaines (roe deer). Although the relative effects of elasticity and temporal variation on ungulate population dynamics have not been previously assessed, most studies suggest an overwhelming importance of the juvenile stage in accounting for between-year variation in population growth rate (32, 35, 80, 106, 123, 183). Exceptions to this pattern may include hunted populations of moose (17), declining populations of caribou (38), and ungulates in Etosha National Park (74), where hunting and predation led to high adult mortality. Elsewhere, we found remarkably similar patterns of variation, despite including in our review populations with (55) and without predators (65) and populations that were introduced or reintroduced (29, 36), feral (34), and semidomestic (7). Because the low elasticity of fitness components during the immature stage is more than compensated for by large temporal variation, for most populations of large herbivores, the immature stage is the critical component of population growth. Therefore, we conclude that temporal variation makes the difference.

PERSPECTIVES

Large herbivores have high economic value; they are often an important source of revenue through sport hunting and ecotourism but can also be agricultural pests or major traffic hazards (119). Consequently, the population dynamics of large herbivores have been the subject of considerable research, and fitness components have been measured in many populations and species. Although new technologies and recent progress in estimation procedures have improved the quality of available data, some problems persist.

Cohort Effects Lead to Interdependence Among Life-History Stages

Contrary to the assumptions of current demographic models, successive life history stages are not independent. Factors affecting fitness components during a cohort's early development may have delayed effects on that cohort's performance later in life (103). Long-term "quality effects" (65) are likely to be pervasive in populations of large herbivores (4, 65, 122, 171, 182) and may lead to an underestimation of the importance of the immature stage in shaping population dynamics.

Partitioning Biological Variability and Sampling Variability

In this review, we did not account for sampling errors that inevitably occur in the estimation of fitness components. The confusion of temporal and sampling variation may bias the assessment of temporal variation and decrease the reliability of comparisons between fitness components (104). To our knowledge, only a study of mule deer reported temporal variation in fitness components after correcting for sampling variation (11). We recommend that future studies attempt to assess more accurately the role of temporal variation in ungulate population dynamics. The available data, however, suggest that the importance of sampling variability is likely greater for adult than for juvenile survival (67). Therefore, if sampling variation were accounted for, the difference in temporal variation between adult and juvenile survival that we reported here would be reinforced.

Data Quality, Modeling, and Population Dynamics: Where Do We Go from Here? More than 20 years ago, Eberhardt (44) suggested that, in marine mammals, an increase of population density would affect fitness components in a predictable order: first juvenile survival, then fecundity of young females, then fecundity of adult females, and last, adult survival. Based on terrestrial large herbivores, our review supports Eberhardt's hypothesis and generalizes it to all sources of temporal variation. For herbivores larger than 50 kg, however, fecundity of young females rather than juvenile survival may be the fitness component most sensitive to environmental perturbations. The production of more realistic and useful population models will require the integration of long-term cohort effects and the partitioning of temporal and sampling variation. It is clear, however, that the greatest obstacle to better understanding the population dynamics of large herbivores is the scarcity of data from long-term field studies of marked individuals. The limited amount of information limits our ability to use a scientific approach for the conservation and management of these ecologically, economically, and socially important animals. A glance at the studies included in our review shows that studies of tropical large herbivores are particularly scarce, as are studies in ecosystems with intact populations of large carnivores. We also found virtually no useful information on kangaroos and wallabies. More than 10 years ago, McNaughton & Georgiadis (118) pointed out that there was a limited amount of data available on population dynamics for the >90 species of large African herbivores. From both a fundamental and an applied viewpoint, the major challenges to improving our knowledge of populations of large herbivores are associated with field ecology. Because studies of large mammals are often costly, require large study areas, and may affect stakeholders such as hunters, farmers, and recreationists, we suggest that increased cooperation between universities and government agencies is required to fully realize the potential of long-term monitoring of marked individuals.

ACKNOWLEDGMENTS

We are grateful to Steve Albon, Mark Boyce, John Byers, Patrick Duncan, Lee Eberhardt, John Fryxell, Stephen Hall, Mark Hewison, Rolf Ims, Jean-Dominique Lebreton, Jim Nichols, Norman Owen-Smith, Dominique Pontier, Bernt-Erik Sæther, and Nils Christian Stenseth for original ideas or comments on previous drafts or for providing us with unpublished data. Financial support was provided by the Programme International de Collaboration Scientifique No. 835 of the Centre

National de Recherche Scientifique (JMG, AL, and NGY) and by the Office National de la Chasse (JMG, AL, and CT). MFB acknowledges the support of the Natural Sciences and Engineering Research Council of Canada; the Alberta Recreation, Parks and Wildlife Fund; the fonds FCAR (Québec); the Foundation for North American Wild Sheep; and the Alberta Conservation Association.

Visit the Annual Reviews home page at www.AnnualReviews.org

LITERATURE CITED

1. Adamczewski JZ, Flood PF, Gunn A. 1997. Seasonal patterns in body composition and reproduction of female muskoxen (*Ovibos moschatus*). *J. Zool.* 241:245–69
2. Adams LG, Singer FJ, Dale BW. 1995. Caribou calf mortality in Denali National Park, Alaska. *J. Wildl. Manage.* 59:584–94
3. Alados CL, Escos J. 1991. Phenotypic and genetic characteristics affecting lifetime reproductive success in female Cuvier's, dama and dorcas gazelles (*Gazella cuvieri, G. dama* and *G. dorcas*). *J. Zool.* 223:307–21
4. Albon SD, Clutton-Brock TH, Guinness FE. 1987. Early development and population dynamics in red deer. II. Density-independent effects of cohort variation. *J. Anim. Ecol.* 56:69–81
5. Albon SD, Mitchell B, Staines BW. 1983. Fertility and body weight in female red deer: a density-dependent relationship. *J. Anim. Ecol.* 52:969–80
6. Anderson JL. 1985. Condition and related mortality of nyala *Tragelaphus angasi* in Zululand, South Africa. *J. Zool.* 207:371–80
7. Asher GW, Adam JL. 1985. Reproduction of farmed red and fallow deer in northern New Zealand. *Bull. R. Soc. N.Z.* 22:217–24
8. Baharav D. 1983. Reproductive strategies in female Mountain and Dorcas gazelles (*Gazella gazella gazella* and *Gazella dorcas*). *J. Zool.* 200:445–53
9. Ballard WB, Miller SM, Whitman JS. 1986. Modeling a south-central Alaskan moose population. *Alces* 22:201–43
10. Barrett MW. 1982. Distribution, behavior, and mortality of pronghorns during a severe winter in Alberta. *J. Wildl. Manage.* 46:991–1002
11. Bartmann RM, White GC, Carpenter LH. 1992. Compensatory mortality in a Colorado mule deer population. *Wildl. Monogr.* 121:1–39
12. Berger J. 1986. *Wild Horses of the Great Basin. Social Competition and Population Size*. Chicago, IL: Univ. Chicago Press
13. Berger J, Cunningham C. 1994. *Bison: Mating and Conservation in Small Populations.* New York: Columbia Univ. Press
14. Berry HH. 1981. Population structure, mortality patterns and a predictive model for estimating future trends in wildebeest numbers in the Etosha National Park. *Madoqua* 12:255–66
15. Bérubé CH, Festa-Bianchet M, Jorgenson JT. 1999. Individual differences, longevity, and reproductive senescence in bighorn ewes. *Ecology* 80:2555–65
16. Blueweiss L, Fox H, Kudzma V, Nakashima D, Peters R, et al. 1978. Relationship between body size and some life history parameters. *Oecologia* 37:257–72
17. Boer AH. 1988. Moose, *Alces alces*, calf mortality in New Brunswick. *Can. Field Natl.* 102:74–75
18. Boer AH. 1992. Fecundity of North American moose (*Alces alces*): a review. *Alces Suppl.* 1:1–10
19. Boyce MS, Tate J. 1982. Pronghorn (*Antilocapra americana*) demography and hunting quotas in the Powder River Basin,

Wyoming. *Trans. Int. Congr. Game Biol.* 14:101–6

20. Byers JA. 1997. *American Pronghorn. Social Adaptations and the Ghosts of Predators Past.* Chicago, IL: Univ. Chicago Press
21. Cameron RD, Smith WT, Fancy SG, Gerhart KL, White RG. 1993. Calving success of female caribou in relation to body weight. *Can. J. Zool.* 71:480–86
22. Catchpole EA, Morgan BJT, Freeman SN, Albon SD, Coulson TN. 1998. *An Integrated Analysis of Soay Sheep Survival Data. Rep. UKC/IMS/98/32.* Canterbury: Univ. Kent
23. Caughley G. 1966. Mortality patterns in mammals. *Ecology* 47:906–18
24. Caughley G. 1970. Eruption of ungulate populations, with emphasis on Himalayan tahr in New Zealand. *Ecology* 51:53–72
25. Challies CN. 1985. Establishment, control, and commercial exploitation of wild deer in New Zealand. *Bull. R. Soc. N.Z.* 22:23–36
26. Choquenot D. 1991. Density-dependent growth, body condition, and demography in feral donkeys: testing the food hypothesis. *Ecology* 72:805–13
27. Clarke CMH, Henderson RJ. 1981. Natural regulation of a non-hunted chamois population. *N.Z. J. Ecol.* 4:126–27
28. Clutton-Brock TH, Albon SD, Guinness FE. 1987. Interactions between population density and maternal characteristics affecting fecundity and juvenile survival in red deer. *J. Anim. Ecol.* 56:857–71
29. Clutton-Brock TH, Guinness FE, Albon SD. 1982. *Red Deer: Behavior and Ecology of Two Sexes.* Chicago, IL: Univ. Chicago Press
30. Clutton-Brock TH, Lonergan ME. 1994. Culling regimes and sex ratio biases in Highland red deer. *J. Appl. Ecol.* 31:521–27
31. Clutton-Brock TH, Major M, Albon SD, Guinness FE. 1987. Early development and population dynamics in red deer. I. Density-dependent effects on juvenile survival. *J. Anim. Ecol.* 56:53–64
32. Clutton-Brock TH, Major M, Guinness FE. 1985. Population regulation in male and female red deer. *J. Anim. Ecol.* 54:831–46
33. Clutton-Brock TH, Price OF, Albon SD, Jewell PA. 1991. Persistent instability and population regulation in Soay sheep. *J. Anim. Ecol.* 60:593–608
34. Clutton-Brock TH, Price OF, Albon SD, Jewell PA. 1992. Early development and population fluctuations in Soay sheep. *J. Anim. Ecol.* 61:381–96
35. Coughenour MB, Singer FJ. 1996. Elk population processes in Yellowstone National Park under the policy of natural regulation. *Ecol. Appl.* 6:573–93
36. Cransac N, Hewison AJM, Gaillard JM, Cugnasse JM, Maublanc ML. 1997. Patterns of mouflon (*Ovis gmelini*) survival under moderate environmental conditions: effects of sex, age and epizootics. *Can. J. Zool.* 75:1867–75
37. Crête M, Courtois R. 1997. Limiting factors might obscure population regulation of moose (Cervidae: *Alces alces*) in unproductive boreal forests. *J. Zool.* 242:765–81
38. Crête M, Couturier S, Hearn BJ, Chubbs TE. 1996. Relative contribution of decreased productivity and survival to recent changes in the demography trend of the Rivière George Caribou Herd. *Rangifer* 9:27–36
39. de Kroon H, Plaisier A, van Groenendael J, Caswell H. 1986. Elasticity: the relative contribution of demographic parameters to population growth rate. *Ecology* 67:1427–31
40. Duncan P. 1992. Horses and grasses: the nutritional ecology of equids and their impact on the Camargue. *Ecol. Stud.* New York: Springer. 87:272 pp.
41. du Plessis SS. 1972. Ecology of blesbok with special reference to productivity. *Wildl. Monogr.* 30:1–71
42. Dusek GL, Mackie RJ, Herriges JD, Compton BB. 1989. Population ecology of

white-tailed deer along the lower Yellowstone River. *Wildl. Monogr.* 104:1–68
43. Dzieciolowski RM, Clarke CMH, Frampton CM. 1992. Reproductive characteristics of feral pigs in New Zealand. *Acta Theriol.* 37:259–70
44. Eberhardt LL. 1977. Optimal policies for conservation of large mammals with special reference to marine ecosystems. *Environ. Conserv.* 4:205–12
45. Eberhardt LL. 1985. Assessing the dynamics of wild populations. *J. Wildl. Manage.* 49:997–1012
46. Eberhardt LL. 1991. Models of ungulate population dynamics. *Rangifer* 7:24–29
47. Epstein MB, Feldhamer GA, Joyner RL. 1983. Predation on white-tailed deer fawns by bobcats, foxes, and alligators: predator assessment. *Proc. Annu. Conf. Southeast. Assoc. Fish Wildl. Agencies, 37th*, pp. 161–72
48. Escos J, Alados CL, Emlen JM. 1994. Application of the stage-projection model with density-dependent fecundity to the population dynamics of Spanish ibex. *Can. J. Zool.* 72:731–37
49. Estes RD, Estes RK. 1979. The birth and survival of wildebeest calves. *Z. Tierpsychol.* 50:45–95
50. Fairbanks WS. 1993. Birthdate, birthweight, and survival in pronghorn fawns. *J. Mamm.* 74:129–35
51. Fandos P. 1989. Reproductive strategies in female Spanish ibex (*Capra pyrenaica*). *J. Zool.* 218:339–43
52. Festa-Bianchet M. 1988. Birthdate and survival in bighorn lambs (*Ovis canadensis*). *J. Zool.* 214:653–61
53. Festa-Bianchet M, Gaillard JM, Jorgenson JT. 1998. Mass and density-dependent reproductive success and reproductive costs in a capital breeder. *Am. Nat.* 152:367–79
54. Festa-Bianchet M, Jorgenson JT. 1998. Selfish mothers: reproductive expenditure and resource availability in bighorn ewes. *Behav. Ecol.* 9:144–50
55. Festa-Bianchet M, Jorgenson JT, Lucherini M, Wishart WD. 1995. Life-history consequences of variation in age of primiparity in bighorn ewes. *Ecology* 76:871–81
56. Festa-Bianchet M, Urquhart M, Smith KG. 1994. Mountain goat recruitment: kid production and survival to breeding age. *Can. J. Zool.* 72:22–27
57. Flueck WT, Smith-Flueck JM. 1996. Kann Energiemangel ein Massensterben unter Cerviden in Sommereinständen der nördlichen Gebirge verursachen? Eine exploratorische Analyse am Schwarzwedelhirsch (*Odocoileus hemionus columbianus*). *Z. Jagdwiss.* 42:85–96
58. Fowler CW. 1987. A review of density dependence in populations of large mammals. In *Current Mammalogy*, ed. HH Genoways, 1:401–41. New York: Plenum
59. Fowler CW, Barmore WJ. 1979. A population model of the northern Yellowstone elk herd. *Proc. Conf. Sci. Res. Natl. Parks, 1st*, ed. RM Linn, 1:427–34. Washington, DC: GPO
60. Fryxell JM. 1987. Food limitation and demography of a migratory antelope, the white-eared kob. *Oecologia* 72:83–91
61. Fryxell JM. 1987. Seasonal reproduction of white-eared kob in Boma National Park, Sudan. *Afr. J. Ecol.* 25:117–24
62. Fuller TK. 1991. Dynamics of a declining white-tailed deer population in north-central Minnesota. *Wildl. Monogr.* 110:1–37
63. Gaillard JM, Andersen R, Delorme D, Linnell JDC. 1998. Family effects on growth and survival of juvenile roe deer. *Ecology* 79:2878–89
64. Gaillard JM, Delorme D, Boutin JM, Laere GV, Boisaubert B, et al. 1993. Roe deer survival patterns: a comparative analysis of contrasting populations. *J. Anim. Ecol.* 62:778–91
65. Gaillard JM, Delorme D, Van Laere G, Duncan P, Lebreton JD. 1997. Early survival in roe deer: causes and consequences

of cohort variation in two contrasted populations. *Oecologia* 112:502–13
66. Gaillard JM, Festa-Bianchet M, Delorme D, Jorgenson JT. 2000. Bodymass and individual fitness in female ungulates: bigger is not always better! *Proc. R. Soc. London Ser. B* 267:471–77
67. Gaillard JM, Festa-Bianchet M, Yoccoz NG. 1998. Population dynamics of large herbivores: variable recruitment with constant adult survival. *Trends Ecol. Evol.* 13:58–63
68. Gaillard JM, Liberg O, Andersen R, Hewison AJM, Cederlund G. 1998. Population dynamics of roe deer. In *The European Roe Deer: The Biology of Success*, ed. R Andersen, P Duncan, JDC Linnell, pp. 309–35. Oslo: Scand. Univ. Press
69. Gaillard JM, Pontier D, Allainé D, Lebreton JD, Trouvilliez J et al. 1989. An analysis of demographic tactics in birds and mammals. *Oikos* 56:59–76
70. Gaillard JM, Sempéré AJ, Boutin JM, Van Laere G, Boisaubert B. 1992. Effects of age and body weight on the proportion of females breeding in a population of roe deer (*Capreolus capreolus*). *Can. J. Zool.* 70:1541–45
71. Gaillard JM, Vassant J, Klein F. 1987. Quelques caractéristiques de la dynamique des populations de sangliers (*Sus scrofa*) en milieu chassé. *Gibier Faune Sauvag.* 4:31–47
72. Garrott RA, Eagle TC, Plotka ED. 1991. Age-specific reproduction in feral horses. *Can. J. Zool.* 69:738–43
73. Gasaway WC, Boertje RD, Grangaard DV, Kelleyhouse DG, Stephenson RO, et al. 1992. The role of predation in limiting moose at low densities in Alaska and Yukon and implications for conservation. *Wildl. Monogr.* 120:1–59
74. Gasaway WC, Gasaway KT, Berry HH. 1996. Persistent low densities of plains ungulates in Etosha National Park, Namibia: testing the food-regulating hypothesis. *Can. J. Zool.* 74:1556–72
75. Gittleman JL, Luh HK. 1992. On comparing comparative methods. *Annu. Rev. Ecol. Syst.* 23:383–404
76. Guinness FE, Clutton-Brock TH, Albon SD. 1978. Factors affecting calf mortality in red deer. *J. Anim. Ecol.* 47:812–32
77. Hamilton WJ, Blaxter KL. 1980. Reproduction in farmed red deer. 1. Hind and stag fertility. *J. Agric. Sci.* 95:261–73
78. Hamlin KL, Mackie RJ. 1989. *Mule deer in the Missouri River Breaks, Montana. A Study of Population Dynamics in a Fluctuating Environment. Rep.* Helena: Mont. Dep. Fish, Wildl. Parks, Study No. BG-1.0
79. Harvey PH, Pagel MD. 1991. *The Comparative Method in Evolutionary Biology.* Oxford, UK: Oxford Univ. Press
80. Hatter IW, Janz DW. 1994. Apparent demographic changes in black-tailed deer associated with wolf control on northern Vancouver Island. *Can. J. Zool.* 72:878–84
81. Hayssen V, van Tienhoven A. 1993. *Asdell's Patterns of Mammalian Reproduction. A Compendium of Species-Specific Data.* Ithaca, NY: Comstock
82. Heard D, Barry S, Watts G, Child K. 1997. Fertility of female moose (*Alces alces*) in relation to age and body composition. *Alces* 33:165–76
83. Heppell SS, Caswell H, Crowder LB. 2000. Life histories and elasticity patterns: perturbation analysis for species with minimal demographic data. *Ecology* 81:654–65
84. Hewison AJM. 1996. Variation in the fecundity of roe deer in Britain: effects of age and body weight. *Acta Theriol.* 41:187–98
85. Houston DB, Stevens V. 1988. Resource limitation in mountain goats: a test by experimental cropping. *Can. J. Zool.* 66:228–38
86. Jingfors KT, Klein DR. 1982. Productivity in recently established muskox populations in Alaska. *J. Wildl. Manage.* 46:1092–96
87. Jorgenson JT, Festa-Bianchet M, Gaillard JM, Wishart WD. 1997. Effects of age, sex, disease and density on survival of bighorn sheep. *Ecology* 78:1019–32

88. Jorgenson JT, Festa-Bianchet M, Lucherini M, Wishart WD. 1993. Effects of body size, population density and maternal characteristics on age of first reproduction in bighorn ewes. *Can. J. Zool.* 71:2509–17
89. Kaji K, Koizumi T, Ohtaishi N. 1988. Effects of resource limitation on the physical and reproductive condition of sika deer on Nakanoshima Island, Hokkaido. *Acta Theriol.* 33:187–208
90. Keiper R, Houpt K. 1984. Reproduction in feral horses: an eight-year study. *Am. J. Vet. Res.* 45:991–95
91. Kie JG, White M. 1985. Population dynamics of white-tailed deer (*Odocoileus virginianus*) on the Welder Wildlife Refuge, Texas. *Southwest. Nat.* 30:105–18
92. Kirkpatrick JF, McCarthy JC, Gudermuth DF, Shideler SE, Lasley BL. 1996. An assessment of the reproductive biology of Yellowstone bison (*Bison bison*) subpopulations using noncapture methods. *Can. J. Zool.* 74:8–14
93. Kirkpatrick JF, Turner JW. 1991. Compensatory reproduction in feral horses. *J. Wildl. Manage.* 55:649–52
94. Klein DR, Olson ST. 1960. Natural mortality patterns of deer in southeast Alaska. *J. Wildl. Manage.* 24:80–88
95. Kojola I. 1993. Early maternal investment and growth in reindeer. *Can. J. Zool.* 71:753–58
96. Kunkel KE, Mech LD. 1994. Wolf and bear predation on white-tailed deer fawns in northeastern Minnesota. *Can. J. Zool.* 72:1557–65
97. Lande R. 1993. Risk of population extinction from demographic and environmental stochasticity and random catastrophes. *Am. Nat.* 142:911–27
98. Langbein J, Putman R. 1992. Reproductive success of female fallow deer in relation to age and condition. In *The Biology of Deer*, ed. RD Brown, pp. 293–99. New York: Springer-Verlag
99. Langenau EE, Lerg JM. 1976. The effects of winter nutritional stress on maternal and neonatal behavior in penned white-tailed deer. *Appl. Anim. Ethol.* 2:207–23
100. Le Bel S, Salas M, Chardonnet P, Bianchi M. 1997. Rusa deer (*Cervus timorensis russa*) farming in New Caledonia: impact of different feed levels on herd breeding rate and performance of new-born fawns. *Aust. Vet. J.* 75:199–203
101. Leader-Williams N. 1980. Population dynamics and mortality of reindeer introduced into south Georgia. *J. Wildl. Manage.* 44:640–57
102. Lent PC, Davis WJ. 1991. Variables influencing survival in four generations of captive-born muskoxen. *Rangifer* 13:137–42
103. Lindström J. 1999. Early development and fitness in birds and mammals. *Trends Ecol. Evol.* 14:343–48
104. Link WA, Nichols JD. 1994. On the importance of sampling variance to investigations of temporal variation in animal population size. *Oikos* 69:539–44
105. Linnell JDC, Aanes R, Andersen R. 1995. Who killed Bambi? The role of predation in the neonatal mortality of temperate ungulates. *Wildl. Biol.* 1:209–23
106. Lloyd PH, Rasa OAE. 1989. Status, reproductive success and fitness in Cape mountain zebra (*Equus zebra zebra*). *Behav. Ecol. Sociobiol.* 25:411–20
107. Logan T. 1973. Study of white-tailed deer fawn mortality on Cookson Hills Deer Refuge in Eastern Oklahoma. *Proc. Southeast. Assoc. Game Fish Comm.* 26:27–35
108. Loison A, Festa-Bianchet M, Gaillard JM, Jorgenson JT, Jullien JM. 1999. Age-specific survival in five populations of ungulates: evidence of senescence. *Ecology* 80:2539–54
109. Loison A, Gaillard JM, Pélabon C, Yoccoz NG. 1999. What factors shape sexual size dimorphism in ungulates? *Evol. Ecol. Res.* 1:611–33
110. Loison A, Langvatn R. 1998. Short and long-term effects of winter and spring

weather on growth and survival of red deer in Norway. *Oecologia* 116:489–500
111. Lycett JE, Henzi SP, Barrett L. 1998. Maternal investment in mountain baboons and the hypothesis of reduced care. *Behav. Ecol. Sociobiol.* 42:49–56
112. Markgren G. 1969. Reproduction of moose in Sweden. *Viltrevy* 6:129–299
113. McCorquodale SM, Eberhardt LL, Eberhardt LE. 1988. Dynamics of a colonizing elk population. *J. Wildl. Manage.* 52:309–13
114. McCullough DR. 1979. *The George Reserve Deer Herd. Population Ecology of a K-Selected Species*. Ann Arbor, MI: Univ. Mich. Press
115. McCullough DR. 1992. Concepts of large herbivore population dynamics. In *Wildlife 2001: Populations*, ed. DR McCullough, RH Barrett, pp. 967–84. London: Elsevier
116. McCullough DR. 1994. What do herd composition counts tell us? *Wildl. Soc. Bull.* 22:295–300
117. McCullough DR. 1997. Breeding by female fawns in black-tailed deer. *Wildl. Soc. Bull.* 25:296–97
118. McNaughton SJ, Georgiadis NJ. 1985. Ecology of African grazing and browsing mammals. *Annu. Rev. Ecol. Syst.* 17:39–65
119. McShea WJ, Underwood HB, Rappole JH. 1997. *The Science of Overabundance. Deer Ecology and Population Management.* Washington, DC: Smithsonian Inst. Press
120. Mduma SAR, Sinclair ARE, Hilborn R. 1999. Food regulates the Serengeti wildebeest: a 40-year record. *J. Anim. Ecol.* 68:1101–22
121. Mech LD, McRoberts RE, Peterson RO, Page RE. 1987. Relationship of deer and moose populations to previous winters' snow. *J. Anim. Ecol.* 56:615–27
122. Mech LD, Nelson ME, McRoberts RE. 1991. Effects of maternal and grandmaternal nutrition on deer mass and vulnerability to wolf predation. *J. Mamm.* 72:146–51
123. Melton DA. 1987. Waterbuck (*Kobus ellipsipyrmnus*) population dynamics: the testing of a hypothesis. *Afr. J. Ecol.* 25:133–45
124. Merrill EH, Boyce MS. 1991. Summer range and elk population dynamics in Yellowstone National Park. In *The Greater Yellowstone Ecosystem*, ed. RB Keiter, MS Boyce, pp. 263–73. New Haven, CT: Yale Univ. Press
125. Millar JS, Zammuto RM. 1983. Life histories of mammals: an analysis of life tables. *Ecology* 64:631–35
126. Miller FL, Broughton E, Gunn A. 1988. Mortality of migratory barren-ground caribou on the calving grounds of the Beverly herd, Northwest Territories, 1981–83. *Can. Wildl. Serv.* 66:1–26
127. Milner JM, Elston DA, Albon SD. 1999. Estimating the contributions of population density and climatic fluctuations to interannual variation in survival of Soay sheep. *J. Anim. Ecol.* 68:1235–47
128. Modafferi RD, Becker EF. 1997. Survival of radiocollared adult moose in lower Susitna River Valley, Southcentral Alaska. *J. Wildl. Manage.* 61:540–49
129. Monard AM, Duncan P, Fritz H, Feh C. 1997. Variations in the birth sex ratio and neonatal mortality in a natural herd of horses. *Behav. Ecol. Sociobiol.* 41:243–49
130. Mundinger JG. 1981. White-tailed deer reproductive biology in the Swan River Valley, Montana. *J. Wildl. Manage.* 45:132–39
131. Murphy DA, Coates JA. 1966. Effects of dietary protein on deer. *N. Am. Wildl. Conf.* 31:129–39
132. Nelson LJ, Peek JM. 1982. Effect of survival and fecundity on rate of increase of elk. *J. Wildl. Manage.* 46:535–40
133. Nichols L. 1978. Dall sheep reproduction. *J. Wildl. Manage.* 42:570–80

134. Nowak. RM. 1991. *Walker's Mammals of the World*, Vol. 2. Baltimore, MD: Johns Hopkins Univ. Press. 5th ed.
135. Orzack SH, Tuljapurkar SD. 1989. Population dynamics in variable environments. VII. The demography and evolution of iteroparity. *Am. Nat.* 133:901–23
136. Owen-Smith N. 1988. *Megaherbivores. The Influence of Very Large Body Size on Ecology*. Cambridge, UK: Cambridge Univ. Press
137. Owen-Smith N. 1990. Demography of a large herbivore, the greater kudu *Tragelaphus strepsiceros*, in relation to rainfall. *J. Anim. Ecol.* 59:893–913
138. Ozoga JJ, Verme LJ. 1984. Effect of family-bond deprivation on reproductive performance in female white-tailed deer. *J. Wildl. Manage.* 48:1326–34
139. Ozoga JJ, Verme LJ. 1986. Relation of maternal age to fawn-rearing success in white-tailed deer. *J. Wildl. Manage.* 50:480–86
140. Pac DF, Mackie RJ, Jorgensen HE. 1991. *Mule Deer Population Organization, Behavior and Dynamics in a Rocky Mountain Environment. Rep.* Mont. Dep. Fish, Wildl. Parks
141. Pascual MA, Hilborn R. 1995. Conservation of harvested populations in fluctuating environments: the case of the Serengeti wildebeest. *J. Appl. Ecol.* 32:468–80
142. Pélabon C, Gaillard JM, Loison A, Portier C. 1995. Is sex-biased maternal care limited by total maternal expenditure in polygynous ungulates? *Behav. Ecol. Sociobiol.* 37:311–19
143. Pemberton JM, Albon SD, Guinness FE, Clutton-Brock TH, Berry RJ. 1988. Genetic variation and juvenile survival in red deer. *Evolution* 42:921–34
144. Peters RH. 1983. *The Ecological Implications of Body Size*. Cambridge, UK: Cambridge Univ. Press
145. Pfister CA. 1998. Patterns of variance in stage-structured populations: evolutionary predictions and ecological implications. *Proc. Natl. Acad. Sci. USA* 95:213–18
146. Portier C, Festa-Bianchet M, Gaillard JM, Yoccoz NG. 1998. Effects of density and weather on survival of bighorn sheep lambs (*Ovis canadensis*). *J. Zool.* 245:271–78
147. Putman RJ, Langbein J, Hewison AJM, Sharma SK. 1996. Relative roles of density-dependent and density-independent factors in population dynamics of British deer. *Mamm. Rev.* 26:81–101
148. Reynolds PE. 1998. Dynamics and range expansion of a reestablished muskox population. *J. Wildl. Manage.* 62:734–44
149. Robinette WL, Baer CH, Pillmore RE, Knittle CE. 1973. Effects of nutritional change on captive mule deer. *J. Wildl. Manage.* 37:312–26
150. Robinette WL, Gashwiler JS. 1950. Breeding season, productivity, and fawning period of the mule deer in Utah. *J. Wildl. Manage.* 14:457–69
151. Roffe TJ. 1993. Perinatal mortality in caribou from the Porcupine herd, Alaska. *J. Wildl. Dis.* 29:295–303
152. Rognmo A, Markussen KA, Jacobsen E, Grav HJ, Blix AS. 1983. Effects of improved nutrition in pregnant reindeer on milk quality, calf birth weight, growth, and mortality. *Rangifer* 3:10–18
153. Rosser AM. 1989. Environmental and reproductive seasonality of puku, *Kobus vardoni*, in Luangwa Valley, Zambia. *Afr. J. Ecol.* 27:77–88
154. Russell DE, Gerhart KL, White RG, van der Wetering D. 1998. Detection of early pregnancy in caribou: evidence for embryonic mortality. *J. Wildl. Manage.* 62:1066–75
155. Rutberg AT. 1987. Adaptive hypothesis of birth synchrony in ruminants: an interspecific test. *Am. Nat.* 130:692–710
156. Sæther BE. 1997. Environmental stochasticity and population dynamics of large

herbivores: a search for mechanisms. *Trends Ecol. Evol.* 12:143–49

157. Sams MG, Lochmiller RL, Qualls CW, Leslie DM, Payton ME. 1996. Physiological correlates of neonatal mortality in an overpopulated herd of white-tailed deer. *J. Mamm.* 77:179–90
158. Sand H. 1996. Life history patterns in female moose (*Alces alces*): the relationship between age, body size, fecundity and environmental conditions. *Oecologia* 106:212–20
159. Sarno RJ, Clark WR, Bank MS, Prexl WS, Behl MJ, et al. 1999. Juvenile guanaco survival: management and conservation implications. *J. Appl. Ecol.* 36:937–45
160. Sarno RJ, Franklin WL. 1999. Maternal expenditure in the polygynous and monomorphic guanaco: suckling behavior, reproductive effort, yearly variation, and influence on juvenile survival. *Behav. Ecol.* 10:41–47
161. Sauer JR, Boyce MS. 1983. Density dependence and survival of elk in northwestern Wyoming. *J. Wildl. Manage.* 47:31–37
162. Sausman KA. 1984. Survival of captive-born *Ovis canadensis* in North American zoos. *Zoo Biol.* 3:111–21
163. Schladweiler P, Stevens DR. 1973. Reproduction of Shiras moose in Montana. *J. Wildl. Manage.* 37:535–44
164. Sinclair ARE. 1974. The natural regulation of buffalo populations in East Africa. III. Population trends and mortality. *E. Afr. Wildl. J.* 12:185–200
165. Sinclair ARE. 1989. Population regulation in animals. In *Ecological Concepts*, ed. JM Cherrett, pp. 197–241. Oxford, UK: Blackwell Sci.
166. Singer FJ, Harting A, Symonds KK, Coughenour MB. 1997. Density-dependence, compensation, and environmental effects on elk calf mortality in Yellowstone National Park. *J. Wildl. Manage.* 61:12–25
167. Skogland T. 1985. The effects of density-dependent resource limitations on the demography of wild reindeer. *J. Anim. Ecol.* 54:359–74
168. Skogland T. 1986. Density-dependent food limitation and maximal production in wild reindeer herds. *J. Wildl. Manage.* 50:314–19
169. Skogland T. 1988. Tooth wear by food limitation and its life history consequences in wild reindeer. *Oikos* 51:238–42
170. Skogland T. 1989. Comparative social organization of wild reindeer in relation to food, mates and predator avoidance. *Adv. Ethol.* 29:1–71
171. Skogland T. 1990. Density dependence in a fluctuating wild reindeer herd: maternal vs. offspring effects. *Oecologia* 84:442–50
172. Smith RH, Lecount A. 1979. Some factors affecting survival of desert mule deer fawns. *J. Wildl. Manage.* 43:657–65
173. Stearns SC, Kawecki TJ. 1994. Fitness sensitivity and the canalization of life-history traits. *Evolution* 48:1438–50
174. Swenson JE. 1985. Compensatory reproduction in an introduced mountain goat population in the Absaroka Mountains, Montana. *J. Wildl. Manage.* 49:837–43
175. Talbot LM, Talbot MH. 1963. The wildebeest in western Masailand, East Africa. *Wildl. Monogr.* 12:1–88
176. Thing H, Klein DR, Jingfors K, Holt S. 1987. Ecology of muskoxen in Jameson Land, northeast Greenland. *Holarct. Ecol.* 10:95–103
177. Thorne ET, Dean RE, Hepworth WG. 1976. Nutrition during gestation in relation to successful reproduction in elk. *J. Wildl. Manage.* 40:330–35
178. Tuljapurkar SD. 1990. *Population Dynamics in Variable Environments.* New York: Springer-Verlag
179. Tuljapurkar S, Caswell H. 1996. *Structured-Population Models in Marine, Terrestrial, and Freshwater Systems.* New York: Chapman & Hall

180. Turner JW, Wolfe ML, Kirkpatrick JF. 1992. Seasonal mountain lion predation on a feral horse population. *Can. J. Zool.* 70:929–34
181. Van Ballenberghe V. 1983. Rate of increase of white-tailed deer on the George Reserve: a re-evaluation. *J. Wildl. Manage.* 47:1245–50
182. Van Ballenberghe V, Ballard WB. 1997. Population dynamics. In *Ecology and Management of the North American Moose*, ed. CC Schwartz, AW Franzmann, pp. 223–45. Washington, DC: Smithsonian Inst. Press
183. Van Sickle J. 1990. Dynamics of African ungulate populations with fluctuating, density-independent calf survival. *Theor. Popul. Biol.* 37:424–37
184. Van Vuren D, Bray MP. 1986. Population dynamics of bison in the Henry Mountains, Utah. *J. Mamm.* 67:503–11
185. Verme LJ. 1965. Reproduction studies on penned white-tailed deer. *J. Wildl. Manage.* 29:74–79
186. Verme LJ. 1991. Decline in doe fawn fertility in southern Michigan deer. *Can. J. Zool.* 69:25–28
187. Walsh NE, Griffith B, McCabe TR. 1995. Evaluating growth of the Porcupine Caribou Herd using a stochastic model. *J. Wildl. Manage.* 59:262–72
188. Wehausen JD. 1996. Effects of mountain lion predation on bighorn sheep in the Sierra Nevada and Granite Mountains of California. *Wildl. Soc. Bull.* 24:471–79
189. Wehausen JD, Bleich VC, Weaver RA. 1987. Mountain sheep in California: a historical perspective on 108 years of full protection. *Trans. West. Sect. Wildl. Soc.* 23:65–74
190. Western D. 1979. Size, life-history and ecology in mammals. *Afr. J. Ecol.* 17:185–204
191. White GC, Bartmann RM. 1998. Effect of density reduction on overwinter survival of free-ranging mule deer fawns. *J. Wildl. Manage.* 62:214–25
192. White GC, Garrott RA, Bartmann RM, Carpenter LH, Alldredge AW. 1987. Survival of mule deer in northwest Colorado. *J. Wildl. Manage.* 51:852–59
193. White RG, Rowell JE, Hauer WE. 1997. The role of nutrition, body condition and lactation on calving success in muskoxen. *J. Zool.* 243:13–20
194. Whitlaw HA, Ballard WB, Sabine DL, Young SJ, Jenkins RA, et al. 1998. Survival and cause-specific mortality rates of adult white-tailed deer in New Brunswick. *J. Wildl. Manage.* 62:1335–41
195. Whitten KR, Garner GW, Mauer FJ, Harris RB. 1992. Productivity and early calf survival in the Porcupine caribou herd. *J. Wildl. Manage.* 56:201–12
196. Williamson DT. 1991. Condition, growth and reproduction in female red lechwe (*Kobus leche leche* Gray 1850). *Afr. J. Ecol.* 29:105–17
197. Wilson DE, Hirst SM. 1977. Ecology and factors limiting Roan and Sable antelope populations in South Africa. *Wildl. Monogr.* 54:1–111
198. Wolfe ML. 1980. Feral horse demography: a preliminary report. *J. Range Manage.* 33:354–60

Annu. Rev. Ecol. Syst. 2000. 31:395–423

IMPACTS OF AIRBORNE POLLUTANTS ON SOIL FAUNA

Josef Rusek

Institute of Soil Biology, Academy of Sciences, Na sádkách 7, 370 05 České Budějovice, Czech Republic; e-mail: rusek@upb.cas.cz

Valin G. Marshall

Applied Research Division, Royal Roads University, Victoria, British Columbia, Canada V9B 5Y2; e-mail: vmarshall@royalroads.ca

Key Words soil biota, soil pH, acid precipitation, bioindicators, litter decomposition

■ **Abstract** The impacts of airborne pollutants have been studied in only a few groups of soil animals, notably protozoans, nematodes, potworms, earthworms, mites, and collembolans. Pollutants in the form of acid depositions, which contain SO_4^{2-}, NO_x, H^+, heavy metals, and some organic compounds, are not homogeneously distributed on the landscape. Deposition patterns depend mainly on landscape configuration and plant cover. Airborne pollutants affect soil animals both directly and indirectly. Direct toxic effects are associated with uptake of free acidic water from the environment by some soil animals and with consumption of polluted food by others. Indirect effects are mediated primarily through disappearance or reduction of the food resources (microflora and microfauna) of soil animals, changes in organic matter content, and modification of microclimate. In the field, changes in competition among species are probably important factors that influence the soil animal community structure as well as the reactions of individual species to soil acidification or liming. The overall effect is a depauperation of soil with an attendant reduction in the rate of organic matter decomposition. We have provided five hypotheses, using soil fauna as indicators, to allow for quick evaluation of environmental changes caused by airborne pollutants.

INTRODUCTION

Ecosystem damage on a global scale has been accelerating during the past few decades. Human activities such as mining, forest cutting, agricultural practices, and land transformation directly affect terrestrial ecosystems. Activities with indirect effects (e.g. emissions from industrial plants) often combine with the direct effects to cause damage. Airborne pollutants affect the soil biota directly through

0066-4162/00/1120-0395$14.00

poisoning and indirectly through changes in vegetation and chemical and physical soil processes.

The soil biota is an important component of all terrestrial ecosystems. It controls many processes in detrital food webs, enhances the decomposition of organic matter, and is essential for humus formation and its further transformation (34). The biota speeds up the mineralization of complex organic compounds into the inorganic salts required by higher plants. Soil animals are also suitable as indicators of the effects of airborne pollutants at the ecosystem level.

Airborne pollutants, chemically very complex compounds, enter the soil via the atmosphere, sometimes as acid deposits that contain H^+, NO_x, SO_3^-, and SO_4^{2-}. Other pollutants, such as CaO dust, can increase soil pH and enhance trapping of heavy metals by organic matter (19). All of these pollutants can be toxic directly to soil animals or can induce pH changes that have complex effects on the soil fauna.

AIRBORNE POLLUTANTS

Pollutants enter the atmosphere from various emission sources. Natural sources include volcanic eruptions, forest fires, and outgassings from anaerobic wetlands (56). However, the major sources are human activities related to the burning of fossil fuels and agricultural activities.

Pollutants from coal-burning plants, which are mainly SO_2 and fly ash, amount to ~900 tons $year^{-1}$ for a 500-MW plant. The Netherlands alone produces ~300,000 tons $year^{-1}$ of fly ash, most of which is deposited on soil. Fly ash consists of noncombustible particles of diverse sizes (<5–100 μm), with surfaces that facilitate adsorption of heavy metals and other chemicals from coal. Inclusions vary depending on the type of coal but, in parts per million, they range as follows: As, 20–40; Cd, 4–<10; Cr, 100–140; Cu, 125–330; Hg, 0–<10; K, 3000–25,000; Na, 3000–15,000; Ni, 150–250; Pb, 100–410; and Zn, 260–920 (42).

Harris (89) classified pollutants entering the environment as point spills, chronic local releases, or widespread releases. The most serious types are the widespread releases. More than 1000 substances are manufactured in quantities capable of polluting the entire globe (30). However, only six of these have caused truly global pollution: radioactive fallout, DDT, PCBs, fluorinated hydrocarbons, carbon dioxide, and oxides and other end products of sulfur and nitrogen, especially acid rain (144). Although DDT and radioactivity can be detrimental to soil fauna, the focus of this review is on pollutants that induce major changes in soil chemistry, especially pH.

Acid depositions contain SO_4^{2-}, NO_x, H^+, heavy metals, and some organic compounds. Good estimates exist for annual bulk deposits of these ions and compounds in parts of Europe and North America (56, 102, 113, 120, 184, 188). For instance, deposition values of 3 g S m^{-2} $year^{-1}$ and 2.5 g N m^{-2} $year^{-1}$ are reached in spruce forests in southern Scandinavia (131). These pollutants are not deposited homogeneously; instead, deposits are influenced by landscape configuration and plant cover. Turbulent diffusion and convection air currents transport fine particles to localized areas within forests.

The mutual influences of forest stands and orography have explained increased levels of deposition on hillsides (165). Forest stand density strongly influences turbulence levels and greatly affects deposition levels, both locally and regionally. Acid deposits are higher in places with increased snow accumulation or in runoff gullies (170). Airborne pollutants carried by stemflow (14, 145, 222) concentrate near deciduous trees. In beech forests, almost 30% of rainfall may enter the soil near the trees as stemflow (121). Consequently, the soil fauna is affected by airborne pollutants in accordance with their distribution pattern in the landscape and in individual ecosystems.

IMPACTS ON VARIOUS TAXONOMIC GROUPS

Protozoa

Estimates of protozoan abundance in soil suggest that numbers may reach 1×10^9 and biomass may be ≤ 20 g m^{-2} (190). Protozoans, being mainly bacterial feeders, have an indirect role in litter decomposition. However, their possible rapid turnover in soil can have a profound effect on the release of plant nutrients. Protozoans are also important pathogens of soil animals.

The distribution of soil protozoa (Rhizopoda, Ciliata, and Flagellata) is influenced by soil pH (189). Amoeban species (naked Rhizopoda) show a wide distribution. Many species of the Testacea (Rhizopoda with "shell"), although widely distributed, prefer acid conditions. The richest ciliate fauna occurs in calcareous soils. Most flagellates will grow over a wide pH range.

Under natural conditions, substrate pH influences the composition of testacean communities (24). The numbers of both species and individuals are generally higher in mosses and sphagnum bogs that have low pHs. As substrate pH reaches neutral and alkaline ranges, protozoan assemblages are less rich and are normally dominated by eurytopic species.

Although high populations of protozoans persist in acid soils, artificial acid rain, which lowers soil pH to ~3, is detrimental to Testacea (215). Such acidification reduces testacean species by about 23% and reduces population density and biomass by two- and four-fold, respectively.

The reaction of soil protozoa to industrial pollution depends on the taxonomic group (196, 197). Acidophilic Testacea predominate in heavily polluted areas, whereas Ciliata and Flagellata persist in less polluted soils. Soil acidification also causes many Flagellata to encyst.

Pollutants (e.g. SO_2 emissions) increase the level of protozoan parasitism in soil invertebrates. Infection rates in heavily polluted forests are high: 60%–80% of Enchytraeidae; 20%–30% of Lumbricidae; 40%–50% of Oribatida; 20%–30% of Collembola; and 5%–10% of Diptera are infected with protozoans. In unpolluted sites, infection rates range from 0 to 20% (161–163). Collembolan infection by

Microsporidia and Coccidia is also high in soils polluted by acid deposits (J Rusek, unpublished observations). We assume that pollutants, like pesticides (160), have a detrimental effect on the immune system of soil invertebrates.

Protozoan responses to soil acidification and their infection of soil invertebrates have led researchers to recommend them as useful bioindicators of airborne pollutants (51, 52, 179, 214).

Nematoda

Nematodes are abundant and ubiquitous in soil, with estimates as high as 30×10^6 nematodes m^{-2} (210) and 20,000 species (26). Like protozoans, their role in organic matter decomposition is indirect. Nevertheless, they contribute significantly to release of plant nutrients (34). There are many nematode trophic groups: bacterivores, fungivores, omnivores, predators (e.g. on protozoans, rotifers, andnematodes), and plant parasites.

Generally, nematode distribution is influenced by pH, with typical acidophilic, neutrophilic, and basiphilic species (28). Many plant parasites occur in soils with high pH. However, the reaction to soil pH is complex. For example, *Tylenchorhynchus dubius* has its highest oviposition rate at pH 4.0, but the greatest number of larvae hatches at pH 7; reproduction is inhibited at pH 3. Although, under field conditions, *T. dubius* is found in sites with soil pH that ranges from 5.1 to 8.5, the highest density of this species occurs at pH 5.6–6.0 (27). On the other hand, the reproduction rate in *Acrobeles complexus* is only slightly influenced by pH within a range of pH 4.0–7.7 (199). Suppression of reproduction in *A. complexus* at pH 4 was attributed to the inhibition of associated bacteria that grew on asparagine-mannitol agar plates (199).

Nematodes usually decrease in density after artificial acidification (59, 60, 75, 105, 106, 128, 186, 200). Nematode density may sometimes increase with soil acidification, but with a concurrent change in the community structure. In an experiment in a spruce forest in south Germany, species numbers were reduced threefold 1.5 years after acid application, whereas acidophilic species increased greatly in numbers (164). The negative effect of artificial acid rain was even more obvious in deeper soil layers (164), where many dominant genera, present throughout the 0- to 6-cm soil layers in natural soil, disappeared in the 4- to 6-cm soil layer with acid treatment.

Furthermore, acidification reduces the colonizers-persisters c-p1 (c-p1 to c-p5 is an *r*- to *K*-strategy continuum, where c-p1 species have the highest and c-p5 the lowest reproduction rates) (see 22) among bacterivore genera such as, *Panagrolaimus, Diploscapter*, and *Rhabditis* s.l. (216). Root feeders tend to decrease with increasing acid concentration (29). Some bacterial feeders (e.g. *Acrobeloides buetschlii*) continue to increase in density at pH 1.0–1.5 (166). We suspect that such increased density of bacteriovores observed at low pH in the laboratory is probably not sustained under field conditions because of eventual suppression of bacterial populations.

Changes in the trophic structure of nematodes after acidification are related to food resources (95). The soil microflora responds to reduction of soil pH with increased dominance of fungi (6, 46). When bacteria decrease after acid treatment, nematode species, notably bacterial feeders, also decrease (60). In a stressed environment with a low soil microbial activity, such as an acidified forest, c-p2 nematodes tend to dominate (35).

Liming at low levels generally increases nematode populations in forest soils (8, 200), but these increases are not sustained because increased biological activity depletes soil organic horizons (136), with an accompanying increase in the numbers of the bacterivorous nematodes (36).

When pollutants (e.g. CaO dust) increased soil pH from 3.5 to 7, nematode populations also changed (8). At lower levels of lime (CaO), species diversity remained unchanged, but bacterivores increased and plant-parasitic and saprophagous genera decreased. However, as soil pH approached 7, the density and number of species of nematodes decreased substantially. In severely damaged sites, higher microbial activity induced an increase in saprophagous and predatory nematodes.

Pollutants that lower pH (e.g. SO_2 and elemental sulfur) are inimical to nematodes (8), producing density reduction and a shift in species dominance (64, 65, 167) characteristic of the changes that result from other forms of acidification mentioned above. This shift can be used to determine the extent of pollution damage and to predict community structure. For example, Háněl (87, 88) found that species of *Malenchus* and *Ecphyadophora* disappeared in heavily polluted forests, whereas the abundance of *Plectus, Metateratocephalus, Rhabditis, Alaimus, Filenchus, Ditylenchus*, and *Eudorylaimus* decreased. The highest *Acrobeloides/Filenchus* ratio occurred in the most polluted sites.

Ohtonen et al (152) did not find nematode numbers correlated with air pollutant gradient (1.2–2.8 mg S g^{-1} of mor humus, the organic layers of a pine forest soil). However, nematode numbers were positively correlated with dehydrogenase activity, mineral N, and soil pH. Because dehydrogenase activity and soil acidity are closely related (151), the results were not surprising. The lack of correlation of nematodes and pollution gradient suggested that S deposition on the soil was not uniform enough to show a clear relationship.

Air pollution has been credited with the success of the hymenopteran *Cephalcia abietis*, a pest of spruce forests (9). The nematode *Steinernema kraussei* is an endoparasite of *C. abietis*. Soil acidity plays a major role in the nematode's ability to parasitize nymphs of the insect (49). There is a positive correlation between soil pH and both the nematode density and number of insects parasitized. Acidic soils (pH <4.0) limit the nematode's ability to find diapausing nymphs of the insect, which may spend more than a year in the mineral soil, where they are exposed to parasitism by *S. kraussei* (147). If nematode density is reduced, the nematode loses its ability to control host density. When soils are amended with lime or magnesium fertilizers to raise pH to 4, there is a significant increase in nematode densities and parasitism rates on the insect pest (58).

Enchytraeidae

Potworms are generally more numerous in boreal forests with acidic mor or moder humus forms than in neutral to basic mull humus. Mor, moder and mull are distinct humus forms. Mors are characterized by a clear separation of organic layers above mineral horizons, are generally acid, and support abundant fungal growth and small invertebrates. Mulls have a more uniform distribution of organic and mineral constituents throughout the soil profile, are generally neutral to slightly alkaline, and support bacterial growth and large invertebrates, notably endogeic and anecic earthworms. Moders are intermediates between mors and mulls. They resemble mors in the distribution of organic and mineral components, and may contain large arthropods and small epigeic earthworms. Other potworm species living in neutral and calcareous (alkaline) soils usually do not reach densities as high as those in acid soils.

Acidophilic species, which tolerate pH ranges 2.5–6.0 (92, 150, 185, 187), usually react negatively to liming, but increase with acidification. Two well-studied acidophilic species are *Cognettia sphagnetorum* (2, 3, 6, 7, 81–83, 111, 129, 132) and *Mesenchytraeus pelicensis* (2, 3, 81–83). Basiphilic species, which generally increase after liming, include *Enchytraeus buchholzi* (2, 3, 81, 82, 111), *Enchytrona parva* (2, 3, 81, 82), *Fridericia paroniana* (83), and *Mesenchytraeus cambrensis* (111).

Because acidophilic potworms may suffer in soils with pH <2.5 (2, 3, 81, 82), attempts have been made to evaluate how such pH reductions might relate to long-range transported pollutants. Sulfur, applied at 98 and 294 kg per hectare (ha), reduced the abundance of *C. sphagnetorum* to 76% and 11% of control values, respectively (6, 7, 129, 132). For comparison, the total deposition of S in the area of Norway most exposed to deposition of long-range transported sulfur is approximately 20 kg ha^{-1} $year^{-1}$ (153). Although an application of 98 and 294 kg ha^{-1} is high compared to long-range transported sulfur, subtle changes in potworm populations have also been observed for the last.

Emissions from an N fertilizer factory eliminated all species but *C. sphagnetorum* (64, 65). When Górny (66, 67) implanted healthy soil in industrially polluted areas, enchytraeids were eliminated within 2 months. Chalupský (32) showed that both the density and dominance relationships of potworms are affected. An undamaged site averaged 51,000 individuals m^{-2} of potworms. It had 10 species dominated by *C. sphagnetorum* ($D = 60\%$), followed by *Marionina cambrensis* ($D = 22\%$). A heavily damaged site harbored 69,000 individuals m^{-2} from 11 species, but the dominance relationship between the two species changed: *C. sphagnetorum* was 76% and *M. cambrensis* was 9%. Chalupský (33) hypothesized that damaged sites will favor *r*-strategists (e.g. *C. sphagnetorum*), whereas stable unpolluted sites will support *K*-strategists (e.g. *M. cambrensis*).

The mechanisms responsible for potworm decline as a result of pollution are not known, but some key factors are most likely changes in food and feeding relationships, increased soil conductivity, and direct poisoning. Acidification (pH 3.1–4.4) and liming (1500 and 4000 kg $CaCO_3$ powder ha^{-1}) affect enzymatic activity in *Fridericia* sp. (202). Acidification decreases enzymatic activities of xylanase and trehalase, but increases those of amylase and cellulase, probably because of the stimulatory effect of acidification on soil amylolytic and cellulolytic microorganisms. Liming decreases both xylanase activity and biomass, the only known negative effects on potworms from liming.

Abundance of *C. sphagnetorum* is positively correlated with soil respiration rate and diversity and production of mycorrhizal fungi, all of which are negatively related to S (1.2–2.8 mg S g^{-1} mor humus) and N concentration in the soil (152). Ohtonen et al (152) attributed the negative correlation to an alteration in the food supply of enchytraeids because of changes in litter quality and amount, as well as changes in species composition of fungi and mycorrhizae.

There seems to be no clear relationship between population densities of enchytraeid species and fundamental chemical properties of soil (83). However, potworms, especially *C. sphagnetorum*, are negatively correlated ($r = -0.82$) with electrical conductivity of the soil solution (94, 96), suggesting that hypertonic soil solutions may upset the water balance of the worms.

Enchytraeid numbers are reduced by deposition of Cu, Zn, and Pb (11). The reasons for such reductions are probably similar to those for earthworms, which are treated below.

Lumbricidae

Boreal coniferous forest soils are generally acid and devoid of many lumbricid species. In acid soils, *Dendrobaena octaedra, Dendrodrillus rubidus*, and *Lumbricus rubellus* may be common, whereas *Aporrectodea rosea, Aporrectodea caliginosa*, and *Lumbricus castaneus* are less frequently encountered and usually at low densities (1, 53, 67, 106). Lumbricids in temperate regions have been classified into three pH-related groups: acid-tolerant species (pH <5), ubiquitous species (pH 4.0–>7.0), and acid-intolerant species (pH ≥4.6) (25, 174).

Experimental acidification decreases numbers of lumbricids (78, 155–158). Except for the epigeic acidophile, *D. octaedra*, species that are common in coniferous forests are all negatively affected. Even *D. octaedra* is adversely affected where soil pH is very low (<3) in localized areas (31). On the other hand, liming usually has a positive effect on lumbricid density (60, 68, 90, 91, 175, 176, 218, 219). Sometimes an increase in total population is accompanied by a decrease in both the density and biomass of individual worms (60). The population structure may also change, with species such as *L. rubellus* being suppressed by increased densities of *Dendrobaena rubida* in limed plots. The density of *L. rubellus* and *A. caliginosa* may increase occasionally, but surprisingly, only that of *D. octaedra*, an acidophile, increases consistently following liming (157).

Industrial pollution may eliminate all species except *D. octaedra* in places where *D. octaedra, D. rubidus, L. rubellus*, and *A. caliginosa* thrived in unpolluted soil (67). Both survival and reproduction of *L. rubellus* are negatively affected by low amounts of fly ash (42). High amounts of Al in fly ash might improve earthworm survival, but Cu is distinctly detrimental. Generally, fresh fly ash is toxic, but aged ash, which contains fewer toxicants, promotes development of soil fauna (38, 39, 42).

The mechanisms responsible for the effects of pollutant stressors on earthworms are poorly understood, but are likely to be many. Earthworms lose large amounts of electrolytes during acute acid stress. Heavy metals affect cellular membrane permeability, and metals such as Cd mimic metal cofactors and affect enzyme activity. Cd, Pb, and H_2SO_4 cause enlargement of epithelial gland cells of *L. terrestris* (72), *A. caliginosa* (69, 70), and *Eisenia fetida* (220, 221). These enlarged gland-cells produce copious secretions that thicken the mucus layer covering the cuticle (70). The quality or quantity of mucus affects integumentary transport and other vital processes (72), including reproduction, lubrication, gas exchange, and cation exchange, and acts as a diffusion barrier. Generally, mucus-cell hyperplasia provides some protection from the action of stressors. However, worms suffer when Cd_2^+, Pb_2^+, H^+, and other toxicants are absorbed from soil or water in large quantities by the integument (50, 220).

Acarina

Mites represent a major arthropod group with an estimated 1 million living species (213). They abound in soil where more than 1×10^6 individuals m^{-2} may occur. The Mesostigmata, Prostigmata, Astigmata, and Cryptostigmata (Oribatida) are common orders of soil mites. As saprovores, fungivores, and predators, they play a vital role in soil decomposition food webs. Some oribatids are important comminutors of coniferous needle litter and wood (see 137); others enhance the recovery of microbial communities after disturbances in soil (134).

Soil mite distribution appears to be pH-related. Many species prefer low-pH sites (44, 45, 73, 119, 146, 154). Others are found in mildly acid soil (pH 4.2–4.8) (119, 146, 154). Some species such as *Nothrus silvestris* tolerate a wider range (pH 3.5–5.6) (118), but very few favor alkaline conditions (73).

Changes in soil pH induced by artificial acidification and liming elicited various responses in acarine orders. Total abundance of Acarina, Mesostigmata, and Oribatida increased significantly after acidification (6, 77, 78, 83, 84). This increase resulted from population explosion of a few species (86). Acidophilic species were favored by acidification or were reduced by liming (6, 83, 84, 86). A few thrived under higher pH conditions, whereas some showed no clear response to pH changes (83, 84).

Generally, increased soil acidity leads to greater dominance of the Oribatida and a reduction in total Mesostigmata and total Prostigmata. In mor forest soils, liming, and to a smaller degree acidification, changes the vertical distribution pattern in Oribatida, increasing dominance in the 3- to 6-cm layer compared to the upper 0–3 cm. On the other hand, liming increases the dominance of Prostigmata in the upper 0- to 3-cm layer (6, 84). Liming is especially detrimental to oribatid mites (54, 55, 98, 135). It increases the decomposition rate and reduces the thickness of the humus layer, thereby creating unfavorable conditions for many oribatid species (136).

Pollution degrades both the quantitative and qualitative structure of oribatid communities (15–17, 65–67, 180–182, 198, 205). These reductions are accompanied by an increase in aggregation of individual species, a decrease in frequency of some species, and a shift in community structure, with recedent (less abundant) and subrecedent (rare) species being replaced by dominant ones (203, 204, 206). Thus three groups of species can be recognized in response to pollution (182): (*a*) sensitive species (e.g. *Adoristes ovatus, Eporibatula rauscheninsis*, and *Oppiella minus*); (*b*) susceptible species (e.g. *Carabodes labyrinthicus* and *Oribatula tibialis*); and (*c*) tolerant species (e.g. *Chamobates schueltzi, Liochthonius* sp., *Tectocepheus velatus, Trichoribates trimaculatus, Zygoribatula exilis*, and *Rhodacarus coronatus*).

Bielska (15–17) obtained somewhat similar results concerning the impacts of pollution and, further, showed that mite response was related to the severity of industrial pollution. She demonstrated a progressive reduction in numbers and species from least- to most-polluted sites. With increasing pollution, the age and dominance structure of the oribatid community changed and species constancy decreased. Larger species had higher densities in less polluted sites. Fly ash is also detrimental to soil mites (143).

Collembola

Collembolans, like mites, are important members of the soil mesofauna. About 6500 species have been described worldwide, and their density in soil reaches up to 300,000 individuals m^{-2} (171). Collembolans play an important role in organic matter decomposition processes in soil. Therefore, negative impacts on their communities affect ecosystem functioning. Two major factors that negatively affect the community relations are acid deposits and liming.

Springtails show special adaptations for regulating their water balance (43). They may experience considerable water loss even with a small deficit in ambient humidity and even in a water vapor-saturated environment. The ventral tube operates as their main organ of extraoral water uptake, allowing springtails to absorb water from moist surfaces (43). However, this ability is pH dependent, as demonstrated by *Tomocerus flavescens*: Liquid absorption is reduced at pH 2–3, increases up to pH 6, and then decreases toward the alkaline range (110).

Both fecundity and longevity in *Folsomia candida, Mesaphorura krausbaueri* s.l., and *Isotoma notabilis* are related to pH (108). Longevity is highest at pH 4, but the greatest egg production takes place at pH 5–7. Fecundity is optimal at pH 5.2. Only *Proisotoma minuta* produces the highest quantity of eggs at pH 7.2. In the laboratory, the parthenogenetic species *F. candida* and *M. krausbaueri* s.l. produce eggs even at pH 3.3, but at only 10% of the normal rate. None of these four species produces eggs at pH 2.5 (108). In the field, the reproduction rates of *F. candida* and another parthenogenetic species, *Heteromurus nitidus*, are closely correlated with soil pH (124). The atmobiotic (i.e. living on vegetation) species *Sminthurus viridis* has optimum growth and fecundity in the weakly acid range (pH 6.0–6.5) (133) and time required for development is doubled at pH 4.1.

Laboratory and field experiments show that pH changes induced by artificial acidification and liming affect abundance (73, 83), behavior (141), community structure (42, 77, 78, 86), and vertical distribution of Collembola (73). The responses between pH 2 and 6 are complex. Although total collembolan population may increase (73, 83), species composition changes, resulting in recognizable pH-related categories. Many species are acidophilic, some are distinctly basiphilic, and others are inconsistent (73, 83, 86), suggesting that pH per se might not be the major controlling factor in their biology.

Liming and acidification can change the vertical distribution of Collembola in the soil profile. Collembolan density increased in the 3- to 6-cm soil layer in comparison to the upper 0- to 3-cm layer of raw humus (mor) (73).

Collembolans were also sensitive to pH changes in the infiltration zone in beech forests where stemflows were often highly acidic and enriched with heavy metals (123–125). Soil in the infiltration zones showed significant chemical changes, with accompanying reduction in the abundance and species diversity of Collembola. With increasing distance from the stem, the dominance of Isotomidae, Onychiurinae, Entomobryidae, and other Collembola increased, whereas the Tullbergiinae decreased and reached typical soil values. Others (6, 12, 74, 83, 85, 104) have observed similar alterations in community structure, resulting from increased soil acidity. When pollution does not alter soil pH, collembolans are not greatly affected (41).

Fly ash (143), as well as fertilizer factory pollution (57), adversely affected soil Collembola. In a polluted transect, epigeic and edaphic collembolan density decreased at sites with both more and less pollution compared with intermediately polluted plots, but the species number of edaphic Collembola decreased significantly closer to the emission source (57). The edaphic collembolan community showed the lowest alpha diversity on the most polluted plots. As many as 50% of individuals from more polluted plots exhibited abnormally higher occurrences of nongenetic purple, reddish-purple, and dark-reddish body colors and decreased occurrences of the whitish and yellowish pigmentation that is normal in animals from unpolluted areas (see also 159).

Even a few years of industrial emissions from an N-fertilizer factory can reduce springtails substantially (64, 65, 67). These results were confirmed by interchanging soil blocks among polluted zones. Within 5 years after fertilizer production began, collembolan population in blocks transferred from polluted zones into healthy areas developed springtail densities similar to healthy soils. On the other hand, Collembola reductions became pronounced when healthy soils were transferred to heavily polluted zones.

Long-term changes in collembolan communities as a result of pollution were confirmed in studies of alpine grassland ecosystems on both limestone and granite (170). The greatest soil pH decreases were observed in the plant communities *Festucetum versicoloris* (pH 7.2–5.7), *Geranio-Alchemilletum crinitae* (pH 6.4–5.1), and *Saxifragetum perdurantis* (pH 7.7–6.4), which dominated places with snow accumulation and water runoff gullies. *Tetracanthella fjellbergi*, previously found in soils on acid bedrock only, became a dominant collembolan species in many ecosystems on limestone. *Folsomia sensibilis, Folsomia penicula*, and *Pseudanurophorus binoculatus*, rare species in the past, became dominant in some ecosystems. *Folsomia alpina*, which was previously dominant on limestone, was replaced by *F. penicula*, and disappeared from the whole territory. The alpine grassland ecosystems reacted earlier to air pollutants than the lower-lying subalpine and mountain forest ecosystems, and Collembola proved to be an excellent bioindicator of the impacts of air pollutants.

Other Groups of Soil Fauna

Macrofauna The soil macrofauna (Diplopoda, Isopoda, Lumbricidae, and the geophilomorph Chilopoda) near tree trunks in beech woods were negatively impacted by acid stemflow (177). Endogeic earthworms and centipedes experienced a dramatic reduction in density near trees compared to soil away from trees. Depth distribution of animals indicated a retreat of most macrofauna species from deeper soil layers near trees. The density of epigeic species inhabiting litter near trees was similar to or higher than that distant from trees. Soil pH was an important factor structuring the soil macrofauna community near trees.

Combining major groups of arthropods as Tracheata (Insecta) and Chelicerata, Vaněk (204) showed that airborne pollution reduced arthropod numbers by 46% compared to their numbers in undamaged forests (194,355 individuals m^{-2}). However, the constituent groups responded differently.

Arachnida Pollutants that caused several changes within a forest ecosystem (18) did not reduce species numbers of Opiliones, but changes occurred in species composition. Within the highly polluted area, a higher harvestman population density and migratory activity were observed. These were probably caused by alteration of microclimate, plant cover, and trophic conditions (18). Spiders have not been studied sufficiently; some are unaffected by pollution (97), whereas others respond negatively (31).

Insecta Many insect groups are negatively impacted by airborne pollutants. Coleoptera, Diptera, Heteroptera, Hymenoptera, and Psocoptera were less abundant in heavily polluted sites (206). Within individual orders, however, response depends on families and on species (31). For example, among staphylinid beetles, although some expansive species increased, relic and rare species decreased their aboveground activity under the influence of airborne pollutants (20, 21, 208).

Liming also decreased some macroarthropod groups, especially saprophagous litter-consuming larvae of Nematocera (Diptera) and *Enoicyla pusilla* (terrestrial Trichoptera) (71). However, Grundmann (71) did not find changes in density of predators (zoophagous Diptera and Coleoptera) after liming. Carabid beetles showed few differences between areas polluted by chemical factories and unpolluted areas (112, 201).

Isopoda Low amounts of fly ash were not inimical to *Trichoniscus pusillus*, but, when this species was given a choice, it preferred a clean substrate to a contaminated one (42).

Myriapoda Symphyla and Diplopoda were depleted by airborne pollutants (206), but Chilopoda were unaffected (209). The diplopod *Glomeris marginata* exhibited the same reaction to fly ash as did the isopod *T. pusillus* mentioned above (42).

Protura Proturans were adversely affected by pollutants (122, 206). Density reduction may be slight from artificial acid rain (194) or dramatic from acid stemflow in beech forests (123). As highly specialized ectomycorrhizal feeders, proturans are probably indirectly influenced through reduction of their fungal food by pollutants. In this way, Protura could be used as early indicators of forest damage because the trees react much later to airborne pollutants than do the proturans feeding on their mycorrhizal roots (59, 61, 62, 194, 195).

DIRECT AND INDIRECT EFFECTS ON SOIL BIOTA

Ecological Groups of Soil Fauna

Two main ecological groups of soil fauna are generally recognized: aquatic and terrestrial. Aquatic soil animals (Protozoa, Turbellaria, Nematoda, Rotatoria, etc) live in soil pores filled with gravity water, in capillary water, and in thin water films covering solid soil particles. Terrestrial soil animals are well represented by the Arthropoda, Pulmonata, and Annelida.

Both ecological groups are susceptible to pollutants. Earthworms, potworms, mollusks, and other invertebrates with permeable skin take up pollutants (e.g. copper) through the skin as well as in their food (191). Even terrestrial soil animals, with a semipermeable cuticle, are directly affected. For example, pH affects water uptake by Collembola (110), and other arthropod groups have morphophysiological systems such as water conducting systems in Isopoda (99), rectal water uptake systems in Diplopoda (142), and eversible bladders in Thysanura, Diplura, Protura, Symphyla, Pauropoda, and others (44) that allow them to absorb water. Therefore, pollutants dissolved in soil water act directly on these animals, as has been shown for Collembola (108, 110, 133). Furthermore, enchytraeid and lumbricid worms will occasionally come into direct contact with heavily polluted soil water that will have a negative effect on them, and this may explain some field observations of both increasing or decreasing population densities in these groups in their reaction to soil acidification.

Soil animals come into direct contact with pollutants through the food they consume. Animals feeding on litter swallow the pollutants and are directly affected when the pollutants are toxic. Sublethal doses indirectly affect their physiology, including reproduction, respiration, length of life cycle, and digestion (85, 115, 202). Through their grazing on contaminated soil microflora, microphytophages especially are indirectly affected by pollutants. Many groups of soil microflora disappear from ecosystems under the pressure of pollutants (148). Soil animals may lose their food resources in this way and consequently die.

Fertilizers and Soil pH

Marshall (136) reviewed the effects of fertilizers on soil fauna. Although general conclusions are difficult because of several unknowns, soil pH is clearly important. Soil pH strongly influences soil microflora, which serves as food for many soil animals. Microfloral activities also render plant detritus more suitable for faunal grazing. Some acidifying nitrogen fertilizers, such as ammonium nitrate, may cause a rapid increase in pH, because lime is added to some commercial formulations to neutralize the acidic action of the main fertilizer.

In the context of fertilization, both pH and nitrogen are important regulatory factors for soil fauna, with changes caused by non-nitrogen fertilizers also being explained by pH alone (4, 130). The responses of the total soil fauna to pH changes induced by fertilization are similar to those mentioned above in relation to acidification experiments (e.g. 6, 83). These data, however, do not show to what extent the effect of pH and fertilizers were direct or indirect via changes in microbial populations or substrate quality. Long-term impacts will vary with soil type (40).

Although faunal responses to pH changes might not be obvious at higher taxonomic levels, this response is seen at the species level (63). Thus studies based only on total numbers of Collembola, Acariformes, Parasitiformes, and other similar higher taxa usually do not exhibit significant changes because the reaction to environmental impact is reflected mostly at the species level. Also, changes might require a long period before they are detected. For example, total abundance of microarthropods showed no apparent changes under the impact of acid deposition and air pollution, but lichen species in a beech forest in western Germany decreased from 33 in 1900 to 7 in 1984 (138, 139).

Influence of Heavy Metals

Most soil animal groups accumulate heavy metals that cause a reduction in abundance and a change in community structure (13, 47, 48, 217, 224, 225, 226). Here we refer mostly to studies that elucidate ecophysiological effects on soil fauna in connection with soil acidification. Blast furnace plants expel large quantities of Fe, Mn, and Zn dust into the environment, where they potentially become incorporated into animal and plant tissues. Near polluted sites, body metal concentration increases with metal concentration in the food, and animals from such sites excrete a greater amount of metal compared to those from unpolluted areas (149). Mn seems to have little effect on soil fauna (115, 149), but Fe and Zn, even at sublethal levels, may adversely affect the soil fauna.

High concentrations of Fe have been found in the midgut cells of isopods (103), without convincing evidence that the increased metal levels are harmless. However, sublethal Fe concentrations decreased feeding activity and growth in Collembola (149) and hampered Mn uptake and increased respiration in isopods, suggesting an energy-demanding detoxication or excretion process (115).

High levels of Zn in the environment are sublethal to invertebrates, depressing reproduction, respiration, and food consumption (115, 117). As a consequence, their role in the decomposition of organic matter is also diminished (169, 192, 193).

Cu (23) and Cd (212a) are inimical to nematodes (23), and both Cu and Pb impair the growth of Collembola (10, 116), even though they have an intriguing method of ridding their bodies of these metals (101). Collembolans store heavy metals in the gut epithelium in the form of sphaerocrystals (107). When the animals molt, they expel the old gut epithelium as a pellet containing the stored metals before the new epithelium develops. By this mechanism, Collembola are able to excrete about 43% of absorbed Pb (114, 116, 207). Unlike Collembola, other soil animals cannot easily rid themselves of heavy metals.

Because some heavy metals play an essential role in many physiological functions, high concentrations of these metals in the environment must always be suspect, especially when the soil pH is changed (184a). As already mentioned, pH affects soil animals both directly and indirectly. Also, the solubility of many heavy metals increases when soil pH is lowered (56).

Influence on Litter Decomposition

Simulated acid rain can reduce CO_2 production and litter decomposition even in mildly acid (pH 5.8) calcareous soil (223). This depression is strong even when the input of protons is 1.5-fold greater than the normal acid deposition. Acid deposition may thus cause an accumulation of primary and secondary C compounds in the litter layer of base-rich soils. The mesofauna significantly reduces the ability of acid rain to inhibit C mineralization (223).

The literature on the effects of soil fauna on litter decomposition is extensive and shows that a healthy faunal population enhances litter decomposition. This positive relationship between soil animals and decomposition rate has been shown for soil animals as a group (65, 66, 93, 124, 125, 173, 224) as well as for individual taxa including Nematoda (216), Lumbricidae (127, 178), Acarina (79), and Collembola (47, 64, 65, 79, 86, 124, 140). In addition, decomposition rate is influenced even by species-specific differences within a taxon (140).

A higher rate of decomposition is also associated with an increase in plant nutrient uptake (34, 183). Therefore, disruption of the natural faunal populations by pH changes can have profound effects on litter decomposition and primary production. The presence of even one species of microarthropods could increase the decomposition rate, also increasing soil pH (79, 85, 86).

Remedial Measures and Bioindicators

A consequence of lowering or raising soil pH is a shift in soil faunal populations, partly brought about through competition (76, 80). In undisturbed areas, soil faunal populations are an integral part of the functioning ecosystem, therefore alterations in these populations have repercussions for litter decomposition and pest status. If the community structure is destroyed, subrecedents could be replaced by dominant species (203, 204, 206). Consequently, the potential species reserve for secondary succession is destroyed, and regeneration of the original fauna becomes a very slow process. Fortunately, damage to soil is sometimes reversible (39, 40).

Liming is widely used for improving forest productivity in central and northern Europe (5). However, liming rarely increases abundance of a species or group of soil fauna (73, 106) on a sustained basis. Even the low earthworm densities in coniferous forests can contribute greatly to the decomposition rate of organic matter (106). Although earthworms are not usually adversely affected by liming, it is important to consider the impacts on other groups of soil biota. Changes in the soil biota community structure may cause long-lasting negative effects on the whole ecosystem. Because broadcast liming has these risks, Funke (60) recommended less drastic ameliorative measures for increasing soil pH in spruce forests, such as patchy liming and adding of compost or litter with allochthonous litter decomposers.

Earthworms have been used to accelerate biological processes in young soils (e.g. newly drained polders) and for restoring some degraded sites (127). Earthworms are also potentially useful for improving the status of soils in which the pH has been lowered, because their activity leads to an increase of soil pH (105). Furthermore, feeding and burrowing activities of some epigeic and endogeic species in calcareous soil counteract the negative effect of acid precipitation on litter decomposition (178). However, not all earthworm species are equally effective, as some epigeic species never penetrate below the F/H (Fermentation/Humus) horizon to incorporate humus with mineral soil.

Techniques are now available for rapid bioassay to determine environmental quality. Most groups of soil animals are useful in this regard and include protozoans (51, 52), nematodes (87, 88, 226), potworms (32, 33), earthworms (25), spiders (31), mites (126, 168, 182), and collembolans (170, 172). Within each group, a different reaction pattern may appear. The total picture thus becomes very complex and emphasizes the necessity of identifying animals to the species level.

The responses of these groups depend on their environment, and the following five hypotheses are offered for further experimentation:

1. Studies on the impact of airborne pollutants on soil fauna must consider the pH tolerance of each species. Acidophilic or acid-tolerant species will

react to acidification and acid pollutants less clearly than would basiphilic ones. This hypothesis can be tested in field microcosms, using the methodology described by Kopeszki (124, 125).

2. Soil animal feeding activity is lower in polluted soil than in unpolluted soil. Litter decomposition rate will be lower in polluted than in unpolluted soil, which can be measured along a pollution gradient in the field, either directly from an emission source or along gradients caused by stemflow near deciduous trees. The methodology applied by Kopeszki (123) or von Törne (211, 212) would be suitable for testing this hypothesis.
3. Less polluted soils in a gradient will exhibit a higher abundance and diversity of soil fauna than more polluted soils. Species missing in the more polluted part of the gradient could serve as indicator species of unpolluted soils. Transect studies of community structure of soil mesofauna near a local pollution source can help to determine indicator species for further testing under hypotheses 4 and 5.
4. Soil animal communities react sensitively to airborne pollutants and often in advance of other components of terrestrial ecosystems. Sites for long-term monitoring of soil should be situated in places with high pollution deposition, such as hillsides with high wind turbulence.
5. Ecotones are more sensitive to disturbance or to changes in the global environment than their adjacent ecosystems. Ecotones (100) in the northern and upper tree-limit zones are assumed to be fragile and therefore very suitable for monitoring.

CONCLUSIONS

The literature dealing with the impacts of airborne pollutants on soil fauna is extensive, but only Nematoda, Enchytraeidae, Lumbricidae, Acarina, and Collembola have been intensively studied. There is a general tendency for all these groups to be reduced by airborne pollutants. This reduction is accompanied by a shift in the community structure brought about by competition and predation. *K*-strategists are replaced by *r*-strategists in the short term.

The few published papers on protozoans suggest that the infestation rate by various parasitic protozoans on soil invertebrates in industrial landscapes is enhanced by airborne pollutants.

Soil nematodes react to different soil pH levels through their reproduction and hatching rates, and the reactions are different for acidophilic, basiphilic, neutrophilic, and acidotolerant species. In field experiments with artificial acid rain, species numbers of soil nematodes decreased drastically; fungivore species increased, whereas bacterivore species decreased. This is a reaction to a reduction in bacterial food resources in soil caused by acidification. The same reaction was found in natural spruce forests damaged by airborne pollutants.

The first reaction of enchytraeids to soil acidification in boreal and mountain coniferous forests is a density increase, probably associated with increased amounts of dead organic matter in soil, which indicates a decreased litter decomposition rate. The density of Enchytraeidae decreased in the following years. The enormous density increase in the first years of forest dieback was associated with *Cognettia sphagnetorum*, whereas other species reached only a low density or became depleted. Most of the species living in boreal forests with mor humus or moder are acidophilic and react negatively to liming.

Earthworms have been treated in relatively few studies. Acidification decreases their number in soil, whereas liming has a positive effect. In heavily polluted industrial areas, parasitic protozoans seriously infest earthworms and other invertebrates.

Microarthropods, especially Collembola and Acarina, have been the most intensely studied soil faunal groups. Both are sensitive to soil acidification caused by airborne pollutants. Many acidophilic species of Collembola and Acarina reach higher density in acidified soils, whereas neutrophilic and basiphilic species suffer, some becoming depleted in such soils. The occurrence of dominant acidophilic species in soils with lower pH results from competition. Acidophilic species have a higher fecundity and longevity at lower pH ranges. Liming after soil acidification affects Collembola and Acarina (and other soil animals) more drastically than acidification alone.

Most of the results on the effects of acid rain and liming on soil animals are from laboratory and field experiments from temperate and boreal Europe. Results from field studies in heavily polluted areas are less numerous, but they confirm and extend the results from the laboratory. The density and number of species of most groups of soil animals become reduced in areas affected by airborne pollutants that cause forest dieback. Soil animal communities in polluted areas change their coenotical parameters. Aggregation of some species increases significantly, whereas the frequencies of others decrease. Dominant species become less abundant and rare species become extirpated. Acidophilic species invade acidified soils on limestone, whereas basiphilic and neutrophilic ones are decimated. Recovery of the population becomes increasingly difficult with the duration of impact. Acute and chronic pollution effects therefore differ. These changes are usually accompanied by a decrease in decomposition rate of soil organic matter.

Soil animals react sensitively to airborne pollutants and are suitable biomonitors for assessing ecological changes caused by such pollutants.

ACKNOWLEDGMENTS

Our work was supported by the Long Range Transport of Atmospheric Pollutants component of Forestry Canada's Green Plan, and grant project 204/93/0276 (Global Changes in Selected Ecosystems) of the Granting Agency of the Czech Republic (GACR). A recent literature review was supported by grant project A6066702 of the Granting Agency of the Academy of Sciences of the Czech

Republic (GA AVCR) and 206/99/1416 of the GACR. Our thanks go to J Addison and T Panesar, Applied Research Division, Royal Roads University, Canada, for comments on an earlier draft of the manuscript, and to David Coleman, Institute of Ecology, The University of Georgia, Athens, GA, for his helpful review.

Visit the Annual Reviews home page at www.AnnualReviews.org

LITERATURE CITED

1. Abrahamsen G. 1972. Ecological study of Lumbricidae (Oligochaeta) in Norwegian coniferous forest soils. *Pedobiologia* 12:267–81
2. Abrahamsen G. 1983. Effects of lime and artificial acid rain on the enchytraeid (Oligochaeta) fauna in coniferous forest. *Holarc. Ecol.* 6:247–54
3. Abrahamsen G, Hovland HY, Hågvar S. 1980. Effects of artificial acid rain and liming on soil organisms and the decomposition of organic matter. In *Effects of Acid Precipitation on Terrestrial Ecosystems*, ed. TC Hutchinson, M Havas, pp. 341–62. New York: Plenum
4. Abrahamsen G, Thompson WN. 1979. A long-term study of the enchytraeid (Oligochaeta) fauna of a mixed coniferous forest and the effects of urea fertilization. *Oikos* 32:318–27
5. Andersson F, Persson T. 1988. *Liming as a Measure to Improve Soil and Tree Condition in Areas Affected by Air Pollution: Results and Experiences of an Ongoing Programme. Rep. 3518*, pp. 1–131. Stockholm: Nat. Swed. Environ. Prot. Board
6. Bååth E, Berg B, Lohm U, Lundgren B, Lundkvist H, et al. 1980. Effects of experimental acidification and liming on soil organisms and decomposition in a Scots pine forest. *Pedobiologia* 20:85–100
7. Bååth E, Berg B, Lohm U, Lundgren B, Lundkvist H, et al. 1980. Soil organisms and litter decomposition in a Scots pine forest: effects of experimental acidification. In *Effects of Acid Precipitation on Terrestrial Ecosystems*, ed. TC Hutchinson, M Havas, pp. 375–80. New York: Plenum
8. Bassus W. 1968. Über Wirkungen von Industrieexhalten auf den Nematodenbesatz im Boden von Kiefernwäldern. *Pedobiologia* 8:289–95
9. Beitensweiler W. 1985. "Waldsterben": forest pests and air pollution. *J. Appl. Entomol.* 99:77–85
10. Bengtsson G, Gunnarsson T, Rundgren S. 1983. Growth changes caused by metal uptake in a population of *Onychiurus armatus* (Collembola) feeding on metal polluted fungi. *Oikos* 40:216–25
11. Bengtsson G, Rundgren S. 1982. Population density and species number of enchytraeids in coniferous forest soils polluted by a brass mill. *Pedobiologia* 24:211–18
12. Bengtsson G, Rundgren S. 1988. The Gusum case: a brass mill and the distribution of soil Collembola. *Can. J. Zool.* 66:1518–26
13. Bengtsson G, Tranvik L. 1989. Critical metal concentrations for forest soil invertebrates. *Water Air Soil Pollut.* 47:381–417
14. Beniamino F, Ponge JF, Arpin P. 1991. Soil acidification under the crown of oak trees. I. Spatial distribution. *For. Ecol. Manage.* 40:221–32
15. Bielska I. 1989. Communities of moss mites (Acari, Oribatei) of grasslands under the pressure of industrial pollution. I. Communities of moss mites of meadows. *Pol. Ecol. Stud.* 15:75–87
16. Bielska I. 1989. Communities of moss mites (Acari, Oribatei) of grasslands under the pressure of industrial pollution. II. Communities of moss mites of pastures. *Pol. Ecol. Stud.* 15:89–99
17. Bielska I. 1989. Communities of moss

mites (Acari, Oribatei) of grasslands under the impact of industrial pollution. III. Communities of moss mites of wastelands. *Pol. Ecol. Stud.* 15:101–10
18. Bliss P, Tietze F. 1984. Die Struktur der epedaphischen Weberknecktfauna (Arachnida, Opiliones) in unterschiedlich immissionsbelasteten Kiefernforsten der Dübener Heide. *Pedobiologia* 26:25–35
19. Bloomfield C, Kelso WJ, Pruden G. 1976. Reactions between metals and humified organic matter *J. Soil Sci.* 27:16–31
20. Boháč J, ed. 1992. *Proc. Int. Conf. Bioidic. Deter. Reg.*, pp. 159–65. České Budějovice, Czechoslov.: Inst. Landsc. Ecol., Czechoslov Acad. Sci.
21. Boháč J, Fuchs R. 1995. The effect of air pollution and forest decline on epigeic staphylinid communities in the Giant Mountains. *Acta Zool. Fenn.* 196:311–13
22. Bongers T. 1990. The maturity index: an ecological measure of environmental disturbance based on nematode species composition. *Oecologia* 83:14–19
23. Bongers T, Bongers M. 1998. Functional diversity of nematodes. *Appl. Soil Ecol.* 10:239–51
24. Bonnet L. 1964. Le peuplement thécamoebien des sols. *Rev. Écol. Biol. Sol.* 1:123–408
25. Bouché MB. 1972. *Lombriciens de France: Écologie et Systématique.* Paris: Inst. Natl. Rech. Agron. 671 pp.
26. Brussaard L, Behan-Pelletier V, Bignell D, Brown V, Didden W, et al. 1997. Biodiversity and ecosystem functioning in soil. *Ambio* 26:563–70
27. Brzeski MW, Dowe A. 1969. Effect of pH on *Tylenchorhynchus dubius* (Nematoda, Tylenchidae). *Nematologica* 15:403–7
28. Brzeski MW, Sandner H. 1974. *Zarys Nematologii.* Warsaw: Pol. Sci. Publ. 400 pp.
29. Burns NC. 1971. Soil pH effects on nematode populations associated with soybeans. *J. Nematol.* 3:238–45
30. Butler GC. 1978. *Principles of Ecotoxicology, SCOPE 12.* Chichester, UK: Wiley & Sons. 350 pp.
31. Cárcamo HA, Parkinson D, Volney JWA. 1998. Effects of sulphur contamination on macroinvertebrates in Canadian pine forests. *Appl. Soil Ecol.* 9:459–64
32. Chalupský J. 1992. Enchytraeids (Annelida, Enchytraeidae) in soils of deteriorated mountain forest sites in the Krkonose and the Beskydy Mts. In *Investigation of the Mountain Forest Ecosystems and of Forest Damage in the Czech Republic. Proc. Workshop, March 17–18*, ed. K Matějka. pp. 81–85. České Budějovice, Czechoslov.: Sci. Pedagog. Publ. (In Czech)
33. Chalupský J. 1995. Long-term study of Enchytraeidae (Oligochaeta) in man-impacted mountain forest soils in the Czech Republic. *Acta Zool. Fenn.* 196: 318–20
34. Coleman DC, Crossley DA Jr. 1996. *Fundamentals of Soil Ecology.* San Diego, CA: Academic. 205 pp.
35. de Goede RGM, Bongers T. 1994. Nematode community structure in relation to soil and vegetation characteristics. *Appl. Soil Ecol.* 1:29–44
36. de Goede RGM, Dekker HH. 1993. Effects of liming and fertilization on nematode communities in coniferous forest soils. *Pedobiologia* 37:193–209
37. Dmowska E. 1995. Influence of simulated acid rain on communities of soil nematodes. *Acta Zool. Fenn.* 196:321–23
38. Dunger W. 1972. Systematische und ökologische Studien an der Apterygotenfauna des Neissetales bei Ostritz/Oberlausitz. *Abh. Ber. Naturkundemus. Görlitz* 47(4):1–42
39. Dunger W. 1991. Langzeitbeobachtungen an der Bodenfauna von Waldstandorten mit steigender Immissions-Belastung. *Rev. Écol. Biol. Sol.* 28:31–39
40. Dunger W, Dunger I, Engelmann H-D, Schneider R. 1972. Untersuchungen

zur Langzeitwirkung von Industrie-Emissionen auf Böden, Vegetation und Bodenfauna des Neissetales bei Ostritz/Oberlausitz. *Abh. Ber. Naturkundemus. Görlitz* 47(3):1–40

41. Dunger W, Schulz H-J. 1995. Long-term observations on the effects of increasing dry deposition on the Collembola fauna of the Neisse valley (Germany). *Acta Zool. Fenn.* 196:324–25
42. Eijsackers H, Lourijsen N, Mentink J. 1983. Effects of fly ash on soil fauna. In *New Trends in Soil Biology*, ed. P Lebrun, HM André, A de Medts, C. Grégoire-Wibo, G Wauthy, pp. 680–81. Louvain-la-Neuve, Belgium: Dieu-Brichard
43. Eisenbeis G. 1982. Physiological absorption of liquid water by Collembola: absorption by the ventral tube at different salinities. *J. Insect Physiol.* 28:11–20
44. Eisenbeis G, Wichard W. 1985. *Atlas zur Bodenbiologie der Bodenarthropoden.* Stuttgart/New York: Gustav Fischer Verlag. 434 pp. [English edition, ed. EA Mole, 1987. Berlin: Springer-Verlag]
45. Engelmann H-D. 1972. Die Oribatidenfauna des Neissetales bei Ostriz/Oberlausitz. *Abh. Ber. Naturkundemus. Görlitz.* 47(5):1–42
46. Feest A, Campbell R. 1986. The microbiology of soils under successive wheat crops in relation to take-all disease. *FEMS Microbiol. Ecol.* 38:99–112
47. Filser J. 1992. *Dynamik der Collembolengesellschaften als Indikatoren für bewirtschaftungsbedingte Bodenbelastungen: Hopfenböden als Beispiel.* Aachen, Germany: Shaker. 136 pp.
48. Filser J, Fromm H, Nagel RF, Winter K. 1995. Effects of previous intensive agricultural management on microorganisms and the biodiversity of soil fauna. *Plant Soil* 170:123–29
49. Fischer P, Führer E. 1990. Effect of soil acidity on the entomophilic nematode *Steinernema kraussei* Steiner. *Biol. Fertil. Soils* 9:174–77
50. Fleming TR, Richards KS. 1982. Localization of adsorbed heavy metals on the earthworm body surface and their retrieval by chelation. *Pedobiologia* 23:415–18
51. Foissner W. 1987. Soil protozoa: fundamental problems, ecological significance, adaptations in ciliates and testaceans, bioindicators, and guide to the literature. *Prog. Protistol.* 2:69–212
52. Foissner W. 1987. Ökologische Bedeutung und bioindikatives Potential der Bodenprotozoen. *Verh. Ges. Okol.* 16:45–52
53. Forsslund K-H. 1944. Studier över det lägre djurlivet. I. Nordsvensk skogsmark. *Medd. Statens Skogsförsöksanst.* 34(1):1–283
54. Franz H. 1959. Das biologische Geschehen im Waldboden und seine Beeinflussung durch die Kalkung. *Allg. Forstzg.* 70:178–81
55. Franz H. 1968. Der Einfluss von Düngemitteln auf die Bodenlebewelt. In *Handbuch Pflanzenernahrung und Düngung.* Band II. *Boden und Düngemittel*, ed. H. Linser, pp. 1715–30. Vienna, Austria: Springer
56. Freedman B. 1995. *Environmental Ecology: The Ecological Effects of Pollution, Disturbance, and Other Stresses.* San Diego, CA: Academic. 606 pp. 2nd ed.
57. Fritzlar F, Dunger W, Schaller G. 1986. Über den Einfluss von Luftverunreinigungen auf Ökosysteme. X. Collembola im Immissionsgebiet eines Phosphat-Düngemittelwerkes. *Pedobiologia* 29: 413–34
58. Führer E, Fischer P. 1991. Towards integrated control of *Cephalcia abietis*, a defoliator of Norway spruce in central Europe. *For. Ecol. Manage.* 39:87–95
59. Funke W. 1986. Tiergesellschaften im Ökosystem "Fichtenforst" (Protozoa, Metazoa-Invertebrata): Indikatoren von Veränderungen in Waldökosystemen. *Proj. Eur. Forsch.zent. Massn. Luftreinhalt. Kernforsch.zent. Karlsr.* 9:1–50
60. Funke W. 1991. Tiergesellschaften in

Waldern: Ihre Eignung als Indikatoren für den Zustand von Ökosystemen. *Proj. Eur. Forsch.zent. Massn. Luftreinhalt. Kernforsch.zent. Karlsr.* 84:1–202

61. Funke W, Bernhard M, Hofer H, Jans W, Lehle E, et al. 1986. Bodentiere als sensitive Indikatoren in Waldökosystemen. *Proj. Eur. Forsch.zent. Massn. Luftreinhalt. Kernforsch.zent. Karlsr.* 4(1):337–46
62. Funke W, Stumpp J, Roth-Holzapfel M. 1987. Bodentiere als Indikatoren von Waldschaden. *Verh. Ges. Ökol.* 15:309–20
63. Gerdsmeier J, Greven H. 1991. Abundanzen und Dominanzen einiger Kleinarthropoden in Buchenwäldern des Eggegebirges, Nordheim-Westfalen. *Acta Biol. Benrodis* 3:1–26
64. Górny M. 1972. Badania zoocenologiczne gleb borów sosnowych w sasiedztwie *Zakładów Azotowych w Puławach. XIX. Ogólnopolski Zjazd Naukowy PTG: Ochrona środowiska Glebowego*, Puławy, pp. 216–18. Krakow, Poland: Pol. Tow. Glebozn.aw.
65. Górny M. 1975. *Zooekologia Gleb Lesnych*. Warsaw: Panstw. Wydaw. Rol. Lesn. 312 pp.
66. Górny M. 1975. Studies on the influence of industrial pollution on soil animals in pine stands, aims and methods of the soil-block model experiment. In *Progress in Soil Zoology*, ed. J Vaněk, pp. 357–62. Prague: Academia
67. Górny M. 1976. Einige Pedo-ökologische Probleme der Wirkung von industriellen Immissionen auf Waldstandorte. *Pedobiologia* 16:27–35
68. Graefe U. 1989. Der Einfluss von sauren Niederschlägen und Bestandeskalkungen auf die Enchytraeidenfauna in Waldböden. *Verh. Ges. Ökol.* 17:597–93
69. Greven H. 1987. Vermehrung epidermaler Schleimzellen als Antwort von Lumbriciden und Gastropoden auf Stresssituationen. *Verh. Ges. Ökol.* 15:321–25
70. Greven H, Bettin C, Reichelt R, Ruther U. 1987. Die Wirkung von Saurestress auf *Lumbricus terrestris* L. (Lumbricidae, Oligochaeta): Methodik und erste Ergebnisse. *Verh. Ges. Ökol.* 15:327–33
71. Grundmann B. 1992. Untersuchungen zur Auswirkung von Kalkungsmassnahmen auf die Arthropodenzönose von Fichten- und Buchenbeständen. See Ref. 20
72. Gunther A, Greven H. 1990. Increase of the number of epidermal gland cells: an unspecific response of *Lumbricus terrestris* L. (Lumbricidae: Oligochaeta) to different environmental stressors. *Zool. Anz.* 225:278–86
73. Hågvar S. 1984. *Ecological studies of microarthropods in forest soils, with emphasis on relations to soil acidity*. PhD thesis. Univ. Oslo. Ås: Nor. Agric. Univ., Nor. For. Res. Inst. 35 pp.
74. Hågvar S. 1984. Effects of liming and artificial acid rain on Collembola and Protura in coniferous forest. *Pedobiologia* 27:341–54
75. Hågvar S. 1987. What is the importance of soil acidity for the soil fauna? *Fauna* 40:64–72 (In Norwegian)
76. Hågvar S. 1987. Why do collemboles and mites react to changes in soil acidity? *Entomol. Medd.* 55:115–19
77. Hågvar S. 1987. Effect of artificial acid precipitation and liming on forest microarthropods. In *Soil Fauna and Soil Fertility*, ed. BR Striganova, pp. 661–67. Moscow: Nauka
78. Hågvar S. 1988. Acid rain and soil fauna. *Biol. Ambient. Proc. World Basque Congr. Bilbao, 2nd*, ed. JC Iturrondobeitia, 1:191–201. Bilbao, Spain: Univ. Pais Vasco
79. Hågvar S. 1988. Decomposition studies in an easily-constructed microcosm: effect of microarthropods and varying soil pH. *Pedobiologia* 31:293–303
80. Hågvar S. 1990. Reactions to soil acidification in microarthropods: Is competition a key factor? *Biol. Fertil. Soils* 9:178–81
81. Hågvar S, Abrahamsen G. 1977. *Acidification Experiments in Conifer Forest. 5.*

Studies on the Soil Fauna. Oslo-Ås: SNSF [Acid Precipitation—Effects on Forest and Fish] Proj. IR 32/77. 47 pp. (In Norwegian)

82. Hågvar S, Abrahamsen G. 1977. Effect of artificial acid rain on Enchytraeidae, Collembola and Acarina in coniferous forest soil, and on Enchytraeidae in *Sphagnum* bog: preliminary results. *Ecol. Bull.* 25:568–70
83. Hågvar S, Abrahamsen G. 1980. Colonisation by Enchytraeidae, Collembola and Acari in sterile soil samples with adjusted pH levels. *Oikos* 34:245–58
84. Hågvar S, Amundsen T. 1981. Effects of liming and artificial acid rain on the mite (Acari) fauna in coniferous forest. *Oikos* 37:7–20
85. Hågvar S, Kjøndal BR. 1981. Decomposition of birch leaves: dry weight loss, chemical changes, and effects of artificial acid rain. *Pedobiologia* 22:232–45
86. Hågvar S, Kjøndal BR. 1981. Effects of artificial acid rain on the microarthropod fauna in decomposing birch leaves. *Pedobiologia* 22:409–22
87. Háněl L. 1992. Soil nematodes in spruce forests with different degree of air pollution damage. See Ref. 20, pp. 166–71
88. Háněl L. 1992. Soil nematodes and disturbed forest ecosystems. See Ref. 32, pp. 76–80
89. Harris RC. 1976. *Suggestions for the Development of a Hazard Evaluation Procedure for Potentially Toxic Chemicals. Rep. 3*, Monit. Assess. Res. Cent. Chelsea Coll., Univ. London. 18 pp.
90. Hartmann P, Fischer R, Scheidler M. 1988. Untersuchungen über den Einfluss von Schadstoffbelastung und Düngungsmassnahmen auf die Wirbellosenfauna oberfrankischer Nadelwälder. *Bayer. Staatsminist. Landesentwickl. Umweltfr.* 169 pp.
91. Hartmann P, Fischer R, Scheidler M. 1989. Auswirkungen der Kalkdüngung auf die Bodenfauna in Fichtenforsten. *Verh. Ges. Ökol.* 17:585–89
92. Healy B. 1980. Distribution of terrestrial Enchytraeidae in Ireland. *Pedobiologia* 20:159–75
93. Heneghan L, Colelman DC, Zou X, Crossley DA Jr, Haines BL. 1998. Soil microarthropod community structure and litter decomposition dynamics: a study of tropical and temperate sites. *Appl. Soil Ecol.* 9:33–38
94. Heungens A. 1980. The influence of salt concentration on an enchytraeid population in pine litter. *Pedobiologia* 20:154–58
95. Heungens A. 1981. Nematode population fluctuations in pine litter after treatment with pH changing compounds. *Med. Fac. Landbouwwet. Rijksuniv. Gent* 46:1267–81
96. Heungens A. 1984. The influence of some acids, bases and salts on an enchytraeid population of a pine litter substrate. *Pedobiologia* 26:137–41
97. Hiebsch H. 1972. Beiträge zur Spinnen- und Weberknechtfauna des Neissetales bei Ostritz. *Abh. Ber. Naturkundemus. Görlitz* 47(6):1–32
98. Hill SB, Metz LJ, Farrier MH. 1975. Soil mesofauna and silvicultural practices. *For. Soils For. Land Manage. Proc. Am. For. Soils Conf., 4th*, ed. B Bernier, CH Winget, pp. 119–35. Québec: Presses Univ. Laval
99. Hoese B. 1981. Morphologie und Funktion des Wasserleitungssystems der terrestrischen Isopoden (Crustacea, Isopoda, Oniscoidea). *Zoomorphology* 98:135–67
100. Holland MM. 1988. SCOPE/MAB technical consultations on landscape boundaries: report of a SCOPE/MAB workshop on ecotones. In *A New Look at Ecotones: Emerging International Projects on Landscape Boundaries*, Biol. Int. Spec. Issue 17, ed. F di Castri, AJ Hansen, MM Holland, pp. 47–106. Paris: Int. Union Biol. Sci.
101. Hopkin SP. 1997. *Biology of the*

Springtails (Insecta: Collembola). Oxford, UK: Oxford Univ. Press. 330 pp.
102. Horsch F, Filby WG, Fund N, Gross S, Hanisch B, et al, eds. 1988. 4. Statuskolloquium des PEF vom 8. bis 10. März 1988 im Kernforschungszentrum Karlsruhe. *Proj. Eur. Forsch.zent. Massn. Luftreinhalt. Kernforsch.zent. Karlsr.* 35: 1–816
103. Hryniewiecka-Szyfter Z. 1972. Ultrastructure of hepatopancreas of *Porcellio scaber* Latr. in relation to function of iron and copper accumulation. *Bull. Soc. Amis Sci. Lett. Pozn.* 113:135–42
104. Huang P. 1986. Belastung und Belastbarkeit streuzersetzender Tiere durch Deposition von Luftverunreinigungen in Waldökosystemen. *UBA Querschnittsemin.* 18:101–12
105. Huhta V, Hyvönen R, Koskenniemi A, Vilkamaa P. 1983. Role of pH in the effect of fertilization on Nematoda, Oligochaeta and microarthropods. See Ref. 42, pp. 61–73
106. Huhta V, Hyvönen R, Koskenniemi A, Vilkamaa P, Kaasalainen P, et al. 1986. Response of soil fauna to fertilization and manipulation of pH in coniferous forests. *Acta For. Fenn.* 195:1–30
107. Humbert W. 1977. The mineral concentrations in the midgut of *Tomocerus minor* (Collembola): microprobe analysis and physiological significance. *Rev. Écol. Biol. Sol.* 14:71–80
108. Hutson BR. 1978. Influence of pH, temperature and salinity on the fecundity and longevity of four species of Collembola. *Pedobiologia* 18:163–79
109. Hyvönen R, Persson T. 1990. Effects of acidification and liming on feeding groups of nematodes in coniferous forest soils. *Biol. Fertil. Soils* 9:205–10
110. Jaeger G, Eisenbeis G. 1984. pH-dependent absorption of solution by the ventral tube of *Tomocerus flavescens* (Tullberg, 1871) (Insecta, Collembola). *Rev. Écol. Biol. Sol.* 21:519–31
111. Jans W, Funke W. 1989. Die Enchytraen (Oligochaeta) von Laub- und Nadelwäldern Süddeutschlands und ihre Reaktion auf substantielle Einflusse. *Verh. Ges. Ökol.* 18:741–46
112. Jarošík V. 1983. A comparison of the diversity of carabid beetles (Col., Carabidae) of two floodplain forests differently affected by emissions. *Věstn. Česk. Spol. Zool.* 47:215–20
113. Jeffries DS. 1990. Snowpack storage of pollutants, release during melting, and impact on receiving waters. In *Soils, Aquatic Processes and Lake Acidification: Acidific Precipitation*, Vol. 4, ed. SA Norton, SE Linberg, AL Page, pp. 107–32. New York: Springer-Verlag
114. Joosse ENG, Buker JB. 1979. Uptake and excretion of lead by litter-dwelling Collembola. *Environ. Pollut.* 18:235–40
115. Joosse ENG, van Vliet LHH. 1984. Iron, manganese and zinc inputs in soil and litter near a blast-furnace plant and the effects on respiration of woodlice. *Pedobiologia* 26:249–55
116. Joosse ENG, Verhoef SC. 1983. Lead tolerance in Collembola. *Pedobiologia* 25:11–18
117. Joosse ENG, Wulffraat KJ, Glas HP. 1981. Tolerance and acclimation to zinc of the isopod *Porcellio scaber* Latr. *Proc. Int. Conf. Heavy Met. Environ., 3rd, Amsterdam*, pp. 425–28. Edinburgh: CEP Consult.
118. Karppinen E. 1955. Ecological and transect survey studies on Finnish camisiids. *Suomen Hyönteistiet. Aikak.* 17(2):1–80
119. Karppinen E. 1958. Über die Oribatiden (Acar.) der Finnishen Waldböden. *Ann. Zool. Soc. Fenn. Vanamo* 19(1)1–43
120. Kauppi P, Anttila P, Kenttamies K. 1990. Acidic precipitation research in Finland. In *Acidic Precipitation*, Vol. 5, *International Overview and Assessment*, ed. AHM Bresser, W Salomons, pp. 281–306. New York: Springer-Verlag

121. Kazda M, Glatzel G, Lindebner L. 1986. Die Belastung von Buchenwald-ökosystemen durch Schadstoffdeposition im Nahbereich stadtischer Ballungsgebiete: Untersuchungen im Wienerwald. *Düsseld. Geobot. Kolloqu.* 3:15–32
122. Kholová H. 1968. Einfluss von Exhalten auf die Bodenfauna. In *Immissionen und Waldzönosen, Jevany u Prahy 1967*, ed. Z RozKošná, pp. 63–67. Prague: ČSAV, ÚTOK
123. Kopeszki H. 1991. Abundanz und Abbauleistung der Mesofauna (Collembolen) als Kriterien für die Bodenzustandsdiagnose im Wiener Buchenwald. *Zool. Anz.* 227:136–59
124. Kopeszki H. 1992. Versuch einer aktiven Bioindikation mit den bodenlebenden Collembolen-Arten *Folsomia candida* (Willem) und *Heteromurus nitidus* (Templeton) in einem Buchenwald-Ökosystem. *Zool. Anz.* 228:82–90
125. Kopeszki H. 1992. Veränderungen der Mesofauna eines Buchenwaldes bei Saurebelastung. *Pedobiologia* 36:295–305
126. Lebrun P. 1979. Soil mite community diversity. In *Recent Advances in Acarology*, ed. JG Rodriguez, 1:603–13. New York: Academic
127. Lee KE. 1985. *Earthworms: Their Ecology and Relationship with Soils and Land Use.* Sydney, Aust.: Academic
128. Lehman PS, Barker KR, Huisingh D. 1971. Effects of pH and inorganic ions on emergence of *Heterodera glycines*. *Nematologica* 17:467–73
129. Lohm U. 1980. Effects of experimental acidification on soil organism populations and decomposition. *Ecol. Impact Acid Precip. Proc. Int. Conf. Impact Acid Precip., Sandefjord, Norway, March 11–14*, ed. D. Drabløs, A Tollan, pp. 178–79. Oslo-Ås: SNSF [Acid Precipitation—Effects on Forest and Fish] Proj.
130. Lohm U, Lundkvist H, Persson T, Wirén A. 1977. Effects of nitrogen fertilization on the abundance of enchytraeids and microarthropods in Scots pine forests. *Studia For. Suec.* 140:1–23
131. Lovblad G, Amann M, Anderson B, Hovmand M, Joffre S, Pedersen U. 1992. Deposition of sulfur and nitrogen in the Nordic countries: present and future. *Ambio* 21:339–47
132. Lundkvist H. 1977. Effects of artificial acidification on the abundance of Enchytraeidae in a Scots pine forest in northern Sweden. *Ecol. Bull.* 25:570–73
133. Maclagan DS. 1932. An ecological study of the “Lucerne flea” (*Sminthurus viridis* L.). *Bull. Entomol. Res.* 23:101–45
134. Maraun M, Visser S, Scheu S. 1998. Oribatid mites enhance the recovery of microbial community after a strong disturbance. *Appl. Soil Ecol.* 9:175–81
135. Märkel K, Bosner R. 1960. Die Bedeutung der Bodentierwelt für den Erfolg von Bestandes-Kalkungen. *Forst Holz Jagd Taschenb.* 10:179–81
136. Marshall VG. 1977. *Effects of Manures and Fertilizers on Soil Fauna: a Review. Spec. Publ. 3*, Commonw. Bur. Soils. Farnham Royal, England: Commonw. Agric. Bur. 79 pp.
137. Marshall VG, Reeves RM, Norton RA. 1987. Catalogue of the Oribatida (Acari) of continental United States and Canada. *Mem. Entomol. Soc. Can.* 139: i-vi + 418 pp.
138. Masuch G. 1985. Flechtenkartierung entlang eines Niederschlagsgradient im Eggegebirge. *Staub Reinhalt. Luft* 45:573–78
139. Masuch G. 1988. Veränderungen der epiphytischen Flechtenflora im Eggegebirge seit 1900. *Acta Biol. Benrodis* 1:7–17
140. Mebes K-H, Filser J. 1998. Does the species composition of Collembola affect nitrogen turnover? *Appl. Soil Ecol.* 9:241–47
141. Mertens J. 1975. L'influence du facteur pH sur le comportement de

Orchesella villosa (Geoffroy, 1764) (Collembola, Insecta). *Soc. R. Zool. Belg.* 105:43–50

142. Meyer E, Eisenbeis G. 1987. The ultrastructural basis for rectal water uptake in *Trimerophorella nivicomes* (Chordeumatida, Diplopoda). *In Soil Fauna and Soil Fertility*, ed. BR Striganova, pp. 128–34. Moscow: Nauka
143. Miko L, Kováč L. 1995. Effects of fly-ash contamination on the microarthropod abundance in East-Slovak agroecosystems. *Acta Zool. Fenn.* 196:342–43
144. Miller DR. 1984. Chemicals in the environment. In *Effects of Pollutants at the Ecosystem Level, SCOPE 22*, ed. PJ Sheehan, DR Miller, GC Butler, P Bourdeau, pp. 7–14. Chichester, UK: Wiley & Sons
145. Mina VN. 1967. Influence of stemflow on soil. *Sov. Soil Sci.* 7:1321–29
146. Moritz M. 1963. Über Oribatidengemeinschaften (Acari: *Oribatei*) norddeutscher Laubwaldböden, unter besonderer Berücksichtigung der Verteilung regelnden Milieubedingungen. *Pedobiologia* 3:142–243
147. Mráček Z. 1986. Nematodes and other factors controlling the sawfly *Cephaleia abietis* (Pamphiliidae: Hymenoptera) in Czechoslovakia. *For. Ecol. Manage.* 15: 75–79
148. Myrold DD, Nason GE. 1992. 3. Effect of acid rain on soil microbial processes. In *Environmental Microbiology*, pp. 59–81. New York: Wiley-Liss
149. Nottrot F, Joosse ENG, van Straalen NM. 1987. Sublethal effects of iron and manganese soil pollution on *Orchesella cincta* (Collembola). *Pedobiologia* 30:45–53
150. Nurminen M. 1967. Ecology of enchytraeids (Oligochaeta) in Finnish coniferous forest soils. *Ann. Zool. Fenn.* 4:147–57
151. Ohtonen R, Markkola AM. 1991. Biological activity and amount of FDA mycelium in mor humus of Scots pine stands (*Pinus sylvestris* L.) in relation to soil properties and degree of pollution. *Biogeochemistry* 13:1–26
152. Ohtonen R, Ohtonen A, Luotonen H, Markkola AM. 1992. Enchytraeid and nematode numbers in urban, polluted Scots pine (*Pinus sylvestris*) stands in relation to other soil biological parameters. *Biol. Fertil. Soils* 13:50–54
153. Overrein LN, Seip HM, Tollan A. 1980. *Acid Precipitation: Effects on Forest and Fish. Final Rep. SNSF Proj. 1972–1980*, pp. 1–175. Oslo-Ås: SNSF [Acid Precipitation—Effects on Forest and Fish] Proj.
154. Persson T. 1975. *Markarthropodernas abundans, biomassa och respiration i ett aldre tallbestand pa Ivantjarnsheden, Gastrikland: en forsta studie. Intern. Rep. 31.* Stockholm: Swed. Conifer. For. Proj.
155. Persson T. 1985. Markorganismerna och forsurningen av skogsmark. In *Vad hander med skogen: skogsdod pa vag*? ed. H Persson, pp. 133–45. Stockholm: Liber
156. Persson T. 1988. *Effects of Liming on the Soil Fauna in Forests: a Literature Review. Rep. 3418*, pp. 1–92. Stockholm: Natl. Swed. Environ. Prot. Board
157. Persson T. 1988. Effects of acidification and liming on soil biology. In *Liming as a Measure to Improve Soil and Tree Condition in Areas Affected by Air Pollution: Results and Experiences of an Ongoing Research Programme*, ed. F Andersson, T Persson. *Rep. 3518*, pp. 53–70. Stockholm: Natl. Swed. Environ. Prot. Board
158. Persson T, Hyvönen R, Lundkvist H. 1987. Influence of acidification and liming on nematodes and oligochaetes in two coniferous forests. In *Soil Fauna and Soil Fertility*, ed. BR Striganova, pp. 191–96. Moscow: Nauka
159. Peter HU. 1984. Über den Einfluss von Luftverunreinigungen auf Ökosysteme. IV. Isopoda, Diplopoda, Collembola und Auchenorrhyncha aus

Bodenfallen in der Umgebung eines Düngemittelwerkes. *Wiss. Z. Univ. Jena Nat.wiss. Rep.* 33:291–307
160. Pižl V. 1985. The effect of the herbicide Zeazin 50 on the earthworm infection by monocystid gregarines. *Pedobiologia* 28:399–402
161. Purrini K. 1981. Protozoa of soil fauna in beech forests, mixed coniferous forests and mixed deciduous forests. *Progr. Protozool. Proc. Int. Congr. Protozool., 6th,* Warsaw: Polish Acad. Sci., p. 298 (Abstr.)
162. Purrini K. 1983. Comparison of pathogenic agents in Collembola (Insecta, Apterygota) from different forests in the Federal Republic of Germany, Austria and Spain. *Pedobiologia* 25:365–71
163. Purrini K. 1983. Soil invertebrates infected by microorganisms. See Ref. 42, pp. 167–78
164. Ratajczak L, Funke W, Zell H. 1989. Die Nematodenfauna eines Fichtenforstes: Auswirkungen anthropogener Einflusse. *Verh. Ges. Ökol.* 17:391–96
165. Ruck B, Adams EW. 1988. Einfluss der Orographie auf die Schadstoffdeposition durch Feinstropfchen. In *4. Statuskolloquium des PEF vom 8. bis 10. März 1988 im Kernforschungszentrum Karlsruhe*, ed. F. Horsch, WG Filby, N Fund, S Gross, B Hanisch, et al. *Proj. Eur. Forsch.zent. Massn. Luftreinhalt. Kernforsch.zent. Karlsr.* 35:627–40
166. Ruess L, Funke W. 1992. Effects of experimental acidification on nematode populations in soil cultures. *Pedobiologia* 36:231–39
167. Ruess L, Funke W. 1995. Nematode fauna of a spruce stand associated with forest decline. *Acta Zool. Fenn.* 196:348–51
167a. Ruess L, Sandbach P, Cudlín P, Dighton J, Crossley A. 1996. Acid deposition in a spruce forest soil: effects on nematodes, mycorrhizas and fungal biomass. *Pedobiologia* 40:51–66
168. Ruf A. 1998. A maturity index for predatory soil mites (Mesostigmata: Gamasina) as an indicator of environmental impacts of pollution on forest soils. *Appl. Soil Ecol.* 9:447–52
169. Ruhling Å, Tyler G. 1973. Heavy metal pollution and decomposition of spruce needle litter. *Oikos* 24:402–16
170. Rusek J. 1993. Air-pollution-mediated changes in alpine ecosystems and ecotones. *Ecol. Appl.* 3:409–16
171. Rusek J. 1998. Biodiversity of Collembola and their functional role in the ecosystem. *Biodivers. Conserv.* 7:1207–19
172. Rusek J. 2000. Soil invertebrate species diversity in natural and disturbed environments. 12. In *Invertebrates as Webmasters in Ecosystems*, ed. DC Coleman, PF Hendrix, pp. 233–52. Wallingford, UK: Commonw. Agric. Bur. Int.
173. Russell DJ, Alberti G. 1998. Effects of long-term, geogenic heavy metal contamination on soil organic matter and microarthropod communities, in particular Collembola. *Appl. Soil Ecol.* 9:483–88
174. Satchell JE. 1955. Some aspects of earthworm ecology. In *Soil Zoology*, ed. DKMcE Kevan, pp. 180–201. London: Butterworths
175. Schauermann J. 1985. Zur Reaktion von Bodentieren nach Düngung von Heinsimsen-Buchen-Wäldern und Siebenstern-Fichtenforsten im Solling. *Allg. Forsteeitschr.* 40:1159–61
176. Schauermann J. 1987. Tiergemeinschaften der Wälder im Solling unter dem Einfluss von Luftschadstoffen und künstlichem Säure- und Düngereintrag. *Verh. Ges. Ökol.* 16:53–62
177. Scheu S, Poser G. 1996. The soil macrofauna (Diplopoda, Isopoda, Lumbricidae and Chilopoda) near tree trunks in a beechwood on limestone: indications for stemflow induced changes in

community structure. *Appl. Soil Ecol.* 3:115–25

178. Scheu S, Wolters V. 1991. Buffering of the effect of acid rain on decomposition of C-14-labelled beech leaf litter by saprophagous invertebrates. *Biol. Fertil. Soils* 11:285–89
179. Schönborn W. 1992. The role of protozoan communities in freshwater and soil ecosystems. *Acta Protozool.* 31:11–18
180. Schwalbe T. 1995. Oribatida in spruce forests influenced by SO_2-immission in the Osterzgebirge Mountains, Germany. *Acta Zool. Fenn.* 196:352–53
181. Seniczak S, Dabrowski J, Klimek A, Kaczmarek S. 1995. Air pollution effects on mites (Acari) in Scots pine forests polluted by a nitrogen fertilizer factory at Włocławek, Poland. *Acta Zool. Fenn.* 196:354–56
182. Seniczak S, Dabrowski J, Klimek A, Kaczmarek S. 1998. Effects of air pollution produced by a nitrogen fertilizer factory on the mites (Acari) associated with young Scots pine forests in Poland. *Appl. Soil Ecol.* 9:453–58
183. Setälä H, Marshall VG, Trofymow JA. 1996. Influence of body size of soil fauna on litter decomposition and ^{15}N uptake by poplar in a pot trial. *Soil Biol. Biochem.* 28:1661–75
184. Sirois A, Vet RJ. 1988. Detailed analysis of sulphate and nitrate atmospheric deposition estimates at the Turkey Lakes watershed. *Can. J. Fish Aquat. Sci.* 45:14–25 (Suppl. 1)

184a. Speir TW, Kettles HA, Percival HJ, Parshotam A. 1999. Is soil acidification the cause of biochemical responses when soils are amended with heavy metal salts? *Soil Biol. Biochem.* 31: 1953–61

185. Springett JA. 1963. The distribution of three species of Enchytraeidae in different soils. In *Soil Organisms*, ed. J Doeksen, J van der Drift, pp. 414–19. Amsterdam: North-Holland
186. Stachurska-Hagen T. 1980. *Effects of Acidification and Liming on Soil Animals. Protozoa, Rotifera, and Nematoda. SNSF Proj. IR 74/80.* Oslo: Ås-NLH, Nor. Inst. For. Res.
187. Standen V, Latter PM. 1977. Distribution of a population of *Cognettia sphagnetorum* (Enchytraeidae) in relation to microhabitats in a blanket bog. *J. Anim. Ecol.* 46:213–29
188. Stokes P. 1984. Clearwater Lake: study of an acidified lake ecosystem. In *Effects of Pollutants at the Ecosystem Level. SCOPE 22*, ed. PJ Sheehan, DR Miller, GC Butler, P Bourdeau, pp. 229–53. Chichester, UK: Wiley & Sons
189. Stout JD. 1968. The significance of the protozoan fauna in distinguishing mull and mor of beech (*Fagus silvatica* L.). *Pedobiologia* 8:387–400
190. Stout JD, Heal OW. 1967. Protozoa. In *Soil Biology*, ed. A Burges, F Raw, pp. 149–95. New York: Academic
191. Streit B. 1984. Effects of high copper concentrations on soil invertebrates (earthworms and oribatid mites): experimental results and a model. *Oecologia* 64:381–88
192. Strojan CL. 1978. Forest leaf litter decomposition in the vicinity of a zinc smelter. *Oecologia* 32:203–12
193. Strojan CL. 1978. The impact of zinc smelter emissions on forest litter arthropods. *Oikos* 31:41–46
194. Stumpp J. 1990. Zur Ökologie einheimischer Proturen (Arthropoda: Insecta) in Fichtenforsten. *Zool. Beitr.* 33(3):345–432
195. Stumpp J, Bernhard M, Funke W, Hofer H, Jans W, et al. 1986. Bodentiere im Fichtenforst: sensitive Indikatoren tiefgreifender Veränderungen in Waldökosystemen. *Verh. Dtsch. Zool. Ges.* 79:403

196. Sztrantowicz H. 1980. Structure and numbers of soil protozoa on industrial areas of Silesia, Poland. *Pol. Ecol. Stud.* 6:607–24
197. Sztrantowicz H. 1984. Protozoa as indicators of environment degradation by industry. *Pol. Ecol. Stud.* 10:67–91
198. Tarman K. 1973. Oribatid fauna in polluted soil. *Biol. Vestn. Ljubl.* 21:153–58
199. Thomas PR. 1965. Biology of *Acrobeles complexus* Thorne cultivated on agar. *Nematologica* 11:395–408
200. Timans PR. 1986. Einfluss der saueren Beregnung und Kalkung auf die Nematodenfauna. *Forstwiss. Centralbl.* 105:335–37
201. Tobisch S, Dunger W. 1974. Carabiden des Neissetales bei Ostritz/Oberlausitz und ihre Reaktion auf Industrie-Emissionen. *Abh. Ber. Nat.kd.Mus. Görlitz* 48(2):1–18
202. Urbášek F, Chalupský J. 1992. Effects of artificial acidification and liming on biomass and on the activity of digestive enzymes in *Enchytraeidae* (Oligochaeta): results of an ongoing study. *Biol. Fertil. Soils* 14:67–70
203. Vaněk J. 1967. Industrieexhalate und Moosmilbengemeinschaften in Nordböhmen. In *Progress in Soil Biology*, ed. O Graff, JE Satchell, pp. 331–39. Amsterdam: North-Holland
204. Vaněk J. 1968. Die Exhalte und die Moosmilbengemeinschaften in dem nordböhmischen Rauchschadengebiete. See Ref. 122, pp. 69–94
205. Vaněk J. 1971. Durch Industrieimmissionen verursachte Veränderungen der Moosmilbengesellschaften. In *Bioindicators of Landscape Deterioration*, pp. 72–77. Prague: ČSAV, ÚTOK
206. Vaněk J. 1974. Změny vyvolané průmyslovými imisemi ve společenstvech pancířníků; (Acarina-Oribatoidea) půd smrkových lesů. *Quest. Geobot.* 14:33–116
207. van Straalen NM, Burgouts TBA, Doornhof MJ. 1985. Dynamics of heavy metals in populations of Collembola in a contaminated forest soil. *Proc. Int. Conf. Heavy Met. Environ., 5th, Athens,* 1:613–15. Edinburgh: CEP Consult.
208. Vogel J. 1980. Ökofaunistische Beobachtungen an der Staphylinidenfauna-des Neissetales bei Ostriz/Oberlausitz. *Abh. Ber. Naturkundemus. Görlitz* 53(4):1–24
209. Voigtlander K, Dunger W. 1992. Long term observations of the effects of increasing dry pollution on the myriapod fauna of the Neisse valley (East Germany). *Ber. Nat.-Med. Ver. Innsbr. Suppl.* 10:251–56
210. Volz P. 1951. Untersuchungen über die Mikrofauna des Waldbodens. *Zool. Jahrb. (Syst.)* 79:514–66
211. von Törne E. 1990. Assessing feeding activities of soil-living animals. I. Bait-lamina-tests. *Pedobiologia* 34:89–101
212. von Törne E. 1990. Schätzung von Fressaktivitäten bodenlebender Tiere. II. Mini-Köder-Tests. *Pedobiologia* 34:269–79
212a. Vranken G, Vanderhaeghen R, Heip C. 1985. Toxicity of cadmium to free-living marine and brackish water nematodes (*Monhystera microphthalma, Monhystera disjuncta, Pellioditis marina*). *Dis. Aquat. Org.* 1:48–58
213. Walter DE, Behan-Pelletier V. 1999. Mites in forest canopies: filling the size distribution shortfall. *Annu. Rev. Entomol.* 44:1–19
214. Wanner M. 1991. Zur Ökologie von Thekamoben (Protozoa, Rhizopoda) in süddeutschen Wäldern. *Arch. Protistenkd.* 140:236–88
215. Wanner M, Funke W. 1989. Zur Mikrofauna von Waldböden: I. Testacea (Protozoa: Rhizopoda). Auswirkungen anthropogener Einflüsse. *Verh. Ges. Ökol.* 17:379–84
216. Wasilewska L. 1998. Changes in the proportions of groups of bacterivorous

soil nematodes with different life strategies in relation to environmental conditions. *Appl. Soil Ecol.* 9:215–20
217. Weigmann G. 1995. Heavy metal burdens in forest soil fauna at polluted sites near Berlin. *Acta Zool. Fenn.* 196:369–70
218. Weigmann G, Kratz W, Heck M, Jaeger-Volmer D, Kielhorn U, et al. 1989. Biologische Dynamik immissionsbelasteter Forsten: Teilprojekt 1.5. In *FE-Vorhaben "Belungsraumnahe Waldökosysteme" (Ball Wos)*. Berlin: UNESCO, MAB pp. 1–205
219. Weigmann G, Renger M, Marschner B. 1989. Untersuchungen zur Belastung und Gefahrung bellungsraumnaher Waldökosysteme. *Verh. Ges. Ökol.* 17:465–72
220. Wielgus-Serafinska E. 1979. Influence of lead poisoning and ultrastructural changes in the body wall of *Eisenia foetida* (Savigny), Oligochaeta. I. Short action of different concentrations of lead and ultrastructural changes in the cells of the body wall. *Folia Histochem. Cytochem.* 17:181–88
221. Wielgus-Serafinska E, Strzelec M. 1983. Influence of lead poisoning and ultrastructural changes in the cells of the body wall of *Eisenia foetida* (Savigny), Oligochaeta. II. Long term action of different concentration of lead and ultrastructural changes in the cells of the body wall. *Folia Histochem. Cytochem.* 21:145–52
222. Wittig R. 1986. Acidification phenomena in beech (*Fagus sylvatica*) forests of Europe. *Water Air Soil Pollut.* 31:317–23
223. Wolters V. 1991. Effects of acid rain on leaf-litter decomposition in a beech forest on calcareous soil. *Biol. Fertil. Soils* 11:151–56
224. Yeates GW, Orchard VA, Speir TW. 1995. Reduction in faunal populations and decomposition following pasture contamination by a Cu-Cr-As based timber preservative. *Acta Zool. Fenn.* 196:297–300
225. Yeates GW, Orchard VA, Speir TW, Hunt JL, Hermans MCC. 1994. Impact of pasture contamination by copper, chromium, arsenic timber preservative on soil biological activity. *Biol. Fertil. Soils* 18:200–8
226. Zullini A, Peretti E. 1986. Lead pollution and moss-inhabiting nematodes of an industrial area. *Water Air Soil Pollut.* 27:403–10

Annu. Rev. Ecol. Syst. 2000. 31:425–39

Ecological Resilience—in Theory and Application

Lance H. Gunderson
Dept. of Environmental Studies, Emory University, Atlanta, Georgia 30322;
e-mail: lgunder@emory.edu

Key Words resilience, stability, stable states, biodiversity, adaptive management

■ **Abstract** In 1973, C. S. Holling introduced the word resilience into the ecological literature as a way of helping to understand the non-linear dynamics observed in ecosystems. Ecological resilience was defined as the amount of disturbance that an ecosystem could withstand without changing self-organized processes and structures (defined as alternative stable states). Other authors consider resilience as a return time to a stable state following a perturbation. A new term, adaptive capacity, is introduced to describe the processes that modify ecological resilience. Two definitions recognize the presence of multiple stable states (or stability domains), and hence resilience is the property that mediates transition among these states. Transitions among stable states have been described for many ecosystems, including semi-arid rangelands, lakes, coral reefs, and forests. In these systems, ecological resilience is maintained by keystone structuring processes across a number of scales, sources of renewal and reformation, and functional biodiversity. In practice, maintaining a capacity for renewal in a dynamic environment provides an ecological buffer that protects the system from the failure of management actions that are taken based upon incomplete understanding, and it allows managers to affordably learn and change.

INTRODUCTION

It has been almost three decades since the term resilience was introduced to the literature by the theoretical ecologist C. S. Holling (22). Since that time, multiple meanings of the concept have appeared (15, 40). Since most management actions are based upon some type of theory, these multiple meanings of resilience can lead to very different sets of policies and actions.

This review is divided into three parts. The first section reviews concepts and multiple meanings of resilience as they have appeared in the literature. That section reviews examples of modeling and field experiments that enrich our understanding of ecological change. The second section includes an assessment of how resilience is related to other key ecosystem properties. The review concludes with a section on how ecological resilience is key to management of complex systems of people and nature.

Resilience, Stability and Adaptive Capacity

Resilience of a system has been defined in the ecological literature in two different ways, each reflecting different aspects of stability. Holling (22) first emphasized these different aspects of stability to draw attention to the distinctions between efficiency and persistence, between constancy and change, and between predictability and unpredictability. Holling (22) characterized stability as persistence of a system near or close to an equilibrium state. By contrast, resilience was introduced to indicate behavior of dynamic systems far from equilibrium, by defining resilience as the amount of disturbance that a system can absorb without changing state. The multiple meanings of resilience are related to assumptions about the presence of either single or multiple equilibria in a system (26), as described in the following sections.

Resilience and Global Equilibrium Many authors define the term resilience as the time required for a system to return to an equilibrium or steady-state following a perturbation (28, 39, 40, 44, 51). Implicit in this definition is that the system exists near a single or global equilibrium condition. Hence the measure of resilience is how far the system has moved from that equilibrium (in time) and how quickly it returns (35).

Other authors (22, 26, 35) consider return times as a measure of stability. Holling (26) described the return time definition of resilience as 'engineering resilience.' The return time definition arises from traditions of engineering, where the motive is to design systems with a single operating objective (9, 41, 54). On one hand, that makes the mathematics tractable, and on the other, it accommodates an engineer's goal to develop optimal designs. There is an implicit assumption of global stability—i.e. there is only one equilibrium or steady state or, if other operating states exist, they should be avoided by applying safeguards. Other fields that use the term resilience, such as physics, control system design, or material engineering, all use this definition.

Resilience and Multiple Equilibrium The second type of resilience emphasizes conditions far from any steady state condition, where instabilities can flip a system into another regime of behavior—i.e. to another stability domain (22). In this case, resilience is measured by the magnitude of disturbance that can be absorbed before the system redefines its structure by changing the variables and processes that control behavior. This has been dubbed ecological resilience in contrast to engineering resilience (22, 57).

One key distinction between these two types of resilience lies in assumptions regarding the existence of multiple stable states. If it is assumed that only one stable state exists or can be designed to so exist, then the only possible definition and measures for resilience are near equilibrium ones—such as characteristic return time as defined above. The concept of ecological resilience presumes the existence of multiple stability domains and the tolerance of the system to perturbations that facilitate transitions among stable states. Hence, ecological resilience refers to the width or limit of a stability domain and is defined by the magnitude of disturbance that a system can absorb before it changes stable states (22, 35).

The presence of multiple stable states and transitions among them have been described in a range of ecological systems. These include transitions from grass-dominated to woody-dominated semi-arid rangelands in Zimbabwe (60) and Australia (36, 59). In these cases the alternative states are described by dominant plant forms, and the disturbance is grazing pressure (59). Other examples include transitions from clear lakes to turbid ones (3, 46); alternative states are indicated by dominant assemblages of primary producers in the water or rooted macrophytes and disturbances include physical variables such as light and temperature. Alternative states are also described in populations levels created by interactions among populations (10, 11, 48, 62).

Carpenter et al. (3, 5) and Scheffer (46) have used the heuristic of a ball and a cup to highlight differences between these types of resilience. The ball represents the system state and the cup represents the stability domain (Figure 1). An equilibrium exists when the ball sits at the bottom of the cup and disturbances shake the marble to a transient position within the cup. Engineering resilience refers to characteristics of the shape of the cup—the slope of the sides dictate the return time of the ball to the bottom. Ecological resilience suggests that more than one cup exists, and resilience is defined as the width at the top of the cup. Implicit in both of these definitions is the assumption that resilience is a static property of

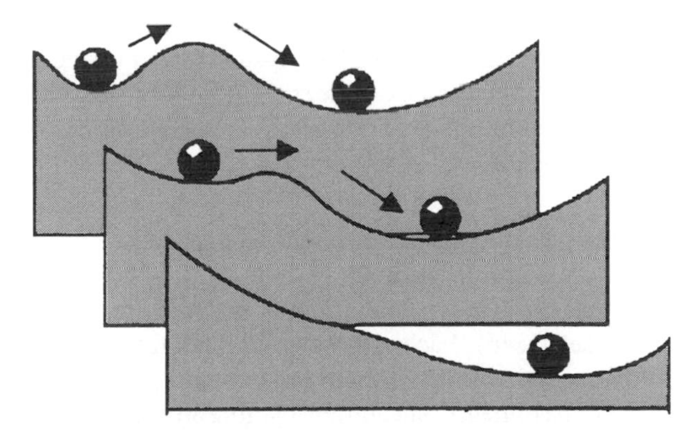

Figure 1 Ball and cup heuristic of system stability. Valleys represent stability domains, balls represent the system, and arrows represent disturbances. Engineering resilience is determined by the slopes in the stability landscapes, whereas ecological resilience is described as the width. Adaptive capacity refers to the ability of the system to remain in a stability domain, as the shape of the domain changes (as shown by the three slices or landscapes).

systems. That is, once defined, the shape of the cup remains fixed over time. But recent work indicates that stability domains are dynamic and variable.

Adaptive Capacity Many of the manifestations of human-induced state changes in ecosystems result from alteration of the key variables that influence the underlying stability domains. The key variables that configure these stability domains change at relatively slow rates (without human intervention). Examples include nutrients in wetlands and lakes (5, 46), species compositions in rangelands (56, 59) or trophic relationships (4). Using the ball in cup heuristic, the shape of the cup is subject to change, altering both stability (return time) and resilience (width of stability domain). Scheffer et al. (46) depict this as multiple stability landscapes (three slices in Figure 1). The property of an ecosystem that describes this change in stability landscapes and resilience is referred to as adaptive capacity (19).

Ecosystem Dynamics and Multiple Stable States

The previous section outlined a contrast among three views of resilience. All describe aspects of change in ecosystems and the degree of that change. But much of the literature over the last 30 years has addressed whether multiple stable states exist in ecosystems, and if so what mediates transition among them.

There is a growing body of literature that documents transitions among stability domains in a variety of ecosystems (4, 21, 35, 38, 59, 60). Many of those systems are influenced by human activities, which has led to a confounding problem around ecological resilience. Some authors (49) suggest that alternative stable states do not exist in systems untouched by humans, while others (10, 47) indicate that these are and have been part of the dynamics of systems with or without humans. Without treading on the question of whether people are or are not natural parts of ecosystems, three examples are presented suggesting that people do change the resilience of system. One involves lake systems, another wetlands, and the other semi-arid rangeland. In each example, the alternative states are discussed, as are the mechanisms that result in the transitions and those processes that contribute or detract from ecological resilience.

Shallow Lakes Limnologists have long recognized the existence of qualitative differences in the state of lakes. In shallow lakes, two alternative states can be characterized as (a) clear water and rooted macrophytes or (b) turbid water with planktonic algae (45, 46). Each of these states is relatively stable due to interactions among nutrients, the types of vegetation, and light penetration (Figure 1). In the clear water state, sediments, and nutrient cycling are stabilized by rooted vegetation (45, 46). The turbid state persists when wind-driven mixing resuspends sediments. The sediments and phytoplankton in the water column decrease light penetration, thereby curtailing establishment of benthic vegetation (45, 46).

Transitions between these two states can be mediated by trophic relationships. Decreasing stocks of planktivorous fish can create a shift from a turbid to a clear lake. As predation on herbivorous zooplankton decreases, their populations increase, leading to an increase in herbivory and a reduction in phytoplankton biomass. Increased light penetration and available nutrients then lead to establishment of rooted vegetation (45). In the other direction, shifts from the clear to turbid state can result from overgrazing of benthic vegetation by fish or waterfowl (45). The shift between stable states is hysteretic—the disturbances that influence change in one direction do not have similar impacts in the opposite direction (45).

Wetlands Nutrient enrichment in the freshwater marshes of the Everglades caused the loss of resilience. The Everglades is an oligotrophic wetland, limited primarily by phosphorus (50). For the past 5000 years or so, the ecosystem effectively self-organized around this low nutrient status, pulsed by annual wet/dry cycles and by decadal recycling associated with fires (17). The resulting landscape mosaic was comprised of sawgrass marshes and wet prairies interspersed with small tree islands (7, 33, 50).

In the late 1970s and early 1980s, large-scale vegetation changes were noticed in the regions downstream from the Everglades agricultural area. Sawgrass marshes and wet prairies had become dominated by a single species—cattail (8). The conversion was attributed to a slow increase in the concentration of soil phosphorus, and a disturbance, such as fire, drought or freeze.

Key ecosystem processes and structures occur at various spatial and temporal scales. The vegetation structures represent the most rapidly changing variables, with plant turnover times on the order of 5 to 10 years (8). Fires operate on return frequencies of 10 to 20 y (20, 53). Other disturbances such as freezes and droughts occur on multiple decade return times (20, 53). The soil phosphorus concentrations are the slowest of the variables, with turnover times on the order of centuries (8).

The resilience of the freshwater marshes is related to the soil nutrient content. The alternative stability domains are characterized by the dominant plant species; sawgrass or wet prairie communities dominate on sites with low nutrients, and cattail dominates on sites with higher soil phosphorus concentrations. Following a disturbance, it is the soil phosphorus level that determines which of these species dominate.

Semi-arid Rangelands Savanna rangelands are found in climatic regimes of hot, rainy summers and mild, dry winters. These systems have high productivity and support a diverse assemblage of perennial and annual grasses and few woody plants. Key biophysical processes in these systems include variation in rainfall, fires, and grazing. Walker (59) and Ludwig et al. (36) identify alternative stable states as either woody/grass coverage, or woody thicket. The transition between these states is mediated by grazing pressures that remove either drought-tolerant or perennial grasses (36, 59). If grazing pressures are high, the perennial grass abundance is decreased, leading to an increased abundance of woody plants. Once the woody community is established, fires burn less frequently (if the thickets burn at all) and the woody community persists for decades. Walker and Ludwig et al (36, 59) suggest that the woody/grass assemblage can be reestablished, either through collapse of matured woody plants or through manipulation of fire and drainage processes (36). The role of functional biodiversity in contributing to the ecological resilience is discussed in the next section.

Resilience in Ecosystems—Patterns and Process

Resilience is an emergent property of ecosystems and is related to self-organized behavior of those ecosystems over time. In this sense, self-organization is the interaction between structure and process that leads to system development, regardless of initial conditions. Self-organization also implies that for certain scale ranges, structure and process are not easily separable and interact in an organic way to generate emergent patterns. The adaptive cycle or four-phase model of Holling (24, 25) describes how patterns and processes change over time in many ecosystems, especially those with disturbance regimes.

Holling (24, 25) describes ecosystem succession in the first two phases of the cycle. The exploitative phase is characterized by rapid colonization of recently disturbed areas. This phase gives way to a conservation phase, as material and energy are accumulated and stored. The exploitative phase is characterized by rapid growth in an arena of scramble competition, while the species that dominate in the later phase tend to have slower growth rates and survive in an arena of exclusive competition. The mature or conservative phase is followed by a phase when a disturbance influences the structure that has accumulated in previous phases. This phase is called creative destruction. Disturbance agents such as forest fires, insect pests, or intense pulses of grazing suddenly release accumulated ecological capital. The system enters the fourth phase, or reorganization. The system passes through the reorganization phase and then enters another exploitative phase.

Resilience is related to the phase of Holling's adaptive cycle. During the exploitation phase, ecological resilience is high—the system can absorb a wide range of disturbances. When the system is reaching the limits to its conservative growth, it becomes increasingly brittle and its accumulated capital is ready to fuel rapid structural changes. The system is very stable, but that stability is local and narrow. A small disturbance can push it out of that stable domain into catastrophe. The nature and timing of the collapse-initiating disturbance determines, within some bounds, the future trajectory of the system. Therefore this brittle state presents the opportunity for a change at a small scale to cascade rapidly through the over-connected system, bringing about its rapid transformation. Either internal conditions or external events can initiate collapse, but typically it is internally induced brittleness (because of high connectivity and accumulated capital) that sets the conditions for collapse. The system becomes "an accident waiting to happen" (24)

During the reorganization phase, a system becomes most vulnerable to changing stability domains. There is little local regulation and stability, so that the system can easily be moved from one state to another. Resources for growth are present, but they are disconnected from the processes that facilitate and control growth. In such a weakly connected state, random seedings can generate multiple trajectories, which then establish the exploitative path along which the system develops. The previous system pattern may reassert itself, or the system may reorganize itself into a novel structure.

Resilience and Biodiversity

The relationship between biological diversity and ecological stability has been an ongoing discussion in ecology (37, 51, 52). Tilman (51, 52) has demonstrated that, over ecologically brief periods, an increase in species number increases the efficiency and stability of some ecosystem functions, but decreases the stability of the populations of the species. While this work is important and interesting, it focuses upon how an ecosystem behaves near some steady state. The role of ecological diversity over a much broader range of variations and especially the relationship between diversity and resilience have only been recently addressed (31, 43, 55, 56).

When grappling with this broader relationship between diversity and resilience Walker (58) developed a driver and passengers analogy. Walker proposed that functional groups of species can be divided into 'drivers' and 'passengers.' Drivers are keystone species that control the future of an ecosystem, while the passengers live in but do not alter significantly this ecosystem. However, as conditions change, endogenously or exogenously, species shift roles. In this model, removing passengers has little effect, but removing drivers can have a large impact. Ecological resilience resides both in the diversity of the drivers, and in the number of passengers who are potential drivers. Walker has more recently shown how the diversity of functional groups also maintains the resilience of ecosystem structure and function (56). Such diversity provides robustness to ecosystem functions and resilience to the system behavior. Moreover, this seems the way many biological processes are regulated—overlapping influences by multiple processes, each one of which is inefficient in its individual effect but together operating in a robust manner.

Recent models indicate that biodiversity provides a cross-scale resilience (43). Species combine to form an overlapping set of reinforcing influences that help spread risks and benefits widely to retain overall consistency in performance independent of wide fluctuations in the individual species. Because of the robustness of this redundancy within functional groups, and the non-linear way behavior suddenly flips from one pattern to another and one set of controls to another, gradual loss of species involved in controlling structure initially would have little perceived effect over a wide range of loss of species. As loss of those species continued, different behavior would emerge more and more frequently in more and more places. To the observer, it would appear as if only the few remaining species were critical when in fact all add to the resilience. Although behavior would change suddenly, resilience measured as the size of stability domains (22) would gradually contract. The system, in gradually losing resilience, would become increasingly vulnerable to perturbations that earlier could be absorbed without change in function, pattern, and controls.

As humans struggle to manage these key ecosystem attributes of function and pattern, there is a growing recognition of the importance of ecological resilience. Ecological resilience is recognized as the property that allows for managers and other actors to learn from and adapt to the unpredictability inherent in these ecosystems. This idea is developed in the final section in which the interaction between ecosystem management and resilience is discussed.

Managing for Resilience in Policy and Practice

A growing number of case histories of large-scale, bureaucratic resource systems (16, 29) and traditional management systems (2) demonstrate patterns of surprise, crises, and reformation. Shifts between alternative conditions are usually signaled as a resource crisis. That is, a crisis occurs when an ecosystem behaves in a surprising manner or when observations of a system are qualitatively different from expectations of that system. Such surprises occur when variation in broad scale processes (such as a hurricane or extreme drought), intersects with internal changes in an ecosystem due to human alteration. Examples of human induced shifts include woody invasion of semi-arid rangelands (59, 60) or algae blooms in freshwater lakes (15) as described earlier. These shifts in stability domains are chronicled as resource crisis. Understanding how and why people chose to react is key to managing for resilience.

When faced with shifting stability domains and resulting crises, management options fall into one of three general classes of response. The first is to do nothing and wait to see if the system will return to some acceptable state. One consequence of this option is that the social benefits of the desired state are foregone while waiting to see if the system will return to the desired state. The second option is to actively manage the system and try to return the system to a desirable stability domain. The third option is to admit that the system is irreversibly changed, and hence the only strategy is to adapt to the new, altered system.

The ecological resilience of the system provides some measure of ease of transition among states and is a key consideration regarding how management actions should or can be structured. This theme is developed in the following two sections, one on how building understanding provides resilience and the second on how to maintain or restore resilience in managed systems.

Uncertainty, Understanding, and Resilience

During most of the 20th century, the goal of technologically-based resource management has been to control the external sources of variability in order to seek a single goal, such as maximization of yield (trees, fish) or controlling levels of pollution. This 'command and control' approach focuses on controlling a target variable, but then slowly changes other parts of the system (27). That is, isolating and controlling the variables of interest (i.e. assuming that the uncertainty of nature can be replaced with the certainty of control) has resulted in erosion of resilience. The manifestation of that erosion is the pattern of policy crisis and reformation as mentioned above and elsewhere (16). Much of subsidized agriculture, where incentives are set up to deal with changes in markets and costs, as well as variability from nature, falls into this category (6).

Much of the 'command and control' resource management that leads to loss of ecological resilience is based upon the presumed predictability of complex ecological systems and driven by the myth that disciplinary science will resolve most uncertainties of management. But there has been a growing sense that traditional scientific approaches are not working, and, indeed, make the problem worse (34). One reason why rigid scientific and technological approaches fail is because they presume a system near equilibrium and a constancy of relationships. In this case, uncertainties arise not from errors in tools or models but from lack of appropriate information for the models. Another reason for failure is that few approaches account for inherent complex relationships among variables that lead to inherent unpredictabilities in ecological systems. Scientific disciplines tend to break the management issue into parts for analysis, and have historically generated piecemeal sets of policies as solutions (16, 61). Yet, recent models by Carpenter and others (5) that integrate ecologic, economic, and social dynamics around a flipping lake system indicate that ecosystem resilience must be continually probed.

Different views of science can contribute to a loss of ecological resilience. One mode of science focuses on parts of the system and deals with experiments that narrow uncertainty to the point of acceptance by peers; it is conservative and unambiguous by being incomplete and fragmentary. The other view is integrative and holistic, searching for simple structures and relationships that explain much of nature's complexity. This view provides the underpinnings for an approach to dealing with resource issues called adaptive management, which assumes surprises are inevitable, that knowledge will always be incomplete, and that human interaction with ecosystems will always be evolving (16, 23, 30, 61).

Adaptive management is an integrated, multidisciplinary method for natural resources management (23, 61). It is adaptive because it acknowledges that the natural resources being managed will always change, so humans must respond by adjusting and conforming as situations change. There is and always will be uncertainty and unpredictability in managed ecosystems, both as humans experience new situations and as these systems change because of management. Surprises are inevitable. Active learning is the way in which this uncertainty is winnowed. Adaptive management acknowledges that policies must satisfy social objectives but also must be continually modified and be flexible for adaptation to these surprises (16, 23, 30, 61). Adaptive management therefore views policies as hypotheses—that is; most policies are really questions masquerading as answers.

Since policies are questions, then management actions become treatments in the experimental sense. The process of adaptive management includes highlighting uncertainties, developing and evaluating hypotheses around a set of desired system outcomes, and structuring actions to evaluate or 'test' these ideas (16, 23, 30, 61). Although learning occurs regardless of the management approach, adaptive management is structured to make that learning more efficient. Trial and error is a default model for learning while managing; people are going to learn and adapt by the simple process of experience. Just as the scientific method promotes efficient learning through articulating hypotheses and testing those hypotheses, adaptive management proposes a similar approach to resolving uncertainties of resource issues.

A unique property of human systems in response to uncertainty is the generation of novelty. Novelty is key to dealing with surprises or crises. Humans are unique in that they create novelty that transforms the future over multiple decades to centuries. Natural evolutionary processes cause the same magnitude of transformation over time spans of millennia. Examples are the creation of new types and arrangements of management institutions after resource crises in the Everglades or Columbia River Basin (30, 32).

Restoration and Maintenance of Resilience

At least two aspects of managing for resilience can be identified; strategies that people employ in order to restore or maintain ecological resilience and properties that contribute to resilience in human organizations. In order to add resilience to managed systems, at least three strategies are employed: increasing the buffering capacity of the system, managing for processes at multiple scales, and nurturing sources of renewal (18). Most activities for buffering tend to address the engineering type of resilience, that is, mitigating the effects of unwanted variation in the system in order to shorten the return time to a desired equilibrium. In many agricultural systems, resistance to change is dealt with by a combination of barriers to outside forces (tariffs, fences, etc.) and internal adjustments such as water or cost control mechanisms (6). Water resource systems can be designed for resilience by increasing the buffering capacity or robustness through redundancy of structures (and flexibility of operations) rather than fewer, larger structures and rigid operational schemes (12). Folke, Berkes and collaborators (1, 2, 13, 14) suggest that traditional approaches (they define as traditional ecological knowledge) buffer systems by allowing smaller scale perturbations to enter the system, thereby lessening the impact of unpredictable or large perturbations. One such example is that the Cree fisherman use a mixed-size mesh net to harvest multiple ages classes, thereby preserving an age class structure that mimics a natural population (1). This age structure helps buffer widely varying reproductive success.

Resource systems that have been sustained over long time periods increase resilience by managing processes at multiple scales. Returning to the example of the Cree in northern Canada, Berkes (1) argues that multiple spatial domains are part of their fishing practices and multiple temporal domains in their hunting practices. While fishing within a season, the Cree monitor catch per unit effort. When they notice the rate dropping, they immediately move to different areas and do not fish those areas for a number of years. Over longer time frames, they rotate fishing effort to more remote sites (1). Similarly, they retain information through belief systems, e.g. that caribou will return for hunting at annual and decadal cycles or periods. Similarly, the Everglades water management system has changed to manage across multiple time frames. In the mid 1970s water deliveries to Everglades Park were based upon a seasonally variable, but annually constant volume of water. This system was changed in the mid 1980s to a statistical formulation that incorporated interannual variation into the volumetric calculation (32).

Folke and Berkes (2) argue that local communities and institutions co-evolve by trial and error at time scales in tune to the key sets of processes that structure ecosystems within which the groups are embedded. Many of the crises chronicled by Gunderson et al. (16) were created by an inherent focus on one scale for management, and reformations of learning recognized the multiple scales by which the ecosystem was functioning. Institutions (defined broadly as the set of rules and structures that allow people to organize for collective action) can add resilience to a system. Young and McCay (63) argue that adding flexibility and renewable structure to property rights regimes will increase ecosystem resilience. They indicate that market-based property right schemes (licenses, leases, quotas or permits) should include termination schemes, with stable arrangements (entitlements, obligations) in the interim years. These principles complement Ostrom's (42) findings that successful institutions allow stakeholders to participate in changing rules that affect them.

A few key institutional ingredients appear necessary to facilitate the movement of systems out of crisis through a reformation. In the review of management histories in western systems (16) these included functions of learning, engagement and trust. Kai Lee (30) calls this "social learning", by combining adaptive management frameworks within a framework of collective choice. Other authors (2, 14) describe this as social capital; comprised of the institutions, traditional knowledge, and common property systems that are the mechanisms by which people link to their environment.

SUMMARY AND CONCLUSIONS

Resilience in engineering systems is defined as a return time to a single, global equilibrium. Resilience in ecological systems is the amount of disturbance that a system can absorb without changing stability domains. Adaptive capacity is described as system robustness to changes in resilience. In ecological systems, resilience lies in the requisite variety of functional groups and the accumulated capital that provide sources for recovery. Resilience within a system is generated by destroying and renewing systems at smaller, faster scales. Ecological resilience is reestablished by the processes that contribute to system 'memory' of those involved in regeneration and renewal that connect that system's present to its past and it to its neighbors.

Many human activities shrink ecological resilience by attempting to control variability in key ecosystem processes. This loss of resilience is often accompanied by a change in system state, signaled as a resource crisis. When a system has shifted into an undesirable stability domain, the management alternatives are to restore the system to a desirable domain, allow the system to return to a desirable domain by itself, or adapt to the changed system because changes are irreversible.

Resilience is maintained by focusing on keystone structuring processes that cross scales, on sources of renewal and reformation, and on multiple sources of capital and skills. No single mechanism can guarantee maintenance of resilience. Strategies that address requisite variety of purposes and concentrate on renewal contribute to resilience. Institutions should focus on learning, and understanding of key cross-scale interactions. Learning, trust and engagement are key components of social resilience. Social learning is facilitated by recognition of uncertainties, monitoring and evaluation by stakeholders. The most difficult issues to deal with are those whose consequences will be realized 10 to 50 years in the future over broad scales.

ACKNOWLEDGMENTS

This work was supported by a grant from the John D. and Catherine T. MacArthur Foundation, and is a contribution of the Resilience Network, a Joint Program of the University of Florida and Beijer International Institute for Ecological Economics. As always, C.S. Holling continues to provide inspiration and novelty in clearly articulating his imaginative theories. Brian Walker, Steve Carpenter, Martin Scheffer have been most helpful in educating me as to the complex dynamics of ecosystems. This work has benefited from discussions with Garry Peterson and Rusty P. Pritchard, Jr.

LITERATURE CITED

1. Berkes F. 1995. Indigenous knowledge and resource management systems: a native Canadian case study from James Bay. In *Property Rights in a Social and Ecological Context*, ed. S Hanna, M Munasinghe, pp. 35–49. Washington, DC: Beijer Int. Inst. & World Bank
2. Berkes F, Folke C. eds. 1998. *Linking Social and Ecological Systems Management Practices and Social Mechanisms for Building Resilience*. New York: Cambridge Univ. Press. 459 pp.
3. Carpenter SR, Cottingham KL. 1997. Resilience and restoration of lakes. *Conservation Ecology.* Vol. 1. (url: http://www.consecol.org)
4. Carpenter SR, Kitchell JF, eds. 1993. *The Trophic Cascade in Lakes*. Cambridge: Cambridge Univ. Press. 385 pp.
5. Carpenter SR, Ludwig D, Brock WA. 1999.

Management of eutrophication for lakes subject to potentially irreversible change. *Ecol. Appl.* 9:751–71
6. Conway G. 1993. Sustainable agriculture: the trade-offs with productivity, stability and equitability. In *Economics and Ecology, New Frontiers and Sustainable Development*, ed. EB Barbier, pp. 57–68. London: Chapman & Hall
7. Craighead FC, Sr. 1971. *The Trees of South Florida*. Vol. I. *The Natural Environments and Their Succession*. Coral Gables, FL: Univ. Miami Press. 212 pp.
8. Davis SM. 1989. Sawgrass and cattail production in relation to nutrient supply in the Everglades. In *Fresh Water Wetlands & Wildlife, 9th Annual Symposium, Savannah River Ecology Laboratory, 24–27 March, 1986*. Charleston, SC: US Dept. Energy
9. DeAngelis DL. 1980. Energy flow, nutrient cycling and ecosystem resilience. *Ecology* 61:764–71
10. Dublin HT, Sinclair ARE, McGlade J. 1990. Elephants and fire as causes of multiple stable states in the Serengeti-mara woodlands. *J. Anthropol. Ecol.* 59:1147–64
11. Estes JA, Duggins DO. 1995. Sea otters and kelp forests in Alaska: Generality and variation in a community ecological paradigm. *Ecol. Monogr.* 65:75–100
12. Fiering MB. 1982. Alternative indices of resilience. *Water Resources Res.* 18:33–39
13. Folke C, Berkes F. 1995. Mechanisms that link property rights to ecological systems. In *Property Rights and the Environment*, ed. S Hanna, M Munasinghe, pp. 50–62. Washington, DC: Beijer Int. Inst. & World Bank
14. Folke C, Holling CS, Perrings C. 1996. Biological diversity, ecosystems, and the human scale. *Ecol. Appl.* 6:1018–24
15. Grimm V, Wissel C. 1997. Babel, or the ecological stability discussions–an inventory and analysis of terminology and a guide for avoiding confusion. *Oecologia* 109:323–34
16. Gunderson LH, Holling C, Light S, eds. 1995. *Barriers & Bridges for the Renewal of Ecosystems and Institutions*. New York: Columbia Univ. Press. 593 pp.
17. Gunderson LH. 1994. *Vegetation: determinants of composition*. In *Everglades: The Ecosystem and Its Restoration*, ed. SM Davis, J Ogden, pp. 323–340. Delray Beach, FL: St. Lucie
18. Gunderson LH. 1999. *Stepping Back: Assessing for Understanding in complex regional systems*. In *Bioregional Assessments: Science at the Crossroads of Management and Policy*, ed. NK Johnson, F Swanson, M Herring, S Greene, pp. 127–40. Washington, DC: Island
19. Gunderson LH, Holling CS, Pritchard L, Peterson G. 1996. *Resilience in Ecosystems, Institutions and Societies, Propositions for a Research Agenda*. Stockholm: Beijer Inst. Discuss. Pap.
20. Gunderson LH, Snyder JR. 1994. Fire patterns in the southern Everglades. In *Everglades: The Ecosystem and Its Restoration*, ed. SM Davis, J Ogden, pp. 291–306. Delray Beach, FL: St. Lucie
21. Hanski I. 1995. Multiple equilibria in metapopulation dynamics. *Nature* 377:618–21
22. Holling CS. 1973. Resilience and stability of ecological systems. *Annu. Rev. Ecol. Syst.* 4:1–23
23. Holling CS, ed. 1978. *Adaptive Environmental Assessment and Management*, London: Wiley & Sons. 377 pp.
24. Holling CS. 1986. The resilience of terrestrial ecosystems: local surprise and global change In *Sustainable Development of the Biosphere*, ed. WC Clark, RE Munn, pp. 292–317. Cambridge: Cambridge Univ. Press.
25. Holling CS. 1992. Cross-scale morphology, geometry and dynamics of ecosystems. *Ecol. Mongr.* 62:447–502
26. Holling CS. 1996. Engineering resilience vs. ecological resilience. In *Engineering Within Ecological Constraints*, ed. PC

Schulze, pp. 31–43. Washington, DC: Natl. Acad.
27. Holling CS, Meffe GK. 1996. Command and control and the pathology of natural resource management. *Conserv. Biol.* 10:328–37
28. Ives AR. 1995. Measuring resilience in stochastic-systems. *Ecol. Monogr.* 65:217–33
29. Johnson KN, Swanson F, Herring M, Greene S. 1999. *Bioregional Assessments.* Washington, DC: Island. 398 pp.
30. Lee KN. 1993. *Compass and Gyroscope.* Washington, DC: Island. 243 pp.
31. Levin S. 1995. Biodiversity: interfacing populations and ecosystems. In *Ecological Perspective of Biodiversity*, ed. M Higashi, T Abe, S Kevin, Kyoto: Kyoto Univ. Press. 295 pp.
32. Light SS, Gunderson LH, Holling CS. 1995. The Everglades: evolution of management in a turbulent ecosystem. In *Barriers and Bridges to the Renewal of Ecosystems and Institutions*, ed. LH Gunderson, CS Holling, SS Light, pp. 103–68. New York: Columbia Univ. Press
33. Loveless CM. 1959. A study of the vegetation of the Florida Everglades. *Ecology* 40:1–9
34. Ludwig D, Hilborn R, Walters C. 1993. Uncertainty, resource exploitation, and conservation: lessons from history. *Science* 260:17–36
35. Ludwig D, Walker B, Holling CS. 1996. Sustainability, stability and resilience. *Conserv. Ecol.* 1:1–27
36. Ludwig J, Tongway D, Freudenberger D, Noble J. 1997. *Landscape Ecology Function and Management Principles from Australia's Rangelands.* Collingwood, Australia: CSIRO. 158 pp.
37. May RM. 1973. *Stability and Complexity in Model Ecosystems.* Princeton, NJ: Princeton Univ. Press. 235 pp.
38. McClanahan TR. 1995. A coral reef ecosystem-fisheries model: impacts of fishing intensity and catch selection on reef structure and processes. *Ecol. Model.* 80:1–19
39. Mittelbach G, Turner A, Hall D, Rettig J. 1995. Perturbation and resilience—a long-term, whole-lake study of predator extinction and reintroduction. *Ecology* 76:2347–60
40. Neubert MG, Caswell H. 1997. Alternatives to resilience for measuring the responses of ecological-systems to perturbations. *Ecology* 78:653–65
41. O'Neill RV, DeAngelis DL, Waide JB, Allen TFH. 1986. *A Hierarchical Concept of Ecosystems. Monogr. Pop. Biol.* Vol. 23, Princeton, NJ: Princeton Univ. Press. 254 pp.
42. Ostrom E. 1995. Designing complexity to govern complexity. In *Property Rights and the Environment*, ed. S Hanna, M Munasinghe, pp. 245–75. Washington, DC: Beijer Intl. Inst. & World Bank
43. Peterson G, Allen C, Holling CS. 1998. Ecological resilience, biodiversity, and scale. *Ecosystems* 1:6–18
44. Pimm SL. 1991. *The Balance of Nature?* Chicago: Univ. Chicago Press. 434 pp.
45. Scheffer M. 1998. *Ecology of Shallow Lakes.* London: Chapman & Hall. 357 pp.
46. Scheffer M, Hosper SH, Meijer ML, Moss B. 1993. Alternative equilibria in shallow lakes. *Trends Evol. Ecol.* 8:275–79
47. Sinclair ARE, ed. 1995. *Serengeti II: Dynamics, Management, and Conservation of an Ecosystem.* Chicago: Univ. Chicago Press. 665 pp.
48. Sinclair ARE, Olsen PD, Redhead TD. 1990. Can predators regulate small mammal populations? Evidence from house mouse outbreaks in Australia. *Oikos* 59:382–92
49. Sousa WP, Connell JH. 1983. On the evidence needed to judge ecological stability or persistence. *Am. Nat.* 121:789–825
50. Steward KK, Ornes WH. 1975. The autecology of sawgrass in the Florida Everglades. *Ecology* 56:162–71

51. Tilman D, Downing JA. 1994. Biodiversity and stability in grasslands. *Nature* 367:363–65
52. Tilman D, Wedin D, Knops J. 1996. Productivity and sustainability influenced by biodiversity in grassland ecosystems. *Nature* 379:718–20
53. Wade DD, Ewel JJ, Hofstetter R. 1980. *Fire in South Florida Ecosystems*. Asheville, NC:SE For. Exp. Stat., US Dep. Agri., For. Serv.
54. Waide JB, Webster JR. 1976. Engineering systems analysis: applicability to ecosystems. In *Systems Analysis and Simulation in Ecology*, ed. BC Patton, pp. 329–71. New York: Academic
55. Walker B. 1995. Conserving biological diversity through ecosystem resilience. *Conserv. Biol.* 9:747–52
56. Walker B, Kinzig A, Langridge J. 1999. Plant attribute diversity, resilience and ecosystem function: the nature and significance of dominant and minor species. *Ecosystems* 2:95–113
57. Walker BH. 1981. Is succession a viable concept in African savanna ecosystems? In *Forest Succession: Concepts and Application*, ed. DC West, HH Shugart, DB Botkin, pp. 431–47. New York: Springer-Verlag
58. Walker BH. 1992. Biological diversity and ecological redundancy. *Conserv. Biol.* 6:18–23
59. Walker BH, Langridge JL, Mcfarlane F. 1997. Resilience of an Australian savanna grassland to selective and nonselective perturbations. *Aust. J. Ecol.* 22:125–35
60. Walker BH, Ludwig D, Holling CS, Peterman RM. 1981. Stability of semi-arid savanna grazing systems. *J. Ecol.* 69:473–98
61. Walters CJ. 1986. *Adaptive Management of Renewable Resources*. New York: McGraw Hill. 374 pp.
62. Weaver J, Paquet PC, Ruggiero L. 1996. Resilience and conservation of large carnivores in the Rocky mountains. *Conserv. Biol.* 10:964–976
63. Young M, McCay BJ. 1995. Building equity, stewardship and resilience into market-based property rights systems. In *Property Rights and the Environment*, ed. S Hanna, M Munasinghe, pp. 210–33. Washington, DC: Beijer Int. Inst. & World Bank

Annu. Rev. Ecol. Syst. 2000. 31:441–80

QUASIREPLICATION AND THE CONTRACT OF ERROR: Lessons from Sex Ratios, Heritabilities and Fluctuating Asymmetry

A. Richard Palmer
Department of Biological Sciences, University of Alberta, Edmonton, Alberta T6G 2E9 Canada, and Bamfield Marine Station, Bamfield, British Columbia V0R 1B0 Canada; e-mail: rich.palmer@ualberta.ca

Key Words publication bias, selective reporting, funnel graph, research synthesis, replication, meta-analysis

■ **Abstract** Selective reporting—e.g., the preferential publication of results that are statistically significant, or consistent with theory or expectation—presents a challenge to meta-analysis and seriously undermines the quest for generalizations. Funnel graphs (scatterplots of effect size vs. sample size) help reveal the extent of selective reporting. They also allow the strength of biological effects to be judged easily, and they reaffirm the value of graphical presentations of data over statistical summaries.

Funnel graphs of published results, including: (*a*) sex-ratio variation in birds, (*b*) field estimates of heritabilities, and (*c*) relations between fluctuating asymmetry and individual attractiveness or fitness, suggest selective reporting is widespread and raise doubts about the true magnitude of these phenomena. Quasireplication—the "replication" of previous studies using different species or systems—has almost completely supplanted replicative research in ecology and evolution. Without incentives for formal replicative studies, which could come from changes to editorial policies, graduate training programs, and research funding priorities, the contract of error will continue to thwart attempts at robust generalizations.

> "For as knowledges are now delivered, there is a kind of contract of error between the deliverer and the receiver: for he that delivereth knowledge desireth to deliver it in such a form as may be best believed, and not as may be best examined; and he that receiveth knowledge desireth rather present satisfaction than expectant inquiry; and so rather not to doubt than not to err: glory making the author not to lay open his weakness, and sloth making the disciple not to know his strength."
>
> *The Advancement of Learning*, Francis Bacon, 1605 (8:170–171)

INTRODUCTION

Little has changed since Bacon penned these perceptive words nearly four hundred years ago. He clearly recognized a fundamental weakness of human nature: We prefer entertainment to challenge. This weakness—when coupled with a deep faith in modern statistics, a general unwillingness to acknowledge that some results will appear significant due to chance, and journal editorial policies that explicitly discourage replication—potentially undermines our quest for a robust understanding of ecological and evolutionary phenomena. Together, these conspire to perpetuate a collective contract of error, where popular beliefs are sanctified by the selective publication of results that are either statistically significant or consistent with theory or expectation, and where a peer-review process discourages the very contributions that are needed most: formal replicate studies.

How do we come to accept that a generalization has been demonstrated scientifically? For the most part, we judge a generalization's validity based on our reading of the scientific literature and our sense of the internal consistency of this literature and its consistency with our own personal observations. But if statistical significance of a result—or its concordance with pre-conceptions—influences the likelihood of publication, or if oft-repeated claims in review papers reinforce belief in a paradigm even where the evidence remains ambiguous (33), then how many emerging generalities reflect biological reality as opposed to collective wishful thinking embroidered with statistical support?

This problem is exacerbated when advocates buttress their strongly held vision of a particular phenomenon with seemingly compelling theoretical and empirical support (42, 73), thereby encouraging others to follow along with their own independent confirmations (2, 86, 96). Eventually, if the claims are exaggerated, either (*a*) the primary protagonists retire or pass on, and without their continued proselytizing others lose interest (as eventually happened to one branch of quantitative genetic methods, (70), or (*b*) the claims trigger a backlash among skeptics, and the weight of evidence declines or swings in the opposite direction (e.g., see Figure 13 below, and Refs. 2, 84, 86, 96, as possible examples). Such oscillations can yield sufficient distrust that even intriguing biological phenomena may be ignored because the waters have become so muddied by contradictory claims.

In fields where repeated tests of specific hypotheses are more commonplace (physics, molecular biology, medicine), an average result seems eventually to emerge from the variation among studies. This happens, in part, because the success of subsequent work depends so much on replication of prior methods and results. In ecology and evolution, where particular outcomes are typically repeated with different species or systems rather than truly repeated, we build up an impression of repeatable patterns by averaging over many heterogeneous studies either qualitatively or else quantitatively via meta-analysis. But, as will become apparent, these tests are as likely to reinforce an illusion as they are to validate a biological pattern.

Quasireplication and Selective Reporting

The practice of repeating studies with different species or systems—but not with the same species or system—seems so entrenched in many biological disciplines, including ecology and evolutionary biology, that I think it deserves a name: *quasireplication*. Quasireplication refers to what others have variously called, in different contexts, "imitative or acquisitive study" (18), "normal science" (60), "encyclopedism" (85), "advocacy science" (120), or "corroborative research" (101). Unfortunately, quasireplication is not true replication. Unless it is combined with true replication, it seems more likely to mislead than to reassure for two reasons.

First, quasireplications seem more vulnerable to selective reporting (the tendency to publish only a subset of studies that were undertaken) and therefore likely to lead to publication bias (deviations in average effect sizes caused by selective reporting) (12, 13). Authors who have explicitly set out to replicate a previous study as closely as possible will likely have more at stake in the outcome, since any outcome—whether confirmatory or not—would be of value. However, authors simply asking whether a popular pattern exists in their favorite organism or system may be less inclined to report negative or contradictory results unless they have unusually great confidence in the power of their test. Quasireplication may seriously compromise the validity of generalizations because repeated reports of popular or trendy results in a variety of different species or systems reinforce belief in their generality, even though much of the support may derive purely from selective reporting.

Second, quasireplication does not provide nearly the same strength of test as does true replication, and therefore it is less effective at resolving differences of opinion about the validity of hypotheses or results. For example, if quasireplicated studies present discrepant results, opponents of a particular hypothesis will highlight them as contradictory evidence, but proponents of the same hypothesis can simply dismiss them as not a true replication of the original claim. The result is bickering, hand waving, and meta-analyses conducted or interpreted in different ways to support one or another cherished belief (74, 80, 105). Such debate is almost entirely deflated by a few well-conducted, fully replicated tests of definitive or classical studies.

Quasireplication nonetheless has an important role. It offers the quickest route to true biological generalizations. However, it should not be used as a substitute for true replication, because of its vulnerability to selective reporting. Clearly, what is needed is a proper balance between the two.

Prior Controversies

Controversies over the validity of prevailing dogma are not new. In the early 1980s, a veritable donnybrook erupted between community ecologists who believed, following Platt (85), that explicit hypothesis testing was the only proper research protocol (95, 101, 102), and others who argued against such a rigid approach (87) or advocated a compromise (92, 93). Elner & Vadas (33) offered a particularly illuminating retrospective of the debate over mechanisms driving the lobster/urchin/kelp system in the western Atlantic, in which "over a period of approximately 20 years, [published] explanations for the phenomenon invoked four separate scenarios, which changed mainly as a consequence of extraneous events rather than experimental testing." Changes in viewpoint were driven as much by sociological as by biological factors. Although much of the debate revolved around how hypotheses are best tested, one recurring cry was for more formal replication of prior research rather than the accumulation of more quasireplicated studies (95, 101).

We seem to have made little progress in the intervening 20 years, but signs of change are increasingly evident. Concerns have recently been raised about selective reporting for several phenomena, including the adaptive significance of enzyme polymorphism (16), correlations with allozyme heterozygosity (52), adaptive sex-ratio variation in mammals (36), and relations between fluctuating asymmetry and sexual selection (80). Concerns have also been raised about selective reporting and publication bias in biological research in general (26). Apparently we are beginning to acknowledge the immensity of the problem.

Unfortunately, while meta-analytic techniques may help reveal biases due to selective reporting, they cannot, at present, reliably correct for them (12, 14, 19, 57). Furthermore, simple graphical techniques like the funnel graph (64) may prove more useful than summary statistics for judging the validity of emerging generalizations.

RESEARCH SYNTHESIS AND PATTERNS OF REPORTING

Few would dispute the value of quantitative approaches to reviews of the literature. But do such quantitative syntheses further exacerbate the contract of error by lending statistical support to collective preconceptions? Disentangling true biological effects from the biases introduced by selective reporting remains one of the most serious unsolved problems of meta-analysis (12, 13, 19, 50, 57).

Meta-Analysis: A Brief Overview

Formal methods of quantitative research synthesis form the domain of meta-analysis (25, 55, 77, 91). In meta-analysis, various statistical results from multiple studies are first converted to a standard statistic—effect size—to allow quantitative summarization. Several effect size statistics are available (25, 77, 91), but the correlation coefficient is perhaps the most popular in ecological and evolutionary studies (7).

Correlation coefficients simply describe the consistency of an association between two variables and have four advantages: (*a*) They are a familiar statistic; (*b*) they range between zero and ± 1.0; (*c*) when squared they yield the coefficient of determination (98), which describes the percent of variation in Y explained by variation in X; and (*d*) statistical significance thresholds provide a useful reference against which published statistics may be compared to test for selective reporting (80). However, correlation coefficients reveal nothing about the strength of an association (i.e. the amount of change in Y for a given change in X), so the familiar—but often ignored—caution about not mistaking statistical significance for biological significance must be repeated.

Meta-analysis has a long history in psychological and educational research (45, 57, 68) and is now widely used to assess the strength and validity of medical research findings (57, 75, 77) [see Becker et al (11) for a particularly useful compact review of books on meta-analytic methods]. In these fields, concerns about publication biases have also been widespread (29, 61) because the validity of a quantitative outcome may have a profound and potentially expensive impact on public policy.

Meta-analysis has also received increasing attention in ecology and evolution following its original application to the effects of age on fecundity in birds (56), a prominent application to the great debate of the mid-1980s over the impact of competition in the field (45), and a convenient, compact review that introduced it to a wide audience (7). More recently, even Bayesian statistical approaches are being applied as an alternative to conventional meta-analytic methods (108). As the popularity of meta-analysis has increased, though, more and more authors have expressed concern about the impact of publication bias on statistical summaries (1, 2, 7, 80), a point that was not always emphasized in earlier meta-analyses (e.g., see 45, 47). Because of the profound bias selective reporting may introduce, great care must be taken to avoid simply re-enforcing the bias by relying on simplified summary statistics within a meta-analysis.

Although not without its shortcomings (27, 30, 39), meta-analysis offers a significant advance over narrative research synthesis. Just as cladistic methods have revolutionized phylogenetics—by forcing all steps in an analysis to be made transparent (including data to be analyzed, characters and character state definitions, weighting protocols, and analytical procedures)—meta-analysis has the potential to revolutionize research synthesis in ecology and evolution. Unfortunately, just as cladistic analyses can be slanted in a preferred direction, so can meta-analysis (31). The exchanges between Givens et al (43) and commentators (14) on the second-hand smoke debate, and between Palmer (80) and Thornhill et al (105) regarding fluctuating asymmetry and sexual selection, are particularly illuminating in this regard.

Potential Causes of Selective Reporting

Many factors influence probability of publication. Some are of little consequence to research synthesis (e.g., loss of funding, loss of motivation or interest unrelated to early results, distraction by other activities) because they are unrelated to a research outcome. However, other causes of underreporting—statistical nonsignificance, inconsistency with expectation, inconsistency with theory—may seriously bias meta-analytic summary statistics. What data exist suggest this problem may be profound.

Statistical Significance of Results Perhaps the most familiar and widespread cause of selective reporting is the statistical significance of results (12, 13, 50). It may influence both an author's willingness or desire to press forward with publication and the willingness of editors and referees to accept a result.

The impact of statistical significance is most easily assessed when the outcomes of both published and unpublished studies may be evaluated retrospectively (12). In retrospective surveys of studies known to have been conducted, those yielding statistically significant results ($P < 0.05$) were more likely to be published, and published sooner (e.g., 24, 100). In addition, among nonsignificant studies, those yielding clearly nonsignificant results ($P > 0.1$) were more likely to be published than those yielding ambiguous results ($0.05 < P < 0.1$) (100). Clearly, not only the statistical significance, but also the statistical clarity of a study's outcome can influence its likelihood of publication.

Consistency with a Preferred Hypothesis Weighted mean effect sizes from a meta-analysis might be trusted if the effects of preconceptions, both positive and negative, averaged out. But what if a preconception is widely shared? Even if it had no biological validity it could still bias the weight of published evidence in its favor, and summary statistics from meta-analyses would serve only to reinforce this bias.

An example from medical research suggests just such a bias. When studies of the effects of acupuncture were compared to other randomized or controlled trials, certain countries reported disproportionately more positive findings than others (112). Those countries reporting more positive findings also happened to be countries in which acupuncture was considered an acceptable treatment. In addition, "no trial published in China or Russia/USSR found [an acupuncture] treatment to be ineffective" (112). Alternatively, in view of the well-known and sizeable placebo effect (49), perhaps the higher incidence of positive findings reflected cultural differences in the belief in acupuncture's effectiveness rather than selective reporting. Patients in China and Russia may, in fact, have shown demonstrably better responses to treatment.

Consistency with Theory Certain results might be under-reported because they make no sense theoretically, even though sampling error dictates that theoretically nonsensical values should arise occasionally due to chance.

Heritability estimates provide a test for such a bias. In the absence of sampling error, theory predicts that heritabilities should range from zero (no resemblance between parents and offspring) to 1.0 (offspring exactly resemble the mean phenotype of their parents) (35). For well-behaved polygenic traits, heritabilities in excess of 1.0 would imply that offspring consistently deviated more from the population mean than the average phenotype of their parents, and heritabilities less than 0.0 would imply that the more parents deviated from the population mean in one direction the more their offspring deviated in the other!

Few evolutionary biologists would seem likely to place much belief in the biological significance of such extreme heritability estimates ($h^2 < 0.0$, $h^2 > 1.0$). However, sampling error dictates that some should arise simply due to chance, particularly when based on small sample sizes (66). The underreporting of negative heritabilities (see Heritability: Impact of Theoretical Preconceptions, below) reveals that consistency with theory clearly influences likelihood of publication.

Even where sampling error may be negligible, inconsistency with a strongly held theory may discourage acceptance of a result. For example (42a), prior to 1956, physicists believed that parity (i.e., mirror image counterparts of all physical phenomena including positive/negative charge, matter/antimatter, right/left spin) was always conserved. This belief was so strong that when three physicists observed a parity violation in the decay of a radium isotope in 1928 and again in 1930, the result was ignored even though it was seen "in all readings in every run, with few exceptions"(42a:218). The result simply did not coincide with any accepted theory. Not until 1956 was theory revised to admit the possibility of parity violations in weak interactions, for which the authors later received the Nobel Prize in physics. Shortly afterwards the first supposedly definitive violation of parity was seen in the beta decay of cobalt 60 (an excess of electrons is emitted from one end of the spinning nucleus) and then quickly confirmed for many other weak interactions. Inconsistency with theory had therefore discouraged the acceptance of repeatable observations of parity violation for over 25 years.

The "Fail-Safe" Number and Its Limitations

The fail-safe number (23, 91) is often invoked to reassure meta-analysts that selective reporting would have to have been severe to account for the overall statistical significance of a particular effect. It estimates the number of studies of zero effect that would have to be published to reduce a weighted-mean effect size to nonsignificance. A fail-safe number of 1000 therefore means that 1000 studies of zero effect would have to have gone unpublished—left in a file drawer (90)—for a particular average effect to have reached statistical significance due to selective reporting.

Although easy to compute and statistically well-defined, the fail-safe number can yield a deceptive impression of how robust a particular meta-analytic result is. This deception arises because the fail-safe computation assumes that all unpublished studies are of zero effect. Clearly, though, some unpublished studies must have yielded results in the opposite direction (e.g., compare Figure 7 to Figure 14 below). So while 1000 studies of zero effect might reduce a meta-analytic mean to nonsignificance, only 100 or fewer studies of zero and opposite effect could reduce a meta-analytic mean to nonsignificance. Therefore, "if the literature is one in which a large number of unreported studies with opposing results may exist, then the usual fail-safe number may add unwarranted confidence to the interpretation of the reported (but potentially biased) results" (10:228).

No simple solution to this problem seems to exist, so conclusions that depend on a seemingly large fail-safe number must be viewed with considerable skepticism.

A GRAPHICAL APPROACH TO RESEARCH SYNTHESIS

The Value of a Graphical Approach

The greatest insurance against being misled by "the pernicious influence of the modern tendency to deify the statistical significance test" (39) is appropriate graphical presentation of research results. If patterns are not apparent in simple graphical form, one may legitimately wonder about their biological significance (67).

Light & Pillemer (64) introduced a particularly attractive graphical approach to the study of selective reporting: the funnel graph (Figure 1; a scatterplot of effect size as a function of sample size). Like many graphical approaches (4), funnel graphs offer several advantages. First, they allow readers to judge for themselves how well behaved the data are: Do published results converge on some average value with increasing sample size? Are data approximately normally distributed so that a mean and standard error are appropriate statistical descriptors? Second, they allow differences between groups to be judged more easily: Are putative differences between groups of interest apparent to the eye or do they depend upon statistical wizardry (4)? Third, a funnel graph also provides a powerful exploratory tool for determining whether statistically significant heterogeneity, due to contrasts of biological or methodological interest, may have been confounded by selective reporting (80): Are studies reporting larger effects disproportionately based on smaller sample sizes?

Several meta-analyses have incorporated funnel graphs to show how effect sizes converged toward an average with increasing sample size (e.g., 6, 28, 45), but rather few have used them to test for selective reporting (32, 80, 114). Why such graphical presentations are not more widely used is hard to understand, except perhaps because, as Magnusson (67:148) wryly reflects, "[scatterplots] are not very scientific. After all, anyone, even a nonscientist, could interpret them."

→

Figure 1 Hypothetical funnel graphs (64)—effect size as a function of sample size (log scale)—as modified in Ref. 80. (*a*) Expected pattern of purely sample size–dependent variation in effect sizes. (*b*) The impact of selective reporting when the true effect size is weak (the classical funnel pattern). (*c*) The impact of selective reporting when the true effect size is moderate (one side of the funnel is missing, and average effect size now depends on sample size). Shaded areas and open circles indicate areas of a reduced likelihood of publication due to selective reporting. Dotted lines indicate the null hypothesis, long-dashed lines indicate overall weighted mean, and curved lines are significance levels for correlation coefficients (P = 0.05) from Table R of Rohlf & Sokal (89). r_{bias} refers to the correlation—sometimes significant statistically—between effect size and sample size (80).

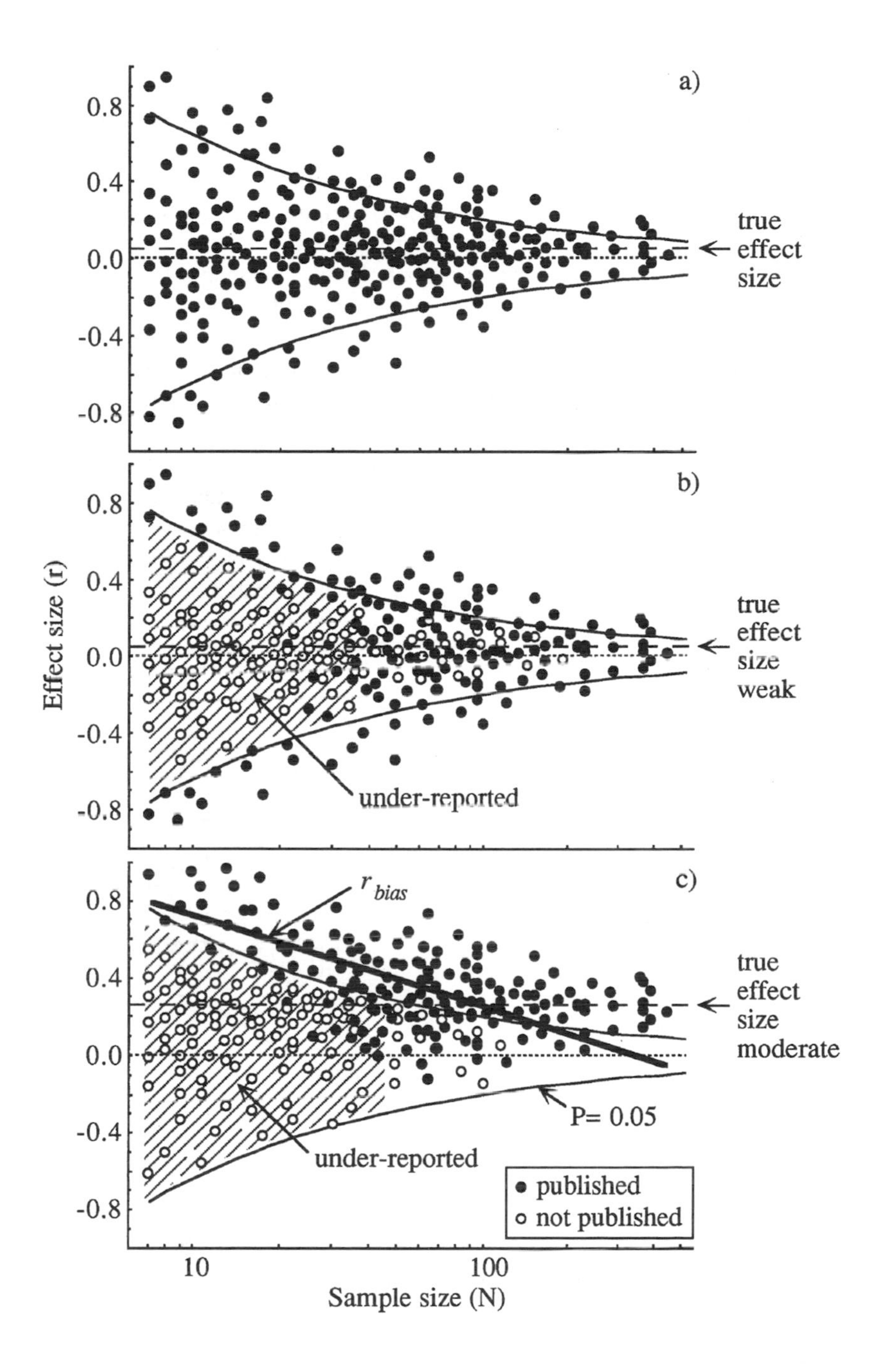
a)
b)
c)
0.8
0.4
0.0
-0.4
-0.8
Effect size (r)
true effect size
true effect size weak
true effect size moderate
under-reported
r_{bias}
P= 0.05
• published
○ not published
10
100
Sample size (N)

Funnel Graphs Showing No Bias

A scatterplot of effect size versus sample size (Figure 1*a*) should exhibit three predictable characteristics if results have not been influenced by selective reporting (80): (*a*) The variance of effect sizes should increase as sample sizes decrease; (*b*) the distribution of effect sizes should be normal for all sample sizes; and (*c*) the mean effect size should be independent of sample size. Any departures from these three criteria suggest nonrandom reporting of results (see Interpreting Apparent Bias in Funnel Graphs below for one alternative explanation).

Where all results come from a single study, and where sample sizes vary considerably, effect sizes should vary as expected in the absence of selective reporting (Figure 1a), unless authors have somehow censored or introduced bias into their own data. Two examples illustrate the expected pattern of purely sample size–dependent variation. In one (Figure 2*a*), the author concluded that no evidence existed for a biased sex ratio in European sparrowhawks (76). In the other (Figure 2*b*), the scatter clearly converged on a value greater than zero, indicating strong statistical support for assortative mating in water striders (6). In both cases, the data behaved as expected in the absence of selective reporting: (*a*) The variance increased with decreasing sample size, (*b*) the data were approximately normally distributed at all sample sizes, and (*c*) the mean effect size did not depend on sample size.

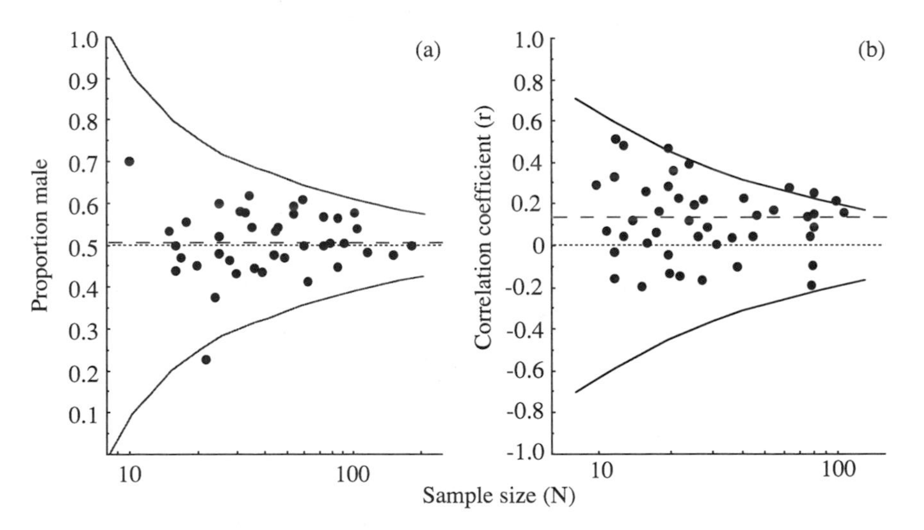

Figure 2 Two examples of within-study variation: (*a*) sex ratio variation among samples of nestlings of the European sparrowhawk (76) and (*b*) estimate of degree of assortative mating among populations of water striders (6). Dotted lines, long-dashed lines, and curved lines in (*b*) as in Figure 1. Curved lines in (*a*) are binomial significance levels (P = 0.05, two tailed) from Table Q of Rohlf & Sokal (89). In (*a*) the scatter converges on the null hypothesis of no sex-ratio bias with increasing sample size, whereas in (*b*) the scatter converges on a weighted mean (long-dashed line) that is different from the null hypothesis (zero).

Two Kinds of Bias Revealed by Funnel Graphs

When combined with statistical significance thresholds, a funnel graph may reveal two different manifestations of selective reporting. First, an underreporting of nonsignificant results based on small sample sizes, where the true underlying effect size is weak, yields a hole in the funnel and therefore a departure from normality (open circles and shaded area, Figure 1*b*). Second, an underreporting of nonsignificant results based on small sample sizes, where the true underlying effect size is moderate, yields a dependence of effect size on sample size (r_{bias} in 80), because results toward one side of the funnel are less likely to reach the threshold of statistical significance (open circles and shaded area, Figure 1*c*).

Interpreting Apparent Bias in Funnel Graphs

Those unfamiliar with the phrases "selective reporting" and "publication bias" might think they imply improper behavior by authors, but such an inference is unwarranted. First, because few scientists publish the results of all studies they undertake, some will inevitably lose interest if a result is not significant statistically, or not clear cut, or somehow doesn't make sense. In this respect, virtually all scientists are guilty to some degree. Second, referees and editors are not inclined to accept negative results based small sample sizes.

But even patterns consistent with selective reporting can arise for legitimate reasons. For example, a significant dependence of effect size on sample size (Figure 1*c*) may also arise for completely rational reasons: Scientists often adjust sample sizes to achieve a desired level of statistical significance. Sample-sizes might be adjusted in two ways. First, if a pilot study reveals a modest effect size, then a biologist may quite legitimately elect to use smaller samples in the final study design, since larger sample sizes are not required to demonstrate the statistical significance of the effect. In other words, sample sizes may be adjusted a priori to the size of the effect via a power analysis (98:260–65). Second, statistical significance may be monitored as data are accumulating, and data collection may be stopped once significance is achieved. Both scenarios would yield a dependence of effect size on sample size (Figure 1*c*), but only if true effect sizes genuinely varied among studies. Therefore, a dependence of effect size on sample size among heterogeneous studies is not unequivocal evidence of selective reporting.

One pattern observed among studies of fluctuating asymmetry and sexual selection—where experimental studies yield larger effect sizes than do observational ones (Figure 3)—illustrates just such a problem. Experimental studies are often based on smaller samples and yield larger effects than observational ones because the investigator has more control over extraneous factors. But if many experimental studies are conducted, and only those that reach statistical significance are published, the pattern apparent in Figure 3 may be entirely an artifact of selective reporting. The conspicuous dependence of effect size on sample size (Figure 3; see also Figures 10, 11, and 12 below) strongly suggests selective reporting.

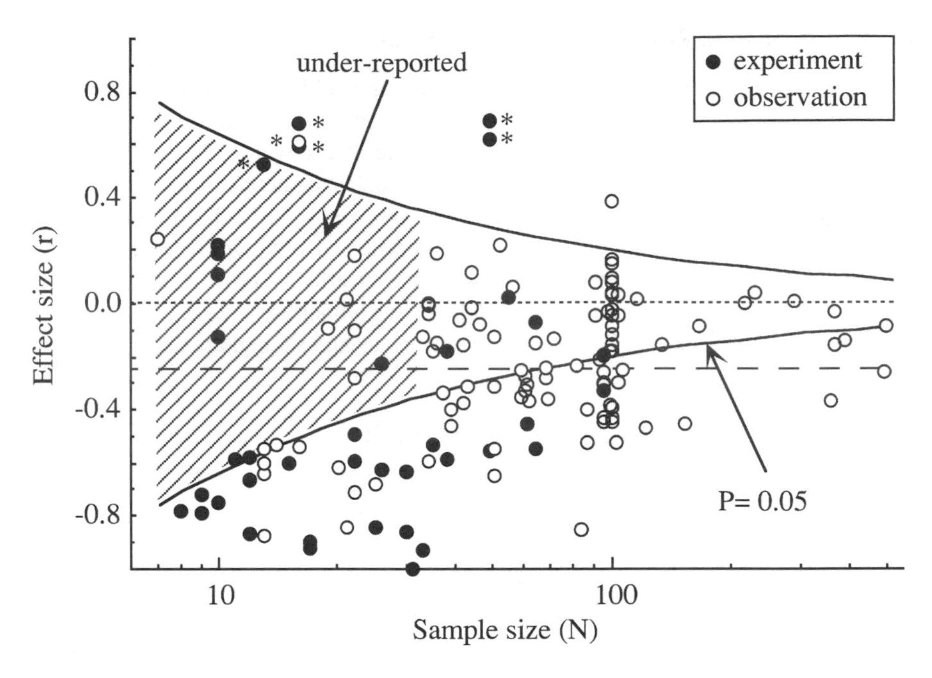

Figure 3 Effect size (correlation coefficient, *r*) as a function of sample size (log scale) for correlations between fluctuating asymmetry and individual attractiveness for experimental (●) as opposed to descriptive (○) studies (74). Dotted line, long-dashed line, curved lines, and shaded region, as in Figure 1. Modified from (80). Asterisked values were excluded from the meta-analysis by the original authors (74). The data on which this figure was based may be obtained from: http://www.biology.ualberta.ca/palmer.hp/DataFiles.htm.

The only way to distinguish between these two hypotheses for a dependence of effect size on sample size (i.e., rational planning vs selective reporting) is to conduct a formal replication of previous studies. Clearly, replicate studies should reveal approximately the same effect size regardless of sample size. If, however, the effect size is significantly lower in a replicated study based on larger sample sizes, then then the original study was biased.

Are studies based on small sample sizes necessarily a biased sample of those that were conducted? Some evidence suggests not. For example, Cappelleri et al (17) compared the results of individual large studies to those of multiple small studies in 61 meta-analyses of pregnancy and childbirth in which at least one large study was included along with a number of studies based on smaller sample sizes. Discrepancies between the single large study, and the average effect of the multiple small studies, were found in 15 of the 61 meta-analyses, but of these, 10 were explained by differences between controls (N = 5), differences in protocol (N = 4), or publication bias (N = 1). In the end, only one discrepancy (of 15) lacked a plausible explanation. For some human studies, where a single common effect size is anticipated, meta-analyses of multiple smaller studies do appear to provide reliable estimates of the outcomes that would be achieved with larger sample sizes.

Limitations to Funnel Graphs Unfortunately, useful as funnel graphs may be for visualizing selective reporting (80), their utility drops dramatically as the number of studies declines (32). Funnel graphs of fewer than 30 studies may still be helpful for judging whether contrasts of biological interest may have been confounded by sample size (see Figure 3), but they will likely lack power adequate to reveal any but the most pronounced selective reporting (32).

Correcting for Bias Due To Selective Reporting

Only by exhaustively uncovering all studies conducted, whether published or not, can the biases due to selective reporting be eliminated. Such a goal would be unachievable in ecology and evolution. Both conventional (12) and Bayesian (19, 57) statistical corrections for publication bias have been proposed, but the limitations of both approaches suggest they offer no quick fix (12, 14, 43).

Ultimately, selective reporting is an inevitable element of the scientific enterprise. The challenge is to determine the extent to which it prejudices our understanding of ecological and evolutionary phenomena.

THREE EXAMPLES OF SELECTIVE REPORTING

Preamble

I have not attempted to conduct formal meta-analyses for any of the examples discussed below, nor have I systematically surveyed the literature for more recent published studies in these areas. Such endeavors lie well outside the scope of this review. In the three sections that follow, my goals are (*a*) to illustrate the value of a graphical approach to data summarization (the funnel graph; 64), and (*b*) to examine recent reviews of biological phenomena for which I could ask specific questions about what motivates biologists to publish their results. Because of the scale of this review, I was obliged to accept values reported in published summary tables as correct, and did not attempt to validate them against the original publications.

The three examples selected for scrutiny were chosen to explore different aspects of reporting patterns: (1) *sex-ratio variation*—how statistically significant associations may arise due to random sampling error and encourage detailed explanations of their putative biological significance, (2) *heritability estimates*—how a priori expectations of what results make sense theoretically influence reporting patterns, and (3) *fluctuating asymmetry and sexual selection*—how both statistical significance of results and consistency with a preferred hypothesis may influence reporting patterns.

1. Adaptive Sex Ratio Variation in Birds: Significant Associations Among Nonsignificant Samples

Studies of sex-ratio variation in birds provide a particularly illuminating example of selective reporting because on closer inspection few, if any, compelling data exist for adaptive departure from a 50:50 sex ratio in any species (20, 119), in spite of numerous published claims to the contrary. In fairness, many studies have reported little or no departure from a 50:50 sex ratio at hatching, and some have specifically drawn attention to the absence of variation greater than expected due to binomial sampling (37, 48). But new papers steadily appear claiming statistical support for adaptive sex-ratio variation (59, 94).

One attractive aspect of sex-ratio variation is explicit alternative hypotheses about which, if either, sex should predominate. Fisher (40) noted that, in the absence of confounding factors, frequency-dependent selection should promote a 50:50 sex ratio: The rarer sex will always have a relatively higher fitness. Even in the absence of selection, a 50:50 sex ratio is expected where sex is determined by the conformation of a single chromosome pair. However, local mate competition or local resource competition (reviewed in 5, 41) may promote an excess of one or the other sex.

Most departures from 50:50 sex ratio are interpreted in terms of these latter two hypotheses (e.g., see 44). However, funnel-graphs of results tabulated in two reviews suggest most, if not all, sex-ratio variation in hatchling birds does not exceed that expected due to binomial sampling variation.

Most authors who compile sex ratios seem aware that sampling error may yield spuriously significant results (e.g., 20, 44), but not all deal with it in the same way. For example, both Cockburn (21) and van Schaik (110) set an arbitrary sample size as large enough ($N = 100$ and 200, respectively), but both include studies of smaller sample sizes in their analyses if results were significant. Clearly, statistical significance in these cases outweighed the authors' a priori belief that large sample sizes were required for adequate confidence. This example also illustrates the widespread tendency to accept results based on small sample sizes if significant but to minimize or dismiss them if not significant.

Clutton-Brock's Review Following a detailed examination of published sex ratio variation in birds, Clutton-Brock (20:326–27) concluded: "Sound evidence for sex ratio variation at hatching is thus scarce. There is some evidence that the sex ratio can vary with position in clutch but trends show no consistency across populations or species. Significant relationships have been found with order of clutch ... and maternal age ... but whether these indicate that birds can vary the hatching sex ratio of their offspring in an adaptive fashion or whether they represent the small proportion of cases where the null hypothesis has been wrongly rejected by chance is as yet not certain." These conclusions echoed an earlier one by Williams (119) that the data for birds did not support any theory of adaptive sex ratio evolution.

Two lines of evidence support Clutton-Brock's (20) suspicion that sampling error alone accounted for the observed sex-ratio variation. First, of 85 separate sex-ratio estimates from 14 studies of 8 species, 11 differed significantly from parity at $0.05 > P > 0.01$ and 2 were significant at $P < 0.01$. These numbers do not differ significantly from those expected due to sampling error ($P = 0.116$; Chi-square test with correction for continuity), although slightly more significant studies appear to have been reported ($P = 0.035$) when they are divided into two groups, those significant at $P < 0.05$ and those not. Second, a funnel-graph of these data clearly reveals the pattern expected for purely random, sample size-dependent variation in sex ratio (Figure 4*a*).

Furthermore, among studies where the title of the original paper stated or implied a biased sex ratio (i.e., where the authors wished to emphasize a significant departure from parity), a closer examination suggests that statistical significance arose due purely to sampling error. First, the sample sizes of these studies were significantly smaller than those of the other studies in Clutton-Brock's review

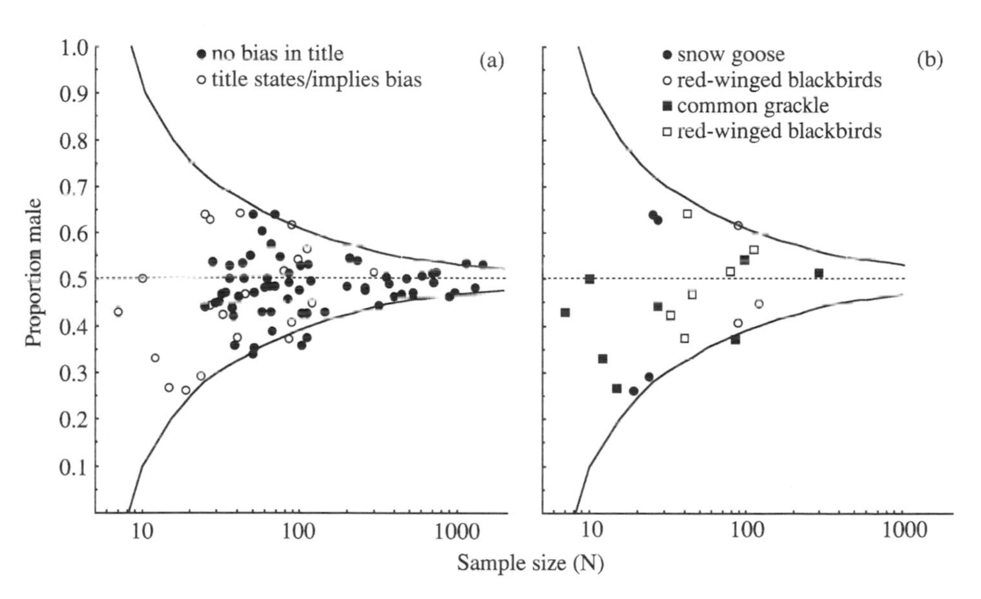

Figure 4 Variation in the sex ratio of samples of birds as a function of sample size (log scale) (all data from Ref. 20). Single species are represented by multiple points and, in some cases, data for single species come from multiple papers. Dotted lines indicate the null hypothesis, and curved lines indicate binomial significance levels ($P = 0.05$, two tailed) from Table Q of Rohlf & Sokal (89). (*a*) samples differentiated by title of paper (does the title state or imply a biased sex ratio was observed?), (*b*) samples for single species exclusively from the four papers with a title stating or implying a biased sex-ratio [●, position in egg-sequence (3); ○, maternal age (15); □, time in laying season (115)]. The data for grackles [■ (53)] are not comparable because the sex-ratio variation tabulated by Clutton-Brock was unrelated to the main point of this paper. The data on which this figure was based may be obtained from: http://www.biology.ualberta.ca/palmer.hp/DataFiles.htm.

(<0.001, Mann-Whitney U-test; Figure 4*a*). Second, although only 2 of the 21 individual samples actually deviated significantly from a 50:50 sex ratio, 3 studies reported significant associations with sex ratio (Figure 4*b*). In other words, random variation among nonsignificant samples yielded statistically significant associations that seemed biologically interesting and could easily be interpreted in light of one or another theory of sex allocation.

Gowaty's Review In contrast to Clutton-Brock (20), Gowaty (44) concluded that small sex-ratio differences between passerine and anseriform birds were significantly correlated with sex-dependent patterns of philopatry: The sex that dispersed further was the sex that tended to be overproduced. Even though sex ratios appeared to depart significantly from parity in only 5 of the 12 passerine species, and in none of the ducks and geese (Figure 5), Gowaty noted that what deviations did exist (whether significant or not) tended toward excess males in ducks and geese (5 of 6 species) and excess females in passerine birds (11 of 12 species) (Figure 5). Gowaty concluded that the consistent directions of these differences were too improbable to be due to chance ($P = 0.004$ at the level of species, and $P = 0.036$ at the level of families; Fisher's exact test), and that "the lack of statistically significant differences from a 50:50 sex ratio may have obscured biologically interesting phenomena associated with sex ratio variation in birds" (42:272).

However, a closer inspection of the data and reasoning raises doubts about these conclusions. First, a funnel graph once again revealed a pattern consistent with simple sampling variation (Figure 5). Second, species in two other orders also exhibited sex-biased philopatry, but in these two orders, either the overproduced sex

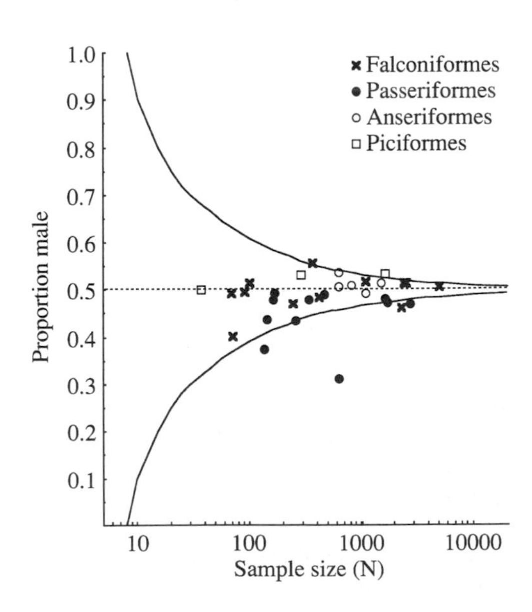

Figure 5 Variation in the sex ratio of nestling or fledgling bird species within each of four orders as a function of log(sample size) (all data from Ref. 44). Dotted line and curved lines as in Figure 4.

TABLE 1 Number of species or families exhibiting an excess of male or female offspring (from Ref. 44)[a]

Order	Philopatric Sex	Number of Species (Families) with Excess Males	Number of Species (Families) with Excess Females	Number of Species (Families) with 50:50 Ratio
Anseriformes	female	5(2)	1(0)	
Passeriformes	male	1(0)	11(6)	
Falconiformes	male	6(2)	6(0)	
Piciformes	male	2(1)	0(0)	1(0)
Male-philopatric pooled		9(3)	17(6)	1(0)

[a]Sex ratios for both species and families were computed using weighted averages (counts of males and females were pooled separately across samples and a new ratio computed from these pooled values).

was uncorrelated with the dispersing sex (Falconiformes) or the overproduced sex dispersed less (two of three woodpecker species) at the level of species or families (Table 1). When all male-philopatric species were pooled, the dependence of sex ratio on philopatry ceased to be significant ($P = 0.09$, contingency table analysis corrected for continuity). Third, sexual size dimorphism—the presumed raison d'etre for the departures from 50:50 sex ratio (44)—is negligible in both passerines and ducks and geese. So the statistical significance of the apparent correlation between sex ratio and philopatry in birds depended on the orders examined.

Conclusion When the bird sex-ratio data are viewed as a whole (Figures 4 and 5), two patterns emerge. First, even though authors may have drawn attention to results that were statistically significant, they also reported results that were not significant. As a consequence, statistically significant samples were only weakly overreported. Second, the pooled data support rather strongly a tightly constrained 50:50 sex ratio, subject to little more than sampling error (Figure 4*a*), as both Williams (119) and Clutton-Brock (20) surmised might be true. The biological significance of the many studies reporting statistically significant departures from parity (44, 59, 94) is therefore questionable.

The advent of modern technologies that allow sexing prior to or shortly after hatching using blood samples (flow cytometry, DNA profiles, microsatellites) has inspired additional studies of sex ratio variation in birds (94) because such techniques can minimize or eliminate effects of sex-biased mortality after hatching. Although some seemingly compelling examples are mentioned (e.g., 58), a more detailed funnel-graph analysis, or explicit replication of some of these results by others, would be required to reject convincingly the possibility that they too are simply examples of selective reporting.

In view of the many reports of statistically significant sex-ratio variation in birds that appear to have arisen simply due to sampling error, I encourage all of those tempted to offer biological explanations for such patterns to restrain their enthusiasm until some of the more striking claims have been replicated independently by others. Sex ratios are too easy to measure incidentally as part of another study, and therefore they would seem particularly prone to selective reporting. Sex ratios obtained with modern techniques (94) may yield values less confounded by differential mortality, but they do not avoid the fundamental problem: Departures from 50:50 require rather large samples to detect reliably (Figures 4 and 5).

2. Heritability: Impact of Theoretical Preconceptions

That theoretical preconceptions alone have an impact on probability of publication is perhaps most clear cut among studies of heritability. Narrow-sense heritability describes the degree to which offspring resemble their parents on average (35). Few biologists would expect offspring to exhibit consistently more extreme phenotypes than their parents (heritability >1.0), or for offspring to deviate consistently from the population mean in the opposite direction from their parents (a negative heritability). But sampling error should yield such heritabilities occasionally, just due to chance (66). In addition, as sample size decreases, the likelihood of a negative or extreme positive heritability increases substantially.

Published estimates of narrow-sense heritabilities (116) reveal what appears to be a clear example of selective reporting (Figure 6). In part, this is influenced by statistical significance: Nonsignificant studies are underrepresented at small sample sizes and average heritability decreases significantly with increasing sample size ($P < 0.001$, Figure 6). More seriously, even at small sample sizes ($N < 50$), where heritabilities are estimated with lower confidence, negative heritabilities are virtually nonexistent, even though several positive estimates exceed the theoretical maximum (Figure 6). At face value, authors appear more comfortable with super-heritability ($h^2 > 1.0$) than with negative heritability ($h^2 < 0.0$).

Studies of the heritability of fluctuating asymmetry reveal the same pattern even more dramatically (Figure 7). The weighted mean heritability is closer to zero than in the previous example (compare to Figure 6), but again, virtually no heritability estimates were less than zero. The absence of any but the slightest negative heritability estimates in both examples (Figures 6 and 7) reveals rather clearly that theoretical expectations influence the likelihood of publication, quite independent of any effects of statistical significance.

Where are all the missing negative heritability estimates? Most likely, they reside in filing cabinets because they made no sense theoretically.

This result is troubling because it implies that even carefully conducted meta-analyses could yield statistical support for a preconception, rather than a genuine biological phenomenon, if the theoretical grounds for that preconception are strongly held (but see Figure 14). For example, consider the following thought experiment. Assume that the true heritability of a particular trait is zero. If 100 biologists independently estimate this heritability, sampling error dictates that half the observations should be negative and half positive. If those biologists who obtain negative heritabilities discard their results, or set them to zero as advised by some

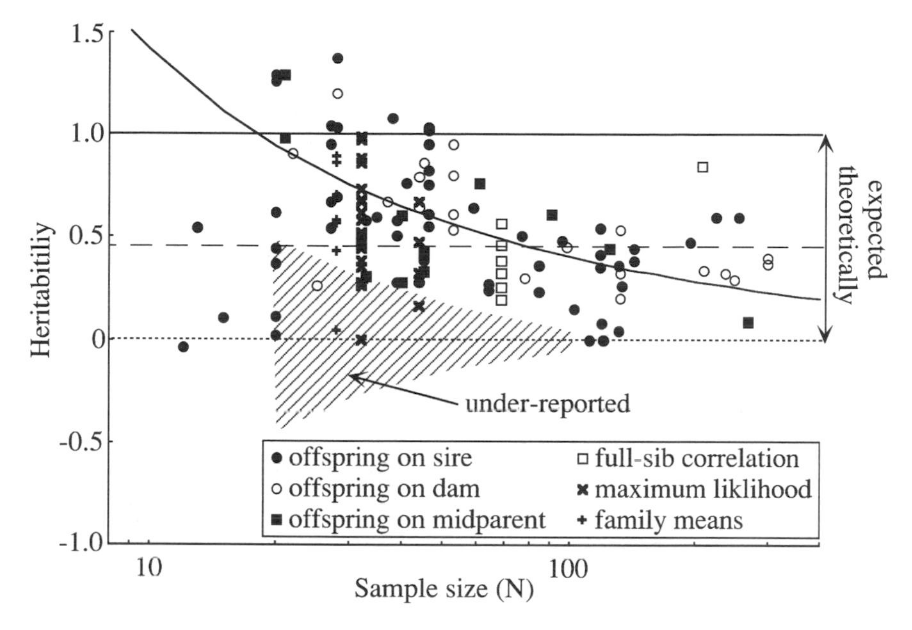

Figure 6 Published heritability estimates (h^2) from field studies as a function of sample size and method of estimation (116). Sample sizes (number of families) were obtained from the original studies. The dotted line indicates the null hypothesis, the long-dashed line indicates the weighted mean heritability across all studies (0.452), and the curved line indicates the upper 95% confidence interval (= 2SE) about zero for heritability estimates according to Falconer (35:166) for parent-offspring regressions based on one offspring per parent ($2/\sqrt{(N-2)}$). This confidence interval is not appropriate for all the methods indicated but does indicate how the significance level varies with sample size. Over all samples, heritability estimates decreased significantly with increasing sample size (Spearman ρ corrected for ties $= -0.33$, $P < 0.001$). The shaded region indicates where observations are expected due to sampling variation but have been underreported. The data on which this figure was based may be obtained from: http://www.biology.ualberta.ca/palmer.hp/DataFiles.htm.

practitioners (35), what is the net result? A meta-analysis would likely yield strong statistical support for a positive heritability that is entirely an artifact of our preconceptions. One can only wonder at how many published heritability estimates have been exaggerated by this preconception.

3. Fluctuating Asymmetry and Fitness: Anatomy of a Bandwagon

Fluctuating asymmetry (FA; small, random departures from perfect symmetry; 65) offers an intuitively appealing measure of developmental precision (the degree to which the right and left sides of a bilaterally symmetrical organism depart from perfect symmetry due to the cumulative effects of developmental noise) (79, 82). It is appealing because of the apparent ease with which it may be measured,

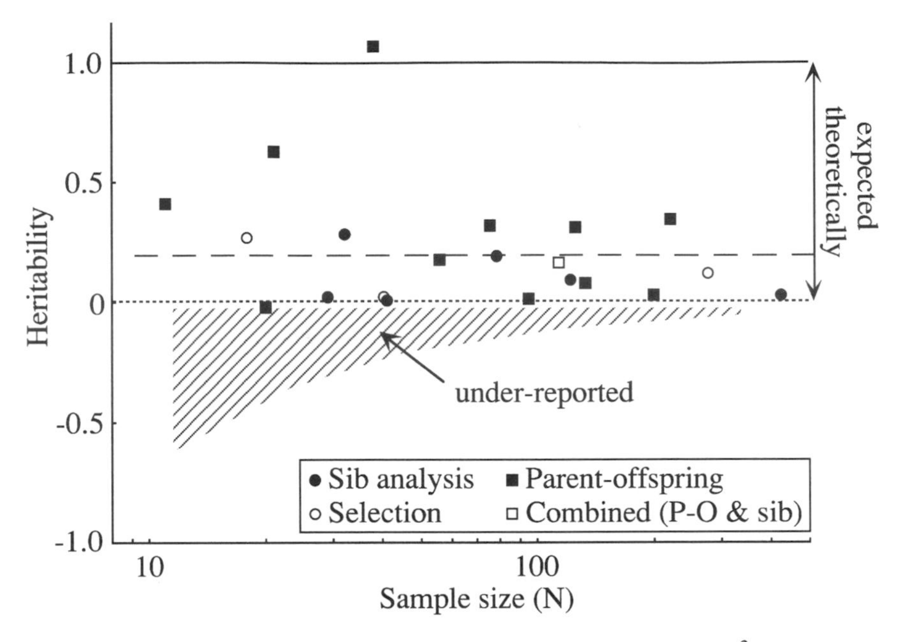

Figure 7 Published estimates of the heritability of fluctuating asymmetry (h^2) for various taxa and methods of estimation as tabulated on the internet site to accompany Ref. 73. Sample sizes cannot be interpreted literally here because very different methods were used in different studies; only the distribution of heritability estimates is relevant. The dotted line indicates the null hypothesis and the long-dashed line indicates the weighted mean heritability across all studies ($h^2 = 0.154$). The shaded region indicates where observations are expected due to sampling variation but have been under-reported. The data on which this figure was based may be obtained from: http://www1.oup.co.uk/MS-asymmetry.

and because of the seemingly sound theoretical grounds for believing that subtle departures from symmetry really should reveal something about underlying developmental stability (69, 82, 111).

Over the last 10 years, interest in FA has increased more than 10-fold (Figure 8). Many biologists have rushed to apply this approach to a variety of questions, since developmental precision is thought (*a*) to be reduced by environmental or genetic stress (see 122 and references therein) and by lowered heterozygosity (113), and (*b*) to correlate negatively with measures of individual fitness (72)—including growth, fecundity, and survival—and with measures of individual attractiveness in studies of sexual selection (74, 104, 105).

Three lines of evidence, however, suggest that selective reporting has greatly exaggerated the apparent strength and generality of one association: the correlation between individual asymmetry and measures of individual quality, fitness, or attractiveness. This evidence includes (*a*) the absence of parallel asymmetry variation among individuals, (*b*) theoretical demonstrations of the limited

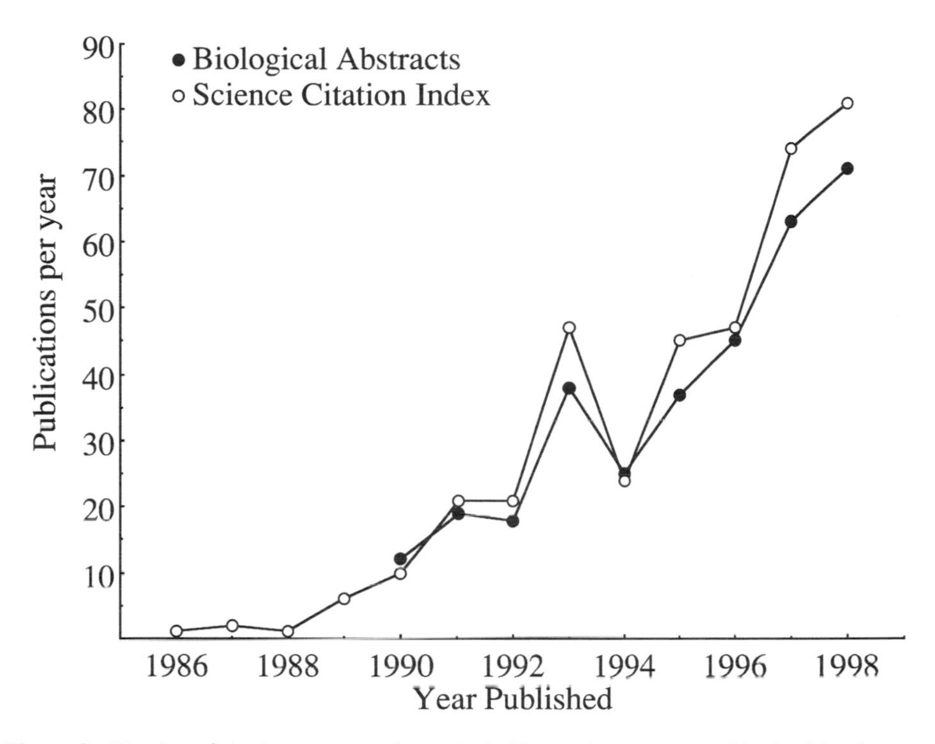

Figure 8 Number of citations per year that include "fluctuating asymmetry" in the title, abstract, or keywords in the electronic database of Biological Abstracts (●), or cite the 1986 review by Palmer and Strobeck (82), as determined from Science Citation Index (○).

statistical power of FA as a measure of underlying developmental instability, and (*c*) patterns of selective reporting among studies of FA and sexual selection that are not observed in other studies of FA variation.

Absence of Parallel Asymmetry Variation Among Individuals If subtle deviation from symmetry is a reliable indicator of underlying developmental stability in an individual, as claimed repeatedly in the FA literature (reviewed in 73), then deviations from symmetry in one trait should correlate with deviations from symmetry in other traits on that same individual. The virtual absence of parallel asymmetry variation among individuals for morphological traits, however, raises serious doubts about whether asymmetries of individual traits should be correlated with any other phenomena of biological interest (52, 78).

An early review of FA variation (82) noted that while asymmetries might be correlated among populations (a population more asymmetrical for one trait was typically more asymmetrical for others—the "population asymmetry parameter" 99), they were rarely correlated among individuals within populations (the "individual asymmetry parameter"; 62). And more recently, an extensive review of

FA variation in many traits in a variety of organisms reached the same conclusion (73:53–55).

Limited Statistical Power The limited ability to detect parallel asymmetry variation among individuals arises from two complementary causes: sampling error and measurement error.

Not until 1994 was a simple statistical explanation advanced for the rarity of parallel asymmetry variation among individuals (78:360): The absolute value of the deviation from symmetry in a single trait of an individual estimates the underlying right-left variance for that trait with only one degree of freedom (see also 107, 117, 118). In other words, the absolute difference between sides in one trait of an individual estimates the underlying developmental instability (the potential right-left variance) of that trait with only one degree of freedom. As most biologists know, the ability to detect differences among means is limited if they are estimated with only one degree of freedom each, and the ability to detect differences among variances is considerably more limited (97).

Simulations confirm this expectation (107). In the absence of measurement error, the observed correlation between asymmetries in two traits was only 0.287 ($r^2 = 0.082$), even where the underlying instability variance differed by 16-fold (Figure 9*a*). So only about 8% of the asymmetry variation in one trait can be predicted by asymmetry variation in a second on the same individual, even under ideal conditions: sizeable variation in underlying developmental stability among individuals and no measurement error.

Measurement error further reduces the expected correlation between asymmetries. It can form a sizeable fraction of the between-sides variation because FA variation is often on the order of 1% of trait size (38, 79) and few biologists measure traits to a precision much greater than 1%. A high percent measurement error significantly reduces the strength of asymmetry correlations. For example, only 3% of the variation in $|R - L|$ in one trait is explained by variation in $|R - L|$ of a second in the same population when measurement error is half of the between-sides variance (Figure 9). In addition, direct evidence suggests that as the repeatability of asymmetry measures increased in published studies, so did the strength of asymmetry correlations (109).

Undaunted, those who believe in the predictive value of FA have invoked this low statistical power to their advantage. For example, Gangestad & Thornhill (42) argued forcefully that the theoretical maximum correlation between individual asymmetry and attractiveness should be −0.27 and therefore that the observed weighted mean effect size of −0.22 implies a high proportion of variation in attractiveness can be attributed to variation in underlying developmental instability (105). That they remained unperturbed by the majority of effect sizes that exceeded this putative theoretical maximum in their own tabulation is a testament to the strength of their convictions.

Clearly, on purely statistical grounds, the rarity of parallel asymmetry variation among individual organisms is not surprising. Even if a true correlation between

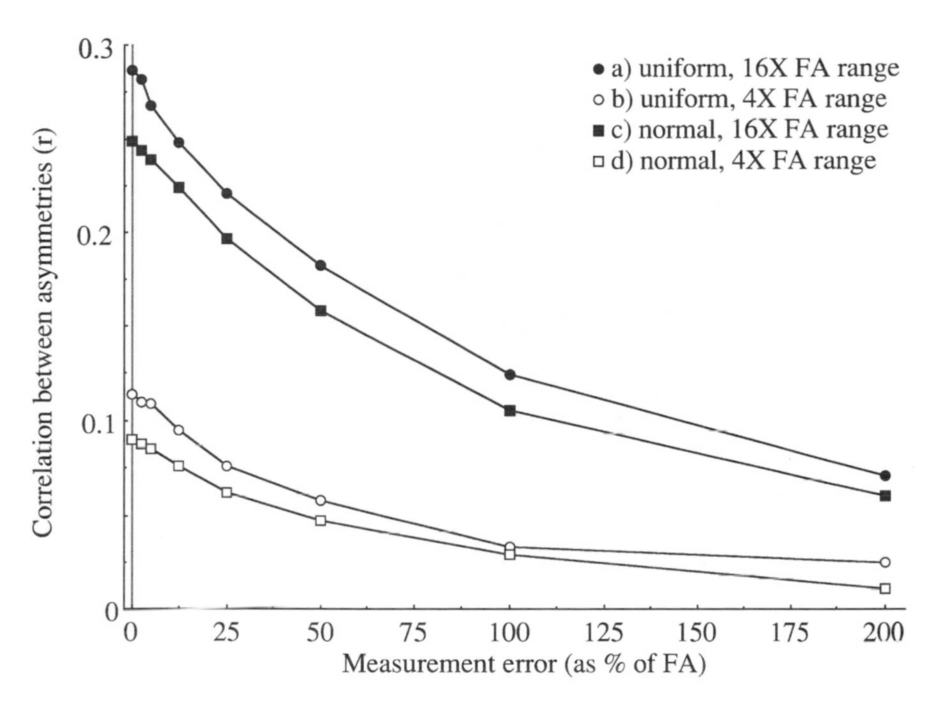

Figure 9 Effect of measurement error on the strength of the correlation (r) between absolute deviations from symmetry (FA) in two traits among individuals in a single sample; kindly simulated by S Van Dongen using the same model as in Ref. (107), but for 100,000 replications. Each population consisted of a mixture of individuals exhibiting three different levels of underlying instability variance: var($R-L$) = FA = $1/x$, 1, x. Two populations were simulated, as were two different distributions of FA variation: (●) x = 4, proportions of all three FA levels equal; (○) x = 2, proportions of all three FA levels equal; (■) x = 4, proportions of FA levels 1:2:1; (□) x = 2, proportions of FA levels 1:2:1. The measurement error variance var(M_1-M_2) is expressed as a percent of the median instability variance (i.e., a value of 100 means the variance of replicate measurements equals the median FA variance between sides).

FA and some trait of interest is 0.2 (close to the putative theoretical maximum), the statistical power of correlation coefficients is low for routine sample sizes (83): With a sample size of N = 40, a significant correlation ($P \leq 0.05$) would be detected only about 20% of the time (power = 0.2), and even with a sample size of N = 100, the power is less than 0.5. In other words, even for a true correlation close to the theoretical maximum, sampling variation should not yield a significant correlation more than 50% of the time.

What accounts for the remarkable number of statistically significant correlations reported between individual asymmetry and other features of animals such as fitness or attractiveness? As Houle (52) noted so pointedly, it is hard to understand how correlations between individual subtle asymmetries and other phenomena of interest can be so common when correlations between asymmetries

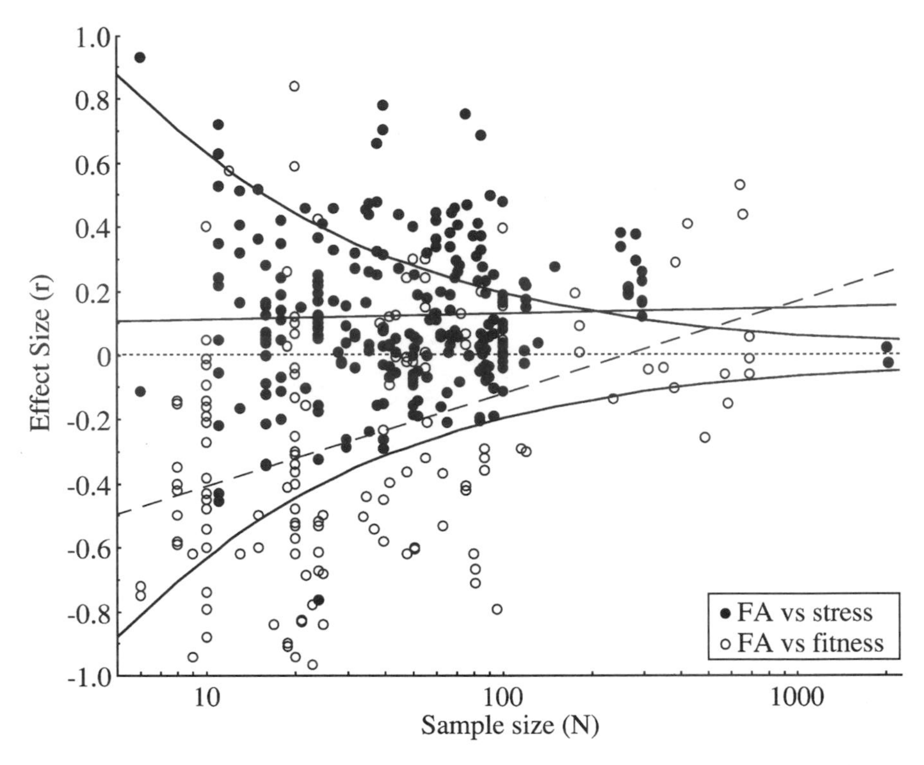

Figure 10 Effect size (correlation coefficient, r) as a function of sample size (log scale) for correlations between fluctuating asymmetry (FA) and stress (●) or various measures of fitness (○) as reported in the meta-analysis by Leung & Forbes (63). Dotted line and curved lines as in Figure 1. The solid line indicates the least-squares linear regression for stress studies ($Y = 0.016\ X + 0.098$, $N = 151$ cases; Spearman's $\rho = 0.031$, $P = 0.63$), and the long-dashed line indicates the regression for fitness studies ($Y = 0.285\ X - 0.691$; Spearman's $\rho = -0.39$, $P < 0.001$). Effect sizes are expected to be positive for FA-stress relations (higher stress results in higher FA), but negative for FA-fitness relations (higher FA is associated with lower fitness). Measures of fitness included traits like body size, mating success, dominance, secondary-sexual trait size, survival, condition, and growth. The data on which this figure was based may be obtained from http://www.biology.ualberta.ca/palmer.hp/DataFiles.htm.

in the same individuals are so rare. Selective reporting seems a likely explanation.

Direct Evidence of Selective Reporting Claims that seem highly improbable (81) certainly raise the possibility of selective reporting (see http://www.biology.ualberta.ca/palmer.hp/asym/Curiosities/Curiosities.htm for some examples). Furthermore, evidence from three meta-analyses strongly suggests that selective reporting of associations between FA and various measures of individual fitness may be a serious problem.

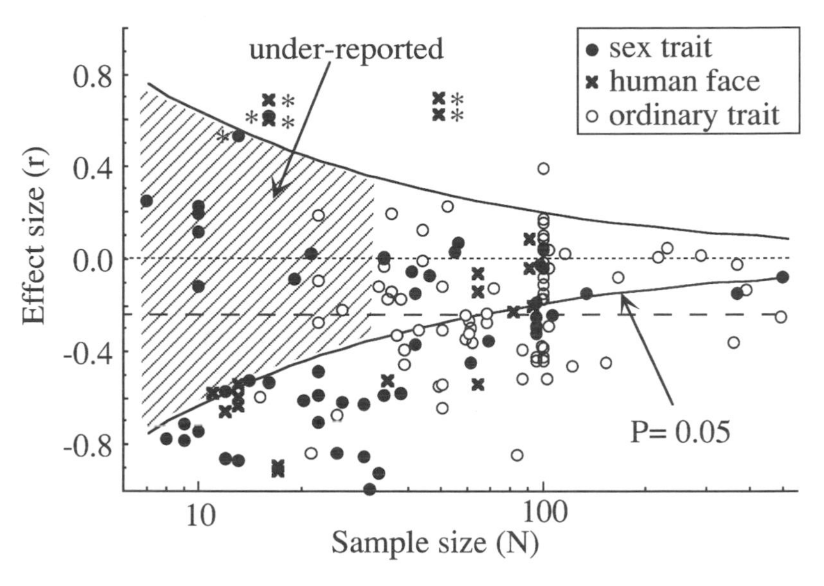

Figure 11 Effect size (correlation coefficient, r) as a function of sample size (log scale) for correlations between fluctuating asymmetry and individual attractiveness for sex traits (●), human faces (x), and ordinary traits (o) (as tabulated in Ref. 74, figure modified from Ref. 80). Dotted line, long-dashed line, curved lines and shaded region, as in Figure 1. Asterisked values were excluded from the original analysis (74) on methodological grounds. The data on which this figure was based may be obtained from http://www.biology.ualberta.ca/palmer.hp/DataFiles.htm.

The dependence of effect size on sample size, r_{bias} (80), was highly significant statistically for signaling traits (sex + human face; Spearman's $\rho = 0.38$, $P = 0.002$, $N = 67$) but not for ordinary or nonsignaling traits (Spearman's $\rho = 0.16$, $P = 0.19$, $N = 73$). The same patterns were apparent when experimental studies were excluded: r_{bias} was highly significant for signaling traits ($\rho = 0.44$, $P = 0.009$, $N = 36$), but not for ordinary traits ($\rho = 0.12$, $P = 0.34$, $N = 69$).

First, Leung & Forbes (63:400), in the earliest meta-analysis of FA variation, concluded that overall correlations between FA and stress, and between FA and various fitness measures, were "non-spurious" but "fairly weak, and highly heterogeneous." A closer examination (Figure 10) reveals that FA-stress correlations were largely independent of sample size ($r_{bias} = 0.031$; $P = 0.63$) and remained more or less centered on the overall weighted mean of $r = 0.17$. A similar non-significant r_{bias} was found in a meta-analysis of FA-heterozygosity correlations (113). FA-fitness correlations, however, revealed a different pattern: As sample size increased, effect size decreased significantly ($r_{bias} = -0.39$, $P < 0.001$). Fully 16% of the overall variation in effect sizes could be attributed to variation in sample size. Average effect sizes appeared moderate (0.3–0.5) (7) when based on sample sizes less than 20, but for sample sizes greater than 50, they were weak (<0.2).

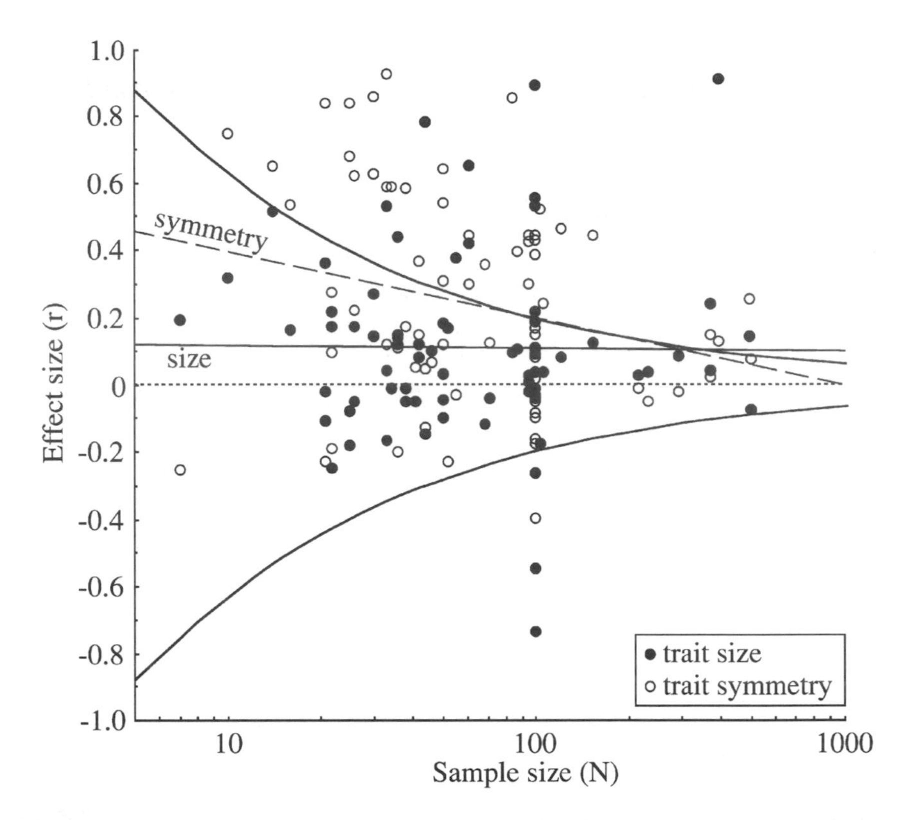

Figure 12 Effect size (correlation coefficient, r) as a function of sample size (log scale) for correlations between trait size and individual attractiveness (●) and correlations between symmetry and individual attractiveness (○) (as tabulated in Ref. 104). Dotted line and curved lines as in Figure 1. The solid line indicates the least-squares linear regression for trait size ($Y = -0.010\,X + 0.130$, N = 73 cases; Spearman's $\rho = -0.054$, $P = 0.65$), and the long-dashed line indicates the regression for fitness studies ($Y = -0.204\,X + 0.609$, N = 73 cases; Spearman's $\rho = -0.260$, $P = 0.027$). Effect sizes are expected to be positive for both relations (larger or more symmetrical traits are more attractive). The trait symmetry data are the same as those in Figure 11 (74), but were limited to studies where both trait size and trait symmetry were examined simultaneously. In addition, the sign of the asymmetry effects was reversed to permit direct comparisons between effects of symmetry and size.

Such a relationship renders statistical summaries of weighted-mean effect sizes virtually meaningless, since average effect size clearly depends on sample size.

Second, a more restricted meta-analysis that specifically examined correlations between asymmetry and attractiveness (74), rather than between asymmetry and a variety of fitness measures, revealed an even stronger suggestion of selective reporting (80): (*a*) The threshold of statistical significance ($P = 0.05$) rather sharply defined the upper boundary to a cluster of published associations based on sample

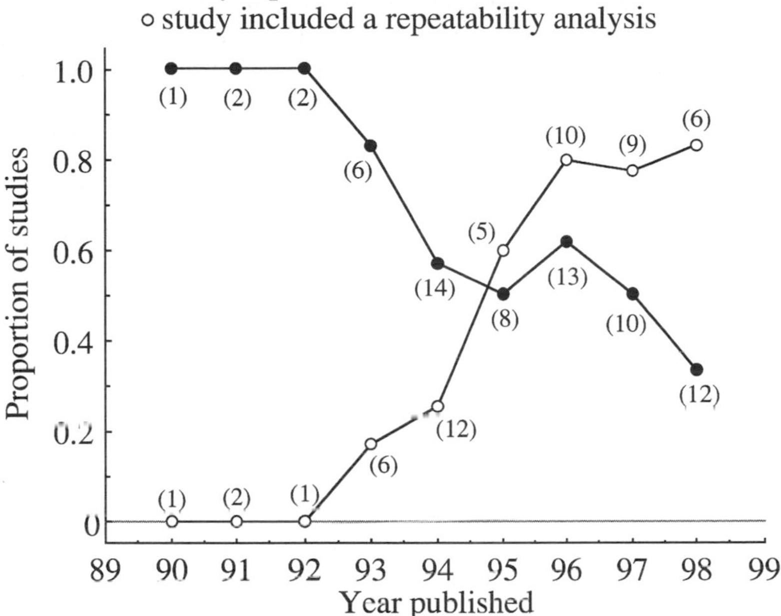

Figure 13 Change over time in the proportion of studies that reported an association between fluctuating asymmetry (FA) and sexual selection and the percent of studies that conducted a test for the repeatability or reliability of the FA signal relative to measurement error (from Ref. 96). Numbers in parentheses indicate number of studies.

sizes less than 20 (lower left portion of Figure 11); (*b*) overall effect size declined significantly with increasing sample size (r_{bias} = Spearman $\rho = 0.39$, $P < 0.001$, $N = 140$; 'included' samples only); (*c*) r_{bias} was highly significant for signaling traits ($P = 0.002$) but not for ordinary or nonsignaling traits ($P = 0.19$). The same patterns were apparent when experimental studies were excluded (r_{bias} was highly significant for signaling traits, $P = 0.009$, but not for ordinary traits, $P = 0.34$), so the disproportionate number of experimental studies at small sample size (Figure 3) was not the cause of the original pattern. Perhaps most seriously of all, r_{bias} was highly significant for signaling traits but not for ordinary traits among the studies conducted by Møller and Thornhill themselves (see Figure 3 of Ref. 80).

Third, another meta-analysis (104) examined cases in which the effects of both trait size and trait symmetry on mating success or attractiveness were studied concurrently. As above (Figure 10), r_{bias} was significant for symmetry variation ($P = 0.027$) but not for variation in trait size ($P = 0.65$) (Figure 12).

Finally, some surprising evidence suggests that many of the initial reports trumpeting the role of FA in sexual selection were actually spurious results that arose from high levels of measurement error coupled with selective reporting. Many early studies did not test whether the putative FA variation being reported exceeded that due simply to measurement error, even though measurement error yields bilateral variation indistinguishable from true FA and was clearly recognized as a serious problem for FA studies (82). As more studies tested for the significance of FA variation relative to measurement error, fewer and fewer reported significant associations with FA (Figure 13). This change over time remains striking, even if the small number of studies prior to 1993 are ignored.

All of the above evidence suggests that studies of FA and individual fitness or attractiveness have been seriously confounded by selective reporting, particularly since a significant r_{bias} was absent among studies of FA and stress (63), FA and heterozygosity (113), and trait size and attractiveness (104). In the end, studies of FA and sexual selection, or FA and individual fitness, will likely reveal more about the sociology of science than about biology.

Alternative Hypotheses to Selective Reporting: Disproof by Replication

The preceding three examples suggest selective reporting may have promoted dubious biological conclusions. This hypothesis is open to disproof. If genuinely replicated studies reveal effect sizes of magnitude and direction similar to those of the original results, then the hypothesis of selective reporting for these cases is rejected.

Sex Ratio Variation Sex-ratio variation in birds, in two formal reviews, appears not to exceed that expected due to sampling error. No doubt this will trouble many who believe otherwise. The hypothesis of purely random variation may be rejected readily by one or two truly replicated studies of published claims of striking departures from parity. For example, if sample sizes exceeding 200 revealed comparable departures from a 50:50 sex ratio in different eggs in the laying sequence, as reported for snow geese based on sample sizes less than 30 (3), then the hypothesis of selective reporting can be rejected. In fact, this replication has already been conducted. Cooke & Harmsen (22), based on a more detailed study (though not on much larger sample sizes), found no statistical support for a dependence of sex-ratio on laying sequence. Seasonal sex-ratio variation in red-winged blackbirds is similarly suspect because two independent studies yielded contradictory results (37, 115). Similarly, if sex ratio of offspring varied by similar amounts among mothers of different ages in red-winged blackbirds based on sample sizes of 300 instead of about 100 (15), then the hypothesis of selective reporting can be rejected.

The only way to determine whether the few significant associations reported by Clutton-Brock (20) and Gowaty (44) were due to chance would be to repeat some of the original studies.

Heritability Negative heritability estimates appear to be greatly underreported in the literature (Figure 6). Among studies of the heritability of FA prior to 1998, such underreporting appears quite pronounced (Figure 7). The claim that FA variation is significantly heritable (73), and the use of that claim to buttress other preferred hypotheses (73), thus seems open to question.

One recent study of the heritability of FA variation reveals just how variable heritability estimates may be (121). Not only does heritability vary by several fold among traits and populations, but as many estimates are negative as are positive (Figure 14). Furthermore, some of the negative estimates are more extreme than the most extreme positive ones. Woods et al (121) are to be commended for reporting such results. Even though negative heritability estimates may make no sense theoretically, they should arise due to sampling error (66). If such negative heritabilities were excluded from prior studies, then the apparent average significant heritability suggested by Figure 7 (73) and elsewhere (108) may be largely or entirely an artifact of selective reporting.

One or two formal replications of earlier studies, particularly of those that reported highly significant heritabilities of FA variation (e.g., $h^2 = 0.63$ in stickleback lateral plate numbers) (46) and ($h^2 = 1.072$ in scorpionfly forewings)

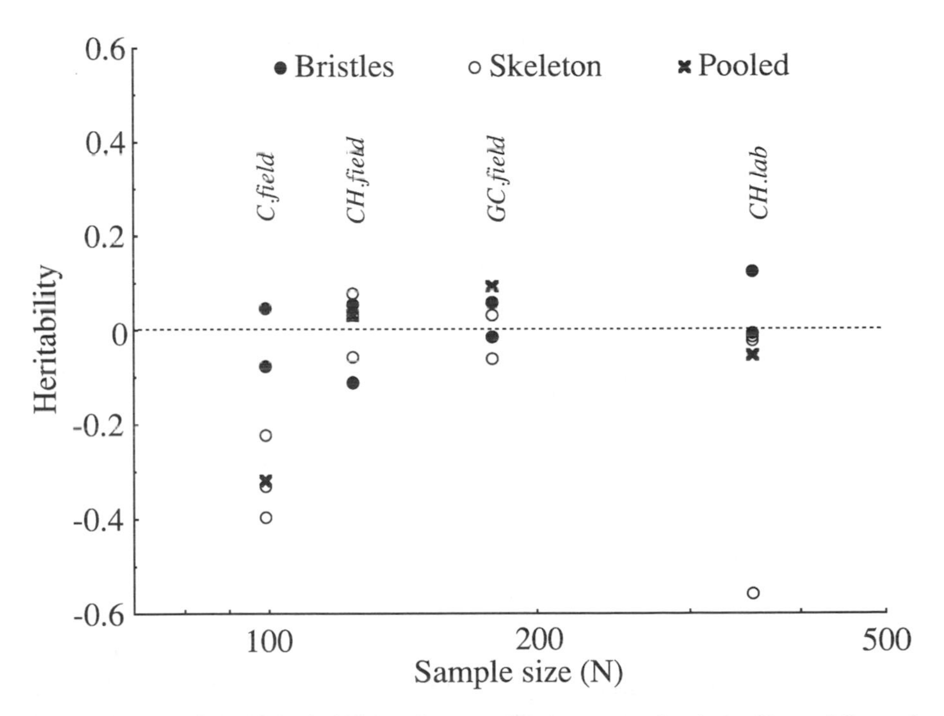

Figure 14 Estimates of heritabilities (parent-offspring regressions) in *Drosophila melanogaster* (121). Estimates were obtained for two bristle and three skeletal traits in three field and one laboratory population. The dotted line indicates the null hypothesis. N refers to the number of families in which "one to two" female offspring were measured per female parent. Abbreviations: C, Cairns; CH, Cherry Hill; GC, Gold Coast. The data on which this figure was based may be obtained from http://www.biology.ualberta.ca/palmer.hp/DataFiles.htm.

(106) would provide a far more robust test of the true heritability of FA variation than additional quasireplications in other taxa or traits. If these high heritabilities could be repeated, then, again, the hypothesis that the overall significant heritability of FA (Figure 7) is due to selective reporting would be rejected.

Fluctuating Asymmetry and Sexual Selection Remarkably, in spite of the huge impact Møller's (71) original study had on interest in asymmetry as a measure of individual attractiveness, I am unaware of any independent published replication of the claim that female barn swallows (*Hirundo rustica*) prefer males with outer tail feathers that are more symmetrical, or of any other claims for correlations between asymmetry and individual attributes in barn swallows or other organisms. In view of the large number of studies of correlations between individual asymmetry and fitness/attractiveness, the absence of replicative studies seems remarkable.

If any claims for correlations between subtle asymmetries and attributes of individuals can be repeated by others with no vested interest in a particular outcome, only then will such claims earn the respect of credibility. Until such time, the hypothesis that correlations between subtle asymmetries and attributes of individuals result predominantly from selective reporting remains a viable one.

DISCUSSION

I have found these results particularly sobering and have gained little pleasure from summarizing them. Clearly all of us—meticulous and conscientious as we may be—are guilty of selective reporting to some degree. Furthermore, those who have tried to publish negative or nonsignificant results may have been discouraged or denied by well-intentioned peers or editors in the review process. As a consequence, what gets published is inevitably not a random sample of studies that were initiated, and we cannot escape the troubling conclusion that some—perhaps many—cherished generalities are at best exaggerated in their biological significance (80) and at worst a collective illusion nurtured by strong a priori beliefs often repeated (33, 73).

Quasireplication likely exacerbates the problem of selective reporting. Authors who explicitly set out to replicate a prior study fully will presumably have a greater desire to publish their results—no matter what the outcome—than those who pursue simply affirmative quasireplication. Quasireplication is still valuable because results and hypotheses ultimately do need to be repeated in other systems to assess their generality. But if quasireplication continues to substitute for formal replication, the pernicious effects of selective reporting will do little more than reinforce the contract of error that entrenches a priori beliefs or perpetuates the unconstructive cycle of bandwagon and backlash (2, 84, 86, 96). Unfortunately, editorial policies may contribute to the problem.

The Impact of Editorial Policy

Editorial policies of many scientific journals appear to reinforce a culture in which original research is valued more highly than replicative research, and in which statistically significant results are favored over nonsignificant ones (26, 50). For example, many journals explicitly discourage replicative studies either in their advice to authors (Table 2a) or in their advice to reviewers (Table 2b). As a consequence, because truly replicative research is so difficult to publish, biologists test the validity of hypotheses or patterns by examining other species or systems.

In fairness, conspicuous examples of replicated studies are sometimes published in premier journals. For example, *Science* reported a study (88) that was unsuccessful at replicating the results of an earlier one, published in *Nature-Genetics*, claiming to have found microsatellite markers linked to human male homosexuality (54). *Science* also drew attention to the great difficulties different labs had at obtaining the same results even in a highly controlled study of mouse behavioral genetics, and concluded by saying "every result should be replicated with a new batch of mice within the same lab, and perhaps elsewhere, before it's published" (34:1599). In these cases, however, conflicting results appeared to make better press than confirmatory ones.

However, even strictly confirmatory results have been deemed of sufficient importance by *Nature* to warrant publication, so long as the subject—Neanderthal

TABLE 2a Portions of instructions to authors from various journals

Journal	Instructions to Authors
American Naturalist	"[*American Naturalist*] welcomes manuscripts that develop new conceptual syntheses, especially those combining verbal or mathematical theory with new empirical information of general significance."
Ecology	*Articles:* "Articles describing significant original research comprise the core of the journal." *Reports:* "Reports are expected to disclose new and exciting work in a concise format. These papers should present results that substantially advance a field or overturn existing ideas." *Notes:* "Present significant new observations and methodological advances."
Evolution	"[W]ell-written papers that represent significant new findings, are of general interest, and are placed in a general context are most likely to be published in *Evolution*."
J. of Animal Ecology	"[P]ublishes the best in original research on any aspect of animal ecology."
Nature	"The initial criteria for a paper to be sent for peer-review are that the results seem novel, arresting (unexpected or surprising), and that the work seems broadly significant outside the field."
Science	"Is your paper one of the best you have ever produced? Will it have a big impact in your field? Will scientists in related fields be interested in the results? Will it surprise the reader? Does it overturn conventional wisdom?"

TABLE 2b Portions of instructions to referees from various journals

Journal	Instructions to Referees
American Naturalist	Will this paper (score on a scale from 1–5): "c. Pose a new and significant problem or introduce a novel subject to the readership. d. Change the way people think about the topic of the manuscript. e. Confirm or refute an unverified theoretical principle or a previously unsupported/ weakly supported generalization."
Ecology	Rating of manuscript (numerical score): "Importance to our readers. Originality of the research."
Evolution	"Does the paper contain new data or new ideas? (yes/no)"
Oecologia	"First, is the science as such sound ... ? Second ... how do you rate its interest value to *Oecologia*? We ... try to select papers which are not merely confirmatory, but make some novel contribution ..."
Oikos	(choose one) "1. Excellent, opens a new and significant area of research; 2. Very good. Makes a conceptual advance in an established field; 3. Good. Adds significantly to knowledge in an established field; 4. Could potentially reach standard 1 or 2 or 3 (indicate which) after revisions; 5. Sound but unexciting routine work that makes no significant contribution to knowledge and no conceptual advance."
Nature	"If the conclusions are not original, it would be very helpful if you could provide relevant references."
Science	"Priority is given to papers that reveal novel concepts of interdisciplinary interest". "In selecting papers for publication, the editors give preference to those papers with novelty and general significance."

DNA sequence (51)—had wide enough appeal. Is the implication here that few data in ecology or evolution are so critical as to require verification?

How To Ameliorate the Impact of Quasireplication and Selective Reporting?

What can be changed to reduce the effects of selective reporting and enhance the stature of formal replicative research? Five changes—in increasing order of difficulty to implement—would help.

(1) Journal Format One minor change to journal formats would have immediate results and be easy to implement. A special category called *Replications*, in which only fully independent replications would be published, offers four advantages. First, it would give greater prominence to truly replicated studies and would presumably encourage more biologists to conduct them. Second, it would ensure that authors, referees, and editors all gave due recognition to the value of formal replications. Third, it would place replicative studies in a defined area of the journal so that readers interested only in novel or original research would not be distracted. Finally, if authors knew that others might attempt to replicate particularly striking claims because replicative studies were actually encouraged by journals, they might be more cautious about rushing flashy but preliminary results through to publication.

(2) Editorial Policy Regarding Replication Journal editorial policies could readily be revised to recognize the value of replicate studies without them becoming a burden to readers. Authors of replicate studies would be obliged to make the case that the study being replicated was central to a developing generalization or dogma, in the same way they are now obliged to make the case that their research is original. Purely pedestrian replication of peripheral studies should garner no more respect than purely pedestrian quasireplications.

(3) Editorial Policy Regarding Statistical Significance Journal editorial policies could include two additional yardsticks for judging the robustness of a result: (*a*) a quantitative sliding scale of statistical significance that depends on sample size, and (*b*) a qualitative nonsignificance test.

The problem of selective reporting is a simple one: As sample size decreases, an author's decision to publish will be more and more influenced by the statistical significance of the result. To discourage the publication of exploratory results that happen to reach statistical significance due to chance, the α level for a result to be considered statistically significant should be more extreme for small than for large samples. For example, for a correlation coefficient, set $\alpha = 0.001$ for sample sizes less than 20, set $\alpha = 0.01$ for sample sizes of 20–100, and retain the usual convention of $\alpha = 0.05$ for sample sizes exceeding 100. Such a rule would go a long way toward reducing the publication of spuriously significant results based on small sample size. What is considered a small sample size would depend on the type of statistical test, but standard meta-analytic techniques (91) allow other statistics to be converted to a common "effect size" (e.g., a correlation coefficient; 7) for easier judgment.

In addition, referees and editors could apply a rough rule of thumb: the nonsignificance test. The nonsignificance test would serve as a kind of litmus test of the strength of a significant result: Is the hypothesis, and sampling or experimental design, sufficiently robust that the results would seem worthy of publication if not significant statistically? In other words, if the primary result was clearly nonsignificant at $P = 0.5$, would the study still seem worthy of publication? If the answer to the nonsignificance test was "yes," then a weakly significant result would be worth reporting. However, if the answer was "no," then a weakly significant result would seem dubious at best.

Both of these suggestions are related to power analysis, but retrospective power analysis suffers from a number of problems (103) and seems unlikely to offer a simple solution. A power analysis asks, If the parametric value of a statistical descriptor for a particular population is truly nonzero, how often would it be found to be significantly different from zero for a given sampling error and sample size?

Unfortunately, retrospective power analyses—those asking about the power of a final result—appear to offer no solution. First, for many statistics a result that is just barely significant (e.g., $P = 0.049$) will always yield a power of approximately 0.5 because 50% of replicate samples from the same population would be higher than this value (and thus significant statistically), and 50% would be lower (and thus nonsignificant), no matter what the sample size. Because it is simply inversely related to the original P value, retrospective estimates of power offer no additional information (103). Second, the confidence intervals on estimates of power can be large (103), thus rendering them uninformative.

(4) Undergraduate and Graduate Training Graduate program coordinators and supervisors should consider encouraging graduate students, as part of a graduate degree program, to conduct at least one formal replicative study. Clearly these would have to be combined with original research so that students could also demonstrate their creativity and ability to tackle novel problems. Here again, a *Replications* section in premier journals would help legitimize and encourage replicative research, and allow students to be recognized for a well-replicated study as much as for a wholly novel one.

(5) Research Funding Priorities Research funding agencies could create a special funding category called *Critical Replications*. Not only would this reward replicative research directly, it would also allow a registry of formally replicated studies to be maintained. As in medical research (100), such registries ensure that reviewers and meta-analysts could directly assess the magnitude of selective reporting. In addition, such funding could be awarded with the condition that it must be published before the same investigators would be eligible to apply for any subsequent funding from the *Critical Replications* fund.

Is "True" Replication Possible?

Some will object that true replication of ecological or evolutionary studies is not possible even in principle. Even where a study is replicated with the same population of the same species using the same protocol, many other factors may not be controllable (weather, genetic makeup of a population, population density of the study organism or of other organisms that might affect the outcome of the study, other historical effects, etc). These factors might also affect replications attempted with a different population of the same species and therefore make interpretation troublesome. This inability to truly replicate a previous study is, undoubtedly, a significant problem.

But is this not precisely why replication is valuable in the first place—to judge just how repeatable (and therefore presumably biologically significant) a result is? Should we not care about how large the effect of uncontrolled variation is on the magnitude or clarity of a particular cherished result? Although it may take different forms, uncontrollable variation is a fact of life for all scientific research.

Surely what matters is either (*a*) how repeatable a result is in the face of uncontrolled variation, or (*b*) how sensitive a result is to specific uncontrollable variables. Clearly, "there is a vast amount of extra information available from repeated experimentation (generality of circumstances; variation in intensity; consistency over seasons, etc.)" (106a). The ineluctable variability of natural systems is a very real and sometimes fascinating aspect of the biology. We ignore it at our peril.

Selective Reporting Among Replicated Studies

Will increased replication eliminate the problem of selective reporting? Of course not. Those with a strong desire to confirm an earlier result will be more inclined to report a positive than a negative outcome. Similarly, those with a vested interested in contradicting a previous claim will be more inclined to publish contrary results. Such biases, even among replicated studies, may be widespread in particularly litigious areas [e.g., see the debate over the effects of second-hand smoke between Givens et al (43) and commentators (14)].

Nonetheless, replicative studies, at the very least, provide some estimate of the among-study variance—due both to genuine sampling error and to selective reporting—that is inevitably present in the scientific enterprise. Quasireplication will never allow the among-study variance to be separated from the among-taxon or among-system variances. Furthermore, if effects are so weak as to yield contradictory results in the hands of different investigators, perhaps it is time to acknowledge that the phenomenon under study may be of little biological significance. For example, the inability of multiple labs to obtain the same results with the same protocol when measuring the anxiety levels of the same strains of laboratory mice (34) should clearly give serious pause to those who wish to study the genes responsible for anxiety.

CONCLUSION

Few would dispute the enviable success that some disciplines in molecular biology have achieved—a success anticipated over 25 years ago (85). Molecular biologists' ability to weed out flawed methods or results via replication has undoubtedly promoted this sustained success. Clearly, ecologists and evolutionary biologists must face an ugly fact: Such success will elude our grasp until formal replication of others' work is embraced as a routine and respected element of research. As Bacon (9) observed, "truth will sooner come out from error than from confusion."

Quasireplication alone will not suffice. It is so vulnerable to selective reporting that it will as likely reinforce trendy notions as it will strengthen genuine biological generalizations. Without true replication we will never know which cherished generalizations are valid and which are the unfortunate consequence of collective wishful thinking re-enforced by an injudicious faith in statistics. Pity, eh?

ACKNOWLEDGMENTS

Many individuals aided me with this review and I am grateful to all for their efforts. Ary Hoffmann, Tim Mousseau, and Derek Roff provided digital or hard copies of their extensive tabulations of heritability estimates though, in the end, space constraints prevented me from incorporating them. I am particularly grateful to Brian Leung for sending me his detailed compilation of effect sizes for pre-1996 FA-stress and FA-fitness relations (63). Stefan Van Dongen and Robert Poulin kindly provided preprints of papers in press. Stefan Van Dongen also generously conducted the simulations that yielded Figure 9. Curt Strobeck, as always, provided a valuable sounding board for statistical issues. Jayson Gillespie carefully entered extensive tables of published results, unearthed sample sizes from original studies, and retrieved many useful papers for me. Many editors or managing editors quickly provided copies of their instructions to referees. Reuben Kaufman, Pete Palmer, Daphne Fautin, and particularly Lois Hammond and Locke Rowe, all provided helpful comments on early drafts of the paper.

LITERATURE CITED

1. Alatalo RV, Kotiaho J, Mappes J, Parri S. 1998. Mate choice for offspring performance: major benefits or minor costs? *Proc. R. Soc. Lond. B* 265:2297–301
2. Alatalo RV, Mappes J, Elgar MA. 1997. Heritabilities and paradigm shifts. *Nature* 385:402–3
3. Ankney CD. 1982. Sex ratio varies with egg sequence in lesser snow geese. *Auk* 99:662–66
4. Anscombe FJ. 1973. Graphs in statistical analysis. *Am. Stat.* 27:17–21
5. Antolin MF. 1993. Genetics of biased sex ratios in subdivided populations: models, assumptions, and evidence. *Oxford Surv. Evol. Biol.* 9:239–81
6. Arnqvist G, Rowe L, Krupa JJ, Sih A. 1996. Assortative mating by size: a meta-analysis of mating patterns in water striders. *Evol. Ecol.* 10:265–84
7. Arnqvist G, Wooster D. 1995. Meta-analysis: synthesizing research findings in ecology and evolution. *Trends Ecol. Evol.* 10:236–40
8. Bacon F. 1891. *The Advancement of Learning*. Oxford: Clarendon. 4th ed.
9. Bacon F. 1960. *The New Organon and Related Writings*. New York: Liberal Arts
10. Becker BJ. 1994. Combining significance levels. See Ref. 25, pp. 215–30
11. Becker BJ. 1998. Mega-review: books on meta-analysis. *J. Educ. Behav. Stat.* 23:77–92
12. Begg CB. 1994. Publication bias. See Ref. 25, pp. 399–409
13. Begg CB, Berlin JA. 1988. Publication bias: a problem in interpreting medical data. *J. R. Stat. Soc.* A151:419–63
14. Begg CB, DuMouchel W, Harris J, Dobson A, Dear K, et al. 1997. Publication bias in meta-analysis: a Bayesian data-augmentation approach to account for issues exemplified in the passive smoking debate—comments and rejoinders. *Stat. Sci.* 12:241–50
15. Blank JL, Nolan V. 1983. Offspring sex ratio in red-winged blackbirds is dependent on maternal age. *Proc. Natl. Acad. Sci. USA* 80:6141–45

16. Brookfield J. 1999. Allozymes again. *Evolution* 53:632–34
17. Cappelleri JC, Ioannidis JPA, Schmid CH, deFerranti SD, Aubert M, et al. 1996. Large trials vs meta-analysis of smaller trials—How do their results compare? *J. Am. Med. Assoc.* 276:1332–38
18. Chamberlin TC. 1897. The method of multiple working hypotheses. *J. Geol.* 5:837–48
19. Cleary RJ, Casella G. 1997. An application of Gibbs sampling to estimation in meta-analysis: accounting for publication bias. *J. Educ. Behav. Stat.* 22:141–54
20. Clutton-Brock TH. 1986. Sex ratio variation in birds. *Ibis* 128:317–29
21. Cockburn A. 1990. Sex ratio variation in Marsupials. *Aust. J. Zool.* 37:467–79
22. Cooke F, Harmsen R. 1983. Does sex ratio vary with egg sequence in lesser snow geese? *Auk* 100:215–17
23. Cooper H. 1979. Statistically combining independent studies: a meta-analysis of sex differences in conformity research. *J. Pers. Soc. Psychol.* 37:131–46
24. Cooper H, DeNeve K, Charlton K. 1997. Finding the missing science: the fate of studies submitted for review by a human subjects committee. *Psychol. Meth.* 2:447–52
25. Cooper H, Hedges LV, eds. 1994. *The Handbook of Research Synthesis*. New York: Russel Sage Found.
26. Csada RD, James PC, Espie RHM. 1996. The 'file drawer problem' of nonsignificant results: Does it apply to biological research? *Oikos* 76:591–93
27. Deeks JJ. 1998. Systematic reviews of published evidence: miracles or minefields? *Ann. Oncol.* 9:703–9
28. Devlin B, Daniels M, Roeder K. 1997. The heritability of IQ. *Nature* 388:468–71
29. Dickersin K. 1997. How important is publication bias? A synthesis of available data. *AIDS Educ. Preven.* 9:15–21
30. Egger M. 1998. Under the metarescope: potential and limits of meta-analyses. *Schweizerische Medizinische Wochenschrift* 128:1893–1901
31. Egger M, Smith GD. 1998. Meta-analysis: bias in location and selection of studies. *Br. Med. J.* 316:61–66
32. Egger M, Smith GD, Schneider M, Minder C. 1997. Bias in meta-analysis detected by a simple, graphical test. *Br. Med. J.* 315:629–34
33. Elner RW, Vadas RLS. 1990. Inference in ecology: the sea-urchin phenomenon in the northwestern Atlantic. *Am. Nat.* 136:108–25
34. Enserink M. 1999. Fickle mice highlight test problems. *Science* 284:1599–600
35. Falconer DS. 1981. *Introduction to Quantitative Genetics*. New York: Longman
36. Festa-Bianchet M. 1996. Offspring sex ratio studies of mammals: Does publication depend upon the quality of the research or the direction of the results? *Ecoscience* 3:42–44
37. Fiala KL. 1981. Sex ratio constancy in the red-winged blackbird. *Evolution* 35:898–910
38. Fields SJ, Spiers M, Herschkovitz I, Livshits G. 1995. Reliability of reliability coefficients in the estimation of asymmetry. *Am. J. Phys. Anthropol.* 96:83–87
39. Finney DJ. 1995. A statistician looks at meta-analysis. *J. Clin. Epidemiol.* 48:87–103
40. Fisher RA. 1958. *The Genetical Theory of Natural Selection*. New York: Dover. 2nd ed.
41. Frank SA. 1990. Sex allocation theory for birds and mammals. *Annu. Rev. Ecol. Syst.* 21:13–55
42. Gangestad SW, Thornhill R. 1999. Individual differences in developmental precision and fluctuating asymmetry: a model and its implications. *J. Evol. Biol.* 12:402–16

42a. Gardner M. 1990. *The New Ambidextrous Universe*. New York: Freeman. 3rd ed.

43. Givens GH, Smith DD, Tweedie RL.

1997. Publication bias in meta-analysis: a Bayesian data-augmentation approach to account for issues exemplified in the passive smoking debate. *Stat. Sci.* 12:221–40

44. Gowaty PA. 1993. Differential dispersal, local resource competition, and sex ratio variation in birds. *Am. Nat.* 141:263–80
45. Gurevitch J, Morrow LL, Wallace A, Walsh JS. 1992. A meta-analysis of competition in field experiments. *Am. Nat.* 140:539–72
46. Hagen DW. 1973. Inheritance of number of lateral plates and gill rakers in *Gasterosteus aculeatus*. *Heredity* 30:303–12
47. Hamilton WJ, Poulin R. 1997. The Hamilton and Zuk hypothesis revisited: a meta-analytical approach. *Behaviour* 134:299–320
48. Harmsen R, Cooke F. 1983. Binomial sex-ratio distribution in the lesser snow goose: a theoretical enigma. *Am. Nat.* 121:1–8
49. Harrington A, ed. 1999. *The Placebo Effect: An Interdisciplinary Exploration*. Cambridge: Harvard Univ. Press
50. Hedges LV. 1992. Modeling publication selection effects in meta-analysis. *Stat. Sci.* 7:246–55
51. Höss M. 2000. Neanderthal population genetics. *Nature* 404:453–54
52. Houle D. 1998. High enthusiasm and low *R*-squared. *Evolution* 52:1872–76
53. Howe HF. 1977. Sex ratio adjustment in the common grackle. *Science* 198:744–46
54. Hu S, Pattatucci AML, Patterson C, Li L, Fulker DW, et al. 1995. Linkage between sexual orientation and chromosome Xq28 in males but not in females. *Nat. Genet.* 11:248–56
55. Hunt M. 1997. *How Science Takes Stock: The Story of Meta-Analysis*. New York: Russell Sage Found.
56. Jarvinen A. 1991. A meta-analytic study of the effects of female age on laying-date and clutch size in the great tit *Parus major* and the pied flycatcher *Ficedula hypoleuca*. *Ibis* 1333:62–66
57. Jones DR. 1995. Meta-analysis: weighing the evidence. *Stat. Med.* 14:137–49
58. Komdeur J, Daan S, Tinbergen J, Mateman C. 1997. Extreme adaptive modification of sex ratio of the Seychelles warbler's eggs. *Nature* 385:522–25
59. Krackow S. 1995. Potential mechanisms for sex ratio adjustment in mammals and birds. *Biol. Rev.* 70:225–41
60. Kuhn TS. 1962. *The Structure of Scientific Revolutions*. Chicago: Univ Chicago Press. 2nd ed.
61. Lau J, Ioannidis JPA, Schmid CH. 1997. Quantitative synthesis in systematic reviews. *Ann. Intern. Med.* 127:820–26
62. Leamy L. 1993. Morphological integration of fluctuating asymmetry in the mouse mandible. *Genetica* 89:139–53
63. Leung B, Forbes MR. 1996. Fluctuating asymmetry in relation to stress and fitness: effects of trait type as revealed by meta-analysis. *Ecoscience* 3:400–13
64. Light RJ, Pillemer DB. 1984. *Summing Up: The Science of Reviewing Research*. Cambridge: Harvard Univ. Press
65. Ludwig W. 1932. *Das Rechts-Links Problem im Teirreich und beim Menschen*. Berlin: Springer
66. Lynch M, Walsh B. 1997. *Genetics Analysis of Quantitative Traits*. Sunderland, MA: Sinauer
67. Magnusson WE. 2000. Error bars: Are they the king's clothes? *Bull. Ecol. Soc. Am.* 147–50
68. Mann C. 1990. Meta-analysis in the breech. *Science* 249:476–80
69. Mather K. 1953. Genetical control of stability in development. *Heredity* 7:297–336
70. Mather K, Jinks JL. 1977. *Introduction to Biometrical Genetics*. New York: Chapman & Hall
71. Møller AP. 1992. Female swallow preference for symmetrical male sexual ornaments. *Nature* 357:238–40
72. Møller AP. 1999. Asymmetry as a predictor of growth, fecundity and survival. *Ecol. Lett.* 2:149–56

73. Møller AP, Swaddle JP. 1997. *Developmental Stability and Evolution*. Oxford: Oxford Univ. Press
74. Møller AP, Thornhill R. 1998. Bilateral symmetry and sexual selection: a meta-analysis. *Am. Nat.* 151:174–92
75. Mosteller F, Colditz GA. 1996. Understanding research synthesis (meta-analysis). *Annu. Rev. Public Health* 17:1–23
76. Newton I, Marquiss M. 1979. Sex ratio among nestlings of the European sparrowhawk. *Am. Nat.* 113:309–15
77. Normand S-LT. 1999. Meta-analysis: formulating, evaluating, combining, and reporting. *Stat. Med.* 18:321–59
78. Palmer AR. 1994. Fluctuating asymmetry analyses: a primer. In *Developmental Instability: Its Origins and Evolutionary Implications*, ed. TA Markow, pp. 335–64. Dordrecht, Netherlands: Kluwer
79. Palmer AR. 1996. Waltzing with asymmetry. *BioScience* 46:518–32
80. Palmer AR. 1999. Detecting publication bias in meta-analyses: a case study of fluctuating asymmetry and sexual selection. *Am. Nat.* 154:220–33
81. Palmer AR, Hammond LM. 2000. The Emperor's codpiece: a post-modern perspective on biological asymmetries. *Int. Soc. Behav. Ecol. Newslett.* 12: In Press
82. Palmer AR, Strobeck C. 1986. Fluctuating asymmetry: measurement, analysis, patterns. *Annu. Rev. Ecol. Syst.* 17:391–421
83. Phillips PC. 1998. Designing experiments to maximize the power of detecting correlations. *Evolution* 52:251–55
84. Pigliucci M, Kaplan J. 2000. The fall and rise of Dr. Pangloss: adaptationism and the Spandrels paper 20 years later. *Trends Ecol. Evol.* 15:66–70
85. Platt JR. 1964. Strong inference. *Science* 146:347–53
86. Poulin R. 2000. Manipulation of host behaviour by parasites: a weakening paradigm? *Proc. R. Soc. Lond. B.* 267:787–92
87. Quinn JF, Dunham AE. 1983. On hypothesis testing in ecology and evolution. *Am. Nat.* 122:602–7
88. Rice G, Anderson C, Risch N, Ebers G. 1999. Male homosexuality: absence of linkage to microsatellite markers at Xq28. *Science* 284:665–67
89. Rohlf FJ, Sokal RR. 1995. *Statistical Tables*. San Francisco: Freeman. 3rd ed.
90. Rosenthal R. 1979. The "file drawer problem" and tolerance for null results. 86:638–41
91. Rosenthal R. 1991. *Meta-Analytic Procedures for Social Research*. Beverly Hills: Sage
92. Roughgarden J. 1983. Competition and theory in community ecology. *Am. Nat.* 122:583–601
93. Salt GW. 1983. Roles: their limits and responsibilities in ecological and evolutionary research. *Am. Nat.* 122:697–705
94. Sheldon BC. 1998. Recent studies of avian sex ratios. *Heredity* 80:397–402
95. Simberloff D. 1983. Competition theory, hypothesis testing, and other community ecological buzzwords. *Am. Nat.* 122:626–35
96. Simmons LW, Tomkins JL, Kotiaho JS, Hunt J. 1999. Fluctuating paradigm. *Proc. R. Soc. Lond. B* 266:593–95
97. Smith BH, Garn SM, Cole PE. 1982. Problems of sampling and inference in the study of fluctuating dental asymmetry. *Am. J. Phys. Anthropol.* 58:281–89
98. Sokal RR, Rohlf FJ. 1995. *Biometry*. New York: Freeman. 3rd ed.
99. Soulé ME, Baker B. 1968. Phenetics of natural populations. IV. The populations asymmetry parameter in the butterfly *Coenonympha tullia*. *Heredity* 23:611–14
100. Stern JM, Simes RJ. 1997. Publication bias: evidence of delayed publication in a cohort study of clinical research projects. *Br. Med. J.* 315:640–45
101. Strong DR. 1980. Null hypotheses in ecology. *Synthese* 43:271–85

102. Strong DR. 1983. Natural variability and the manifold mechanisms of ecological communities. *Am. Nat.* 122:636–60
103. Thomas L. 1997. Retrospective power analysis. *Conserv. Biol.* 11:276–80
104. Thornhill R, Møller AP. 1998. The relative importance of size and asymmetry in sexual selection. *Behav. Ecol.* 9:546–51
105. Thornhill R, Møller AP, Gangestad SW. 1999. The biological significance of fluctuating asymmetry and sexual selection: a reply to Palmer. *Am. Nat.* 154:234–41
106. Thornhill R, Sauer P. 1992. Genetic sire effects on the fighting ability of sons and daughters and mating success of sons in a scorpionfly. *Anim. Behav.* 43:255–64
106a. Underwood AJ. 1999. Publication of so-called 'negative' results in marine ecology. *Mar. Ecol. Prog. Ser.* 191:307–9
107. Van Dongen S. 1998. How repeatable is the estimation of developmental stability by fluctuating asymmetry? *Proc. R. Soc. Lond. B* 265:1423–27
108. Van Dongen S. 2000. The heritability of fluctuating asymmetry: a Bayesian hierarchical model. *Ann. Zool. Fennici.* 37:15–23
109. Van Dongen S, Lens L. 2000. The evolutionary potential of developmental stability. *J. Evol. Biol.* 13:326–35
110. van Schaik CP, Hrdy SB. 1991. Intensity of local resource competition shapes the relationship between maternal rank and sex ratios at birth in cercopithecine primates. *Am. Nat.* 138:1555–62
111. Van Valen L. 1962. A study of fluctuating asymmetry. *Evolution* 16:125–42
112. Vickers A, Goyal N, Harland R, Rees R. 1998. Do certain countries produce only positive results? A systematic review of controlled trials. *Contr. Clin. Trials* 19:159–66
113. Vollestad LA, Hindar K, Moller AP. 1999. A meta-analysis of fluctuating asymmetry in relation to heterozygosity. *Heredity* 83:206–18
114. Wang MC, Bushman BJ. 1998. Using the normal quantile plot to explore meta-analytic data sets. *Psychol. Meth.* 3:46–54
115. Weatherhead P. 1983. Secondary sex ratio adjustment in red-winged blackbirds (*Agelaius phoeniceus*). *Behav. Ecol. Sociobiol.* 12:57–61
116. Weigensberg I, Roff DA. 1996. Natural heritabilities: Can they be reliably estimated in the laboratory? *Evolution* 50: 2149–57
117. Whitlock M. 1996. The heritability of fluctuating asymmetry and the genetic control of developmental stability. *Proc. R. Soc. Lond. B* 263:849–53
118. Whitlock M. 1998. The repeatability of fluctuating asymmetry: a revision and extension. *Proc. R. Soc. Lond. B* 265:1429–31
119. Williams GC. 1979. On the question of adaptive sex ratio in outcrossed vertebrates. *Proc. R. Soc. Lond. B* 205:567–80
120. Wilson EE. 1975. *Sociobiology*. Cambridge, MA: Harvard Univ. Press
121. Woods RE, Hercus MJ, Hoffmann AA. 1998. Estimating the heritability of fluctuating asymmetry in field *Drosophila*. *Evolution* 52:816–24
122. Woods RE, Sgro CM, Hercus MJ, Hoffmann AA. 1999. The association between fluctuating asymmetry, trait variability, trait heritability, and stress: a multiply replicated experiment on combined stresses in *Drosophila melanogaster*. *Evolution* 53:493–505

Annu. Rev. Ecol. Syst. 2000. 31:481–531

INVASION OF COASTAL MARINE COMMUNITIES IN NORTH AMERICA: Apparent Patterns, Processes, and Biases

Gregory M. Ruiz and Paul W. Fofonoff
Smithsonian Environmental Research Center, Edgewater, Maryland 21037; e-mail: ruiz@serc.si.edu; fofonoff@serc.si.edu

James T. Carlton
Maritime Studies Program, Williams College—Mystic Seaport, Mystic, Connecticut 06355; e-mail: jcarlton@williams.edu

Marjorie J. Wonham
Department of Zoology, University of Washington, Seattle, Washington 98915; e-mail: mwonham@u.washington.edu

Anson H. Hines
Smithsonian Environmental Research Center, Edgewater, Maryland 21037; e-mail: hines@serc.si.edu

Key Words nonindigenous, invasion resistance, disturbance, propagule supply, introduced species

■ **Abstract** Biological invasions of marine habitats have been common, and many patterns emerge from the existing literature. In North America, we identify 298 nonindigenous species (NIS) of invertebrates and algae that are established in marine and estuarine waters, generating many "apparent patterns" of invasion: (*a*) The rate of reported invasions has increased exponentially over the past 200 years; (*b*) Most NIS are crustaceans and molluscs, while NIS in taxonomic groups dominated by small organisms are rare; (*c*) Most invasions have resulted from shipping; (*d*) More NIS are present along the Pacific coast than the Atlantic and Gulf coasts; (*e*) Native and source regions of NIS differ among coasts, corresponding to trade patterns. The validity of these apparent patterns remains to be tested, because strong bias exists in the data. Overall, the emergent patterns reflect interactive effects of propagule supply, invasion resistance, and sampling bias. Understanding the relative contribution of each component remains a major challenge for invasion ecology and requires standardized, quantitative measures in space and time that we now lack.

INTRODUCTION

Biological invasions, or the establishment of species beyond their historical range, have long been of great interest to ecologists, evolutionary biologists, and paleontologists (40, 41, 50, 72, 84, 85, 135, 148). The establishment and study of small populations has generated a wide range of opportunities to understand fundamental population, community, and ecosystem processes across many taxonomic groups (38, 130, 136, 137). In recent years, invasion research has focused especially on the patterns and process of invasions themselves (25, 37, 81, 101, 108, 129), and we have witnessed a virtual explosion in the quantity and diversity of research in this topic area (104, 115, 144).

The recent growth of invasion research has been stimulated largely by an apparent increase in the rate of nonindigenous species (NIS) invasions and their effects on native populations and communities, ecosystem function, and economies, as well as human health (6, 10, 38, 61, 62, 132, 140). As a result, new information and shifting perspectives have emerged rapidly. This emergence is particularly striking for invasions of marine environments, which had historically received little attention compared to terrestrial or freshwater systems (13).

In this article we explore patterns, mechanisms and hypotheses associated with marine invasions. Although new data on marine invasions have increased rapidly, they have never been summarized, beyond analysis for single bays or estuaries (11, 29, 30, 69, 117) (JT Carlton 2000, unpublished checklist; Carlton & Wonham 2000, unpublished manuscript). Moreover, the complexities and potential biases of these data, and inferences that can be drawn from the data, have not been evaluated critically. Here, we provide such a synthesis for marine invasions of North America and begin to evaluate some of the emergent patterns and underlying mechanisms. More specifically, we wish to summarize spatial and temporal patterns of invasion and to identify (*a*) key gaps in data, (*b*) hypotheses about mechanisms, and (*c*) future directions for research. Although our analysis is specific to marine and estuarine invasions, we explore issues and approaches that are relevant generally for both invasion biology and invasion management.

PATTERNS OF INVASION

Classification and Analysis

We characterized patterns of invasion for marine (including estuarine) invertebrates and algae on multiple spatial scales, focusing primarily on North America. Our focus on invertebrates and algae is intended to illustrate general issues, using a relatively large group of NIS known for North America across many phyla. Although vascular plants and fish were excluded from our analysis, these groups have contributed hundreds of additional NIS that are established in coastal bays and estuaries of North America (30, 55, 117; P Fofonoff, GM Ruiz & AH Hines, submitted; GM Ruiz, P Fofonoff, AH Hines & JT Carlton, unpublished data). Their exclusion from our analysis was pragmatic, to reduce the complexity of patterns and analyses. Furthermore, there are also many fundamental differences for vascular plants and fish, compared to invertebrates and algae, with respect to invasion patterns (e.g. transfer mechanisms, habitat distributions, dates of arrival, and biology) that warrant separate analyses.

For each of 298 NIS that are reported to be established in coastal waters of North America, we summarized available information about the distribution and invasion history for the Atlantic, Pacific, and Gulf coasts. We defined NIS as those organisms transported by human activities to coastal regions where they did not previously occur. We omitted species that underwent range expansions attributed to natural dispersal, even if some resulted from anthropogenic changes in environmental conditions (19).

We considered the marine and estuarine waters of North America to extend from outer coastlines to the limit of tidal waters within bays and estuaries, including the oligohaline and tidal freshwater reaches of estuaries. Within this coastal zone, we included all species that occurred below the mean monthly limit of spring tides. Our list therefore includes marine organisms but also some species found commonly in salt marshes and strand-lines of beaches as well as species reported from estuarine freshwater. We also included insects released for biocontrol when their host plants were reported as occurring in tidal waters. However, we excluded some "boundary species" that appeared occasionally or rarely within our study area but were found primarily in terrestrial habitats and inland freshwater (see 117 for additional discussion).

Our review and synthesis relied on four main sources of information. The primary source was published information, including especially some existing analyses of NIS for particular bays (11, 29, 30, 69, 117; JT Carlton 2000, unpublished checklist; Carlton & Wonham 2000, unpublished manuscript) as well as a diffuse collection of literature. We also reviewed unpublished reports, theses, and records from long-term monitoring efforts along each coast. In addition, we corresponded with many scientists who were expert either in particular taxonomic groups or the biota of particular geographic regions. Finally, we also conducted some limited field surveys at selected sites.

As a minimum, we sought to characterize the following attributes for each species: (*a*) Date of First Record; (*b*) Native Geographic Region; (*c*) Source Region of invasion; (*d*) Vector (mechanism) of introduction; (*e*) Salinity Distribution; (*f*) Geographic Distribution; (*g*) Invasion Status; and (*h*) Population Status. The information was collected separately for each coast, since some species have invaded multiple coasts and these attributes may differ among coasts. Where multiple sites of invasion for a species existed within a coast, data were always collected for the first site and date of successful introduction (as attributes such as Source Region and Vector may differ among sites). The details of this classification scheme and subsequent analyses are described below.

Invasion Status

To assess the invasion status of species (as below), we used a graded set of criteria, relying on the historical record, paleontological record, archaeological record, biogeographic distribution, dispersal mechanisms, documented introductions, and a suite of ecological and biological characteristics (26, 27, 30, 138). We assigned species to one of three categories of invasion status, reflecting the degree of certainty that a species was introduced or native:

Introduced species Native and introduced ranges of these species were well established and provided a clear invasion history, in most cases. We considered a few additional species to be introductions where the evidence was very convincing; included here are a few intracoast invasions, for which natural dispersal is possible but highly unlikely (see below).

Cryptogenic species (Possible Introductions)—No definitive evidence of either native or introduced status [*sensu* Carlton (16)]. For some of these species, introduced status has been suggested or appears likely.

Native species Native range of these species was well established and provided clear evidence of native status.

Owing to intracoastal invasions, it was possible for a species to have a compound assignment to two or more categories of species status. For example, the hooked mussel (*Ischadium recurvum*) is native to the southeastern United States but introduced to the northeast. Thus, this species is considered native, cryptogenic, and introduced along different regions of the Atlantic coast. In our analyses, all information about this species along the Atlantic coast refers to the introduced populations. Although such intracoastal invasions possibly occurred for many species, we included only those that were clearly documented; all others were considered cryptogenic and thus excluded from our present analyses.

Date of First Record

For date of first record, we used the first date of collection, sighting, or documented deliberate release. If these were not reported, dates of written documents or publications were used; however, we recognize that these later dates may be many years after the date a species was first collected or sighted.

Vector

We evaluated plausible mechanisms (or vectors) for each introduction, using information about the first date of record, life history, habitat utilization, and ecological attributes. We assigned each species invasion to one of eight broad vector categories: Shipping; Fisheries; Biocontrol; Ornamental escape; Agricultural escape; Research escape; Canals, created by humans, as a corridor for dispersal; or Multiple. Several of the broad categories are composed of subcategories. For example, the Shipping vector included organisms moved on the hull, in ballast water or dry ballast, in or on cargo, on deck, on anchors, etc. Fisheries introductions involved both intentional and unintentional release, including those that resulted from aquaculture. Both Fisheries and Ornamental introductions also included species associated with the target species (e.g. fouling organisms on oysters). Although some of the subcategories are discussed further in this review, this higher resolution is the focus of a separate analysis (GM Ruiz, JT Carlton, P Fofonoff & AH Hines, submitted).

Several simultaneous mechanisms of introduction were clearly possible for many species, creating some uncertainty about the vector responsible for each invasion. In these cases, we simply classified the vector as Multiple and indicated the plausible mechanisms. For example, the green crab (*Carcinus maenas*) was recently to introduced western North America, and multiple mechanisms of introduction exist for this invasion: Shipping and Fisheries.

Sequential mechanisms of introduction also existed for some species, where the first introduction can be ascribed to a particular vector but subsequent introductions may have occurred due to additional mechanisms. To recognize this, there must exist a clear chronological sequence in the operation of the respective vectors, such that one predates any additional vectors. For example, the barnacle (*Balanus improvisus*) was introduced to western North America in the mid 19th century by Shipping, but the latter movement of oysters (Fisheries) represents an additional vector that was active afterwards. In such cases, we identified both the initial vector and additional vectors.

Native and Probable Source Region

Native Region identifies the range of each species before human transport, and Source Region identifies the likely area from which an invasion occurred. The Source Region may differ from Native Region for various reasons. First, a species may have a wide native distribution, whereas an introduction may have been most likely to occur from a particular region, based upon the prevalent trade patterns (and vectors) in operation. Second, there are many "stepping stone" invasions, where a species may invade secondarily from a previously invaded region that is outside the native range.

We assigned a probable Source Region, based upon the extent of available transfer mechanisms, known association with those mechanisms, and proximity to site of invasion. Identification of Source Region has some degree of uncertainty. This was particularly problematic for some widespread species, where many potential source regions (with operating transfer mechanisms) exist. For these species, we have indicated "Unknown" for Source Region.

For the purposes of our analysis, we classified Native and Source regions in terms of broad oceanic and continental regions. Ocean regions were used for species with strong marine affinities, whereas continental regions were used for those with primarily fresh water (or continental) distributions. The categories included: Western Atlantic, Eastern Atlantic, Amphi- (both Western and Eastern) Atlantic, Indian Ocean, Indo-West Pacific, Western Pacific, Eastern Pacific,

Amphi-Pacific, North America, Eurasia, South America, Africa, and Australia. As indicated above, Unknown was used when Source Region remained unresolved, and was also used as necessary for unresolved Native Region.

Native and Source Region categories were selected to accommodate most species and highlight general patterns. For each species, we identified distribution according to commonly used biogeographic regions (8, 134). These were then combined into our broad categories to simplify analyses (e.g. Northwestern Atlantic and Southwestern Atlantic became Western Atlantic) or to reflect species distributions that traversed boundaries (e.g. Indian Ocean and Western Pacific became Indo-West Pacific). Additional data on the known native and introduced range of each species, providing much finer resolution than presented here, are available upon request; these can also be found in associated references listed in the supplemental Appendix 1 on the Annual Reviews online website.

Population Status

To distinguish between introductions with persistent populations and those that may have failed, we classified the Population Status of each species as one of the following:

1. Established species have been documented as present and reproducing within the last 30 y. Multiple records were required for a species to be considered established. Furthermore, for species detected in the past 10 y, occurrence was necessary in at least two locations or in two consecutive years.
2. Population status was considered Unknown for introductions with no records within the past 20–30 y or for recent introductions with too few records (as above).
3. We recognized two categories of species introductions that do not appear to be established. Failed introductions are species that were reported but for which there is no evidence of establishment. In contrast, *Extinct* introductions survived and reproduced for many years before disappearing.

We did not include the extensive literature that exists on failed introductions. There are literally hundreds of species that have been released but apparently never established (see 11, 21, 69, 123, 146). Our goal was to document the history and fate of established populations, accounting also for those that were extinct or of unknown population status. Since population status can vary along a coastline, just as invasion status (above), some species were assigned to multiple categories. It is only in this context that we refer to failed introductions in our analyses.

Salinity Range

The salinity distribution of each species was classified by the Venice system of salinity zones. A species could occur in one or more of the following salinity zones: Freshwater, Limnetic (tidal freshwater, 0–0.5 ppt), Oligohaline (0.5–5 ppt salinity); Mesohaline (5–18 ppt); Polyhaline (18–30 ppt); and Euhaline (30–35 ppt).

All species present in nontidal freshwater also occurred within tidal waters, occurring sometimes across a broad range of salinity. Throughout, salinity ranges were considered to be the sum of ranges for all life-stages reported for a species.

Geographic Distribution

For each species, we characterized the reported geographic distribution within North America, allowing a comparison of invasion patterns among coasts. We also compared patterns of invasion among relatively large estuaries, which included commercial ports. For comparisons among estuaries, we focused most of our attention on five locations: Prince William Sound (Alaska), Puget Sound (Washington), Coos Bay (Oregon), San Francisco Bay (California), and Chesapeake Bay (Maryland and Virginia). Most marine invasions have been reported in bay and estuarine environments (116), and these selected estuaries have been the foci of intensive analyses on the patterns and extent of invasion (11, 29, 30, 69, 117; JT Carlton 2000, unpublished checklist; Carlton & Wonham 2000, unpublished manuscript). As a result, the five estuaries offer the most complete data on spatial variation in the extent of invasions and species overlap within North America.

We also included in our comparison of NIS among estuaries data from Port Philip Bay, Australia (67). As with the focal estuaries in North America (above), patterns of NIS invasions at this site have recently received intensive analysis, providing the opportunity for an initial comparison with the North American sites.

Analyses

We used the resulting database (Appendix 1) to examine patterns of marine invasion in North America by taxonomic group, date of first record, vector, source region, native region, and salinity distribution. We included all introductions that were considered established (as above). We excluded from this analysis cryptogenic species as well as boundary residents, but we discuss the importance of cryptogenic species further below. Due to the size of Appendix 1, it does not appear in this article but is available at the Annual Reviews website repository in the Supplemental Materials section.

Our primary goal is to examine patterns of invasion for the entire continent. We have also included a comparison of invasion patterns on two additional spatial scales. First, we examine concordance of patterns among the three separate coasts of North America. Second, we describe the number and overlap of NIS among estuaries, including five in North America (as above) and one in Australia (Port Philip Bay) for which invasions have been well analyzed.

Although we have characterized the current knowledge on marine invasion patterns for North America, it is important to recognize the sources and limitations of the data from which these patterns emerge. We therefore consider our analysis to outline the apparent patterns of invasion from the literature. We address both the limitations and underlying assumptions that must be tested to adequately interpret these patterns.

Extent of Marine Invasions in North America

We identified 298 NIS of invertebrates and algae that are established in coastal waters of North America (Appendix 1). The 76 established species that have successfully colonized more than one coast we designate as "repeat invaders" (also "repeat" or "secondary" invasions) in our subsequent analyses. Thus, among all three coasts of North America, there have been a total of 374 successful invasion events (= 298 initial invasions +76 repeat invasions).

An additional three species are classified as extinct invasions, and the success of another 33 species is unknown (Appendix 1). In all subsequent analyses, we have restricted our focus to species known to have successfully invaded, which are classified as established invasions.

Our data provide only a minimum estimate for established invasions of marine invertebrates and algae. We have excluded consideration of boundary residents and cryptogenic species from our estimates, and the latter group may include hundreds of NIS that have gone unrecognized as such. Furthermore, many sites and taxa within North America have received little scrutiny. Below, we discuss the potential consequences of these limitations to the overall patterns.

Although our analysis is restricted to invertebrates and algae, it is noteworthy that at least 100 species of nonindigenous fish and 200 species of nonindigenous vascular plants are known to be established within this coastal area (55, 30; P Fofonoff, GM Ruiz & AH Hines, submitted; GM Ruiz, P Fofonoff, AH Hines & JT Carlton, unpublished data). In general, the identification and knowledge of established populations is better for these groups than for invertebrates and algae, due to both the size of the organisms and the extent of research and monitoring programs. However, a relatively large proportion of boundary residents exist among the nonindigenous fish and plants, and the tendency of species to occur within estuaries can vary geographically, complicating numerical estimates of NIS (P Fofonoff, GM Ruiz & AH Hines, submitted).

Taxonomic Distribution of Marine Invasions in North America

The NIS in our analysis were distributed among 11 phyla, with a significant difference in the contribution of each phylum to the 298 species (Figure 1*A*; Appendix 1; $X^2 = 224.6$, df $= 10$, $P < 0.001$). Half of all species were crustaceans or molluscs, accounting respectively for 28% and 22% of the initial invasions.

→

Figure 1 Total number of established nonindigenous species of invertebrates and algae reported in marine waters of North America shown by: (*A*) Taxonomic group, (*B*) Vector, (*C*) Date of First Record, (*D*) Rate of Invasion, (*E*) Native Region, and (*F*) Salinity zone. *Filled bar*, number of unique or initial species invasions (n = 298); *open bar*, number of repeat invasions among coasts (n = 76; see text for description). Rate of invasion was estimated for 30 y intervals, with number of new invasions shown for the first year of each interval since 1790.

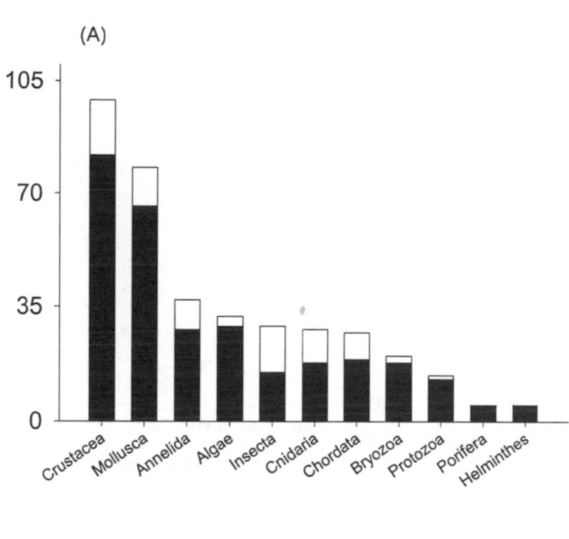
(A)
105
70
35
0
Crustacea
Mollusca
Annelida
Algae
Insecta
Cnidaria
Chordata
Bryozoa
Protozoa
Porifera
Helminthes
Number of Invasions

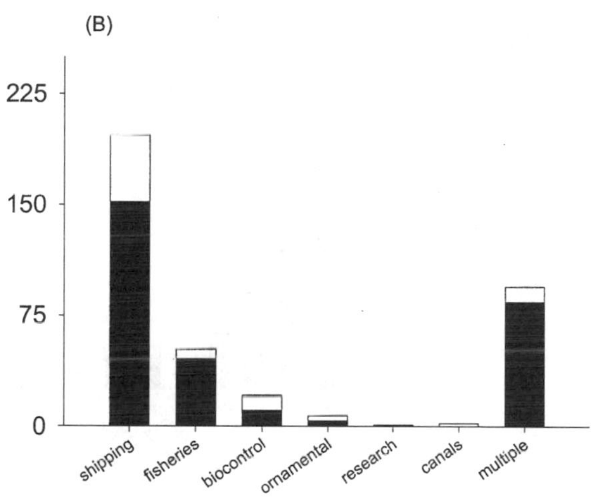
(B)
225
150
75
0
shipping
fisheries
biocontrol
ornamental
research
canals
multiple

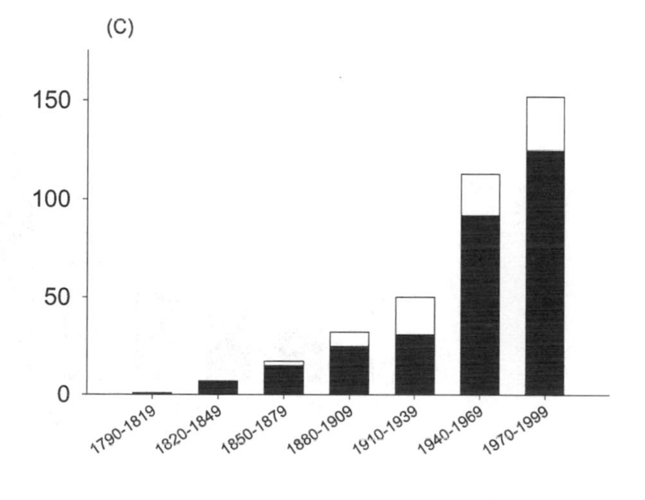
(C)
150
100
50
0
1790-1819
1820-1849
1850-1879
1880-1909
1910-1939
1940-1969
1970-1999

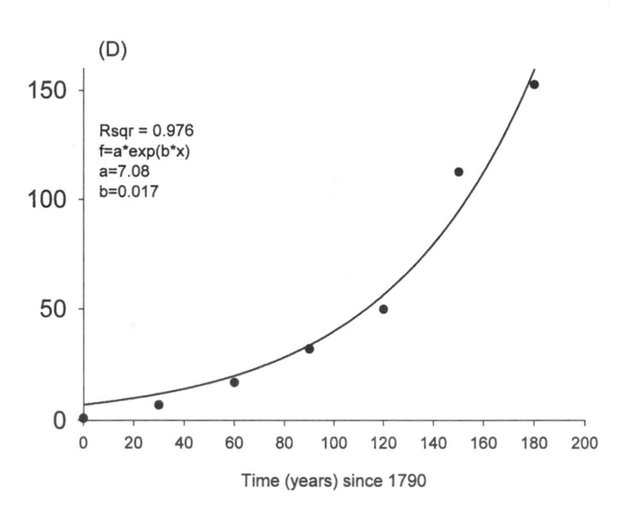
(D)
150
100
50
0
Rsqr = 0.976
f=a*exp(b*x)
a=7.08
b=0.017
0
20
40
60
80
100
120
140
160
180
200
Time (years) since 1790

Number of Invasions

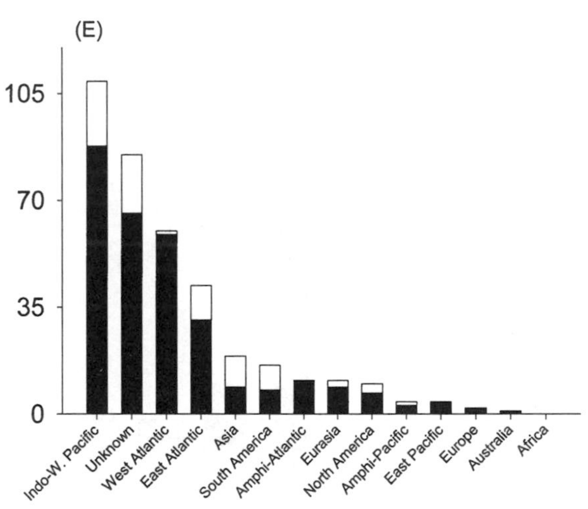
(E)
105
70
35
0
Indo-W. Pacific
Unknown
West Atlantic
East Atlantic
Asia
South America
Amphi-Atlantic
Eurasia
North America
Amphi-Pacific
East Pacific
Europe
Australia
Africa

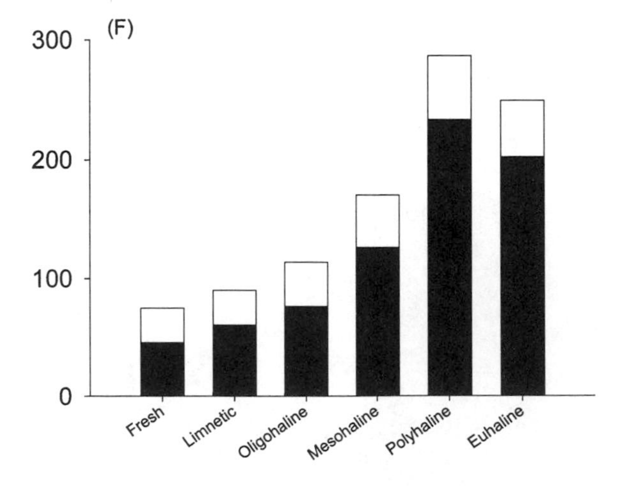
(F)
300
200
100
0
Fresh
Limnetic
Oligohaline
Mesohaline
Polyhaline
Euhaline

These two phyla together also accounted for 38% of the 76 repeat invasions (22% of crustaceans and 16% of molluscs), which was high relative to the other phyla (mean = 7%, range = 0–18%).

Perhaps most striking is the low number of relatively small organisms recognized as NIS. A few species of protists and diatoms are recognized as NIS, but many groups of microorganisms (including bacteria, viruses, fungi, microsporidia, coccidia, etc.) are absent from our database (Appendix 1). For Chesapeake Bay alone, the number of known NIS declines significantly with size of the organism, using maximum size for all taxa other than plants and fish (117). Although there may be something fundamentally different about invasion opportunities or success for small taxa, we hypothesize that this pattern results from bias in the data (see below).

Mechanisms of Introduction for Marine Invasions in North America

Shipping and fisheries have been the dominant vectors for marine invasions in North America. Shipping was the sole vector for 51% of the 298 initial invasions, and fisheries were responsible for another 15% (Figure 1*B*; Appendix 1). Although multiple vectors were plausible for 29% of all initial invasions, 78% of these 85 invasions were attributed to shipping and fisheries as the only plausible mechanisms. Shipping and fisheries together accounted for 89% of all 298 initial invasions (=51% shipping +15% fisheries +22% shipping and fisheries as multiple vectors).

Shipping and fisheries were also responsible for most (74%) of the 76 secondary or repeat invasions, occurring on coasts other than the initial coast of invasion (Figure 1*B*, Appendix 1). Shipping alone accounted for 59% of the repeat invasions, and the remaining 15% were attributed to fisheries or multiple vectors for which shipping and fisheries were the only vectors.

Despite the predominance of shipping and fisheries as vectors, there remains a great deal of uncertainty about the relative contribution or importance of each mechanism individually. This is underscored by the frequency of invasions attributed to multiple vectors, creating a wide range of importance for shipping and fisheries vectors. For example, 51% of 298 initial invasions are attributed to shipping as the sole vector, but shipping may be involved in an additional 27% of the invasions (as a possible mechanism in 94% of the 85 invasions attributed to multiple vectors; Appendix 1). It is possible to weight each vector, based upon their spatial and temporal pattern of operation, suggesting a probable vector in many cases. However, this cannot reliably exclude the other possible vectors as a mechanism for introduction. Furthermore, the multiple vectors are all in operation and may each contribute propagules to the initial or subsequent introduction of a species

A further analysis divided each vector category into subcomponents (GM Ruiz, JT Carlton, P Fofonoff & AH Hines, submitted), indicating that most invasions from shipping resulted from ballast water and hull fouling communities, and those from fisheries were dominated by translocations of organisms associated with oysters (see 14 for description). However, the relative importance of these shipping subcategories (hull fouling or ballast water) to the overall number of invasions remains poorly resolved, due to the existence of multiple vectors (as discussed above), as well as multiple subcategories that were plausible within the shipping vector, in many cases.

Rate of Marine Invasions in North America

The rate of reported invasions has increased over the past 2 centuries, using the date of first record for all 374 initial and secondary invasions of North America (Figure 1*C*; Appendix 1). The increase of initial invasions is best described by an exponential function (Figure 1*D*; $y = 7.08^{(0.017x)}$, $r^2 = 0.976$; where y is the number of new invasions and x is time in 30y intervals, indicated as the first year of the 30y interval). In contrast, although the rate of known repeat invasions is also increasing in North America, this is best described by a linear function ($y = 0.135x$, $r^2 = 0.874$).

The relative contribution of shipping to reported invasions has also increased over time (GM Ruiz, JT Carlton, P Fofonoff & AH Hines, submitted). The rate of invasions attributed solely to shipping has increased over the past 200 y, accounting for 62% of initial invasions in the past 30 y (Figure 2; see also Appendix 1). This increase is best described by an exponential function ($y = 1.127^{(0.024x)}$, $r^2 = 0.992$). In contrast, the rate of reported fisheries invasions is not increasing consistently over time and may be declining in recent years. Only 8% of reported invasions were attributed solely to fisheries in the past 30y, and the rate of fisheries invasions since 1790 is best described by a slightly positive linear relationship (Figure 2; $y = 0.083x - 0.929$, $r^2 = 0.669$).

Finally, our temporal data indicate that 20% of initial invasions in the last 30y interval are attributed to multiple vectors, usually the combination of shipping and fisheries (Appendix 1). Although lower than the prevalence of multiple vectors for all time periods (29%), uncertainty exists about particular vector responsible for even many of the most recent invasions.

Native and Source Region of Marine Invasions in North America

Approximately half of all initial invasions were classified as native to western ocean margins: the Indo-West Pacific and West Atlantic regions (30% and 20%, respectively; Figure 1*E*; Appendix 1). Indo-West Pacific species were either from the West Pacific (69%) or shared between the Indian Ocean and West Pacific (31%; as shown in Appendix 1). In contrast, 12% of all initial invaders were considered native to the eastern ocean margins of the Atlantic (10%) and the Pacific (1%). Continents were the native region for 12% of initial invaders, including primarily species of freshwater origin. Roughly 5% of the initial invaders were classified as

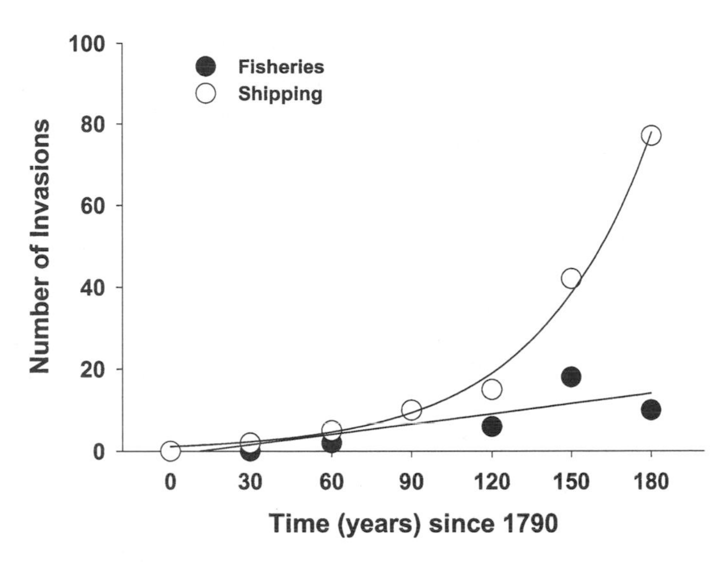

Figure 2 Rate of reported marine invasions of invertbrates and algae that were attributed to shipping and fisheries in North America since 1790. *Open circles* indicate number of new invasions associated with shipping per 30 y interval; *closed circles* indicate number of new invasions associated with fisheries per 30 y interval. Data are plotted as in Figure 1.

Amphi-Atlantic or Amphi-Pacific for native region. Surprisingly, the native region for 22% of all initial invaders was considered unknown, although the taxonomic identity was also uncertain for 33 (50%) of these 66 species.

Addition of the 76 repeat invasions has little effect on the overall prevalence of native regions (Figure 1*E*). However, it was notable that few repeat invaders were classified as native to the West Atlantic (2% of total for that region) and Amphi-Atlantic (0%), compared to the other regions with relatively large numbers of initial invaders (Indo-West Pacific—19%; Unknown—22%; East Atlantic—26%; Eurasia—18%). This is mostly an artifact of West Atlantic species being native to eastern North America, making it impossible to invade this coast (in our analysis) and reducing the opportunity for repeat invasions to occur.

We estimate that source and native regions were different for approximately 22% of all initial and repeat invasions (combined), excluding species classified as unknown for either region (Appendix 1). The source and native regions differed for all species with Amphi-Atlantic or Amphi-Pacific native regions, indicating that the invasion of North American sites occurred often from a limited portion of the native range. However, for all species from other native ranges, this implies a "stepping-stone" mode of invasion, where invasions are occurring from secondary populations outside the native range. Most of these cases involved secondary source populations in the East Atlantic (12 species), West Atlantic (8 species, excluding those that were Amphi-Atlantic), and North America (9 species). Furthermore, we estimate that the proportion of stepping-stone invasions was relatively high for NIS with native regions of Indo-West Pacific (25% of 109 species) and Eurasia, Europe and Asia (31% of 32 species).

Although native region is well defined for 77% of species, source region should be considered a rough approximation. Strong evidence often underlies the choice of source region for each invasion (e.g. operation and relative strength of vectors, established populations, date of first record, etc). For example, the European green crab, *Carcinus maenas*, may have invaded western North America from multiple source regions, including: Europe, eastern North America, Australia, and South Africa. Recent genetic analyses indicate that this newly established population derived from eastern North America (3). However, in most cases, alternate sources (in and outside of the native region) cannot easily be excluded.

Salinity Distribution of Marine Invasions in North America

Significantly more invasions are known in high (polyhaline and euhaline) than low salinity zones, for initial invasions as well as for all invasions combined (Figure 1*F*; initial invasions: $X^2 = 245$, df $= 5$, $p < 0.001$; all invasions: $X^2 = 234$, df $= 5$, $P < 0.001$). However, repeat invasions were evenly distributed across salinity zones. Importantly, many species were reported to be euryhaline and to occur in several salinity zones. Only 18 species were considered to be restricted to one salinity category.

Geographic Variation in the Extent and Patterns of Marine Invasions

Variation Among Coasts of North America

The largest number of initial invaders are known from the West Coast (187 NIS), compared to the East Coast (108 NIS) and Gulf Coast (7 NIS), representing a significant difference among coasts (Figure 3; $X^2 = 161$, df $= 2$, $P < 0.001$); the total number of initial invasions (302) exceeds the total number of species (298), because the initial invasion for some species occurred on more than one coast).

In contrast, repeat invasions were proportionally greatest for the Gulf Coast (82% of all invasions) compared to the West Coast and East Coast (18% and 0%, respectively), resulting also in a significant difference among coasts ($X^2 = 120$, df $= 2$, $P < 0.001$). The lack of known repeat invaders for the East Coast is striking and underscores the asymmetry in the sequence (or direction) of repeat invasions among coasts. The high prevalence of repeat invaders for the Gulf Coast results primarily from high overlap in NIS composition with the East Coast. Of the 39 NIS on the Gulf Coast, 87% are also known invaders on the East Coast, compared to 51% on the West Coast.

The intercoast overlap appears to be much greater for NIS present on the East Coast (38% overlap with West Coast; 32% overlap with Gulf Coast) than the West Coast (18% overlap with East Coast; 9% overlap with Gulf Coast).

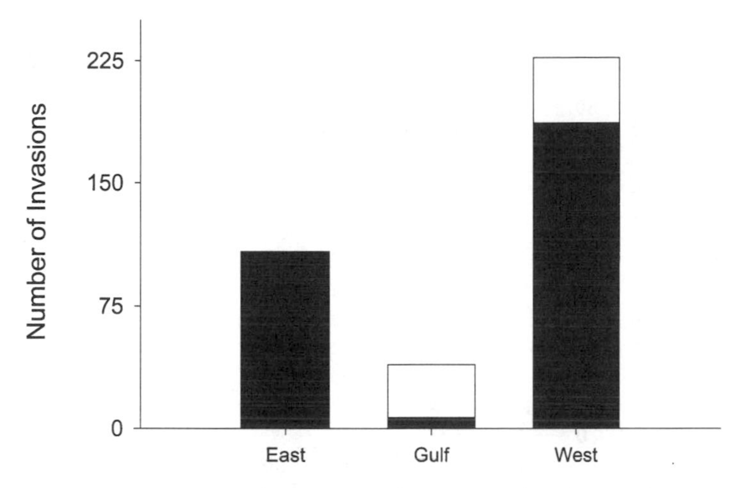

Figure 3 Total number of established nonindigenous species of invertebrates and algae reported among three coasts of North America. *Filled bar*, number of unique or initial species invasions (n = 298); *open bar*, number of repeat invasions, or the number of additional invasion events among coasts that involved a subset of initial invaders (n = 76; see text for further explanation).

However, the actual overlap of species among coasts is much greater than appears above. We have indicated overlap in a narrow sense, including only species known to be invaders to each coast. Many additional species that are native or cryptogenic to one coastal region (especially the East Coast) have invaded another coast (Appendix 1; see also Variation Among Estuaries and Data Bias Hypotheses sections for further discussion). Although these cases are excluded from our comparisons, they increase the overall overlap of biotas among regions.

Many similarities and differences exist in invasion patterns among the three coasts (GM Ruiz, JT Carlton, P Fofonoff & AH Hines, submitted; see also Appendix 1). The rate of reported invasions has increased exponentially on each coast over the past 200 y (Figure 4; West Coast, $y = 4.951^{0.0165x}$, $r^2 = 0.968$; East Coast, $y = 1.40^{0.0195x}$, $r^2 = 0.916$; Gulf Coast, $y = 0.940^{0.0160x}$, $r^2 = 0.855$). The NIS on each coast are dominated by crustaceans and molluscs, accounting together for 41–50% of the total. Shipping is the sole vector for approximately half of the known invasions on each coast: East Coast (60%), West Coast (48%), Gulf Coast (64%). However, the relative importance of fisheries as the sole vector was greater on the West Coast (19% of the total) compared to the East Coast and Gulf Coast (7% and 5%, respectively).

The native and source regions of NIS differs among coasts (GM Ruiz, JT Carlton, P Fofonoff & AH Hines, submitted; see also Appendix 1). Most West Coast NIS (53%) were native to the Indo-West Pacific and the Western Atlantic, and a smaller proportion (7%) were native to the Eastern Atlantic. In contrast, the first two of these native regions accounted for only 33% of the NIS known

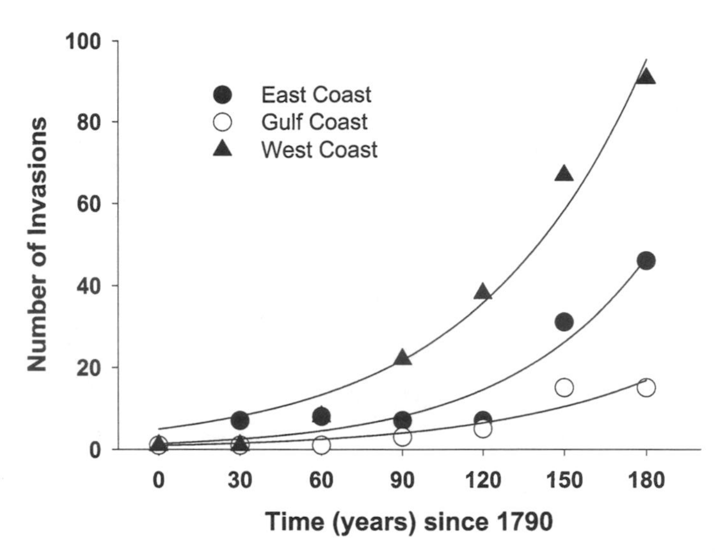

Figure 4 Rate of reported marine invasions of invertebrates and algae for each coast of North America since 1790. *Filled triangles* indicate the number of new invasions for the West Coast; those for the East Coast and Gulf Coast are indicated by *filled circles* and *open circles*, respectively. Data are plotted for 30y intervals as in Figure 1.

from the East Coast, where 19% were native to the Eastern Atlantic. Compared to the West Coast, a larger number of NIS on the East Coast were "within-coast" invasions, being introduced from one portion of the coast to another (10% of East Coast NIS compared to 1% of West Coast NIS). Finally, the disparity between native and source region was greater for the East Coast (28% of NIS had different native and source regions) than for the West Coast (15%). This results primarily from secondary invasion of the East Coast by western Pacific natives, which first invaded Europe and North America. The disparity between native and source regions was also relatively high for the Gulf Coast (38%), but finer analyses were not performed for this coast due to low sample size.

For each coast, there was a significant difference in the number of reported invasions among salinity zones (West: $X^2 = 254$, df = 5, $P < 0.001$; East: $X^2 = 32$, df = 5, $P < 0.001$; Gulf: $X^2 = 1.9$, df = 5, $P = 0.05$; see Appendix 1). A much greater proportion of NIS occurred in the polyhaline and euhaline zones of the West Coast (63%) compared to the East and Gulf Coasts (47% and 23%, respectively; number of NIS per salinity zone, with increasing salinity as follows: West Coast—24, 37, 54, 97, 187, 169; East Coast—33, 35, 38, 55, 75, 68; Gulf Coast—18, 18, 22, 19, 25, 22).

Variation Among Estuaries

Considerable variation exists in the number and overlap of known NIS among estuaries (11, 29, 30, 67, 69, 117; JT Carlton 2000, unpublished checklist; Carlton & Wonham 2000, unpublished manuscript; Table 1 and 2). The number of known

TABLE 1 Checklist of nonindigenous species reported as established in each of six different bays and estuaries

Phylum	Site[a]					
Species	SFB	CB	PS	PWS	ChB	PPB
Dinoflagellata						
Alexandrium catenella						X
Bacillariophyta						
Attheya armata		X				
Coscinodiscus wailesii					X	
Odontella sinensis					X	
Thalassiosira punctigera					X	
Phaeophyta						
Asperococcus compressus						X
Fucus cottoni				X		
Microspongium globosum				X		
Sargassum muticum	X	X	X			
Sorocarpus micromorus						X
Stictyosiphon soriferus						X
Striaria attenuata					X	
Undaria pinnatifida						X
Chlorophyta						
Bryopsis sp.	X					
Cladophora prolifera						X
Codium fragile ssp. *tomentosoides*	X				X	X
Ulva fasciata						X
Rhodophyta						
Antithamnionella spirographidis						X
Bonnemaisonia hamifera					X	
Callithamnion byssoides	X					
Ceramium sinicola				X		
Chondria arcuata						X
Chroodactylon ramosum				X		
Deucalion levringii						X
Gelidium vagum			X			
Gymnogongrus crenulatus						X
Lomentaria hakodatensis			X			
Medeiothamnion lyalli						X
Polysiphonia brodiaei						X
Polysiphonia denudata	X					
Polysiphonia harveyi					X	
Polysiphonia senticulosa (pungens)						X
Schottera nicaeensis						X
Solieria filiformis						X

(*Continued*)

TABLE 1 (*Continued*)

Phylum	Site[a]					
Species	**SFB**	**CB**	**PS**	**PWS**	**ChB**	**PPB**
Foraminifera						
Trochammina hadai	X		X	X		
Ciliophora						
Ancistrocoma pelseneeri	X					
Ancistrum cyclidioides	X					
Boveria teredinis	X					
Cothurnia limnoriae	X					
Lobochona prorates	X					
Mirofolliculina limnoriae	X	X				
Sphenophyra dosiniae	X					
Haplosporidia						
Haplosporidium nelsoni					X	
Porifera						
Aplysilla rosea						X
Cliona sp.	X	X	X	X		
Corticium candelabrum						X
Dysidea avara						X
Dysidea fragilis						X
Halichondria bowerbankii	X	X				
Haliclona heterofibrosa						X
Haliclona loosanoffi	X	X				
Halisarca dujardini						X
Microciona prolifera	X					
Prosuberites sp.	X					
Cnidaria						
Amphisbetia operculata						X
Antennella secundaria						X
Blackfordia virginica	X	X			X	
Bougainvillea muscus (ramosa)						X
Cladonema radiatum			X			
Cladonema uchidae	X					
Clava multicornis	X					
Clytia hemisphaerica						X
Clytia paulensis						X
Cordylophora capsia	X	X	X		X	
Corymorpha sp.	X					
Diadumena "cincta"	X					
Diadumene franciscana	X					

TABLE 1 (*Continued*)

Phylum	Site[a]					
Species	**SFB**	**CB**	**PS**	**PWS**	**ChB**	**PPB**
Diadumene leucolena	X	X				
Diadumene lineata	X	X	X		X	
Ectopleura crocea	X	X				X
Filellum serpens						X
Garveia franciscana	X				X	
Halecium delicatulum						X
Maeotias inexspectata	X				X	
Moerisia lyonsi					X	
Moerisia sp.	X					
Monotheca obliqua						X
Obelia dichotoma (australis)						X
Phialella quadrata						X
Plumularia setacea						X
Sarsia eximia						X
Turritopsis nutricula						X
Platyhelminthes						
Pseudostylochus ostreophagus			X			
Kamptozoa						
Barentsia benedeni	X	X	X		X	
Loxosomatoides laevis					X	
Urnatella gracilis	X					
Nematoda						
Anguillicola crassus					X	
Bryozoa						
Aetea anguina						X
Alcyonidium sp.	X	X				
Amathia distans						X
Anguinella palmata	X					
Bowerbankia gracilis	X	X	X			
Bowerbankia spp.						X
Bugula "neritina"	X	X				X
Bugula calathus						X
Bugula flabellata						X
Bugula simplex						X
Bugula sp. 1			X			
Bugula sp. 2			X			
Bugula stolonifera	X		X			X
Celleporella hyalina						X
Conopeum reticulum						X
Conopeum tenuissimum	X	X				

(*Continued*)

TABLE 1 (*Continued*)

Phylum	Site[a]					
Species	**SFB**	**CB**	**PS**	**PWS**	**ChB**	**PPB**
Cryptosula pallasiana	X	X	X			X
Electra pilosa						X
Fenestrulina malusii						X
Membranipora membranacea						X
Microporella ciliata						X
Schizoporella unicornis	X	X	X	X		X
Scruparia ambigua						X
Scrupocellaria bertholetti						X
Scrupocellaria scrupea						X
Scrupocellaria scruposa						X
Tricellaria occidentalis						X
Victorella pavida	X					
Watersipora "subtorquata"	X	X				X
Watersipora arcuata						X
Zoobotryon verticillatum	X					
Mollusca						
Aplysiopsis formosa						X
Batillaria attramentaria			X			
Bithynia tentaculata					X	
Boonea bisuturalis	X					
Busycotypus canaliculatus	X					
Catriona rickettsi	X					
Cecina manchurica			X			
Cipangopaludina chinensis	X				X	
Corbicula fluminea	X	X			X	
Corbula gibba						X
Crassostrea virginica			X			
Crepidula convexa	X					
Crepidula fornicata			X			
Crepidula plana	X		X			
Cuthona perca	X				X	
Cyrenoida floridana					X	
Eubranchus misakiensis	X					
Gemma gemma	X					
Geukensia demissa	X					
Ilyanassa obsoleta	X		X			
Janolus hyalinus						X
Littorina littorea					X	
Littornia saxatilis	X					
Lyrodus pedicellatus	X					
Macoma petalum	X					
Melanoides tuberculata	X					

TABLE 1 (*Continued*)

Phylum			**Site**[a]			
Species	**SFB**	**CB**	**PS**	**PWS**	**ChB**	**PPB**
Mercenaria mercenaria			X			
Musculista senhousia	X		X			X
Mya arenaria	X	X	X	X		
Myosotella myosotis	X	X	X		X	
Mytilus galloprovinicialis	X	X	X			
Nassarius fraterculus			X			
Nuttallia obscurata			X			
Ocinebrellus inornatus			X			
Okenia plana	X					
Petricolaria pholadiformis	X					
Philine auriformis	X	X				
Potomocorbula amurensis	X					
Raeta pulchella						X
Rangia cuneata					X	
Rapana venosa					X	
Sukuraeolis enosimesis	X					
Stramonita haemostoma					X	
Tenellia adspersa	X	X				
Teredo navalis	X	X			X	
Theora lubrica (fragilis)	X					X
Urosalpinx cinerea	X		X			
Venerupis (Ruditapes) philippinarum	X	X	X			
Viviparus georgianus					X	
Annelida						
Boccardia proboscidea						X
Boccardiella ligerica	X					
Branchiura sowerbyi	X					X
Demonax leucaspis						X
Eteone sp.		X				
Euchone limnicola						X
Ficopomatus engimaticus	X				X	X
Heteromastus filiformis	X	X		X		
Hydroides elegans						X
Limnodrilus monothecus	X					
Manayunkia speciosa	X					
Marenzellaria viridis	X					
Marphysa "sanguinea"	X					
Myxicola infundibulum						X
Nereis succinea	X	X	X			X
Paranais frici	X					
Potamilla sp.	X					
Potamothrix bavaricus	X					

(*Continued*)

TABLE 1 (*Continued*)

Phylum	**Site[a]**					
Species	**SFB**	**CB**	**PS**	**PWS**	**ChB**	**PPB**
Pseudopolydora paucibranchiata						X
Sabaco elongatus	X					
Sabella spallanzanii						X
Streblospio benedicti	X	X	X			
Tubificoides apectinatus	X					
Tubificoides brownae	X	X				
Tubificoides diazi		X	X			
Tubificoides wasselli	X					
Variachaetadrilus angustipenis	X					
Crustacea						
Acanthomysis aspera	X					
Acanthomysis bowmani	X					
Acartiella sinensis	X					
Ampelisca abdita	X					
Ampithoe valida	X	X	X			
Argulus japonicus					X	
Balanus amphitrite	X				X	X
Balanus improvisus	X	X				
Caprella mutica	X	X	X			
Carcinus maenas	X	X			X	X
Chelura terebrans	X					
Cirolana harfordi						X
Corophium acherusicum	X	X	X			X
Corophium alienense	X					
Corophium heteroceratum	X					
Corophium insidiosum	X	X	X			X
Corophium sextonae						X
Deltamysis holmquistae	X					
Dynoides dentisinus	X					
Eobrolgus spinosus		X	X			
Eochelidium sp.			X			
Epinebalia sp.	X					
Eriocheir sinensis	X					
Eurylana arcuata	X					
Eusarsiella zostericola	X					
Gammarus daiberi	X					
Gitanopsis sp.						X
Grandierella japonica	X	X	X			
Hemigrapsus sanguineus						X
Iais californica	X	X				
Ilyocryptus agilis						X
Jassa marmorata	X	X	X			X

TABLE 1 (*Continued*)

Phylum	**Site[a]**					
Species	**SFB**	**CB**	**PS**	**PWS**	**ChB**	**PPB**
Lernaea cyprinacea	X					
Leucothoe sp.	X					
Ligia exotica					X	
Limnoithona sinensis	X					
Limnoithona tetraspina	X					
Limnoria quadripunctata	X					
Limnoria tripunctata	X	X	X			
Loxothylacus panopaei					X	
Melita nitida	X	X	X			
Melita sp.	X					
Mytilicola orientalis	X		X			
Nippoleucon hinumensis	X	X	X			
Oithona davisae	X					
Orconectes virilis	X					X
Pacifastcus leniusculus	X					
Palaemon macrodactylus	X	X				
Paracerceis sculpta						X
Paradexamine sp.	X					
Paranthura sp.	X					
Parapleustes derzhavini	X	X	X			
Procambarus clarkii	X				X	
Pseudodiaptomus forbesi	X					
Pseudodiaptomus inopinus		X	X			
Pseudodiaptomus marinus	X		X			
Pyromaia tuberculata						X
Rhithropanopeus harrisii	X	X				
Sinelobus sp.	X	X				
Sinocalanus doerri	X					
Sphaeroma quoyanum	X	X				
Stenothoe valida	X					
Stephos pacificus			X			
Stephos sp.			X			
Synidotea laevidorsalis	X					
Tortanus dextrilobatus	X					
Transorchestia engimatica	X					
Hexapoda (Insecta)						
Anisolabis maritima	X				X	
Conchopus borealis	X					
Galerucella calmariensis					X	
Galerucella pusilla					X	
Hylobius transversovittatus					X	
Neochetina bruchi	X					
Neochetina eichornia	X					

(*Continued*)

TABLE 1 (*Continued*)

Phylum	Site[a]					
Species	SFB	CB	PS	PWS	ChB	PPB
Procancace diannae						X
Trignotylus uhleri	X					
Echinodermata						
Asterias amurensis						X
Chordata						
Ascidia sp.	X					
Ascidiella aspersa						X
Botrylloides leachi						X
Botrylloides sp.	X					
Botrylloides violaceus	X	X	X	X		
Botryllus schlosseri	X	X	X		X	X
Ciona intestinalis	X					X
Ciona savignyi	X		X			
Diplosoma listerianum		X				
Ecteinascidia turbinata					X	
Molgula manhattensis	X	X	X			X
Styela clava	X	X	X			X
Styela plicata						X
Totals	**157**	**55**	**57**	**10**	**49**	**91**

[a]Abbreviations and primary literature sources for each site are as follows:

San Francisco Bay	(SF; 30)
Coos Bay	(CB; JT Carlton 2000, unpublished checklist)
Puget Sound	(PS; JT Carlton & M Wonham, unpublished manuscript)
Prince William Sound	(PWS; 69)
Chesapeake Bay	(ChB; 117)
Port Philip Bay	(PPB; 67)

NIS per estuary ranges from 10 to 157, and the percent overlap of NIS between pairs of estuaries varies between 0% and 90%. The percent overlap was often asymmetrical between estuary pairs, especially where a disparity existed in total number of NIS. Estuaries with relatively few NIS had the greatest overlap with estuaries with a larger number of invaders. As discussed above, the total number of species shared among estuaries is much greater than it appears, when native and cryptogenic species are included.

Although it may not be surprising to find relatively high NIS overlap among West Coast estuaries from California to Washington, the degree of overlap among the more distant estuaries (e.g. across oceans or continents) is notable. The pairwise overlap ranges from 6% to 41% among San Francisco Bay, Chesapeake Bay, and Port Philip Bay (Australia), as shown in Table 2. Five species are known as invaders in all three estuaries, and twenty NIS are established in both San Francisco Bay and Chesapeake Bay (Appendix 1). An additional 43 established

TABLE 2 Overlap of established nonindigenous species reported among six different estuaries[a], shown as number and percent (parentheses)[b]

	SFB	CB	PS	PWS	ChB	PPB
SFB	157 (100)	51 (32.5)	37 (23.6)	6 (3.8)	20 (12.7)	20 (12.7)
CB	51 (89.5	57 (100)	31 (54.4)	5 (8.8)	9 (15.8)	13 (22.8)
PS	37 (64.9)	31 (54.4)	57 (100)	5 (8.8)	5 (8.8)	11 (19.3)
PWS	6 (60.0)	5 (50.0)	5 (50.0)	10 (100)	0 (0.0)	1 (10.0)
ChB	20 (40.8)	9 (18.4)	5 (10.2)	0 (0.0)	49 (100)	5 (10.2)
PPB	20 (22)	13 (14.3)	11 (12.1)	1 (1.1)	5 (5.5)	91 (100)

[a] See Table 1 for abbreviations.

[b] Diagonal (in black) indicates the number of nonindigenous species in each estuary. Numbers below diagonal (gray) indicate the overlap as the percentage of species at the row site that also occur at the column sites. Numbers above the diagonal (clear) indicate overlap as the percentage of species at the column site that also occur at the row site.

species introduced to San Francisco Bay occur as natives in Chesapeake Bay, and at least 6 other species (*Boccardiella ligerica, Bowerbankia gracilis, Halichondria bowerbankia, Molgula manhattensis*, and *Tenellia adspersa*) are cryptogenic in Chesapeake Bay and introduced to San Francisco Bay; both of these groups are not included in our estimates for Table 2 (which requires the species be introduced in both systems). However, there are only two cases of species cryptogenic or native on the west coast, including San Francisco Bay, and introduced into Chesapeake Bay (the diatoms *Coscinodiscus wailesii* and *Thalassiosira punctigera*).

The data for West Coast estuaries show an intriguing latitudinal pattern, with number of NIS increasing significantly from north to south (for Prince William Sound, Puget Sound, Coos Bay, and San Francisco Bay: $y = -5.51x + 332.7$, $r^2 = 0.762$, $P < 0.01$, where degrees latitude is the independent variable; see Tables 1 and 2). Fewer invasions are, however, known from southern California ($n = 35$) than San Francisco Bay ($n = 157$) (35) (Table 2). Importantly, intensive analysis of NIS invasions is not yet available for any bay in Southern California, similar to that for other estuaries in Table 1.

Recent analysis of Pearl Harbor, Hawaii, indicates that a relatively high number of that NIS are established in this low latitude embayment. Although not yet complete for direct comparison in our analyses, Carlton and Eldridge (20) report 156 NIS of invertebrates and algae are established at this site, increasing an earlier initial estimate (32). This result appears consistent with the latitudinal pattern above. However, island sites may differ from continental ones in many respects (83, 91) that could confound interpretation of latitudinal effects. Analysis of additional island sites, both tropical and temperate, is necessary to partition the relative effects of latitude versus land area (i.e. mainland versus island sites) on invasion number.

The five focal estuaries (Table 1) display both similarities and differences with respect to invasion patterns, as described in recent analyses (30, 31, 67, 117; Carlton & Wonham 2000, unpublished manuscript). Most display a strong increase in the reported rate of invasions. The NIS among sites are dominated by crustaceans and molluscs, and most invasions are attributed to shipping or the combination of shipping and fisheries. Native and source regions of the NIS are highly variable, particularly across continents or oceans. Although some sites do not exhibit a large range in salinity (e.g. Port Philip Bay, Puget Sound), the other estuaries exhibit contrasting salinity patterns of invasion. For example, the total number of NIS increases three- to fourfold from low to high salinity in San Francisco Bay, peaking in the polyhaline zone; whereas in Chesapeake Bay, the total number varies much less among zones (up to twofold) and peaks in the mesohaline zone. In contrast, the number of NIS in the fouling community increases with salinity in Chesapeake Bay but decreases with salinity in Coos Bay (46, 117; GM Ruiz, AH Hines, LD McCann & JA Crooks, unpublished data).

Most data about the extent and patterns of marine invasion in North America and elsewhere come from protected bays and estuaries, and we have presented data from a small number of sites. Although data on marine invasion patterns exist for other global regions, (e.g. 2, 33, 51, 59, 67, 73, 94, 102, 103, 105, 109, 150; G Pauley, pers. comm.), these are not directly comparable to the sites evaluated in North America, differing substantially in area (e.g. entire seas), habitat type, or intensity of analysis. Our analyses therefore represent only a subset of existing latitudes, habitat types, as well as continents. A clear next step is to test the robustness or generality of emerging patterns across these additional scales.

UNDERSTANDING PATTERNS OF INVASION

Hypotheses that could explain the observed patterns of invasion may be grouped into three general categories related to: (*a*) variation in propagule supply characteristics; (*b*) variation among recipient regions in susceptibility or resistance to invasion; or (*c*) bias in the quantity or quality of existing data. These hypotheses are not mutually exclusive and have been advanced in various forms and combinations to account for invasion patterns by space, time, taxonomic group, habitat type, and donor region (34, 50, 81, 82, 84, 101, 116, 121, 129, 143). Below, we review these hypotheses in more detail and evaluate existing support for them in marine communities, discussing ways in which these hypotheses may operate to generate observed patterns.

Another theme in invasion ecology has been the invasion potential or capacity of a species to invade (5, 39, 43, 52, 66, 93, 101, 107, 120, 121, 142). This theme focuses on the applicant pool of species, examining a range of questions, such as which species will invade, what makes a good invader, and which attributes of species contribute to differential invasion success? A general view is that life history attributes such as the intrinsic rate of population increase, *r*, play a significant role in determining which species succeed, although empirical data in support of this hypothesis are often confounded by other variables (52, 120). Geographic range, which integrates environmental tolerance and a variety of population characteristics, has also emerged as a predictor of invasion success and has some empirical support (143). In this review, variation in invasion potential has been considered as a component of the above three hypotheses, because our focus is on patterns of NIS richness, not individual species characteristics. Thus, for our purposes, we consider invasion potential as a component of propagule supply (as described below under Different Donor Regions's, and Condition of Propagules), which may modify the relationship between supply and invasion success.

Progagule Supply Hypotheses

Supply hypotheses propose that variation in propagule delivery accounts for variation in invasion patterns. In its basic form, propagule supply is portrayed as the quantity of arriving propagules (propagule pressure), with the number of invasions increasing as a function of total propagule quantity (84, 119, 120, 143). Supply hypotheses by themselves predict that the same propagule pressure in space or time would result in approximately the same number of invasions. Thus, the relationship between propagule supply and invasions would be described by a single function across spatial and temporal scales (82, 143).

Propagule supply can be broken down into component parts that each may affect the invasion outcome, including the following:

Total Quantity (Propagule Pressure)

The quantity of propagules released may be correlated significantly to invasion success. Studies of propagule pressure have (*a*) correlated estimates of the rate of propagule arrival and invasion or (*b*) measured the success of intentional introductions as a function of propagule number (10, 84, 114, 119, 120, 143).

Inoculation Density, Frequency, and Duration

The spatial dispersion and tempo of supply may have important consequences. The same quantity of propagules can be distributed differently in space and time, affecting invasion success (112, 121, 142). In particular, inoculation characteristics that result in consistently low propagule densities may produce different success rates compared to those resulting in high local propagule densities (1, 100).

Different Donor Regions

The source of propagules may influence invasion success, due to a number of differences in the donor region. These include different species pools, differing genotypes of the same species, as well as differences in density and condition (i.e. "inoculation density" and "condition" as discussed elsewhere). Propagules from two donor regions may therefore differ in their physiological, life-history, and ecological characteristics that can modify capacity to invade the same recipient environment (37, 43, 101, 135). Discussions of such variation among donor regions are often accompanied by consideration of "environmental matching" or "biotic resistance" in the recipient region (see below).

Condition of Propagules

The physiological condition of propagules upon arrival to various recipient regions may vary, depending upon the individual organisms (e.g. life stage or age) and the history of transfer (e.g. vector, food, temperature, season, length of journey, etc). In turn, this may affect performance of propagules and their invasion success (69, 96, 97, 123, 147).

For marine and estuarine habitats, these aspects of propagule supply exhibit considerable spatial and temporal variation that may contribute to observed invasion patterns. The strongest evidence for supply hypotheses derives from increased propagule pressure and a corresponding increase in the rate of reported invasions. In general, the transfer rate of marine organisms is thought to have increased, especially during the twentieth century, due to changes in the size, speed, and operation of global shipping traffic (17, 116). Ship size is correlated positively to the volume of ballast water (within vessel class) and the surface area of hulls and seachests (24; GM Ruiz, AW Miller, B Steves, RA Everett & AH Hines, unpublished data). The increase in ship sizes could result in increasing propagule pressure. The increased speed of vessels over time may have further increased the number and condition of arriving propagules, especially in ballast water where survivorship of organisms during transit is often time-dependent (58, 77, 123, 147; LD Smith, GM Ruiz, AH Hines, BSS Galil & JT Carlton, unpublished data). With expanding global trade, both the number of arriving vessels and the number of source regions (i.e. last ports of call) have increased at many recipient ports. This combination of factors may increase the overall number of propagules arriving to ports over time as well as the diversity of species and genotypes involved. Furthermore, as invasions continue to accrue at the source ports, the diversity of exported propagules may further expand and promote a positive feedback of "stepping-stone" invasions (17, 69, 77).

Working against this presumptive increase in transfer rate, however, is the probable decline in ship fouling communities that characterized wooden vessels for many centuries, as well as steel and iron vessels up to the mid-twentieth century (12, 22). Several factors are thought to be involved in a decline, including the development of anti-fouling paints, lower port residencies (leading to reduced settlement of fouling organisms), and greater speeds at sea (leading to more species being washed away by shear forces, while also facilitating potential survivorship for those organisms that remain, as noted above). The balance among these various processes, operating to both enhance and depress transfer rates by ships over time, remains to be quantified.

A large portion of observed spatial variation in the origin (i.e. native and source regions) and vectors of NIS clearly reflects qualitative differences in propagule supply. This is most evident in comparison of West Coast and East Coast patterns (Appendix 1). Most marine invasions to the West Coast originate from the Indo-West Pacific (including Western Pacific) and Western Atlantic, having the same native and source regions. In contrast, most invasions to the East Coast originate from Eastern Atlantic, although many of these species are native to the Pacific. These patterns correspond directly to the dominant trade corridors for the respective coasts in historical time (12, 24, 30, 117; US Maritime Administration, unpublished data). Furthermore, although shipping was identified as the probable vector for most invasions along each coast of North America, many more invertebrate species arrived via fisheries to the West Coast compared to the East Coast (30, 117; Appendix 1). This difference in introductions via fisheries corresponds to a marked difference in supply of oysters between coasts: the West Coast received extensive shipments of oysters from the western Atlantic and the western Pacific, whereas oyster transfers to East Coast fisheries have been minor by comparison and primarily intracoastal (14, 117).

Any quantitative relationships between propagule supply and spatial invasion patterns are much less evident. For example, despite the disparity in number of NIS between the East and West coasts, or between San Francisco Bay and Chesapeake Bay (Figure 3, Table 2), there is no evidence that propagule supply has been greater for the West Coast compared to the East Coast, or for San Francisco Bay compared to Chesapeake Bay. Estimates of the present number of ship arrivals and the total volume of ships' ballast water arriving from overseas are negatively associated with apparent invasion patterns. For 1997–1999, more vessels arrived from overseas to the East Coast, as well as the Gulf Coast, than the West Coast (U.S. Maritine Administration, unpublished data; GM Ruiz, AW Miller, B Steves, RA Everett & AH Hines, unpublished data). Carlton et al (24) estimated that the largest volumes of ballast water discharged at selected ports from foreign arrivals in 1991 occurred on the Gulf Coast (New Orleans) and East Coast (Chesapeake Bay), whereas discharge volumes to West Coast ports (including San Francisco Bay) were relatively low; however, total volume estimates are not yet available for the entire coasts. Furthermore, the cumulative supply of exotic propagules over historic time (which is unknown) may also be greatest for the East Coast, reflecting temporal differences in the development of extensive European colonization and oceanic trade among North American coasts (GM Ruiz, JT Carlton, P Fofonoff & AH Hines, submitted).

Although it appears that total propagule supply in these pairwise comparisons may run counter to the number of known invasions, ships now arriving at the West Coast (relative to the East Coast) may include fewer donor ports (24; Ruiz et al., unpublished data). A relatively limited number of donor sites would perhaps result in repeated inoculations of the same species more frequently to the West Coast than to the East Coast, and this could increase invasion success (24, 112, 123). Furthermore, the donor ports for ships arriving to the West Coast are from different regions than those arriving to the East Coast (US Maritime Administration, unpublished data), and may also differ in the diversity and densities of arriving propagules. We do not know the extent to which the spatial and temporal patterns of these progagule supply characteristics differ among coasts.

The relationship between supply and taxonomic distribution of known marine invasions has received little consideration (however see 146). In general, we expect that the number and diversity of propagules released into marine environments by humans is inversely correlated with organism size, reflecting the general availability of organisms in their natural environment as well as the abundance of organisms measured in ballast water of ships (9, 56, 60). For example, it is not unusual to detect bacteria in the range of 10^7–10^8 cells per liter in ballast water, compared to densities of 10^2–10^3 crustaceans per liter (47, 69, 123; GM Ruiz, FC Dobbs, TK Rawlings, LA Drake, TH Mullady, A Huq & RR Colwell, unpublished data). However, the relative number of reported invasions for small organisms, and taxonomic groups dominated by small organisms, is perhaps not surprisingly counter to this expectation (as discussed in Data Bias Hypotheses section, below).

The relationship between supply and salinity distribution of invaders is also poorly resolved. It is clear that propagules have arrived frequently from high salinity zones of donor regions (14, 21, 30, 69, 123; GM Ruiz, LD Smith, AH Hines & JT Carlton, unpublished data). We speculate that most ballast water (and entrained propagules) arriving in North America from overseas is of relatively high salinity, as described for Chesapeake Bay (123) and Coos Bay (21), but this is not at all clear. Since species richness often increases with salinity in estuaries (7, 42), this may increase the species pool arriving in ballast of higher salinity. For marine invertebrates and algae, we hypothesize that both species richness and absolute number of human-transferred propagules have generally been greatest from high salinity zones of donor regions, corresponding to the pattern of invasion for North America. If true, however, this would not explain the observed difference in salinity distribution of NIS between the East Coast and West Coast, or between Chesapeake Bay and San Francisco Bay.

So far, we have limited our discussion of the supply-invasion relationship to species richness and patterns of delivery, but variation in invasion potential may exist among donor regions, taxonomic groups, and time periods that can modify this relationship. There is reason to believe that invasion potential differs among donor regions and taxonomic groups, corresponding to size, life history characteristics, and environmental requirements of organisms (43, 66, 100, 101, 135, 143; see also below). Furthermore, the condition of propagules at the donor region or during transfer may change over time and influence their capacity to invade, independent of invasion resistance among recipient regions (17). Although an intriguing possibility, we presently lack the data to critically evaluate variation in invasion potential or its importance to patterns of marine invasion.

We have identified some possible associations between propagule supply and invasion patterns, based primarily upon qualitative data. Some quantitative data exist on propagule supply in particular marine systems (21, 28, 56, 58, 61, 63, 64, 69, 76,

86, 99, 123, 127, 141, 146, 149). However, these data have not been collected in a standard fashion or compared directly to invasion patterns to test for specific associations. Comparative data on propagule supply among sites largely remain to be collected and should include measures of diversity, frequency, density, and condition of propagules (as above). Until such quantitative measures of propagule supply are available, we cannot adequately and formally test the various relationships between supply and invasion in marine systems (82, 143).

Invasion Resistance Hypotheses

Resistance hypotheses hold that invasion patterns result from variation in characteristics of recipient environments that prevent (or facilitate) survival and establishment of NIS. Lonsdale (82) has effectively illustrated this concept with a simple equation: $E = IS$. Here, the number of successful invasions at a site (E) is the product of number of exotic species that are introduced (I) and the survival rate of these species at this site (S); we assume that the term I controls for density, frequency, and tempo of introductions (as discussed above). Supply hypotheses assume that S is approximately constant, whereas resistance hypotheses do not. Instead, the latter predict that given the same supply characteristics (I), the resulting number of invasions (E) will differ among sites or times due to variation in survivorship (S). When controlling for supply, such differences in the number of successful invasions are considered to result from differences in invasion resistance (*i.e.*, susceptibility or invasibility). Invasion resistance is therefore a relative term, which arises from variation in S and can only be defined by measuring residuals from the relationship between I and S (82, 143). As a practical matter, survivorship is often equated with invasibility (the inverse of resistance), since survivorship is the response variable to resistance that is difficult to quantify. Unfortunately, equating survivorship with invasibility may obscure rigorous analysis of resistance as an independent attribute of the recipient system, which regulates the survivorship response.

As with propagule supply, resistance hypotheses can be divided among categories based upon factors causing variation in survivorship. Lonsdale (82) suggested survivorship (S) is a product of multiple survivorship functions: $S = S_1S_2S_3...S_n$, where S_{1-n} represents survivorship due to different attributes (1...n) of the recipient environment (e.g. environmental conditions, predators, pathogens, etc.). Thus, resistance can result from any one or a combination of attributes that affect survivorship of propagules differently among recipient environments.

For our discussion of resistance hypotheses in marine habitats, we divide resistance into two general components: abiotic and biotic. The abiotic component includes environmental conditions, such as habitat distribution and availability, that affect mortality; this is roughly equivalent to mortality resulting from maladaptation (82). For example, tropical species arriving in polar ecosystems may experience very poor survival compared to those arriving in temperate or tropical ecosystems. We consider this difference in survivorship and invasion success to result from differences in abiotic resistance. The biotic component includes variation in mortality due to biological interactions (e.g. competition, predation, disease, parasitism, etc), and differences in the strength of interactions among sites may result in differences in biotic resistance. The roles of biotic and abiotic resistance to invasion success have been explored using a variety of quantitative and theoretical approaches, providing strong support for both (4, 25, 101, 106, 111, 128–130, 133; see also 79, 110). In addition, an extensive literature since Elton (50) also indicates that disturbance can significantly affect invasion resistance (54, 70, 71, 100; but see also 126). We consider disturbance to either facilitate or inhibit invasion through changes in biological and environmental conditions, affecting biotic and abiotic resistance respectively.

Considering known differences in the biological and environmental attributes among marine systems throughout the world (8, 48, 134), variation in resistance to invasion is virtually certain. Furthermore, temporal variation in invasion resistance is an expected outcome of the broadscale changes and disturbance in recipient coastal ecosystems resulting from habitat alteration, freshwater diversion, eutrophication, fisheries exploitation, sedimentation, anoxia, chemical pollution, and invasion (25, 45, 54, 70, 101, 112, 118, 122). Many, if not most, considerations of disturbance suggest that invasion resistance should diminish with increasing magnitude and frequency of change over time (6, 62, 89, 90, 95).

Despite these predictions, tests of biotic or abiotic resistance to invasion are extremely rare in marine systems compared to terrestrial systems. Smith et al (123) showed that most propagules released in Chesapeake Bay with the ballast water of ships derive from high salinity environments and are faced with low salinity conditions upon release, suggesting that abiotic resistance may be relatively high and limit many potential invasions. In addition, an experimental study has shown that species richness had a significant effect on establishment and survivorship of exotic marine species, using pre-assembled fouling communities exposed to natural rates of recruitment at field sites (125). In this case, it appears that the outcome was mediated by resource (space) competition, as space became more limiting with increasing species richness. A similar outcome may result from disturbance events that reduce diversity and space occupation, allowing rare species to become established and persist (98, 124; but see also 79). Despite some support for effects of species richness on invasibility in freshwater and terrestrial systems, the interaction between species richness and invasibility has been variable among communities, suggesting that the interactions are often complex and any generalizations are premature (78, 79, 82, 110, 145; see also 80 for review).

Climatic differences between the respective coasts may also contribute to differences in abiotic resistance (8, 134). Although there is a growing literature about use of environmental matching to predict invasibility, or abiotic resistance to invasion (59, 68, 143), we urge caution. To date, there are no clear demonstrations of this approach as a predictive tool in marine systems. For example, it is tempting to suggest that the paucity of NIS known from Prince William Sound, Alaska, is in large part a result of environmental resistance, which may be intrinsic to high latitude ecosystems. We surmise that relative propagule pressure to this region historically has been low. Propagule supply to Prince William Sound increased markedly in the latter half of the twentieth century, as oil tankers have delivered annually since 1977 an estimated 20 million metric tons of ballast water to the area (69). Most of this ballast water and associated plankton originates from western U.S. ports, including San Francisco Bay, that are invaded by NIS. Recent field surveys in Prince William Sound have failed to detect many new invasions (69). Although the low number of NIS is consistent with an environmental resistance hypothesis, there may be significant time lags in detection of recent invasions. More fundamentally, we lack comparisons of invasion success among sites that control for propagule supply, providing the necessary reference point(s) to estimate resistance.

Despite the current lack of data to evaluate invasion resistance, we suggest that variation in invasion success among sites is probably the rule rather than the exception. In our view, the question is not whether biotic or abiotic resistance exists but how much variation exists in space and time. Furthermore, given the plethora of missing data and potential confounding factors in analysis of correlative field data (82; see also above), we advocate an experimental and theoretical approach to explore both the variation in invasion resistance and its role in observed invasion patterns.

Data Bias Hypotheses

Many potential biases exist in the observed patterns of marine invasions. Potential biases may result from three fundamental aspects of the present data:

1. The search effort for NIS is unevenly distributed spatially, temporally, and taxonomically. All of the data used in our analyses, and those of existing compilations for the well-studied estuaries, are derived primarily from literature-based syntheses (11, 20, 30, 117; JT Carlton 2000, unpublished checklist; Carlton & Wonham 2000, unpublished manuscript; but see 67). These data represent "by-catch" from a wide spectrum of research, surveys, and observations. The focus (habitats, salinity zones, taxonomic groups), resolution (taxonomic expertise and level of identification), and extent (sampling effort, areal coverage, number) of analyses are therefore uneven in space and time. Thus, data for analyses are not directly comparable (especially among sites, times, or taxonomic groups), as they result primarily from the accumulation of historical analyses that were conducted for a diverse variety of reasons.
2. The quality of systematic and biogeographic information is unevenly distributed taxonomically. There is an inherent bias in the quality of information available among taxonomic groups. Some organisms are relatively large and conspicuous, with hard parts that are preserved in the fossil record (e.g., molluscs, crustaceans). In general, the systematics and biogeography of these groups are well known relative to smaller organisms

(e.g. nematodes, annelids, dinoflagellates, and microorganisms) for which the historical surveys and paleontological records are much more limited and the systematics are often poorly resolved.

3. The quality of biogeographic information is unevenly distributed among sites and regions. The extent and timing of search efforts differ relative to the onset of intensive propagule pressure. For example, intensive shipping began in Chesapeake Bay approximately four centuries ago, predating surveys of many taxonomic groups by decades to centuries. In contrast, intensive propagule pressure in San Francisco Bay commenced about 1850, and major biological surveys of this region commenced within approximately 60 y (rather than the 300 y lag-time evident for many groups in the Chesapeake). The relative timing of these activities may have affected our ability to differentiate early invaders from native species (18, 53, 117).

Although we are confident about the information presented for the 298 NIS (Appendix 1), providing a minimum level of invasions on the coasts, the uneven quality of data may cause invasions to go undetected and thereby influence observed invasion patterns. It is for this reason that we have characterized our analysis of existing data as "apparent" patterns. Here, we evaluate further some biases that may exist for many of the patterns discussed above.

Spatial Patterns of Invasion

We hypothesize that significant bias exists in the apparent spatial patterns of invasion in our analyses. As suggested above, this may result primarily from two sources. First, the search effort among sites is uneven (#1 above). For example, the research effort for San Francisco Bay is undoubtedly greater than that for Prince William Sound, and effort for individual taxonomic groups also varies among sites. Second, and possibly more significant, the quality of historical baseline information on biotic communities is highly variable among sites (#3 above). We expect both the search effort and quality of baseline to affect the number of NIS detected within each site.

There are no standardized, quantitative measures across sites to evaluate (or control for) the effect of variable search effort on spatial invasion patterns. It is also difficult to compare or normalize prior search effort among sites, because the historical information results from a variety of studies and methods. We presently know of no approach to control for these missing data, short of conducting surveys.

On the level of coasts, we hypothesize that the relatively low number of invaders known for the Gulf coast results from bias in search effort: (*a*) there has been no "case study" of invasions for a Gulf Coast estuary, similar to those in San Francisco Bay or Chesapeake Bay, and (*b*) the extent of historical research on marine invertebrate communities is lowest for the Gulf coast. On the level of estuaries, similar potential for strong bias clearly exists, especially for sites like Prince William Sound compared to other estuaries. A recent survey of Prince William Sound resulted in 20 new species records for the region, including 3 NIS and many additional species that were native or cryptogenic (69). This suggests that the biota remains poorly described. Although the poor records of local biota may result in bias, this survey also did not detect many of the NIS that have been evident at other West Coast estuaries. In contrast, a similar set of surveys in Chesapeake Bay detected five new NIS for that region (117). We therefore hypothesize that the observed numerical differences in NIS among estuaries in Table 2 does not result from differences in search effort.

While it is possible to test for bias due to search effort by implementing standardized surveys, the issue of uneven historical baseline information among sites is more difficult to resolve. For each estuary and coast, there are cryptogenic species that may be either native or non-native (30, 69, 117; Carlton & Wonham 2000, unpublished manuscript). Some cryptogenic marine species are conspicuous, structurally and functionally, but the historical record is ambiguous about their distribution prior to ocean trade (Table 3). There is strong reason to believe that the extent of cryptogenic species is unevenly distributed among sites and coasts, corresponding to the extent of biotic surveys prior to the onset of intense propagule supply (18, 117). We therefore predict that the number of invasions that are cryptogenic is greater for the East and Gulf Coasts compared to the West Coast, and for Prince William Sound compared to other West Coast estuaries.

To explore the potential magnitude of cryptogenic species, Fofonoff et al (53) found that approximately 34% of 780 species from the Chesapeake Bay also occur in Europe. Although a few (5%) of these species are known invaders to the Chesapeake or Europe, the invasion status of most have never been evaluated, suggesting that 30% of 739 species should be considered cryptogenic at present. The first records for many of these taxa follow by decades to centuries the initiation of extensive commerce with Europe. Fofonoff et al surmise that many NIS may have arrived with early trade and are now included in this unevaluated group of species. We suggest that the extent of overlap between West Coast estuaries and the western Pacific may be lower, due in part to the timing of trade and surveys. The extent of cryptogenic species may also be lower along the West than East Coast. It would be instructive to quantify and test for such asymmetry among coasts with comparable data sets.

To further explore this possible bias in the observed pattern of invasion, we examined spatial patterns of invasion for molluscs, which are relatively conspicuous and well studied, have a fossil record, and are presumably less prone to be missed as invaders. Despite our predictions about bias due to cryptogenic species, the distribution of invasions for molluscs shows the same general spatial patterns that we reported across all taxonomic groups (15). Specifically, the data indicate that the largest number of mollusc invasions are known from the West Coast (47 NIS) compared to the East and Gulf Coasts (28 and 8 NIS, respectively; Appendix 1).

TABLE 3 Examples of cryptogenic marine and estuarine species present in North America[a]

Species[b]	Phylum	Present World Distribution	Date of 1st North American Record	Distribution[c]
Pathogens & Parasites				
Vibrio cholerae	Omnibacteria	Cosmopolitan	Unknown	East, West, Gulf
Labyrinthula zosterae	Labyrinthulamycota	N Atlantic, NW Pacific	1930s	East
Perkinsus marinus	Apicomplexa	NW Atlantic	1920s	East[1]
Minchinia teredinis	Haplosporida	NW Atlantic	1976	East
Phytoplankton				
Pseudo-nitzschia australis	Bacillariophyta	SW Pacific, NE Pacific	1930s	West
Gyrodinium "aureolum"	Dinophyta	Cosmopolitan	1957	East
Pfiesteria piscicida	" "	NW Atlantic	1991	East, Gulf
Heterosigma akashiwo	Raphidophyta	Pacific, Atlantic	1950s	East, West
Fibrocapsa japonica	Raphidophyta	W Pacific, N Atlantic	1980s	East
Aureococcus anophagefferens	Chrysophyta	NW Atlantic	1985	East
Macroalgae				
Enteromorpha "intestinalis"	Chlorophyta	Cosmopolitan	1858	East, Gulf, West
Ulva "lactuca"	" "	Cosmopolitan	1858	East, Gulf, West
Cladophora spp.	" "	Cosmopolitan	1858	East, Gulf, West
Myriocladia loveni	Phaeophyta	N Atlantic	Unknown	East

Zooplankton				
Tintinnopsis "corniger"	Ciliophora	NW Pacific, Gulf of Mexico	1968	Gulf, West[2]
Eurytemora "affinis"	Crustacea	Circumboreal	1906	East, Gulf, West
Americamysis almyra	" "	Gulf of Mexico, NW Atlantic	1977	Gulf, East[3]
Benthic Invertebrates				
Ectopleura dumortieri	Cnidaria	NE Atlantic, NW Atlantic	1862	East
Obelia spp.	" "	Cosmopolitan	1857	East, West
Protohydra leuckarti	" "	Cosmopolitan	1939	East, West
Nematostella vectensis	" "	N Atlantic, NE Pacific	1939	East, Gulf, West
Limnodrilus hoffmeisteri	Annelida	Cosmopolitan	Unknown	East, West
Capitella "capitata"	" "	Cosmopolitan	Unknown	East, Gulf, West
Harmothoe "imbricata"	" "	Circumboreal	1881	East, West
Namanereis "littoralis"	" "	Cosmopolitan	1942	East, West
Polydora "cornuta"	" "	Cosmopolitan	1820	East, Gulf, West

(*Continued*)

TABLE 3 (*Continued*)

Species[b]	Phylum	Present World Distribution	Date of 1st North American Record	Distribution[c]
Pygospio "elegans"	" "	Circumpolar	Unknown	East, West
Spiophanes "bombyx"	" "	Cosmopolitan	1881	East, West
Siphonaria pectinata	Mollusca	N Atlantic, S Atlantic	1841	East, Gulf
Alderia modesta	" "	Cosmopolitan	1848	East, Gulf, West
Teredo clappi	" "	Cosmopolitan	1923	East, Gulf
Leptochelia "dubia"	Crustacea	Cosmopolitan	1901	East, Gulf, West
Limnoria lignorum	" "	Circumboreal	1841	East, West
Caprella "equilbria"	" "	Cosmopolitan	1818	East, Gulf, West
Caprella "pennantis"	" "	Cosmopolitan	1818	East, Gulf, West
Platorchestia platensis	" "	Cosmopolitan	1873	East, Gulf, West
Aetea anguina	Bryozoa	Cosmopolitan	1891	East, West
Alcyonidium parasiticum	Bryozoa	N Atlantic, N Pacific	1873	East, West
Molgula manhattensis	Chordata	Cosmopolitan	1843	East, West[4]

[a] References for each species published in reference 117.

[b] Quotation marks indicate possible species complexes.

[c] North American Distribution.

[1] Cryptogenic in Chesapeake Bay and southward, but introduced to Delaware-Maine.

[2] Collected from balast water only on the West Coast.

[3] We consider this species native on the Gulf Coast, where it was described from specimens collected in 1953, but cryptogenic on the Atlantic Coast (Florida-Maryland), where it was first reported in 1977.

[4] Cryptogenic on the Atlantic Coast, but introduced on West Coast.

Temporal Patterns of Invasion

The number of established marine NIS is increasing over time, but the actual rate of invasions warrants further scrutiny and discussion. Based upon current data, the rate of invasion appears to have increased over the past 200 y for North America as well as for multiple estuaries (31, 67, 117). A similar pattern has also been observed for freshwater and terrestrial ecosystems, across many taxonomic groups (55, 87, 88; GM Ruiz & JT Carlton, submitted). Although the signal is consistent and appears robust, two sources of bias may contribute to this temporal pattern. First, many early invasions may simply be undetected and are considered cryptogenic species. Second, the search effort has increased over time, due to (*a*) to increased level of research activity and publications, both generally and specifically on invasions, (*b*) increased public interest and search effort, and (*c*) improved tools for systematic analyses (e.g. molecular techniques).

These three sources of bias would serve to increase the apparent rate of invasions, but their relative importance has not been explored (115). It would be useful to standardize rate of past detection against search effort, since number of species detected clearly increases with both temporal and spatial components of effort (113). However, this is not possible because the effort is so uneven among the various information sources and is poorly documented (115; but see 31). Furthermore, with the advent of new molecular approaches, we are now detecting invasions that previously could not be discerned (3, 57, 115). For example, the recent discovery of two bivalves, *Macoma petulum* and *Mytilus galloprovincialis*, along the West Coast reflects use of molecular tools to identify NIS that clearly arrived many decades earlier (Appendix 1). Although few marine NIS in North America have been identified with such molecular techniques, this underscores the general issue of increasing search effort.

If we examine the temporal pattern for molluscs, following the rationale above, the rate of reported marine NIS known for North America increased significantly over time (Figure 5; $y = 0.1417x - 1.6$, $r^2 = 0.956$; where y is number of new invasions and x is time in 30 year intervals, indicated as the first year of each interval). However, the increase is linear in contrast to the exponential function observed for all taxonomic groups combined (Figure 5). Although we suggest that molluscs provide a good proxy measure to remove temporal bias, the extent to which they represent other taxonomic groups with respect to invasion rates has not been tested.

We hypothesize that the rate of marine invasions is increasing over time, driven by the combination of increasing propagule supply and decreasing invasion resistance. However, we also predict that the apparent rates of invasion are inflated, due to the prevalence of undetected early invasions and increasing search effort over time. Lag times in population increase of invaders may also serve to inflate the apparent rate of invasion, as detection probably depends upon both density and search effort (36, 65, 113). This could be especially important if the population dynamics of invaders has changed over time, perhaps in response to anthropogenic

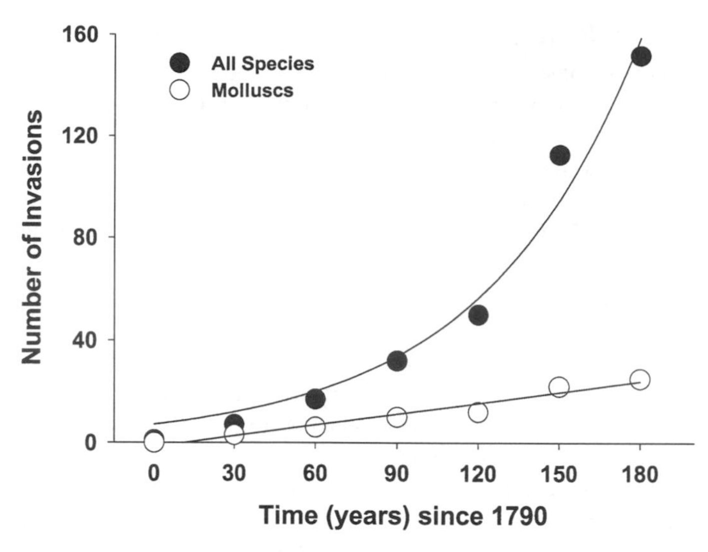

Figure 5 Rate of reported marine invasions of North America since 1790 for (a) molluscs and (b) all species of invertebrates and algae. *Open circles* indicate the number of new invasions for molluscs; *filled circles* indicate those for all taxa. Data are plotted for 30 y intervals as in Figure 1.

changes to estuarine habitats. These hypotheses remain to be tested and are best examined with temporally replicated, standardized measures that we now lack.

Taxonomic Patterns of Invasion

We hypothesize that strong bias also exists for the taxonomic distribution of invasions, and that the prevalence of small invading organisms is grossly underrepresented by current measures. In general, the quality of systematic and biogeographic information diminishes with organism size, and the available baseline information for small organisms and microorganisms is poor relative to large invertebrates (49). With the exception of a few dinoflagellates (61), there is a conspicuous lack of surveys or baseline studies to evaluate the extent of marine microorganism invasions. If invasions were occurring, even at a high rate, how would we know? Furthermore, invasion biology in these groups is even more complicated than for other taxa, due to the occurrence of gene transfer that is reported in the field (74). At present, invasion by microorganisms, including parasites and pathogens that cause disease (62), is a fundamental gap in our understanding of patterns and consequences of marine invasions.

Salinity and Habitat Patterns of Invasion

Sites vary significantly in the relative size of salinity zones and habitats that may strongly influence some of the spatial, temporal, taxonomic, and salinity patterns observed. Perhaps most striking is the extent of variable freshwater and oligohaline zones in estuaries. As noted previously, several bays have only small areas of low salinity, restricting the opportunity for colonization by species of freshwater origin. This is especially so for Puget Sound and Port Philip Bay in our analyses. Consequently, the search effort for low salinity taxa is uneven among bays. If we remove the freshwater and oligohaline species from our analysis (Appendix 1 and Table 1), many of the same patterns are evident: (*a*) more NIS are known for the West Coast (217) compared to the East Coast (95) and Gulf Coast (30), and (*b*) more NIS are known for San Francisco Bay compared to other bays. However, attempting to standardize for salinity generally reduces the observed spatial variation in the extent of invasions.

It would be instructive to standardize all data for salinity, habitat, and area. Each of these attributes should have a strong influence on both the taxonomic distribution and number, and perhaps rate, of successful invasions. Effects of habitat type and area on species richness have been well documented (84, 113). We surmise that such species-area relationships cannot explain the overall patterns observed (*e.g.*, East Coast and West Coast are roughly similar in size, Chesapeake Bay and Prince William Sound are larger than San Francisco Bay), but it would be informative to examine invasion patterns with a multivariate approach to control for size, habitat, and salinity characteristics. This approach should be especially valuable as data become available from a broader array of sites, increasing overall sample size and statistical power.

Although it is possible to control for the size and habitat/salinity composition of sites, the uneven nature of existing data remains a problem due to the difficulty in assessing variation in search effort. As a first step toward standardizing the present data, it is possible to document the distribution and types of existing information sources (by habitat, salinity zone, area, and taxonomic group), which can serve to identify studies with the most similar methods as well as to identify conspicuous gaps. It may even be possible to find similar studies for comparisons among bays. Although such approaches can provide useful insights, interpreting the data gathered from non-standardized methods is often an insurmountable problem, which is inherent to studies that were not designed for direct comparison.

CONCLUSIONS AND FUTURE DIRECTIONS

In the past decade, the ecology of marine invasions has come into focus, as knowledge of extent and history of NIS invasions has expanded rapidly for coastal ecosystems of North America and elsewhere. The existing literature for North American coasts shows that scores to hundreds of exotic species have invaded each coast. We have summarized for the first time many of the apparent patterns that emerge from a comprehensive analysis of these North American data. Some patterns are clear, particularly that shipping has been the most important vector of introduction historically and at present, although an array of other vectors have also operated to introduce many marine and estuarine species into North American waters. Limitations of the present data confound interpretation of other patterns, especially spatial variation in both the extent of invasion and taxonomic distribution of NIS. Geographic variation in the number of invasions is particularly difficult to sort out. Whereas San Francisco Bay clearly is more invaded than Prince William Sound and many other West Coast sites, the greater number of NIS reported in San Francisco Bay than Chesapeake Bay may be an artifact of the Chesapeake's longer history of active vectors (and relative paucity of early biotic surveys coincident with the onset of European trade and colonization), resulting in a greater number of unrecognized invasions. The observed exponential rate of increase for invasions may also be an artifact of data biases. Although we remain uncertain about the actual rate, we are more confident that a striking increase is not artifact, because it holds for subsets of better quality data, including (a) molluscs, a large and well-studied taxon and (b) San Francisco Bay, a site at which early biological studies were nearly contemporaneous with the operation of major vectors.

Despite the value of the existing data for marine invasions of North America, as elsewhere, it is important to recognize their limitations. Most data are derived from literature-based analyses, rather than direct field surveys designed to detect NIS. In fact, contemporary field surveys are lacking for many taxonomic groups and habitats at each site, where the most recent assessments may be decades old, and the quantity and quality of data on marine invasions is highly variable among sites. As a result, this "by-catch" approach to data collection has resulted in inherent biases that confound the interpretation of invasion patterns and processes.

Thus, modifying Lonsdale (82), we consider that the observed patterns of marine invasions reflect the interactive effects of propagule supply (PS), invasion resistance (R) of the recipient system, and bias (B) of the data, such that

$$I = \sum_{i=1}^{n} (PS_i)\,(R_i)\,(B_i)$$

where I is the number of established NIS summed across species from $i = 1$ to n at a location and time. Each component may vary spatially and temporally within a single bay. Rigorous interpretation of invasion patterns is confounded in most ecosystems, because fluctuations of the independent variables of PS and R are not controlled, and because B creates so many interaction terms (PS*B, R*B, PS*R*B) that the main effects are obscured. Clearly, priorities for invasion research include the collection of quantitative and experimental data that allow controlled analysis of the independent variables and elimination of data bias.

Testing many of the invasion hypotheses about propagule supply and invasion resistance requires standardized, quantitative measures of community composition in space and time (75, 115). At present, no program or framework exists to implement such quantitative measures of NIS invasions in North America. We therefore underscore the need to establish standardized ecological surveys of NIS across major regions of the North American shoreline and elsewhere (115; see also 67 for discussion of such a program). Furthermore, we advocate repeated quantitative measures at multiple sites, as well as across taxa and habitats, to avoid conclusions based upon a single site (taxa or habitat) that may not be broadly representative. Indeed, it is measures of variation in space and time that are necessary to test hypotheses about supply and resistance. Although a proposition for long-term temporal data, only this approach can provide data free of many potential biases that presently confound our interpretation of invasion patterns and processes.

Spatial and temporal measures of propagule supply are also fundamental to understanding of invasion mechanisms. To a large extent, we presently lack standardized measures of propagule supply. Despite existing measures of propagule supply, the data derive from a diverse variety of methods and often only focus on a single vector for a short (1–2 y) time period. As with ecological surveys, we must establish quantitative measures of propagule supply (or vector strength) that are collected in a standard fashion in space and time, if we are to test many key hypotheses about supply and invasion resistance.

Multiple approaches are clearly required to understand the patterns and underlying processes of marine invasions. Standardized measures of variation in vector strength and invasion success (as above) will allow us describe extant patterns and test invasion mechanisms, particularly at large spatial and temporal scales. Although this approach is necessary to understand invasions, it is not sufficient. For example, it will be difficult to control for all of the differences in propagule supply characteristics to different sites (e.g., source region, taxonomic composition, density and tempo of delivery, and capacity of different organisms to invade; for further discussion see Propagule Supply Hypotheses). However, manipulative laboratory and field experiments can provide an effective and unambiguous approach to address such a complex suite of variables at smaller scales, and show a great deal of promise for invasion ecology (70, 79, 125, 129). Thus, a combined strategy of mensurative and experimental approaches to invasion ecology is both complementary and most desirable, given their respective strengths and weaknesses, particularly in addressing issues of scale and complexity (44, 144).

Using these approaches, measures of invasion resistance are urgently needed. Resistance, the independent variable that defines propagule survivorship (the dependent variable), remains vague and poorly measured, especially in marine communities. When propagule supply is known, resistance can be estimated as a relative trait by comparing survivorship (or number of invasions) among locations, habitats, or time periods (82, 144). Comparing estimates of propagule supply and the resulting invasions can be used both to estimate resistance and to test for correlation to particular environmental or biological attributes of the recipient community. Thus, we advocate use of these estimates to advance and test predictions about specific attributes that underlie resistance and mediate the patterns of invasion.

Finally, we emphasize the dual value of these approaches, providing information that is key to both basic and applied areas of invasion ecology. Our emphasis throughout has been on the fundamental science, and we have argued for rigorous, quantitative and experimental data to test relationships between propagule supply and invasion patterns. However, there are many management and policy initiatives now underway at regional, national and international levels of government to reduce the risk and impacts of coastal invasions (23, 92, 131, 139). The success of fundamental science to guide and evaluate invasion management actions also depends on accurate measures of the basic relationship between propagule supply and invasions.

ACKNOWLEDGMENTS

We thank Brian Steves, Kelly Lion, and Marco Sigala for their assistance and insights in preparation of the manuscript. We also thank many of our colleagues for valuable discussions and comments on various aspects of invasion patterns: John Chapman, Jeff Cordell, Jeff Crooks, Richard Everett, Jon Geller, Ted Grosholz, Chad Hewitt, Whitman Miller, and David Smith. We also wish to acknowledge support for our invasion research from the Maryland and Connecticut Sea Grant Programs, National Sea Grant Program, NOAA, Pew Fellowship in the Environment and Conservation (The Pew Foundation and Pew Charitable Trusts), Smithsonian Institution, U.S. Department of Defense, and U.S. Fish and Wildlife Service.

LITERATURE CITED

1. Allee WC. 1931. *Animal Aggregations: A Study in General Sociology*. Chicago: Univ. Chicago Press. 431 pp.
2. Asakura A. 1992. Recent introductions of marine benthos into Tokyo Bay (review): process of invasion into an urban ecosystem with discussion on the factors inducing their successful introduction. *J. Nat. Hist. Mus. Inst., Chiba* 2(1):1–14
3. Bagley M, Geller JB. 2000. Sources for worldwide invasions of the European shore crab: Inferences from microsatellite DNA variation. *Proceedings of the First Marine Bioinvasions Conference*. In press.
4. Baltz DM, Moyle PB. 1993. Invasion resistance to introduced species by a native assemblage of California stream fishes. *Ecol. Appl.* 3(2):246–55
5. Barrett SCH, Richardson BJ. 1986. Genetic attributes of invading species. *In Ecology of Biological Invasions*, ed. RH Groves, JJ Burdon, pp. 21–33 Cambridge: Cambridge University Press
6. Baskin Y. 1998. Winners and losers in a changing world: global changes may promote invasions and alter the fate of invasive species. *BioScience* 48(10):788–92
7. Boesch DF. 1977. A new look at zonation of benthos along the estuarine gradient. In *Ecology of Marine Benthos*, ed. BC Coull, pp. 245–66. Columbia, SC: Univ. S. C. Press
8. Briggs JC. 1974. *Marine Zoogeography*. New York: McGraw-Hill
9. Brock TD, Madigan MT, Matinko JM, Parker J. 1993. *Biology of Microorganisms*. Englewood Cliffs, NJ: Prentice Hall
10. Bryan RT. 1999. Alien species and emerging infectious diseases: past lessons and future implications. In *Invasive Species and Biodiversity Management*, ed. OT

Sandlund, PJ Schei, Å Viken, pp. 163–75. Dordrecht, The Netherlands: Kluwer Academic

11. Carlton JT. 1979. *History, biogeography, and ecology of the introduced marine and estuarine invertebrates of the Pacific Coast of North America*. PhD thesis. Univ. Calif., Davis
12. Carlton JT. 1985. Transoceanic and interoceanic dispersal of coastal marine organisms: the biology of ballast water. *Oceanogr. Mar. Biol., Ann. Rev.* 23:313–71
13. Carlton JT. 1989. Man's role in changing the face of the ocean: biological invasions and implications for conservation of near-shore environments. *Conserv. Biol.* 3(3):265–73
14. Carlton JT. 1992. Dispersal of living organisms into aquatic ecosystems as mediated by aquaculture and fisheries activities. In *Dispersal of Living Organisms into Aquatic Ecosystems*, ed. A Rosenfield, R Mann, pp. 13–45. College Park, MD.: MD. Sea Grant
15. Carlton JT. 1993. Introduced marine and estuarine mollusks of North America: an end-of-the-20th-century perspective, part 2. *J. ShellFish Res.* 11:489–505
16. Carlton JT. 1996. Biological invasions and cryptogenic species. *Ecology* 77(6):1653–55
17. Carlton JT. 1996. Patterns, process, and prediction in marine invasion ecology. *Biol. Cons.* 78:97–106
18. Carlton JT. 1999. The scale and ecological consequences of biological invasions in the World's oceans. In *Invasive Species and Biodiversity Management*, ed. OT Sandlund, PJ Schei, Å Viken, pp. 195–212. Dordrecht, The Netherlands: Kluwer Academic
19. Carlton JT. 2000. Global change and biological invasions in the oceans. In *The Impact of Global Change on Invasive Species*, ed. HL Mooney, R Hobbs. Covelo, CA: Island Press
20. Carlton JT, Eldredge L. *Marine Bioinvasions of the Hawaiian Archipelago*. Honolulu, HI: Bernice P. Bishop Museum Press. In press
21. Carlton JT, Geller JB. 1993. Ecological roulette: the global transport of nonindigenous marine organisms. *Science* 261:78–82
22. Carlton JT, Hodder J. 1995. Biogeography and dispersal of coastal marine organisms: experimental studies of a replica of a 16th century sailing vessel. *Mar. Biol.* 121:721–30
23. Carlton JT, Kelly J. 1998. Foreword. In *Ballast Water: Ecological and Fisheries Implications*, 1–4, Int. Counc. Explor. Sea (ICES), Denmark
24. Carlton JT, Reid DM, van Leeuwen H. 1995. *The role of shipping in the introduction of nonindigenous aquatic organisms to the coastal waters of the United States (other than the Great Lakes) and an analysis of control options*, Report to U. S. Coast Guard, Washington D.C.
25. Case TJ. 1990. Invasion resistance arises in strongly interacting species-rich model competition communities. *Proc. Natl. Acad. Sci. USA* 87:9610–14
26. Chapman JW, Carlton JT. 1991. A test of the criteria for introduced species: the global invasion by the isopod *Synidotea laevidorsalis* (Miers, 1881). *J. Crustac. Biol.* 11(3):386–400
27. Chapman JW, Carlton JT. 1994. Predicted discoveries of the introduced isopod *Synidotea laevidorsalis*. *J. Crustac. Biol* 14(4):700–14
28. Chu KH, Tam PF, Fung CH, Chen QC. 1997. A biological survey of ballast water in container ships entering Hong Kong. *Hydrobiologia* 352:201–6
29. Cohen A, Mills C, Berry H, Wonham M, Bingham B, et al. 1998. *Puget Sound Expedition: A Rapid Assessment Survey of Non-Indigenous Species in the Shallow Waters of Puget Sound.* Olympia, WA: *Wash. State Dept. Nat. Resour.*

30. Cohen AN, Carlton JT. 1995. *Nonindigenous Species in a United States Estuary: a Case Study of the Biological Invasions of the San Francisco Bay and Delta*, U.S. Fish and Wildlife Service and National Sea Grant College Program (Connecticut Sea Grant)
31. Cohen AN, Carlton JT. 1998. Accelerating invasion rate in a highly invaded estuary. *Science* 279:555–58
32. Coles SL, DeFelice RC, Eldredge LG, Carlton JT. 1999. Historical and recent introductions of non-indigenous marine species into Pearl Harbor, Oahu, Hawaiian Islands. *Mar. Biol.* 135:147–58
33. Cranfield HJ, Gordon DP, Willan RC, Marshall BA, Battershill CN, et al. 1998. *Adventive Marine Species in New Zealand, 48 pp,* The National Institute of Water and Atmospheric Research, New Zealand
34. Crawley MJ. 1987. What makes a community invasible? In *Colonization, Succession and Stability*, ed. AJ Gray, MJ Crawley, PJ Edwards, pp. 424–53. Oxford: Blackwell Scientific Publications
35. Crooks JA. 1998. *Effects of the introduced mussel, Musculista senhousia, and other anthropogenic agents on benthic ecosystems of Mission Bay, San Diego*. PhD thesis. Univ. Calif., San Diego
36. Crooks JA, Soulé ME. 1999. Lag times in population explosions of invasive species: causes and implications. In *Invasive Species and Biodiversity Management*, ed. OT Sandlund, PJ Schei, Å Viken, pp. 103–25. The Netherlands: Kluwer Academic Publishers
37. Crosby AW. 1986. Weeds. *Ecological Imperialism: The Biological Expansion of Europe, 900–1900*, pp. 145–336. London: Cambridge University Press
38. D'Antonio CM, Vitousek PM. 1992. Biological invasions by exotic grasses, the grass/fire cycle, and global change. *Ann. Rev. Ecol. Syst.* 23:63–87
39. Daehler CC, Strong DR. 1997. Hybridization between introduced smooth cordgrass (*Spartina alterniflora*; Poaceae) and native California cordgrass (*S. foliosa*) in San Francisco Bay (California). *Am. J. Bot.* 84(5):607–11
40. Darwin C. 1854. *A Monograph on the Sub-Class Cirripedia*. London: The Bay Society
41. Darwin C. 1859. *The Origin of Species*. London: J. Murray
42. Deaton LE, Greenberg MJ. 1986. There is no horohalinicum. *Estuaries* 9:20–30
43. di Castri F. 1989. History of biological invasions with special emphasis on the old world. In *Biological Invasions: a Global Perspective*, ed. JA Drake, HA Mooney, F di Castri, RH Groves, FJ Kruger, M Rejmánek, M Williamson, pp. 1–30. Chichester, UK: John Wiley & Sons, Ltd.
44. Diamond J. 1986. Overview: laboratory experiments, field experiments, and natural experiments. In *Community Ecology*, ed. J Diamond, TJ Case, pp. 3–23. New York: Harper & Row
45. Dickerson JE Jr, Robinson JV. 1986. The controlled assembly of microcosmic communities: the selective extinction hypothesis. *Oecologia* 71:12–17
46. Drake JA, Huxel GR, Hewitt CL. 1996. Microcosms as models for generating and testing community theory. *Ecology* 77(3):670–77
47. Drake LA, Dobbs FC, Choi KH, Ruiz GM, McCann LD, Mullady TL. 1999. *Inventory of microbes in ballast water of ships arriving in Chesapeake Bay.* Presented at Natl. Conf. Mar. Bioinvasions, 1st, Mass. Inst. Technol., Cambridge, Mass.
48. Ekman S. 1953. *Zoogeography of the Sea*. London: Sidgwick & Jackson, Ltd.
49. Elbrächter M. 1999. Exotic flagellates of coastal North Sea waters. *Helgoländer Meeresunters.* 52:235–42
50. Elton CS. 1958. *The Ecology of Invasions by Animals and Plants*. London: Methuen & Co. Ltd
51. Eno NC. 1996. Non-native marine species

in British waters: effects and controls. *Aquat. Cons.: Mar. Freshwater Ecosyst.* 6:215–28

52. Erlich PR. 1986. Which animal will invade? In *Ecology of Biological Invasions of North America and Hawaii*, ed. HA Mooney, JA Drake, pp. 79–95. New York: Springer-Verlag
53. Fofonoff P, Ruiz GM, Hines AH, Carlton JT. 2000. *Assessing the importance of cryptogenic species in estuaries*. American Society of Limnology and Oceanography, Aquatic Sciences Meeting, Copenhagen; Abstract SS21-23
54. Fox MD, Fox BJ. 1986. The susceptibility of natural communities to invasion. In *Ecology of Biological Invasions*, ed. RH Groves, JJ Burdon, pp. 57–66. Cambridge: Cambridge University Press
55. Fuller PM, Nico LG, Williams JD. 1999. *Nonindigenous Fishes Introduced into Inland Waters of the United States*. Bethesda, MD: American Fisheries Society
56. Galil BS, Hulsmann N. 1997. Protist transport via ballast water—biological classification of ballast tanks by food web interactions. *Europ. J. Protis.* 33:244–53
57. Geller, JB. 1996. Molecular approaches to the study of marine biological invasions. In *Molecular Zoology: Advances, Strategies and Protocols*, ed. J Ferraris, S Palumbi, pp. 119–32. New York: Wiley-Liss
58. Gollasch S, Dammer M, Lenz J, Andres HG. 1998. Non-indigenous organisms introduced via ships into German waters. In *Ballast Water: Ecological and Fisheries Implications 50–64*, Int. Counc. Explor. Sea (ICES), Denmark
59. Gollasch S, Leppäkoski E. 1999. *Initial Risk Assessment of Alien Species in Nordic Coastal Waters*. Copenhagen, Denmark: Nordic Council of Ministers
60. Grimes DJ. 1991. Ecology of estuarine bacteria capable of causing human disease: a review. *Estuaries* 14(4):345–60
61. Hallegraeff GM. 1998. Transport of toxic dinoflagellates via ships' ballast water: bioeconomic risk assessment and efficacy of possible ballast water management strategies. *Mar. Ecol. Prog. Ser.* 168:297–309
62. Harvell CD, Kim K, Burkholder JM, Colwell RR, Epstein PR, et al. 1999. Emerging marine diseases—climate links and anthropogenic factors. *Science* 285:1505–10
63. Harvey M, Gilbert M, Gauthier D, Reid D. 1999. *A preliminary assessment of risks for the ballast water–mediated introduction of nonindigenous marine organisms in the estuary and Gulf of St. Lawrence*, Canadian Technical Report of Fisheries and Aquatic Sciences
64. Hay C, Handley S, Dodgshun T, Taylor M, Gibbs W. 1997. *Cawthron's Ballast Water Research Programme Final Report 1996-1997*, Cawthron Institute, Nelson, New Zealand
65. Hayek L-AC, Buzas MA. 1997. *Surveying Natural Populations*. New York: Columbia University Press
66. Hengeveld R. 1988. Mechanisms of biological invasions. *J. Biogeogr.* 15:819–28
67. Hewitt CL, Campbell ML, Thresher RE, Martin RB, ed. 1999. *Marine Biological Invasions of Port Phillip Bay, Victoria*, Technical Report No.20, Centre for Research on Introduced Marine Pests, Hobart
68. Hillman SP, ed. 1999. The ballast water problem–where to from here? *Proceedings of a Workshop Held 5-6 May 1999, Brisbane, Australia.* EcoPorts Monograph Series No. 19, Brisbane, Australia
69. Hines AH, Ruiz GM. 2000. *Biological invasions at cold-water coastal ecosystems: ballast- mediated introductions in Port Valdez/Prince William Sound*, Final Report to Regional Citizens Advisory Council of Prince William Sound
70. Hobbs RJ. 1989. The nature and effects of disturbance relative to invasions. In *Biological Invasions: a Global Perspective*, ed. JA Drake, HA Mooney, F Di Castri, RH Groves, FJ Kruger, M Rejmánek, M

Williamson, pp. 389–405. Chichester, UK: John Wiley & Sons, Ltd.

71. Horvitz CC. 1997. The impact of natural disturbances. In *Strangers in Paradise: Impact and Management of Nonindigenous Species in Florida*, ed. D Simberloff, DC Schmitz, TC Brown, pp. 63–74. Washington D. C.: Island Press
72. Jablonski D. 1998. Geographic variation in the molluscan recovery from the end-cretaceous extinction. *Science* 279:1327–30
73. Jansson K. 1994. *Alien Species in the Marine Environment—Introductions to the Baltic Sea and the Swedish West Coast 68 pp.*, Swedish Environmental Protection Agency, Sweden
74. Jiang SC, Paul JH. 1998. Gene transfer by transduction in the marine environment. *Appl. Environ. Microbiol.* 64(8):2780–87
75. Kareiva P. 1996. Developing a predictive ecology for non-indigenous species and ecological invasions. *Ecology* 77:1651–52
76. Kelly JM. 1993. Ballast water and sediments as mechanisms for unwanted species introductions into Washington State. *J. Shellfish Res.* 12(2):405–10
77. Lavoie DM, Smith LD, Ruiz GM. 1999. The potential for intracoastal transfer of non-indigenous species in the ballast water of ships. *Estuarine Coast. Shelf Sci.* 48:551–64
78. Law R, Weatherby AJ, Warren PH. 2000. On the invasibility of persistent protist communities. *Oikos* 88:319–26
79. Levine JM. 2000. Species diversity and biological invasions: relating local process to community pattern. *Science* 288:852–54
80. Levine JM, D'Antonio CM. 1999. Elton revisited: a review of evidence linking diversity and invasibility. *Oikos* 87:15–26
81. Lodge DM. 1993. Biological invasions: lessons for ecology. *Trends Ecol. Evol.* 8(4):133–36
82. Lonsdale WM. 1999. Global patterns of plant invasions and the concept of invasibility. *Ecology* 80(5):1522–36
83. Loope LL, Mueller-Dombois D. 1989. Characteristics of invaded islands, with special reference to Hawaii. In *Biological Invasions: a Global Perspective*, ed. JA Drake, F DiCastri, RH Groves, FJ Kruger, HA Mooney, M Rejmánek, MH Williamson, pp. 257–80. Chichester, UK: John Wiley & Sons, Ltd.
84. MacArthur R, Wilson E. 1967. *The Theory of Island Biogeography*, Princeton: Princeton University Press
85. Mayr E. 1963. *Animal Species and Evolution*. Cambridge, MA: The Belknap Press of Harvard University Press
86. Medcof JC. 1975. Living marine animals in a ship's ballast water. *Proc. Natl. Shellfish. Ass.* 65:11–12
87. Mills EL, Leach JH, Carlton JT, Secor CL. 1993. Exotic species in the Great Lakes: a history of biotic crises and anthropogenic introductions. *J. Great Lakes Res.* 19 (1):1–54
88. Mills EL, Scheuerell MD, Carlton JT, Strayer D. 1997. Biological invasions in the Hudson River: an inventory and historical analysis. *New York State Mus. Circ.* 57:1–51
89. Mooney HA, Hobbs R, ed. 2000. *The Impact of Global Change on Invasive Species*. Covelo, CA: Island Press
90. Mooney HA, Hofgaard A. 1999. Biological invasions and global change. In *Invasive Species and Biodiversity Management*, ed. OT Sandlund, PJ Schei, Å Viken, pp. 139–48. The Netherlands: Kluwer Academic Publishers
91. Moulton MP, Pimm SL. 1986. Species introductions to Hawaii. In *Ecology of Biological Invasions in North America and Hawaii,* ed. HA Mooney, JA Drake, pp. 231–49. New York: Springer-Verlag
92. National Research Council. 1996. *Stemming the Tide: Controlling Introductions of Nonindigenous Species by Ships' Ballast Water*, ed. Marine Board Commission on Engineering and Technical Systems. Washington, D.C.: National Academy Press

93. Newsome AE, Noble IR. 1986. Ecological and physiological characters of invading species. In *Ecology of Biological Invasions*, ed. RH Groves, JJ Burdon, pp. 1–20. Cambridge: Cambridge University Press
94. Olenin S, Leppäkoski E. 2000. *Inventory of Baltic Sea alien species*. http://www.ku.lt/nemo/species.htm
95. Patz JA, Epstein PR, Burke TA, Balbus JM. 1996. Global climate change and emerging infectious diseases. *J. Am. Med. Ass.* 275(3):217–23
96. Pechenik JA. 1990. Delayed metamorphosis by larvae of benthic marine invertebrates: Does it occur? Is there a price to pay? *Ophelia* 32:63–94
97. Pechenik JA, Wendt DE, Jarrett JN. 1998. Metamorphosis is not a new beginning—larval experience influences juvenile performance. *BioScience* 11:901–10
98. Pickett STA, White PS. 1985. *The Ecology of Natural Disturbance and Patch Dynamics*. New York: Academic Press
99. Pierce RW, Carlton JT, Carlton DA, Geller JB. 1997. Ballast water as a vector for tintinnid transport. *Mar. Ecol. Prog. Ser.* 149:295–97
100. Pimm SL. 1989. Theories of predicting success and impact of introduced species. In *Biological Invasions: a Global Perspective*, ed. JA Drake, F. Di Castri, RH Groves, FJ Kruger, HA Mooney, M Rejmánek, MH Williamson, pp. 351–67. Chichester, UK: John Wiley & Sons, Ltd.
101. Pimm SL 1991. *The Balance of Nature?* Chicago, Illinois: University of Chicago Press
102. Por FD. 1978. *Lessepsian Migration: The Influx of Red Sea Biota into the Mediterranean by Way of the Suez Canal*. Heidelberg: Springer-Verlag
103. Por FD. 1990. Lessepsian migration. An appraisal and new data. *Bull. Inst. Océanogr. Monaco Spec. Vol.* 7:1–10
104. Pysek P. 1995. On the terminology used in plant invasion studies. In *Plant Invasions - General Aspects and Special Problems*, ed. P Pysek, K Prach, M Rejmánek, M Wade, pp. 71–81. Amsterdam: SPB Academic
105. Reise K, Gollasch S, Wolff WJ. 1999. Introduced marine species of the North Sea coasts. *Helgoländer Meeresunters.* 52:219–34
106. Rejmánek M. 1989. Invasibility of plant communities. In *Biological Invasions: a Global Perspective*, ed. JA Drake, F DiCastri, RH Groves, FJ Kruger, HA Mooney, M Rejmánek, MH Williamson, pp. 369–88. Chichester, UK: John Wiley & Sons, Ltd.
107. Rejmánek M. 1996. Theory of seed plant invasiveness: The first sketch. *Biol. Conserv.* 78:171–81
108. Rejmánek M, Richardson DM. 1996. What attributes make some plant species more invasive? *Ecology* 77(6):1655–61
109. Ribera MA, Boudouresque C-F. 1995. Introduced marine plants, with special reference to macroalgae: mechanisms and impact. *Prog. Phycol. Res.* 11:187–268
110. Robinson GR, Quinn JF, Stanton ML. 1995. Invasibility of experimental habitat islands in a California winter annual grassland. *Ecology* 76:786–94
111. Robinson JV, Dickerson JE Jr. 1984. Testing the invulnerability of laboratory island communities to invasion. *Oecologia* 61:169–74
112. Robinson JV, Edgemon MA. 1988. An experimental evaluation of the effect of invasion history on community structure. *Ecology* 69(5):1410–17
113. Rosenzweig ML. 1995. *Species Diversity in Space and Time*. Cambridge: Cambridge University Press
114. Roughgarden J. 1986. Predicting invasions and rates of spread. In *Ecology of Biological Invasions of North America and Hawaii*, ed. HA Mooney, JA Drake, pp. 179–90. New York: Springer-Verlag
115. Ruiz GM, Carlton JT, Fofonoff P, Strayer D, Mills E, et al. 2000. *Interpreting*

invasion patterns from ecological surveys, Report to the U.S. Fish and Wildlife Service

116. Ruiz GM, Carlton JT, Grosholz ED, Hines AH. 1997. Global invasions of marine and estuarine habitats by non-indigenous species: mechanisms, extent, and consequences. *Am. Zool.* 37:621–32
117. Ruiz GM, Fofonoff P, Carlton JT, Hines AH. 2000. *Invasion History of Chesapeake Bay*, Report to U.S. Fish and Wildlife Service Washington, D.C.
118. Ruiz GM, Fofonoff P, Hines AH. 1999. Non-indigenous species as stressors in estuarine and marine communities: assessing invasion impacts and interactions. *Limnol. Oceanogr.* 44(3, part 2):950–72
119. Schoener TW, Spiller DA. 1995. Effect of predators and area invasion: An experiment with island spiders. *Science* 267:1811–13
120. Simberloff D. 1986. Introduced insects: a biogeographic and systematic perspective. In *Ecology of Biological Invasions of North America and Hawaii*, ed. HA Mooney, JA Drake, pp. 3–26. New York: Springer-Verlag
121. Simberloff D. 1989. Which insect introductions succeed and which fail? In *Biological Invasions: a Global Perspective*, ed. JA Drake, F DiCastri, RH Groves, FJ Kruger, HA Mooney, M Rejmánek, MH Williamson, pp. 61–75. Chichester, UK: John Wiley & Sons, Ltd.
122. Simberloff D, Von Holle B. 1999. Positive interactions of nonindigenous species: invasional meltdown? *Biol. Invasions* 1: 21–32
123. Smith LD, Wonham MJ, McCann LD, Ruiz GM, Hines AH, Carlton JT. 1999. Invasion pressure to a ballast-flooded estuary and an assessment of inoculant survival. *Biol. Invasions* 1:67–87
124. Sousa WP. 1984. The role of disturbance in natural communities. *Ann. Rev. Ecol. Syst.* 15:353–91
125. Stachowicz JJ, Whitlach RB, Osman RW. 1999. Species diversity and invasion resistance in a marine ecosystem. *Science* 286:1577–79
126. Stohlgren TJ, Binkley D, Chong GW, Kalkhan MA, Schell LD, et al. 1999. Exotic plant species invade hot spots of native plant diversity. *Ecol. Monogr.* 69(1):25–46
127. Subba Rao DV, Sprules WG, Locke A, Carlton JT. 1994. *Exotic phytoplankton from ships' ballast waters: risk of potential spread to mariculture sites on Canada's east coast*, Canadian Data Report of Fisheries and Aquatic Sciences
128. Symstad AJ. 2000. A test of the effects of functional group richness and composition on grassland invasibility. *Ecology* 81(1):99–109
129. Tilman D. 1997. Community invasibility, recruitment limitation, and grassland biodiversity. *Ecology* 78(1):81–92
130. Tilman D. 1999. The ecological consequences of changes in biodiversity: a search for general principles. *Ecology* 80(5):1455–74
131. U.S. Congress. 1996. National Invasive Species Act, Public Law 104–332. In *Congressional Record*, 142. Washington, D.C.: U.S. Government Printing Office
132. U.S. Congress Office of Technology Assessment. 1993. *Harmful Non-Indigenous Species in the United States, OTF-F-565*, Washington, D.C.: U.S. Government Printing Office
133. Usher MB. 1988. Biological invasions of nature reserves: a search for generalizations. *Biol. Conserv.* 44:119–35
134. Vermeij GJ. 1978. *Biogeography and Adaptation: Patterns of Marine Life*. Cambridge: Harvard University Press
135. Vermeij GJ. 1991. When biotas meet: Understanding biotic interchange. *Science* 253 (5024):1099–104
136. Vitousek PM. 1990. Biological invasions and ecosystem processes: toward an integration of population biology and ecosystem studies. *Oikos* 57:7–13

137. Vitousek PM, Walker LR. 1989. Biological invasion by *Myrica faya* in Hawai'i: plant demography, nitrogen fixation, ecosystem effects. *Ecol. Monogr.* 59:247–65
138. Webb DA. 1985. What are the criteria for presuming native status? *Watsonia* 15:231–365
139. White House. 1999. *Executive Order: Invasive Species.* http://www.pub.whitehouse.gov/urires/IZR?urn:pdi://oma.eop.gov.us/1999/2/3/14.text.1
140. Wilcove DS, Rothstein D, Dubow J, Phillips A, Losos E. 1998. Quantifying threats to imperiled species in the United States. *BioScience* 48(8):607–15
141. Williams RJ, Griffiths FB, van der Wal EJ, Kelly J. 1988. Cargo vessel ballast water as a vector for the transport of non-indigenous marine species. *Estuarine Coast. Shelf Sci.* 26(4):409–20
142. Williamson M. 1989. Mathematical models of invasion. In *Biological Invasions: a Global Perspective*, ed. JA Drake, HA Mooney, F di Castri, RH Groves, FJ Kruger, M Rejmánek, M Williamson, pp. 329–50. Chichester, UK: John Wiley & Sons, Ltd.
143. Williamson M. 1996. *Biological Invasions*. London: Chapman & Hall
144. Williamson M. 1999. Invasions. *Ecography* 22:5–12
145. Wiser SK, Allen RB, Clinton PW, Platt KH. 1998. Community structure and forest invasion by an exotic herb over 23 years. *Ecology* 79:2071–81
146. Wonham MJ, Carlton JT, Ruiz GM, Smith LD. 2000. Fish and ships: relating dispersal frequency to success in biological invasions. *Mar. Biol.* 136:1111–21
147. Wonham MJ, Walton WC, Frese AM, Ruiz GM. 1996. *Transoceanic transport of ballast water: Biological and physical dynamics of ballasted communities and the effectiveness of mid-ocean exchange*, Final Report to the U.S. Fish and Wildlife Service and the Compton Foundation
148. Wright S. 1978. *Evolution and the Genetics of Populations Volume 4–Variability Within and Among Natural Populations.* Chicago: University of Chicago Press
149. Zhang F, Dickman M. 1999. Mid-ocean exchange of container vessel ballast water. 1: Seasonal factors affecting the transport of harmful diatoms and dinoflagellates. *Mar. Ecol. Prog. Ser.* 176:243–51
150. Zolotarev V. 1996. Black Sea ecosystem changes related to the introduction of new mollusc species. *Mar. Ecol.* 17(1–3):227–36

Annu. Rev. Ecol. Syst. 2000. 31:533–63

DIVERSIFICATION OF RAINFOREST FAUNAS: An Integrated Molecular Approach

C. Moritz,[1] J. L. Patton,[2] C. J. Schneider,[3] and T. B. Smith[4]

[1]*Department of Zoology and Entomology and Cooperative Research Centre for Tropical Rainforest Ecology and Management, University of Queensland, Brisbane, Queensland 4072, Australia; e-mail: cmoritz@zoology.uq.edu.au*

[2]*Museum of Vertebrate Zoology, University of California, Berkeley, California 94720; e-mail: patton@uclink4.berkeley.edu*

[3]*Department of Biology and Center for Ecology and Conservation Biology, Boston University, Boston, Massachusetts 02215; e-mail: cschneid@bio.bu.edu*

[4]*Department of Biology and Center for Tropical Research, San Francisco State University, San Francisco, California 94132; e-mail: tsmith@sfsu.edu*

Key Words molecular systematics, evolution, speciation, tropical diversity, biogeography

■ **Abstract** Understanding the evolutionary processes that generate and sustain diversity in tropical faunas has challenged biologists for over a century and should underpin conservation strategies. Molecular studies of diversity within species and relationships among species, when integrated with more traditional approaches of biogeography and paleoecology, have much to contribute to this challenge. Here we outline the current major hypotheses, develop predictions relevant to integrated molecular approaches, and evaluate the current evidence, focusing on central African, Australian, and South American systems. The available data are sparse relative to the scale of the questions. However, the following conclusions can be drawn: (*a*) in most cases, the divergence of extant sister taxa predates the Pleistocene; (*b*) areas with high habitat heterogeneity and recent climatic or geological instability appear to harbor more species of recent origin; (*c*) there is support for both allopatric and gradient models of diversification and more attention should be given to the role of diversifying selection regardless of geographic context; and (*d*) conservation strategies should seek to protect heterogeneous landscapes within and adjacent to large rainforest areas, rather than rainforests alone.

SCOPE AND ISSUES

The search for the basis of high species diversity in tropical forests has occupied biologists for over a century (91, 115, 143, 151). What biologists seek to understand is how there come to be so many species in a single place (alpha diversity) and so much turnover of species between habitats (beta diversity) and

0066-4162/00/1120-0533$14.00

regions (gamma diversity; 27). Traditionally, this question has been subdivided into defining processes that maintain high species diversity, primarily within areas (ecological processes), and those that generate high diversity through time and across regions (historical evolutionary and biogeographical processes). However, this dichotomy between ecological and evolutionary processes is being blurred through the recognition that history has played a substantial role in shaping both regional and local species diversity (71, 118, 154) and that ecological factors can influence or drive speciation (83, 102, 121, 122).

Here we focus on the processes, both ecological and geographic, that have generated species diversity in tropical forests and affected the balance between the origin and extinction of species. In so doing, we need to be clear about what we mean by "species." There is a plethora of species definitions in the literature (70, 88, 103). Our approach is shaped by the phenomenon that we seek to explain—functional diversity in complex ecosystems. Thus, we recognize as species geographically bounded sets of populations that are distinct for morphological traits or are reproductively isolated from congeners, with or without corresponding molecular divergence. This definition encompasses situations in which there are concordant molecular and phenotypic differences, but allows for circumstances in which the progenitor of a recently evolved, phenotypically distinct lineage is paraphyletic with respect to the derived species (105). This definition excludes allopatric populations that are divergent for presumed neutral molecular genetic characters but not clearly distinguishable using behavioral or morphological traits. While such populations represent independent evolutionary lineages and would qualify as species under an Evolutionary or Phylogenetic Species Concept (31), they do not represent the type of diversity typically considered by ecologists.

The purpose of this review is to explore the contribution that analyses of molecular phylogenetics and population genetics are making to the understanding of the origin of faunal diversity in tropical forests. Examining patterns of molecular diversity provides the potential to link anagenetic evolution and speciation, to shed light on historical biogeography, and to test alternative hypotheses about mechanisms of diversification (40, 41). One finding that emerges from this and previous reviews (3) is the need to integrate molecular systematics with analyses of phenotypic variation and reproductive isolation and, where possible, with independent evidence of landscape history such as that derived from geology or paleoclimatology.

The importance of testing alternative hypotheses about speciation mechanisms was emphasized by Endler (40, 41), who demonstrated that the same geographic pattern of species abundance and congruence of subspecies and species boundaries can be explained by both parapatric (gradient) and allopatric (refuge) mechanisms. We review major current hypotheses about the cause of high species diversity in tropical faunas and develop contrasting predictions about patterns of molecular and phenotypic variation for alternative speciation models. We then evaluate molecular evidence from recent studies spanning three continents, starting with ages of rainforest species inferred from interspecific comparisons and proceeding through case studies that illustrate the use of molecular and, where available, morphological and ecological information to test predictions of specific hypotheses about mechanisms of diversification. We conclude that while geographic context is important, the role of ecology and divergent selection in determining functional diversity within and among species in rainforests deserves greater attention. This has important implications for conservation.

The molecular analysis of rainforest faunas is in its infancy, yet already it is clear that it will contribute to understanding the history of species diversification (52) and historical biogeography of particular regions. This, in turn, should inform the study of community ecology (118) by providing the historical framework with which to interpret patterns of species distribution and community composition (124) and, perhaps, by estimating critical variables such as rates of speciation, extinction and immigration (71, 98).

Insights from molecular systematics and population genetics, when combined with analyses of species distributions, phenotypic variation, and landscape history should also lead to improved strategies for conservation. Given that we should seek to maintain evolutionary processes (7, 42, 94, 140), including the requisite ecological viability of systems, it is important to understand how historical processes have shaped genetic and species diversity in whole communities of organisms (2, 95) and how current evolutionary processes are maintaining phenotypic diversity. While the details of these processes will differ among systems, it should be possible to devise conservation strategies that protect both the (irreplaceable) genetic diversity attributable to historical isolation and the landscape features that promote phenotypic diversity through a balance between gene flow and selection (94). Such process-oriented strategies for conservation are likely to be more effective in a changing world than those that assume a static distribution of diversity (20).

CURRENT HYPOTHESES OF DIVERSIFICATION

The major hypotheses concerning high species diversity in tropical systems can be placed into two nonexclusive categories: those that invoke low rates of extinction versus those focused on high rates of speciation. The former argue that tropical biotas have accumulated large numbers of species over long time periods because, relative to temperate or boreal systems, the tropics are old and have been stable climatically (143). An extension is that ecologically stable rainforest areas either accumulate paleoendemics or combine low extinction rates with high speciation rates to accumulate both paleo- and neoendemic species (44, 45). In principle, the hypothesis that high species richness is partly attributable to low extinction and/or high speciation rates is testable by comparing sister groups between tropics and temperate zones (24, 43) or between tropical areas of different stability (e.g. stable lowland tropics versus volatile Andes; see below).

Hypotheses concerning factors that promote speciation in tropical faunas are too numerous to review in detail here (for recent reviews, see 19, 62). By and large,

TABLE 1 Brief description of major models of evolutionary processes that promote diversification of rainforest faunas[a]

Name—Geographic Mode[b]	Isolating Barrier	Evolutionary Mechanisms	Key Reference(s)
Refugia model—allopatric	Dry forests, savanna	Isolation, drift, selection	37, 57, 62, 148
Riverine model—allopatric	Major rivers	Isolation, drift?[c], selection?[c]	6, 151
Vanishing refuges—allopatric	Dry forests, savanna	Isolation, drift, directional selection	148
Disturbance—vicariance—allopatric	Heterogeneous forest structure	Competition, directional selection?[c]	19, 28
Gradient model—parapatric or allopatric	None necessary	Divergent selection, with or without gene flow	39, 117, 141

[a]See Figure 1 and text for further description.

[b]Several variations on these themes are reviewed by Haffer (62). These include a river-refuge model that combines major rivers and dry biomes between rainforest refugia as agents promoting geographic isolation within Amazonia and various paleogeography models that invoke tectonic and mountain-building events barriers.

[c]?, Refers to processes for which operation is uncertain.

these hypotheses have been preoccupied with the geographic context of speciation and, for allopatric models, the physical cause of isolation (Table 1). The refugia model has been the most widely discussed (85, 115, 153) and rests on the premise that climatic change caused rainforests to contract to refugia separated by dry forests and savanna and that this isolation promoted speciation. Initial discussions focused on Pleistocene events, particularly those for the last glacial cycle or two (37, 57), although more recently the model has been extended to Tertiary events on the assumption that climatic oscillations driven by Milankovitch cycles through this period were of sufficient amplitude and duration to promote speciation (61, 62).

Criticisms of the refuge model include uncertainty about whether Amazonian rainforests contracted or just changed in composition (29), concerns about sampling bias in identifying locations of refugia (100), debate over whether contact zones between presumed sister-taxa are appropriately located (40, 92), increased complexity (and reduced testability) as additional refugia are proposed (85), and the argument that alternative speciation mechanisms provide equally good explanations for the biogeographic patterns observed (40, 41, 87). Moreover, with few exceptions, the evolutionary mechanisms supposed to promote morphological or reproductive divergence among the isolated populations are rarely explicit. Finally, only rarely can the paleoecological and biogeographic evidence specify either the size or ecological characteristics of putative refugia.

One interesting variation on this theme is the vanishing refuge model (148), which posits that some populations differentiated to species through directional selection toward tolerance of ecotones or dry habitats as rainforest patches became too small to retain viable populations. This idea was prompted by the impression that sister taxa commonly occur in geographically adjacent but distinct habitats, a pattern that is also predicted by gradient models (see below). Another refuge-based model that explicitly discussed the mechanisms involved in population divergence was developed to account for the extraordinary diversity of mimetic forms of butterflies in the neotropics (17, 18, 89). Turner (146; but see also 15, 87) proposed that random loss of model species among refugia was a major selective agent that resulted in diversification of mimetic species.

The second major class of allopatric model (Table 1) invokes substantial river systems as barriers to gene flow, such that populations on either side gradually diverge to form separate species (151). Empirical support for this "riverine barrier" model comes from observation that the boundaries of closely related species or subspecies often coincide with the major rivers of Amazonia [e.g. tamarins and marmosets (64); various birds (30, 58–60, 63); Amazonian lizards (1)]. Difficulties with this model arise because the strength of the barrier to gene flow in widely distributed species diminishes toward the upper reaches of a river (6, 114), and the location of many rivers is highly dynamic on both short and long time scales (84). A problem shared with the refuge model is that distributional data alone do not distinguish between the rivers as current meeting points for species that diverged elsewhere versus locations of primary diversification (60, 62, 128).

Models based on divergent selection across strong environmental gradients (gradient model, Table 1) differ fundamentally from allopatric models in that complete suppression of gene flow is not a prerequisite for phenotypic divergence and speciation (39, 102, 117). For rainforests, the gradient model suggests that strong environmental (e.g. habitat) gradients resulted in adaptive divergence and speciation. This is expected to result in sister species adapted to adjacent but distinct environments (e.g. rainforest—dry forest). Evidence consistent with this model in rainforest faunas comes from the frequent location of hybrid zones in ecotones (41, 42) and observations of species-level phenotypic differentiation between populations in rainforest and adjacent habitats (126, 141). Recent emphasis on current (145) and, possibly, historical (28) heterogeneity of vegetation structure within Amazonia suggests additional possibilities for this mechanism to operate. However, once again, distributional data alone are open to multiple interpretations (e.g. 85, 92).

Aside from the inherent ambiguity of current distribution patterns, a further difficulty with all of the above models is the spatio-temporal dynamics of key variables such as the structure, location, and contiguity of rainforests and the concomitant strength and location of selection gradients. These are poorly resolved or contentious for the late Pleistocene, let alone earlier periods. Any discussion of refuge or other models should bear this in mind and critically evaluate evidence from paleopalynology or climate modeling to confirm that such habitat mosaics did, in fact, exist (18).

PREDICTIONS FROM MOLECULAR SYSTEMATICS

Molecular data, when combined with information on phenotypic variation, species distributions, and landscape ecology and history can provide four types of information relevant to hypotheses about speciation in rainforests:

1. Relationships among species and historical (phylogeographic) lineages within species, in particular identification of sister groups in relation to geography and habitat/ecological attributes (32, 85) although nonmolecular data are also informative.
2. Approximate estimates of the timing of divergence events (4, 5, 14, 35, 66, 124) and, potentially, examination of the long-term balance between speciation and extinction within monophyletic lineages (98);
3. Estimates of current and historical gene flow rates among populations (129), against which patterns of morphological variation or reproductive isolation can be compared (39, 102);
4. Tests for historical founder events and range expansions (130, 144), especially in relation to the location of putative refugia and expansion zones.

The predictions of three major hypotheses, the refugia, riverine, and gradient models, are illustrated in Figure 1 and Table 1. Briefly, the refugia model predicts that recently evolved sister taxa occur in adjacent refugia (as defined a priori by substantial paleoecological evidence) and that there should be evidence from intraspecific gene trees for range expansions and secondary contact between refugia and, if drift is invoked, for population restriction within refugia. These patterns should be broadly congruent among species with similar ecological requirements and vagility (3). An important caveat is that a rainforest refugium itself may be

---

Figure 1 Complementary approaches to discriminating among major hypotheses concerning mechanisms that promote diversity in rainforest faunas (see Table 1). For both Riverine (allopatric) vs Gradient (panel *A1*) and Refugia (allopatric) vs Gradient (panel *B1*) comparisons, expected outcomes in relation to phenotypic divergence vs gene flow in multipopulation comparisons (panels *A2*, *B2*), and phylogeny of taxa (panels *A3*, *B3*). In the Riverine vs Gradient test, populations are distributed among habitats (*A vs B* and *A′ vs B′*) that may result in divergent selection pressures and, on either side of a river, are hypothesized to suppress gene flow (*A and B vs A′ and B′*). The Refugia vs Gradient test is illustrated for populations occupying adjacent elevational zones (*A* and *a vs B* and *b*) across which a selection gradient is hypothesized and geographically disjunct habitats (*A* and *B* vs *a* and *b*) are hypothesized to represent separate historical refugia (e.g. 104). In both cases, the Gradient model predicts an inverse relationship between phenotypic divergence and gene flow across habitats but not within habitats, and that recently derived sister-taxa should occupy distinct environments. Conversely, the allopatric (riverine or refugia) models predict that sister taxa should be separated by the barrier to gene flow. For further discussion and predictions, see text.

A. Riverine vs. Gradient

Habitat 1 Habitat 2

1\.

A

X X X X X X

River

X X X X X X

A' B'

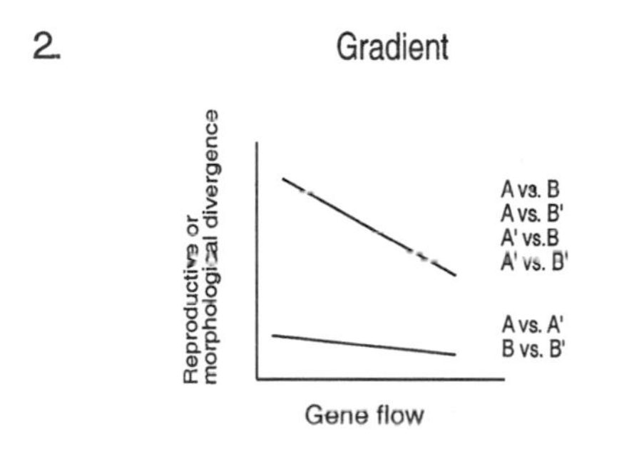

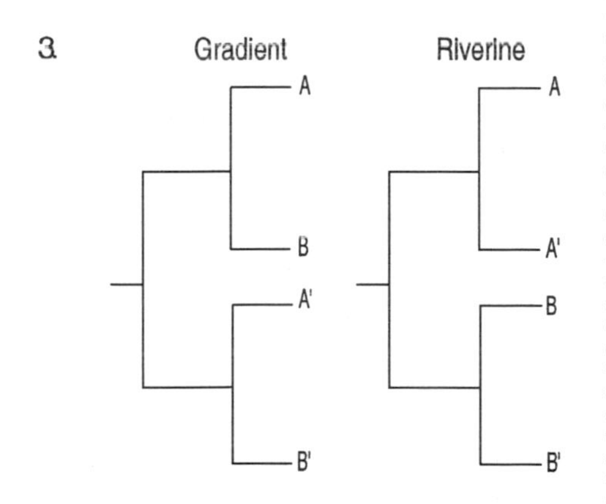

B. Refugia vs. Gradient

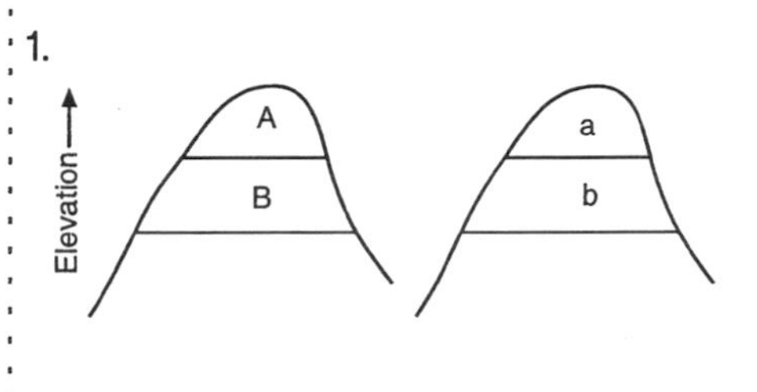

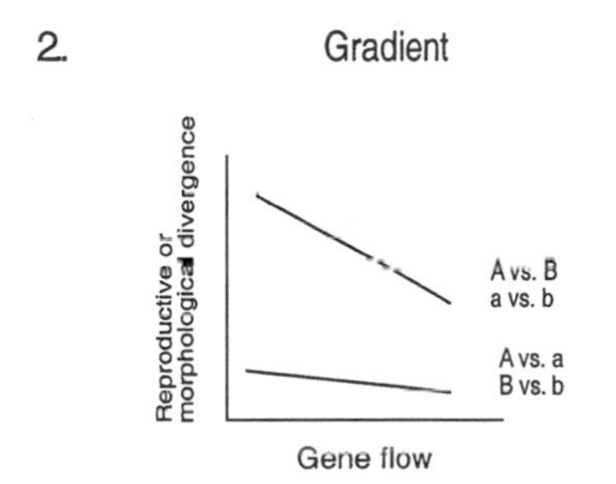

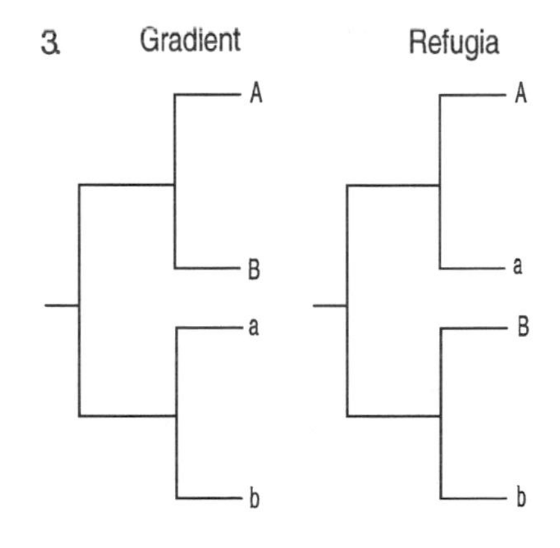

dissected by dry habitats (68), in which case the refugial population may retain multiple genetic lineages within a metapopulation structure, potentially confounding tests for population bottlenecks and range expansion. Another limitation is that refuge hypotheses do not predict a unique phylogenetic structure for sister species (36) unless the putative refugia are linearly arrayed (124) or the sequence of isolation can be predicted with confidence from independent paleoecological or geological evidence (85).

The riverine model predicts that sister species, and distinct phylogeographic clades within species, should occur across major rivers rather than being segregated within fluvial areas or across gradients (Figure 1). In contrast to the refugia model, there is no expectation for genetic signatures of range expansion. Again, the vicariant pattern should be congruent among species with similar ecological requirements and capacity to disperse across rivers. Importantly, phylogenies within and among closely related species can distinguish between primary divergence across rivers versus secondary contact between nonsister lineages that diversified elsewhere (110). Complications arise for species that occupy flood plains across which the actual location of the river channel varies in time, potentially relocating populations of the adjacent lineages onto opposite banks due to meander cutoffs. As a consequence, specialists of floodplain forest are less likely to exhibit riverine diversification than are upland ("terra firma") forest specialists (109, 110).

For both of the above "allopatric" models (and the river-refugia combination; 62), the insights into historical biogeography provided by intra- and interspecific gene trees provide an opportunity to ask what happens when vicariant populations come into secondary contact. Is there evidence for reproductive isolation, for example, linkage disequilibrium or pre- or postmating isolation (9, 10)? In the context of determining whether geographic isolation alone is sufficient to promote speciation, these are important questions regardless of the level of phenotypic differentiation between the sister lineages.

The gradient (or "divergence with gene flow") model makes unique predictions about the relationship between gene flow and morphological or reproductive divergence within species and also predicts that sister taxa should occupy distinct but adjacent habitats (Figure 1). Within species, this model predicts that, for a given level of gene flow, phenotypic and reproductive divergence among populations should be much greater among habitats than within habitats. Morphological divergence should be strongly inversely correlated with gene flow because of contrasting selection pressures in different habitats (Figure 1; 102). Divergence for neutral quantitative traits should also be inversely correlated with gene flow due to drift, but no difference is expected for comparisons among versus within habitats (Figure 1; 86, 102). The prediction that sister species should occupy distinct habitats is shared with the vanishing refugia model (148), but the latter also requires severe population bottlenecks and range expansion, whereas the gradient model does not.

Tests of speciation models will be most informative for recent to middle-aged taxa. Lineages resulting from ancient divergences are likely to have undergone species extinctions, diverged in ecology, or changed distributions to the point where the original geographic/ecological context of speciation is obscured. On the other hand, very recent lineages may not have achieved a genetic signature sufficient to pinpoint their mode of origin. This issue is relevant to both phylogenetic and biogeographic approaches. Although analytical methods to distinguish between allopatric and sympatric speciation mechanisms from species phylogenies have been proposed (8), it seems unlikely that they can distinguish parapatric models (e.g. gradient model) from other alternatives.

TEMPO OF SPECIATION

Molecular phylogenies of extant taxa offer the opportunity to examine the tempo and mode of speciation for groups in which fossil data are lacking (99) as is the case for most tropical rainforest taxa. Several questions regarding the evolution of the high species diversity of tropical regions may be addressed as comprehensive molecular phylogenies for diverse groups become available. First, is there evidence for recent bursts of speciation consistent with Pleistocene Refuge models? Second, how do the relative divergence times of tropical and temperate lineages compare—are tropical taxa older than temperate taxa? Third, how do estimates of extinction and speciation rates in tropical clades compare to those of nontropical sister groups—does the high species diversity in tropical groups result from higher speciation rates, lower extinction rates, or a combination of the two? Fourth, how does the tempo of speciation, inferred from molecular divergence, differ among rainforest regions with different ecological and/or geological histories?

Is there evidence for recent Pleistocene speciation? Given uncertainties about rates of molecular evolution and the typically low precision of most estimates (67), the following compilation of genetic divergence among sister species of rainforest vertebrates seeks only to examine broad patterns. Comparisons of relative divergence estimates among sister taxa in tropical and temperate regions involve few assumptions, other than that of similar rates of substitution across taxa and among groups. Bearing the above caveats in mind, molecular phylogenies of birds, mammals, lizards, frogs, and salamanders from tropical regions of Central and South America, Australia, and Africa suggest that most speciation events in tropical rainforests predate the Pleistocene (Table 2). For example, only 7 of 125 speciation events in 22 genera of tropical South American small mammals occurred in a time frame consistent with Pleistocene divergence (<about 4% sequence divergence for mtDNA cytochrome b sequences; Figure 2) and most of those occurred in a single rodent genus, *Oecomys*. In birds, 18 of 64 speciation events show sister groups differing by >4% in mitochondrial protein coding genes (cyt-b and/or ND2). Most of these (10) are concentrated in a single genus of Spinetails (*Cranioleuca*; 48). In total, the available data indicate that Pleistocene climate oscillations played little role in generating vertebrate species diversity in tropical rainforests.

Are tropical taxa older than temperate taxa? We do not know of any studies that directly compare ages of species from temperate and tropical regions. However, a

TABLE 1 Summary of genetic distance among sister species and/or sister groups from molecular systematic studies of tropical vertebrate taxa[a]

Taxon	Data Type	Mean (range)	Time (My)	Reference(s)
Caudata				
Oedipina	Allozymes Nei's D	0.73 (0.35–1.60)	4.87 (2.3–10.7)	53
Anura				
Physalaemus	12S rRNA	10.12 (2.24–18.71)	10.12 (2.24–18.71)	21
Bufo (neotropical species)	16S rRNA	15.14 (6.17–24.89)	15.14 (6.17–24.89)	55
Cophixalis	16S rRNA	9.66 (4.82–14.47)	9.66 (4.82–14.47)	C Hoskins & C Moritz, unpublished data
Cyclorhamphus	MCF[b]	33.95 (3.0–102.0)	16.98 (1.5–61.0)	66
Squamata				
Australian carphodactyline geckos (*Carphodactylus*, *Saltuarius*, and *Phyllurus*)	12SrRNA	8.55 (3.7–14.35)	8.55 (3.7–14.35)	CJ Schneider, unpublished data
Mammalia				
Australasian ring-tail possums	ND2, Cyt-b	15.24 (4.11–23.29)	7.62 (2.06–12.65)	96, MC Lara & C Moritz, unpublished data
Amazonian possums	Cyt-b	9.36 (2.30–19.90)	4.68 (1.65–9.95)	36, 97, 106
Amazonian rodents	Cyt-b	12.51 (0.90–19.63)	6.75 (0.45–9.82)	35, 36, 79, 108–110, 131, 132

Andean *Akodon*	Cyt-b	7.95 (5.40–12.20)	3.98 (2.70–6.10)	104, 111
Atlantic Forest marsupials	Cyt-b	13.99 (5.80–19.30)	7.00 (2.90–9.65)	97
Atlantic Forest rodents	Cyt-b	13.30 (5.84–17.40)	6.65 (2.92–8.70)	78, 79
Saguinus monkeys	308bp Cyt-b, 600bp control region.	11.85 (6.79–13.90)[c]	5.93 (3.40–7.95)	72
Aves				
Tanagers (*Rhamphocelus*)	Cyt-b	4.69 (1.60–8.32)	2.35 (0.80–4.16)	56
Nyctibiid Potoos	Cyt-b	14.41 (11.10–16.20)	7.21 (5.55–8.10)	90
Bul–buls (*Andropadus*)	Cyt-b	10.60 (4.59–18.24)	5.30 (2.30–9.12)	119
Scrubwrens (*Sericornis*)	Cyt-b	8.27 (5.1–13.1)	4.18 (2.55–6.51)	73
Chat tyrants (*Silvicultrix*)	ND2	7.40 (5.80–9.70)	3.7 (2.90–4.85)	47
Spinetails (*Cranioleuca*)	ND2, Cyt-b	1.9 (0.50–4.00)	0.95 (0.25–2.00)	48
Tit–tyrants (*Anairetes*)	ND2, Cyt-b	3.3 (1.10–5.30)	1.65 (0.50–2.65)	120

[a]Studies were selected based on their relatively complete representation of known species and use of mitochondrial DNA in most cases. In all cases, Kitsch trees were produced from a distance matrix as listed under Data Type, using Phylip version 3.5. Divergence values represent the branch length between phylogenetically independent pairs of species or sister groups on the Kitsch tree and, with the exception of the *Saguinus* monkeys (uncorrected p distance) are based on Kimurs 2-parameter estimates. Approximate divergence times were estimated by using 0.15 D/My for allozyme data and 1% and 2%/My sequence divergence for mitochondrial 12S and 16S rDNA and protein-coding genes (ND2 and Cyt-b), respectively (see text for caveats).

[b]MCF, Micro-Complement Fixation.

[c]Sequence unavailable in Genbank.

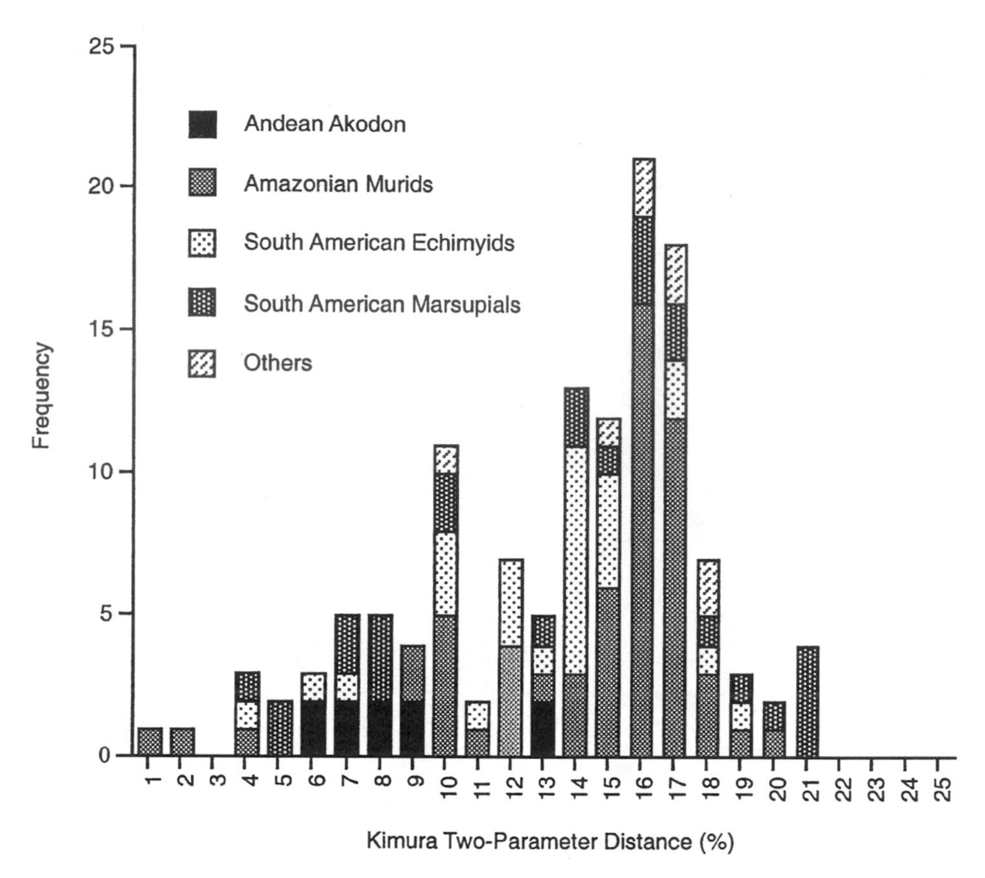

Figure 2 Histogram of Kimura two-parameter distances between mammalian sister species, as shown in Table 2, grouped by taxonomic affiliation and geographic distribution.

limited sampling of tropical bird taxa (Table 2) allows a preliminary comparison. Klicka & Zink (77) found the mean sequence divergence among temperate zone sister species to be 5.1% $\pm$ 3.00%. Among the tropical bird species listed in Table 2, the average divergence is 7.2% $\pm$ 4.37%. This difference is not statistically significant (t-test, $p = 0.12$) but additional data are needed for a robust analysis. Avian tribes at low latitudes tend to be older (as estimated from DNA hybridization data) than those at high latitudes (51). Further, avian sister taxa (tribes to families) at low latitudes are also more species rich, implying a higher rate of diversification (speciation less extinction; 24). As molecular phylogenies accumulate for species-rich genera and families, it should be possible to analyze more precisely the tempo of diversification in tropical vs temperate clades.

Does the tempo of speciation differ among rainforest regions with different ecological and/or geological histories? Again, data are lacking for most taxa and regions but evidence for two groups from South America (birds and mammals) is informative. Based on the DNA hybridization data of Sibley & Alquist (127), Fjeldså (44) suggested that the tropical Andes contain a greater proportion of young bird species than do the Amazonian lowlands. This view is supported by low levels of mtDNA sequence divergence among species within genera that have a primarily Andean distribution (48, 49, 120; but see 47). For South American mammals, there are sufficient data for murid rodents (104, 113, 131, 132) to examine the relative depth of divergence among species from the Andes and lowland Amazonian rainforests. For species in the genus *Akodon* and closely related genera, mitochondrial cytochrome-b sequence divergence indicates that species in the Andes are significantly younger than those in the Amazonian lowlands. The mean distance between sister taxa in Andean genera is 7.9% ± 2.45%, whereas the mean distance between sister taxa within Amazonia is 13.8% ± 3.76% (t-test, $p < 0.001$).

Although these comparisons are based on few taxa, the pattern of relatively old species in tropical rainforests may reflect a general pattern worldwide. Roy (119) found that lowland species of a widespread African rainforest bird genus (*Andropadus*) were older than species in the Rift Mountains of East Africa. In Australia, species of rainforest-endemic vertebrates appear to be extremely old, as inferred from the large genetic distance among them (Table 2; 96), and may represent the remnants of an ancient, previously more widespread rainforest fauna. In addition, mtDNA divergence of small mammal species from the Amazon basin and the Atlantic Forest of eastern Brazil reveal similarly ancient speciation events in many taxa (see below and 78, 97, 132), consistent with earlier indications of pre-Pleistocene divergence of Amazonian and Atlantic forest amphibians (66). As better calibrations of molecular evolutionary rates become available and paleoecological reconstructions of habitat distributions are extended further back in time, comparisons of absolute time of divergence among species may allow us to infer the geographic and climatic context of speciation in these ancient lineages.

MOLECULAR PERSPECTIVES ON SPECIATION MECHANISMS

Refugia Models

While rainforest refugia have been postulated for most rainforest areas of the world, the molecular data necessary to test whether such refugia existed and, if so, whether their existence is associated with speciation have begun to be collected for only a few systems. We summarize findings from analyses of low-moderate elevation areas in Africa, South America, and eastern Australia as a series of case studies. We exclude the high Andes because, while the fauna of these paramo and grassland communities were likely the subject of vicariance (150), they are not germane to the rainforest refuge model.

Africa Molecular analyses have been applied to avian fauna of montane rainforests of east Africa. Using published DNA hybridization data (127), Fjeldså and colleagues (44, 46) demonstrated that these, along with other mountains and the forest/savanna ecotone surrounding the Guineo-Congolian lowland forest, contain concentrations of both recent and old species, whereas the lowland area is dominated by old species. The mountains of east Africa have been proposed as refugia that provided relatively stable montane forest environments during the Quaternary and as a site of recent diversification (37). Phylogenetic analyses of mtDNA sequences from various species and subspecies of bulbuls (*Andropadus*) indicated that the montane taxa were a monophyletic group relative to those in the lowlands and that some of the former had diverged recently (<3 million years (My), 119), whereas the latter were mid-early Miocene in age. These observations were interpreted as supporting a montane refuge model of speciation through genetic drift in populations of reduced size (119).

South America The forested Andean slopes, the western Amazon Basin, and the Atlantic Forest of coastal Brazil are all major areas of biotic endemism, with refugia proposed for each (17, 18). Despite the extraordinarily rich biota of these regions, however, few molecular data are available, and phylogeographic tests of proposed refuges have yet to be developed (15, 36).

Using the DNA hybridization data of Sibley & Ahlquist (127), Fjeldså (44) identified the tropical Andes and the lower eastern Andean fringe as areas with the highest concentration of recently evolved endemic bird species. In contrast, phyletic relics and medium-old taxa characterize the forests of lowland Amazonia. Fjeldså (45) described those areas with concentrations of old lineages as "ecologically extremely stable" (EESA) regions, ones that remained stable through the shifting climatic periods in the Pleistocene. "Stability" here refers to ecological conditions that maintained species pools despite local habitat shifts and changing community compositions, rather than constancy of specific extant communities over long time periods.

Fjeldså (44, 45) related the high proportion of young species in the Andes (by a factor of three or more relative to Amazonia) to the topographic and climatic complexity created by uplift of the central and northern Andes since the Miocene. These analyses also suggested that most young species within the Atlantic Forest of coastal Brazil occur in the central region in the states of Rio de Janeiro and Espirito Santo, another area where faulting has generated relatively recent escarpments and coastal ranges. Importantly, this region is actually located between postulated Pleistocene refuges.

These conclusions accord with the few available phylogeographic studies of bird lineages distributed from Amazonian lowlands to Andean highlands. For example, the Amazonian species of the flycatcher genus *Leptopogon* is basal to all others in the complex, and the altitudinally zoned Andean species become increasingly younger from low to high elevations (11). Hence, speciation in this genus is consistent with old lowland forest taxa diversifying into habitats of successively higher elevation in the Andes as the mountains rose. Creighton (33) had proposed a similar model for lowland and Andean taxa of mouse opossums of the genera *Marmosops*, *Marmosa*, and *Gracilinanus*, although without molecular data.

The same pattern of older lowland and younger high-elevation species is apparent in small-bodied mammals for which molecular data are available (35, 36, 104, 132). As noted above, sister-species of lowland forest mice are significantly more divergent than are those of Andean taxa. Moreover, divergences among species within lowland lineages of both rodents and marsupials are not only deep but relatively uniform (mtDNA distances 13.7% $\pm$ 0.16%). These data suggest that speciation episodes in each were approximately concordant at some time in the late Tertiary (assuming similar rates of molecular divergence).

In coastal Brazil, those few mammalian taxa for which molecular data are available (78, 97) exhibit distribution patterns inconsistent with diversification between the proposed refuges along the Atlantic escarpment. For example, the two species of the slender mouse opossum *Marmosops* in the Atlantic Forest are elevationally segregated in the mountains of southern coastal Brazil, rather than replacing one another between proposed refugees.

Replacement, with possible contact at a midpoint between refuges, would be expected if refugial isolation had promoted their divergence. The substantial molecular difference between these two species (mtDNA distance of 16.7%) also supports ancient differentiation. Finally, this pair of species cannot be confirmed as phyletic sisters in relation to the five lowland Amazonian species, as would be required if their divergence within the Atlantic Forest were refuge-related. The combination of distribution pattern, depth of molecular divergence, and nonsister relationship all point to differentiation at a time earlier than, and by a process different from, Pleistocene refuge formation. The distribution pattern of elevational segregation, with a geographically limited higher-elevation species in the southern Atlantic Forest and much more broadly distributed lower-elevation one also characterizes other mammals [e.g. rodents of the genera *Delomys* (149) and *Akodon* (26)] as well as some birds [ant wrens of the genus *Drymophila* (155)]. Although molecular estimates of divergence in these taxa have not been published, the shared distribution pattern with *Marmosops* suggests the possibility of a common history.

The association between recent diversification and areas of tectonic, as well as climatic, activity is supported by recent reconstructions of the geological history of the central and northern Andes and adjacent Amazonia. The last (and continuing) major episode of Andean uplift began in early to mid-Miocene (10–12 MYA). By the end of the Pliocene, the eastern cordillera was sufficiently high to alter the flow of the proto-Amazon from north into the Caribbean to its present exit into the Atlantic Ocean to the east (82, 84). Andean uplift also generated substantial subsidence of foreland basins and forebulge formation in western Amazonia, producing a series of subbasins now filled with sediments of Andean origin. Phylogeographic signatures concordant with these underlying paleogeophysical features remain evident in the small mammal fauna of that region today (36, 81, 107, 110).

A general conclusion reached from the molecular studies is that species diversification in the lowland tropical forests is older than previously thought, with a substantial proportion of the extant fauna initiating divergence prior to the Pleistocene. In the case of Amazonia, the high species richness and endemism of the proposed refugia may have resulted from greater age and, perhaps, longer-term stability. These regions have thus served as reservoirs for speciation in the geologically active areas around the periphery of Amazonia, but not as centers of active and recent speciation.

Australian Wet Tropics The rainforests of the Australian wet tropics occupy a small (<1 million hectare) area in northeast Queensland but have the high species diversity and endemicity (relative to temperate/arid regions in Australia) typical of rainforests elsewhere. Geologically, the major topographic features—peaks of the Great Dividing Range—are ancient eroded remains of an uplift that is perhaps >150 Mya in age; superimposed on this has been localized volcanic activity in the late Cenozoic to as recently as 10 kya. From paleopalynology, there is strong evidence for contractions of complex angiosperm rainforests during glacial cycles, so that the forests were restricted to mesic (mostly high elevation) areas for much of the Quaternary and perhaps earlier (76). High-elevation Pleistocene refugia have been proposed by plant biogeographers (152) although analyses of charcoal indicate that these were extensively dissected by dry forests (68). Paleoclimate modeling (101) infers a major disjunction of the cool complex rainforests, which harbored the majority of endemic vertebrates, to either side of a relatively low, dry saddle termed the Black Mountain Corridor (BMC) during the last glacial maximum. This was followed by a rapid expansion of these forests at 8 thousand years ago (kya) (68, 101) reaching a maximum extent in a cool wet phase between 7.5 and 6 kya, which would have facilitated migration between areas of upland rainforest, even those that are now isolated.

Molecular comparisons of endemic species indicate that very few are likely to have arisen within the Pleistocene, with most probably representing Miocene divergences (see above). This, together with evidence from geographic patterns of species richness (154), indicates that the major effect of mid-late Pleistocene climate fluctuations has been local extinction rather than speciation. In the absence of recent radiations within genera so far examined, we focus on the patterns of molecular and morphological variation within widely distributed endemic species.

Phylogeography of mtDNA has been examined for 13 species of vertebrate (five birds, four frogs, four lizards): all but three birds (two of which are not rainforest specialists) show a congruent phylogeographic break located at or adjacent to the BMC (74, 124, 126; M Cunningham, unpublished data). Levels of sequence divergence across the BMC range from <1% to 12%, with a mean of 0.84% for birds, 7.58% for frogs, and 7.35% for lizards. For the herpetofauna, most (7/8) divergence values across the BMC are >4%, indicating largely pre-Pleistocene isolation of these high-elevation communities. Within areas, there is evidence from the gene trees for recent population reductions followed by (presumably

Holocene) range expansions (124, 125) as required by the refuge model (Table 1). Notably, it appears that the community of rainforest-dependent vertebrates from the southernmost block has been largely or wholly reconstituted by migration from the north during the Holocene expansion events (124). Two rainforest species examined, a high-elevation bird (*Sericornis keri*, 74) and a low-mid elevation *Drosophila* (*D. birchii*, 75), showed exceptionally low diversity and no geographic structure across the wet tropics, suggestive of recent restriction to a single region.

The evidence for both refugial isolation, in some cases extending back millions of years, and recent range expansions provides an unusual opportunity to evaluate critically the effect of isolation combined with population bottlenecks on morphological differentiation within species. The result is straightforward—little or no differentiation in ecomorphological traits was observed in the lizards (125, 126) or frogs (M Cunningham, unpublished data). Detailed morphometric comparisons have not yet been made for the birds, but it is notable that there are no pairs of subspecies separated by the BMC. Experimental tests for reproductive isolation between the isolated populations have not been carried out. However, genetic analyses of secondary contact zones in three taxa (a frog, skink, and mammal; M Cunningham, B Phillips, L Pope, respectively, unpublished data) did not reveal linkage disequilibrium or consistent heterozygote deficiency, as would be expected with strong reproductive isolation.

The Wet Tropics system provides an ideal test case for the refuge model, even allowing for pre-Pleistocene events. The paleoecological and molecular evidence for long-term isolation combined with late Pleistocene contractions and subsequent range expansions is compelling, yet there is little detectable effect on morphological or reproductive divergence where this has been examined. While not all species are expected to diversify in response to refugial isolation and drift (85), the consistency of these observations across taxa must cast doubt on the general validity of the refugia hypothesis. More detailed studies of zones of secondary contact between the northern and southern lineages are needed to substantiate our conclusion.

Riverine Barrier Model

Few studies have used molecular approaches to investigate the role of rivers as boundaries demarcating diversified taxa within Amazonia, and none, to our knowledge, have been concerned with such a role in other lowland tropical forest regions of the globe. Results of the few molecular studies designed to test explicitly for riverine isolation (22, 23, 50, 56, 81, 108, 110) tend to be equivocal, as might be expected, since the question must be asked for each individual taxon for each of the major river systems of Amazonia. Capparella's research on birds (22, 23) shows a riverine effect, although its extent is unknown given the short section of the Rio Amazonas examined. Broad-scale phylogeographic (mtDNA) analyses of felids (ocelots, margays) revealed a congruent division across the Rio Amazonas (38), although sample sizes were small.

Divergence patterns for both frogs (50, 81) and small mammals (108–110) have been examined for the Rio Juruá, western Brazil, with the same sampling design, using allozymes and/or mtDNA sequences. Of the 29 species or species pairs of small mammals (7 marsupials, 1 primate, and 21 rodents), a clear riverine effect was observed only for subspecies of the saddle-back tamarin (*Saguinus fuscicollis*; 114). However, phylogenetic analyses of the entire species complex (72) suggest that the Rio Juruá is only a point of secondary contact and was not involved in primary divergence (107). A riverine effect is plausible for two other species (*Oecomys* sp. and *Proechimys echirothrix*, but the geographic extent of sampling for these is inadequate. In all other cases, no riverine effect is apparent. In 15 of these taxa, little molecular divergence is present, and individual haplotypes are broadly distributed on both banks and along the entire length. In the remaining 11 taxa, deep divergences between sister lineages are concordant with their geographic placements along the river. The turnover point is perpendicular to the river, not along its bank, and is coincident with both paleogeographic features and a transition between present forest formations (110). The Rio Juruá also fails to form a barrier among any of the frog taxa examined to date, although the same phylogeographic break related to paleogeography and/or forest type is apparent (50, 81).

Gradient Models

In Central Africa, Smith et al (141, 142) evaluated the Gradient (divergence with gene flow) model in several species of birds found both in central rainforest and the vast (often greater than 1,000 km wide) forest-savanna mosaic that surrounds it. Central African rainforest contrasts sharply with ecotone habitats in both climate and ecology (93). Ecotonal forest patches experience both lower annual rainfall and greater annual variation in rainfall than the main rainforest (80). In addition to these climatic differences, there are also large differences in community structure, involving both fauna and flora. For example, mosaic habitats include fewer numbers of forests species in addition to savanna species absent from the forest. These physical and ecological differences have resulted in contrasting selective environments and corresponding differences in morphology (25) of the organisms that live in them. Endler (40) estimated that 52% of the contact zones between sister species in Africa occurred between forest and savanna, not between historical refugia as predicted by the forest refugia hypothesis. More recently, Fjeldså (44) concluded that newly evolved species were concentrated in the transitional ecotones and montane environments that surround the main central African rainforest.

Smith and colleagues (139, 141, 142) compared morphological divergence within and between habitats for three widely distributed and phylogenetically distinct species of birds in Central Africa. These include the little greenbul (*Andropadus virens*), black-bellied seedcracker (*Pyrenestes ostrinus*), and olive sunbird (*Nectarinia olivacea*). Genetic divergence as a measure of gene flow was also examined within two of the species (cf Figure 1). In the little greenbul (141), the relationship between gene flow and morphologic divergence in heritable characters was found to support a central prediction of the Gradient model (39, 117). Ecotone and forest populations diverged more morphologically per unit gene flow than did populations from the same habitat. Finally, the magnitude of morphological differences between ecotone and forest populations were similar to that seen between species, suggesting that differential selection was resulting in species level differences in morphology.

The agent of selection causing the morphological differences between the two habitats was not identified. However, the direction of morphological change, particularly the finding that ecotone populations of little greenbuls have longer wings, is consistent with the need for greater aerodynamic efficiency in more open habitats (141) because of increased predation by aerial predators (12, 123).

Causes of habitat-associated morphological shifts in the black-bellied seedcracker are clearer. Seedcrackers show a polymorphism in bill size (133, 138). Small- and large-billed morphs interbreed throughout their range. However, they differ in feeding ecologies (134, 136, 137), in diet, and in feeding efficiencies on sedges, which results in disruptive selection on bill width (138). In contrast to rainforest populations in which only small- and large-billed morphs occur (135), a "mega-billed" morph is found in ecotones (139). Mega-billed morphs specialize on a species of sedge found only in ecotonal areas that has seeds too hard for either small- or large-billed morphs to crack (139). Despite high levels of gene flow with central rainforest populations (mtDNA control region Nm = 7), the mega-billed morph is maintained because of its selective advantage for feeding on these extremely hard seeds (139).

Examination of morphological and geographic distance in populations of the olive sunbird revealed a similar pattern of divergence between forest and ecotones (142). As in both other species, morphological divergence is greater between than within habitats, regardless of the geographic distance separating populations. Although estimates of gene flow are unavailable, populations of ecotones and forests separated by only 200 km are more morphologically divergent than forest populations separated by more than 600 km.

None of these studies examined levels of reproductive isolation, but Orr & Smith (102) suggested substituting morphological divergence with an index of reproductive divergence and comparing levels of gene flow within and between habitats to examine whether reproductive divergence is higher for between-habitat comparisons per unit gene flow than within-habitat comparison. Indices of reproductive divergence could be obtained from mate choice experiments or analyses of genetically based signaling characters important in mate recognition (102).

A similar pattern of morphological divergence between habitats was found in a species of ground-dwelling rainforest skink (*Carlia*) in the Wet Tropics of Australia (126). Despite substantial gene flow, adult males and females were found to be about 15% smaller in body size in open (wet sclerophyll) forests than in adjacent rainforest and also to differ in head and limb proportions independent of size. Although this variation could partially result from phenotypic plasticity, it is notable that life history also differs, with adults reproducing at an earlier age in open habitats. Like other endemic rainforest vertebrates in the Wet Tropics (see above), *C. rubrigularis* shows a major phylogeographic break in mtDNA variation across the BMC (>12% sequence divergence). Differences in ecomorphological traits across the BMC were trivial relative to those between habitats, despite the long period of isolation between regions. These observations of intraspecific variation appear more consistent with the gradient model than the refuge model. Furthermore, using model lizards in the two habitats, Schneider et al (126) identified increased predation by birds as a possible agent of selection for smaller body size in the open forest habitat.

Explicit tests of the role of elevational gradients in speciation are scarce. Patton and colleagues (104, 111) applied the phylogenetic approach (Figure 1) to species of akodontine and oryzomyine rodents distributed across geographically adjacent elevational transects. The results rejected the gradient model and supported a vicariance hypothesis, as sister-species were distributed between geographically adjacent areas of similar altitude rather than among different elevations along each montane transect.

For African greenbuls, Smith et al (142) found that morphological divergence per unit geographic distance was as high between central lowland rainforest and mountain populations as it was between lowland forest and ecotone populations. In contrast, other comparisons between habitats, and those among mountain populations, showed much less morphological divergence. Therefore, isolation on mountains alone, at least for greenbuls, does not appear to cause morphological divergence. These results suggest that elevational gradients may be an important factor leading to morphological divergence.

SYNTHESIS

Tempo and Causes of Speciation

Molecular data, when appropriately integrated with paleoecology, biogeography, and studies of phenotypic variation and reproductive divergence (13), have great potential to inform theories about the factors that generate species diversity in tropical rainforest faunas and how these faunas have changed through time. Given that such investigations have only recently begun in earnest, it is no surprise that this potential has not yet been realized. Despite their limitations, the results to date have sharpened the questions and suggested improved ways of addressing previously intractable problems.

The analyses of intraspecific diversity described above provide general support for the effectiveness of selection gradients in generating phenotypic diversity and are consistent with several requirements of the gradient (divergence with gene flow) model. Further, they suggest that geographic isolation per se, even when combined with population bottlenecks and subsequent range expansions, does not generate substantial diversity in phenotypes. There are few cases where the gradient and refuge models have been contrasted directly for the same system.

However, juxtaposition of the first two observations above suggests that differential selection, such as that occurring among populations inhabiting different forest types, is a primary determinant of phenotypic variation within species of rainforest fauna, with its effectiveness depending on the balance between selection pressure and gene flow. This conclusion is hardly novel (39) but is worth stating given the prominence accorded to genetic drift in many recent discussions (119). Given that selection-driven divergence can occur without complete suppression of gene flow, we suggest that greater attention should be given to ecological factors that promote selection gradients, with less attention on whether the geographic context is allopatric, parapatric, or sympatric (102, 121, 141). The role of sexual selection in promoting speciation (116) in rainforest faunas has yet to be explored.

The link between factors driving intraspecific diversity and those causing speciation is still to be established. While the divergence with gene flow model underlying the gradient hypothesis has a sound theoretical basis (39) and is supported by experimental studies of *Drosophila* (39, 117), it remains to be demonstrated unequivocally that selection across ecological gradients is a major cause of speciation in tropical faunas. Whereas the intraspecific analyses tend to support the gradient model, the few tests employing interspecific phylogeny did not (104). Data from both approaches are too sparse to warrant further speculation, but one thing is clear—we need to integrate studies of intraspecific molecular, phenotypic, and reproductive variation with those of interspecies phylogenies in order to rigorously distinguish between the alternative hypotheses about speciation mechanisms (Table 1). Such studies need to be extended to taxa that can be manipulated experimentally.

The foregoing review does provide one clear result—the great majority of vertebrate species so far examined appear to have diverged from their extant sister species well before the onset of the climatic perturbations of the Pleistocene. Further, there is tantalizing evidence that relatively young taxa are geographically concentrated, occurring predominantly in areas that have been recently active geologically and in ecotonal regions. Whether these patterns reflect geographic differences in rates of extinction, speciation, or both remains to be seen, but we note that higher speciation rates in areas with strong habitat heterogeneity is consistent with the potential for speciation via diversifying selection. That species within most genera examined to date tend to be very old has limited the inferences that can be made about speciation mechanisms. In this context, it would be rewarding to focus more attention on species-rich lineages occupying geographic regions with concentrations of young species.

The substantial age of most species examined raises one intriguing question—given that any one of the mechanisms suggested has the potential to cause rapid speciation, particularly if connected with shifts in mate choice, why are there not more young species? We expect that the rapid shifts in the composition and distribution of vegetation and associated climates during the last two glacial cycles would have generated abundant opportunity for speciation via selection gradients or differential selection among isolates, but there is little evidence for this in tropical systems. By contrast, there are numerous examples of recent adaptive radiations or speciation events in postglacial temperate environments, on islands, or in recently formed lakes (52, 54, 65). There are several possible, nonexclusive reasons why this might be so. First, it may be that late Pleistocene fluctuations did cause numerous speciation events, but that the derived taxa were subject to a high extinction rate relative to ancestral or unaffected taxa. Second, it may be that most rainforest (and adjacent) environments lacked the ecological opportunity for survival of new species that is characteristic of islands, recently formed lakes, etc. Third, perhaps species responded to rapid climate change predominantly by migration rather than adaptive differentiation (112), so that populations remained in structurally similar environments with morphology under stabilizing selection. Further progress on this question will require close integration of molecular and morphological information with information from the fossil record and paleoecology.

While some progress is being made with rainforest vertebrates, comparable studies of invertebrates and plants are mostly lacking. Given the prominence of tropical arthropods and plants in accounts of global and tropical species diversity, this is clearly a major omission. Neotropical butterflies are an exception (reviewed by 15, 18, 89). Almost no attention has been paid by to the evolutionary dynamics of species interactions in tropical systems, a dominant theme in ecological perspectives on tropical rainforest diversity. Molecular analyses of interacting species [e.g. host plants and phytophagous arthropods (43), host-parasite systems, mutualisms, etc.] would be extremely rewarding in this respect and are becoming more plausible as taxonomic and biogeographic data are improving for some groups.

Implications for Conservation Planning

That tropical rainforests and the species they sustain are under threat from habitat destruction and degradation has been emphasized repeatedly, although not all accounts are uniformly negative (16). In response to this and the recommendations of the 1992 International Convention on Biological Diversity, there is an effort by global and national agencies to prioritize and set aside areas for conservation (e.g. reserves) and sustainable development. How can a better understanding of the tempo and causes of diversification of rainforest taxa inform this process?

One important contribution is to identify geographic areas that represent concentrations of old endemic species versus those where recently derived species are concentrated. The work of Fjeldså and colleagues (44, 46) is notable in this respect. While areas of both types might be rich in endemic species (i.e. hotspots), the reasons for conserving them and the optimal strategies differ. The former are important as reservoirs of phylogenetic diversity (34, 147) and their protection should focus on maintaining sufficiently large tracts of intact rainforest to preserve the endemics. Areas with concentrations of young species are important from the context of maintaining the evolutionary processes that generate species and phenotypic diversity (42, 94, 140, 141). Several such areas, for example, the central African ecotones, East African rift mountains, Andean slopes, and their parallels elsewhere warrant increased attention from conservation agencies, with particular attention to maintaining the viability, representation, and connectedness of habitats that contribute to these heterogeneous landscapes.

Protection of intraspecific genetic diversity is also prominent in conventions and conservation policies. In this context, we can distinguish between historical and adaptive components of diversity (94)—the former focusing on representation of historically isolated areas, which can be identified by comparative molecular phylogeography combined with paleoecology (2, 95), and the latter focusing on maintaining viable populations within heterogeneous landscapes (141). Conservation priorities based on complementarity of species distributions do not necessarily predict those based on identification of congruent genetic divisions within species (95; T Smith, unpublished data). Moritz (94, 95) suggested an iterative approach whereby areas of high values from species irreplaceability and historical isolation within species are combined. Within such areas, it is essential to maintain the integrity and heterogeneity of habitats in order to protect the adaptive component of intraspecific diversity.

Finally, along with other evidence, comparative molecular phylogeography can reveal how historical fluctuations have shaped the current distribution and composition of communities (124). These studies, along with paleoecological data (68, 69), are providing unambiguous evidence that individual species and ecological communities are highly dynamic in space in time. By contrast, most conservation assessments are essentially static, seeking to identify areas that maximize the representation of species, community, and environmental diversity under present conditions. While these efforts are important and must continue, planners need to ensure that such areas are extensive enough to permit natural dynamic processes to continue. Again, maintaining heterogeneous landscapes on the margins as well as interior of rainforests is essential. Further, as the necessary geographic scale will often be greater than can be reasonably be accommodated within conservation reserves and there is extensive human pressure on landscapes adjacent to rainforests, there is urgent need to focus on conservation strategies and sustainable development outside of protected areas.

LITERATURE CITED

1. Avila-Pires TCS. 1995. Lizards of Brazilian Amazonia (Reptilia: Squamata). *Zool. Verh.* 299:1–706
2. Avise JC. 1992. Molecular population structure and the biogeographic history of a regional fauna: a case history with lessons for conservation biology. *Oikos* 63:62–76
3. Avise JC. 2000. *Phylogeography: The History and Formation of Species*. Cambridge, MA: Harvard Univ. Press
4. Avise JC, Walker D. 1998. Pleistocene phylogeographic effects on avian population and the speciation process. *Proc. R. Soc. London Ser. B* 1998:547–63
5. Avise JC, Walker D, Johns GC. 1998. Speciation durations and Pleistocene effects on

vertebrate phylogeography. *Proc. R. Soc. London Ser. B* 265:1707–12
6. Ayres JM, Clutton-Brock TH. 1992. River boundaries and species range size in Amazonian primates. *Am. Nat.* 140:531–37
7. Balmford A, Mace GM, Ginsgerg JR. 1999. The challenges to conservation in a changing world: putting processes on the map. In *Conservation in a Changing World*, ed. GM Mace, A Balmford, JR Ginsburg. Cambridge, UK: Cambridge Univ. Press
8. Barrowclough TG, Vogler AP, Harvey PH. 1998. Revealing the factors that promote speciation. *Philos. Trans. R. Soc. London Ser. B* 353:241–49
9. Barton N, Gale K. 1993. Genetic analysis of hybrid zones. In *Hybrid Zones and the Evolutionary Process*, ed. RG Harrison, Vol. 2, pp. 13–45. Oxford, UK: Oxford Univ. Press
10. Barton NH, Hewitt GM. 1985. Analysis of hybrid zones. *Annu. Rev. Ecol. Syst.* 16:113–48
11. Bates JM, Zink RM. 1994. Evolution into the Andes: molecular evidence for species relationships in the genus *Letopogon*. *Auk* 111:507–15
12. Benkman CW. 1991. Predation, seed size partitioning and the evolution of body size in seed-eating finches. *Evol. Ecol.* 5:118–27
13. Brawn JD, Collins TM, Medina M, Bermingham E. 1996. Associations between physical isolation and geographical variation within three species of neotropical birds. *Mol. Ecol.* 5:33–46
14. Brower AVZ. 1994. Rapid morphological radiation and convergence among races of the butterfly *Heliconius erato* inferred from patterns of mitochondrial DNA evolution. *Proc. Natl. Acad. Sci. USA* 91:6491–95
15. Brower AVZ. 1996. Parallel race formation and the evolution of mimicry in *Heliconius* butterflies: a phylogenetic hypothesis from mitochondrial DNA sequences. *Evolution* 50:195–221
16. Brown KS, Brown GG. 1992. Habitat alteration and species loss in Brazilian forests. In *Tropical Deforestation and Species Extinction*, ed. T Whitmore, JA Sayer, pp. 119–42. London: Chapman & Hall
17. Brown KS Jr. 1982. Historical and ecological factors in the biogeography of aposematic neotropical butterflies. *Am. Zool.* 22:453–71
18. Brown KS Jr. 1987. Conclusions, synthesis, and alternative hypotheses. In *Biogeography and Quaternary History of Tropical America*, ed. TC Whitmore, KS Brown, pp. 175–96. New York: Oxford Univ. Press
19. Bush M. 1994. Amazonian speciation: a necessarily complex model. *J. Biogeogr.* 21:5–17
20. Bush MB. 1996. Amazonian conservation in a changing world. *Biol. Conserv.* 76:219–28
21. Cannatella DC, Hillis DM, Chippindale PT, Weigt L, Rand AS, Ryan MJ. 1998. Phylogeny of frogs of the *Physalaemus pustolosus* species group, with an examination of data incongruence. *Syst. Biol.* 47:311–35
22. Capparella AP. 1988. Genetic variation in neotropical birds: implications for the speciation process. *Acta Congr. Int. Ornithol.* 19:1658–64
23. Capparella AP. 1991. Neotropical avian diversity and riverine barriers. *Acta Congr. Int. Ornithol.* 20:307–16
24. Cardillo M. 1999. Latitude and rates of diversification in birds and butterflies. *Proc. R. Soc. London Ser. B.* 266:1–5
25. Chapin JP. 1932. The birds of the Belgium Congo. *Bull. Am. Mus. Nat. Hist.*
26. Christoff AU. 1997. *Contribucao a sistematica das especies do genero Akodon (Rodentia: Sigmodontinae) do leste do Brasil: estudos anatomicos, citogeneticos e de distribuicao geographica*. DSc thesis. Univ. Sao Paulo
27. Cody ML. 1996. Introduction to neotropical diversity. In *Neotropical Diversity and*

Conservation, ed. AC Gibson, pp. 1–20. Los Angeles: Univ. Calif.
28. Colinvaux P. 1993. Pleistocene biogeography and diversity in tropical forests of South America. In *Biological Relationships Between Africa and South America*, ed. P Goldblatt, 16:473–99. New Haven, CT: Yale Univ. Press
29. Colinvaux P, De Oliveira P, Moreno J, Miller M, Bush M. 1996. A long pollen record from lowland Amazonia: forest and cooling in glacial times. *Science* 274:85–88
30. Cracraft J. 1985. Historical biogeography and patterns of differentiation within the South American avifauna: areas of endemism. In *Neotropical Ornithology, Ornith. Monogr. 36*, ed. P Buckley, MS Foster, ES Morton, RS Ridgely, FG Buckley, pp. 49–84. Washington DC: Am. Ornithol. Union
31. Cracraft J. 1989. Speciation and its ontology: the empirical consequences of alternative species concepts for understanding patterns and processes of differentiation. In *Speciation and Its Consequences*, ed. D Otte, JA Endler, pp. 28–59. Sunderland, MA: Sinauer
32. Cracraft J, Prum RO. 1988. Patterns and processes of diversification: speciation and historical congruence in some neotropical birds. *Evolution* 42:603–20
33. Creighton GK. 1985. Phylogenetic inference, biogeographic interpretations, and the pattern of speciation in *Marmosa* (Marsupialia: Didelphidae). *Acta Zool. Fenn.* 170:121–24
34. Crozier RH. 1997. Preserving the information content of species: genetic diversity, phylogeny, and conservation worth. *Annu. Rev. Ecol. Syst.* 28:243–68
35. da Silva M, Patton J. 1993. Amazonian phylogeography: mtDNA sequence variation in arboreal echimyid rodents (Caviomorpha). *Mol. Phylogenet. Evol.* 2:243–55
36. da Silva MNF, Patton JL. 1998. Molecular phylogeography and the evolution and conservation of Amazonian mammals. *Mol. Ecol.* 7:475–86
37. Diamond AW, Hamilton AC. 1980. The distribution of forest passerine birds and quaternary climate change in tropical Africa. *J. Zool.* 191:379–402
38. Eizirik E, Bonatto SL, Johnson WE, Crawshaw PG, Vie JC, et al. 1998. Phylogeographic patterns and evolution of the mitochondrial DNA control region in two neotropical cats (Mammalia, Felidae). *J. Mol. Evol.* 47:613–24
39. Endler JA. 1977. *Geographic Variation, Speciation, and Clines*. Princeton, NJ: Princeton Univ. Press
40. Endler JA. 1982. Pleistocene forest refuges: fact or fancy. In *Biological Diversification in the Tropics*, ed. GT Prance. New York: Columbia Univ. Press
41. Endler JA. 1982. Problems in distinguishing historical from ecological factors in biogeography. *Am. Zool.* 22:441–52
42. Erwin TL. 1991. An evolutionary basis for conservation strategies. *Science* 253:750–52
43. Farrell BD, Mitter C. 1993. Phylogenetic determinants of insect/plant community diversity. In *Species Diversity in Ecological Communities*, ed. RE Ricklefs, D Schluter, pp. 253–66. Chicago, IL: Univ. Chicago Press
44. Fjeldså J. 1994. Geographical patterns for relict and young species of birds in Africa and South America and implications for conservation priorities. *Biodivers. Conserv.* 3:207–26
45. Fjeldså J. 1995. Geographic patterns of neoendemic and older relic species of Andean forest birds: the significance of ecologically stable areas. In *Biodiversity and Conservation of Neotropics Forests*, eds. SP Churchill, H Balslev, E Forero, JL Luteyn, pp. 89–102. New York: NY Bot. Gard.
46. Fjeldså J, Lovett JC. 1997. Geographical

patterns of old and young species in African forest biota: the significance of specific montane areas as evolutionary centres. *Biodivers. Conserv.* 6:325–46

47. Garcia-Moreno J, Arctander P, Fjeldså J. 1998. Pre-Pleistocene differentiation among chat-tyrants. *Condor* 100:629–40
48. Garcia-Moreno J, Arctander P, Fjeldså J. 1999. A case of rapid diversification in the neotropics: phylogenetic relationships among *Cranioleuca* spinetails (Aves, Furnariidae). *Mol. Phylogenet. Evol.* 12:273–81
49. Garcia-Moreno J, Arctander P, Fjeldså J. 1999. Strong diversification at the treeline among *Metallura* hummingbirds. *Auk* 116:702–11
50. Gascon CS, Lougheed C, Bogart JP. 1996. Genetic and morphological variation in *Vanzollinius discodactylus*: a test of the river hypothesis of speciation. *Biotropica* 28:376–87
51. Gaston KJ, Blackburn TM. 1996. The tropics as a museum of biological diversity: analysis of the New World avifauna. *Proc. R. Soc. London Ser. B* 263:63–68
52. Givnish TJ, Systma KJ. 1997. *Molecular Evolution and Adaptive Radiations.* Cambridge, UK: Cambridge Univ. Press
53. Good D, Wake D 1997. Phylogenetic and taxonomic implications of protein variation in the Mesoamerican salamander genus *Oedipina* (Caudata: Plethodontidae). *Rev. Biol. Trop.* 45:1185–208
54. Grant PR. 1998. *Evolution on Islands.* Oxford, UK: Oxford Univ. Press
55. Graybeal A. 1997. Phylogenetic relationships of bufonid frogs and tests of alternate macroevolutionary hypotheses characterizing their radiation. *Zool. J. Linn. Soc.* 119:297–338
56. Hackett SJ. 1996. Molecular phylogenetics and biogeography of tanagers in the genus *Ramphocelus* (Aves). *Mol. Phylogenet. Evol.* 5:368–82
57. Haffer J. 1969. Speciation in Amazonian forest birds. *Science* 165:131–37
58. Haffer J. 1974. *Avian speciation in tropical South America. Nuttall Ornithol. Club Publ.* 14:1–390. Cambridge MA: NuHall Ornithol. Club
59. Haffer J. 1978. Distribution of Amazonian forest birds. *Bonner Zool. Beitr.* 38:38–78
60. Haffer J. 1992. On the "river effect" in some forest birds of southern Amazonia. *Bol. Mus. Para. E. Goldi Sér. Zool.* 8:217–45
61. Haffer J. 1993. Time's cycle and time's arrow in the history of Amazonia. *Biogeographica* 69:15–45
62. Haffer J. 1997. Alternative models of vertebrate speciation in Amazonia: an overview. *Biodivers. Conserv.* 6:451–77
63. Haffer J. 1997. Contact zones between birds of southern Amazonia. *Ornithol. Monogr.* 48:281–305
64. Hershkovitz P. 1977. *Living New World Monkeys* (*Platyrhini*), Vol. 1. Chicago, IL: Univ. Chicago Press
65. Hewitt GM. 1996. Some genetic consequences of ice ages, and their role in divergence and speciation. *Biol. J. Linn. Soc.* 58:247–76
66. Heyer WR, Maxson LR. 1983. Relationships, zoogeography, and speciation mechanisms of frogs of the genus *Cycloramphus* (Amphibia, Leptodactylidae). *Arq. Zool. São Paulo* 30:341–73
67. Hillis D, Mable B, Moritz C. 1996. Applications of molecular systematics. In *Molecular Systematics*, ed. DM Hillis, C Moritz, BK Mable, Vol. 12, pp. 515–43. Sunderland, MA: Sinauer
68. Hopkins MS, Ash J, Graham AW, Head J, Hewett RK. 1993. Charcoal evidence of the spatial extent of the *Eucalyptus* woodland expansions and rainforest contractions in north Queensland during the late Pleistocene. *J. Biogeogr.* 20:59–74
69. Hopkins MS, Head J, Ash JE, Hewett RK, Graham AW. 1996. Evidence of a Holocene and continuing expansion of lowland rainforest in humid, tropical North Queensland. *J. Biogeogr.* 23:737–45

70. Howard DJ, Berlocher SH, eds. 1998. *Endless Forms: Species and Speciation.* New York: Oxford Univ. Press
71. Hubbell SP. 2001. *The Unified Theory of Biodiversity and Biogeography.* Princeton, NJ: Princeton Univ. Press
72. Jacobs S, Larson A, Cheverud JM. 1995. Phylogenetic relationships and orthogenetic evolution of coat colour among tamarins (genus *Saguinus*). *Syst. Biol.* 44:515–32
73. Joseph L. 1994. *A molecular approach to species diversity and evolution in eastern Australian rainforest birds.* PhD thesis. Univ. Queensland
74. Joseph L, Moritz C, Hugall A. 1995. Molecular support for vicariance as a source of diversity in rainforest. *Proc. R. Soc. London Ser. B* 260:177–82
75. Kelemen L, Moritz C. 1999. Comparative phylogeography of a sibling pair of rainforest Drosophila species (*Drosophila serrata* and *Drosophila birchii*). *Evolution.* 53:1306–11
76. Kershaw AP. 1994. Pleistocene vegetation of the humid tropics of northeastern Queensland, Australia. *Palaeogeogr. Palaeoclimatol. Palaeoecol.* 109:399–412
77. Klicka J, Zink RM. 1997. The importance of recent ice ages in speciation: a failed paradigm. *Science* 277:1666–69
78. Lara MC, Patton JL. 2000. Phylogeography of terrestrial spiny rats (*Trinomys*, Echimyidae) of the Atlantic forests of Brazil. *Zool. J. Linn. Soc.* In press
79. Lara MC, Patton JL, da Silva MNF. 1996. The simultaneous diversification of South American Echimyid rodents (Hystricognathi) based on complete cytochrome b sequences. *Mol. Phylogenet. Evol.* 5:403–13
80. Longman KA, Jenik J. 1992. Forest-savanna boundaries: general considerations. In *Nature and Dynamics of Forest-Savanna Boundaries*, ed. PA Furley, J Proctor, JA Ratter, pp. 3–20. New York: Chapman & Hall
81. Lougheed SC, Gascon C, Jones DA, Bogart JP, Boag PT. 1999. Ridges and rivers: a test of competing hypotheses of Amazonian diversification using a dart-poison frog (*Epipedobates femoralis*). *Proc. R. Soc. London Ser. B* 266:1829–35
82. Lovejoy NRE, Bermingham E, Martin AP. 1998. Marine incursion in South America. *Nature* 396:421–22
83. Lu G, Bernatchez L. 1999. Correlated trophic specialisation and genetic divergence in sympatric lake whitefish ecotypes (*Coregonus clupeaformis*): support for the ecological speciation hypothesis. *Evolution* 53:1491–505
84. Lundberg JG, Marshall LG, Guerrero J, Horton B, Malabarba MCSL, Wesselingh F. 1998. The stage for neotropical fish diversification: a history of South American rivers. In *Phylogeny and Classification of Neotropical Fishes*, ed. MCSL Malabarba, RE Reis, RP Vari, ZM Lucena, CAS Lucena, pp. 13–48. Porto Alegre, Brazil: Edipucrs
85. Lynch JD. 1988. Refugia. In *Analytical Biogeography*, ed. AA Myers, PS Giller, pp. 311–42. New York: Chapman & Hall
86. Lynch M, Pfender M, Spitze K, Lehman N, Hicks J, et al. 1999. The quantitative and molecular genetic architecture of a subdivided species. *Evolution* 53:100–10
87. Mallet JLB. 1993. Speciation, raciation and color pattern evolution in *Heliconius* butterflies: evidence from hybrid zones. In *Hybrid Zones and the Evolutionary Process*, ed. RG Harrison, pp. 226–60. New York: Oxford Univ. Press
88. Mallet JLB. 1995. A species definition for the modern synthesis. *Trends Ecol. Evol.* 10:294–99
89. Mallet JLB, Turner JRG. 1998. Biotic drift or the shifting balance—did forest islands drive the diversity of warningly coloured butterflies. In *Evolution on Islands*, ed. PR Grant, pp. 262–80. Oxford, UK: Oxford Univ. Press
90. Mariaux J, Braun MJ. 1996. A molecular phylogenetic survey of the nightjars and

allies (Caprimulgiformes) with special emphasis on the potoos (Nyctibiidae). *Mol. Phylogenet. Evol.* 6:228–44

91. Mayr E. 1963. *Animal Species and Evolution*. Cambridge MA: Belknap
92. Mayr E, O'Hara RJ. 1986. The biogeographical evidence supporting the Pleistocene forest refuge hypothesis. *Evolution* 40:55–67
93. Millington AC, Styles PJ, Critchley RW. 1992. Mapping forests and savannas in sub-Saharan Africa from advanced very high resolution radiometer (AVHRR) imagery. In *Nature and Dynamics of Forest-Savanna Boundaries*, ed. PA Furley, J Proctor, JA Ratter, pp. 37–62. New York: Chapman & Hall
94. Moritz C. 1999. A molecular perspective on the conservation of diversity. In *The-Biology of Biodiversity*, ed. M Kato, pp. 21–34. Tokyo: Springer-Verlag
95. Moritz C, Faith D. 1998. Comparative phylogeography and the identification of genetically divergent areas for conservation. *Mol. Ecol.* 7:419–29
96. Moritz C, Joseph L, Cunningham M, Schneider CJ. 1997. Molecular perspectives on historical fragmentation of Australian tropical and subtropical rainforest: implications for conservation. In *Tropical Rainforest Remnants: Ecology, Management and Conservation of Fragmented Communities*, ed. WF Laurance, RO Bieregard, pp. 442–54. Chicago: Chicago Univ. Press
97. Mustrangi MA, Patton JL. 1997. Phylogeography and systematics of the slender mouse opossum *Marmosops* (Marsupialia: Didelphidae). *Univ. Calif. Publ. Zool.* 130:1–86
98. Nee S, Holmes EC, May RM, Harvey PH. 1994. Extinction rates can be estimated from molecular phylogenies. *Philos. Trans. R. Soc. London Ser. B* 344:77–82
99. Nee S, Mooers A, Harvey P. 1992. Tempo and mode of evolution revealed from molecular phylogenies. *Proc. Natl. Acad. Sci. USA* 89:8322–26
100. Nelson BW, Ferreira CAC, da Silva MF, Kawaski ML. 1990. Endemism centres, refugia and botanical collection density in Brazilian Amazonia. *Nature* 345: 714–16
101. Nix HA. 1991. Biogeography: patterns and process. In *Rainforest Animals: Atlas of Vertebrates Endemic to Australia's Wet Tropics*, ed. HA Nix, M Switzer, pp. 11–39. Canberra: Commonwealth Aust.
102. Orr MR, Smith TB. 1998. Ecology and speciation. *Trends Ecol. Evol.* 13: 502–6
103. Otte D, Endler JA, eds. 1987. *Speciation and Its Consequences*. Sunderland, MA: Sinauer
104. Patton J, Smith MF. 1992. MtDNA phylogeny of Andean mice: a test of diversification across ecological gradients. *Evolution* 46:174–83
105. Patton J, Smith M. 1994. Paraphyly, polyphyly, and the nature of species boundaries in pocket gophers (genus *Thomomys*). *Syst. Biol.* 43:11–26
106. Patton JL, da Silva MNF. 1997. Definition of species of pouched four-eyed opossums (Didelphidae, Philander). *J. Mammal.* 78:90–102
107. Patton JL, da Silva MNF. 1998. Rivers, refuges and ridges: the geography of speciation of Amazonian mammals. In *Endless Forms: Species and Speciation*, ed. DJ Howard, SH Berlocher, pp. 203–13. New York: Oxford Univ. Press
108. Patton JL, da Silva MNF, Malcolm JR. 1994. Gene genealogy and differentiation among arboreal spiny rats (Rodentia: Echimyidae) of the Amazon basin: a test of the riverine barrier hypothesis. *Evolution* 48:1314–23
109. Patton JL, da Silva MNF, Malcolm JR. 1996. Hierarchical genetic structure and gene flow in three sympatric species of Amazonian rodents. *Mol. Ecol.* 5: 229–38

110. Patton JL, da Silva MNF, Malcolm JR. 2000. Mammals of the Rio Juruá and the evolutionary and ecological diversification of Amazonia. *Bull. Am. Mus. Nat. Hist.* 244:1–306
111. Patton JL, Myers, P, Smith MF. 1990. Vicariant versus gradient models of diversification: the small mammal fauna of eastern Andean slopes of Peru. In *Vertebrates in the Tropics*, ed. G Peters, R Hutterer, pp. 355–71. Bonn, Germany: Mus. Alexander Koenig
112. Pearse CM, Lande R, Bull JJ. 1989. A model of population growth, dispersal and evolution in a changing environment. *Ecology* 70:1657–64
113. Pearson OP, Smith MF. 1999. Genetic similarity between *Akadon olivaceus* and *Akadon xanthorhinus* (Rodentia: Muridae) in Argentina. *J. Zool.* 247:43–52
114. Peres CA, Patton JL, da Silva MNF. 1996. Riverine barriers and gene flow in Amazonian saddle-back tamarins. *Folia Primatol.* 67:113–24
115. Prance GT, ed. 1982. *Biological Diversification in the Tropics.* New York: Columbia Univ. Press
116. Price T. 1998. Sexual selection and natural selection in bird speciation. *Philos. Trans. R. Soc. London Ser. B* 353:1–12
117. Rice WR, Hostert EE. 1993. Perspective: laboratory experiments on speciation: what have we learned in forty years. *Evolution* 47:1637–53
118. Ricklefs RE, Schluter D, eds. 1993. *Species Diversity in Ecological Communities*. Chicago: Univ. Chicago Press
119. Roy M. 1997. Recent diversification in African greenbuls (Pycnonotidae: Andropadus) supports a montane speciation model. *Proc. R. Soc. London Ser. B* 264:1337–44
120. Roy MS, Torres-Mura JC, Hertel F. 1999. Molecular phylogeny and evolutionary history of the tit- tyrants (Aves: Tyrannidae). *Mol. Phylogenet. Evol.* 11:67–76
121. Schluter D. 1996. Ecological causes of adaptive radiation. *Am. Nat.* 148:S40–S64
122. Schluter D. 1998. Ecological causes of speciation. In *Endless Forms: Species and Speciation*, ed. DJ Howard, p. 114–29. Oxford, UK: Oxford Univ. Press
123. Schluter D, Repasky RR. 1991. Worldwide limitation of finch densities by food and other factors. *Ecology* 72:1763–74
124. Schneider CJ, Cunningham M, Moritz C. 1998. Comparative phylogeography and the history of endemic vertebrates in the wet tropics rainforests of Australia. *Mol. Ecol.* 7:487–98
125. Schneider CJ, Moritz C. 1999. Refugial isolation and evolution in the wet tropics rainforests of Australia. *Proc. R. Soc. London Ser. B* 266:191–96
126. Schneider CJS, Smith TB, Larison B, Moritz C. 1999. A test of alternative models of diversification in tropical rainforests: ecological gradients vs. refugia. *Proc. Natl. Acad. Sci. USA* 96:13869–73
127. Sibley CG, Ahlquist JE. 1990. *Phylogeny and Classification of Birds: A Study in Molecular Evolution*. New Haven, CT: Yale Univ. Press
128. Simpson BB, Haffer J. 1978. Speciation patterns in the Amazonian forest biota. *Annu. Rev. Ecol. Syst.* 9:497–518
129. Slatkin M. 1994 Gene flow and population structure. In *Ecological Genetics*, ed. L Real, pp. 3–17. Princeton, NJ: Princeton Univ. Press
130. Slatkin M, Hudson RR. 1991. Pairwise comparisons of mitochondrial DNA sequences in stable and exponentially growing populations. *Genetics* 129:555–62
131. Smith MF, Patton JL. 1993. The diversification of South-American murid rodents: evidence from mitochondrial-DNA sequence data for the Akodontine tribe. *Biol. J. Linn. Soc.* 50:149–77
132. Smith MF, Patton JL. 1999. Phylogenetic

relationships and the radiation of sigmodontine rodents in South America: evidence from cytochrome b. *J. Mammal. Evol.* 6:89–128

133. Smith TB. 1987. Bill size polymorphism and intraspecific niche utilization in an African finch. *Nature* 329:717–19
134. Smith TB. 1990. Natural selection on bill characters in the two bill morphs of the African finch *Pyrenestes ostrinus. Evolution* 44:832–42
135. Smith TB. 1990. Patterns of morphological and geographic variation in trophic bill morphs of the African finch *Pyrenestes. Biol. J. Linn. Soc.* 41:381–414
136. Smith TB. 1990. Resource use by bill morphs of an African finch: evidence for intraspecific competition. *Ecology* 71:1246–57
137. Smith TB. 1991. Inter- and intra-specific diet overlap during lean times between *Quelea erythrops* and bill morphs of *Pyrenestes ostrinus. Oikos* 60:76–82
138. Smith TB. 1993. Disruptive selection and the genetic basis of bill size polymorphism in the African finch. *Pyrenestes. Nature* 363:618–20
139. Smith TB. 1997. Adaptive significance of the mega-billed form in the polymophic black-bellied seedcracker *Pyrnestes ostrinus. Ibis* 139:382–87
140. Smith TB, Bruford MW, Wayne RK. 1993. The preservation of process: the missing element of conservation programs. *Biodivers. Lett.* 1:164–67
141. Smith TB, Wayne RK, Girman D, Bruford M. 1997. A role for ecotones in generating rainforest biodiversity. *Science* 276:1855–57
142. Smith TB, Wayne RK, Girman DJ, Bruford MW. 2000. Evaluating the divergence-with-gene-flow model in natural populations: the importance of ecotones in rainforests: speciation. In *Rainforests: Past and Future*, ed. C Moritz, E Bermingham, C Dick. Chicago: Univ. Chicago Press
143. Stebbins GL. 1974. *Flowering Plants, Evolution Above the Species Level.* Cambridge, MA: Harvard Univ. Press
144. Templeton AR. 1998. Nested clade analyses of phylogeographic data: testing hypotheses about gene flow and population history. *Mol. Ecol.* 7:381–97
145. Tuomisto H, Ruokolainen K, Kalliola R, Linna A, Danjoy W, Rodriguez Z. 1995. Dissecting Amazonian biodiversity. *Science* 269:63–66
146. Turner JRG. 1982. How do refuges produce biological diversity? Allopatry and parapatry, extinction and gene flow in mimetic butterflies. In *Biological Diversification in the Tropics*, ed. GT Prance, pp. 309–34. New York: Plenum.
147. Vane-Wright RI, Humphries CJ, Williams PH. 1991. What to protect: systematics and the agony of choice. *Biol. Conserv.* 55:235–54
148. Vanzolini PE, Williams EE. 1981. The vanishing refuge: a mechanism for ecogeographic speciation. *Pap. Avulsos. Zool.* 34:251–55
149. Voss RS. 1993. A revision of the Brazilian muroid rodent genus *Delomys* with remarks on thomasomyine characters. *Am. Mus. Novit.* 3073:1–44
150. Vuilleumier F, Simberloff D. 1980. Ecology versus history as determinants of patchy and insular distributions in high Andean birds. In *Evolutionary Biology*, ed. MK Hecht, WC Steere, B Wallace, Vol. 12, pp. 235–379. New York: Plenum
151. Wallace AR. 1852. On the monkeys of the Amazon. *Proc. Zool. Soc. London* 20:107–10
152. Webb L, Tracey J. 1981. Australian rainforests: pattern and change. In *Ecological Biogeography of Australia*, ed. JA Keast, pp. 605–94. The Hague: Junk
153. Whitmore TC, Prance GT, eds. 1987. *Biogeography and Quaternary History in*

Tropical America. New York: Oxford Univ. Press

154. Williams S, Pearson R. 1997. Historical rainforest contractions, localized extinctions and patterns of vertebrate endemism in the rainforests of Australia's wet tropics. *Proc. R. Soc. London Ser. B* 264:709–16
155. Willis EO. 1988. *Drymophila rubicolis* (Bertoni 1901) is a valid species (Aves, Formicariidae). *Rev. Bras. Biol.* 48:431–38

Annu. Rev. Ecol. Syst. 2000. 31:565–95

THE EVOLUTIONARY ECOLOGY OF TOLERANCE TO CONSUMER DAMAGE

Kirk A. Stowe,[1] Robert J. Marquis,[2] Cris G. Hochwender,[3] and Ellen L. Simms[1]

[1]*Department of Integrative Biology, University of California, Berkeley, California 94720-3140; e-mail kstowe@socrates.berkeley.edu*

[2]*Department of Biology, University of Missouri—St. Louis, St. Louis, Missouri 63121-4499*

[3]*Department of Biology, Vassar College, Poughkeepsie, New York 12604*

Key Words phenotypic plasticity, evolutionary constraints, compensatory growth, herbivory, plant pathogens

■ **Abstract** Recent theoretical studies suggest that the ability to tolerate consumer damage can be an important adaptive response by plants to selection imposed by consumers. Empirical studies have also found that tolerance is a common response to consumers among plants. Currently recognized mechanisms underlying tolerance include several general sets of traits: allocation patterns; plant architecture; and various other traits that may respond to consumer damage, e.g., photosynthetic rate. Theoretical studies suggest that tolerance to consumer damage may be favored under a range of conditions, even when the risk and intensity of damage varies. However, most of these models assume that the evolution of tolerance is constrained by internal resource allocation trade-offs. While there is some empirical evidence for such trade-offs, it is also clear that external constraints such as pollinator abundance or nutrient availability may also limit the evolution of tolerance. Current research also suggests that a full understanding of plant adaptation to consumers can only be achieved by investigating the joint evolution of tolerance and resistance. While tolerance to consumer damage has just recently received significant attention in the ecological literature, our understanding of it is rapidly increasing as its profound ecological and evolutionary implications become better appreciated.

INTRODUCTION

Plant tissue damage caused by consumers is an important selective force molding plant phenotypes (3, 79, 92, 108, 109). Until recently, most studies of plant adaptation to consumers focused exclusively on the evolution of traits that prevent or reduce tissue damage, i.e. resistance (3, 36, 108, 109). However, consumers may also select for traits that allow plants to maintain fitness in the face of tissue loss (92, 108, 109, 127, 140, 153a). Plant genotypes that can sustain tissue loss with little or no decrease in fitness relative to that in the undamaged state are termed tolerant of damage (108, 109). The term tolerance has also been used to describe the ability of plants to cope with other stresses (e.g. salinity, drought, heavy metals). In this paper, we drop the qualifier unless we are referring to an environmental stress other than consumer damage.

Among ecologists and evolutionary biologists, initial interest in tolerance was stimulated by several empirical studies reporting that consumer damage may increase, rather than decrease, plant productivity (89, 90, 103, 104). This remarkable observation was initially dismissed as the result of reallocation of below-ground resources to above-ground structures in perennial plants, which would eventually entail a net fitness decrement (13, 153a). However, when Paige & Whitham (107) discovered that grazed individuals of an Arizona population of *Ipomopsis aggregata* exhibited higher lifetime fitness than their ungrazed neighbors, the apparently paradoxical phenomenon of overcompensation could no longer be summarily dismissed.

Subsequent research stimulated by this seminal paper (107) focused largely on overcompensating tolerance. For example, considerable effort was devoted to determining whether other populations of *Ipomopsis* exhibit the same phenomenon (14, 15). Initially, these studies did not find overcompensating tolerance (14–16), leading to skepticism about the original result. Nevertheless, further work suggested that while overcompensating tolerance may be unusual in *Ipomopsis aggregata*, it does exist in some populations (47, 105, 106).

It is best to view overcompensating tolerance as one extreme along a continuum of plant responses to consumer damage (83). Even in populations for which the mean response to damage is incomplete tolerance, there is evidence that genetic variation in tolerance exists. Specifically, some families exhibit overcompensating tolerance, whereas others express incomplete tolerance (16, 56, 63, 137, 146). Moreover, recent studies comparing historically grazed and ungrazed plant populations indicate that repeatedly grazed populations can evolve overcompensating tolerance, even while other populations remain incompletely tolerant (74).

The ecological and evolutionary effects of tolerance differ from those of resistance in several important ways. For example, plant resistance traits may have a selective impact on herbivore traits, whereas tolerance will not (123). Further, evidence is mounting that a comprehensive theory of plant adaptation to consumption requires joint consideration of both resistance and tolerance traits (3, 38, 87, 137, 140, 146). Here, we synthesize recent literature concerning the evolutionary implications of tolerance to consumer damage and highlight numerous ecological and evolutionary questions therein.

DEFINING TOLERANCE

While resistance was originally defined in the agricultural literature as an umbrella term including both tolerance and defense (26, 108, 109, 127, 138), we follow the convention often used in the ecological literature. They use defense as the blanket term with resistance and tolerance as subcategories (3, 4, 37, 38, 64, but see 140). Thus, we consider defense as the umbrella term which includes both tolerance and resistance.

The distinction between tolerance and resistance was first described by agricultural scientists (26, 109, 108, 127), who determined that these composite traits are comprised of different sets of underlying characteristics, often controlled by different sets of genes (6, 19, 39, 44, 76, 88, 110, 116). As early as 1894, Cobb (26) distinguished between the ability to endure disease yet still "mature a far crop of grain" from the ability to resist disease attack. Painter (108) first described herbivore tolerant plants as those "surviving under levels of infestation that would kill or severely injure susceptible plants," but he subsequently expanded his definition to include a plant that "shows an ability to grow and reproduce itself or to repair injury ... in spite of supporting a population of herbivores approximately equal to that damaging a susceptible host" (109). In a review of tolerance to plant disease, Schafer (127) defined tolerance as "that capacity of a cultivar resulting in less yield or quality loss relative to disease severity or pathogen development when compared with other cultivars or crops." These definitions highlight the steps involved in measuring tolerance. First, individuals must be classified by their genetic relationships. Then, within each genetic class, fitness (or yield) must be measured at different levels of damage, disease, or pest population density. The fitness (or yield) responses of the individuals in a genetic class, across the damage, disease, or pest density gradient, is then an estimate of the tolerance of that group.

Thus, we can define tolerance as the reaction norm of fitness across a damage gradient, and it can be treated as a phenotypically plastic trait (137). As with other phenotypically plastic traits, tolerance to damage can be modeled by a mathematical function (129, 130, 154). The fitness function is probably modeled most accurately as a complex polynomial equation (3, 111, 112, 115, 146). For example, some potato varieties maintain yield at low damage levels (compensating tolerance), but experience decreasing yield with further increases in damage (incomplete tolerance) (95). Other varieties increase yield in response to damage (overcompensating tolerance), but undercompensate for heavier damage (95; also see 77, 78, 94, 133). However, for simplicity, a linear function is frequently used to describe tolerance (3, 56, 87, 137, 140, 146) and is often a good approximation (11).

The tolerance of a genotype can then be described by the linear function, $Y = a + bX$, where Y indicates fitness and X indicates damage level (Figure 1). The Y-intercept, a, denotes fitness when undamaged and describes the genotype's ability to tolerate all environmental stresses other than consumer damage. The mean height of the line, $\overline{Y}$, describes fitness averaged across all damage levels and is a measure of general vigor (42, 41, 43). Finally, the slope of the reaction norm, b, describes the change in fitness in response to consumer damage, or tolerance. A completely tolerant genotype has a flat reaction norm ($b = 0$) and experiences no fitness impact of damage. A negative slope ($b < 0$) indicates undercompensating tolerance; a positive slope ($b > 0$) describes overcompensating tolerance.

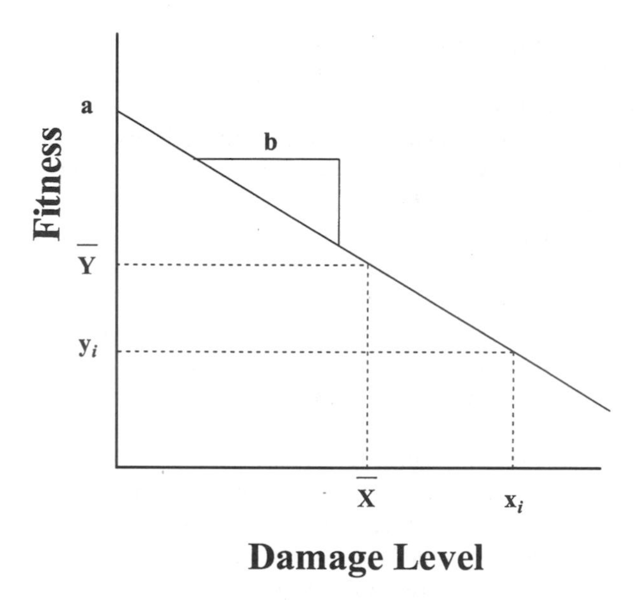

Figure 1 Hypothetical fitness reaction norm of a genotype, obtained by regressing fitnesses of clonal replicates of the genotype on the damage level that each replicate sustained. The estimated slope, *b*, is an operational measure of tolerance to damage; the influence that damage has on fitness. The *y*-intercept, *a*, indicates the influence that other environmental variables have on fitness. This value can be obtained by extrapolation but is best measured on clonal replicates experimentally protected from damage. In this example, the slope is negative, indicating incomplete compensation for damage.

Some authors have defined tolerance to damage simply as the fitness of an individual at a particular level of damage (108, 109, 127). However, our model illustrates that defining tolerance in terms of fitness in a single damage environment fails to distinguish the effects of damage on fitness from those of other uncontrolled environmental factors. Therefore, defining tolerance as fitness at a definite level of damage (e.g. 108, 109, 127) obscures the action of traits that specifically allow plants to tolerate damage by consumers (see 56). To illustrate this with our simple model, we define fitness of a genotype X experiencing the *ith* level of damage as Y_i. This value is predicted by three factors: (i) the known level of damage, X_i, (ii) the slope of the genotype's fitness reaction norm to damage, b, and (iii) the genotype's fitness in the absence of damage, a. In the absence of damage, fitness is determined by the interaction of the genotype with all other environmental factors. Thus, the only component of the model that is not determined by the level of damage or the genotype's fitness response to damage is the intercept, which is determined by the genotype's fitness response to all other environmental variables. Further, the only way to determine whether a component trait contributes to tolerance of consumer damage rather than to tolerance of some other environmental stress is to measure how it affects fitness along a damage gradient.

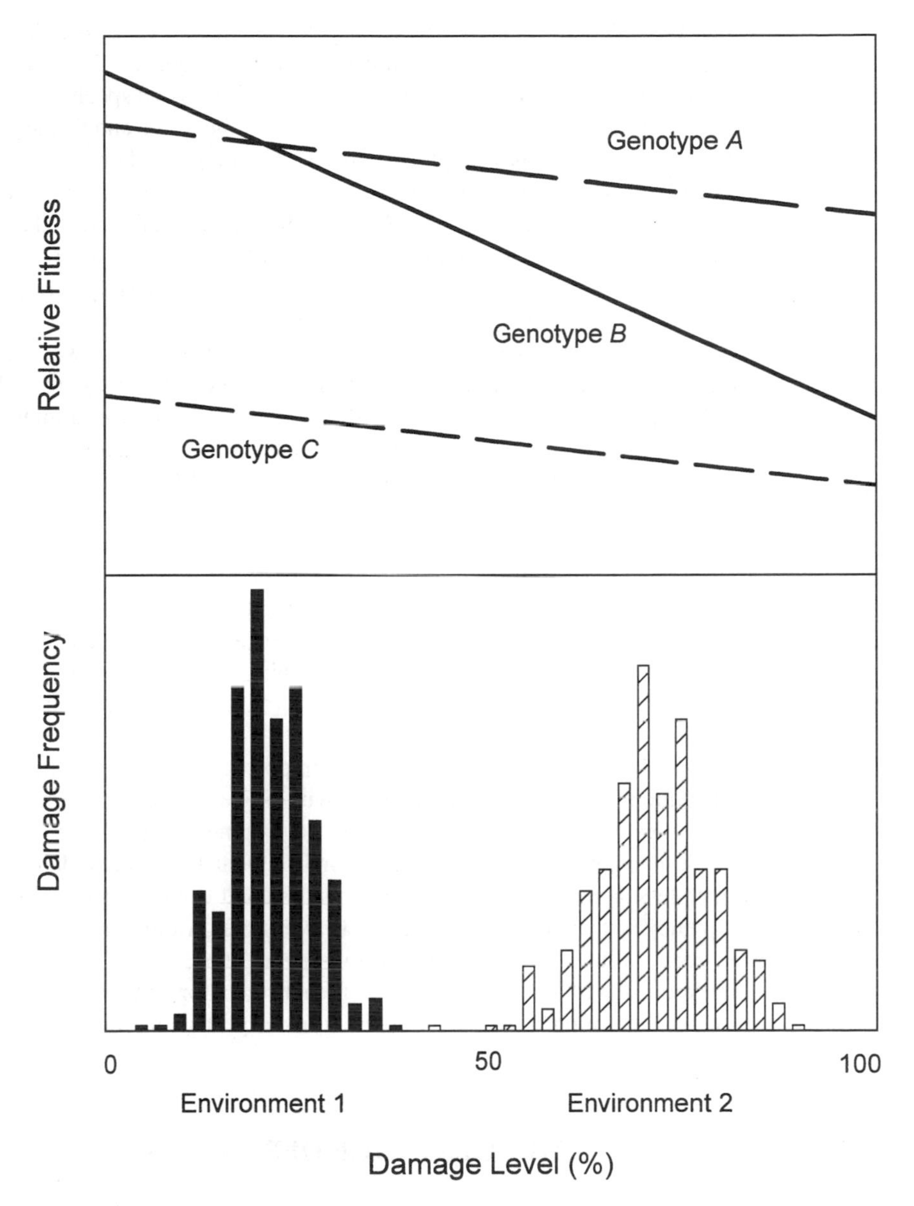

Figure 2 Fitness reaction norms of three plant genotypes exposed to two environments that differ in mean levels of damage. In the upper panel, the relative fitnesses of the three genotypes are plotted against damage level; their tolerance phenotypes are indicated by their reaction norm slopes whereas the heights of their reaction norms indicate their general vigor in these environments. The lower panel indicates the relative frequencies of damage levels in each environment. Environment 1 has a lower mean level of damage than does environment 2.

It is also possible for genotypes to be highly tolerant of consumer damage (i.e. b close to, or greater than, 0), but to have low general vigor ($\overline{Y}$). Thus, the relative contributions of general vigor versus tolerance to fitness of a species, variety, or genotype in any specific environment depends upon the range and frequencies of damage levels to which they may be exposed. The range of damage can also determine the detectability of fitness trade-offs (see below, artificial vs natural damage). These points are illustrated in Figure 2. In this figure, the mean damage level in environment 2 is relatively high and genotype A is always more fit than genotype B. Further, although genotype C is more tolerant than genotype B, genotype C always has the lowest fitness because it has low general vigor. Genotype A has the overall fitness advantage in environment 2 because it is always more vigorous than genotype C, as well as being both more tolerant and vigorous than genotype B. In this environment, selection should favor both increased tolerance and vigor, i.e. genotype A. However in environment 1, which has a lower average level of damage, the norms of reaction of genotypes A and B cross. Within this environment, genotypes A and B have approximately equal general vigor, but their crossing reaction norms reveal a cost of tolerance. Genotype A is more tolerant and therefore has higher fitness during times of moderate damage. But because of the cost of tolerance, genotype B is more fit during times of low damage. Both genotypes are more fit than genotype C, which still suffers from low general vigor. These hypothetical data demonstrate that distinguishing between general vigor and tolerance may be important in understanding the evolutionary response of plants to consumer-imposed selection.

The example in Figure 2 also illustrates the importance of knowing the probability distribution of different damage levels within plant populations (56, 140). When damage levels are uniformly high, as in environment 2, genotype A is always more fit than genotype B. However, in environment 1, in which plants tend to experience less damage, the relative fitnesses of A and B will depend critically on the frequency distribution of damage levels over time. An important question to answer, then, is how often these conditions occur (56, 140). Clearly, determining the frequency of damage environments that select for or against tolerance is essential for determining the evolutionary trajectory of plant-consumer interactions.

OPERATIONALIZING THE TOLERANCE DEFINITION

While in theory the tolerance of an individual can be determined, in practice tolerance cannot be quantified as a property of an individual. Instead, it must be estimated as the property of a group (i.e. individuals of a species, variety, cultivar, population, or genotype). Thus, characterizing tolerance entails measuring the fitnesses of replicate individuals from the group that have experienced a range of damage levels. These individual values are then summarized statistically to estimate the tolerance of the group. Recent empirical studies have employed variants of this method to evaluate genetic variance for tolerance to both artificial and natural damage (56, 87, 135, 137, 140, 146). Despite the seeming simplicity of this protocol, however, there are many potential pitfalls involved in measuring tolerance.

Natural versus Artificial Damage

While it has been argued that to completely characterize the tolerance function, all damage levels should be investigated (3), others have argued that a focus on natural damage is more effective when estimating tolerance (145). Thus, controversy exists over whether tolerance should be evaluated using natural variation in damage or experimentally imposed damage levels (56, 140, 145). This controversy is composed of several issues that should be considered individually. The first issue involves statistical biases. Natural levels of damage are determined by a multiplicity of environmental and genetic factors that may also directly influence plant fitness. These confounding factors may create a bias if natural damage is used to estimate tolerance (56, 140, 146). For example, natural damage levels undoubtedly reflect individual variation in resistance. If resistance is costly, then individuals with less damage (i.e. more resistance) may exhibit lower fitness, thereby creating a downward bias in the estimate of tolerance (140).

Concerns about confounding factors have prompted several investigators to impose controlled levels of damage that can be randomly assigned among individuals (e.g. 56, 140). However, the use of artificial damage entails its own problems (145). Recent studies clearly indicate that the nature of plant resistance traits induced in response to tissue damage can be strongly influenced by the identity of the causal agent (4, 10a, 88a). While the mechanisms by which different responses are induced are not always clear (10b), it seems ever more likely that the same unit of damage might impose different fitness effects, depending upon its causal agent.

Even when damage is imposed using natural agents, achieving controlled levels of damage often requires some artificiality. For example, insects may be confined to small cages attached to individual leaves (4, 72a). Such constraints will alter the dispersion, timing, and duration of damage. Several studies now suggest that these factors influence the fitness impact of damage (79a, 85). These concerns have led some investigators to argue that the biases caused by natural damage are relatively less important than the loss of precision caused by using artificial damage to estimate tolerance (145).

Difficulty Estimating Fitness

In most organisms, it is virtually impossible to measure *total* fitness. In many plants, however, it is even more complicated. Fitness is determined not only by the number of offspring produced by an individual, but also by the number of offspring it sired, and the survival and fecundity of offspring. Thus, estimates of tolerance may depend on which fitness components are considered.

Most empirical studies have estimated tolerance using female components of fitness (e.g. seed production: 49, 50, 87, 137, 140, 146). However, damage may also impact male fitness components (40, 47, 93, 141, 143). For example, foliar damage can alter attractiveness to pollinators (141, 143) and thus reduce a plant's success at siring seeds. Mechanisms that mitigate the impact of damage on female fitness components may differ from those that reduce the impact on male fitness. Indeed, tolerance measured via the female fitness components might even be achieved by shifting resources from male to female reproductive structures (50, 156a). However, if siring seeds consumes less resources than filling seeds, tolerance may be achieved by shifting resources from female to male reproduction. Thus, a better measure of tolerance would consider both female and male components of fitness (47, 140).

Even when growth and/or reproduction of damaged individuals appear equivalent to that of undamaged plants, subtle fitness trade-offs may cause new tissues (i.e. leaves) or replacement progeny to differ in quality (i.e. seed viability, seedling survivorship, and/or seed output). When damaged individuals produce the same number of offspring, yet of lower quality, tolerance may be overestimated. For example, *Pastinaca sativa* appeared tolerant to *Depressaria pastinacella* caterpillars because damaged plants produced the same number of seeds as undamaged plants. However, seeds from damaged and undamaged plants were not always equally large or viable (50). Consumer damage may also affect flowering phenology (50). Depending on the length of the growing season, changes in phenology may affect seed quality or threaten the possibility of reproduction altogether (63, 146). Biotic factors such as pollinator availability may also constrain phenology and limit tolerance via reproductive delay (62). Such constraints may not always be manifested. For example, although seeds produced by browsed *Ipomopsis aggregata* are produced later in the growing season, existing studies have shown that delayed seed maturation does not reduce germination or subsequent growth of progeny (107). Of course, this outcome may vary with the date of first frost.

Difficulty Measuring Damage

Finally, when consumer damage is difficult to quantify, resistance and tolerance may appear indistinguishable. It is particularly difficult to quantify systemic damage, such as that imposed by sap feeders or systemic diseases (see 125). Agricultural scientists have developed some methods to deal with these types of problems (76, 110, 138). For example, rather than measuring leaf area damaged, they use the biomass gain of the consumer or its population growth as an estimate of resistance. Thus, plants that produce less consumer biomass are considered more resistant. Further, regression methods have been used to distinguish between tolerance and resistance in crop varieties, and may be usefully applied to natural populations (76, 110, 138). This is done by regressing plant fitness on damage level, or consumer biomass. The intersection of a line marking the mean damage level and the regression line then forms four quadrants that indicate different combinations of tolerance and resistance.

MECHANISMS OF TOLERANCE

Attempts to predict the evolution of tolerance will benefit from a more detailed understanding of the genetic mechanisms controlling plant responses to damage. Despite extensive efforts devoted to elucidating how increases in defense are selected for by consumer feeding, there is still only cursory understanding of these processes (3, 4, 27, 36, 56, 80, 87, 125, 140, 141). Total fitness represents a lifetime integration of phenotypic interactions with the ever-changing environment. Thus, mechanisms underlying the fitness response to damage (tolerance) are likely to be far more complex than those underlying resistance. Consequently, current understanding of the fundamental mechanisms of tolerance resides at a very gross level (147).

Traits currently known to provide tolerance to damage are involved in two general roles: one, resource reallocation; and two, plant architecture. Clearly, replacing tissue or progeny lost to consumers involves allocation of mobile resources (49, 50). Yet, even when resources are adequate, tolerance may be limited by the number of available meristems (45). Further, patterns of vascular architecture can restrict the flow of resources among plant parts (157), limiting the use of existing resources to tolerate tissue loss (81).

Resource Allocation Patterns

Resource allocation patterns prior to and/or following damage may contribute to tolerance. Patterns of resource allocation are characterized by (*a*) relative allocation to storage, growth, or reproductive organs, (*b*) qualitative aspects of the organs (e.g. thick vs thin leaves, storage vs feeder roots), and (*c*) the timing of allocation.

Allocation decisions prior to damage may condition tolerance. In *Asclepias syriaca*, for example, genotypes that stored more resources in rhizomes were more tolerant of losing leaf area to consumers (56).

Changes in allocation following damage may also influence tolerance. In undamaged individuals, source-sink relationships direct resource allocation among organs (157), with the relative strength of sinks determining which will accumulate more resources (54). Consumer damage often removes sinks and/or sources, thereby altering source-sink relationships and modifying allocation patterns (147). The existence of sinks is determined by hormones, and the way that allocation patterns shift in response to damage are under hormonal control as well. Consequently, tolerance may depend in part on hormonal control of meristem release and differentiation. For example, removal of the apical meristem in *Ipomopsis aggregata* releases lateral meristems, which would otherwise remain dormant (105–107).

The branches formed from these lateral meristems together can produce as many, or even more, seeds as the lost apical shoot (105, 107).

Damage-induced changes in hormonal meristem control may also alter allocation to sexual functions. For example, in undamaged wild parsnip (*Pastinaca sativa*) (49, 50), early inflorescences are composed mainly of hermaphroditic flowers, which will become sinks if they set seed, whereas later maturing inflorescences produce mostly male flowers. However, floral damage on early inflorescences alters sex expression of later maturing inflorescences, allowing them to produce more hermaphroditic flowers. As a consequence, damaged plants produce as many seeds as undamaged plants (although, as described earlier, seeds produced on the later maturing inflorescences are smaller than those that would have been borne on the earlier maturing inflorescences).

Tolerance may also depend on qualitative differences among plant modules made or modified following damage (5, 30, 97, 126, 155). Much of the research concerning physiological changes has focused on changes in photosynthetic rates, which often increase in damaged leaves and/or in neighboring undamaged leaves (90a). The pathways to increased photosynthetic rates are numerous. For example, photosynthetic rates of *Phaseolus vulgaris* increased after herbivory because the remaining leaves had higher levels of RUBISCO (155). In contrast, in *Agropyron* species, increased photosynthetic rates were due to delays in leaf senescence (97). Further, *Solidago* leaves produced following damage showed higher photosynthetic rates than leaves on control plants because they had both high specific leaf area and delayed senescence (90a). Future experimentation should take into account how both the pattern of damage within a leaf (90b) and within the plant canopy (90a) may influence photosynthetic rates. However, the question of whether increased photosynthetic rates are of sufficient magnitude or duration to compensate for lost tissue is still open.

The timing of allocation to growth, storage, and reproduction can also be critical to tolerance. In *Isomeris arborea*, floral damage by a pollen-feeding beetle occurs during the first half of the flowering season (68, 69). Some genotypes achieved tolerance by prolonging flowering in response to damage, allowing them to set seed after the beetles stopped feeding. This example highlights how tolerance may depend critically on the degree of overlap in plant and consumer phenology. It also illustrates the close relationship between resistance and tolerance. Genotypes of *I. arborea* that delay flowering until after the beetle has pupated are generally described as resistant (escape in time). In contrast, genotypes that flower while the beetles are active, but then reallocate resources to flowering post-damage are termed tolerant, yet they can only be tolerant if they escape damage in the second half of the season, i.e. are resistant.

Plant Architecture

As described above, allocation patterns contribute to plant architecture (45). In turn, plant architecture influences resistance (80, 82) and may affect plant tolerance in several different ways (81). First, architecture may influence tolerance through its impact on resource capture. For example, wild tomatoes are more tolerant than their domesticated kin, largely because the canopy structure of wild tomatoes allows them to better exploit increased light availability following damage (159).

The number and distribution of meristems can also influence tolerance (20, 81, 100). For example, the extent of growth following damage may depend on the number and distribution of meristems surviving damage, the pattern of meristem release from dormancy, and the number of new meristems produced after damage. Plants may be tolerant by having more meristems prior to damage. For example, two wild relatives of maize have a greater number of tillers than the domesticated species and are more tolerant of stem borer damage (124). Palms provide a classic example of meristem limitation of tolerance. Because they lack basal adventitious shoots, palms are completely intolerant of meristem herbivory. Damage to the single meristem kills the tree (31).

Sectoriality imposed by vascular architecture may be another important constraint on tolerance to damage (81, 156, 157). Since plants are modular in their construction, resource movement among modules is not completely independent (156, 157). Vascular piping does not connect every plant part to all other parts, and resources from distant modules may be unavailable for allocation to sites of localized damage. Two factors may limit the ability of plants to respond to localized damage (81, 157). First, plants may be unable to respond to damage if sources cannot detect the demand from distant sinks (156, 156a, 157). Second, inadequate vascular connections may impede resource transfer to distant locations, thereby hampering the ability to tolerate consumer damage (81, 157). In many cases, plant resources appear sufficient to compensate for localized leaf area loss, but vascular constraints on resource allocation limit compensatory growth, leading to localized decreases in growth and reproduction (81). Further, vascular constraints on reallocation responses can also decrease overall plant fitness (79, 85, 86). In some plants, it appears that new vascular connections form after damage (134). It seems reasonable that genotypes better able to form such connections would be more tolerant, yet this idea remains unexplored.

Although increased vascular integration may improve plant tolerance to foliar damage (131), it might also increase resources available to internal feeding consumers such as systemic pathogens and xylem- or phloem-feeding herbivores. Consequently, selection on vascular architecture by one consumer guild might be opposed by selection imposed by other consumer guilds. For example, stem-boring and sucking consumers may have a smaller fitness impact on clonal plants because their feeding is restricted to a particular ramet (124). However, individual ramets may sacrifice tolerance to leaf chewers by giving up the ability to share resources (7, 131). Because so little is known about the effects of plant architecture on tolerance to damage, future studies are needed to test these predictions. Further comparative and experimental studies examining plant architecture effects on tolerance may be a promising avenue for investigation.

EVOLUTION OF TOLERANCE

A large body of theory (1, 60, 61, 96, 136, 148, 150, 149, 151) has been developed to elucidate the conditions under which overcompensating tolerance may evolve. These models all make specific assumptions about how patterns of resource allocation constrain fitness under different consumption regimes. Consequently, in our review below, we refer to them as *reallocation models*. More recent theoretical work has addressed tolerance that is not necessarily overcompensating (3, 125, 146). These models also consider the *joint evolution of tolerance and resistance*. Below, we summarize the predictions and evaluate the assumptions of both types of models.

Models

Reallocation Models Several models were designed to investigate the implications of internal resource trade-offs for the evolution of overcompensating tolerance to meristem damage. As a group, they explore how the pattern of meristem removal, both within and among growing seasons, influences the evolution of tolerance (1, 60, 61, 96, 132, 136, 150, 149, 151). Some models examine the effects of individual selection (96, 148, 150), whereas others explore selection among lineages (61, 136).

Individual Selection

Simple mathematical models invoking individual selection suggest that overcompensating tolerance can evolve when the chance of being damaged is greater than the chance of escaping (148, 150, 152). When the probability of being damaged is greater than 50%, individual fitness is maximized by withholding investment in reproduction until after damage (150, 151). In contrast, when there is less than a 50% probability of being damaged, individual fitness is maximized by committing all resources to reproduction prior to damage, leaving none available for reallocation (63, 151, 152).

These models assume that meristem-removal occurs randomly among individuals, individuals are damaged only once, and damage occurs at the same life stage every generation. Thus, damage is the environmental cue that indicates that it is safe to allocate stored reserves to reproduction. However, plants that rely on environmental cues for their allocation decisions may allocate resources suboptimally (33, 63). Further, plants that delay reproduction regardless of damage can invade a population of overcompensating individuals unless damage is universal (84, 144). Thus, individuals that allocate resources to reproduction late in the season, i.e. after the danger of damage, will be more fit than those that wait until damage occurs (61, 63). This highlights the important influence of allocation patterns on the evolution of tolerance.

These simple allocation models seem unrealistic because they predict the evolution of incomplete tolerance only when all allocation strategies are equally fit, i.e. in the limited case when being damaged or undamaged are equally likely (150). Further, their assumptions about the pattern and timing of damage are quite restrictive and maybe unrealistic (79a, 85). Relaxing these assumptions produces predictions that appear more realistic. For example, tolerance can evolve if consumers preferentially damage larger individuals that have invested more of their total resource pool into growth or reproduction rather than storage (150). Further, as consumers become choosier (i.e. express stronger preference for larger individuals), the amount that plants should invest in growth/reproduction prior to damage decreases, relaxing the conditions for the evolution of overcompensating tolerance. However, introducing a trade-off between active and future meristem production favors intermediate allocation strategies (148), which leads to incomplete tolerance. Such a trade-off can be included in the model by assuming a cost of currently active buds versus maintaining dormant buds. This added cost increases the intensity and risk of damage needed to favor bud dormancy, thereby restricting the conditions under which tolerance in any form can evolve.

Other assumptions have been explored in a series of models developed by Tuomi and colleagues. They first (148) relaxed the assumption that heavy damage is required to activate dormant meristems. This model predicts that a high probability of experiencing a single grazing episode favors meristems that are activated by small amounts of damage. Next, they relaxed the twin assumptions that plants experience only a short window of vulnerability prior to flowering and never suffer secondary damage (96). By incorporating repeated grazing events within a season, they found that optimal meristem reactivity to damage declined with increases in the number of grazing events, grazing risk, or intensity of damage. Thus, when a population is likely to experience multiple grazing events, grazers must remove a relatively large proportion of active meristems from an individual before its dormant buds are stimulated to grow. However, the model also predicts that repeated within-season grazing events relax the conditions that favor tolerance. Specifically, as the number of grazing events increases, overcompensating tolerance may evolve at lower risks of and levels of damage than when damage occurs only once during a growing season (96).

Does Overcompensating Tolerance Equal Plant-Consumer Mutualism?

Several authors have suggested that plants with overcompensating tolerance have evolved a mutualistic relationship with consumers (for review, see 13, 90, 84, 153a). However, theory makes clear that simply observing overcompensation in response to artificial damage (or even to natural damage that is heavier than normal) is not sufficient evidence with which to conclude that plants have evolved a mutualistic relationship with consumers. A mutualism requires that the two parties perform better when acting in concert than when separate (17a). However, models that predict evolution of overcompensation in the presence of consumers usually predict higher fitness in populations that are never damaged (84). Moreover, even when selection does not explicitly favor overcompensating tolerance, plants that have evolved incomplete compensation may exhibit overcompensation after experiencing heavier than typical meristem loss (96). This phenomenon arises because adaptation to repeated grazing creates a hump-shaped fitness function in which a small number of intense grazing episodes can produce startlingly large overcompensation. These theoretical outcomes challenge the notion that plants expressing overcompensating tolerance participate in a mutualism with consumers.

Lineage Selection Selection among lineages may act directly on the reaction norm of phenotypically plastic traits that are expressed only once per lifetime. In this case, the optimal norm of reaction is found in the lineage with the greatest among-generation geometric mean fitness (113, 132). Models incorporating this form of selection predict that tolerance can evolve under less restricted conditions than needed under individual selection (61, 136). For example, they predict evolution of overcompensating tolerance under lower probabilities of within-season damage without requiring that damage be size-dependent (136). Neither do such models require that the risk of damage be predictable among growing seasons (61).

One such lineage selection model explored how variation in the timing and probability of damage among growing seasons affects selection on the relationship between tolerance and flowering time (61). This model found that less predictable damage favors plants that avoid damage by flowering early, i.e. escape in time, but are less tolerant. Damage that is more predictable favors plants that achieve tolerance by delaying investment of resources in reproduction until late in the season.

Models of the Joint Evolution of Tolerance and Resistance Because resistance and tolerance may both evolve in response to the same selection pressure, i.e. consumer damage, their evolution is most appropriately considered jointly (3, 27, 38, 87, 125, 140, 144, 146, 153). Van der Meijden et al (153) first considered the possibility that tolerance and resistance might be negatively genetically correlated and therefore evolve antagonistically. A negative correlation could arise from pleiotropic effects of genes involved in an allocation trade-off between resistance and tolerance or could arise from linkage disequilibrium produced by correlated selection on these traits (37, 71, 137, 140, 153). Correlated selection occurs because as resistance increases, damage declines. Thus, the fitness advantage of tolerance declines with increasing resistance. Likewise, because the fitness decrement due to damage declines as tolerance increases, the fitness advantage of resistance decreases as tolerance increases (3, 38, 125, 144, 153).

The first mathematical model to examine the joint evolution of resistance and tolerance was developed by Fineblum & Rausher (38). Like van der Meijden et al (153), these authors assumed a trade-off between resistance and tolerance. Their model predicts two fitness peaks: one fixed for high resistance with no tolerance, the other for high tolerance but no resistance. However, the model does not consider the strongly divergent effects that resistance and tolerance may have on consumer populations, which could feed back as very different selection regimes on the two traits.

Like the models discussed above, this model assumes that changes in plant damage are independent of both plant and consumer traits. Such an assumption may be unrealistic. Of course, plants may evolve increased defense, which may include phenological traits allowing them to escape damage. Similarly, consumers may evolve the ability to overcome plant resistance, to detect, and/or find suitable host plants. Consumers may also evolve phenological changes that allow them to use a previously unavailable host (36, 59). Further, evolution of consumers in response to unrelated selective factors might change levels of plant damage. For example, damage may increase as consumers escape regulation by natural enemies through the evolution of resistance to these enemies (17). Finally, consumer populations may exhibit numerical responses to the frequency of resistant or tolerant plant genotypes, causing changes in the fitness consequences of possessing such traits. Many aspects of this complexity have yet to be incorporated into theoretical treatments of tolerance evolution. The models reviewed below (3, 125, 144) begin the challenging task of integrating some of these ideas into our understanding of the evolution of plant-consumer interactions.

Feedback among resistance and tolerance alleles has recently been addressed by three models (3, 125, 144). These models assume that consumer fitness declines as resistant genotypes increase in frequency, which then decreases damage, and thereby, the fitness advantage of resistance. This negative feedback between resistance alleles and selection imposed on them prevents fixation of alleles conferring complete resistance (3, 125, 144). Instead, plant populations remain polymorphic for resistance (125, 144), a classic prediction of antagonistic optimality models, which are reviewed by Burdon (23).

In contrast to resistance, tolerance alleles do not reduce consumer fitness or population size (3, 125). Thus, the possibility of damage remains the same or increases as tolerance alleles increase in frequency. This positive feedback allows the benefits of tolerance alleles, and thus selection for such alleles, to increase with their frequency, ultimately resulting in their fixation (125, 144). Further, as the level of tolerance conferred by an allele increases, its probability of becoming fixed also increases (125). Similarly, alleles coding for resistance through avoidance of damage should also become fixed within a population, so long as the consumer is not truly monophagous (144). As with tolerance, this outcome is based on the premise that avoidance alleles have no impact on consumer fitness. However, generalist and specialist consumers may create different selection pressures on plant populations (140, 144), resulting in the maintenance of both tolerant and resistant genotypes within plant populations. For example, some phytochemicals may defend against generalists but attract specialists, which use them as feeding or oviposition cues (32, 48, 57, 101, 117). This may create a situation in which generalists select for increased chemical resistance, whereas specialists that are attracted to these compounds may select for increased tolerance. Thus, whether tolerance becomes fixed within a population may depend in part on the composition of the consumer community (140, 144).

The evolutionary dynamics of tolerance and resistance alleles are slightly different if the genes are linked (125, 144). Consider, for example, a tolerance mutation linked to a resistance gene. When the benefit of tolerance is greater than that of resistance, tolerance will invade the population and increase to fixation. If the cost of tolerance is greater than that of resistance, however, tolerant and susceptible alleles will instead achieve a stable polymorphism.

EVALUATION OF MODEL ASSUMPTIONS

Selection Factors

Models investigating the evolution of tolerance share several key assumptions. Primary among these is the assumption that tolerance evolves in response to consumer damage (37, 38, 146, 147). Several lines of evidence support the validity of this assumption (87, 146). For example, wild plants which probably experience more intense damage from consumers than do domesticated plants, which are protected by pesticides, are more tolerant of damage (124, 159).

However, several authors contend that rather than being a specific adaptation to consumer damage, tolerance of damage is a by-product of selection for the ability to tolerate other environmental stresses (1, 2, 153). Thus, an alternative explanation for the wild–domestic comparison is that wild plants have evolved tolerance to competition because they experience relatively more intense competition than do their domesticated kin. This idea may stem from definitions of tolerance that tie the ability to outgrow damage primarily to intrinsic growth rate (24, 27, 51). However, rather than being directly associated with intrinsic growth rate, tolerance may also depend on traits associated with plant architecture and internal resource allocation. Further, if tolerance of damage were the by-product of selection for tolerance of competition, then these two traits would be positively correlated genetically. Instead, tolerance and competitive ability may be negatively correlated, each constraining the evolution of the other. For example, the most damage-tolerant genotypes of the common milkweed, *Asclepias syrica*, allocate more resources to root tissue than do less tolerant genotypes (56), possibly sacrificing their ability to compete for light.

The best evidence that tolerance to damage has evolved in response to selection by consumers comes from studies comparing multiple populations with divergent grazing histories. In several of these studies, individuals originating from predictably grazed (35) or more frequently damaged (114) populations exhibit greater levels of tolerance. A particularly compelling study compared managed and unmanaged populations of field gentian, *Gentianella campestris* (74). Some populations have been either mown or grazed every year for at least 100 years (and possibly 1000 years), whereas unmanaged populations growing along roadsides or (more recently) in electric power line clearings, have not experienced such consistent damage. Individuals from managed populations exhibited greater tolerance to damage than those from unmanaged populations (74).

Safe Storage of Resources

Another implicit assumption of reallocation models is that plants can store resources/meristems in a manner that makes them unavailable to consumers. For example, if the probability of foliar damage is high and there is a relatively low risk of root damage, plants can defend resources (through avoidance) by allocating them to roots. Plants that do so may be more tolerant because these resources are available for subsequent reallocation following foliar damage (56). Thus, the best tolerance strategy may be to resist damage to tissues that contribute most significantly to fitness (95a), as predicted by *optimal defense theory* (67, 162). For example, in acorns the apical area adjacent to the developing embryo is highly resistant to consumers due to the level of tannins (139). Thus, consumer damage is concentrated in the basal portion of acorns (139). This allocation to resistance increases the probability that the embryo will survive, and thus tolerate, consumer damage to the acorn. In cases where tolerance is achieved by allocating resources to highly resistant plant organs (e.g. 95a), then tolerance and resistance may be positively genetically correlated. This expectation contrasts with the trade-off between these two traits generally predicted from theory (38, 136a, 153).

Herbivory Limits Plant Fitness

Another unstated assumption in tolerance theory is that plant fitness is limited solely by consumers. However, individual fitness is often limited by processes independent of damage. Such factors may restrict the expression of the phenotypic variation in tolerance among genotypes, thereby constraining the evolutionary response to damage.

Abiotic conditions may alter the phenotypic expression of tolerance (8, 21, 40, 46, 52, 56, 58, 98, 99). For example, while *Brassica napus* plants grown at field capacity (well watered) were more fit overall, they were less tolerant of damage than those grown under drought conditions (98). This observation suggests that water stress may limit the expression of genetic variance in tolerance. Further, in *Asclepias syriaca*, genetic variation in the ability to tolerate damage was greater under high compared to low nutrient conditions (56).

Biotic factors other than damage may also limit fitness and thereby limit the expression of variation in tolerance. For example, under pollen-limiting conditions, pollinator availability can influence the expression of tolerance (62). Pollinator availability significantly affected the ability of *Ipomopsis aggregata* individuals to tolerate damage (62). Damaged individuals produced more flowers than undamaged individuals, but exhibited little tolerance in terms of seed production. Hand-pollination greatly increases tolerance as measured by seed production. Such a population could evolve greater tolerance to damage by losing its dependence on pollinators, i.e. evolving selfing. Evolving selfing would remove pollen limitation and restore the dependence of fitness on damage. Selfing would also provide tolerance to the detrimental effects of folivory on pollinator attraction (90c, 141, 143).

Semelparous Reproduction

Most existing models of the evolution of tolerance assume semelparous reproduction (but see 125). Specifically, resources are acquired and stored until maturity, at which time all available resources are committed to reproduction and the plant subsequently dies. However, tolerance is also expressed by iteroparous plant species (for a review, see 147). Because damage in one growing season can have an impact on survivorship and reproduction through many reproductive seasons (34, 122), iteroparity adds significant complexity to both modeling and measuring tolerance. Tackling this added complexity is an important challenge for future theoretical and empirical studies.

CONSTRAINTS ON THE EVOLUTION OF TOLERANCE

Allocation Trade-off

The most prominent assumption of evolutionary models is that tolerance is costly (37, 38, 125, 137, 140, 144), although in some cases these models require vanishingly small costs (125). As discussed earlier, tolerance can be described as the reaction norm of fitness across a damage gradient, which can be treated as a phenotypically plastic trait (137). Schlicting (130) and Scheiner (129) provide general insights into the evolution of phenotypically plastic traits. It is important to note, however, that methods for detecting costs of most phenotypically plastic traits (33) may not be applicable to examining costs of tolerance. This is because tolerance has been defined as the reaction norm of fitness, rather than that of a specific trait.

While a cost of tolerance may be manifested in various forms, reallocation models assume that tolerant individuals store resources in some form that is available for allocation only after damage. Thus, plants that withhold allocation of resources to growth or reproduction until after damage will be more tolerant of damage but will be less fit in the absence of damage. This assumption has rarely been explicitly tested (but see 35, 56).

Because an allocation trade-off dictates that the optimal allocation pattern in an environment with natural damage differs from that in an environment without damage, its existence might be inferred from a negative fitness correlation across damage environments (41, 42, 56, 137). However, a negative fitness correlation across damage environments could also be caused by a cost of resistance (136a). Thus, a better method for detecting an allocation trade-off is to examine the genetic correlation between the slope (tolerance) and the intercept (fitness in the absence of damage) of the fitness reaction norm. If the slope and the intercept are estimated from measurements made on the same individuals, however, the two variables are not independent (87, 146). This problem can be circumvented in two ways. First, fitness in the absence of damage can be measured on a different set of replicates than that used to measure tolerance (87, 137). Using this method, Simms & Triplett (137) found a trade-off in *Ipomea purpurea* between tolerance to damage imposed by a fungal pathogen and fitness in the absence of the pathogen. It is also possible to apply a statistical correction to deal with the lack of independence between the slope and intercept of the reaction norm in response to damage (87, 146). One of the two studies that have used this statistical correction has detected a cost of tolerance (87, 146).

A cost of tolerance can also be detected by comparing the levels of tolerance exhibited by ancestral populations that are subjected to damage with those descendent populations protected from damage. If tolerance is costly, it is likely to be lost in populations not experiencing damage because its costs will no longer be offset by the fitness benefit it provides in the presence of damage. For example, domesticated populations of agricultural species, which are commonly protected from consumers by pesticides, typically express lower levels of tolerance than do their wild ancestors (124, 159), suggesting that tolerance declined due to its fitness cost. However, domestication is a complex process, and this decline in tolerance may be due to selection on other correlated traits. In a more natural example, recently established Pacific Coast populations of *Spartina alterniflora* exhibit less tolerance than their Atlantic Coast ancestors (28). While this decline in tolerance might be due to genetic drift during the founding of new populations, it suggests that tolerance is lost in the absence of damage. This is further supported by comparisons among introduced Pacific Coast populations in habitats differing in herbivory. Populations in which herbivores are present are more tolerant (28); however, which Pacific Coast populations are ancestral or derived is unknown. Thus, determining if tolerance is costly in this case requires more investigation. While these cases suggest that tolerance is costly (28, 124, 159), whether the costs of tolerance are due to allocation trade-offs or trade-offs with other traits is unknown.

Tolerance and Resistance Trade-off

Models of the joint evolution of tolerance and resistance have either assumed (38) or predicted (3, 125, 144) a trade-off between these traits. Some studies have detected a negative correlation between resistance and tolerance across species (125, 153), and others have found evidence within populations for a negative genetic correlation between tolerance and constitutive resistance (38, 140, 146). Tolerance may also be negatively correlated with induced resistance. For example, induction of glucosinolates in *Raphanus raphanistrum* decreases the ability of individuals to tolerate damage by consumers (4).

Two genetic mechanisms may cause a negative correlation between resistance and tolerance: (*a*) antagonistic pleiotropy, caused by the same sets of genes affecting both traits, and (*b*) linkage disequilibrium, caused by correlational selection (71) favoring individuals that possess one or the other trait, but not both (37, 137, 140). Selection experiments can be used to disentangle these two mechanisms. In one such experiment, selection for increasing resistance produced a decrease in tolerance in *Brassica rapa* (140), suggesting that this trade-off was due to antagonistic pleiotropy rather than linkage disequilibrium.

Trade-offs between tolerance and resistance may not be universal, however (87). For example, across species of *Zea*, constitutive resistance to a stem borer was positively correlated with tolerance of its damage (124). Similarly, many crop varieties are both resistant and tolerant (76, 102, 110, 138).

If tolerance and resistance are negatively correlated genetically, then selection may maintain both traits at intermediate levels (140). For example, in *Brassica rapa*, the fitnesses of undamaged, highly defended individuals are statistically indistinguishable from those of damaged, poorly defended individuals (140). In contrast to some recent theoretical predictions (38), this empirical result suggests that, highly defended, poorly tolerant individuals can coexist with poorly defended, highly tolerant individuals.

Ontogenetic Trade-offs

Another common assumption of tolerance evolution models is that the expression of tolerance does not differ among growth stages. However, growth stages may vary in tolerance (19, 160). For example, young seedlings of the woody shrub *Prosopis glandulosa* tolerate damage better than do older seedlings (160). Although unexamined, ontogenetic variation in allocation to above- and below-ground tissues may explain observed age differences in tolerance in this species. Thus, depending on when selection occurs, ontogenetic changes in the expression of tolerance may significantly influence its evolution. Specifically, the evolution of tolerance may be constrained if selection operates in opposite directions at different ontogenetic stages, if selection acts only at specific stages, or if there are genetic correlations among stage-specific levels of tolerance.

Similar to "safe storage," plants may invest more resources to resist consumers at certain life stages that contribute most to fitness (67, 162) and thus be able to tolerate damage. Life history theory and empirical evidence suggest that the life stage with the greatest probability of mortality should be the most defended (118–120). The fact that juvenile plants often allocate more resources to resistance than do older plants (12, 18, 22, 72, 128) has been used to argue that plants defend tissues in an optimal manner (162). This hypothesis depends upon the assumption that juveniles have the highest mortality, which appears to be a common observation (9a, 9b, 25b, 27a, 64a). This suggests that greater understanding of the evolution of plant responses to damage requires further exploration of the timing of consumer damage relative to plant life history stages.

Architectural Constraints

Storing meristems in case of damage may entail a cost to immediate growth and reproduction (132). Further, maintaining meristems in a quiescent state may also involve costs associated with the synthesis and transport of growth inhibitors. These costs, due to meristem storage, are similar in concept to allocation costs associated with storing resources (carbohydrates, proteins, nutrients) (45, 156, 156a, 157; see below) in protected or unavailable organs that can later be reallocated to replace tissues lost to consumer attack. These potential costs vis-à-vis growth and reproduction may constrain the evolution of tolerance.

Vascular architecture, and the degree to which it limits flow of resources within and among plant parts (79a, 85, 86, 156, 157), may also constrain a plant's ability to tolerate tissue loss. Marquis (80, 81) predicted that individuals with more vascular connections among plant parts would show less impact of spatially restricted damage. He suggested that this is due to the greater possibility of resources to flow from undamaged to damaged portions of the plant. Currently, however, we are lacking the empirical studies concerning the range of phenotypes available among plant species and genotypes upon which selection might act. Thus, evaluation of this hypothesis may be premature.

However, a response to selection for more resource communication among plant parts might be constrained by the opposing selective effects of different consumer guilds. Increased vascular connections, allowing greater resource movement from undamaged to damaged portions of plants, which may contribute to tolerance of foliar damage, may simultaneously increase the potential resource base available to phloem- and xylem-tapping consumers. This constraint should apply to plants with multiple ramets (e.g. tillering grasses), as well as to nonclonal species.

ECOLOGICAL IMPLICATIONS OF TOLERANCE

Once a plant responds to tissue loss, its value as a food resource and habitable environment may change (4, 64). These changes may influence the community of plant consumers, as well as that of the natural enemies of those consumers (65). While plant responses to damage, such as the production of new tissues, may affect consumers either positively or negatively (29, 53, 64, 121, 161, Hochwender and Fritz, unpublish. data), traits contributing directly to tolerance are generally thought to have no negative impact on consumer population size (3). In fact, they may actually allow consumer populations to increase (125). However, tolerance may also be accompanied by induced resistance (4), which may negatively affect consumer populations.

The best-studied examples of the effects of plant tolerance on consumers are found in studies of the effect of vertebrate consumers on later abundances of insect consumers (29, 53, 121; Hochwender & Fritz, unpublished data). Most research has found that insect abundance typically increases on plants following natural browsing or clipping (29, 52, 53, 121, Hochwender & Fritz, unpublished data). Changes in plant chemical or physical traits following damage may also affect higher trophic levels. For example, if consumer populations increase due to tolerance, this provides a larger resource for their predators and parasitoids. This increase in resources, i.e. herbivore population size, may also result in increased population densities at higher trophic levels (65). However, not all species in the consumer community increase in response to prior damage (29, 53, 56). Thus, plant tolerance may change not only the structure of the consumer community, but also that of the predators and parasitoids.

Consumers can have profound effects on the structure of plant communities in which they feed, altering both succession (10, 23a, 73, 75, 91) and regional differences in plant diversity (25a). Such changes are commonly attributed to the combined effects of host preference by consumers (i.e., interspecific differences in plant resistance) and interspecific differences in the competitive and colonizing ability of plants (52, 99a). However, differential plant tolerance among species may also influence community structure (9). One study, specifically designed to differentiate between the relative importance of plant tolerance and resistance for explaining consumer impacts on plant communities (9), suggests that the effects of consumers on plant community composition are more influenced by plant resistance than tolerance to damage. That is, replaced species were as tolerant or more tolerant than those that were replacing them. Thus, it appeared that levels of damage due to plant resistance were causing the species replacement. However, it is still too early to make general conclusions concerning the effects of tolerance on plant diversity.

CONCLUSION

Since Ehrlich & Raven's (36) seminal paper describing coevolution between consumers and their host plants, our knowledge of the evolutionary responses of plants to damage has dramatically increased. With the incorporation of tolerance as a possible response to consumer-imposed selection, we have begun to form a more comprehensive understanding of the evolution of plant-consumer interactions. While significant advances concerning the evolution of tolerance have been made, important questions still need to be answered, including the following:

- What are the specific traits underlying tolerance to damage?
- Are there mechanistic distinctions between resistance and tolerance?
- How does consumer-imposed selection differ between generalist and specialist consumers?
- Is tolerance evolution constrained by environmental variation?
- What implications does perenniality have for the measurement and evolution of tolerance?

- What role does ontogenetic constraints play in the evolution of tolerance?
- Are costs of tolerance a general phenomenon?
- What influence does plant tolerance have on higher trophic levels?
- What is the role of plant tolerance in structuring biotic communities?

These questions point to promising areas of future research. Further, when we have answers to them, our understanding of the evolution of plant-consumer interactions will be greatly enhanced.

ACKNOWLEDGMENTS

We would like to thank Robert Fritz, Tom Juenger, John Lill, Kristine Mothershead, Diana Pilson, and Art Weis for the many discussions that helped clarify our ideas concerning tolerance. Nels Holmberg, Damond Kyllo, John Lill, Kristine Mothershead, Bitty Roy, and Peter Tiffin provided valuable comments on the manuscript. This work was supported by Dropkin (University of Chicago) and a National Research Council (Ford Foundation) Fellowships to K.A.S., DEB-9527900 and DEB-9815550 to E.L.S., and BSR 96-15038 to R.S.F.

LITERATURE CITED

1. Aarssen L. 1995. Hypotheses for the evolution of apical dominance in plants: implications for the interpretation of overcompensation. *Oikos* 74:49–156
2. Aarssen L, Irwin D. 1991. What selection: herbivory or competition? *Oikos* 60:261–62
3. Abrahamson W, Weis A. 1997. *Evolutionary Ecology Across Three Trophic Levels: Goldenrods, Gallmakers, and Natural Enemies*. Princeton, NJ: Princeton Univ. Press
4. Agrawal A, Strauss S, Stout M. 1999. Cost of induced responses and tolerance to herbivory in male and female fitness components of wild radish. *Evolution* 53:1093–104
5. Alderfer R, Eagle C. 1976. The effect of partial defoliation on the growth and photosynthesis of bean leaves. *Bot. Gaz.* 137:351–55
6. Allsopp PG, McGill NG. 1997. Variation in resistance to *Eumargarodes laingi* (Hemiptera: Margarodidae) in Australian sugarcane. *J. Econ. Entomol.* 90:1702–9
7. Alpert P, Mooney H. 1986. Resource sharing among ramets in the clonal herb, Fragaria chilonesis. *Oecologia* 70:227–33
8. Alward R, Joern A. 1993. Plasticity and overcompensation in grass responses to herbivory. *Oecologia* 95:358–64
9. Anderson V, Briske D. 1995. Herbivore-induced species replacement in grasslands: Is it driven by herbivory tolerance or avoidance. *Ecol. Appl.* 5:1014–24

9a. Augspurger CK, Kelly CK. 1984. Pathogen mortality of tropical tree seedlings: experimental studies of the dispersal distance, seedling density, and light conditions. *Oecologia* 61:211–17

9b. Augspurger CK, Kitajima K. 1992. Experimental studies of seedling recruitment from contrasting seed distributions. *Ecology* 73:1270–84

10. Augustine D, McNaughton S. 1998. Ungulate effects on the functional species composition of plant communities: herbivore selectivity and plant tolerance. *J. Wildl. Manage.* 62:1165–83

10a. Baldwin IT. 1990. Herbivory simulations in ecological research. *Trends Ecol. Evol.* 5:91–93
10b. Baldwin IT, Preston CA. 1999. The eco-physiological complexity of plant responses to insect herbivores. *Planta* 208:137–45
11. Bardner R, Fletcher K. 1974. Insect infestations and their effects on the growth and yield of field crops: a review. *Bull. Entomol. Res.* 64:141–60
12. Basey J, Jenkins S, Busher P. 1988. Optimal central-place foraging by beavers: tree size selection in relation to defensive chemicals of quaking aspen. *Oecologia* 76:278–82
13. Belsky A. 1986. Does herbivory benefit a plant? A review of the evidence. *Am. Nat.* 127:870–92
14. Bergelson J, Crawley MJ. 1992. The effects of grazers on the performance of individuals and populations of scarlet gilia, Ipomopsis aggregata. *Oecologia* 90:435–44
15. Bergelson J, Crawley MJ. 1992. Herbivory and Ipomopsis aggregata: the disadvantages of being eaten. *Am. Nat.* 139:870–82
16. Bergelson J, Juenger T, Crawley MJ. 1996. Regrowth following herbivory in Ipomopsis aggregata: compensation but not overcompensation. *Am. Nat.* 148:744–55
17. Bernays E, Graham M. 1988. On the evolution of host specificity in phytophagous arthropods. *Ecology* 69:886–92
17a. Boucher DH. 1985. *The Biology of Mutualism: Ecology and Evolution*. New York, NY: Oxford Univ. Press
18. Bowers M, Stamp N. 1993. Effects of plant-age, genotype, and herbivory on plantago performance and chemistry. *Ecology* 74:1778–91
19. Brandt R, Lamb R. 1994. Importance of tolerance and growth rate in the resistance of oilseed rapes and mustards to flea beetle, Phyllotreta cruciferae (Goez) (Coleoptera: Chrysomelidae). *Can. J. Plant Sci.* 74:169–76
20. Briske DD, Boutton TW, Wang Z. 1996. Contribution of flexible allocation priorities to herbivory tolerance in C-4 perennial grasses: an evaluation with C-13 labeling. *Oecologia* 105:151–59
21. Brook KD, Hearn AB, Kelly CF. 1992. Response of cotton, Gossypium hirsutum L., to damage by insect pests in Australia: manual simulation of damage. *J. Econ. Entomol.* 85:1368–77
22. Bryant J, Julkunentiitto R. 1995. Ontogenetic development of chemical defense by seedling resin birch: energy-cost of defense production. *J. Chem. Ecol.* 21:883–96
23. Burdon J. 1987. *Diseases and Plant Population Biology*. Cambridge, UK: Cambridge Univ. Press
23a. Carson WP, Root RB. 1999. Top-down effects of insect herbivores during early succession: influence on biomass and plant dominance. *Oecologia* 121:260–72
24. Chapin FI. 1980. The mineral nutrition of wild plants. *Annu. Rev. Ecol. Syst.* 11:233–60
25. Charlesworth B. 1994. *Evolution in Age Structured Populations*. Cambridge, UK: Cambridge Univ. Press
25a. Chase JM, Leibold MA, Downing AL, Shurin JB. 2000. Grassland foodwebs: a review of the effects of productivity, herbivory, and plant species compositional turnover. *Ecology*. In press
25b. Clark DB, Clark DA. 1985. Seedling dynamics of a tropical tree: impacts of herbivory and meristem damage. *Ecology* 66:1884–92
26. Cobb N. 1894. Contributions to an economic knowledge of Australian rusts (Uredineae). *Agric. Gaz. N.S.W.* 5:239–50
27. Coley P, Bryant J, Chapin F. 1985. Resource availability and plant antiherbiovre defense. *Science* 230:895–99

27a. Connell JH, Tracey JG, Webb TJ. 1984. Compensatory recruitment, growth, and mortality as factors maintaining rain forest tree diversity. *Ecol. Mono.* 54:141–64

28. Daehler CC, Strong DR. 1995. Lower tolerance to herbivory in *Spartina alterniflora* after a century of herbivore-free growth. *Bull. Ecol. Soc. Am.* 76:59

29. Danell K, Huss-Danell K. 1985. Feeding by insects and hares on birches earlier affected by moose browsing. *Oikos* 44:75–81

30. Danell K, Huss-Danell K, Bergstrom R. 1985. Interactions between browsing moose and two species of birch in Sweden. *Ecology* 68:1867–78

31. del Campo M, Renwick J. 1999. Dependence of host constituents controlling food acceptancc by Manduca sexta larvae. *Entomol. Exp. Appl.* 93:209–15

32. De Stevens D, Putz F. 1984. Mortality rate of some rain forest palms in Panama. *Principes* 29:162–65

33. DeWitt T, Sih A, Wilson D. 1998. Costs and limits of phenotypic plasticity. *Trends Ecol. Evol.* 13:77–81

34. Doak DF. 1992. Lifetime impacts of herbivory for a perennial plant. *Ecology* 73:2086–99

35. Dyer M, Acra M, Wang G, Coleman D, Freckman D, et al. 1991. Source-sink carbon relations in 2 *Panicum coloratum* ecotoypes in response to herbivory. *Ecology* 72:1472–83

36. Ehrlich P, Rave P. 1964. Butterflies and plants: a study in coevolution. *Evolution* 18:586–608

37. Fineblum W. 1991. *Genetic constraints on the evolution of resistance and tolerance to herbivore damage in morning glory*. PhD thesis. Duke Univ., Durham, NC

38. Fineblum WL, Rausher MD. 1995. Tradeoff between resistance and tolerance to herbivore damage in a morning glory. *Nature* 377:517–20

39. Flynn JL, Reagan TE. 1988. Corn phenology in relation to natural and simulated infestations of the sugarcane borer [Diatraea saccharalis] (Lepidoptera: Pyralidae). *J. Econ. Entomol.* 77:1524–29

40. Frazee J, Marquis R. 1994. Environmental contributions to floral trait variation in Chamaecrista-fasciculata (Fabaceae, Caesalpinoideae). *Am. J. Bot.* 81:206–15

41. Fry J. 1992. The mixed-model analysis of variance applied to quantitative genetics: biological meaning of the parameters. *Evolution* 46:540–50

42. Fry J. 1993. The "general vigor" problem: can antagonistic pleiotropy be detected when genetic covariances are positive. *Evolution* 47:327–33

43. Futuyma D, Philippi T. 1987. Genetic variation and covariation in responses to host plants by Alsophila pometaria (Lepidoptera: Gemetridae). *Evolution* 41:269–79

44. Gao H, Beckman CH, Mueller WC. 1995. The nature of tolerance to *Fusarium oxysporum* f. sp. *lycopersici* in polygenically field-resistant Marglobe tomato plants. *Physiol. Mol. Plant Pathol.* 46:401–12

45. Geber M. 1990. The cost of meristem limitation in Polygonum arenastrum: negative genetic correlations between fecundity and growth. *Evolution* 44:799–819

46. Gertz AK, Bach CE. 1995. Effects of light and nutrients on tomato plant compensation for herbivory by *Manduca sexta* (Lepidoptera, Sphingidae). *Gt. Lakes Entomol.* 27:217–22

47. Gronemeyer P, Dilger B, Bouzat J, Paige K. 1997. The effects of herbivory on paternal fitness in scarlet gilia: better moms make better pops. *Am. Nat.* 150:592–602

48. Haribal M, Renwick J. 1998. Identification and distribution of oviposition stimulants for monarch butterflies in hosts and nonhost plants. *J. Chem. Ecol.* 24:891–904

49. Hendrix SD. 1979. Compensatory reproduction in a biennial herb Pastinaca sativa

following insect Depressaria pastinacella defloration. *Oecologia* 42:107–18
50. Hendrix SD, Trapp EJ. 1989. Floral herbivory in Pastinaca-Sativa: do compensatory responses offset reductions in fitness. *Evolution* 43:891–95
51. Herms D, Mattson W. 1992. The dilemma of plants: to grow or defend. *Q. Rev. Biol.* 67:283–335
52. Hjalten J, Danell K, Ericson L. 1993. Effects of simulated herbivory and intraspecific competition on the compensatory ability of birches. *Ecol. Publ. Ecol. Soc. Am.* 74:1136–42
53. Hjalten J, Price P. 1996. The effect of pruning on willow growth and sawfly population densities. *Oikos* 77:549–55
54. Ho L. 1988. Metabolism and compartmentalization of imported sugars in sink organs in relation to sink strength. *Annu. Rev. Plant Physiol. Plant Mol. Biol.* 39:355–78
55. Deleted in proof
56. Hochwender C, Marquis R, Stowe K. 2000. The potential for and constraints on the evolution of compensatory ability in Asclepias syriaca. *Oecologia* 122:361–70
57. Hughes P, Renwick J, Lopez K. 1997. New oviposition stimulants for they diamondback moth in cabbage. *Entomol. Exp. Appl.* 85:281–83
58. Jacobs KA, Johnson GR. 1996. Ornamental cherry tolerance of flooding and phytophthora root rot. *HortScience* 31:988–91
59. Janzen D. 1980. When is it coevolution? *Evolution* 34:611–12
60. Jaremo J, Nilsson P, Tuomi J. 1996. Plant compensatory growth: herbivory or competition. *Oikos* 77:238–47
61. Jaremo J, Ripa J, Nilsson P. Flee or fight uncertainty: plant strategies in relation to anticipated damage. *Ecol. Lett.* In press
62. Juenger T, Bergelson J. 1997. Pollen and resource limitation of compensation to herbivory in scarlet gilia, Ipomopsis aggregata. *Ecology* 78:1684–95
63. Juenger T, Bergelson J. 2000. The evolution of compensation to herbivory in scarlet gilia, Ipomopsis aggregata: herbivore-imposed natural selection and the quantitative genetics of tolerance. *Evolution* 54:79–92
64. Karban R, Baldwin I. 1997. *Induced Responses to Herbivory*. Chicago, Ill.: Univ. Chicago Press
65. Kitajima K, Augspuger CK. 1989. Seed and seedling ecology of a monocarpic tropical tree, *Tachigilia versicolor*. *Ecology* 70:1102–14
66. Krause S, Raffa K. 1996. Defoliation tolerance affects the spatial and temporal distribution of larch sawfly and natural enemy populations. *Ecol. Entomol.* 21:259–69
67. Krischik V, Denno R. 1983. Individual, population, and geographic patterns in plant defense. In *Variable Plants and Herbivores in Natural and Managed System*, ed. R Denno, M McClure, pp. 463–512. New York, NY: Academic
68. Krupnick G, Weis A. 1999. The effect of floral herbivory on male and female reproductive success in Isomeris arborea. *Ecology* 80:135–49
69. Krupnick G, Weis A, Campbell D. 1999. The consequences of floral herbivory for pollinator service to Isomeris arborea. *Ecology* 80:125–34
70. Lambers H, Poorter H. 1992. Inherent variation in growth-rate between higher plants: a search for physiological causes and ecological consequences. *Adv. Ecol. Res.* 23:187–261
71. Lande R, Arnold S. 1983. The measurement of selection on correlated characters. *Evolution* 37:1210–26
72. Langenheim J, Stubblebine W. 1983. Variation in leaf resin composition between parent tree and progeny in Hymenea: implication for herbivory in the humid tropics. *Biochem. Syst. Ecol.* 11:97–106
72a. Lehtila K, Strauss SY. 1999. Effects of foliar herbivory on male and female

reproductive traits of wild radish, *Raphanus raphanistrum*. Ecology 80: 116–24
73. Leibold M. 1996. A graphical model of keystone predators in food webs: trophic regulation of abundance, incidence, and diversity. *Am. Nat.* 147:784–812
74. Lennartsson T, Tuomi J, Nilsson P. 1997. Evidence for an evolutionary history of overcompensation in the grassland biennial *Gentianella campestris* (Gentianaceae). *Am. Nat.* 149:1147–55
75. Louda S, Holt R, Keeler K. 1990. Herbivore influences on plant performance and competitive interactions. In *Perspectives on Plant Competition*, ed. J Grace, D Tilman, pp. 413–44. San Diego, CA: Academic
76. Lye B, Smith C. 1988. Evaluation of rice cultivars for antibiosis and tolerance to fall armyworm (Lepidoptera: Noctuidea). *Fla. Entomol.* 71:254–61
77. Mailloux G, Binns M, Bostanian N. 1991. Density-yield relationships and economic injury models for Colorado potato beetle larvae on potatoes. *Res. Popul. Ecol.* 33:101–13
78. Mailloux G, Bostanian N, Binns M. 1995. Density-yield relationships for Colorado potato beetle adults on potato. *Phytoparasitica* 23:101–18
79. Marquis R. 1992. The selective impact of herbivores. In *Plant Resistance to Herbivores and Pathogens*, ed. R Fritz, E Simms, pp. 301–25. Chicago, IL: Univ. Chicago Press
79a. Marquis, RJ. 1992. A bite is a bite? Constraints on responses to folivory in *Piper arieianum* (Piperaceae). *Ecology* 73:143–152
80. Marquis R, Whelan C. 1996. Plant morphology, and recruitment of the third tropic level: subtle and little-recognized defenses? *Oikos* 75:330–34
81. Marquis RJ. 1996. Plant architecture, sectoriality and plant tolerance to herbivores. *Vegetatio* 127:85–97
82. Marquis RJ, Whelan CJ. 1994. Insectivorous birds increase growth of white oak through consumption of leaf-chewing insects. *Ecology* 75:2007–14
83. Maschinski J, Whitham TG. 1989. The continuum of plant responses to herbivory: the influence of plant association, nutrient availability, and timing. *Am. Nat.* 134:1–19
84. Mathews J. 1994. The benefits of overcompensation and herbivory: the difference between coping with herbivores and liking them. *Am. Nat.* 144:528–33
85. Mauricio R, Bowers M. 1990. Do caterpillars disperse their damage: larval foraging behavior of 2 specialist herbivores, Euphydry-phaeton (Nymphalidae) and Pieris–rapae (Pieridae). *Ecol. Entomol.* 15:153–61
86. Mauricio R, Bowers M, Bazzaz F. 1993. Pattern of leaf damage affects fitness of the annual plant Raphanus-sativa (Brassicaceae). *Ecology* 74:2066–71
87. Mauricio R, Rausher MD, Burdick DS. 1997. Variation in the defense strategies of plants: Are resistance and tolerance mutually exclusive? *Ecology* 78:1301–11
88. McBlain BA, Zimmerly MM, Schmitthenner AF, Hacker JK. 1991. Tolerance to phytophthora rot in soybean. I. Studies of the cross 'Ripley' × 'Harper'. *Crop Sci.* 31:1405–11
88a. McCloud, ES and Baldwin, IT. 1997. Herbivory and caterpillar regurgitants amplify the wound-induced increases in jasmonic acid but not nicotine in *Nicotiana* sylvesteris. *Planta* (Heidelberg). 203:430–35
89. McNaughton S. 1979. Grazing as an optimization process: grass ungulate relationships in the Serengeti. *Am. Nat.* 113:691–703
90. McNaughton S. 1983. Compensatory plant-growth as a response to herbivory. *Oikos* 40:329–36

90a. Meyer GA. 1998. Patterns of defoliation and its effects on photosynthesis and growth of goldenrod. *Funct. Ecol.* 12:270–79
90b. Morrison KD, Reekie, EG. 1995. Pattern of defoliation and its effect on photosynthesis capacity in *Oenothera biennis*. *J. Ecol.* 83:759–67
90c. Mothershead K, Marquis, RJ. 2000. Fitness impacts of herbivory through indirect effects on plant-pollinator interactions in Oenothera macrocarpa. *Ecology* 81:30–40
91. Mudler C, Koricheva J, Huss-Danell K, Hogberg P, Joshi J. 1999. Insects affect relationships between plant species richness and ecosystem processes. *Ecol. Lett.* 2:237–46
92. Mussell H. 1980. Tolerance to disease host plant-pathogen interactions. *Plant Dis. Adv. Treatise* 5:39–52
93. Mutikainen P, Delph L. 1996. Effects of herbivory on male reproductive success in plants. *Oikos* 75:353–58
94. Nault B, Follet P, Gould F, Kennedy G. 1995. Assessing compensation for insect damage in mixed plantings of resistant and susceptible potatoes. *Am. Potato J.* 72:157–76
95. Nault B, Kennedy G. 1998. Limitations of using regression and mean separation analyses for describing the response of crop yield to defoliation: a case study of the Colorado potato beetle (Coleoptera: Chrysomelidae) on potato. *J. Econ. Entomol.* 91:7–19
95a. Newman RM, Kerfoot CW, Hanscom III, Z. 1996. Watercress allelochemical defends high-nitrogen foliage against consumption: Effects on freshwater invertebrate herbivores. *Ecology* 77:2312–23
96. Nilsson P, Tuomi J, Astrom M. 1996. Even repeated grazing may select for overcompensation. *Ecology* 77:1942–46
97. Nowak RS, Caldwell MM. 1984. A test of compensatory photosynthesis in the field: implications for herbivory tolerance [Agropyron, grazing pressure]. *Oecologia* 61:311–18
98. Nowatzki T, Weiss M. 1997. Effects of simulated and flea beetle injury to cotyledons on growth of drought-stressed oilseed rape, Brassica napus L. *Can. J. Plant Sci.* 77:475–81
99. Oesterheld M, McNaughton SJ. 1991. Effect of stress and time for recovery on the amount of compensatory growth after grazing. *Oecologia* 85:305–13
99a. Olff H, Ritchie, ME. 1998. Effects of herbivores on grassland plant diversity. *Trends Ecol. Evol.* 13:261–65
100. Olson BE, Richards JH. 1988. Tussock regrowth after grazing: intercalary meristem and axillary bud activity of tillers of Agropyron desertorum. *Oikos* 51:374–82
101. Orians C, Huang C, Wild A, Zee P, Dao P, Fritz R. 1997. Causes of variable beetle responses to hybrid willows. *Entomol. Exp. Appl.* 83:285–94
102. Ortega A, Vasal S, Mihm J, Hershey C. 1980. Breeding for insect resistance in maize. In *Breeding Plants Resistant to Insects*, ed. F Maxwell, P Jennings, pp. 371–417. New York, NY: Wiley & Sons
103. Owen D. 1980. How plants benefit from the animals that eat them. *Oikos* 35:230–35
104. Owen D, Wiegert R. 1976. Do consumers maximize plant fitness? *Oikos* 27:488–92
105. Paige K. 1999. Regrowth following ungulate herbivory in Ipomopsis aggregata: geographic evidence for overcompensation. *Oecologia* 118:316–23
106. Paige KN. 1992. Overcompensation in response to mammalian herbivory from mutualistic to antagonistic interactions. *Ecology* 73:2076–85
107. Paige KN, Whitham TG. 1987. Overcompensation in response to mammalian herbivory: the advantage of being eaten. *Am. Nat.* 129:407–16
108. Painter R. 1951. *Insect Resistance in Crop Plants*. New York, NY: Wiley & Sons

109. Painter R. 1958. Resistance of plants to insects. *Annu. Rev. Entomol.* 3:267–90
110. Panda N, Heinrichs E. 1983. Levels of tolerance and antibiosis in rice varieties having moderate resistance to the brown planthopper, Nilaparvata lugens (Stal) (Hemiptera: Delphacidae). *Environ. Entomol.* 12:1204–14
111. Pedigo L. 1989. *Entomology and Pest Management*. New York, NY: Macmillan
112. Pedigo L, Hutchins S, Higley L. 1986. Economic injury levels in theory and practice. *Annu. Rev. Entomol.* 31:341–68
113. Philippi T, Seger J. 1989. Hedging one's evolutionary bets, revisited. *Trends Ecol. Evol.* 4:41–44
114. Polley HW, Detling JK. 1988. Herbivory tolerance of Agropyron smithii populations with different grazing histories. *Oecologia* 77:261–67
115. Poston F, Pedigo L, Welch S. 1983. Economic injury levels: reality and practicality. *Bull. Ecol. Soc. Am.* 29:49–53
116. Radcliffe D, Hussey R, McClendon R. 1990. Cyst nematode vs tolerant and intolerant soybean cultivars. *Agron. J.* 82:855–60
117. Renwick J, Lopez K. 1999. Experience-based food consumption by larvae of Pieris rapae: addiction to glucosinolates? *Entomol. Exp. Appl.* 91:51–58
118. Reznick D. 1982. The impact of predation on life-history evolution in Trinidadian guppies: genetic-basis of observed life-history patterns. *Evolution* 36:1236–50
119. Reznick D, Butler M, Rodd F, Ross P. 1996. Life-history evolution in guppies (Poecilia reticulata). 6. Differential mortality as a mechanism for natural selection. *Evolution* 50:1651–60
120. Reznick D, Endler J. 1982a. The impact of predation on life-history evolution in Trinidadian guppies (Poecilla-Reticulata). *Evolution* 36:160–77
121. Roininen H, Price P, Bryant J. 1997. Response of galling insects to natural browsing by mammals in Alaska. *Oikos* 80:481–86
122. Root R. 1996. Herbivore pressure on goldenrods (Solidago altissima): its variation and cumulative effects. *Ecology* 77:1074–87
123. Rosenthal JP, Kotanen PM. 1994. Terrestrial plant tolerance to herbivory. *Trends Ecol. Evol.* 9:145–48
124. Rosenthal JP, Welter SC. 1995. Tolerance to herbivory by a stemboring caterpillar in architecturally distinct maizes and wild relatives. *Oecologia* 102:146–55
125. Roy B, Kirchner J. 2000. Evolutionary dynamics of pathogen resistance and tolerance. *Evolution.* 54:51–63
126. Ruess R, McNaughton S, Coughenour M. 1983. The effects of clipping, nitrogen source and nitrogen concentration on the growth responses and nitrogen uptake of an East African sedge. *Oecologia* 59:253–61
127. Schafer J. 1971. Tolerance to plant disease. *Annu. Rev. Phytopathol.* 9:235–52
128. Schappert P, Shores J. 1995. Cyanogenesis in Turnera ulmifolia L. (Turneraceae) 1. Phenotypic distribution and genetic variation for cyanogenesis on Jamaica. *Heredity* 74:392–404
129. Scheiner S. 1993. Genetics and evolution of phenotypic plasticity. *Annu. Rev. Ecol. Syst.* 24:35–68
130. Schlichting C. 1986. The evolution of phenotypic plasticity. *Annu. Rev. Ecol. Syst.* 17:667–93
131. Schmid B, Puttick G, Burgess K, Bazzaz F. 1988. Clonal integration and effects of simulated herbivory in old-field perennials. *Oecologia* 75:465–71
132. Seger J, Brockman J. 1987. What is bet hedging? *Oxford Surv. Evol. Biol.* 4:182–211
133. Senanayake D, Holliday N. 1990. Economic injury levels for Colorado potato beetle (Coleoptera: Chrysomelidae) on 'Norland' potatoes in Manitoba. *J. Econ. Entomol.* 83:2058–64

134. Shea P. 1989. Phytophagous insect complex associated with cone of white fir, Abies concolor (Gord and Glend) Lindl and its impact on seed production. *Can. Entomol.* 121:699–708
135. Shen CS, Bach CE. 1997. Genetic variation in resistance and tolerance to insect herbivory in Salix cordata. *Ecol. Entomol.* 22:335–42
136. Simmons A, Johnston M. 1999. The cost of compensation. *Am. Nat.* 153:683–87
136a. Simms EL. 1992. Costs of plant resistance to herbivory. In *Plant Resistance to Herbivores and Pathogens*, ed. R Fritz, E Simms, pp. 392–425. Chicago, IL: Univ. Chicago Press
137. Simms E, Triplett J. 1994. Costs and benefits of plant responses to disease: resistance and tolerance. *Evolution* 48:1973–85
138. Smith C. 1989. *Plant Resistance to Insects: A Fundamental Approach.* New York, NY: Wiley & Sons
139. Steele M, Knowles T, Bridle K, Simms E. 1993. Tannins and partial consumption of acorns: implications for dispersal of oaks by seed predators. *Am. Midl. Nat.* 130:229–38
140. Stowe KA. 1998. Experimental evolution of resistance in Brassica rapa: correlated response of tolerance in lines selected for glucosinolate content. *Evolution* 52:703–12
141. Strauss S. 1997. Floral characters link herbivores, pollinators, and plant fitness. *Ecology* 78:1640–45
142. Strauss S, Agrawal A. 1999. The ecology and evolution of tolerance to herbivory. *Trends Ecol. Evol.* 14:179–85
143. Strauss S, Conner J, Rush S. 1996. Foliar herbivory affects floral characters and plant attractiveness to pollinators: implications for male and female plant fitness. *Am. Nat.* 147:1098–1107
144. Tiffin P. 2000. Are tolerance, avoidance, and antibiosis evolutionarily and ecologically equivalent responses of plants to herbivores? *Am. Nat.* 155: 128–138
145. Tiffin P, Inouye B. 2000. Measuring tolerance to herbivory: accuracy and precision of estimates using natural versus imposed damage. *Evolution.* In press
146. Tiffin P, Rausher M. 1999. Genetic constraints and selection acting on tolerance to herbivory in the common morning glory, Ipomoea purpurea. *Am. Nat.* 154: 700–16
147. Trumble JT, Kolodnyhirsch DM, Ting IP. 1993. Plant compensation for arthropod herbivory. *Annu. Rev. Entomol.* 38:93–119
148. Tuomi J, Nilsson P, Astrom M. 1994. Plant compensatory responses: bud dormancy as an adaptation to herbivory. *Ecology* 75:1429–36
149. Vail S. 1992. Selection for overcompensatory plant responses to herbivory: a mechanism for the evolution of plant-herbivore mutualism. *Am. Nat.* 139:1–8
150. Vail S. 1993. Overcompensation as a life-history phenomenon. *Evol. Ecol.* 7:122–23
151. Vail SG. 1994. Overcompensation, plant-herbivore mutualism, and mutalistic coevolution: a reply to Mathews. *Am. Nat.* 144:534–36
152. van der Meijden E. 1990. Herbivory as a trigger for growth. *Funct. Ecol.* 4:121–29
153. van der Meijden E, Wijn M, Verkaar H. 1988. Defence and regrowth, alternative plant strategies in the struggle against herbivores. *Oikos* 51:355–63
153a. Vekaar HJ. 1986. When does grazing benefit plants? *Trends Ecol. Evol.* 1: 168–169
154. Via S, Lande R. 1985. Genotype-environment interaction and the evolution of phenotypic plasticity. *Evolution* 39:505–22
155. Wareing P, Khalifa M, Terharne K. 1968. Rate-limiting processes in photosynthesis at saturating light intensities. *Nature* 220:453–57

156. Watson M. 1986. Integrated physiological units in plants. *Trends Ecol. Evol.* 1:119–23
156a. Watson MA. 1995. Sexual differences in plant development phenology affect plant-herbivore interactions. *Trends Ecol. Evol.* 10:180–182
157. Watson M, Casper B. 1984. Morphogenetic constraints on patterns of carbon distribution in plants. *Annu. Rev. Ecol. Syst.* 15:233–58
158. Weis A, Gorman W. 1990. Measuring natural selection on reaction norms: an exploration of the Erustoga-Solidago system. *Evolution* 44:820–31
159. Welter SC, Steggall JW. 1993. Contrasting the tolerance of wild and domesticated tomatoes to herbivory: agroecological implications. *Ecol. Appl.* 3:271–78
160. Weltzin JE, Archer SR, Heitschmidt RK. 1998. Defoliation and woody plant (Prosopis glandulosa) seedling regeneration: potential vs realized herbivory tolerance. *Plant Ecol.* 138:127–35
161. Wold E, Marquis R. 1997. Induced defense in white oak: effects on herbivores and consequences for the plant. *Ecology* 78:1356–69
162. Zangerl A, Bazzaz F. 1992. Theory and pattern in plant defense allocation. In *Plant Resistance to Herbivores and Pathogens*, ed. R Fritz, E Simms, pp. 363–91. Chicago, IL: Univ. Chicago Press

Subject Index

B

F

G

J

K

Q

S

CUMULATIVE INDEXES

CONTRIBUTING AUTHORS, VOLUMES 27–31

CHAPTER TITLES, VOLUMES 27–31

Volume 27 (1996)

Volume 28 (1997)

Volume 29 (1998)

Volume 30 (1999)

Volume 31 (2000)